AN INTRODUCTION TO GENERAL RELATIVITY AND COSMOLOGY

Experts Plebański and Krasiński provide a thorough introduction to the tools of general relativity and relativistic cosmology. Assuming familiarity with advanced calculus, classical mechanics, electrodynamics and special relativity, the text begins with a short course on differential geometry, taking a unique top-down approach. Starting with general manifolds on which only tensors are defined, the covariant derivative and affine connection are introduced before moving on to geodesics and curvature. Only then is the metric tensor and the (pseudo)-Riemannian geometry introduced, specialising the general results to this case. The main text describes relativity as a physical theory, with applications to astrophysics and cosmology. It takes the reader beyond traditional courses on relativity through in-depth descriptions of inhomogeneous cosmological models and the Kerr metric. Emphasis is given to complete and clear derivations of the results, enabling readers to access research articles published in relativity journals.

JERZY PLEBAŃSKI (1928–2005) was a Polish theoretical physicist best known for his extensive research into general relativity, nonlinear electrodynamics and mathematical physics. He split his time between Warsaw, Poland and Mexico, his permanent residence from the mid-1970s onwards. He is remembered, among other things, for defining the algebraic classification of the tensor of matter, for finding new solutions of the Einstein equations (for example, the Plebański–Demiański metric), formulation of the heavenly equations and the effective field theory relating GR and supergravity, known as Plebański action. The first part of the book is developed from Plebański's lecture notes.

ANDRZEJ KRASIŃSKI is Emeritus Professor at the Nicolaus Copernicus Astronomical Center, Polish Academy of Sciences in Warsaw. He served for many years on the Editorial Board of the journal *General Relativity and Gravitation* and has acted as Poland's elected representative on the International Committee of the International Society for General Relativity and Gravitation. He is author of *Inhomogeneous Cosmological Models* (Cambridge, 1997), co-author of *Structures in the Universe by Exact Methods* (Cambridge, 2009) and co-editor of *Golden Oldies in General Relativity* (Springer, 2013). His research record includes many papers on the interpretation of inhomogeneous cosmological models. He is also a co-author of the computer program, Ortocartan, for algebraic calculations in general relativity.

AN INTRODUCTION TO GENERAL RELATIVITY AND COSMOLOGY

JERZY PLEBAŃSKI

National Polytechnic Institute of Mexico

ANDRZEJ KRASIŃSKI

Nicolaus Copernicus Astronomical Center, Polish Academy of Sciences

CAMBRIDGE
UNIVERSITY PRESS

Shaftesbury Road, Cambridge CB2 8EA, United Kingdom

One Liberty Plaza, 20th Floor, New York, NY 10006, USA

477 Williamstown Road, Port Melbourne, VIC 3207, Australia

314–321, 3rd Floor, Plot 3, Splendor Forum, Jasola District Centre,
New Delhi – 110025, India

103 Penang Road, #05–06/07, Visioncrest Commercial, Singapore 238467

Cambridge University Press is part of Cambridge University Press & Assessment,
a department of the University of Cambridge.

We share the University's mission to contribute to society through the pursuit of
education, learning and research at the highest international levels of excellence.

www.cambridge.org
Information on this title: www.cambridge.org/9781009415620

DOI: 10.1017/9781009415651

First published 2006
Second edition published 2024

A catalogue record for this publication is available from the British Library

A Cataloging-in-Publication data record for this book is available from the Library of Congress

ISBN 978-1-009-41562-0 Hardback

Contents

The scope of this text	*page* xiii	
Preface to the second edition	xv	
Acknowledgements	xviii	
1	**How the theory of relativity came into being (a brief historical sketch)**	1
	1.1 Special versus general relativity	1
	1.2 Space and inertia in Newtonian physics	1
	1.3 Newton's theory and the orbits of planets	2
	1.4 The basic assumptions of general relativity	3
	Part I Elements of differential geometry	7
2	**A short sketch of 2-dimensional differential geometry**	9
	2.1 Constructing parallel straight lines in a flat space	9
	2.2 Generalisation of the notion of parallelism to curved surfaces	10
3	**Tensors, tensor densities**	12
	3.1 What are tensors good for?	12
	3.2 Differentiable manifolds	12
	3.3 Scalars	14
	3.4 Contravariant vectors	14
	3.5 Covariant vectors	14
	3.6 Tensors of second rank	15
	3.7 Tensor densities	16
	3.8 Tensor densities of arbitrary rank	16
	3.9 Algebraic properties of tensor densities	16
	3.10 Mappings between manifolds	18
	3.11 The Levi-Civita symbol	20
	3.12 Multidimensional Kronecker deltas	21
	3.13 Examples of applications of the Levi-Civita symbol and of the multidimensional Kronecker delta	22
	3.14 Exercises	23
4	**Covariant derivatives**	24
	4.1 Differentiation of tensors	24

	4.2	Axioms of the covariant derivative	26
	4.3	A field of bases on a manifold and scalar components of tensors	27
	4.4	The affine connection	28
	4.5	The explicit formula for the covariant derivative of tensor density fields	28
	4.6	Exercises	30
5	**Parallel transport and geodesic lines**		**31**
	5.1	Parallel transport	31
	5.2	Geodesic lines	32
	5.3	Exercises	33
6	**The curvature of a manifold; flat manifolds**		**34**
	6.1	The commutator of second covariant derivatives	34
	6.2	The commutator of directional covariant derivatives	35
	6.3	The relation between curvature and parallel transport	36
	6.4	Covariantly constant fields of vector bases	41
	6.5	A torsion-free flat manifold	41
	6.6	Parallel transport in a flat manifold	41
	6.7	Geodesic deviation	42
	6.8	Algebraic and differential identities obeyed by the curvature tensor	43
	6.9	Exercises	44
7	**Riemannian geometry**		**45**
	7.1	The metric tensor	45
	7.2	Riemann spaces	45
	7.3	The signature of a metric, degenerate metrics	46
	7.4	Christoffel symbols	47
	7.5	The curvature of a Riemann space	48
	7.6	Flat Riemann spaces	49
	7.7	Subspaces of a Riemann space	49
	7.8	Flat Riemann spaces that are globally non-Euclidean	50
	7.9	The Riemann curvature versus the normal curvature of a surface	51
	7.10	The geodesic line as the line of extremal distance	51
	7.11	Mappings between Riemann spaces	52
	7.12	Conformally related Riemann spaces	53
	7.13	Conformal curvature	54
	7.14	Timelike, null and spacelike intervals in a 4-dimensional spacetime	56
	7.15	Embeddings of Riemann spaces in Riemann spaces of higher dimension	58
	7.16	The Petrov classification	65
	7.17	Exercises	67
8	**Symmetries of Riemann spaces, invariance of tensors**		**69**
	8.1	Symmetry transformations	69
	8.2	The Killing equations	69
	8.3	The connection between generators and the invariance transformations	72
	8.4	Finding the Killing vector fields	72
	8.5	A Killing vector field along a geodesic is a geodesic deviation field	74

8.6	Invariance of other tensor fields	74
8.7	The Lie derivative	75
8.8	The algebra of Killing vector fields	76
8.9	Surface-forming vector fields	76
8.10	Spherically symmetric 4-dimensional Riemann spaces	77
8.11	* Conformal Killing fields and their finite basis	81
8.12	* The maximal dimension of an invariance group	83
8.13	Exercises	85
9	**Methods to calculate the curvature quickly: differential forms and algebraic computer programs**	**89**
9.1	The basis of differential forms	89
9.2	The connection forms	90
9.3	The Riemann tensor	91
9.4	Using computers to calculate the curvature	92
9.5	Exercises	93
10	**The spatially homogeneous Bianchi-type spacetimes**	**94**
10.1	The Bianchi classification of 3-dimensional Lie algebras	94
10.2	The dimension of the group versus the dimension of the orbit	99
10.3	Action of a group on a manifold	99
10.4	Groups acting transitively, homogeneous spaces	100
10.5	Invariant vector fields	100
10.6	The metrics of the Bianchi-type spacetimes	102
10.7	The isotropic Bianchi-type (Robertson–Walker) spacetimes	103
10.8	Exercises	106
11	***The Petrov classification by the spinor method**	**108**
11.1	What is a spinor?	108
11.2	Translating spinors to tensors and vice versa	109
11.3	The spinor image of the Weyl tensor	111
11.4	The Petrov classification in the spinor representation	111
11.5	The Weyl spinor represented as a 3×3 complex matrix	112
11.6	The equivalence of the Penrose classes to the Petrov classes	114
11.7	The Petrov classification by the Debever method	115
11.8	Exercises	116
	Part II The theory of gravitation	**121**
12	**The Einstein equations and the sources of a gravitational field**	**123**
12.1	Why Riemannian geometry?	123
12.2	Local inertial frames	123
12.3	Trajectories of free motion in Einstein's theory	124
12.4	Special relativity versus gravitation theory	127
12.5	The Newtonian limit of general relativity	127
12.6	Sources of the gravitational field	128
12.7	The Einstein equations	128

12.8   Hilbert's derivation of the Einstein equations   129
12.9   The Palatini variational principle   133
12.10 The asymptotically Cartesian coordinates and the asymptotically flat
       spacetime   133
12.11 The Newtonian limit of Einstein's equations   133
12.12 Examples of sources in the Einstein equations: perfect fluid and dust   137
12.13 Equations of motion of a perfect fluid   139
12.14 The cosmological constant   140
12.15 An example of an exact solution of Einstein's equations: a Bianchi type
       I spacetime with dust source   141
12.16 * Other gravitation theories   144
       12.16.1 The Brans–Dicke theory   144
       12.16.2 The Bergmann–Wagoner theory   145
       12.16.3 The Einstein–Cartan theory   145
       12.16.4 The bi-metric Rosen theory   146
12.17 Matching solutions of Einstein's equations   146
12.18 The weak-field approximation to general relativity   148
12.19 Exercises   154

**13 The Maxwell and Einstein–Maxwell equations
     and the Kaluza–Klein theory**   156
13.1   The Lorentz-covariant description of electromagnetic field   156
13.2   The covariant form of the Maxwell equations   156
13.3   The energy-momentum tensor of electromagnetic field   157
13.4   The Einstein–Maxwell equations   158
13.5   * The variational principle for the Maxwell and Einstein–Maxwell
       equations   158
13.6   * The Kaluza–Klein theory   159
13.7   Exercises   161

14 Spherically symmetric gravitational fields of isolated objects   162
14.1   The curvature coordinates   162
14.2   Symmetry inheritance   165
14.3   Spherically symmetric electromagnetic field in vacuum   166
14.4   The Schwarzschild and Reissner–Nordström solutions   166
14.5   Orbits of planets in the gravitational field of the Sun   169
14.6   Deflection of light rays in the Schwarzschild field   175
14.7   Measuring the deflection of light rays   178
14.8   Gravitational lenses   180
14.9   The spurious singularity of the Schwarzschild solution at $r = 2m$ 182
14.10 * Embedding the Schwarzschild spacetime in a flat Riemannian space   187
14.11 Interpretation of the spurious singularity at $r = 2m$; black holes   191
14.12 The Schwarzschild metric in other coordinate systems   193
14.13 The equation of hydrostatic equilibrium   194
14.14 The 'interior Schwarzschild solution'   196

14.15 * The maximal analytic extension of the Reissner–Nordström metric   197

14.16 Motion of particles in the Reissner–Nordström spacetime with $e^2 < m^2$ 205

14.17 Exercises   209

15 Relativistic hydrodynamics and thermodynamics   211

15.1 Motion of a continuous medium in Newtonian hydrodynamics   211

15.2 Motion of a continuous medium in relativistic hydrodynamics   213

15.3 The equations of evolution of θ, $\sigma_{\alpha\beta}$, $\omega_{\alpha\beta}$ and $\dot{u}^\alpha$; the Raychaudhuri equation   216

15.4 Singularities and singularity theorems   218

15.5 Relativistic thermodynamics   219

15.6 Exercises   222

16 Relativistic cosmology I: general geometry   223

16.1 A continuous medium as a model of the Universe   223

16.2 The geometric optics approximation   224

16.3 The redshift   226

16.4 The optical tensors   228

16.5 The apparent horizon   230

16.6 * The double-null tetrad   231

16.7 * The equations of propagation of the optical scalars   233

16.8 * The Goldberg–Sachs theorem   235

16.9 * The area distance   240

16.10 * The reciprocity theorem   243

16.11 Other observable quantities   246

16.12 The Fermi–Walker transport   246

16.13 Position drift of light sources   248

16.14 Exercises   253

17 Relativistic cosmology II: the Robertson–Walker geometry   255

17.1 The Robertson–Walker metrics as models of the Universe   255

17.2 Optical observations in an R–W universe   257

     17.2.1 The redshift   257

     17.2.2 The redshift–distance relation   258

     17.2.3 Number counts   259

17.3 The Friedmann equation   260

17.4 The Friedmann solutions with $\Lambda = 0$ 262

17.5 The redshift–distance relation in the $\Lambda = 0$ Friedmann models   263

17.6 The Newtonian cosmology   264

17.7 The Friedmann solutions with the cosmological constant   266

17.8 The ΛCDM model   269

17.9 The redshift–distance relation in the $\Lambda \neq 0$ Friedmann models   271

17.10 The redshift drift: a test for accelerating expansion   272

17.11 Horizons in the Robertson–Walker models   273

17.12 The inflationary models and the 'problems' they solved   277

17.13 The value of the cosmological constant   282

17.14   The 'history of the Universe'   282
17.15   Invariant definitions of the Robertson–Walker models   285
17.16   Different representations of the R–W metrics   286
17.17   Exercises   287

18   Relativistic cosmology III: the Lemaître–Tolman geometry   289
18.1   The comoving-synchronous coordinates   289
18.2   The spherically symmetric inhomogeneous models   289
18.3   The Lemaître–Tolman model   291
18.4   Conditions of regularity at the centre   295
18.5   Formation of voids in the Universe   296
18.6   Formation of other structures in the Universe   297
        18.6.1   Density to density evolution   298
        18.6.2   Velocity to density evolution   300
        18.6.3   Velocity to velocity evolution   302
18.7   The influence of cosmic expansion on planetary orbits   303
18.8   * The apparent horizons for a central observer in L–T models   305
18.9   * Black holes in the evolving Universe   309
18.10   * Shell crossings and necks/wormholes   314
        18.10.1 $E < 0$ 318
        18.10.2 $E = 0$ 319
        18.10.3 $E > 0$ 320
        18.10.4 Final comment   320
18.11   The redshift along radial rays   320
18.12   The blueshift   322
18.13   * Apparent horizons for noncentral observers   326
18.14   The influence of inhomogeneities in matter distribution on the cosmic
        microwave background radiation   327
18.15   Matching the L–T models to the Schwarzschild and Friedmann solutions   330
18.16   * The shell focussing singularity   331
18.17   * Extending an L–T spacetime through a shell crossing singularity   334
18.18   * Singularities and cosmic censorship   336
18.19   Solving the 'horizon problem' without inflation   341
18.20   * The evolution of $R(t, M)$ versus the evolution of $\rho(t, M)$ 343
18.21   * Increasing and decreasing density perturbations   344
18.22   Mimicking accelerating expansion of the Universe by inhomogeneities
        in matter distribution   347
18.23   Drift of light rays   349
18.24   * L&T curio shop   351
        18.24.1 Lagging cores of the Big Bang   351
        18.24.2 Strange or nonintuitive properties of the L–T model   352
        18.24.3 Chances to fit an L–T model to observations   355
        18.24.4 An 'in one ear and out the other' Universe   355
        18.24.5 A 'string of beads' Universe   357

18.24.6 Uncertainties in inferring the spatial distribution of matter 358

18.24.7 Is the distribution of matter in our Universe fractal? 360

18.24.8 General results related to the L–T models 360

18.25 Exercises 361

19 Relativistic cosmology IV: simple generalisations of L–T and related geometries 366

19.1 The plane- and hyperbolically symmetric spacetimes 366

19.2 G_3/S_2-symmetric dust solutions with $R_{,r} \neq 0$ 367

19.3 Plane symmetric dust solutions with $R_{,r} \neq 0$ 368

19.4 G_3/S_2-symmetric dust in electromagnetic field, the case $R_{,r} \neq 0$ 370

19.4.1 Integrals of the field equations 370

19.4.2 Matching the charged dust metric to the Reissner–Nordström metric 374

19.4.3 Prevention of the Big Crunch singularity by electric charge 375

19.4.4 * Charged dust in curvature and mass-curvature coordinates 378

19.4.5 Regularity conditions at the centre 380

19.4.6 * Shell crossings in charged dust 381

19.5 The Datt–Ruban solution 383

19.6 Exercises 390

20 Relativistic cosmology V: the Szekeres geometries 391

20.1 The Szekeres–Szafron family of metrics 391

20.1.1 The $\beta_{,z} = 0$ subfamily 392

20.1.2 The $\beta_{,z} \neq 0$ subfamily 396

20.1.3 Interpretation of the Szekeres–Szafron coordinates 397

20.1.4 Common properties of the two subfamilies 399

20.1.5 * The invariant definitions of the Szekeres–Szafron metrics 401

20.2 The Szekeres solutions and their properties 402

20.2.1 The $\beta_{,z} = 0$ subfamily 402

20.2.2 The $\beta_{,z} \neq 0$ subfamily 403

20.2.3 * The $\beta_{,z} = 0$ family as a limit of the $\beta_{,z} \neq 0$ family 404

20.3 Properties of the quasi-spherical Szekeres solutions with $\beta_{,z} \neq 0 = \Lambda$ 405

20.3.1 Basic physical restrictions 406

20.3.2 The significance of $\mathcal{E}$ 406

20.3.3 Conditions of regularity at the origin 409

20.3.4 Shell crossings 412

20.3.5 Regular maxima and minima 415

20.3.6 The mass dipole 416

20.3.7 * The absolute apparent horizon 418

20.3.8 * The apparent horizon and its relation to the AAH 422

20.3.9 * Which is the true horizon – the AH or the AAH? 427

20.4 * The Goode–Wainwright representation of the Szekeres solutions 429

20.5 Selected interesting subcases of the Szekeres–Szafron family 433

20.5.1 The Szafron–Wainwright model 433

	20.5.2 The toroidal universe of Senin	435
20.6	Selected further reading on the Szekeres models	438
20.7	Exercises	440

21   The Kerr metric   442
21.1   The Kerr–Schild metrics   442
21.2   The derivation of the Kerr metric by the original method   444
21.3   Basic properties   449
21.4   * Derivation of the Kerr metric by Carter's method – from the
         separability of the Klein–Gordon equation   454
21.5   The event horizons and the stationary limit hypersurfaces   459
21.6   The Hamiltonian and the Poisson bracket   464
21.7   General geodesics   465
21.8   Geodesics in the equatorial plane   468
21.9   *The maximal analytic extension of the Kerr metric   475
21.10 * The Penrose process   485
21.11 Stationary–axisymmetric spacetimes and locally nonrotating observers 487
21.12 * Ellipsoidal spacetimes   490
21.13 A Newtonian analogue of the Kerr solution   492
21.14 A source of the Kerr field?   493
21.15 Exercises   493

22   Relativity enters technology: the Global Positioning System   499
22.1   Purpose and setup   499
22.2   The principle of position determination   500
22.3   The reference frames and the Sagnac effect   500
22.4   Earth's gravitation and the SI time units   502
22.5   Selected corrections of the orbits of the GPS satellites   504
         22.5.1 Corrections for gravity and velocity   504
         22.5.2 The eccentricity correction   507
22.6   The 9 largest relativistic effects in the GPS   507
22.7   Exercises   509

23   Subjects omitted from this book   511

24   Comments to selected exercises and calculations   514
24.1   Exercise 1 to Chapter 14   514
24.2   Exercise 14 to Chapter 14   515
24.3   Verifying Eqs. (19.35) with (19.31) and (19.32) with (19.28)   516
24.4   Verifying the Einstein equations (20.2), (20.9) and (20.11)   519
24.5   Equation (20.179) defines η at the AAH uniquely   521
24.6   The four curves in Fig. 20.4 meet at one point   521
24.7   The discarded case in Eqs. (20.2)–(20.11)   522
24.8   Hints for verifying Eq. (21.28)   526
References   527
Index   546

The scope of this text

General relativity is the currently accepted theory of gravitation. Under this heading one could include a huge amount of material. For the needs of this theory an elaborate mathematical apparatus was created. It has partly become a self-standing sub-discipline of mathematics and physics, and it keeps developing, providing input or inspiration to physical theories that are being newly created (such as gauge field theories, supergravitation and the brane-world theories). From the gravitation theory, descriptions of astronomical phenomena taking place in strong gravitational fields and in large-scale sub-volumes of the Universe are derived. This part of gravitation theory develops in connection with results of astronomical observations. For the needs of this area, another sophisticated formalism was created (the Parametrised Post-Newtonian, PPN, formalism). Finally, some tests of the gravitation theory can be carried out in laboratories, either terrestrial or orbital. These tests, their improvements and projects of further tests have led to developments in mathematical methods and in technology that are by now an almost separate branch of science – as an example, one can mention here the search for gravitational waves and the calculations of properties of the wave signals to be expected.

In this situation, no single textbook can attempt to present the whole of gravitation theory, and the present text is no exception. We made the working assumption that relativity is part of physics (this view is not universally accepted!). The purpose of this course is to present those results that are most interesting from the point of view of a physicist, and were historically the most important. We are going to lead the reader through the mathematical part of the theory by a rather short route, but in such a way that he/she does not have to take anything on our word, is able to verify every detail and, after reading the whole text, will be prepared to solve several problems by him/herself. Further help in this should be provided by the exercises in the text and the literature recommended for further reading.

The introductory part (Chapters 1–7), although assembled by J. Plebański long ago, has never been published in book form.[1] It differs from other courses on relativity in that it introduces differential geometry by a top-down method. It begins with general manifolds, on which no structures except tensors are defined, and discusses their basic properties. Then it adds the notion of the covariant derivative and affine connection,

---

[1] A part of that material had been semi-published as copies of typewritten notes (Plebański, 1964).

without introducing the metric yet, and again proceeds as far as possible. At that level it defines geodesics via parallel displacement and presents the properties of curvature. Only at this point it introduces the metric tensor and the (pseudo) Riemannian geometry and specialises the results derived earlier to this case. Then it proceeds to the presentation of more detailed topics, such as symmetries, the Bianchi classification and the Petrov classification.

Some of the chapters on classical relativistic topics contain material that, to the best of our knowledge, has never been published in any textbook. In particular, this applies to Chapter 8 (on symmetries) and to Chapter 16 (on cosmology with general geometry). Chapters 18–20 (on inhomogeneous cosmologies) are entirely based on original papers. Parts of Chapters 18, 19 and 20 cover the material introduced in A. K.'s monograph on inhomogeneous cosmological models (Krasiński, 1997). However, the presentation here was thoroughly rearranged, extended and brought up to date. We no longer intended to briefly mention all contributions to the subject; rather, we have placed the emphasis on complete and clear derivations of the most important results. That material has so far existed only in scattered journal papers and has been assembled into a textbook for the first time (the monograph by Krasiński, 1997, was only a concise review). Taken together, this collection of knowledge constitutes an important and interesting part of relativistic cosmology whose meaning has, unfortunately, not yet been appreciated properly by the astronomical community.

Most figures for this text, even when they look the same as the corresponding figures in the published papers, were newly generated by A. K. using the program Gnuplot, sometimes on the basis of numerical calculations programmed in Fortran 90. The only figures taken verbatim from other sources are those that illustrated the joint papers of A. K. with C. Hellaby and K. Bolejko, and of N. Ashby in Chapter 22. The latter are reproduced here with the permission of the author and of N. Dadhich as a representative of the publisher, Inter-University Centre for Astronomy and Astrophysics in Pune, India.

J. Plebański kindly agreed to be included as a co-author of this text – having done his part of the job long ago. Unfortunately, he was not able to participate in the writing up and proofreading. He died while the first edition of this book was being edited. Therefore, the second author (A. K.) is exclusively responsible for any errors.

Note for the reader. Some parts of this book may be skipped on first reading, since they are not necessary for understanding the material that follows. They are marked by asterisks. Chapters 18–20 are expected to be the highlights of this book. However, they go far beyond standard courses of relativity and may be skipped by those readers who wish to remain on the well-beaten track. Hesitating readers may read on, but can skip the sections marked by asterisks.

Andrzej Krasiński
Warsaw, September 2005 and May 2023

Preface to the second edition

The main reason why a new edition of this book was needed is that during the 17 years after the first edition several little errors have been discovered in it. Mostly, they are innocent typos that can be corrected by a careful reader (most of them were discovered by my then-student, Mr Przemysław Jacewicz, to whom I express my gratitude). However, a few required significant changes. The list of those changes is given further below.

By this opportunity, a few sections and a new chapter were added. The added sections describe the developments in understanding the cosmological implications of the Lemaître–Tolman and Szekeres models that occurred after 2006. The added Chapter 22, on relativistic effects in the Global Positioning System, is a thought-provoking demonstration that general relativity became a necessary component in today's down-to-Earth technology.

A few reviewers of the first edition complained that this book is not really an introduction to relativity and cosmology, but rather an advanced-level monograph, so its title is misleading. Still, together with the publisher, we decided to keep the 'introduction' in the title, for the following reasons:

1. No prior familiarity of the reader with general relativity and differential geometry is assumed in this book. It takes a careful reader to some height of advancement, but it begins at a very basic level.
2. The book omits several topics (see Chapter 23). For example, the reviewers of the first edition complained about the omission of perturbative methods in cosmology. Consequently, this is not a complete monograph. To become an expert in relativity the reader will have to continue his/her reading with other books. The reason for the omissions was that we intended to make our readers familiar with the mathematical and physical basics of relativity, leaving out those applications that require extended additional study. *Ergo*, our book is not an encyclopaedia of the subject, but ... well, an introduction.

My co-author and former master, Jerzy Plebański, died in 2005. But the part of this book that is directly based on his works and on my notes to his lectures is still a solid fundament of the whole.

The list of significant changes from the first edition

Several figures were redrawn to make them graphically clearer or to make the symbols in them consistent with the text.

Section 9.4 (on algebraic computing) was shortened because algebraic computing is no longer a novelty.

Section 10.5 (on invariant vector fields) was re-edited for better clarity.

Chapter 11 (on spinors) was re-edited and rearranged to improve the clarity of presentation, and extended hints were added to most exercises.

Section 12.16 (on other theories of gravitation) was re-edited.

Section 12.18 (on the weak-field approximation to general relativity) was re-edited to improve the clarity of presentation, and the information about the Gravity Probe B experiment was updated.

Sections 14.6 (on measuring the deflection of light rays) and 14.8 (on gravitational lenses) were re-edited to make them more up to date, and photographs of two gravitational lenses were added at the end of Sec. 14.8.

A part of Chapter 16 (on cosmology in general geometry) was reshuffled to achieve a more logical ordering of the exposition.

Chapter 17 (on cosmology in Robertson–Walker geometry) was substantially re-edited and updated in several places.

Chapter 18 (on cosmology in Lemaître–Tolman geometry) was re-edited in several places. In particular,

- Section 18.18 (on singularities and cosmic censorship) was re-edited and shortened.
- The old Subsection 18.20.2 (now 18.24.2) had to be partly corrected, but the conclusion remains unchanged.

The old Chapter 19 was split into two. The old Sections 19.1–19.5 became the new Chapter 19 under the title 'Relativistic cosmology IV: simple generalisations of L–T and related geometries'. By this opportunity, it was extended for results obtained after the first edition (see below). The remaining part of the old Chapter 19 became Chapter 20 with the title: 'Relativistic cosmology V: the Szekeres geometries'.

Consequently, the old Chapter 20 (on the Kerr metric) became Chapter 21.

The old Sec. 19.5.3 (now Sec. 20.1.3) was corrected at several places.

The old Subsection 19.6.3 (now Subsection 20.2.3) was completely rewritten to get rid of several errors.

The old Subsection 19.7.6 (now Subsection 20.3.7) was completely rewritten to put its terminology into agreement with the existing literature.

The old Subsection 19.7.7 (on nonexistent Szekeres wormholes) was removed.

The old Section 19.10, containing the proof that one subcase in solving the Einstein equations for the Szekeres-type metrics leads to a contradiction, was transferred to the new Chapter 24 and is now Section 24.7. It contains a large amount of calculations with a rather disappointing end result.

Chapter 21 (old number 20) was re-edited in several places to improve the clarity of presentation.

The following new material was added:

(i) A new Section 8.5 on the relation between the Killing vectors and geodesic deviation.

(ii) New Exercises 9 and 14 at the end of Chapter 8.

(iii) Several new exercises at the end of Chapter 12.

(iv) At the end of Chapter 16:

- A derivation of the formula for the Fermi–Walker transport and references to the original Fermi and Walker papers.
- A derivation of the formula for the position drift of a light source in a general geometry, following the paper by M. Korzyński and J. Kopiński (2018).

(v) In Chapter 17, the following three new sections:

- Section 17.8, introducing the ΛCDM model.
- Section 17.9, with the derivation of the distance–redshift relation in the Friedmann models with Λ.
- Section 17.10, on the redshift drift in Robertson–Walker models.

(vi) Several new exercises at the end of Chapter 17.

(vii) Section 18.12, explaining the reasons why some objects in the Universe may display blueshift rather than redshift.

(viii) Section 18.13, describing how apparent horizons in L–T models are radically different for noncentral observers.

(ix) At the end of Chapter 18 – a new section that demonstrates how the accelerating expansion of the Universe can be mimicked by inhomogeneities in mass distribution.

(x) Also at the end of Chapter 18 – an exemplary illustration of the rate of change of position of a distant galaxy on the sky caused by the drift described in the newly added section at the end of Chapter 16.

(xi) A new Section 19.3, in which it is argued that spaces of constant time in plane symmetric spacetimes are most naturally interpreted as flat tori.

(xii) At the end of the new Sec. 19.4.6 – the demonstration how a charged dust ball can be evolved through the minimal size, only to hit a shell crossing singularity after that.

(xiii) Second half of the new Section 19.5 – a discussion of evolution of the charged Ruban (1972, 1983) solution matched to Reissner–Nordström (RN).

(xiv) At the end of the new Section 19.5 – spacetime diagrams of the configuration mentioned above and of the neutral Ruban (1968, 1969) model matched to Schwarzschild.

(xv) The new Subsections 20.3.8 and 20.3.9, in which the 'absolute apparent horizon' (AAH) and the ordinary apparent horizon (AH) of the central observer in the quasi-spherical Szekeres spacetimes are compared.

(xvi) At the end of the new Chapter 20, a new Section 20.6 with short descriptions of a few papers published after the first edition of this book that contain instructive contributions to the Szekeres geometries.

(xvii) Several new exercises to Chapter 21 (old Chapter 20).

(xviii) A new Chapter 22 on the relativistic effects in the Global Positioning System.

(xix) Another new Chapter 24 that contains detailed hints on how to solve the more difficult exercises and verify complicated calculations.

Andrzej Krasiński
Warsaw, 2023

Acknowledgements

We thank Charles Hellaby for comments on the various properties of the Lemaître–Tolman models and for providing copies of his unpublished works on this subject. Some of the figures used in this text were copied from C. Hellaby's files, with his permission. We are grateful to Pankaj S. Joshi for helpful comments on cosmic censorship and singularities, and to Amos Ori for clarifying the matter of shell crossings in charged dust. The correspondence with Amos significantly contributed to clarifying several points in Section 19.4. We are also grateful to George Ellis for his very useful comments on the first draft of this book. We thank Bogdan Mielnik and Maciej Przanowski, who were of great help in the difficult communication between one of the authors residing in Poland and the other in Mexico. M. Przanowski had carefully proofread a large part of this text and caught several errors. So did Krzysztof Bolejko, who was the first reader of this text even before it was typed into a computer file. J. P. acknowledges the support from the Consejo Nacional de Ciencia y Teconología projects 32427E and 41993F.

In preparing the second edition, the second author (A. K.) received valuable help from Leszek Sokołowski concerning the variational principle for the Einstein–Maxwell equations. Mauro Carfora and Bob Jantzen helped with locating the original papers on the Fermi–Walker transport. Marie-Noelle Célérier helped with finding references to Michel Chasles papers. Neil Ashby kindly sent me the eps files of the figures illustrating his papers and allowed me to use them in this book. Consultations by Kayll Lake, George Ellis, Roy Maartens and Mikołaj Korzyński are also gratefully acknowledged.

1

How the theory of relativity came into being (a brief historical sketch)

1.1 Special versus general relativity

The name 'relativity' covers two physical theories. The older one, called special relativity, published in 1905, is a theory of electromagnetic and mechanical phenomena taking place in reference systems that move with large velocities relative to an observer, but are not influenced by gravitation. It is considered to be a closed theory. Its parts had entered the basic courses of classical mechanics, quantum mechanics and electrodynamics. Students of physics study these subjects before they begin to learn general relativity. Therefore, we shall not deal with special relativity here. Familiarity with it is, however, necessary for understanding the general theory. The latter was published in 1915. It describes the properties of time and space, and mechanical and electromagnetic phenomena in the presence of gravitational field.

1.2 Space and inertia in Newtonian physics

In the Newtonian mechanics and gravitation theory the space was just a background – a room to be filled with matter. It was considered obvious that the space is Euclidean. The masses of matter particles were considered their internal properties independent of any interactions with the remaining matter. However, from time to time it was suggested that not all of the phenomena in the Universe can be explained using such an approach. The best known among those concepts was the so-called Mach principle. This approach was made known by Ernst Mach in the second half of the nineteenth century, but had been originated by the English philosopher Bishop George Berkeley, in 1710, while Newton was still alive. Mach began with the following observation: in the Newtonian mechanics a seemingly obvious assumption is tacitly made, namely that all the space points can be labelled, for example by assigning Cartesian coordinates to them. One can then observe the motion of matter by finding in which point of space a given particle is located at a given instant. However, this is not actually possible. If we accept another basic assumption of Newton, namely that the space is Euclidean, then its points do not differ from one another in any way. They can be labelled only by matter being present in the space. In truth, we thus can observe only the motion of one portion of matter relative to another portion of matter. Hence, a correctly formulated theory should speak only about relative motion

1

(of matter relative to matter), not about absolute motion (of matter relative to space). If this is so, then the motion of a single particle in a totally empty Universe would not be detectable. Without any other matter we could not establish whether the lone particle is at rest, or is moving or experiencing acceleration. But the reaction of matter to acceleration is the only way to measure its inertia. Hence, that lone particle would have zero inertia. It follows then that inertia is, likewise, not an absolute property of matter, but is induced by the remaining matter in the Universe, supposedly via the gravitational interaction.

One can question this principle in several ways. No-one will ever be able to find him/herself in an empty Universe, so any theorems on such an example cannot be verified. It is possible that the inertia of matter is a 'stronger' property than the homogeneity of space, and would still exist in an empty Universe, thus making it possible to measure absolute acceleration. Criticism of Mach's principle is made easier by the fact that it has never been formulated as a precise physical theory. It is just a collection of critical remarks and suggestions, partly based on calculations. It happens sometimes, though, that a new way of looking at an old theory, even if not sufficiently justified, becomes a starting point for meaningful discoveries. This was the case with Mach's principle that inspired Einstein at the starting point of his work.

1.3 Newton's theory and the orbits of planets

In addition to the above-mentioned theoretical problem, Newton's theory had a serious empirical problem. It was known already in the first half of the nineteenth century that the planets revolve around the Sun in orbits that are not exactly elliptic. The real orbits are rosettes – curves that can be imagined as follows: let a point go around an ellipse, but at the same time let the ellipse rotate slowly around its focus in the same direction (see Fig. 1.1). Newton's theory explained this as follows: an orbit of a planet is an exact ellipse only if we assume that the Sun has just one planet.[1] Since the Sun has several planets, they interact gravitationally and mutually perturb their orbits. When these perturbations are taken into account, the effect is *qualitatively* the same as observed.

However, in 1859, Urbain J. LeVerrier (the same person who, a few years earlier, had predicted the existence of Neptune on the basis of similar calculations) verified whether the calculated and observed motions of Mercury's perihelion agree. It turned out that they do not – and the discrepancy was much larger than the observation error. The calculated rate of perihelion shift was smaller than the one observed by 43 arc seconds per century (the modern value is $43.11 \pm 0.45''$ per century (Will, 2018)). Astronomers and physicists tried to explain this effect in various simple ways, e.g. by assuming that yet another planet, called Vulcan, revolves around the Sun inside Mercury's orbit and perturbs it; by allowing for gravitational interaction of Mercury with the interplanetary dust; or by assuming that the Sun is flattened in consequence of its rotation. In the last case, the gravitational field of the Sun would not be spherically symmetric, and a sufficiently large flattening would

---

[1] More assumptions were actually made, but the other ones seemed so obvious at that time that they were not even mentioned: that the Sun is exactly spherical, and that the space around the Sun is exactly empty. None of these is strictly correct, but the departures of observations from theory caused by the nonsphericity of the Sun and by the interplanetary matter are insignificant.

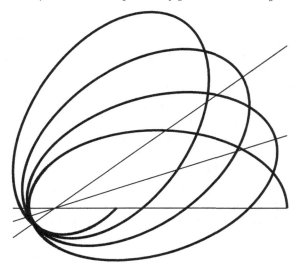

Fig. 1.1 Real planetary orbits, in consequence of various perturbations, are not ellipses but non-closed curves. The perihelion shift in this figure is greatly exaggerated. In reality, the greatest angle of perihelion shift observed in the Solar System, for Mercury, equals approximately 1.5° per 100 years.

explain the additional rotation of Mercury's perihelion. All these hypotheses did not pass the observational tests. The hypothetical planet Vulcan would have to be so massive that it would be visible in telescopes, but wasn't. There was not enough interplanetary dust to cause the observed effect. The Sun, if it were sufficiently flattened to explain Mercury's motion, would cause yet another effect: the planes of the planetary orbits would swing periodically around their mean positions with an amplitude of about 43″ per century, and that motion would have been observed, but wasn't (Dicke, 1964).

In spite of these difficulties, nobody doubted the correctness of Newton's theory. The general opinion was that Mach's critique would be answered by formal corrections in the theory, and the anomalous perihelion motion of Mercury would be explained by new observational discoveries. Nobody expected that any other gravitation theory could replace Newton's, which had been going from one success to another for over 200 years. General relativity was not created in response to experimental or observational needs. It resulted from speculation, it preceded all but one of the experiments and observations that confirmed it, and it became broadly testable only about 50 years after it had been created, in the 1960s. So much time did technology need to catch up and go beyond the opportunities provided by astronomical phenomena.

1.4 The basic assumptions of general relativity

It is interesting to follow the development of relativistic ideas in the same order as that in which they actually appeared in the literature. However, this was not a straight and smooth road. Einstein made a few mistakes and put forward a few hypotheses that he had to revoke later. He had been constructing the theory gradually, while at the same time learning the Riemannian geometry – the mathematical basis of relativity. If we followed that gradual progress, we would have to take into account not only some blind paths but also competitors of Einstein, some of whom questioned the need for the (then) new theory, while some others tried to get ahead of Einstein, but without success (Mehra, 1974). Learning relativity in this way would not be efficient, so we will take a shortcut. We shall begin by justifying the need for this theory, then we shall present the basic elements of Riemann's geometry and then we will present Einstein's theory in its final shape. The history of relativity's taking shape is presented in Mehra's book (Mehra, 1974), and its original presentation is to be found in the collection of classic papers (Einstein et al., 1923).

Einstein's starting point was a critique of Newton's theory based on Mach's ideas. Newtonian physics said that in a space free of any interactions, material bodies would either remain at rest or would move by uniform rectilinear motion. Since, however, the real Universe is permeated by gravitational fields that cannot be shielded, all bodies in the Universe move on curved trajectories in consequence of gravitational interactions.

There is a problem here. When we say that a trajectory is curved, we assume that we can define a straight line. But how can we do this when no real body follows a straight line? The terrestrial standards of straight lines are useful only because no distances on the Earth are truly great, and at short distances the deformation of 'rigid' bodies due to gravitation is unmeasurably small. Maybe then the trajectory of a light ray would be a good model of a straight line?

To see whether this could be the case, consider two Cartesian reference systems K and K', whose axes (x, y, z) and (x', y', z') are, respectively, parallel. Let K be inertial, and let K' move with respect to K along the z-axis with acceleration $g(t, x, y, z)$. Let the origins of both systems coincide at $t = 0$. Then

$$x' = x, \qquad y' = y, \qquad z' = z - \int_0^t \mathrm{d}\tau \int_0^\tau \mathrm{d}s g(s, x, y, z).$$

Hence, the equations of motion of a free particle, that in K are

$$\frac{\mathrm{d}^2 x}{\mathrm{d}t^2} = \frac{\mathrm{d}^2 y}{\mathrm{d}t^2} = \frac{\mathrm{d}^2 z}{\mathrm{d}t^2} = 0,$$

in K' assume the form

$$\frac{\mathrm{d}^2 x'}{\mathrm{d}t^2} = \frac{\mathrm{d}^2 y'}{\mathrm{d}t^2} = 0, \qquad \frac{\mathrm{d}^2 z'}{\mathrm{d}t^2} = -g(t, x, y, z).$$

The quantity that we interpreted in K as acceleration would be interpreted in K' as the intensity of a gravitational field (with opposite sign). The gravitational field can thus be simulated by accelerated motion, or, more exactly, the gravitational force is simulated by the force of inertia. If so, then light in a gravitational field should behave similarly to when it is observed from an accelerated reference system.

How would we see a light ray in such a system? Imagine a space vehicle that flies across a light ray. Let the light ray enter through the window W and fall on a screen on the other side of the vehicle (see Fig. 1.2). If the vehicle were at rest, then the light ray entering at W would hit the screen at the point A. Since the vehicle keeps flying, it will move a bit

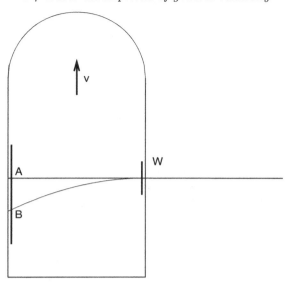

Fig. 1.2 A space vehicle flying across a light ray. See the explanation in the text.

before the ray hits the screen, and the bright spot will appear at the point B. Now assume that the light ray indeed moves in a straight line when observed by an observer who is at rest. Then it is easy to see that the path WB will be straight when the vehicle moves with a constant velocity, whereas it will be curved when the vehicle moves with acceleration. Hence, if the gravitational field behaves analogously to the field of inertial forces, then the light ray should be deflected also by gravitation. Consequently, it cannot be the standard of a straight line.

If we are unable to provide a physical model of a fundamental notion of Newton's physics, let us try to do without it. Let us assume that no such thing exists as 'gravitational forces' that curve the trajectories of celestial bodies, but that the geometry of space is modified by gravitation in such a way that the observed trajectories are paths of free motion. Such a theory might be more complicated than Newton's in practical instances, but it will use only such notions as are related to actual observations, without an unobservable background of the Euclidean space.

A modified geometry means non-Euclidean geometry. A theory created in order to deal with broad classes of non-Euclidean geometries is differential geometry. It is the mathematical basis of general relativity, and we will begin by studying it.

Part I
Elements of differential geometry

2

A short sketch of 2-dimensional differential geometry

2.1 Constructing parallel straight lines in a flat space

The classical Greek geometric constructions, with the help of rulers and compasses, fail over large distances. For example, if we wish to construct a straight line parallel to the momentary velocity of the Earth that passes through a given point on the Moon, compasses and rulers do not help. What method might work in such a situation? For the beginning, let us assume that great distance is our only problem – that we live in a space without gravitation, so we can use a light ray or the trajectory of a stone shot from a sling as a model of a straight line.

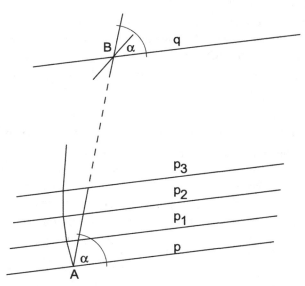

Fig. 2.1 Constructing parallel straight lines at a distance in a flat space. See the explanation in the text.

Assume that an observer is at the point A (see Fig. 2.1) on the straight line p, and wants to construct a straight line parallel to p through the point B. The following programme is 'technically realistic': we first determine the straight line passing through both A and B (for example, by directing a telescope towards B), then we measure the angle α between

9

the lines p and AB, then, from B, we construct a straight line q that is inclined to AB at the same angle α *and lies in the same plane as p and AB*. The second condition requires that we can control points of q other than B, and it can pose some problems. However, if our observer is able to construct parallel straight lines that are not too distant from the given one, he/she can carry out the following operation: the observer moves from A to A_1, constructs a straight line $p_1 \parallel p$, then moves on to A_2, constructs a straight line $p_2 \parallel p_1$, etc., until, in the n-th step, he/she reaches B and constructs $q = p_n \parallel p_{n-1}$ there.

This construction can be generalised. The observer does not have to move from A to B on a straight line. He/she can start from A in an arbitrary direction and, at a point A_1, construct a straight line parallel to p; it has to lie in the plane pAA_1 and be inclined to AA_1 at the same angle as p. Then, from A_1 the observer can continue in still another direction and at a point A_2 repeat the construction: a straight line p_2 has to lie in the plane $p_1A_1A_2$ and be inclined to A_1A_2 at the same angle as p_1 was. When the broken line he/she is following reaches B, the last straight line will be the desired one.

We can imagine broken lines whose straight segments are becoming still shorter. In the limit, we conclude that we would be able to carry out this construction along an arbitrary differentiable curve. The plane needed in the construction will be in each step determined by the tangent vector of the curve and the last straight line we had constructed.

In this way, we arrived at the idea of constructing parallel straight lines by parallelly transporting directions. Note that a straight line is privileged in this construction: it is the only one to which the parallelly transported direction is inclined always at the same angle. In particular, a parallelly transported vector tangent to a straight line remains tangent to it at every point. A straight line can be defined by this property, provided we are able to independently define what it means to be parallel. One possible definition is this: a vector field $\mathbf{v}(x)$ along a curve $C \subset \mathbb{R}^n$ consists of parallel vectors when there exists a coordinate system such that $\partial v^i / \partial x^j \equiv 0$.

2.2 Generalisation of the notion of parallelism to curved surfaces

On a curved surface, the analogue of a straight line is a geodesic line. This is a curve whose arc PQ (see Fig. 2.2) is the shortest among all curved arcs connecting P and Q. Note that the vector tangent to a curve on a curved surface S is not a subset of this surface. The collection of all vectors tangent to S at a point $P \in S$ spans a plane tangent to S at P.

On a curved surface S, parallel transport is defined as follows. Suppose that we are given the pair of points P and Q, an arc of a curve C connecting P and Q and a vector tangent to S at P that we plan to parallelly transport to Q. If C is a geodesic, then we transport the vector v along it in such a way that it is everywhere inclined to the tangent vector of C at the same angle. If C is not a geodesic, then we proceed as follows:

1. We divide the arc PQ into n segments.
2. We connect the ends of each arc by a geodesic.
3. We transport $\mathbf{v}$ parallelly along each geodesic arc.
4. We calculate the result of this operation as $n \to \infty$.

It is easy to note that the parallel transport thus defined depends on the curve along which the transport was carried out. For example, consider a sphere, its pole C and two

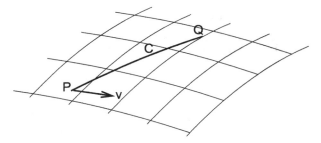

Fig. 2.2 Parallel transport of vectors on a curved surface. See the explanation in the text.

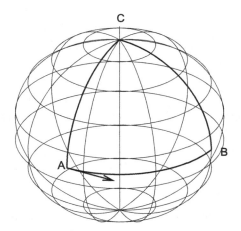

Fig. 2.3 Parallel transport of vectors on a sphere. See the explanation in the text.

points A and B lying on the equator, 90° away from each other (Fig. 2.3). Let $\mathbf{v}$ be the vector tangent to the equator at A. Transport $\mathbf{v}$ parallely to C along the arc AC, and then again along the arcs AB and BC. All three arcs are parts of great circles, which are geodesics, so $\mathbf{v}$ makes always the same angle with the tangent vectors of the arcs. The first transport will yield a vector at C that is tangent to BC, while the second one will yield a vector at C perpendicular to BC. In consequence, if we transport (in differential geometry one says 'drag') a vector along a closed loop, we will not obtain the same vector that we started with. The curvature of the surface is responsible for this. The connection between the initial vector, the final vector and the curvature is rather complicated; we will come to it further on.

We have discussed 2-dimensional surfaces in this chapter in order to visualise things more easily. However, this gave us an unfair advantage: on a 2-dimensional surface, the direction inclined to a given tangent vector at a given angle is uniquely determined. In spaces of higher dimension we will need a definition of 'parallelism at a distance' that will be analogous to $\partial v^i / \partial x^j = 0$ that we used in a flat space.

3
Tensors, tensor densities

3.1 What are tensors good for?

In Newtonian physics, a preferred class of reference systems is used. They are the inertial systems – those in which the three Newtonian principles of dynamics hold true. However, it may be difficult to identify them in practice. As we have seen in Chapter 1, it may not be easy to decide whether a given object moves with acceleration or remains at rest in a gravitational field. Hence, laws of physics should be formulated in such a way that no reference system is privileged. The choice of a reference system, even when it is evidently convenient (like, e.g., the centre of mass system), is an act of human will, while the laws of physics should not depend on our decisions.

Tensors are objects defined so that no reference system is privileged. For the beginning, we will settle for a vague definition that we will make precise later. Suppose we change the coordinate system in an n-dimensional space from $\{x^\alpha\}$, $\alpha = 1, 2, \ldots, n$, to $\{x^{\alpha'}\}$, $\alpha' = 1, 2, \ldots, n$. A tensor is a collection of functions on that space that changes in a specific way under such a coordinate transformation. The appropriate class of spaces and the 'specific way' in which the functions change will be defined in subsequent sections.

3.2 Differentiable manifolds

As already stated, in relativity we will be using non-Euclidean spaces. The most general class of spaces that we will consider are **differentiable manifolds**. This is a generalisation of the notion of a curved surface for which a tangent plane exists at every point of it. An n-dimensional differentiable manifold of class p is a space M_n in which every point x has a neighbourhood $\mathcal{O}_x$ such that the following conditions hold:

1. There exists a one-to-one mapping κ_x of $\mathcal{O}_x$ onto a subset of $\mathbb{R}^n$, called a **map** of $\mathcal{O}_x$. The coordinates of the image $\kappa_x(x)$ are called the coordinates of $x \in M_n$.

2. If the neighbourhoods $\mathcal{O}_x$ and $\mathcal{O}_y$ of $x, y \in M_n$ have a nonempty intersection ($\mathcal{O}_x \cap \mathcal{O}_y \neq \emptyset$), κ_x is a map of $\mathcal{O}_x$ and κ_y is a map of $\mathcal{O}_y$, then the mappings $(\kappa_y \circ \kappa_x^{-1})$ and $(\kappa_y \circ \kappa_x^{-1})^{-1}$ are mappings of class p of $\mathbb{R}^n$ into itself.

A **tangent space** to the manifold M_n at the point x is a vector space spanned by vectors tangent at x to curves in M_n that pass through x.

If $\kappa_x(x) = \{x^1, \ldots, x^n\}$ are the coordinates of the point x, then the equation $x^i =$ constant, where $i \in \{1, \ldots, n\}$ is a fixed index, defines a hypersurface in $\mathbb{R}^n$, and thus also a hypersurface H in M_n. A coordinate system is thus a set of n one-parameter families of hypersurfaces; x^i being the parameter in the i-th family.

Now let the manifold be $\mathbb{R}^n$. Each hypersurface H defines then a family of vectors: if $\Phi(x) = C =$ constant is the equation of H, then $\partial\Phi/\partial x^\alpha$, where $\{x^\alpha\}$ are the coordinates of the point x, is an equation of a family of vectors attached to H and orthogonal to it. Taking all the hypersurfaces of the $\Phi = C$ family, we obtain a family of curves tangent to the vectors $\partial\Phi/\partial x^\alpha$. Thus, each coordinate system in $\mathbb{R}^n$ defines a family of curves.

The converse is not true: an n-parameter family of curves C_x in $\mathbb{R}^n$ defines a family of hypersurfaces orthogonal to C_x only when the vectors tangent to C_x have zero rotation (to be defined in Chapter 15).

The reason why, for this example, we had to take the special case of $M_n = \mathbb{R}^n$ is that, as we shall see later in this section, in a general vector space, vectors like the gradient of a function (called *covariant vectors*) and vectors like a tangent vector to a curve (called *contravariant vectors*) are unrelated objects. A relation between them exists only in spaces tangent to such manifolds in which a *metric* is defined (see Chapter 7); $\mathbb{R}^n$ is one of them. Without a metric, a covariant vector cannot be converted into a contravariant one, and a curve tangent to a field of covariant vectors cannot be constructed.

Let $U \subset M_n$ be an open subset. Suppose we are given a collection of n families of curves, each of $(n-1)$ parameters, such that n curves pass through each point $x \in U$. Suppose that the tangent vectors to these curves are linearly independent at every $x \in U$. Then the tangent vectors to these curves at the point x, $e_a(x)|_{a=1,\ldots,n}$ are a basis of the space tangent to M_n at x. Let $v(x)$ be an arbitrary vector tangent to M_n at x. Then

$$v = \sum_{a=1}^{n} v^a(x)e_a(x).$$

The coefficients $\{v^a\}$ are called the components of the vector v in the basis $e_a(x)$. The mapping $x \to v(x)$ assigns to each point $x \in U \subset M_n$ a vector tangent to M_n at x and is called a **vector field** on U.

Note that the vectors of a vector field are defined on tangent spaces to the manifold, while the components of vector fields are functions on the manifold. The vectors $e_a(x)$ can be identified with directional derivatives: $e_a(x)\Phi(x) = (\partial/\partial x^a)\Phi(x)$ (that is, the a-th vector in the basis is the directional derivative down the a-th family of hypersurfaces in the coordinate system).

We adopt three conventions:

1. If any index in a formula appears twice, a sum over all of its values is implied. Hence, if α changes from 1 to n, while $\{V_\alpha\}$ and $\{U^\alpha\}$, $\alpha = 1, \ldots, n$ are collections of functions labelled by α, then

$$U^\alpha V_\alpha \equiv \sum_{\alpha=1}^{n} U^\alpha V_\alpha.$$

2. The collection of all coordinates $\{x^\alpha\}$, $\alpha = 1, \ldots, n$ will be denoted $\{x\}$, and a function $f(\{x\})$ will be denoted $f(x)$.

3. The derivative of a function $f(x)$ with respect to x^α will be denoted as follows:

$$f_{,\alpha} \equiv \frac{\partial f}{\partial x^\alpha}.$$

3.3 Scalars

The simplest tensor is a **scalar**. It is a function on a manifold whose value, when the coordinates are transformed, changes just by substituting the transformation in its argument:

$$\varphi'(x'(x)) = \varphi(x). \tag{3.1}$$

Examples of scalars are physical constants, rest masses of elementary particles, their electric charges and a density distribution in a continuous medium. The coordinates of a point in a manifold are scalars, too, since their values transform by the simple law (3.1) when coordinates are changed from $\{x'\}$ to $\{x''\}$: $x^\alpha(x''(x')) = x^\alpha(x')$.

3.4 Contravariant vectors

The functions $v^\alpha(x)$, $\alpha = 1, \ldots, n$ are said to be coordinates of a **contravariant vector field** when, by a change of coordinates on M_n, they transform by the law

$$v^{\alpha'}(x'(x)) = x^{\alpha'},_\alpha \, v^\alpha(x). \tag{3.2}$$

Examples are vectors tangent to curves. Suppose that a curve C is given by the parametric equations $t \to x^\alpha(t)$, $\alpha = 1, \ldots, n$, $t \in [a, b] \subset \mathbb{R}^1$, where x^α is the value of a coordinate of a point on the curve. An arbitrary field of vectors tangent to C is then given by

$$t \to v^\alpha(t) = f(x)g(t)\frac{\mathrm{d}x^\alpha}{\mathrm{d}t},$$

where $f(x)$ is an arbitrary function of the coordinates and $g(t)$ is an arbitrary function of the parameter. When the coordinate system is changed from (x) to (x'), we have

$$v^{\alpha'}(t) = f(x'(x))g(t)\frac{\mathrm{d}x^{\alpha'}}{\mathrm{d}t} = f(x)g(t)x^{\alpha'},_\alpha\frac{\mathrm{d}x^\alpha}{\mathrm{d}t},$$

in agreement with (3.2).

3.5 Covariant vectors

The functions $u_\alpha(x)$, $\alpha = 1, \ldots, n$ are said to be coordinates of a **covariant vector field** when, by a change of coordinates on M_n, they transform by the law

$$u_{\alpha'}(x'(x)) = x^\alpha,_{\alpha'} \, u_\alpha(x). \tag{3.3}$$

By convention, the indices of contravariant vectors are placed as superscripts, whereas those of covariant vectors are placed as subscripts. An example of a covariant vector is the gradient of a scalar function $\varphi(x)$. For such a function, we have, in agreement with (3.3),

$$\varphi_{,\alpha'}(x'(x)) = x^\alpha,_{\alpha'} \, \varphi_{,\alpha}(x).$$

Note that the quantity $v^\alpha u_\alpha$ is a scalar, since we have

$$v^{\alpha'} u_{\alpha'} = x^{\alpha'},_\alpha v^\alpha x^\beta,_{\alpha'} u_\beta = \left(x^\beta,_{\alpha'} x^{\alpha'},_\alpha \right) v^\alpha u_\beta$$
$$= x^\beta,_\alpha v^\alpha u_\beta = \delta^\beta,_\alpha v^\alpha u_\beta = v^\alpha u_\alpha.$$

Another example of a scalar field is the directional derivative of a scalar field along a contravariant vector field, $v(\varphi) \overset{\text{def}}{=} v^\alpha \varphi,_\alpha$.

3.6 Tensors of second rank

Scalars are sometimes called tensors of rank zero, to emphasise that they have no indices. The contravariant and covariant vectors are collectively called tensors of rank 1. The tensors of rank 2 are objects whose components are labelled by two indices. There are three kinds of them:

1. Doubly contravariant tensors. Their components $T^{\alpha\beta}(x)$ transform under a coordinate transformation $x \to x'$ on M_n as follows:

$$T^{\alpha'\beta'}(x'(x)) = x^{\alpha'},_\alpha x^{\beta'},_\beta T^{\alpha\beta}(x). \tag{3.4}$$

2. Doubly covariant tensors. Their components $T_{\alpha\beta}(x)$ transform by the rule

$$T_{\alpha'\beta'}(x'(x)) = x^\alpha,_{\alpha'} x^\beta,_{\beta'} T_{\alpha\beta}(x). \tag{3.5}$$

3. Mixed tensors. Their components transform by the rule

$$T_{\alpha'}{}^{\beta'}(x'(x)) = x^\alpha,_{\alpha'} x^{\beta'},_\beta T_\alpha{}^\beta(x). \tag{3.6}$$

The components of a second-rank tensor form a square matrix that transforms in a prescribed way. An example of a doubly covariant tensor is a matrix of a quadratic form

$$\Phi(A) = \Phi_{\alpha\beta} A^\alpha A^\beta,$$

where A^α are components of a contravariant vector, and the value of $\Phi(A)$ is a scalar.

An example of a mixed tensor is a matrix of a mapping of one vector space into another

$$V^\alpha = B^\alpha{}_\beta W^\beta,$$

where V^α and W^β are contravariant vectors in different vector spaces.

An example of a doubly contravariant tensor is the inverse matrix to a matrix of a quadratic form, but we will be able to prove this after we learn about some other objects.

The quantity $T^\alpha{}_\alpha$ for a mixed tensor (the sum of its diagonal components) is called the **trace** of T and is a scalar. Quantities like $\sum_\alpha T^{\alpha\alpha}$ for a contravariant second-rank tensor and $\sum_\alpha T_{\alpha\alpha}$ for a covariant second-rank tensor are not tensorial objects. Summations over indices standing on the same level occur only exceptionally in differential geometry – for example, when a calculation is done in a chosen coordinate system.

3.7 Tensor densities

A **tensor density** differs from the corresponding tensor in that, when transformed from one coordinate system to another, it gets multiplied by a certain power of the Jacobian of the transformation. The exponent of the Jacobian is called the **weight** of the density. For example, a scalar density of weight w transforms as follows

$$\Phi'(x') = \left[\frac{\partial(x')}{\partial(x)}\right]^w \Phi(x), \tag{3.7}$$

and a contravariant vector density of weight w transforms by the rule

$$v^{\alpha'}(x') = \left[\frac{\partial(x')}{\partial(x)}\right]^w x^{\alpha'}{}_{,\alpha}\, v^\alpha(x). \tag{3.8}$$

An example of a scalar density is the element of volume in a multidimensional integral. It transforms by the law

$$\mathrm{d}_n x = \frac{\partial(x)}{\partial(x')}\, \mathrm{d}_n x', \tag{3.9}$$

so it is a scalar density of weight $+1$.

An arbitrary tensor is by definition a tensor density of weight zero.

3.8 Tensor densities of arbitrary rank

The components of a tensor density of weight w, k times contravariant and l times covariant, transform by the law

$$\begin{aligned}
T^{\alpha'_1 \alpha'_2 \ldots \alpha'_k}_{\beta'_1 \beta'_2 \ldots \beta'_l}(x'(x)) &= \left[\frac{\partial(x')}{\partial(x)}\right]^w x^{\alpha'_1}{}_{,\alpha_1} x^{\alpha'_2}{}_{,\alpha_2} \ldots x^{\alpha'_k}{}_{,\alpha_k} \\
&\times x^{\beta_1}{}_{,\beta'_1} x^{\beta_2}{}_{,\beta'_2} \ldots x^{\beta_l}{}_{,\beta'_l}\, T^{\alpha_1 \alpha_2 \ldots \alpha_k}_{\beta_1 \beta_2 \ldots \beta_l}(x).
\end{aligned} \tag{3.10}$$

Such an object is called a tensor density of type $[w, k, l]$.

For general tensor densities one can carry out an operation that is analogous to finding the trace of a mixed tensor of rank 2. This operation is called **contraction**. It consists in making an upper index equal to a lower index, and summing over all its allowed values. The resulting density is of type $[w, k-1, l-1]$, thus

$$[\text{contracted } T]^{\alpha_1 \ldots \alpha_{i-1} \alpha_{i+1} \ldots \alpha_k}_{\beta_1 \ldots \beta_{j-1} \beta_{j+1} \ldots \beta_l}(x) = T^{\alpha_1 \ldots \alpha_{i-1} \rho \alpha_{i+1} \ldots \alpha_k}_{\beta_1 \ldots \beta_{j-1} \rho \beta_{j+1} \ldots \beta_l}(x) \tag{3.11}$$

(note that sum over all values of ρ is implied above). The indices over which the summation is carried out are called 'dummy indices', since they do not show up in the transformation law of the contracted density.

The contraction may be done over several pairs of indices at the same time. Then, one must take care to give different names to each pair of dummy indices, to avoid confusion.

3.9 Algebraic properties of tensor densities

Here is a list of the most basic properties of tensor densities.

(i) If $T^{\cdots\cdots}_{\cdots\cdots\cdots} \equiv 0$ in one coordinate system, then $T^{\cdots\cdots}_{\cdots\cdots\cdots} \equiv 0$ in all coordinate systems (this follows easily from the transformation law).

(ii) A linear combination of two tensor densities of type $[w, k, l]$ is a tensor density of the same type. (Adding tensor densities of different types makes no sense.)

(iii) The collection of quantities obtained when each component of one tensor density (of type $[w, k, l]$) is multiplied by each component of another tensor density (of type $[w', k', l']$) is called a **tensor product** of the two densities, and is a tensor density of type $[w + w', k + k', l + l']$. For example, out of u_α and v^α one can form such tensor products as $v^\alpha v^\beta$, $v^\alpha u_\beta$, $u_\alpha u_\beta$, $v^\alpha v^\beta v^\gamma$, $v^\alpha u_\beta v^\gamma$, $v^\alpha u_\alpha v^\beta u_\gamma v^\delta$. The tensor product is denoted by $\otimes$, thus, for example, $v^\alpha u_\beta = (v \otimes u)^\alpha{}_\beta$.

(iv) If a tensor density does not change its value when two indices (either both upper or both lower) are interchanged, then it is called **symmetric** with respect to this pair of indices. If it only changes sign, then it is called **antisymmetric** in this pair of indices. The property of being symmetric or antisymmetric with respect to a given pair of indices is preserved under transformations of coordinates.

(v) This last property allows us to define the **symmetrisation** and the **antisymmetrisation**. The **symmetric part** of a tensor density with respect to the indices $\{\alpha_1, \ldots, \alpha_k\}$ is the quantity

$$T_{(\alpha_1 \ldots \alpha_k)} \stackrel{\text{def}}{=} \frac{1}{k!} \sum_{\substack{\text{over all permutations} \\ i_1, \ldots, i_k}} T_{\alpha_{i_1} \ldots \alpha_{i_k}}. \tag{3.12}$$

The **antisymmetric part** of a tensor density with respect to the indices $\{\alpha_1, \ldots, \alpha_k\}$ is the quantity

$$T_{[\alpha_1 \ldots \alpha_k]} \stackrel{\text{def}}{=} \frac{1}{k!} \sum_{\substack{\text{over all permutations} \\ i_1, \ldots, i_k}} (\text{sign of the permutation}) \, T_{\alpha_{i_1} \ldots \alpha_{i_k}}. \tag{3.13}$$

One can carry out the symmetrisation or antisymmetrisation (i.e. respectively, *symmetrise* or *antisymmetrise*) also with respect to the upper indices, and, in each case, over only a subset of all the indices that a tensor density has. For example, a tensor of rank 4 $T_{\alpha\beta\gamma\delta}$ can be symmetrised over only two of its indices

$$T_{(\alpha\beta)\gamma\delta} = \frac{1}{2} \left(T_{\alpha\beta\gamma\delta} + T_{\beta\alpha\gamma\delta} \right),$$

or over three indices

$$T_{(\alpha\beta\gamma)\delta} = \frac{1}{6} \left(T_{\alpha\beta\gamma\delta} + T_{\alpha\gamma\beta\delta} + T_{\beta\alpha\gamma\delta} + T_{\beta\gamma\alpha\delta} + T_{\gamma\alpha\beta\delta} + T_{\gamma\beta\alpha\delta} \right).$$

If we want to emphasise that a certain index is excluded from symmetrisation or antisymmetrisation, then we put it between vertical strokes. For example, the antisymmetrisation of $T_{\alpha\beta\gamma\delta}$ with respect to α and γ only would be denoted $T_{[\alpha|\beta|\gamma]\delta}$.

The symmetrisation and the antisymmetrisation with respect to just two indices are complementary operations, as can easily be verified, thus

$$T_{(\alpha\beta)\gamma\delta\ldots} + T_{[\alpha\beta]\gamma\delta\ldots} \equiv T_{\alpha\beta\gamma\delta\ldots}.$$

This is no longer true for symmetrisations/antisymmetrisations over larger numbers of indices. The other parts of $T_{\alpha\beta\gamma\delta...}$ that enter neither $T_{(\alpha\beta\gamma)\delta...}$ nor $T_{[\alpha\beta\gamma]\delta...}$ are obtained when different rules of assigning signs to the different terms in (3.12) and (3.13) are chosen. We will not encounter those operations in this book.

If the antisymmetrisation is done with respect to a larger number of indices than the dimension of the manifold, then the result is a tensor density that identically equals zero. This is because, with n possible different values of each index, there must be repetitions of values in a set of $m > n$ indices. Then, each term in (3.13) has its counterpart that is identical, but enters the sum with an opposite sign.

3.10 Mappings between manifolds

The following notation will be used.

M_n and P_m are two arbitrary differentiable manifolds, $\dim M_n = n$, $\dim P_m = m$; $\{x^\alpha\}$, $\alpha = 1,\ldots,n$ and $\{y^a\}$, $a = 1,\ldots,m$ are the coordinate systems on M_n and P_m, respectively. (Recall that several coordinate neighbourhoods may be needed to cover each manifold, but here we consider a single coordinate patch on each one.)

The $F : M_n \to P_m$ is a mapping of class C^1 represented by the functions $y^a = F^a(\{x\})$.

Now consider a function $f : P_m \to \mathbb{R}^1$. For those points of P_m that are images of some points of M_n, the mapping F and the function f automatically define a function acting on M_n. Let $P_m \ni q = F(p)$, where $p \in M_n$, and let $f(q) = r \in \mathbb{R}^1$. Then $(f \circ F)(p) = r$, and so $f \circ F : M_n \to \mathbb{R}^1$. We can thus say that the mapping F that takes points of M_n to points of P_m defines an **associated mapping** of functions on P_m to functions on M_n. This associated mapping will be denoted F_0^*. The zero denotes that this is a mapping of functions, i.e. tensors with zero indices, and the asterisk placed as a superscript denotes that the functions are sent in the opposite direction to points of the manifold M_n. An associated mapping that sends objects backwards with respect to the main mapping is called a **pullback**. The function associated to f will thus be denoted $F_0^* f$, so

$$F_0^* f : M_n \to \mathbb{R}^1, \qquad (F_0^* f)(p) \stackrel{\text{def}}{=} f(F(p)). \tag{3.14}$$

Now consider a contravariant continuous vector field v^α on M_n. This field defines a family of curves $x^\alpha(\tau)$ tangent to the vectors of this field by

$$v^\alpha = \frac{\mathrm{d}x^\alpha}{\mathrm{d}\tau}. \tag{3.15}$$

Suppose that an arc of the curve $x^\alpha(\tau)$ is in the domain of the mapping F. Then points of this curve have their images in P_m, $y^a(\tau) = F^a(x^\alpha(\tau))$. Hence, the arc of the image-curve in P_m is automatically parametrised by the same parameter τ. Since the functions $x^\alpha(\tau)$ are of class C^1 (as integrals of continuous functions), and F is of the same class by assumption, we can differentiate the functions $y^a(\tau)$ by τ. The derivatives are components of the vector field w^a tangent to the curve $y^a(\tau)$. The mapping F thus defines an associated mapping of vector fields on M_n to vector fields on P_m:

$$w^a(F(x)) = \frac{\mathrm{d}}{\mathrm{d}\tau}(F^a(x)) = \frac{\partial F^a}{\partial x^\alpha}\frac{\mathrm{d}x^\alpha}{\mathrm{d}\tau} = \frac{\partial F^a}{\partial x^\alpha}v^\alpha. \tag{3.16}$$

If $\dim P_m < \dim M_n$, then some curves in M_n will be mapped on single points of P_m, and then, according to (3.16), the image of the vector field v^α tangent to such a curve will be a zero vector on P_m (because the image of such a curve, $y^a(\tau)$, will in fact be independent of τ). Note the similarity of (3.16) to the transformation law of contravariant vectors; we will come back to it at the end of this section.

We have thus found that the mapping of manifolds $F : M_n \to P_m$ defines an associated mapping of vector fields on M_n to vector fields on P_m. We will denote this associated mapping F_{1*}. This time it is a mapping of tensors with one index, and it takes objects in the same direction as the main mapping. Such a mapping is called a **pushforward**. If v is a vector field on M_n, then $F_{1*}(v)$ is a vector field on P_m. The mapping F_{1*} is a linear mapping between vector spaces, determined via (3.16) by the matrix $||F^a{}_{,\alpha}||$.

Vector fields v^α on M_n and w^a on P_m can be uniquely represented by directional derivatives of functions. Let $g : M_n \to \mathbb{R}^1$ and $f : P_m \to \mathbb{R}^1$, then

$$v^\alpha(x) \quad \longleftrightarrow \quad v(g) \overset{\text{def}}{=} v^\alpha \frac{\partial g}{\partial x^\alpha},$$
$$w^a(y) \quad \longleftrightarrow \quad w(f) \overset{\text{def}}{=} w^a \frac{\partial f}{\partial y^a}. \tag{3.17}$$

Then the mapping F_{1*} can be defined in another, equivalent way:

$$v\left(F_0^* f\right) = \left(F_{1*}v\right)\left(f\right), \tag{3.18}$$

where v is a vector field defined on M_n.

A field of covariant vectors ω_a on P_m can be understood as a linear form ω that maps vector fields on P_m to $\mathbb{R}^1$:

$$\omega(w) \overset{\text{def}}{=} \omega_a w^a. \tag{3.19}$$

If a field of forms ω is defined on P_m, then, for those points of P_m that are images of points of M_n, we can define a field of forms on M_n by

$$\left(F_1^*\omega\right)(v) = \omega\left(F_{1*}v\right), \tag{3.20}$$

where v is a vector field on M_n. This is an associated mapping of tensors with one index that is again a pullback. In terms of coordinates, this can be written as follows:

$$\left(F_1^*\omega\right)_\alpha v^\alpha = \omega_a F^a{}_{,\alpha}\, v^\alpha, \tag{3.21}$$

or equivalently

$$\left(F_1^*\omega\right)_\alpha = \omega_a F^a{}_{,\alpha}. \tag{3.22}$$

Note the similarity of (3.22) to the transformation law of covariant vectors.

To sum up: a mapping $F : M_n \to P_m$ between manifolds defines the mapping F_{1*} of vector fields on M_n to vector fields on P_m and the mappings F_0^* and F_1^* of, respectively, functions and forms on P_m to functions and forms on M_n. These definitions can be extended, in a rather obvious way, to mappings of contravariant tensor fields of arbitrary rank from M_n to P_m and of covariant tensor fields from P_m to M_n. Note that tensor densities are defined only for nonsingular transformations, while F can be singular. Therefore, no definition of an associated mapping of general tensor densities can be given.

If $n = m$ and F is a diffeomorphism of class C^1, then F^{-1} exists and is also of class C^1. In this case, tensors of arbitrary rank, including mixed tensors, can be transported in both directions between M_n and P_m. If $F : M_n \to P_m$ is not reversible, then mixed tensors cannot be transported in any direction.

Now observe the following. Coordinate transformations are in fact nothing else than mappings of $\mathbb{R}^n$ into itself – because coordinate patches are subsets of $\mathbb{R}^n$ by definition. Hence, if such a mapping occurs within the image of a single map, a mapping of the manifold into itself is associated with it. Consequently, coordinate transformations can be interpreted as mappings of the underlying manifold into itself.

Let us now specialise the considerations of this section to the case when $n = m$, and M_n and P_m are subsets of the same manifold. Then it follows from (3.14), (3.16) and (3.22), and from their extensions to tensor fields of higher ranks, that the transformation laws of tensors are consistent with the interpretation of a coordinate transformation as a mapping of the manifold into itself. We shall make use of this analogy in Chapter 8.

3.11 The Levi-Civita symbol

Let $\epsilon_{\alpha_1 \ldots \alpha_n}$ be a tensor density (whose weight is as yet unknown) defined on a differentiable manifold M_n, which is antisymmetric with respect to any pair of its indices. Because of the antisymmetry, each component that has at least a pair of identical indices will be equal to zero. Only those components can be different from zero for which the set $\{\alpha_1, \ldots, \alpha_n\}$ is a permutation of the set $\{1, \ldots, n\}$. For even permutations $\epsilon_{\alpha_1 \ldots \alpha_n} = \epsilon_{1 \ldots n}$, for odd permutations $\epsilon_{\alpha_1 \ldots \alpha_n} = -\epsilon_{1 \ldots n}$. Hence, defining the component $\epsilon_{1 \ldots n}$ suffices to determine the whole tensor density $\epsilon_{\alpha_1 \ldots \alpha_n}$. We thus define

$$\epsilon_{1 \ldots n} = +1. \tag{3.23}$$

The quantity $\epsilon_{\alpha_1 \ldots \alpha_n}$ is called the **Levi-Civita symbol**.

Let $A^\alpha{}_\beta$ be an arbitrary matrix. Let us investigate the quantity

$$[D(A)]_{\beta_1 \ldots \beta_n} \overset{\text{def}}{=} \epsilon_{\alpha_1 \ldots \alpha_n} A^{\alpha_1}{}_{\beta_1} \ldots A^{\alpha_n}{}_{\beta_n}. \tag{3.24}$$

Let α_i label the columns and β_j the rows of the matrix. The expression above is a sum of n-tuple products of elements of the matrix $A^\alpha{}_\beta$. In each product, each factor comes from a different column (because if two indices α_i and α_j are equal, then $\epsilon_{\alpha_1 \ldots \alpha_n} = 0$). The i-th factor in each product is always from the same (i-th) row, but each time from a different column, and, in the whole sum, runs over all columns. For even permutations $\alpha_1 \ldots \alpha_n$, the product enters the sum with a plus sign, for odd permutations – with a minus sign. The quantity (3.24) is also antisymmetric in all the β_i indices (see Exercise 1). Hence, $[D(A)]_{\beta_1 \ldots \beta_n}$ vanishes if any two indices β_i, β_j are equal (i.e. if the same row of the matrix A appears in the positions (i, j)). Thus $[D(A)]_{\beta_1 \ldots \beta_n}$ has all the properties of the determinant of A. It will equal $+\det(A)$ when the permutation $\{1, \ldots, n\} \to \{\beta_1, \ldots, \beta_n\}$ is even, and $[-\det(A)]$ when the permutation is odd. Hence,

$$\epsilon_{\alpha_1 \ldots \alpha_n} A^{\alpha_1}{}_{\beta_1} \ldots A^{\alpha_n}{}_{\beta_n} = \epsilon_{\beta_1 \ldots \beta_n} \det(A). \tag{3.25}$$

Now let us apply this formula to the matrix of derivatives (the Jacobi matrix) $x^\alpha{}_{,\alpha'}$ of the coordinate transformation $\{x\} \to \{x'\}$. We have

$$\epsilon_{\alpha'_1...\alpha'_n} = \frac{\partial(x')}{\partial(x)} x^{\alpha_1}{}_{,\alpha'_1} \cdots x^{\alpha_n}{}_{,\alpha'_n} \epsilon_{\alpha_1...\alpha_n}. \tag{3.26}$$

This shows that $\epsilon_{\alpha_1...\alpha_n}$ is a tensor density of type $[1, 0, n]$.

By a similar method one can verify that $\epsilon^{\alpha_1...\alpha_n}$ is a tensor density of type $[-1, n, 0]$.

3.12 Multidimensional Kronecker deltas

Let us recall the definition of the ordinary Kronecker delta:

$$\delta^\alpha{}_\beta = \begin{cases} 1 & \text{when} \quad \alpha = \beta \\ 0 & \text{when} \quad \alpha \neq \beta \end{cases} \tag{3.27}$$

(this is the unit matrix). This is a tensor, since, on the one hand

$$\delta'^\alpha{}_\beta = \delta^\alpha{}_\beta \tag{3.28}$$

by definition, and on the other hand

$$\delta'^{\alpha'}{}_{\beta'} = x^{\alpha'}{}_{,\beta'} = x^{\alpha'}{}_{,\rho} x^\rho{}_{,\beta'} = x^{\alpha'}{}_{,\rho} x^\sigma{}_{,\beta'} \delta^\rho{}_\sigma, \tag{3.29}$$

which is the transformation law of a mixed tensor of rank 2.

A multidimensional Kronecker delta is defined as follows:

$$\delta^{\alpha_1...\alpha_k}_{\beta_1...\beta_k} = \begin{vmatrix} \delta^{\alpha_1}{}_{\beta_1} & \cdots & \cdots & \delta^{\alpha_1}{}_{\beta_k} \\ \vdots & & & \vdots \\ \vdots & & & \vdots \\ \delta^{\alpha_k}{}_{\beta_1} & \cdots & \cdots & \delta^{\alpha_k}{}_{\beta_k} \end{vmatrix}. \tag{3.30}$$

From the definition we have at once

$$\delta^{\alpha_1...\alpha_k}_{\beta_1...\beta_k} = \delta^{[\alpha_1...\alpha_k]}_{\beta_1...\beta_k} = \delta^{\alpha_1...\alpha_k}_{[\beta_1...\beta_k]}, \tag{3.31}$$

and it follows that k must not be greater than the dimension of the manifold, n, or else $\delta^{\alpha_1...\alpha_k}_{\beta_1...\beta_k} \equiv 0$. Other than this, k is unrelated to n.

Now let us consider the special case $k = n$. Since an object that is antisymmetric in all n indices has just one independent component, it must be proportional to the Levi-Civita symbol with the same indices. For the upper indices of the multidimensional delta we thus have $\delta^{\alpha_1...\alpha_n}_{\beta_1...\beta_n} = T_{\beta_1...\beta_n} \epsilon^{\alpha_1...\alpha_n}$. But $T_{\beta_1...\beta_n}$ is antisymmetric in $\{\beta_1, \ldots, \beta_n\}$, so it must be proportional to $\epsilon_{\beta_1...\beta_n}$, thus $\delta^{\alpha_1...\alpha_n}_{\beta_1...\beta_n} = \lambda \epsilon_{\beta_1...\beta_n} \epsilon^{\alpha_1...\alpha_n}$. To calculate λ it now suffices to substitute in the last formula any sets of indices for which both sides are nonzero. We substitute $\{\alpha_1, \ldots, \alpha_n\} = \{\beta_1, \ldots, \beta_n\} = \{1, \ldots, n\}$, and we see that $\lambda = 1$. Hence, finally

$$\delta^{\alpha_1...\alpha_n}_{\beta_1...\beta_n} = \epsilon_{\beta_1...\beta_n} \epsilon^{\alpha_1...\alpha_n}. \tag{3.32}$$

This is a tensor of weight zero.

From (3.30) it follows that $\delta^{\alpha_1\ldots\alpha_k}_{\beta_1\ldots\beta_k} \neq 0$ only when $\{\alpha_1,\ldots,\alpha_k\}$ are all different and are a permutation of $\{\beta_1\ldots\beta_k\}$. If any pair of upper indices has equal values, or if any pair of lower indices has equal values, then the determinant in (3.30) has two identical rows or two identical columns and is zero. If any index β_i of the set $\{\beta_1\ldots\beta_k\}$ is different from all the α-s in $\{\alpha_1,\ldots,\alpha_k\}$, the determinant in (3.30) has only zeros in the i-th column and is zero again. If $\delta^{\alpha_1\ldots\alpha_k}_{\beta_1\ldots\beta_k} \neq 0$, then $\delta^{\alpha_1\ldots\alpha_k}_{\beta_1\ldots\beta_k} = +1$ when the lower indices are an even permutation of the upper ones, and $\delta^{\alpha_1\ldots\alpha_k}_{\beta_1\ldots\beta_k} = -1$ when they are an odd permutation.

Using these properties one can verify that

$$\delta^{\alpha_1\ldots\alpha_k\rho}_{\beta_1\ldots\beta_k\rho} = (n-k)\delta^{\alpha_1\ldots\alpha_k}_{\beta_1\ldots\beta_k}. \tag{3.33}$$

This is seen as follows. If $\{\beta_1,\ldots,\beta_k\}$ are not a permutation of $\{\alpha_1,\ldots,\alpha_k\}$, or if any index in either of the two sets is repeated, then both sides of (3.33) are zero and the equation holds. When all $\{\alpha_1,\ldots,\alpha_k\}$ are different, while $\{\beta_1,\ldots,\beta_k\}$ are their permutation, each term in the sum on the left-hand side is equal to $\delta^{\alpha_1\ldots\alpha_k}_{\beta_1\ldots\beta_k}$ when $\rho \notin \{\alpha_1,\ldots,\alpha_k\}$, and is zero when $\rho \in \{\alpha_1,\ldots,\alpha_k\}$. Consequently, there are $(n-k)$ values of ρ with which there are nonzero contributions on the left-hand side, and each contribution is equal to the delta on the right. $\square$

The following equations are simple consequences of (3.33):

$$\delta^{\alpha_1\ldots\alpha_{n-1}\rho}_{\beta_1\ldots\beta_{n-1}\rho} = \delta^{\alpha_1\ldots\alpha_{n-1}}_{\beta_1\ldots\beta_{n-1}}, \tag{3.34}$$

$$\delta^{\rho_1\ldots\rho_s}_{\rho_1\ldots\rho_s} = (n-s+1)(n-s+2)\ldots n = \frac{n!}{(n-s)!}, \tag{3.35}$$

$$\delta^{\rho_1\ldots\rho_n}_{\rho_1\ldots\rho_n} = n!, \tag{3.36}$$

$$\delta^{\alpha_1\ldots\alpha_k\rho_{k+1}\ldots\rho_n}_{\beta_1\ldots\beta_k\rho_{k+1}\ldots\rho_n} = (n-k)!\,\delta^{\alpha_1\ldots\alpha_k}_{\beta_1\ldots\beta_k}. \tag{3.37}$$

3.13 Examples of applications of the Levi-Civita symbol and of the multidimensional Kronecker delta

The Levi-Civita symbols and the multidimensional deltas are useful in calculations with determinants or antisymmetrisations: they allow us to replace tricky reasonings with simple computational rules.

With the help of (3.25), (3.32) and (3.36) we can verify that

$$\det(A^{\cdot}_{\cdot}) = \frac{1}{n!}\delta^{\beta_1\ldots\beta_n}_{\alpha_1\ldots\alpha_n} A^{\alpha_1}{}_{\beta_1} \ldots A^{\alpha_n}{}_{\beta_n}, \tag{3.38}$$

so the determinant of a mixed tensor is a scalar;

$$\det(B_{..}) = \frac{1}{n!}\epsilon^{\alpha_1\ldots\alpha_n}\epsilon^{\beta_1\ldots\beta_n} B_{\alpha_1\beta_1} \ldots B_{\alpha_n\beta_n}, \tag{3.39}$$

so the determinant of a doubly covariant tensor is a scalar density of weight (-2);

$$\det(C^{..}) = \frac{1}{n!}\epsilon_{\alpha_1\ldots\alpha_n}\epsilon_{\beta_1\ldots\beta_n} C^{\alpha_1\beta_1} \ldots C^{\alpha_n\beta_n}, \tag{3.40}$$

so the determinant of a doubly contravariant tensor is a scalar density of weight $(+2)$.

One can also verify that the antisymmetrisation with respect to any set of indices can be written as follows:

$$T_{[\alpha_1\ldots\alpha_k]} = \frac{1}{k!}\delta^{\rho_1\ldots\rho_k}_{\alpha_1\ldots\alpha_k}T_{\rho_1\ldots\rho_k};\tag{3.41}$$

and similarly for upper indices.

3.14 Exercises

1. Verify that the $D(A)$ defined in (3.24) is antisymmetric with respect to the β-indices.

Hint. Interchange the names of any two α-s in the sum and then move the new α-s to their old positions by transposing the indices of ϵ and interchanging the A-s in the product.

2. Prove that the cofactor of the element $A^\alpha{}_\beta$ in a mixed tensor $\{A^\alpha{}_\beta\}$ is given by

$$M^\beta{}_\alpha = \frac{1}{(n-1)!}\delta^{\beta\beta_1\ldots\beta_{n-1}}_{\alpha\alpha_1\ldots\alpha_{n-1}}A^{\alpha_1}{}_{\beta_1}\ldots A^{\alpha_{n-1}}{}_{\beta_{n-1}}.\tag{3.42}$$

3. Find the formulae for the cofactors of the elements $B_{\alpha\beta}$ and $C^{\alpha\beta}$ in a doubly covariant and a doubly contravariant tensor, respectively. Note that the cofactor has in each case its indices positioned opposite to its corresponding element.

4. Find the formula for the coefficient of λ^i in the characteristic equation for a matrix $M^\mu{}_\nu$:

$$\det\left(M^\mu{}_\nu - \lambda\delta^\mu{}_\nu\right) = 0.$$

Prove that all the coefficients of this polynomial are scalars. Note the coefficients of λ^0 and of λ^{n-1} – what functions of the matrix are they?

4

Covariant derivatives

4.1 Differentiation of tensors

Let us calculate the derivative of a contravariant vector field v^α after it has been transformed from the $\{x\}$-coordinates to the $\{x'\}$-coordinates:

$$v^{\alpha'}{}_{,\beta'} = \left(x^{\alpha'}{}_{,\alpha}\, v^\alpha\right){}_{,\beta'} = x^{\alpha'}{}_{,\alpha\beta}\, x^\beta{}_{,\beta'}\, v^\alpha + x^{\alpha'}{}_{,\alpha}\, x^\beta{}_{,\beta'}\, v^\alpha{}_{,\beta} . \tag{4.1}$$

This is not a tensor field, in consequence of the term $x^{\alpha'}{}_{,\alpha\beta}\, x^\beta{}_{,\beta'}\, v^\alpha$. The same is true for most other tensor fields. There are only a few cases in which the derivatives of tensor fields are themselves tensor fields. One example is the derivative of a scalar field, which is a covariant vector field. Other examples are the following:

1. The derivatives of the Levi-Civita symbols and of all the Kronecker deltas are identically equal to zero, and hence are tensors.

2. If $T_{\alpha_1 \ldots \alpha_k}$ is a tensor field (of weight 0), then $T_{[\alpha_1 \ldots \alpha_k, \alpha_{k+1}]}$ (a generalisation of rotation) is a tensor field, too.

3. If $T^{\alpha_1 \ldots \alpha_k}$ is a tensor density field of weight -1, and is antisymmetric in all the indices, then $T^{\alpha_1 \ldots \alpha_k}{}_{,\alpha_k}$ (a generalisation of the divergence of a vector field) is also a tensor density field of weight -1.

We will verify the third example. By assumption, when the coordinates are transformed from $\{x\}$ to $\{x'\}$, $T^{\alpha_1 \ldots \alpha_k}$ transforms as follows:

$$T^{\alpha'_1 \ldots \alpha'_k} = \left(\frac{\partial(x')}{\partial(x)}\right)^{-1} x^{\alpha'_1}{}_{,\alpha_1} \ldots x^{\alpha'_k}{}_{,\alpha_k}\, T^{\alpha_1 \ldots \alpha_k} . \tag{4.2}$$

Let us differentiate this by $x^{\alpha'_k}$ and contract the result by α'_k. In that term, in which the differentiation acts on the determinant $(\partial(x')/\partial(x))^{-1} = \partial(x)/\partial(x')$, we will develop the determinant according to (3.38). We have

$$T^{\alpha'_1 \ldots \alpha'_k}{}_{,\alpha'_k} = \frac{1}{n!} \left(\sum_{i=1}^{n} \delta^{\rho'_1 \ldots \rho'_n}_{\sigma_1 \ldots \sigma_n} x^{\sigma_1}{}_{,\rho'_1} \ldots x^{\sigma_i}{}_{,\rho'_i \alpha'_k} \ldots x^{\sigma_n}{}_{,\rho'_n}\right)$$

$$\times x^{\alpha'_1}{}_{,\alpha_1} \ldots x^{\alpha'_k}{}_{,\alpha_k}\, T^{\alpha_1 \ldots \alpha_k}$$

$$+ \frac{\partial(x)}{\partial(x')} \left(\sum_{i=1}^{k-1} x^{\alpha'_1}{}_{,\alpha_1} \ldots x^{\alpha'_i}{}_{,\alpha_i \alpha'_k} \ldots x^{\alpha'_k}{}_{,\alpha_k}\right) T^{\alpha_1 \ldots \alpha_k}$$

$$+ \frac{\partial(x)}{\partial(x')} x^{\alpha'_1}{}_{,\alpha_1} \ldots x^{\alpha'_{k-1}}{}_{,\alpha_{k-1}} x^{\alpha'_k}{}_{,\alpha_k \alpha'_k} T^{\alpha_1 \ldots \alpha_k}$$

$$+ \frac{\partial(x)}{\partial(x')} x^{\alpha'_1}{}_{,\alpha_1} \ldots x^{\alpha'_k}{}_{,\alpha_k} T^{\alpha_1 \ldots \alpha_k}{}_{,\rho} x^\rho{}_{,\alpha'_k} . \tag{4.3}$$

In the first term we recall that δ is antisymmetric with respect to both upper and lower indices, so interchanging simultaneously a pair of upper indices and a pair of lower indices does not change it. Hence, the products in the first term can be ordered so that the factor with the second derivative is contracted with the last indices of δ. After such an ordering we see that all the n components of the sum are identical, so

$$\text{first term} = \frac{1}{(n-1)!} \left(\delta^{\rho'_1 \ldots \rho'_n}_{\sigma_1 \ldots \sigma_n} x^{\sigma_1}{}_{,\rho'_1} \ldots x^{\sigma_{n-1}}{}_{,\rho'_{n-1}} x^{\sigma_n}{}_{,\rho'_n \alpha'_k} \right)$$

$$\times x^{\alpha'_1}{}_{,\alpha_1} \ldots x^{\alpha'_k}{}_{,\alpha_k} T^{\alpha_1 \ldots \alpha_k} . \tag{4.4}$$

In the second term of (4.3), each component of the sum contains $x^{\alpha'_i}{}_{,\alpha_i \alpha'_k} x^{\alpha'_k}{}_{,\alpha_k} = x^{\alpha'_i}{}_{,\alpha_i \alpha_k}$, which is symmetric in (α_i, α_k). However, it is then contracted with respect to both (α_i, α_k) with the $T^{\alpha_1 \ldots \alpha_k}$, which is antisymmetric in these two indices. Such a contraction is always identically equal to zero, hence the second term is zero.

In the third term, we note that $x^{\alpha'_k}{}_{,\alpha_k}$ is an element of the inverse matrix to $[x^\alpha{}_{,\alpha'}]$. Thus, $x^{\alpha'_k}{}_{,\alpha_k}$ is equal to the cofactor of the element transposed to $\left({}^{\alpha'_k}{}_{\alpha_k} \right)$ in the matrix $[x^\alpha{}_{,\alpha'}]$, divided by the determinant of $[x^\alpha{}_{,\alpha'}]$. The element transposed to $\left({}^{\alpha'_k}{}_{\alpha_k} \right)$ in $[x^\alpha{}_{,\alpha'}]$ is $x^{\alpha_k}{}_{,\alpha'_k}$, while $\det [x^\alpha{}_{,\alpha'}] = \partial(x)/\partial(x')$. Hence, using (3.42), we have

$$x^{\alpha'_k}{}_{,\alpha_k \alpha'_k} = \left[\left(\frac{\partial(x)}{\partial(x')} \right)^{-1} \frac{1}{(n-1)!} \delta^{\alpha'_k \mu'_1 \ldots \mu'_{n-1}}_{\alpha_k \nu_1 \ldots \nu_{n-1}} x^{\nu_1}{}_{,\mu'_1} \ldots x^{\nu_{n-1}}{}_{,\mu'_{n-1}} \right]_{,\alpha'_k} . \tag{4.5}$$

We see at once that the differentiation of the factors $x^{\nu_i}{}_{,\mu'_i}$ by $x^{\alpha'_k}$ will give zero contributions because the results of these differentiations, $x^{\nu_i}{}_{,\mu'_i \alpha'_k}$, are symmetric in (μ'_i, α'_k) and will be contracted with the delta which is antisymmetric in the same indices. The only nonzero contribution will thus be from the derivative of the determinant, so

$$\text{third term} = - \left(\frac{\partial(x)}{\partial(x')} \right)^{-2} \times \frac{1}{(n-1)!} \cdot \delta^{\rho'_1 \ldots \rho'_n}_{\sigma_1 \ldots \sigma_n} x^{\sigma_1}{}_{,\rho'_1} \ldots x^{\sigma_{n-1}}{}_{,\rho'_{n-1}} x^{\sigma_n}{}_{,\rho'_n \alpha'_k}$$

$$\times \frac{1}{(n-1)!} \delta^{\alpha'_k \mu'_1 \ldots \mu'_{n-1}}_{\alpha_k \nu_1 \ldots \nu_{n-1}} x^{\nu_1}{}_{,\mu'_1} \ldots x^{\nu_{n-1}}{}_{,\mu'_{n-1}} . \tag{4.6}$$

Note that the last line above is just

$$\left(\frac{\partial(x)}{\partial(x')} x^{\alpha'_k}{}_{,\alpha_k} \right) .$$

Using this we see that the third term in (4.3) has the same absolute value as the first one, but is of opposite sign. Hence, the first and third terms cancel each other, and the final result in (4.3) is the last term which equals

$$T^{\alpha'_1 \ldots \alpha'_k}{}_{,\alpha'_k} = \frac{\partial(x)}{\partial(x')} x^{\alpha'_1}{}_{,\alpha_1} \ldots x^{\alpha'_{k-1}}{}_{,\alpha_{k-1}} T^{\alpha_1 \ldots \alpha_k}{}_{,\alpha_k} , \tag{4.7}$$

where we have used $x^{\alpha'_k}{}_{,\alpha_k} x^\rho{}_{,\alpha'_k} = x^\rho{}_{,\alpha_k} = \delta^\rho{}_{\alpha_k}$. Hence, $T^{\alpha_1\ldots\alpha_k}{}_{,\alpha_k}$ is indeed a tensor density field of weight -1.

The statement proved above also holds for the case when T^α has just one index and is a contravariant vector density field. Then, the first and third terms still cancel each other while the second term in (4.3) simply does not exist.

The fact that partial derivatives of tensor fields are not tensor fields themselves is unfortunate because the laws of physics are usually formulated as differential equations. We would like these equations to have the form (a tensor) $= 0$, since this would hold in all coordinate systems. This suggests the following idea: let us define a 'generalised differentiation', which will yield tensor fields when acting on tensor fields, and will coincide with ordinary differentiation when acting on scalars and the Kronecker deltas, for which partial derivatives are tensors. Then, we will replace the partial derivatives with the generalised derivatives in the basic equations. We guess that this generalised differentiation, called **covariant differentiation**, will reduce to ordinary differentiation in certain privileged coordinate systems.

4.2 Axioms of the covariant derivative

We want the covariant differentiation to have all the algebraic properties of an ordinary differentiation, and in addition to yield tensor densities when acting on tensor densities. We will denote the covariant derivative by ∇_α or ${}_{|\alpha}$ or $\mathrm{D}/\partial x^\alpha$. The symbols $T_i[w, k, l]$ will denote tensor densities (of weight w, k times contravariant and l times covariant) whose explicit indices are irrelevant.

Specifically, we want ∇_α to have the following properties:

1. To be distributive with respect to addition:

$$\nabla_\alpha \left(T_1[w, k, l] + T_2[w, k, l]\right) = \nabla_\alpha \left(T_1[w, k, l]\right) + \nabla_\alpha \left(T_2[w, k, l]\right). \tag{4.8}$$

2. To obey the Leibniz rule when acting on a tensor product:

$$\nabla_\alpha \left(T_1 \otimes T_2\right) = \left(\nabla_\alpha T_1\right) \otimes T_2 + T_1 \otimes \left(\nabla_\alpha T_2\right). \tag{4.9}$$

3. To reduce to the partial derivative when acting on a scalar field:

$$\nabla_\alpha \Phi = \Phi_{,\alpha}. \tag{4.10}$$

4. To yield zero when acting on the Levi-Civita symbols and Kronecker deltas:

$$\nabla_\alpha \epsilon^{\alpha_1\ldots\alpha_n} = 0, \qquad \nabla_\alpha \epsilon_{\alpha_1\ldots\alpha_n} = 0, \qquad \nabla_\alpha \delta^\alpha{}_\beta = 0. \tag{4.11}$$

The last equation implies at once that

$$\nabla_\alpha \delta^{\alpha_1\ldots\alpha_k}_{\beta_1\ldots\beta_k} = 0 \tag{4.12}$$

for any k. It also implies that ∇_α commutes with contraction.

5. When acting on a tensor density field of type $[w, k, l]$, it should produce a tensor density field of type $[w, k, l + 1]$, thus

$$\nabla_\alpha \left(T_1[w, k, l]\right) = T_2[w, k, l + 1].$$

Only the last property does not hold for partial derivatives.

From these postulated properties we will now derive an operational formula for the covariant derivative.

4.3 A field of bases on a manifold and scalar components of tensors

In every tangent space to an n-dimensional manifold M_n we can choose a set of n linearly independent contravariant vectors, $\{e_1{}^\alpha, \ldots, e_n{}^\alpha\}$. The indices a, b, c, $\ldots = 1, \ldots, n$ will label vectors (while Greek indices label coordinate components of tensors). After such a basis is chosen at every $x \in M_n$, we consider the set of n vector *fields*:

$$x \to e_a{}^\alpha(x), \qquad a = 1, \ldots, n.$$

The collection of quantities $\{e_a{}^\alpha(x)\}$, $\alpha = 1, \ldots, n$, $a = 1, \ldots, n$ forms a matrix whose elements are functions on the manifold. Since all the vectors are linearly independent at every x, the matrix is nonsingular, so there exists an inverse matrix $e^a{}_\alpha$ that obeys

$$e^a{}_\alpha e_a{}^\beta = \delta^\beta{}_\alpha. \tag{4.13}$$

A subset of the matrix $\|e^a{}_\alpha\|$ defined by a fixed a is then a covariant vector field. The set of fields corresponding to all values of a forms a **dual basis** to $\{e_a{}^\alpha(x)\}$, $a = 1, \ldots, n$. By virtue of the $\{e_a{}^\alpha(x)\}$ being linearly independent, (4.13) implies the following:

$$e^a{}_\alpha e_b{}^\alpha = \delta^a{}_b. \tag{4.14}$$

For any tensor field (of weight 0) $T^{\alpha_1 \ldots \alpha_k}_{\beta_1 \ldots \beta_l}$, the collection of quantities

$$T^{a_1 \ldots a_k}_{b_1 \ldots b_l} \stackrel{\text{def}}{=} e^{a_1}{}_{\alpha_1} \ldots e^{a_k}{}_{\alpha_k} e_{b_1}{}^{\beta_1} \ldots e_{b_l}{}^{\beta_l} T^{\alpha_1 \ldots \alpha_k}_{\beta_1 \ldots \beta_l}, \tag{4.15}$$

labelled by the indices $a_1, \ldots, a_k, b_1, \ldots, b_l = 1, \ldots, n$, is a set of n^{k+l} scalar fields that uniquely represents the set of n^{k+l} coordinate components of the tensor field $T^{\alpha_1 \ldots \alpha_k}_{\beta_1 \ldots \beta_l}$. This is so because, in consequence of (4.13) – (4.14), an inverse formula to (4.15) exists that allows one to calculate $T^{\alpha_1 \ldots \alpha_k}_{\beta_1 \ldots \beta_l}$ when $T^{a_1 \ldots a_k}_{b_1 \ldots b_l}$ are given:

$$T^{\alpha_1 \ldots \alpha_k}_{\beta_1 \ldots \beta_l} = e_{a_1}{}^{\alpha_1} \ldots e_{a_k}{}^{\alpha_k} e^{b_1}{}_{\beta_1} \ldots e^{b_l}{}_{\beta_l} T^{a_1 \ldots a_k}_{b_1 \ldots b_l}. \tag{4.16}$$

Let us denote

$$e := \det \|e_a{}^\alpha\| = \frac{1}{n!} \epsilon^{a_1 \ldots a_n} \epsilon_{\alpha_1 \ldots \alpha_n} e_{a_1}{}^{\alpha_1} \ldots e_{a_n}{}^{\alpha_n}. \tag{4.17}$$

Now, $\epsilon_{\alpha_1 \ldots \alpha_n}$ is a tensor density of weight $+1$, whereas $\epsilon^{a_1 \ldots a_n}$ is a set of scalars because it depends only on the basis in the vector space, and not on the coordinate system. Hence, e is a scalar density of weight $+1$.

The quantity e, together with $\{e_a{}^\alpha\}$ and $\{e^a{}_\alpha\}$, can be used to represent tensor density fields by sets of scalar fields. Let $T^{\alpha_1 \ldots \alpha_k}_{\beta_1 \ldots \beta_l}$ now be a tensor density field of type $[w, k, l]$; then each element of the set

$$T^{a_1 \ldots a_k}_{b_1 \ldots b_l} := e^{-w} e^{a_1}{}_{\alpha_1} \ldots e^{a_k}{}_{\alpha_k} e_{b_1}{}^{\beta_1} \ldots e_{b_l}{}^{\beta_l} T^{\alpha_1 \ldots \alpha_k}_{\beta_1 \ldots \beta_l} \tag{4.18}$$

is a scalar field. The set $T^{a_1 \ldots a_k}_{b_1 \ldots b_l}$ then uniquely defines $T^{\alpha_1 \ldots \alpha_k}_{\beta_1 \ldots \beta_l}$ via a formula analogous to (4.16), with the factor e^{+w} added. The weight w is an extra bit of information.

4.4 The affine connection

We now define the set of quantities

$$\Gamma^\alpha{}_{\beta\gamma} = -e_s{}^\alpha \left(\nabla_\gamma - \partial_\gamma\right) e^s{}_\beta. \tag{4.19}$$

The elements of this set are called the coefficients of **affine connection**. When specified explicitly, they tell us how the covariant derivative acts on the basis vector fields. Later we will consider manifolds in which these coefficients can be calculated from more basic objects (see Chapter 7). For now, we consider manifolds in which the $\Gamma^\alpha{}_{\beta\gamma}$ are just given.

The $\Gamma^\alpha{}_{\beta\gamma}$ do not depend on the choice of basis. Let us assume that $\{e_a{}^\alpha\}$ and $\{e_{a'}{}^\alpha\}$ are two different bases. The vector fields of the second basis can then be decomposed in the first basis

$$e_{a'}{}^\alpha = A^b{}_{a'} e_b{}^\alpha, \tag{4.20}$$

and then the elements of the transformation matrix

$$A^b{}_{a'} = e^b{}_\alpha e_{a'}{}^\alpha \tag{4.21}$$

are scalar fields. Hence, $A^b{}_{a'|\alpha} = A^b{}_{a',\alpha}$ and $\left(A^{-1}\right)^{c'}{}_{d|\alpha} = \left(A^{-1}\right)^{c'}{}_{d,\alpha}$. Then, calculating the $\Gamma^\alpha{}_{\beta\gamma}$ in the basis $\{e_{a'}{}^\alpha\}$ we have

$$\begin{aligned}
\left(\Gamma^\alpha{}_{\beta\gamma}\right)_{e'} &= -A^r{}_{s'} e_r{}^\alpha \left(\nabla_\gamma - \partial_\gamma\right) \left[\left(A^{-1}\right)^{s'}{}_s e^s{}_\beta\right] \\
&= -\delta^r{}_s e_r{}^\alpha \left(\nabla_\gamma - \partial_\gamma\right) e^s{}_\beta = \left(\Gamma^\alpha{}_{\beta\gamma}\right)_e.
\end{aligned} \tag{4.22}$$

Now let us note that the $\Gamma^\alpha{}_{\beta\gamma}$ *are not tensor fields*. When coordinates are transformed, these coefficients change as follows:

$$\Gamma^{\alpha'}{}_{\beta'\gamma'} = x^{\alpha'}{}_{,\alpha}\, x^\beta{}_{,\beta'}\, x^\gamma{}_{,\gamma'}\, \Gamma^\alpha{}_{\beta\gamma} + x^{\alpha'}{}_{,\rho}\, x^\rho{}_{,\beta'\gamma'}. \tag{4.23}$$

However, the antisymmetric part

$$\Omega^\alpha{}_{\beta\gamma} \overset{\text{def}}{=} \Gamma^\alpha{}_{[\beta\gamma]} \tag{4.24}$$

is a tensor, since $x^\rho{}_{,[\beta'\gamma']} = 0$. It is called the **torsion tensor**.

4.5 The explicit formula for the covariant derivative of tensor density fields

In order to obtain an explicit formula for the covariant derivative, we need to know two other properties of the connection coefficients:

$$\text{(I)} \qquad \Gamma^\alpha{}_{\beta\gamma} = e^s{}_\beta \left(\nabla_\gamma - \partial_\gamma\right) e_s{}^\alpha. \tag{4.25}$$

$$\text{(II)} \qquad \nabla_\alpha \left(e^w\right) = w e^{w-1} \nabla_\alpha e. \tag{4.26}$$

The verification of (I) is easy. To verify (II), consider the quantity

$$F_\alpha(w) \overset{\text{def}}{=} e^{-w} \nabla_\alpha \left(e^w\right). \tag{4.27}$$

Using the postulated properties of the covariant derivative, we obtain

$$F_\alpha\left(w_1 + w_2\right) = F_\alpha(w_1) + F_\alpha(w_2). \tag{4.28}$$

Every continuous function that has the property $f(w_1 + w_2) = f(w_1) + f(w_2)$ for all real w_1 and w_2 also has the property $f(w) = f(1)w$. Hence $e^{-w}\nabla_\alpha(e^w) = we^{-1}\nabla_\alpha e$, which is equivalent to (4.26).

Equation (4.26) holds also for partial derivatives, so

$$e^{-w}(\nabla_\gamma - \partial_\gamma)(e^w) = we^{-1}(\nabla_\gamma - \partial_\gamma)e. \tag{4.29}$$

Now, using (3.38) and (3.42), we obtain

$$(\nabla_\gamma - \partial_\gamma)e = ee^{a_n}{}_{\rho_n}(\nabla_\gamma - \partial_\gamma)e_{a_n}{}^{\rho_n} = e\Gamma^\rho{}_{\rho\gamma}. \tag{4.30}$$

At this point, we can derive the formula for the covariant derivative of an arbitrary tensor density field. Let us convert the tensor density field to the set of scalar fields given by (4.18). Axiom 3 implies then

$$(\nabla_\gamma - \partial_\gamma)T^{a_1 \ldots a_k}_{b_1 \ldots b_l} = 0. \tag{4.31}$$

On the other hand, using (4.18) and axiom 2, we obtain

$$(\nabla_\gamma - \partial_\gamma)T^{a_1 \ldots a_k}_{b_1 \ldots b_l} = -we^{-w}\Gamma^\rho{}_{\rho\gamma}e^{a_1}{}_{\alpha_1}\cdots e^{a_k}{}_{\alpha_k}e_{b_1}{}^{\beta_1}\cdots e_{b_l}{}^{\beta_l}T^{\alpha_1 \ldots \alpha_k}_{\beta_1 \ldots \beta_l}$$

$$+ e^{-w}\sum_{i=1}^{k}e^{a_1}{}_{\alpha_1}\cdots(\nabla_\gamma - \partial_\gamma)e^{a_i}{}_{\alpha_i}\cdots e^{a_k}{}_{\alpha_k}e_{b_1}{}^{\beta_1}\cdots e_{b_l}{}^{\beta_l}T^{\alpha_1 \ldots \alpha_k}_{\beta_1 \ldots \beta_l}$$

$$+ e^{-w}\sum_{j=1}^{l}e^{a_1}{}_{\alpha_1}\cdots e^{a_k}{}_{\alpha_k}e_{b_1}{}^{\beta_1}\cdots(\nabla_\gamma - \partial_\gamma)e_{b_j}{}^{\beta_j}\cdots e_{b_l}{}^{\beta_l}T^{\alpha_1 \ldots \alpha_k}_{\beta_1 \ldots \beta_l}$$

$$+ e^{-w}e^{a_1}{}_{\alpha_1}\cdots e^{a_k}{}_{\alpha_k}e_{b_1}{}^{\beta_1}\cdots e_{b_l}{}^{\beta_l}(\nabla_\gamma - \partial_\gamma)T^{\alpha_1 \ldots \alpha_k}_{\beta_1 \ldots \beta_l}. \tag{4.32}$$

Now we convert this back to coordinate components, by contracting it with $e^w e_{a_1}{}^{\alpha_1}\cdots e_{a_k}{}^{\alpha_k}e^{b_1}{}_{\beta_1}\cdots e^{b_l}{}_{\beta_l}$ and using (4.19) and (4.25):

$$0 = -w\Gamma^\rho{}_{\rho\gamma}T^{\alpha_1 \ldots \alpha_k}_{\beta_1 \ldots \beta_l} - \sum_{i=1}^{k}\Gamma^{\alpha_i}{}_{\rho_i\gamma}T^{\alpha_1 \ldots \rho_i \ldots \alpha_k}_{\beta_1 \ldots \beta_l} + \sum_{j=1}^{l}\Gamma^{\rho_j}{}_{\beta_j\gamma}T^{\alpha_1 \ldots \alpha_k}_{\beta_1 \ldots \rho_j \ldots \beta_l}$$

$$+ (\nabla_\gamma - \partial_\gamma)T^{\alpha_1 \ldots \alpha_k}_{\beta_1 \ldots \beta_l}. \tag{4.33}$$

From this, finally

$$\nabla_\gamma T^{\alpha_1 \ldots \alpha_k}_{\beta_1 \ldots \beta_l} = \partial_\gamma T^{\alpha_1 \ldots \alpha_k}_{\beta_1 \ldots \beta_l} + w\Gamma^\rho{}_{\rho\gamma}T^{\alpha_1 \ldots \alpha_k}_{\beta_1 \ldots \beta_l} + \sum_{i=1}^{k}\Gamma^{\alpha_i}{}_{\rho_i\gamma}T^{\alpha_1 \ldots \rho_i \ldots \alpha_k}_{\beta_1 \ldots \beta_l}$$

$$- \sum_{j=1}^{l}\Gamma^{\rho_j}{}_{\beta_j\gamma}T^{\alpha_1 \ldots \alpha_k}_{\beta_1 \ldots \rho_j \ldots \beta_l}. \tag{4.34}$$

The following special cases of (4.34) occur frequently:

$$A^\alpha{}_{|\gamma} = A^\alpha{}_{,\gamma} + \Gamma^\alpha{}_{\rho\gamma}A^\rho; \tag{4.35}$$

$$a_{\alpha|\gamma} = a_{\alpha,\gamma} - \Gamma^\rho{}_{\alpha\gamma}a_\rho; \tag{4.36}$$

$$T^{\alpha\beta}{}_{|\gamma} = T^{\alpha\beta}{}_{,\gamma} + \Gamma^\alpha{}_{\rho\gamma}T^{\rho\beta} + \Gamma^\beta{}_{\rho\gamma}T^{\alpha\rho}; \tag{4.37}$$

$$T_{\alpha\beta|\gamma} = T_{\alpha\beta,\gamma} - \Gamma^\rho{}_{\alpha\gamma}T_{\rho\beta} - \Gamma^\rho{}_{\beta\gamma}T_{\alpha\rho}; \tag{4.38}$$

$$T^\alpha{}_{\beta|\gamma} = T^\alpha{}_{\beta,\gamma} + \Gamma^\alpha{}_{\rho\gamma}T^\rho{}_\beta - \Gamma^\rho{}_{\beta\gamma}T^\alpha{}_\rho. \tag{4.39}$$

Note that, unlike the partial derivative, the covariant derivative does not act on single components of tensor densities. It is an operator that acts on the whole tensor density and produces another tensor density.

4.6 Exercises

1. What is the condition for the 'covariant rotation' $T_{[\alpha|\beta]}$ of a covariant vector field T_α to coincide with the ordinary rotation $T_{[\alpha,\beta]}$?

2. Let $g_{\alpha\beta} = g_{(\alpha\beta)}$ be a doubly covariant tensor that is nonsingular, i.e. $\det||g_{\alpha\beta}|| \neq 0$. Let $g^{\alpha\beta}$ be its inverse matrix, i.e.

$$g^{\alpha\rho}g_{\rho\beta} = \delta^\alpha{}_\beta.$$

Show that the object defined as

$$\left\{ \begin{matrix} \alpha \\ \beta\gamma \end{matrix} \right\} = \frac{1}{2}g^{\alpha\rho}\left(g_{\beta\rho,\gamma} + g_{\gamma\rho,\beta} - g_{\beta\gamma,\rho}\right)$$

transforms under coordinate transformations by the same law as the coefficients of affine connection. What is the torsion in this case?

3. Verify that (4.28) implies $F_\alpha(w) = wF_\alpha(1)$.

5

Parallel transport and geodesic lines

5.1 Parallel transport

Let a curve C be given on a manifold with affine connection, and let v be a field of vectors along the curve, $C \ni x \to v^\alpha(x)$. In a Euclidean space, in Cartesian coordinates, the vectors $v(x)$ are parallel when

$$\frac{\mathrm{d}v^\alpha}{\mathrm{d}\tau} = \frac{\partial v^\alpha}{\partial x^\beta}\frac{\mathrm{d}x^\beta}{\mathrm{d}\tau} = 0, \tag{5.1}$$

where τ is a parameter along C, while $x^\alpha(\tau)$ are coordinates of a point on C. Then, for two points on C corresponding to τ_1 and τ_2:

$$v^\alpha(\tau_1)\underset{*}{=}v^\alpha(\tau_2) \tag{5.2}$$

(the asterisk means that the equation holds only in some coordinate systems; for example, (5.2) does not hold in polar coordinates on a plane). We will generalise the definition of parallelism so that it does not depend on the coordinates. The following definition suggests itself:

$$\frac{\mathrm{D}v^\alpha}{\mathrm{d}\tau} \overset{\mathrm{def}}{=} (\nabla_\rho v^\alpha)\frac{\mathrm{d}x^\rho}{\mathrm{d}\tau} = 0, \tag{5.3}$$

which is at the same time the definition of a covariant derivative along a curve. Using (4.35) we obtain from this

$$v^\alpha{}_{,\rho}\frac{\mathrm{d}x^\rho}{\mathrm{d}\tau} + \Gamma^\alpha{}_{\sigma\rho}v^\sigma\frac{\mathrm{d}x^\rho}{\mathrm{d}\tau} = 0. \tag{5.4}$$

This can be written equivalently as

$$\frac{\mathrm{d}v^\alpha}{\mathrm{d}\tau} = -\Gamma^\alpha{}_{\sigma\rho}v^\sigma\frac{\mathrm{d}x^\rho}{\mathrm{d}\tau}. \tag{5.5}$$

Thus, the vector $v^\alpha(\tau_1)$, while being parallely transported from the point $x^\alpha(\tau_1)$ to the point $x^\alpha(\tau_2)$, changes in the following way:

$$v^\alpha_\parallel(\tau_2) = v^\alpha(\tau_1) - \int_{x(\tau_1)}^{x(\tau_2)} \Gamma^\alpha{}_{\sigma\rho}(x)v^\sigma(x)\mathrm{d}x^\rho. \tag{5.6}$$

The expression under the integral may not be a perfect differential. Then, the result of integration will depend on the curve C. Thus, in general, the result of the parallel transport so defined depends on the path of transport, and for a closed curve

$$v^\alpha_\parallel(\tau_1) \stackrel{\text{def}}{=} v^\alpha(\tau_1) - \oint_C \Gamma^\alpha{}_{\sigma\rho}(\tau)v^\sigma(\tau)\frac{\mathrm{d}x^\rho}{\mathrm{d}\tau}\mathrm{d}\tau \neq v^\alpha(\tau_1). \tag{5.7}$$

The conditions under which parallel transport is independent of the path, i.e. under which (5.7) does not occur, will be given in Section 6.3.

The parallel transport of an arbitrary tensor density field $T^{\alpha_1\dots\alpha_k}_{\beta_1\dots\beta_l}$ is defined by

$$\frac{\mathrm{D}}{\mathrm{d}\tau}T^{\alpha_1\dots\alpha_k}_{\beta_1\dots\beta_l} = T^{\alpha_1\dots\alpha_k}_{\beta_1\dots\beta_l|\rho}\frac{\mathrm{d}x^\rho}{\mathrm{d}\tau} = 0. \tag{5.8}$$

5.2 Geodesic lines

A **geodesic line** is a curve G whose tangent vector $v^\alpha = \mathrm{d}x^\alpha/\mathrm{d}\tau$, after being parallelly transported along it from $x^\alpha(\tau_0) \in G$ to $x^\alpha(\tau) \in G$, is collinear with the tangent vector that is defined at $x(\tau)$. Hence

$$v^\alpha_\parallel(\tau)\Big|_{\tau_0 \to \tau} = \lambda(\tau)v^\alpha(\tau), \tag{5.9}$$

or, according to (5.6)

$$v^\alpha_\parallel(\tau)\Big|_{\tau_0 \to \tau} = v^\alpha(\tau_0) - \int_{\tau_0 \atop G}^{\tau} \Gamma^\alpha{}_{\sigma\rho}(t)\lambda(t)v^\sigma(t)v^\rho(t)\mathrm{d}t = \lambda(\tau)v^\alpha(\tau). \tag{5.10}$$

An example is a straight line in a Euclidean space. In that case, the integral in (5.10) is zero and, *if the length of arc is chosen as the parameter τ, $\lambda(\tau) = 1$*. A geodesic line is a generalisation of the notion of a straight line to any manifold with affine connection.

Let us differentiate (5.10) by τ. Since $v^\alpha = \mathrm{d}x^\alpha/\mathrm{d}\tau$, the result is

$$\lambda\left(\frac{\mathrm{d}^2x^\alpha}{\mathrm{d}\tau^2} + \Gamma^\alpha{}_{\sigma\rho}(\tau)\frac{\mathrm{d}x^\sigma}{\mathrm{d}\tau}\frac{\mathrm{d}x^\rho}{\mathrm{d}\tau}\right) = -\frac{\mathrm{d}\lambda}{\mathrm{d}\tau}\frac{\mathrm{d}x^\alpha}{\mathrm{d}\tau}. \tag{5.11}$$

Now let us change the parameter as follows:

$$\tau \to s(\tau) \stackrel{\text{def}}{=} \int_{\tau_0}^\tau \frac{c}{\lambda(t)}\mathrm{d}t, \tag{5.12}$$

where c and τ_0 are arbitrary constants. Then, (5.11) becomes

$$\frac{\mathrm{d}^2x^\alpha}{\mathrm{d}s^2} + \Gamma^\alpha{}_{\sigma\rho}(s)\frac{\mathrm{d}x^\sigma}{\mathrm{d}s}\frac{\mathrm{d}x^\rho}{\mathrm{d}s} = 0, \tag{5.13}$$

or, equivalently

$$\frac{\mathrm{D}}{\mathrm{d}s}\left(\frac{\mathrm{d}x^\alpha}{\mathrm{d}s}\right) = 0. \tag{5.14}$$

The form (5.13) of the **geodesic equation** is privileged in that, in the parametrisation (5.12), the tangent vector transported parallelly along the geodesic is not only collinear with the locally defined tangent vector but also coincides with it. The parameter s that has this property exists for any $\lambda(\tau)$ such that $\lambda(\tau) \neq 0$ for every τ, and it is called the **affine parameter**. It is defined up to the linear transformations

$$s' = as + b, \qquad a, b = \text{constant}. \tag{5.15}$$

Equation (5.13) allows us to prove the following theorem:

Theorem 5.1 *In a manifold M_n with an affine connection, for every point x and for every vector v tangent to M_n at x, there exists a geodesic line passing through x that is tangent to v.*

This is so because a solution of a second-order differential equation is uniquely determined by its value at one point and its first derivative at that point.

However, two points of a manifold with affine connection cannot always be connected by one and only one geodesic. A geodesic joining two points may not exist in a nonconnected manifold, like a two-sheeted hyperboloid. On the other hand, any two points on the surface of a cylinder (where the geodesic lines are straight lines, circles and screw-lines) can be connected by an infinite number of geodesics. (Just imagine a screw-line that connects two points p and q by the shortest path, then another one that runs one extra time around the cylinder between p and q, then another one that runs two times around the cylinder, and so on.)

Note that only the symmetric part of the affine connection gives a nonzero contribution to the geodesic equation.

5.3 Exercises

1. Consider a vector on a Euclidean plane being transported parallely along a straight line. Find how its components change when they are given in polar coordinates.

2. Do the same for a vector in a 3-dimensional Euclidean space when its components are given in spherical coordinates. From the result, read out the connection coefficients of the Euclidean space in spherical coordinates.

6

The curvature of a manifold; flat manifolds

6.1 The commutator of second covariant derivatives

The covariant derivative was introduced in order that derivatives of tensor densities are still tensor densities. This advantage has a few inconvenient consequences. One of these we already know: parallel transport defined via covariant differentiation depends on the path. Another one will appear now: the second covariant derivatives do not commute.

For a proper scalar field (with weight $w = 0$) we have

$$(\nabla_\delta \nabla_\gamma - \nabla_\gamma \nabla_\delta) \, T = \nabla_\delta \, (T_{,\gamma}) - \nabla_\gamma \, (T_{,\delta}) = -2\Omega^\rho{}_{\gamma\delta} T_{,\rho} \, . \tag{6.1}$$

This is zero only in torsion-free manifolds.

For a scalar density field of weight w we have

$$\begin{aligned} (\nabla_\delta \nabla_\gamma - \nabla_\gamma \nabla_\delta) \, \check{T} &= \left(\nabla_\gamma \check{T}\right)_{,\delta} + w \Gamma^\rho{}_{\rho\delta} \nabla_\gamma \check{T} - \Gamma^\rho{}_{\gamma\delta} \nabla_\rho \check{T} \\ &\quad - \left(\nabla_\delta \check{T}\right)_{,\gamma} - w \Gamma^\rho{}_{\rho\gamma} \nabla_\delta \check{T} + \Gamma^\rho{}_{\delta\gamma} \nabla_\rho \check{T} \\ &= w \left(\Gamma^\rho{}_{\rho\gamma,\delta} - \Gamma^\rho{}_{\rho\delta,\gamma}\right) \check{T} - 2\Omega^\rho{}_{\gamma\delta} \nabla_\rho \check{T}. \end{aligned} \tag{6.2}$$

Let us leave this formula without comment for a while and let us note the analogous result for a covariant vector field:

$$\begin{aligned} (\nabla_\delta \nabla_\gamma - \nabla_\gamma \nabla_\delta) \, T_\beta &= - \left(\Gamma^\rho{}_{\beta\gamma,\delta} - \Gamma^\rho{}_{\beta\delta,\gamma}\right) T_\rho \\ &\quad + \left(\Gamma^\rho{}_{\sigma\gamma} \Gamma^\sigma{}_{\beta\delta} - \Gamma^\rho{}_{\sigma\delta} \Gamma^\sigma{}_{\beta\gamma}\right) T_\rho - 2\Omega^\sigma{}_{\gamma\delta} T_{\beta|\sigma}. \end{aligned} \tag{6.3}$$

Now let us define

$$\begin{aligned} B^\alpha{}_{\beta\gamma\delta} &\overset{\text{def}}{=} -\Gamma^\alpha{}_{\beta\gamma,\delta} + \Gamma^\alpha{}_{\beta\delta,\gamma} + \Gamma^\alpha{}_{\sigma\gamma} \Gamma^\sigma{}_{\beta\delta} - \Gamma^\alpha{}_{\sigma\delta} \Gamma^\sigma{}_{\beta\gamma} \\ &\equiv -2\Gamma^\alpha{}_{\beta[\gamma,\delta]} + 2\Gamma^\alpha{}_{\sigma[\gamma} \Gamma^\sigma{}_{|\beta|\delta]}. \end{aligned} \tag{6.4}$$

The quantity $B^\alpha{}_{\beta\gamma\delta}$ is called the **curvature tensor**. In order to see that it is indeed a tensor, note that the left-hand side of (6.3) is a tensor by definition. The quantities $\Omega^\sigma{}_{\gamma\delta}$ and $T_{\beta|\sigma}$ are tensors, too, hence the last term on the right-hand side is a tensor. It follows that $B^\rho{}_{\beta\gamma\delta} T_\rho$ is a tensor. Now, since T_ρ is an arbitrary covariant vector, $B^\rho{}_{\beta\gamma\delta}$ must be a tensor itself. (This can also be verified directly using (6.4) and (4.23).)

Now let us observe that

$$B^\rho{}_{\rho\gamma\delta} = -\Gamma^\rho{}_{\rho\gamma,\delta} + \Gamma^\rho{}_{\rho\delta,\gamma}. \tag{6.5}$$

Hence, (6.2) can be rewritten as follows:

$$(\nabla_\delta \nabla_\gamma - \nabla_\gamma \nabla_\delta) \, \check{T} = -w B^\rho{}_{\rho\gamma\delta} \check{T} - 2\Omega^\rho{}_{\gamma\delta} \check{T}_{|\rho}. \tag{6.6}$$

Thus, the same curvature tensor determines also the commutator of the covariant derivatives of a scalar density field.

Finally, for the commutator acting on a contravariant vector field

$$(\nabla_\delta \nabla_\gamma - \nabla_\gamma \nabla_\delta) \, T^\alpha = -B^\alpha{}_{\rho\gamma\delta} T^\rho - 2\Omega^\sigma{}_{\gamma\delta} T^\alpha{}_{|\sigma}. \tag{6.7}$$

It can be verified that the commutator of covariant derivatives acts on the tensor product of two tensor density fields in the following way:

$$\begin{aligned}
&(\nabla_\delta \nabla_\gamma - \nabla_\gamma \nabla_\delta) \, (T_1 \otimes T_2) \\
&= \quad [(\nabla_\delta \nabla_\gamma - \nabla_\gamma \nabla_\delta) T_1] \otimes T_2 + T_1 \otimes [(\nabla_\delta \nabla_\gamma - \nabla_\gamma \nabla_\delta) T_2]
\end{aligned} \tag{6.8}$$

(the weights and indices of T_1 and T_2 are irrelevant here). Thus, the operator $(\nabla_\delta \nabla_\gamma - \nabla_\gamma \nabla_\delta)$ has the property of an ordinary differentiation. Then, an arbitrary tensor density field of type $[w, k, l]$ behaves under covariant differentiation like a tensor product of a single scalar density of weight w, k contravariant vectors and l covariant vectors – see (4.34). Hence, we can guess the formula for the commutator of second covariant derivatives acting on an arbitrary tensor density field:

$$(\nabla_\delta \nabla_\gamma - \nabla_\gamma \nabla_\delta) \, T^{\alpha_1 \ldots \alpha_k}_{\beta_1 \ldots \beta_l} = -w B^\rho{}_{\rho\gamma\delta} T^{\alpha_1 \ldots \alpha_k}_{\beta_1 \ldots \beta_l} - \sum_{i=1}^{k} B^{\alpha_i}{}_{\rho_i \gamma\delta} T^{\alpha_1 \ldots \rho_i \ldots \alpha_k}_{\beta_1 \ldots \beta_l}$$

$$+ \sum_{j=1}^{l} B^{\rho_j}{}_{\beta_j \gamma\delta} T^{\alpha_1 \ldots \alpha_k}_{\beta_1 \ldots \rho_j \ldots \beta_l} - 2\Omega^\rho{}_{\gamma\delta} T^{\alpha_1 \ldots \alpha_k}_{\beta_1 \ldots \beta_l | \rho}. \tag{6.9}$$

(This can be formally derived by projecting $T^{\alpha_1 \ldots \alpha_k}_{\beta_1 \ldots \beta_l}$ on the basis fields and calculating the commutator for the projection in two ways, just like we did with (4.31) – (4.32).) Equation (6.9) is called the **Ricci formula**. Let us note, from (6.4)

$$B^\alpha{}_{\beta\gamma\delta} = B^\alpha{}_{\beta[\gamma\delta]}. \tag{6.10}$$

We will see in Section 7.9 that the name 'curvature tensor' evokes the right associations.

6.2 The commutator of directional covariant derivatives

Let k^α and l^α be two linearly independent contravariant vector fields determined on a manifold with affine connection, M_n. Let us calculate, for an arbitrary tensor density field, the commutator of second directional covariant derivatives along k and l, thus

$$\begin{aligned}
[\nabla_k, \nabla_l] \, T^{\alpha_1 \ldots \alpha_k}_{\beta_1 \ldots \beta_l} &\stackrel{\text{def}}{=} k^\mu \nabla_\mu \left(l^\rho \nabla_\rho T^{\alpha_1 \ldots \alpha_k}_{\beta_1 \ldots \beta_l} \right) - l^\mu \nabla_\mu \left(k^\rho \nabla_\rho T^{\alpha_1 \ldots \alpha_k}_{\beta_1 \ldots \beta_l} \right) \\
&= (k^\mu l^\rho{}_{,\mu} - l^\mu k^\rho{}_{,\mu}) \, T^{\alpha_1 \ldots \alpha_k}_{\beta_1 \ldots \beta_l | \rho} - 2 k^\mu l^\nu \Omega^\rho{}_{\mu\nu} T^{\alpha_1 \ldots \alpha_k}_{\beta_1 \ldots \beta_l | \rho} \\
&\quad + k^\mu l^\rho \left(\nabla_\mu \nabla_\rho - \nabla_\rho \nabla_\mu \right) T^{\alpha_1 \ldots \alpha_k}_{\beta_1 \ldots \beta_l}.
\end{aligned} \tag{6.11}$$

The quantity

$$[k, l]^\rho \stackrel{\text{def}}{=} k^\mu l^\rho{}_{,\mu} - l^\mu k^\rho{}_{,\mu} \tag{6.12}$$

is called the **commutator** of k and l. It is a tensor field, since

$$[k, l]^\rho \equiv k^\mu l^\rho{}_{|\mu} - l^\mu k^\rho{}_{|\mu} + 2\Omega^\rho{}_{\mu\nu} k^\mu l^\nu. \tag{6.13}$$

The term with $\Omega^\rho{}_{\mu\nu}$ in (6.11) cancels the term with torsion that will appear from (6.9). In the end, we obtain

$$[\nabla_k, \nabla_l] T^{\alpha_1 \ldots \alpha_k}_{\beta_1 \ldots \beta_l} = [k, l]^\rho T^{\alpha_1 \ldots \alpha_k}_{\beta_1 \ldots \beta_l | \rho} - w B^\rho{}_{\rho\nu\mu} k^\mu l^\nu T^{\alpha_1 \ldots \alpha_k}_{\beta_1 \ldots \beta_l}$$

$$- \sum_{i=1}^{k} B^{\alpha_i}{}_{\rho_i \nu\mu} k^\mu l^\nu T^{\alpha_1 \ldots \rho_i \ldots \alpha_k}_{\beta_1 \ldots \beta_l} + \sum_{j=1}^{l} B^{\rho_j}{}_{\beta_j \nu\mu} k^\mu l^\nu T^{\alpha_1 \ldots \alpha_k}_{\beta_1 \ldots \rho_j \ldots \beta_l}. \tag{6.14}$$

Suppose in addition that $[k, l]^\alpha = 0$. This is a necessary and sufficient condition for the existence of coordinates **adapted** simultaneously to both k and l, call them τ^a and τ^b:

$$k^\alpha = \frac{\partial x^\alpha}{\partial \tau^a}, \qquad l^\alpha = \frac{\partial x^\alpha}{\partial \tau^b}, \tag{6.15}$$

so that, if τ^a and τ^b are chosen as two of the coordinates, then, with $x'^1 = \tau^a$ and $x'^2 = \tau^b$, we have $k'^\alpha = \delta^\alpha_1$, $l'^\alpha = \delta^\alpha_2$. (For the proof see Exercise 2.) For such fields, (6.14) becomes

$$[\nabla_{\tau_a}, \nabla_{\tau_b}] T^{\alpha_1 \ldots \alpha_k}_{\beta_1 \ldots \beta_l} = -w B^\rho{}_{\rho\nu\mu} \frac{\partial x^\nu}{\partial \tau^b} \frac{\partial x^\mu}{\partial \tau^a} T^{\alpha_1 \ldots \alpha_k}_{\beta_1 \ldots \beta_l}$$

$$+ \frac{\partial x^\nu}{\partial \tau^b} \frac{\partial x^\mu}{\partial \tau^a} \left(- \sum_{i=1}^{k} B^{\alpha_i}{}_{\rho_i \nu\mu} T^{\alpha_1 \ldots \rho_i \ldots \alpha_k}_{\beta_1 \ldots \beta_l} + \sum_{j=1}^{l} B^{\rho_j}{}_{\beta_j \nu\mu} T^{\alpha_1 \ldots \alpha_k}_{\beta_1 \ldots \rho_j \ldots \beta_l} \right). \tag{6.16}$$

6.3 The relation between curvature and parallel transport

The definition of parallel transport, (5.8), is a linear homogeneous set of first-order differential equations for the functions $T^{\alpha_1 \ldots \alpha_k}_{\beta_1 \ldots \beta_l}(\tau)$. The solutions of such equations are linear functions of the initial conditions. Let the initial values of $T^{\alpha_1 \ldots \alpha_k}_{\beta_1 \ldots \beta_l}(\tau)$ be given at $\tau = 0$. Then the solution of (5.8) can be represented as

$$T^{\alpha_1 \ldots \alpha_k}_{\beta_1 \ldots \beta_l}(\tau) = P^{\alpha_1 \ldots \alpha_k}_{\overline{\alpha_1} \ldots \overline{\alpha_k}} {}^{\overline{\beta_1} \ldots \overline{\beta_l}}_{\beta_1 \ldots \beta_l}(\tau) T^{\overline{\alpha_1} \ldots \overline{\alpha_k}}_{\overline{\beta_1} \ldots \overline{\beta_l}}(0), \tag{6.17}$$

where $P^{\alpha_1 \ldots \alpha_k}_{\overline{\alpha_1} \ldots \overline{\alpha_k}} {}^{\overline{\beta_1} \ldots \overline{\beta_l}}_{\beta_1 \ldots \beta_l}(\tau)$ is a linear operator that maps the initial value $T^{\overline{\alpha_1} \ldots \overline{\alpha_k}}_{\overline{\beta_1} \ldots \overline{\beta_l}}(0)$ into the running value $T^{\alpha_1 \ldots \alpha_k}_{\beta_1 \ldots \beta_l}(\tau)$. It has two sets of indices, those with the bar correspond to the initial point $x^\alpha(0)$, those without a bar correspond to the running point $x^\alpha(\tau)$. We will call it the **propagator of parallel transport**. It depends on the two points and, apart from exceptional cases, on the curve along which the transport occurs. It transforms like a tensor density at the point $x^\alpha(\tau)$, with the indices ${}^{\alpha_1 \ldots \alpha_k}_{\beta_1 \ldots \beta_l}$, multiplied in the sense of a tensor product by a tensor density at the point $x^\alpha(0)$, with the indices ${}^{\overline{\beta_1} \ldots \overline{\beta_l}}_{\overline{\alpha_1} \ldots \overline{\alpha_k}}$. Such an object is called a **bi-tensor**. In consequence of (5.8), the parallel propagator must obey

$$\frac{\mathrm{D}}{\mathrm{d}\tau} P^{\alpha_1 \ldots \alpha_k}_{\overline{\alpha_1} \ldots \overline{\alpha_k}} {}^{\overline{\beta_1} \ldots \overline{\beta_l}}_{\beta_1 \ldots \beta_l}(\tau) = 0, \tag{6.18}$$

and, in consequence of (6.17), it must obey the initial condition

$$P^{\alpha_1 \ldots \alpha_k}_{\overline{\alpha_1} \ldots \overline{\alpha_k}} {}^{\overline{\beta_1} \ldots \overline{\beta_l}}_{\beta_1 \ldots \beta_l}(0) = \delta^{\alpha_1}_{\overline{\alpha_1}} \ldots \delta^{\alpha_k}_{\overline{\alpha_k}} \delta^{\overline{\beta_1}}_{\beta_1} \ldots \delta^{\overline{\beta_l}}_{\beta_l}. \tag{6.19}$$

In (6.18), the covariant differentiation applies only to the indices without a bar because the initial point does not depend on τ.

We will give a special name to the propagator of parallel transport along a closed curve. Let $0 \leq \tau \leq 1$ be a parameter on a closed curve and let $x^\alpha(0) = x^\alpha(1)$ be its initial-final point. We denote

$$P^{\alpha_1...\alpha_k \; \overline{\beta_1}...\overline{\beta_l}}_{\overline{\alpha_1}...\overline{\alpha_k} \; \beta_1...\beta_l}(1) = S^{\alpha_1...\alpha_k \; \overline{\beta_1}...\overline{\beta_l}}_{\overline{\alpha_1}...\overline{\alpha_k} \; \beta_1...\beta_l}. \tag{6.20}$$

In the symbol $S^{.....}_{.....}$ all the indices refer to the point $x^\alpha(0) = x^\alpha(1)$, hence $S^{.....}_{.....}$ is an ordinary tensor density field of type $[w, k + l, k + l]$.

Now let us define elementary propagators: P_w to transport scalar densities, $P^\alpha_{\overline{\alpha}}$ to transport contravariant vectors and $P^{\overline{\beta}}_\beta$ to transport covariant vectors. All of them obey (6.18) and the initial conditions

$$P_w(0) = 1, \qquad P^\alpha_{\overline{\alpha}}(0) = \delta^\alpha_{\overline{\alpha}}, \qquad P^{\overline{\beta}}_\beta(0) = \delta^{\overline{\beta}}_\beta. \tag{6.21}$$

The following equation holds:

$$P^{\alpha_1...\alpha_k \; \overline{\beta_1}...\overline{\beta_l}}_{\overline{\alpha_1}...\overline{\alpha_k} \; \beta_1...\beta_l} = P_w P^{\alpha_1}_{\overline{\alpha_1}} \dots P^{\alpha_k}_{\overline{\alpha_k}} P^{\overline{\beta_1}}_{\beta_1} \dots P^{\overline{\beta_l}}_{\beta_l} \tag{6.22}$$

because both its sides obey the same set of differential equations (6.18) and the same set of initial conditions (6.19).

Similarly, the general propagator of transport along a closed curve, $S^{.....}_{.....}$, is just the tensor product of elementary propagators:

$$S^{\alpha_1...\alpha_k \; \overline{\beta_1}...\overline{\beta_l}}_{\overline{\alpha_1}...\overline{\alpha_k} \; \beta_1...\beta_l} = S_w S^{\alpha_1}_{\overline{\alpha_1}} \dots S^{\alpha_k}_{\overline{\alpha_k}} S^{\overline{\beta_1}}_{\beta_1} \dots S^{\overline{\beta_l}}_{\beta_l}. \tag{6.23}$$

Hence, in order to investigate the properties of general propagators, it suffices to deal with the elementary propagators.

Let us consider the parallel transport along a curve $x^\alpha(\tau)$ from the point $x^\alpha(0)$ to an arbitrary point $x^\alpha(\tau)$, of the following scalar field:

$$V_\alpha(\tau) W^\alpha(\tau) \equiv P^{\overline{\alpha}}_\alpha(\tau) P^\alpha_{\overline{\beta}}(\tau) V_{\overline{\alpha}}(0) W^{\overline{\beta}}(0). \tag{6.24}$$

Let us take, for the beginning, the special case when the above scalar is constant along the whole curve. Then

$$V_\alpha(\tau) W^\alpha(\tau) = V_{\overline{\alpha}}(0) W^{\overline{\alpha}}(0) = \delta^{\overline{\alpha}}_{\overline{\beta}} V_{\overline{\alpha}}(0) W^{\overline{\beta}}(0). \tag{6.25}$$

Comparing this with the previous equation, we see that

$$\left(P^{\overline{\alpha}}_\alpha(\tau) P^\alpha_{\overline{\beta}}(\tau) - \delta^{\overline{\alpha}}_{\overline{\beta}} \right) V_{\overline{\alpha}}(0) W^{\overline{\beta}}(0) = 0. \tag{6.26}$$

However, the vectors $V_{\overline{\alpha}}(0)$ and $W^{\overline{\beta}}(0)$ are arbitrary, so

$$P^{\overline{\alpha}}_\rho(\tau) P^\rho_{\overline{\beta}}(\tau) = \delta^{\overline{\alpha}}_{\overline{\beta}}. \tag{6.27}$$

Now let us contract this equation with $P^{\overline{\beta}}_\gamma(\tau) V_{\overline{\alpha}}(0)$. The result is

$$\left(P^\rho_{\overline{\beta}} P^{\overline{\beta}}_\gamma - \delta^\rho_\gamma \right) P^{\overline{\alpha}}_\rho(\tau) V_{\overline{\alpha}}(0) = \left(P^\rho_{\overline{\beta}} P^{\overline{\beta}}_\gamma - \delta^\rho_\gamma \right) V_\rho(\tau) = 0. \tag{6.28}$$

Since $V_\rho(\tau)$ is an arbitrary vector, we have as a consequence

$$P_{\overline{\beta}}^\alpha P_\gamma^{\overline{\beta}} = \delta_\gamma^\alpha. \tag{6.29}$$

Equations (6.27) and (6.29) show that the operators $P_\alpha^{\overline{\alpha}}$ and $P_{\overline{\beta}}^\beta$ are inverse to each other: $P_{\overline{\beta}}^\beta$ is not only the propagator of parallel transport of a contravariant vector from $x^\alpha(0)$ to $x^\alpha(\tau)$ but at the same time also the propagator of parallel transport of a covariant vector from $x^\alpha(\tau)$ to $x^\alpha(0)$ *along the same curve.* A similar duality exists for $P_\alpha^{\overline{\alpha}}$.

In consequence of (6.27) and (6.29), the following also holds

$$S_{\overline{\rho}}^\alpha S_\beta^{\overline{\rho}} = \delta_\beta^\alpha, \qquad S_\sigma^{\overline{\gamma}} S_{\overline{\delta}}^\sigma = \delta_{\overline{\delta}}^{\overline{\gamma}}. \tag{6.30}$$

Therefore, we shall consider only the propagators P_w, S_w, $P_{\overline{\beta}}^{\overline{\alpha}}$ and $S_{\overline{\beta}}^{\overline{\alpha}}$, because $P_{\overline{\beta}}^\alpha$ and $S_{\overline{\beta}}^\alpha$ are algebraically determined by the first four.

Equation (6.18) for $P_w(\tau)$ becomes

$$\frac{\mathrm{d}P_w}{\mathrm{d}\tau} + w\Gamma^\rho{}_{\rho\sigma}\frac{\mathrm{d}x^\sigma}{\mathrm{d}\tau}P_w = 0. \tag{6.31}$$

With the initial condition $P_w(0) = 1$, this has the following solution:

$$P_w(\tau) = \exp\left(-w\int_{\substack{0\\C}}^\tau \Gamma^\rho{}_{\rho\sigma}(t)\frac{\mathrm{d}x^\sigma}{\mathrm{d}t}(t)\mathrm{d}t\right). \tag{6.32}$$

Hence,

$$S_w = \exp\left(-w\oint_C \Gamma^\rho{}_{\rho\sigma}(x)\mathrm{d}x^\sigma\right). \tag{6.33}$$

Now let us assume that the loop C can be contracted to a point (this assumption will be made in all that follows). Then we can use the Stokes theorem and express the integral along C through the integral over an arbitrary 2-surface leaf S_C spanned on C:

$$\oint_C \Gamma^\rho{}_{\rho\sigma}(x)\mathrm{d}x^\sigma = -\int_{S_C} \Gamma^\rho{}_{\rho[\gamma,\delta]}\mathrm{d}x^\gamma \wedge \mathrm{d}x^\delta = \frac{1}{2}\int_{S_C} B^\rho{}_{\rho\gamma\delta}\mathrm{d}x^\gamma \wedge \mathrm{d}x^\delta. \tag{6.34}$$

Hence, in (6.33)

$$S_w = \exp\left(-\frac{1}{2}w\int_{S_C} B^\rho{}_{\rho\gamma\delta}\mathrm{d}_2 x^{\gamma\delta}\right), \tag{6.35}$$

where $\mathrm{d}_2 x^{\gamma\delta}$ denotes the surface element $\mathrm{d}x^\gamma \wedge \mathrm{d}x^\delta$.

To solve (6.18) for $P_{\overline{\beta}}^\beta$, we span a 2-surface leaf S_C on the loop C, and then we embed C in a one-parameter family of loops $C(\varepsilon)$ such that $C(0) = x^\alpha(0) = x^\alpha(1)$ is the single initial/final point of C, and $C(1) \equiv C$ (see Fig. 6.1). Points on S_C will be labelled by τ and ε in such a way that $x^\alpha(0,\varepsilon) = x^\alpha(1,\varepsilon)$ and $x^\alpha(\tau_1,0) = x^\alpha(\tau_2,0)$ for all τ_1 and τ_2.

Then, the surface element under the integral is

$$\mathrm{d}x^\gamma \wedge \mathrm{d}x^\delta = \left(\frac{\partial x^\gamma}{\partial \tau}\frac{\partial x^\delta}{\partial \varepsilon} - \frac{\partial x^\gamma}{\partial \varepsilon}\frac{\partial x^\delta}{\partial \tau}\right)\mathrm{d}\tau\mathrm{d}\varepsilon, \tag{6.36}$$

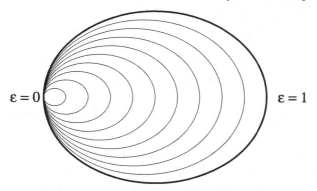

$\varepsilon = 0$ $\varepsilon = 1$

Fig. 6.1 Embedding a loop in a one-parameter family of loops. The thicker line is the loop along which we consider the parallel transport, it corresponds to $\varepsilon = 1$. As ε goes down to 0 through all the values in $[0, 1]$, the loop becomes shorter and, in the limit $\varepsilon \to 0$, degenerates to the single point where all the loops are tangent. The loops in this figure are ellipses given by the parametric equations $x = \varepsilon a[1 - \cos(2\pi t)], y = \varepsilon b \sin(2\pi t)$.

so, for an arbitrary function $F(\tau, \varepsilon)$

$$\int_{S_C} F(\tau, \varepsilon) \mathrm{d}x^\gamma \wedge \mathrm{d}x^\delta = \int_0^1 \mathrm{d}\tau \int_0^1 \mathrm{d}\varepsilon \; F(\tau, \varepsilon) \left(\frac{\partial x^\gamma}{\partial \tau} \frac{\partial x^\delta}{\partial \varepsilon} - \frac{\partial x^\gamma}{\partial \varepsilon} \frac{\partial x^\delta}{\partial \tau} \right). \tag{6.37}$$

Along each loop $C(\varepsilon)$ the propagator $P_\beta^{\overline{\beta}}(\tau, \varepsilon)$ is defined by the equation

$$\frac{\mathrm{D}}{\partial \tau} P_\beta^{\overline{\beta}}(\tau, \varepsilon) \underset{\varepsilon}{\equiv} 0, \tag{6.38}$$

with the initial condition

$$P_\beta^{\overline{\beta}}(0, \varepsilon) \underset{\varepsilon}{\equiv} \delta_\beta^{\overline{\beta}}. \tag{6.39}$$

We also have

$$S_\beta^{\overline{\beta}}(\varepsilon) = P_\beta^{\overline{\beta}}(1, \varepsilon) \tag{6.40}$$

(note that $S_\beta^{\overline{\beta}}$ in general depends on the curve, i.e. is a function of ε).

On the leaf S_C we can consider also the covariant derivatives by the parameter ε (i.e. along the curves given by $\tau = $ constant). We have

$$\frac{\mathrm{D}}{\partial \varepsilon} S_\beta^{\overline{\beta}}(\varepsilon) = \frac{\mathrm{d}}{\mathrm{d}\varepsilon} S_\beta^{\overline{\beta}}(\varepsilon) \tag{6.41}$$

because under a transformation of the parameter ε the propagator $S_\beta^{\overline{\beta}}(\varepsilon)$ transforms like a scalar (by substitution only).

Now let us use (6.16) to calculate the commutator of the second covariant derivatives by τ and ε acting on the propagator $P_\beta^{\overline{\beta}}(\tau, \varepsilon)$:

$$\left(\frac{\mathrm{D}}{\partial \tau}\frac{\mathrm{D}}{\partial \varepsilon} - \frac{\mathrm{D}}{\partial \varepsilon}\frac{\mathrm{D}}{\partial \tau}\right) P^{\overline{\beta}}_{\beta}(\tau,\varepsilon) = \frac{\partial x^{\gamma}}{\partial \tau}\frac{\partial x^{\delta}}{\partial \varepsilon}B^{\rho}{}_{\beta\gamma\delta}P^{\overline{\beta}}_{\rho}(\tau,\varepsilon). \tag{6.42}$$

But the second term on the left-hand side is zero by virtue of (6.38). Knowing this, we contract (6.42) with $P^{\beta}_{\overline{\alpha}}$. Using (6.38) again we obtain

$$\frac{\mathrm{D}}{\partial \tau}\left(P^{\beta}_{\overline{\alpha}}(\tau,\varepsilon)\frac{\mathrm{D}}{\partial \varepsilon}P^{\overline{\beta}}_{\beta}(\tau,\varepsilon)\right) = \frac{\partial x^{\gamma}}{\partial \tau}\frac{\partial x^{\delta}}{\partial \varepsilon}B^{\rho}{}_{\beta\gamma\delta}P^{\overline{\beta}}_{\rho}(\tau,\varepsilon)P^{\beta}_{\overline{\alpha}}(\tau,\varepsilon). \tag{6.43}$$

The expression in parentheses on the left-hand side is a scalar with respect to the non-barred indices. For covariant differentiation with respect to τ, those quantities that carry only barred indices are scalars (the barred indices refer to the initial points of curves, where $\tau \equiv 0$). Hence, the covariant derivative by τ on the left-hand side of (6.43) reduces to the partial derivative by τ. Then, we integrate the resulting equation by τ from 0 to 1, we make use of (6.39), of the covariant constancy of $\delta^{\overline{\beta}}_{\beta}$ (axiom 4 of the covariant derivative) and of (6.40). The result is

$$S^{\beta}_{\overline{\alpha}}(\varepsilon)\frac{\mathrm{D}}{\partial \varepsilon}S^{\overline{\beta}}_{\beta}(\varepsilon) = \int_{0}^{1}\mathrm{d}\tau\,\frac{\partial x^{\gamma}}{\partial \tau}\frac{\partial x^{\delta}}{\partial \varepsilon}B^{\rho}{}_{\beta\gamma\delta}P^{\overline{\beta}}_{\rho}(\tau,\varepsilon)P^{\beta}_{\overline{\alpha}}(\tau,\varepsilon). \tag{6.44}$$

Now we contract both sides of the above equation with $S^{\overline{\alpha}}_{\alpha}$, we make use of (6.41) and integrate the result by ε from 0 to 1, obtaining

$$\begin{aligned}
S^{\overline{\beta}}_{\alpha} - \delta^{\overline{\beta}}_{\alpha} &= \int_{0}^{1}\mathrm{d}\varepsilon\oint_{0}^{1}\mathrm{d}\tau\,\frac{\partial x^{\gamma}}{\partial \tau}\frac{\partial x^{\delta}}{\partial \partial\varepsilon}B^{\rho}{}_{\beta\gamma\delta}P^{\overline{\beta}}_{\rho}(\tau,\varepsilon)P^{\beta}_{\overline{\alpha}}(\tau,\varepsilon)S^{\overline{\alpha}}_{\alpha}(\varepsilon) \\
&= \frac{1}{2}\int_{S_C}B^{\rho}{}_{\beta\gamma\delta}P^{\overline{\beta}}_{\rho}(\tau,\varepsilon)P^{\beta}_{\overline{\alpha}}(\tau,\varepsilon)S^{\overline{\alpha}}_{\alpha}(\varepsilon)\mathrm{d}_2x^{\gamma\delta}. \tag{6.45}
\end{aligned}$$

Now it is seen that if $B^{\alpha}{}_{\beta\gamma\delta} = 0$, then $S^{\overline{\beta}}_{\alpha} = \delta^{\overline{\beta}}_{\alpha}$ – the parallel transport of an arbitrary covariant vector along an arbitrary loop that can be contracted to a point reproduces the initial vector. In consequence of (6.30), the same is true for an arbitrary contravariant vector and, in consequence of (6.35), for any scalar density. Then, in consequence of (6.23), with $B^{\alpha}{}_{\beta\gamma\delta} = 0$, *any* tensor density will return to its initial value when transported around any loop that can be contracted to a point.

The converse theorem also holds: if the parallel transport reproduces the initial tensor density *for every closed loop that can be contracted to a point*, then $B^{\alpha}{}_{\beta\gamma\delta} = 0$. This is because the integral in (6.45) is then zero for any arbitrary surface leaf S_C. We have thus

Theorem 6.1 *Parallel transport along any closed loop that can be contracted to a point reproduces the initial value of every tensor density if and only if the curvature tensor is zero.*

This is at the same time a necessary and sufficient condition for the parallel transport to be independent of the path.

The manifolds for which $B^{\alpha}{}_{\beta\gamma\delta} = 0$ are called **flat**.

6.4 Covariantly constant fields of vector bases

Suppose that a set of n vector fields exists on a manifold M_n such that at every point of M_n the vectors defined by these fields are linearly independent, and moreover all the fields are covariantly constant: $e_a{}^\alpha{}_{|\beta} \underset{a}{\equiv} 0$, $a = 1, \ldots, n$. Then we have

$$0 = e_a{}^\alpha{}_{|\gamma\delta} - e_a{}^\alpha{}_{|\delta\gamma} = B^\alpha{}_{\rho\gamma\delta} e_a{}^\rho. \tag{6.46}$$

Contracting both sides of this with $e^a{}_\beta$ we obtain $B^\alpha{}_{\beta\gamma\delta} = 0$ – the necessary condition for the existence of such a field of bases.

We will show that this is a sufficient condition, too. Assume that $B^\alpha{}_{\beta\gamma\delta} = 0$. Then the result of parallel transport does not depend on the path. Hence, with a fixed initial point and fixed initial data at that point, this is a well-defined operation with a unique result at any other point of the manifold. Choose then an arbitrary point $x_0 \in M_n$, and in the tangent space to M_n at x_0 choose an arbitrary basis of vectors $e_a{}^\alpha(x_0)$. Then define bases in spaces tangent to M_n at other points as sets of vectors obtained from $e_a{}^\alpha(x_0)$ by parallel transport. Such a set of vector fields will be covariantly constant on M_n.

The field of bases $e^a{}_\alpha$, dual to $e_a{}^\alpha$, is then also covariantly constant. Choosing such bases to calculate the connection coefficients we obtain

$$\Gamma^\alpha{}_{\beta\gamma} = e_s{}^\alpha e^s{}_{\beta,\gamma} = -e^s{}_\beta e_s{}^\alpha{}_{,\gamma}. \tag{6.47}$$

6.5 A torsion-free flat manifold

If a manifold is not only flat but also torsion-free, then for a covariantly constant field of bases we obtain from (6.47)

$$0 = -e_s{}^\alpha \left(e^s{}_{\beta,\gamma} - e^s{}_{\gamma,\beta} \right). \tag{6.48}$$

Contracting this with $e^a{}_\alpha$ we obtain $e^a{}_{\beta,\gamma} - e^a{}_{\gamma,\beta} \underset{a}{\equiv} 0$. This equation implies that, in a neighbourhood of every point of the manifold, a set of scalar functions $\phi^a(x)$ exists such that $e^a{}_\beta = \phi^a{}_{,\beta}$, $a = 1, \ldots, n$. Since we assumed that the vectors $e^a{}_\alpha$ are linearly independent at every point, i.e. $\det \|e^a{}_\alpha\| \neq 0$, this now implies $\partial(\phi^1, \ldots, \phi^n)/\partial(x^1, \ldots, x^n) \neq 0$. This means that the connection between the functions $\{\phi^a\}_{a=1,\ldots,n}$ and the coordinates $\{x^\alpha\}$ is unique and thus reversible, so the functions $\{\phi^a\}$ can be chosen as new coordinates. Then, let $x^{\alpha'}(x) = \phi^{\alpha'}(x)$. In these coordinates, $e^a{}_{\alpha'} = \phi^a{}_{,\alpha'} = \partial\phi^a/\partial\phi^{\alpha'} = \delta^a{}_{\alpha'}$ and $e_a{}^{\alpha'} = \partial\phi^{\alpha'}/\partial\phi^a = \delta^{\alpha'}{}_a$. Hence, in (6.47) we have $\Gamma^{\alpha'}{}_{\beta'\gamma'} = 0$.

In a flat torsion-free manifold one can thus choose such coordinates in which the connection coefficients are zero. In these coordinates, the covariant derivatives reduce to ordinary partial derivatives; we shall call these coordinates Cartesian. A special case (but not the only one) of a flat torsion-free manifold is a Euclidean space (of any dimension).

6.6 Parallel transport in a flat manifold

Since the parallel transport in a flat manifold does not depend on the path, the propagators of parallel transport are functions that depend only on the initial point $\bar{x}$ and on the final point x, but not on the curve that is chosen to join $\overline{x}$ to x. For covariantly constant fields of bases $\{e_A{}^\alpha\}$ and $\{e^A{}_\alpha\}$ we then have

$$P_w = \left[\frac{e(x)}{e(\overline{x})}\right]^w, \qquad P^\alpha{}_{\overline{\alpha}} = e_S{}^\alpha(x)e^S{}_{\overline{\alpha}}(\overline{x}), \qquad P^{\overline{\beta}}{}_\beta = e^S{}_\beta(x)e_S{}^{\overline{\beta}}(\overline{x}). \tag{6.49}$$

These equations follow because their right-hand sides obey the differential equations for propagators and the corresponding initial conditions.

If the manifold is torsion-free in addition, then $e^A{}_\alpha$ are gradients of scalar functions. Choosing these functions as new coordinates we obtain

$$P_w(x,\overline{x}) = 1, \qquad P^\alpha{}_{\overline{\alpha}} = \delta^\alpha{}_{\overline{\alpha}}, \qquad P^{\overline{\beta}}{}_\beta = \delta_\beta{}^{\overline{\beta}}. \tag{6.50}$$

Thus, in a flat torsion-free manifold, in Cartesian coordinates, a vector that is transported parallelly has the same components at every point and the point to which the vector is attached becomes irrelevant.

6.7 Geodesic deviation

As we will see in Section 12.3, the inertial forces that act on a body moving along a geodesic exactly cancel the gravitational forces. On the other hand, the geodesic lines are the only curves privileged by the geometry of a manifold with connection, hence only they can be used as a privileged reference system. Since a gravitational field cannot be detected on a geodesic, one can only observe neighbouring geodesics from the given one and draw conclusions about the gravitational field on the basis of those observations. The vector field of geodesic deviation is a measure of the position of a particle on a neighbouring geodesic with respect to a particle on a given geodesic.

Let $U \subset M_n$ be an open subset of the manifold M_n with connection. Choose in U a one-parameter family of geodesics labelled by the parameter ε. Let $\varepsilon = 0$ correspond to that geodesic from which we will be observing the neighbourhood. Every point on any of the geodesics of the family may be identified by the values of the parameter ε (whose value identifies the geodesic) and the affine parameter s on that geodesic, $x^\alpha|_U = x^\alpha(s,\varepsilon)$. The **geodesic deviation** is the vector field

$$\delta x^\alpha(s) \stackrel{\text{def}}{=} \left.\frac{\partial x^\alpha}{\partial\varepsilon}\right|_{\varepsilon=0} \tag{6.51}$$

defined along our geodesic $\varepsilon = 0$. We shall find how the deviation changes as we move down this geodesic. Since the set $\{x^\alpha(s,\varepsilon)\}$ with fixed ε and changing s is a geodesic, the following is true:

$$\left.\frac{\mathrm{D}}{\mathrm{d}s}\frac{\partial x^\alpha}{\partial s}\right|_{\varepsilon=\text{const}} = 0. \tag{6.52}$$

Then, from (6.16)

$$\left(\frac{\mathrm{D}}{\mathrm{d}s}\frac{\mathrm{D}}{\mathrm{d}\varepsilon} - \frac{\mathrm{D}}{\mathrm{d}\varepsilon}\frac{\mathrm{D}}{\mathrm{d}s}\right)\frac{\partial x^\alpha}{\partial s} = B^\alpha{}_{\rho\mu\nu}\frac{\partial x^\rho}{\partial s}\frac{\partial x^\mu}{\partial s}\frac{\partial x^\nu}{\partial\varepsilon}. \tag{6.53}$$

But, by virtue of (6.52), the second term on the left-hand side is zero. Now note that

$$
\begin{aligned}
\frac{\mathrm{D}}{\mathrm{d}\varepsilon}\frac{\partial x^\alpha}{\partial s} &= \frac{\partial^2 x^\alpha}{\partial\varepsilon\partial s} + \Gamma^\alpha{}_{\rho\sigma}\frac{\partial x^\rho}{\partial s}\frac{\partial x^\sigma}{\partial\varepsilon} \\
&= \frac{\partial^2 x^\alpha}{\partial s\partial\varepsilon} + \Gamma^\alpha{}_{\sigma\rho}\frac{\partial x^\sigma}{\partial\varepsilon}\frac{\partial x^\rho}{\partial s} + 2\Omega^\alpha{}_{\rho\sigma}\frac{\partial x^\rho}{\partial s}\frac{\partial x^\sigma}{\partial\varepsilon} \\
&= \frac{\mathrm{D}}{\mathrm{d}s}\frac{\partial x^\alpha}{\partial\varepsilon} + 2\Omega^\alpha{}_{\rho\sigma}\frac{\partial x^\rho}{\partial s}\frac{\partial x^\sigma}{\partial\varepsilon}.
\end{aligned}
\tag{6.54}
$$

Making use of (6.54) in (6.53) we obtain

$$
\frac{\mathrm{D}^2}{\mathrm{d}s^2}\frac{\partial x^\alpha}{\partial\varepsilon} + 2\frac{\partial x^\rho}{\partial s}\frac{\mathrm{D}}{\mathrm{d}s}\left(\Omega^\alpha{}_{\rho\sigma}\frac{\partial x^\sigma}{\partial\varepsilon}\right) = B^\alpha{}_{\rho\mu\nu}\frac{\partial x^\rho}{\partial s}\frac{\partial x^\mu}{\partial s}\frac{\partial x^\nu}{\partial\varepsilon}.
\tag{6.55}
$$

Denoting $k^\alpha = \partial x^\alpha/\partial s$ (the vector field tangent to the geodesic) and δx^α as in (6.51), then substituting $\varepsilon = 0$ in the above, we finally obtain

$$
\frac{\mathrm{D}^2}{\mathrm{d}s^2}\delta x^\alpha + 2k^\rho\frac{\mathrm{D}}{\mathrm{d}s}\left(\Omega^\alpha{}_{\rho\sigma}\delta x^\sigma\right) - B^\alpha{}_{\rho\mu\nu}k^\rho k^\mu\delta x^\nu = 0.
\tag{6.56}
$$

This is the **geodesic deviation equation**. It is useful in experimental tests of general relativity, since it allows one (in principle) to calculate the curvature by measuring relative displacements of bodies moving along neighbouring geodesics.

In a flat torsion-free manifold and in the Cartesian coordinates, (6.56) simplifies to $\mathrm{d}^2\left(\delta x^\alpha\right)/\mathrm{d}s^2 = 0$, so its solution is $\delta x^\alpha = A^\alpha s + B^\alpha$, where A^α and B^α are constant vectors. From (6.51) then, the position of the point under observation, $x^\alpha(s,\varepsilon_0)$, relative to the corresponding point $x^\alpha(s,0)$ of the geodesic $\varepsilon = 0$ is given by

$$
x^\alpha(s,\varepsilon_0) = x^\alpha(s,0) + \varepsilon_0\delta x^\alpha.
\tag{6.57}
$$

Hence, in a flat torsion-free manifold two neighbouring geodesics either diverge or converge with a constant velocity, or remain parallel (when $A^\alpha = 0$). In curved manifolds and in manifolds with torsion, solutions of the geodesic deviation equation are more complicated. (Actually, it is rarely possible to find the solution explicitly. Usually, solutions are found numerically or by perturbations.)

Note that the form of (6.54) does not change when the parameter ε is transformed by $\varepsilon = \varepsilon(\varepsilon')$. The derivatives $\mathrm{d}\varepsilon'/\mathrm{d}\varepsilon$ are constant along geodesics (because ε itself is constant), hence the factor $\mathrm{d}\varepsilon'/\mathrm{d}\varepsilon$ will cancel in (6.56). Changing the parametrisation means merely choosing a different basis in the space of solutions of (6.56).

6.8 Algebraic and differential identities obeyed by the curvature tensor

The curvature tensor obeys two other sets of identities in addition to (6.10). One follows from

$$
\begin{aligned}
B^\alpha{}_{[\beta\gamma\delta]} &= \frac{1}{3!}\delta^{\rho\sigma\mu}_{\beta\gamma\delta}B^\alpha{}_{\rho\sigma\mu} = \frac{1}{3!}\delta^{\rho\sigma\mu}_{\beta\gamma\delta}\left(-2\Gamma^\alpha{}_{\rho[\sigma,\mu]} + 2\Gamma^\alpha{}_{\nu[\sigma}\Gamma^\nu{}_{|\rho|\mu]}\right) \\
&= \frac{1}{3}\delta^{\rho\sigma\mu}_{\beta\gamma\delta}\left(-\Gamma^\alpha{}_{\rho\sigma,\mu} + \Gamma^\alpha{}_{\nu\sigma}\Gamma^\nu{}_{\rho\mu}\right) = \frac{1}{3}\delta^{\rho\sigma\mu}_{\beta\gamma\delta}\left(-\Omega^\alpha{}_{\rho\sigma,\mu} + \Gamma^\alpha{}_{\nu\sigma}\Omega^\nu{}_{\rho\mu}\right) \\
&= \frac{1}{3}\delta^{\rho\sigma\mu}_{\beta\gamma\delta}\left(-\Omega^\alpha{}_{\rho\sigma|\mu} + \Gamma^\alpha{}_{\nu\mu}\Omega^\nu{}_{\rho\sigma} - \Gamma^\nu{}_{\rho\mu}\Omega^\alpha{}_{\nu\sigma} - \Gamma^\nu{}_{\sigma\mu}\Omega^\alpha{}_{\rho\nu} + \Gamma^\alpha{}_{\nu\sigma}\Omega^\nu{}_{\rho\mu}\right).
\end{aligned}
\tag{6.58}
$$

We used the fact that in the contraction over three indices with the antisymmetric $\delta^{\rho\sigma\mu}_{\beta\gamma\delta}$ the antisymmetrisation with respect to $\sigma\mu$ occurs automatically, and we expressed the partial derivative of $\Omega^\alpha{}_{\rho\sigma}$ through the corresponding covariant derivative. Now let us note that the sum of the second term and the last one is symmetric in $\sigma\mu$, so gives zero when contracted with the $\delta^{\rho\sigma\mu}_{\beta\gamma\delta}$. Thus

$$B^\alpha{}_{[\beta\gamma\delta]} = \frac{1}{3}\delta^{\rho\sigma\mu}_{\beta\gamma\delta}\left(-\Omega^\alpha{}_{\rho\sigma|\mu} - 2\Omega^\alpha{}_{\nu\sigma}\Omega^\nu{}_{\rho\mu}\right) = -2\Omega^\alpha{}_{[\beta\gamma|\delta]} - 4\Omega^\alpha{}_{\nu[\gamma}\Omega^\nu{}_{\beta\delta]}. \tag{6.59}$$

The identities of the other set are differential:

$$B^\alpha{}_{\beta[\gamma\delta|\varepsilon]} = -2\Omega^\nu{}_{[\gamma\delta}B^\alpha{}_{|\beta|\varepsilon]\nu}, \tag{6.60}$$

they are called the **Bianchi identities**. For their derivation see Exercise 3.

Recall that the curvature tensor arose by calculating the commutator of second covariant derivatives. Commutators of linear operators obey the well-known Jacobi identity. The Bianchi identities arise from the commutators of second covariant derivatives in the same way as the Jacobi identity arises from commutators of linear operators.

The Bianchi identities are important for the physical interpretation of relativity. They ensure that equations of motion of material media follow from the field equations and need not be postulated separately.

6.9 Exercises

1. Observe that coordinates can always be adapted to *one* vector field k^α so that, in the new coordinates, $k^\alpha = \delta^\alpha{}_1$.

2. Using the result of the previous exercise show that coordinates can be simultaneously adapted to two vector fields k^α and l^α so that $k'^\alpha = \delta^\alpha_1$ and $l'^\alpha = \delta^\alpha_2$ if and only if the two vector fields commute, $[k,l]^\alpha = 0$. It follows that if i vector fields all commute, then coordinates can be adapted simultaneously to all of them.

Hint. To show that this is sufficient, adapt the coordinates to k^α, then find the transformations of coordinates that preserve the property $k'^\alpha = \delta^\alpha_1$, and, using only these transformations, try to adapt the coordinates to l^α.

3. Derive the Bianchi identities (6.60).

Hint. Use Kronecker 3-deltas to calculate antisymmetrisations, in the same way in which they were used in deriving (6.59).

7

Riemannian geometry

The most detailed sources for the subject of this chapter are Eisenhart (1940) – a thorough presentation of differential geometry of curves and 2-dimensional surfaces in a Euclidean space, and Eisenhart (1964) – a textbook on Riemannian geometry in n dimensions.

7.1 The metric tensor

Up to here, we have dealt with manifolds on which the only additional objects were the affine connection coefficients and the curvature tensor calculated from them. That structure allowed us to define the parallel transport and thereby to compare directions of vectors attached at different points. However, we could not calculate distances between points or angles between vectors. Now we will add a new object that will allow for that – a symmetric second-rank covariant tensor, $g_{\alpha\beta} = g_{(\alpha\beta)}$, called the **metric tensor**. Using it, we can define the **metric form**, also called the **metric**:

$$\mathrm{d}s^2 = g_{\alpha\beta}\mathrm{d}x^\alpha\mathrm{d}x^\beta. \tag{7.1}$$

This expression (a scalar) in fact makes sense only under an integral. The length of arc of a curve $x^\alpha(\lambda)$ between the points $x^\alpha(\lambda_0)$ and $x^\alpha(\lambda_1)$ is

$$l_{\lambda_0\lambda_1} = \int_{\lambda_0}^{\lambda_1} \mathrm{d}s = \int_{\lambda_0}^{\lambda_1} \left| g_{\alpha\beta}(\lambda)\frac{\mathrm{d}x^\alpha}{\mathrm{d}\lambda}\frac{\mathrm{d}x^\beta}{\mathrm{d}\lambda}\right|^{1/2} \mathrm{d}\lambda. \tag{7.2}$$

We take the absolute value above because the form need not be positive-definite. We will explain later the meaning of $g_{\alpha\beta}\mathrm{d}x^\alpha\mathrm{d}x^\beta \leq 0$.

An example of a manifold with a metric tensor is a Euclidean space (of arbitrary dimension). Its metric form in rectangular Cartesian coordinates is

$$\mathrm{d}s^2 = \left(\mathrm{d}x^1\right)^2 + \left(\mathrm{d}x^2\right)^2 + \cdots + \left(\mathrm{d}x^n\right)^2. \tag{7.3}$$

7.2 Riemann spaces

It is logical to require that a vector $k^\alpha(\lambda_0)$ (attached at a point on the curve $x^\alpha(\lambda)$ given by $\lambda = \lambda_0$), when parallely transported along that curve in the affine parametrisation, preserves its length. What condition does such a requirement impose on the metric tensor?

Let $\lambda = s$ be an affine parameter. The length of the vector $k^\alpha(s)$, parallely transported along the curve $x^\alpha(s)$, will not change when

$$0 = \frac{\mathrm{d}}{\mathrm{d}s}\left(g_{\alpha\beta}k^\alpha k^\beta\right) = \frac{\mathrm{D}}{\mathrm{d}s}\left(g_{\alpha\beta}k^\alpha k^\beta\right) = \frac{\mathrm{D}g_{\alpha\beta}}{\mathrm{d}s}k^\alpha k^\beta \tag{7.4}$$

(see (5.3)). If this equation is to hold for every vector field k^α, then

$$0 = \frac{\mathrm{D}g_{\alpha\beta}}{\mathrm{d}s} = \frac{\mathrm{d}x^\rho}{\mathrm{d}s}g_{\alpha\beta|\rho}. \tag{7.5}$$

Now, this should hold for every curve, which implies in the end

$$g_{\alpha\beta|\rho} = 0 \Longrightarrow g_{\alpha\beta,\rho} - \Gamma^\sigma{}_{\alpha\rho}g_{\sigma\beta} - \Gamma^\sigma{}_{\beta\rho}g_{\alpha\sigma} = 0. \tag{7.6}$$

Those manifolds, on which a metric tensor obeying (7.6) is defined, and in addition

$$\Omega^\alpha{}_{\beta\gamma} = 0 \Longrightarrow \Gamma^\alpha{}_{\beta\gamma} = \Gamma^\alpha{}_{(\beta\gamma)} \tag{7.7}$$

are called **Riemann spaces**. We shall use the same name also for those manifolds on which the metric tensor is not positive-definite, although this does not agree with the habit of mathematicians. In mathematics, the manifolds on which $g_{\alpha\beta}$ is not positive-definite are called *pseudo-Riemannian* in general, and sometimes have special names, e.g. those used in relativity are called **Lorentzian**. The statements we will make further in this book apply to both the proper Riemannian and pseudo-Riemannian geometries. In order to be precise, we would thus have to say 'Riemannian and pseudo-Riemannian' every time.

Riemann spaces are the mathematical basis of relativity. Manifolds with a metric tensor that in addition have nonzero torsion are sometimes considered. However, torsion has not so far been related to any observable effects, so we shall keep the assumption (7.7).

7.3 The signature of a metric, degenerate metrics

The expressions $g_{\alpha\beta}\mathrm{d}x^\alpha\mathrm{d}x^\beta$ and $g_{\alpha\beta}k^\alpha k^\beta$ are quadratic forms. A coordinate transformation $\{x\} \to \{x'\}$ changes $g_{\alpha\beta}k^\alpha k^\beta$ to $g_{\alpha'\beta'}k^{\alpha'}k^{\beta'}$, where $k^{\alpha'} = x^{\alpha'}{}_{,\alpha}k^\alpha$ and $g_{\alpha'\beta'} = x^\alpha{}_{,\alpha'}x^\beta{}_{,\beta'}g_{\alpha\beta}$. Let X denote the matrix $\left\|x^{\alpha'}{}_{,\alpha}\right\|$, let g denote the matrix $\|g_{\alpha\beta}\|$ and let k denote the vector k^α. In the terminology of quadratic forms, the transition $k \to k' = Xk$, $g \to g' = \left(X^{-1}\right)^T gX^{-1}$ is called a transformation of the basis. For matrices of quadratic forms, the Sylvester *theorem on the inertia of forms* holds. It says that whatever method is used to diagonalise a real symmetric matrix, the numbers of diagonal elements that are positive, zero and negative are always the same. This set of signs is denoted by the symbol

$$(\underbrace{+,\ldots,+}_{n_1 \text{ times}},\underbrace{-,\ldots,-}_{n_2 \text{ times}},\underbrace{0,\ldots,0}_{n_3 \text{ times}}), \tag{7.8}$$

where n_1 is the number of positive elements, n_2 is the number of negative elements and n_3 is the number of zero elements, $n_1 + n_2 + n_3 = n = \dim V_n$. The symbol (7.8), called the **signature** of the metric tensor, is a basis-independent property of every matrix, and so is a coordinate-independent characteristic of a point of a Riemann space V_n. When $n_3 > 0$, the metric is called degenerate. When $n_2 = n_3 = 0$, the metric is positive-definite.

In relativity, 4-dimensional Riemann spaces are used, with the signature $(+ - - -)$ or $(- + + +)$ or $(+ + + -)$, depending on the convention; in this text, the signature will always be $(+ - - -)$. Since such a metric is not positive-definite, the Riemann spaces of relativity are not metric spaces. If the distance $d(x, y)$ between the points $x \in V_n$ and $y \in V_n$ is defined by the integral (7.2) along any curve (e.g. a geodesic), then $d(x, y) = 0$ does not imply $x = y$; the integral can vanish also for such pairs of points that do not coincide. The geometrical and physical meaning of $l_{\lambda_0 \lambda_1} = 0$ in (7.2) will be explained in Sec. 7.14.

Coordinate transformations preserve the signature of the metric tensor at each single point. No theorem guarantees that the signature must be the same at different points of the manifold. However, a region with a signature different from $(+ - - -)$ would have no physical interpretation. Usually, the subset on which the signature changes to an unphysical one is, in one sense or another, an 'edge' of the manifold or of the allowed coordinate patch. However, transitions of the type $(+ - - -) \to (- + - -)$ do occur, in which two coordinates interchange their roles and another becomes the privileged one. This happens, for example, at the horizon of a black hole (see Section 14.11).

For a degenerate metric, its determinant is zero and the inverse matrix to $g_{\alpha\beta}$ does not exist. With such a metric, the mapping $k^\alpha \to g_{\alpha\beta} k^\beta$ is analogous to projecting a vector on a subspace of lower dimension, so no inverse mapping exists. For nondegenerate metrics, the matrix inverse to $g_{\alpha\beta}$ does exist, is denoted $g^{\alpha\beta}$ and obeys $g^{\alpha\rho} g_{\rho\beta} = \delta^\alpha{}_\beta$. Thus, in a Riemann space with a nondegenerate metric, for every contravariant vector k^α there exists the corresponding covariant vector $g_{\alpha\beta} k^\beta$ denoted k_α, and for every covariant vector m_α there exists the corresponding contravariant vector $g^{\alpha\beta} m_\beta$ denoted m^α. The mapping $k^\alpha \to k_\alpha = g_{\alpha\beta} k^\beta$ is called **lowering the index**, the inverse mapping $m_\alpha \to m^\alpha = g^{\alpha\beta} m_\beta$ is called **raising the index**. In such a Riemann space, we can consider contravariant and covariant *components* of the same vector. In a general manifold, with no metric defined on it, there is no relation between covariant and contravariant vectors.

7.4 Christoffel symbols

Now we shall solve (7.6) for $\Gamma^\alpha{}_{\beta\gamma}$ assuming that the metric tensor is nondegenerate. Let us rewrite (7.6) three times, each time with a different permutation of the indices:

$$g_{\alpha\beta,\rho} - \Gamma^\sigma{}_{\alpha\rho} g_{\sigma\beta} - \Gamma^\sigma{}_{\beta\rho} g_{\alpha\sigma} = 0, \tag{7.9}$$

$$g_{\beta\rho,\alpha} - \Gamma^\sigma{}_{\beta\alpha} g_{\sigma\rho} - \Gamma^\sigma{}_{\rho\alpha} g_{\beta\sigma} = 0, \tag{7.10}$$

$$g_{\rho\alpha,\beta} - \Gamma^\sigma{}_{\rho\beta} g_{\sigma\alpha} - \Gamma^\sigma{}_{\alpha\beta} g_{\rho\sigma} = 0. \tag{7.11}$$

Let us add the last two equations, and subtract the first one from the result. Using (7.7) and contracting with $g^{\gamma\rho}$ we obtain

$$\Gamma^\gamma{}_{\alpha\beta} = \frac{1}{2} g^{\gamma\rho} \left(g_{\alpha\rho,\beta} + g_{\beta\rho,\alpha} - g_{\alpha\beta,\rho} \right). \tag{7.12}$$

The affine connection coefficients that are built of the metric tensor in this way are called the **Christoffel symbols** and are denoted $\left\{ {\gamma \atop \alpha\beta} \right\}$.

7.5 The curvature of a Riemann space

The curvature tensor of a Riemann space, built of the Christoffel symbols standing in place of connection coefficients, is called the **Riemann tensor** and denoted $R^{\alpha}{}_{\beta\gamma\delta}$. Since Riemann spaces are torsion-free, the identities (6.59) and (6.60) now reduce to

$$R^{\alpha}{}_{[\beta\gamma\delta]} = 0, \tag{7.13}$$

$$R^{\alpha}{}_{\beta[\gamma\delta;\epsilon]} = 0. \tag{7.14}$$

The covariant derivative in a Riemann space is denoted by a semi-colon.

The Riemann tensor obeys one more set of identities. Applying (6.9) and recalling that a metric is covariantly constant, we obtain

$$0 = g_{\alpha\beta;\gamma\delta} - g_{\alpha\beta;\delta\gamma} = R^{\rho}{}_{\alpha\gamma\delta}g_{\rho\beta} + R^{\rho}{}_{\beta\gamma\delta}g_{\alpha\rho}. \tag{7.15}$$

This can be written as

$$R_{\beta\alpha\gamma\delta} + R_{\alpha\beta\gamma\delta} = 0 \iff R_{\alpha\beta\gamma\delta} = R_{[\alpha\beta]\gamma\delta}. \tag{7.16}$$

From this, one more identity can be derived. Equation (7.13), in consequence of $R_{\alpha\beta\gamma\delta} = R_{\alpha\beta[\gamma\delta]}$, can be rewritten in the form

$$R_{\alpha\beta\gamma\delta} + R_{\alpha\gamma\delta\beta} + R_{\alpha\delta\beta\gamma} = 0. \tag{7.17}$$

Let us rewrite this equation three more times taking cyclic permutations of all four indices:

$$R_{\beta\gamma\delta\alpha} + R_{\beta\delta\alpha\gamma} + R_{\beta\alpha\gamma\delta} = 0, \tag{7.18}$$

$$R_{\gamma\delta\alpha\beta} + R_{\gamma\alpha\beta\delta} + R_{\gamma\beta\delta\alpha} = 0, \tag{7.19}$$

$$R_{\delta\alpha\beta\gamma} + R_{\delta\beta\gamma\alpha} + R_{\delta\gamma\alpha\beta} = 0. \tag{7.20}$$

Now add the first and fourth equation, and subtract the second and third from the result. Taking into account (7.16) and (6.10) we obtain

$$R_{\alpha\beta\gamma\delta} = R_{\gamma\delta\alpha\beta}, \tag{7.21}$$

which means that the Riemann tensor is symmetric with respect to the interchange of the first pair of indices with the second.

Since $R_{\alpha\beta\gamma\delta}$ is a tensor of rank 4, it has n^4 components, i.e. 256 in a 4-dimensional space. However, in consequence of the symmetries, only some of the components are algebraically independent. Antisymmetry in (α, β) provides $n(n+1)/2$ equations $R_{(\alpha\beta)\gamma\delta} = 0$ for each set (γ, δ), which leaves us with at most $n^4 - n^2 \cdot n(n+1)/2 = n^2 \cdot n(n-1)/2$ independent components. Because of antisymmetry in (γ, δ), we obtain $[n(n+1)/2] \times [n(n-1)/2]$ additional equations $R_{\alpha\beta(\gamma\delta)} = 0$, which leaves us with at most $[n(n-1)/2]^2$ independent components. From this, we have to subtract $n\begin{pmatrix} n \\ 3 \end{pmatrix}$ equations (7.13). This gives finally

$$\left[\frac{n(n-1)}{2}\right]^2 - n\begin{pmatrix} n \\ 3 \end{pmatrix} = \frac{1}{12}n^2\left(n^2 - 1\right) \tag{7.22}$$

independent components, i.e. only 20 when $n = 4$.

7.6 Flat Riemann spaces

If $R^\alpha{}_{\beta\gamma\delta} = 0$ in a Riemann space, then the results of Section 6.5 apply – one can choose coordinates so that $\left\{{\alpha \atop \beta\gamma}\right\}_* = 0$. This equation is preserved by all linear coordinate transformations. In such coordinates, as seen from (7.6), the metric has constant components. Then, linear coordinate transformations can be used so that the metric tensor becomes diagonal and all its diagonal elements are either $+1$, -1 or 0, thus

$$\mathrm{d}s^2 = \sum_{i=1}^n \varepsilon_i \left(\mathrm{d}x^i\right)^2, \tag{7.23}$$

where $\varepsilon_i = +1, 0$ or -1.

7.7 Subspaces of a Riemann space

Let a subspace S_m of a Riemann space V_n be given by the parametric equations $x^\alpha = f^\alpha\left(\tau^1, \tau^2, \ldots, \tau^m\right)$, $\alpha = 1, \ldots, n$. The metric of the space V_n for a pair of points $\{\tau_0^a\} \in S_m$ and $\{\tau_0^a + \mathrm{d}\tau^a\} \in S_m$ is then

$$g_{\alpha\beta}\mathrm{d}x^\alpha\mathrm{d}x^\beta = g_{\alpha\beta}\frac{\partial x^\alpha}{\partial \tau^a}\frac{\partial x^\beta}{\partial \tau^b}\mathrm{d}\tau^a\mathrm{d}\tau^b. \tag{7.24}$$

Hence, $g_{\alpha\beta}\left(\partial x^\alpha/\partial \tau^a\right)\left(\partial x^\beta/\partial \tau^b\right)$ plays the role of a metric tensor in S_m. We say that the metric $g_{\alpha\beta}$ of V_n **induces** the metric $h_{\alpha\beta}$ in S_m by the formula

$$h_{ab} = g_{\alpha\beta}\frac{\partial x^\alpha}{\partial \tau^a}\frac{\partial x^\beta}{\partial \tau^b}. \tag{7.25}$$

In the 3-dimensional Euclidean space, in rectangular Cartesian coordinates, the length of a curve is

$$l_{\lambda_0, \lambda_1} = \int_{\lambda_0}^{\lambda_1} \left[\left(\frac{\mathrm{d}x}{\mathrm{d}\lambda}\right)^2 + \left(\frac{\mathrm{d}y}{\mathrm{d}\lambda}\right)^2 + \left(\frac{\mathrm{d}z}{\mathrm{d}\lambda}\right)^2\right]^{1/2} \mathrm{d}\lambda. \tag{7.26}$$

Hence, the metric form here is $\mathrm{d}s^2 = \mathrm{d}x^2 + \mathrm{d}y^2 + \mathrm{d}z^2$, and the metric tensor is the unit matrix. Knowing this, we can use (7.25) to find the metric tensor of an arbitrary 2-dimensional surface in the Euclidean space. For example, for a sphere of radius a, with the centre at $x = y = z = 0$, the parametric equations in spherical coordinates are:

$$x = a\sin\vartheta\cos\varphi, \qquad y = a\sin\vartheta\sin\varphi, \qquad z = a\cos\vartheta. \tag{7.27}$$

Hence,

$$\mathrm{d}s^2 = \mathrm{d}x^2 + \mathrm{d}y^2 + \mathrm{d}z^2 = a^2\left(\mathrm{d}\vartheta^2 + \sin^2\vartheta\mathrm{d}\varphi^2\right), \tag{7.28}$$

and the metric tensor of the sphere is the matrix

$$||h_{ab}|| = \begin{pmatrix} a^2 & 0 \\ 0 & a^2\sin^2\vartheta \end{pmatrix}. \tag{7.29}$$

7.8 Flat Riemann spaces that are globally non-Euclidean

A flat space is not necessarily Euclidean. As an example, take a cylinder of radius a. It has the parametric equations

$$x = a\cos\varphi, \qquad y = a\sin\varphi, \qquad z = z, \tag{7.30}$$

i.e. z will be used as the second parameter. Hence,

$$\mathrm{d}x^2 + \mathrm{d}y^2 + \mathrm{d}z^2 = a^2\mathrm{d}\varphi^2 + \mathrm{d}z^2. \tag{7.31}$$

Thus, the metric tensor of a cylinder has constant coefficients in these coordinates, so the Christoffel symbols will all be zero, hence, the Riemann tensor will be zero – in these coordinates, and so, being a tensor, also in all other coordinates. It follows that a cylinder is flat in the sense of Riemann geometry. However, it is not Euclidean – because one can travel along a geodesic (one of the circles that are perpendicular to generators) still in the same direction and arrive back at the starting point. The path of this journey will be a closed curve that is not continuously contractible to a point. In a Euclidean plane such curves do not exist. Hence, the metric tensor determines the local geometry, but does not determine the topology. Every 2-dimensional subset of a cylinder that is continuously contractible to a point is isometric to a certain subset of a Euclidean plane.

The surface of a torus in a Euclidean 3-space has nonzero curvature (see Exercise 1). But a torus can be flat – provided it is embedded in a 4-dimensional space. To see this, let us introduce polar coordinates in the plane $\mathbb{R}^2$:

$$x = r\cos\varphi, \qquad y = r\sin\varphi.$$

In these coordinates, the metric form of the plane becomes

$$\mathrm{d}s_2{}^2 = \mathrm{d}x^2 + \mathrm{d}y^2 = \mathrm{d}r^2 + r^2\mathrm{d}\varphi^2,$$

where $\varphi \in [0, 2\pi]$, so the point (r, φ) is identical to the point $(r, \varphi + 2\pi)$ and the curves $r = $ constant are circles. Now take the 4-dimensional Euclidean space with the metric

$$\mathrm{d}s_4{}^2 = \mathrm{d}x^2 + \mathrm{d}y^2 + \mathrm{d}z^2 + \mathrm{d}u^2$$

and introduce the polar coordinates in the planes ($z = $ constant, $u = $ constant) and in the planes ($x = $ constant, $y = $ constant): $x = r\cos\varphi$, $y = r\sin\varphi$, $z = \rho\cos\psi$, $u = \rho\sin\psi$, where $\varphi \in [0, 2\pi]$, $\psi \in [0, 2\pi]$. The following points are identical: $(r, \varphi, \rho, \psi) \equiv (r, \varphi, \rho, \psi + 2\pi) \equiv (r, \varphi + 2\pi, \rho, \psi)$. The curves on which (r, ρ, ψ) are constant and the curves on which (r, φ, ρ) are constant are all circles. Now choose the 2-surface given by $r = r_0 = $ constant, $\rho = \rho_0 = $ constant. The metric form of this surface is

$$\mathrm{d}s_2{}^2 = r_0{}^2\mathrm{d}\varphi^2 + \rho_0{}^2\mathrm{d}\psi^2.$$

This is a flat surface (because the metric tensor has constant coefficients), but it has all the topological properties of a torus.

7.9 The Riemann curvature versus the normal curvature of a surface

In an n-dimensional Riemann space, the curvature tensor has $n^2(n^2 - 1)/12$ algebraically independent components, which makes 1 when $n = 2$. Hence, the curvature of a 2-dimensional surface is determined by just one function, for example $R = g^{\alpha\beta}R^\rho{}_{\alpha\rho\beta}$. This quantity turns out to be equal to the Gauss curvature, whose description is given below. For the full reasoning see Eisenhart (1940).

We assume that the reader knows the definition of the curvature of a curve. Now consider a 2-dimensional surface S and a point $P \in S$, at which we want to calculate the curvature of S. Draw the straight line L through P that is perpendicular to S, and then consider an arbitrary plane F containing L. The plane intersects the surface along a curve C called the **normal section** of S. Now imagine F being rotated around L, and consider the curvatures at P of the resulting normal sections. It can happen that all these curvatures will be equal – this is the case at every point of a sphere, or when L is a symmetry axis of S. But in general the curvature of C changes when F is rotated, and in the collection of all curvatures there will be the largest and the smallest value. (If the point P is nonsingular, then all the curvatures will be finite.) The Gauss curvature of the surface S at P is the product of the largest curvature of C by the smallest one. When all curvatures are equal, the curvature of the surface is the square of the curvature of normal sections. Knowing this, one can verify that the Gauss curvature is equal to $g^{\alpha\beta}R^\rho{}_{\alpha\rho\beta}$. This shows that the name 'curvature tensor' evokes the right association.

Now it can be seen why the curvature of a cylinder is equal to zero: at any point of the cylinder one of the normal sections is a straight line whose curvature is zero, while all the other normal sections have positive curvatures. Hence, the smallest curvature of a normal section is zero. The curvature of a one-sheeted hyperboloid is negative (see Exercise 1). The reason: even though a straight line is among its normal sections, the other normal sections have positive and negative curvatures. So, the zero curvature is neither maximum nor minimum and does not enter the product contained in the Gauss curvature.

7.10 The geodesic line as the line of extremal distance

If the length of a curve arc is defined on a manifold, then in the collection of all arcs joining two given points we can look for the arc of extremal (i.e. greatest or smallest) length. Thus, we look for a curve on which, with fixed λ_0 and λ_1, the quantity

$$\int_{\lambda_0}^{\lambda_1} \left| g_{\alpha\beta}(x)\frac{\mathrm{d}x^\alpha}{\mathrm{d}\lambda}\frac{\mathrm{d}x^\beta}{\mathrm{d}\lambda} \right|^{1/2} \mathrm{d}\lambda \tag{7.32}$$

takes the extremal value. It must obey the Euler–Lagrange equations

$$\frac{\mathrm{d}}{\mathrm{d}\lambda}\left(\frac{\partial f}{\partial \dot{x}^\gamma} \right) - \frac{\partial f}{\partial x^\gamma} = 0, \tag{7.33}$$

where f is the integrand in (7.32), and $\dot{x}^\gamma \stackrel{\text{def}}{=} \mathrm{d}x^\gamma/\mathrm{d}\lambda$. For convenience, let us write

$$\left| g_{\alpha\beta}\dot{x}^\alpha \dot{x}^\beta \right| = \varepsilon g_{\alpha\beta}\dot{x}^\alpha \dot{x}^\beta, \tag{7.34}$$

where $\varepsilon = +1$ or $\varepsilon = -1$, as appropriate. Then, from (7.33)

$$\frac{\mathrm{d}}{\mathrm{d}\lambda}\left(\frac{g_{\gamma\beta}\dot{x}^\beta}{\sqrt{\varepsilon g_{\mu\nu}\dot{x}^\mu\dot{x}^\nu}}\right) - \frac{1}{2}\frac{g_{\alpha\beta,\gamma}\dot{x}^\alpha\dot{x}^\beta}{\sqrt{\varepsilon g_{\mu\nu}\dot{x}^\mu\dot{x}^\nu}} = 0. \tag{7.35}$$

The case $g_{\mu\nu}\dot{x}^\mu\dot{x}^\nu = 0$ requires separate treatment. We shall do this further on, but for now we assume that $g_{\mu\nu}\dot{x}^\mu\dot{x}^\nu \neq 0$. Let us introduce the new parameter $s(\lambda)$ defined by

$$\frac{\mathrm{d}s}{\mathrm{d}\lambda} = \sqrt{\varepsilon g_{\mu\nu}\dot{x}^\mu\dot{x}^\nu}. \tag{7.36}$$

Then we have in (7.35):

$$\frac{\mathrm{d}s}{\mathrm{d}\lambda}\frac{\mathrm{d}}{\mathrm{d}s}\left(g_{\gamma\beta}\frac{\mathrm{d}x^\beta}{\mathrm{d}s}\right) - \frac{1}{2}g_{\alpha\beta,\gamma}\frac{\mathrm{d}s}{\mathrm{d}\lambda}\frac{\mathrm{d}x^\alpha}{\mathrm{d}s}\frac{\mathrm{d}x^\beta}{\mathrm{d}s} = 0. \tag{7.37}$$

Hence,

$$g_{\gamma\beta,\alpha}\frac{\mathrm{d}x^\alpha}{\mathrm{d}s}\frac{\mathrm{d}x^\beta}{\mathrm{d}s} + g_{\gamma\beta}\frac{\mathrm{d}^2x^\beta}{\mathrm{d}s^2} - \frac{1}{2}g_{\alpha\beta,\gamma}\frac{\mathrm{d}x^\alpha}{\mathrm{d}s}\frac{\mathrm{d}x^\beta}{\mathrm{d}s} = 0 \tag{7.38}$$

and, after contracting this with $g^{\alpha\gamma}$,

$$\frac{\mathrm{d}^2x^\alpha}{\mathrm{d}s^2} + \left\{\begin{matrix}\alpha\\\mu\nu\end{matrix}\right\}\frac{\mathrm{d}x^\mu}{\mathrm{d}s}\frac{\mathrm{d}x^\nu}{\mathrm{d}s} = 0. \tag{7.39}$$

This is the geodesic equation, (5.13). Thus, in Riemann spaces, a geodesic has another characteristic property: it extremises the distance.

Note that the geodesic extremises not only (7.32) but also the functional $\int_{\lambda_0}^{\lambda_1} g_{\mu\nu}\dot{x}^\mu\dot{x}^\nu\mathrm{d}\lambda$. In this case, the assumption $g_{\mu\nu}\dot{x}^\mu\dot{x}^\nu \neq 0$ is not necessary, and a curve of zero length can also be the extremal.

7.11 Mappings between Riemann spaces

Now let us apply the considerations of Section 3.10 to the Riemann spaces.

If a Riemann space P_m is an image of another manifold M_n (or of a subset thereof) under the mapping $F : M_n \to P_m$, then the metric tensor of P_m can be pulled back to M_n by the mapping $F_2{}^*$. However, M_n can itself be a Riemann space and have its own metric tensor, in general different from the one pulled back. Hence, more than one metric may be defined on the same manifold.

An example of such a situation is a geographic map. It is a projection of a subset of the surface of the Earth into the flat sheet of paper. Within the domain of the map, the projection is invertible, i.e. the inverse mapping is defined. This inverse mapping can then be used to pull back the metric of the Earth surface into the plane of the map. Thus, the surface of a page in an atlas has its own Euclidean metric, and also the pulled back metric of the surface of the Earth. In past centuries, navigators who travelled over large areas were in fact reading out the metric of a sphere from their maps.

7.12 Conformally related Riemann spaces

Let V_n and U_n be Riemann spaces of the same dimension, and let $\{x^\alpha\}$ and $\{y^a\}$ be the coordinates in V_n and U_n, respectively, where $\alpha, a = 1, \ldots, n$. Let $F: V_n \to U_n$ be a diffeomorphism of class C^1, and let $g_{\alpha\beta}$ and h_{ab} be the metric tensors on V_n and U_n. On each of the spaces we then have two metric tensors. For example on U_n there is its own metric tensor h_{ab}, and the metric pulled back from V_n by $\left[(F_2)^{-1} \right]^*$ as follows:

$$\left\{ \left[(F_2)^{-1} \right]^* g \right\}_{ab} (y) = \frac{\partial x^\alpha}{\partial y^a}(y) \, \frac{\partial x^\beta}{\partial y^b}(y) \, g_{\alpha\beta}(x(y)). \tag{7.40}$$

If there exists a scalar function $\phi_F: U_n \to \mathbb{R}^1$ such that for any $y \in U_n$

$$\left\{ \left[(F_2)^{-1} \right]^* g \right\}_{ab} (y) = (\phi_F(y))^{-2} \, h_{ab}(y), \tag{7.41}$$

or, equivalently,

$$\frac{\partial x^\alpha}{\partial y^a}(y) \, \frac{\partial x^\beta}{\partial y^b}(y) \, g_{\alpha\beta}(x(y)) = (\phi_F(y))^{-2} \, h_{ab}(y), \tag{7.42}$$

then we call F a **conformal mapping**.[1] The inverse mapping F^{-1} is then conformal, too, and $\phi_{F^{-1}} = 1/\phi_F$. Riemann spaces that can be related by a conformal mapping are called **conformally related**. A Riemann space that is conformally related to a flat Riemann space is called **conformally flat**.

A conformal mapping does not change the angles between vectors. Take two vectors $k^\alpha(x_0)$ and $l^\alpha(x_0)$ of nonzero length at $x_0 \in V_n$, then the angle between them is

$$\cos \alpha^{(V)} = \frac{g_{\alpha\beta} k^\alpha l^\beta}{\sqrt{|g_{\rho\sigma} k^\rho k^\sigma|} \, \sqrt{|g_{\mu\nu} l^\mu l^\nu|}}. \tag{7.43}$$

Now, using (3.16), the analogue of (3.22) for covariant tensors of rank 2 and (7.42), we obtain for the images of the vectors k and l in the tangent space to U_n at $y_0 = F(x_0)$:

$$h_{rs} \, (F_{1*}k)^r \, (F_{1*}k)^s = \phi_F{}^2 \frac{\partial x^\rho}{\partial y^r} \frac{\partial x^\sigma}{\partial y^s} \frac{\partial y^r}{\partial x^\lambda} \frac{\partial y^s}{\partial x^\tau} g_{\rho\sigma} k^\lambda k^\tau$$

$$= \phi_F{}^2 \delta^\rho{}_\lambda \delta^\sigma{}_\tau g_{\rho\sigma} k^\lambda k^\tau = \phi_F{}^2 g_{\rho\sigma} k^\rho k^\sigma, \tag{7.44}$$

and similarly for the remaining scalar products in (7.43). Hence, the angle $\alpha^{(U)}$ between the images of k and l is determined by

$$\cos \alpha^{(U)} = \frac{h_{ab} \, (F_1^* k)^a \, (F_1^* l)^b}{\sqrt{|h_{rs} \, (F_1^* k)^r \, (F_1^* k)^s|} \, \sqrt{|h_{mn} \, (F_1^* l)^m \, (F_1^* l)^n|}}$$

$$= \frac{g_{\alpha\beta} k^\alpha l^\beta}{\sqrt{|g_{\rho\sigma} k^\rho k^\sigma|} \, \sqrt{|g_{\mu\nu} l^\mu l^\nu|}} = \cos \alpha^{(V)}. \tag{7.45}$$

Equation (7.42) also shows that a conformal mapping maps vectors of zero length on V_n to vectors of zero length on U_n.

---

[1] The $\phi_F{}^{-2}$ is just a convention that simplifies some of the further formulae; the important point is that the two metrics differ only by a scalar factor.

7.13 Conformal curvature

A second-rank tensor can be constructed from the Riemann tensor

$$R_{\alpha\beta} \overset{\text{def}}{=} R^{\rho}{}_{\alpha\rho\beta}, \tag{7.46}$$

called the **Ricci tensor**. It is symmetric in $(\alpha\beta)$. In addition, we calculate its trace

$$R \overset{\text{def}}{=} g^{\alpha\beta} R_{\alpha\beta} \equiv R^{\alpha}{}_{\alpha}, \tag{7.47}$$

and its trace-free part

$$\overline{R}^{\alpha}{}_{\beta} \overset{\text{def}}{=} R^{\alpha}{}_{\beta} - \frac{1}{n}\delta^{\alpha}{}_{\beta}R. \tag{7.48}$$

Using these quantities we can now define a new tensor of rank 4 called the **conformal curvature tensor** or the **Weyl tensor**:

$$C^{\alpha\beta}{}_{\gamma\delta} \overset{\text{def}}{=} R^{\alpha\beta}{}_{\gamma\delta} + \frac{1}{n-2}\delta^{\alpha\beta\rho}_{\gamma\delta\sigma}\overline{R}^{\sigma}{}_{\rho} - \frac{1}{n(n-1)}\delta^{\alpha\beta}_{\gamma\delta}R. \tag{7.49}$$

This definition makes sense only for $n > 2$. We shall deal with the case $n = 2$ separately.

The Weyl tensor has all the same symmetries in indices as the Riemann tensor, and in addition all of its traces are zero. In relativity, the Weyl tensor describes that part of the gravitational field that propagates into vacuum and is detectable outside the sources, gravitational waves among other things. It is the same for conformally related Riemann spaces: if $\widetilde{g}_{\alpha\beta} = \phi^{-2}g_{\alpha\beta}$, then $\widetilde{C}^{\alpha}{}_{\beta\gamma\delta} = C^{\alpha}{}_{\beta\gamma\delta}$ (note the positions of indices; with other positions the two tensors are proportional, but not equal). This fact will be verified in the course of proving the following theorem:

Theorem 7.1 *If in a Riemann space V_n the Weyl tensor is zero, then the metric of V_n is conformally flat, i.e. there exists a function ϕ such that $g_{\alpha\beta} = \phi^{-2}\eta_{\alpha\beta}$, where $R^{\alpha\beta}{}_{\gamma\delta}(\eta) = 0$. When $n = 3$, $C^{\alpha\beta}{}_{\gamma\delta} \equiv 0$, but not every 3-dimensional metric is conformally related to a flat one. The necessary and sufficient condition for a 3-dimensional Riemann space to be conformally flat is the vanishing of the* **Cotton–York tensor**

$$C^{\alpha\beta} \overset{\text{def}}{=} 2\epsilon^{\alpha\gamma\delta}\left(R^{\beta}{}_{\gamma;\delta} - \frac{1}{4}\delta^{\beta}{}_{\gamma}R_{,\delta}\right), \tag{7.50}$$

where $\epsilon^{\alpha\gamma\delta}$ is the Levi-Civita symbol. In two dimensions, the Weyl tensor is undetermined, but every 2-dimensional metric is conformally flat.

Proof: (adapted from Raszewski (1958, pp. 516 – 521)):

Part I: $n > 3$.

We shall show that if $C^{\alpha\beta}{}_{\gamma\delta} = 0$, then the equations determining the function ψ such that $\psi^2 g_{\alpha\beta} = \eta_{\alpha\beta}$ do have a solution. The quantities without a tilde will be those determined by $g_{\alpha\beta}$, those with a tilde will be determined by $\widetilde{g}_{\alpha\beta} \overset{\text{def}}{=} \psi^{-2}g_{\alpha\beta}$. We introduce the following conventions: all displayed covariant derivatives will be with respect to $g_{\alpha\beta}$, indices of objects without a tilde will be manipulated by $g_{\alpha\beta}$ and indices of objects with a tilde will be manipulated by $\widetilde{g}_{\alpha\beta}$.

We first find that

$$\widetilde{\left\{ \begin{matrix} \alpha \\ \beta\gamma \end{matrix} \right\}} = \left\{ \begin{matrix} \alpha \\ \beta\gamma \end{matrix} \right\} - \psi^{-1} \left(\delta^\alpha{}_\beta \psi_{,\gamma} + \delta^\alpha{}_\gamma \psi_{,\beta} - g_{\beta\gamma} \psi^{;\alpha} \right). \tag{7.51}$$

The last term is in fact a new symbol: the derivative of ψ, which is a covariant vector, has its index raised. Then we find the corresponding relations between the other quantities:

$$\widetilde{R}^{\alpha\beta}{}_{\gamma\delta} = \psi^2 \left(R^{\alpha\beta}{}_{\gamma\delta} - \psi^{-1} \delta^{\rho\lambda}_{\gamma\delta} \delta^{\alpha\beta}_{\sigma\rho} \psi_{;\lambda}{}^{;\sigma} - \psi^{-2} \delta^{\alpha\beta}_{\gamma\delta} \psi_{,\rho} \psi^{;\rho} \right), \tag{7.52}$$

$$\widetilde{R}_{\alpha\beta} = R_{\alpha\beta} + (n-2)\psi^{-1}\psi_{;\alpha\beta} + g_{\alpha\beta}\psi^{-1}\psi_{;\rho}{}^{;\rho} - (n-1)g_{\alpha\beta}\psi^{-2}\psi_{,\rho}\psi^{;\rho}, \tag{7.53}$$

$$\widetilde{R} = \psi^2 R + (n-1)\left(2\psi\psi_{;\rho}{}^{;\rho} - n\psi_{,\rho}\psi^{;\rho}\right). \tag{7.54}$$

Using these equations one can now verify that $\widetilde{C}^\alpha{}_{\beta\gamma\delta} = C^\alpha{}_{\beta\gamma\delta}$. Then we require $\widetilde{R}^\alpha{}_{\beta\gamma\delta} = 0$ (which automatically implies $\widetilde{C}^\alpha{}_{\beta\gamma\delta} = C^\alpha{}_{\beta\gamma\delta} = 0$). By virtue of (7.52) this means

$$R^{\alpha\beta}{}_{\gamma\delta} = \psi^{-1} \delta^{\rho\lambda}_{\gamma\delta} \delta^{\alpha\beta}_{\sigma\rho} \psi_{;\lambda}{}^{;\sigma} + \psi^{-2} \delta^{\alpha\beta}_{\gamma\delta} \psi_\rho \psi^{;\rho}. \tag{7.55}$$

If $\widetilde{R}^\alpha{}_{\beta\gamma\delta} = 0$, then also $\widetilde{R}_{\alpha\beta} = 0$ and $\widetilde{R} = 0$, so, from (7.54):

$$\psi_{;\rho}{}^{;\rho} = -\frac{1}{2(n-1)}\psi R + \frac{1}{2}n\psi^{-1}\psi_{,\rho}\psi^{;\rho}, \tag{7.56}$$

and, from (7.53) with use of (7.56):

$$\psi_{;\alpha\beta} = -\frac{1}{n-2}\psi R_{\alpha\beta} + \frac{1}{2(n-1)(n-2)}g_{\alpha\beta}\psi R + \frac{1}{2}g_{\alpha\beta}\psi^{-1}\psi_{,\rho}\psi^{;\rho}. \tag{7.57}$$

This equation can be rewritten in the form $\psi_{,\alpha\beta} = $ [the appropriate expression]. Hence, if such a ψ exists, then the integrability conditions $\psi_{,\alpha\beta\gamma} - \psi_{,\alpha\gamma\beta} = 0$ should be fulfilled by the right-hand sides. They are equivalent to $\psi_{;\alpha\beta\gamma} - \psi_{;\alpha\gamma\beta} = R^\rho{}_{\alpha\beta\gamma}\psi_{,\rho}$. In this last equation, we must substitute for $R^\rho{}_{\alpha\beta\gamma}$ from (7.55) and for all second covariant derivatives of ψ from (7.57). After a long calculation, the following result emerges:

$$- R_{\alpha\beta;\gamma} + R_{\alpha\gamma;\beta} + \frac{1}{2(n-1)}\left(g_{\alpha\beta}R_{,\gamma} - g_{\alpha\gamma}R_{,\beta}\right) = 0. \tag{7.58}$$

The Bianchi identities $R^\alpha{}_{\beta[\gamma\delta;\epsilon]} = 0$, with (7.49) and $C^{\alpha\beta}{}_{\gamma\delta} = 0$ substituted in them, when contracted over (α, γ), become

$$(n-3)\left(-R_{\alpha\beta;\gamma} + R_{\alpha\gamma;\beta}\right) + g_{\alpha\beta}R^\rho{}_{\gamma;\rho} - g_{\alpha\gamma}R^\rho{}_{\beta;\rho}$$
$$+ \frac{1}{n-1}\left(g_{\alpha\gamma}R_{,\beta} - g_{\alpha\beta}R_{,\gamma}\right) = 0. \tag{7.59}$$

This, when contracted with $g^{\alpha\beta}$, is equivalent to

$$R^\rho{}_{\gamma;\rho} - \frac{1}{2}R_{,\gamma} = 0. \tag{7.60}$$

With $n > 3$, Eq. (7.58) follows from (7.59) and (7.60), and so is fulfilled in consequence of the Bianchi identities. This means that (7.57) is then integrable, so a function ψ exists such that $g_{\alpha\beta} = \psi^{-2}\eta_{\alpha\beta}$, where $R^\alpha{}_{\beta\gamma\delta}(\eta) = 0$. $\square$

Part II: $n = 3$.

When $n = 3$, Eq. (7.59) follows from (7.60) and does not determine $(-R_{\alpha\beta;\gamma} + R_{\alpha\gamma;\beta})$, so it cannot be equivalent to (7.58). In this case, (7.58) is an additional condition that must be obeyed by the curvature in order that the metric is conformally flat. This is equivalent to $C^{\alpha\beta} = 0$, where $C^{\alpha\beta}$ is the Cotton–York tensor of (7.50). However, with $n = 3$, the Weyl tensor is identically zero, as can be verified by substituting consecutively all the values of all indices in (7.49). $\square$

Part III: $n = 2$.

With $n = 2$, the second term in (7.49) becomes undetermined. However, then

$$R^{\alpha\beta}{}_{\gamma\delta} = \frac{1}{2}\delta^{\alpha\beta}_{\gamma\delta}R, \tag{7.61}$$

because any tensor antisymmetric in $(^{\alpha\beta})$ in two dimensions must be proportional to $\epsilon^{\alpha\beta}$, and similarly for the lower indices. Equations (7.52) and (7.54) still apply, but (7.53) – (7.54) follow from (7.52). Consequently, the only limitation on ψ is in this case the equation $\widetilde{R} = 0$, i.e.

$$\psi_{;\rho}{}^{;\rho} = -\frac{1}{2}\psi R + \psi^{-1}\psi_{,\rho}\,\psi^{;\rho}.$$

This can be equivalently rewritten as

$$(\ln\psi)_{;\rho}{}^{;\rho} = -\frac{1}{2}R,$$

which is a linear inhomogeneous equation of the type of the Poisson equation (for positive-definite signature) or of the wave equation (for indefinite signature). Hence, in every case it will have a solution, so every 2-dimensional metric is conformally flat. $\square$

7.14 Timelike, null and spacelike intervals in a 4-dimensional spacetime

The physically important 4-dimensional Riemann spaces are those with the signature $(+---)$. They are called **spacetimes**. Consider the following equation in a spacetime:

$$ds^2 = g_{\alpha\beta}dx^\alpha dx^\beta = 0. \tag{7.62}$$

Let us choose a point $P_0 \in V_4$. We can choose coordinates so that at P_0

$$g_{\alpha\beta}(P_0) = \begin{bmatrix} +1 & 0 & 0 & 0 \\ 0 & -1 & 0 & 0 \\ 0 & 0 & -1 & 0 \\ 0 & 0 & 0 & -1 \end{bmatrix},$$

i.e. the metric becomes the Minkowski metric of special relativity. (Note: it is exactly Minkowskian only at P_0, and approximately Minkowskian in a small neighbourhood of P_0.) In these coordinates, called **locally Cartesian**, (7.62) taken at P_0 becomes

$$ds^2 = (dx^0)^2 - (dx^1)^2 - (dx^2)^2 - (dx^3)^2 = 0. \tag{7.63}$$

In a 3-dimensional Euclidean space, the equation $(z - z_0)^2 - (x - x_0)^2 - (y - y_0)^2 = 0$ describes a cone with the vertex at (x_0, y_0, z_0) and the axis at $z = z_0$. By analogy, the hypersurface in spacetime determined by (7.62) is called a **light cone** (see Fig. 7.1). The coordinate x^0 is called the **time coordinate**, the remaining ones are **space coordinates**. The light cone divides the neighbourhood of P_0 into three disjoint regions. Each point lying on the light cone can be connected to P_0 by a geodesic arc of zero length. All curves of zero length, geodesic or not, are called **null curves**, and the points on the cone are said to be in a **null relation** to P_0. The tangent vector to a null curve at any point has zero length, such vectors are called **null vectors**. Each point in the regions F and P inside the cone can be connected to P_0 by a curve $x^\alpha(\lambda)$ on which $g_{\alpha\beta} \frac{\mathrm{d}x^\alpha}{\mathrm{d}\lambda} \frac{\mathrm{d}x^\beta}{\mathrm{d}\lambda} > 0$ everywhere. These points are said to be in a **timelike relation** with P_0, and the vectors v^α for which $g_{\alpha\beta} v^\alpha v^\beta > 0$ are called **timelike vectors**. Finally, each point in the region E outside the light cone can be connected to P_0 by a curve on which $g_{\alpha\beta} \frac{\mathrm{d}x^\alpha}{\mathrm{d}\lambda} \frac{\mathrm{d}x^\beta}{\mathrm{d}\lambda} < 0$ everywhere. These points are said to be in a **spacelike relation** with P_0, and vectors v^α for which $g_{\alpha\beta} v^\alpha v^\beta < 0$ are called **spacelike vectors**. Justifications of these names come from special relativity. The region F inside the cone, in which the x^0-coordinates of all the points are greater than x^0 of P_0, is called the **future** of P_0. The region P is called the **past** of P_0. The region E does not have a name and is usually called 'elsewhere' with respect to P_0.[2]

Not every curve lying on the light cone (7.62) has zero length. Only the generators of the cone, which are **null geodesics**, have this property. Other curves on the cone, for example spirals winding on its surface towards P_0, are spacelike curves whose tangent vectors v^α have $g_{\alpha\beta} v^\alpha v^\beta < 0$ all along. Curves that are null but not geodesic are at each point Q tangent to the light cone of Q, but veer from one cone to another. Figure 7.2 shows a broken null line whose straight segments are null geodesic arcs, but at the corners the line goes from one cone to another. A general nongeodesic null line can be imagined as a limit of a sequence of such broken null lines as the extent of each geodesic segment (as measured, for example, by the range of the affine parameter) goes to zero. Such null lines through P_0 enter the region F, or reach P_0 from within P.

Similarly, not every curve arc in F or P is timelike. These regions contain also null nongeodesic and spacelike curves. The characteristic property of the region $F \cup P$ is that for each point there a timelike curve joining it to P_0 *exists*. Such curves do not exist on the light cone and in E. Curves on the light cone that reach P_0 are either null or spacelike, and every curve arc in E that reaches P_0 must have at least a segment that is spacelike.

It should now be clear that a light cone exists at every point of a spacetime, and the analysis above applies to every point. Since $\mathrm{d}s^2$ is a scalar, Eq. (7.62) is covariant, so the light cone is a geometric object, it does not depend on the choice of coordinates.

---

[2] The whole of Minkowski spacetime can be divided into F, P, E and the light cone relative to every P_0. In curved spacetimes, the light cones can be complicated hypersurfaces that neither are axially symmetric nor have straight generators, and can have self-intersections, see Perlick (2004). The latter necessarily happens in a sufficiently large neighbourhood of a black hole (see Section 14.11). The division of a curved spacetime by a light cone is usually well defined only in a finite neighbourhood of each P_0.

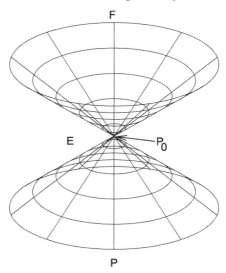

Fig. 7.1 Equation (7.62) determines the light cone that divides the neighbourhood of P_0 into the future F of P_0, the past P of P_0 and 'elsewhere' E. A light cone looks so simple only in the flat spacetime in which the coordinates with the property (7.63) can be introduced globally. See Perlick (2004) for examples of light cones having highly nontrivial geometries and topologies.

7.15 Embeddings of Riemann spaces in Riemann spaces of higher dimension

Let W_m be a subspace of the Riemann space V_n, and let the metric tensor of V_n be $g_{\alpha\beta}$. Let W_m be defined by the parametric equations $x^\alpha = f^\alpha \left(\tau^1, \ldots, \tau^m\right)$. Then, as we observed in Section 7.7, W_m is itself a Riemann space of m dimensions, with the metric tensor (7.25).

Now we ask a question reciprocal to the one answered in Section 7.7: when can a given Riemann space V_n be a subspace of another Riemann space? We do not consider *null subspaces* (i.e. subspaces whose tangent spaces contain null vectors) because in them the determinant of the metric is zero, and they require a separate treatment. Also, we will investigate only the question of *local embeddings*, i.e. whether an open subset of V_n can be embedded in another Riemann space. Global embeddings pose additional problems that we will not discuss. If V_n is a subspace of U_N of dimension $N > n$, then a set of functions

$$Y^A = f^A \left(x^1, \ldots, x^n\right), \qquad A = 1, \ldots, N, \tag{7.64}$$

should exist on U_N such that

$$g_{\alpha\beta} = G_{AB} Y^A{}_{,\alpha} Y^B{}_{,\beta}, \tag{7.65}$$

where G_{AB} is the metric tensor of U_N. Note that Y^A and G_{AB} are scalars with respect to coordinate transformations in V_n.

The question of whether a given V_n can be embedded in a U_N can be answered with the help of the reasoning presented below (see Eisenhart (1964) for a more detailed exposition).

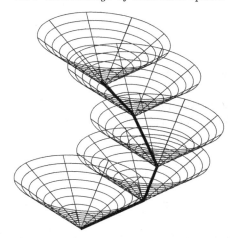

Fig. 7.2 Future light cones along a nongeodesic null line with geodesic segments (the thicker line). The line is tangent to each light cone at its vertex, but passes from one cone to another as it proceeds.

Let $X_{\widehat{A}}{}^A$ be a set of vector fields on U_N (where $\widehat{A} = n+1,\ldots,N$ labels vectors and $A = 1,\ldots,N$ labels their components) that are orthogonal to V_n and orthogonal to each other, with each field being normalised to ± 1:

$$G_{AB}X_{\widehat{A}}{}^A X_{\widehat{A}}{}^B = \varepsilon_{\widehat{A}} = \pm 1 \qquad \text{(no sum over } \widehat{A}\text{)}, \tag{7.66}$$

$$G_{AB}X_{\widehat{A}}{}^A X_{\widehat{B}}{}^B = 0 \qquad \text{for } \widehat{A} \neq \widehat{B}. \tag{7.67}$$

Since $Y^A{}_{,\alpha}$ are tangent to V_n and $X_{\widehat{A}}{}^A$ are orthogonal to V_n, we have for all α and $\widehat{A}$

$$G_{AB}Y^A{}_{,\alpha} X_{\widehat{A}}{}^B = 0 \tag{7.68}$$

We differentiate (7.65) covariantly by x^γ. Since Y^A and G_{AB} are scalars in V_n, we obtain

$$\frac{\partial G_{AB}}{\partial Y^C}Y^A{}_{,\alpha}\,Y^B{}_{,\beta}\,Y^C{}_{,\gamma} + G_{AB}\left(Y^A{}_{;\alpha\gamma}\,Y^B{}_{,\beta} + Y^A{}_{,\alpha}\,Y^B{}_{;\beta\gamma}\right) = 0. \tag{7.69}$$

Rewriting this equation with indices and signs permuted as $-\{\alpha\beta\gamma\} + \{\alpha\gamma\beta\} + \{\beta\gamma\alpha\}$ and adding all three equations we obtain

$$G_{AB}Y^B{}_{,\gamma}\left(Y^A{}_{;\alpha\beta} + \left\{\begin{matrix} A \\ MN \end{matrix}\right\}_G Y^M{}_{,\alpha}\,Y^N{}_{,\beta}\right) = 0. \tag{7.70}$$

For a fixed B, $Y^B{}_{,\gamma}$ is a vector in U_N tangent to V_n. Equation (7.70) means that the object in parentheses is orthogonal to all the n tangent vectors to V_n, so must be spanned on the $(N-n)$ vectors $X_{\widehat{A}}{}^A$ that are orthogonal to V_n:

$$Y^A{}_{;\alpha\beta} + \left\{\begin{matrix} A \\ MN \end{matrix}\right\}_G Y^M{}_{,\alpha}\,Y^N{}_{,\beta} = \sum_{\widehat{S}=n+1}^{N} \varepsilon_{\widehat{S}}\Omega_{(\widehat{S})\alpha\beta}X_{\widehat{S}}{}^A, \tag{7.71}$$

where $\Omega_{(\widehat{S})\alpha\beta}$ are the coefficients to be determined below. For a fixed $\widehat{S}$, $\Omega_{(\widehat{S})\alpha\beta}$ is a tensor in V_n, symmetric in $(\alpha\beta)$, and a scalar in U_N; the index $\widehat{S}$ labels different Ω-s. The $\varepsilon_{\widehat{S}}$ are the same as in (7.66) – each $\varepsilon_{\widehat{S}}$ tells whether its associated $X_{\widehat{S}}{}^A$ is normalised to $+1$ or -1. Using (7.65)–(7.68), we find:

$$\Omega_{(\widehat{B})\alpha\beta} = G_{AB} Y^A{}_{;\alpha\beta} X_{\widehat{B}}{}^B + G_{AB} \left\{ {A \atop MN} \right\}_G Y^M{}_{,\alpha} Y^N{}_{,\beta} X_{\widehat{B}}{}^B. \tag{7.72}$$

In the case $N = n + 1$, the single quantity $\Omega_{\alpha\beta}$ defined by (7.72) is called the **second fundamental form** of the subspace V_n.

Note that the term arising from $Y^A{}_{;\alpha\beta}$ in (7.72):

$$G_{AB} Y^A{}_{;\rho} \left\{ {\rho \atop \alpha\beta} \right\} X_{\widehat{B}}{}^B = 0$$

in virtue of (7.68). In what remains of (7.72), the quantity

$$Y^A{}_{,\alpha\beta} + \left\{ {A \atop MN} \right\}_G Y^M{}_{,\alpha} Y^N{}_{,\beta} \equiv \left(Y^A{}_{,\alpha} \right)_{|N} Y^N{}_{,\beta}$$

is the directional covariant derivative of $Y^A{}_{,\alpha}$ along $Y^N{}_{,\beta}$ (the subscript $|N$ denotes a covariant derivative in U_N). Consequently, this quantity measures the rate of change of the α-th tangent vector $Y^A{}_{,\alpha}$ as we move along the β-th tangent vector field $Y^N{}_{,\beta}$. Then, $\Omega_{(\widehat{B})\alpha\beta}$ is the projection of this rate of change on the $\widehat{B}$-th normal vector field to V_n. For this reason, the $\Omega_{(\widehat{B})\alpha\beta}$ are sometimes called the **extrinsic curvatures** of the subspace V_n embedded in U_N. They allow us to 'view' the geometry of V_n from an enveloping space and to see a difference between Riemann spaces that have the same intrinsic geometry (i.e., equivalent metric forms[3]). For example, a plane and a cylinder in a Euclidean 3-space that have the same intrinsic geometries have different second fundamental forms.

In order to know whether a given V_n can be embedded in a given U_N, we have to find out whether the functions Y^A obeying (7.65) exist. They must obey (7.71), which determine the second covariant derivatives of Y^A in V_n. These equations will be solvable if the integrability condition – the Ricci formula (6.9) – is fulfilled:

$$Y^A{}_{;\alpha\beta\gamma} - Y^A{}_{;\alpha\gamma\beta} = R^\rho{}_{\alpha\beta\gamma}(g) Y^A{}_{,\rho}, \tag{7.73}$$

where $R^\rho{}_{\alpha\beta\gamma}(g)$ is the Riemann tensor of V_n. In calculating the third covariant derivatives of Y^A from (7.71), we will encounter the first derivatives of $X_{\widehat{S}}{}^A$, so we need to know more about them.

Differentiating (7.68) covariantly by x^β, eliminating $G_{AB} Y^A{}_{;\alpha\beta} X_{\widehat{B}}{}^B$ with use of (7.72) and making use of covariant constancy of G_{AB}:

$$G_{AB,\beta} \equiv G_{AB,C} Y^C{}_{,\beta} = \left(\left\{ {R \atop AC} \right\}_G G_{RB} + \left\{ {R \atop BC} \right\}_G G_{AR} \right) Y^C{}_{,\beta} \tag{7.74}$$

we obtain another expression for $\Omega_{(\widehat{B})\alpha\beta}$, equivalent to (7.72):

$$\Omega_{(\widehat{A})\alpha\beta} = -G_{AB} Y^A{}_{,\alpha} X_{\widehat{A}}{}^B{}_{,\beta} - G_{AR} \left\{ {R \atop BC} \right\}_G Y^A{}_{,\alpha} Y^C{}_{,\beta} X_{\widehat{A}}{}^B. \tag{7.75}$$

---

[3] Equivalent means either the same or one being a coordinate transform of the other.

This can be equivalently written as

$$-\Omega_{(\widehat{A})\alpha\beta} = G_{AB}Y^A{}_{,\alpha}\left(X_{\widehat{A}}{}^B{}_{|C}\right)Y^C{}_{,\beta}, \tag{7.76}$$

which is a covariant derivative of $X_{\widehat{A}}{}^B$ in U_N projected onto vectors tangent to V_n. This provides another interpretation of the second fundamental forms: $\left(X_{\widehat{A}}{}^B{}_{|C}\right)Y^C{}_{,\beta}$ is the rate of change of the $\widehat{A}$-th normal vector to V_n as we move along the β-th tangent field $Y^C{}_{,\beta}$, and $\Omega_{(\widehat{A})\alpha\beta}$ is the projection of that rate on the α-th tangent vector field $Y^A{}_{,\alpha}$. Equation (7.76) clearly shows that $\Omega_{(\widehat{A})\alpha\beta}$ are scalars in U_N.

Now we define the following set of vector fields on V_n:

$$\mu_{[\widehat{R}\,\widehat{S}]\beta} \overset{\text{def}}{=} G_{AB}X_{\widehat{R}}{}^A\left(X_{\widehat{S}}{}^B\right)_{|C}Y^C{}_{,\beta} \equiv G_{AB}X_{\widehat{R}}{}^A X_{\widehat{S}}{}^B{}_{|\beta}, \tag{7.77}$$

where $\widehat{R}, \widehat{S} = n+1, \ldots, N$, and make use of the identity

$$G_{AB}X_{\widehat{R}}{}^A\left(X_{\widehat{S}}{}^B\right)_{|C} = \left(G_{AB}X_{\widehat{R}}{}^A X_{\widehat{S}}{}^B\right)_{|C} - G_{AB|C}X_{\widehat{R}}{}^A X_{\widehat{S}}{}^B - G_{AB}\left(X_{\widehat{R}}{}^A\right)_{|C}X_{\widehat{S}}{}^B.$$

The first term is zero by virtue of (7.66) – (7.67), the second one is zero because $G_{AB|C} = 0$. What remains shows, on comparison with (7.77), that the $\mu_{[\widehat{R}\,\widehat{S}]\beta}$ are antisymmetric in $[\widehat{R}\,\widehat{S}]$:

$$\mu_{[\widehat{R}\,\widehat{S}]\beta} = -\mu_{[\widehat{S}\,\widehat{R}]\beta}, \tag{7.78}$$

so they vanish identically when $N = n+1$.

The derivatives $X_{\widehat{S}}{}^B{}_{,\beta}$ are not tensors in U_N. Still, they are objects with one contravariant index in U_N, so in every fixed coordinate system they can be decomposed in the vector basis $\{Y^B{}_{,\beta}, X_{\widehat{S}}{}^B\}$; only the coefficients of the decomposition will not be scalars in U_N:

$$X_{\widehat{S}}{}^B{}_{,\beta} = A_{(\widehat{S})\beta}{}^\gamma Y^B{}_{,\gamma} + \sum_{\widehat{P}=n+1}^{N} B_{(\widehat{P}\,\widehat{S})\beta}X_{\widehat{P}}{}^B, \tag{7.79}$$

where the coefficients $A_{(\widehat{S})\beta}{}^\gamma$ and $B_{(\widehat{P}\widehat{S})\beta}$ are determined by substituting (7.79) in (7.75) and (7.77). Making use of (7.65) – (7.68) and contracting the result of substitution in (7.75) with $g^{\alpha\gamma}$ we obtain:

$$A_{(\widehat{A})\beta}{}^\gamma = -g^{\alpha\gamma}\Omega_{(\widehat{A})\alpha\beta} - g^{\alpha\gamma}G_{AR}\begin{Bmatrix} R \\ MN \end{Bmatrix}_G X_{\widehat{A}}{}^M Y^N{}_{,\beta}Y^A{}_{,\alpha}, \tag{7.80}$$

$$B_{(\widehat{R}\,\widehat{S})\beta} = \varepsilon_{\widehat{R}}\mu_{[\widehat{R}\,\widehat{S}]\beta} - \varepsilon_{\widehat{R}}G_{BR}\begin{Bmatrix} R \\ MN \end{Bmatrix}_G Y^M{}_{,\beta}X_{\widehat{S}}{}^N X_{\widehat{R}}{}^B. \tag{7.81}$$

Note now that the set $\{Y^A{}_{,\alpha}, X_{\widehat{B}}{}^A\}$, $\alpha = 1, \ldots, n$; $\widehat{B} = n+1, \ldots, N$ is a field of vector bases on U_N, of exactly the kind we used in Section 4.3. In agreement with our considerations there, the metric tensor G_{AB} in U_N can be represented through its scalar components $\widehat{G}_{\widehat{A}\widehat{B}}$ in these bases:

$$\begin{aligned} \widehat{G}_{\alpha\beta} &= Y^A{}_{,\alpha}Y^B{}_{,\beta}G_{AB} = g_{\alpha\beta}, & \widehat{G}_{\widehat{A}\beta} &= X_{\widehat{A}}{}^A Y^B{}_{,\beta}G_{AB}, \\ \widehat{G}_{\widehat{A}\widehat{B}} &= X_{\widehat{A}}{}^A X_{\widehat{B}}{}^B G_{AB}, & \widehat{A}, \widehat{B} &= n+1, \ldots, N. \end{aligned} \tag{7.82}$$

Using (7.66) – (7.68) we obtain for $\widehat{A}, \widehat{B} = n+1, \ldots, N$

$$\widehat{G}_{\widehat{A}\beta} = 0, \qquad \widehat{G}_{\widehat{A}\,\widehat{B}} = \varepsilon_{\widehat{A}}\delta_{\widehat{A}\,\widehat{B}}. \tag{7.83}$$

Therefore, the inverse metric $\widehat{G}^{AB}$ has the same block-diagonal form

$$\widehat{G}^{\alpha\beta} = g^{\alpha\beta}, \qquad \widehat{G}^{\widehat{A}\beta} = 0, \qquad \widehat{G}^{\widehat{A}\,\widehat{B}} = \varepsilon_{\widehat{A}}\delta^{\widehat{A}\,\widehat{B}}. \tag{7.84}$$

The same coefficients $\{Y^A{}_{,\alpha}, X_{\widehat{B}}{}^A\}$ can then be used to represent the inverse metric G^{AB} via the $\widehat{G}^{\alpha\beta}$, $\widehat{G}^{\widehat{A}\beta}$ and $\widehat{G}^{\widehat{A}\,\widehat{B}}$. Adapting (4.16) to our present notation, we have

$$G^{AB} = Y^A{}_{,\alpha}\,Y^B{}_{,\beta}\,g^{\alpha\beta} + \sum_{\widehat{P}=n+1}^{N} \varepsilon_{\widehat{P}}X_{\widehat{P}}{}^A X_{\widehat{P}}{}^B. \tag{7.85}$$

Using this to eliminate $Y^A{}_{,\alpha}\,Y^B{}_{,\gamma}\,g^{\alpha\gamma}$, we find from (7.80):

$$
A_{(\widehat{S})\beta}{}^{\gamma}Y^B{}_{,\gamma} = -g^{\alpha\gamma}\Omega_{(\widehat{S})\alpha\beta}Y^B{}_{,\gamma} - \left\{{B \atop MN}\right\}_G X_{\widehat{S}}{}^M Y^N{}_{,\beta}
$$
$$
+ G_{AR}\left\{{R \atop MN}\right\}_G X_{\widehat{S}}{}^M Y^N{}_{,\beta} \sum_{\widehat{P}=n+1}^{N} \varepsilon_{\widehat{P}}X_{\widehat{P}}{}^A X_{\widehat{P}}{}^B. \tag{7.86}
$$

Using this and (7.81) in (7.79) we see that two sums over $\widehat{P}$ cancel out and the result is

$$
X_{\widehat{S}}{}^B{}_{,\beta} = -g^{\mu\rho}\Omega_{(\widehat{S})\mu\beta}Y^B{}_{,\rho} - \left\{{B \atop MN}\right\}_G X_{\widehat{S}}{}^M Y^N{}_{,\beta} + \sum_{\widehat{P}=n+1}^{N} \varepsilon_{\widehat{P}}\mu_{[\widehat{P}\,\widehat{S}]\beta}X_{\widehat{P}}{}^B. \tag{7.87}
$$

Now we can employ the integrability condition (7.73). Substituting for $Y^A{}_{;\alpha\beta}$ from (7.71), then using (7.71) and (7.87) to eliminate the second derivatives of Y^A and the derivatives of $X_{\widehat{S}}{}^A$, we obtain:

$$
R^A{}_{MNP}(G)Y^M{}_{,\alpha}\,Y^N{}_{,\beta}\,Y^P{}_{,\gamma} + \sum_{\widehat{S}=n+1}^{N} \varepsilon_{\widehat{S}}X_{\widehat{S}}{}^A\left(\Omega_{(\widehat{S})\alpha\beta;\gamma} - \Omega_{(\widehat{S})\alpha\gamma;\beta}\right)
$$
$$
+ \sum_{\widehat{S}=n+1}^{N} \varepsilon_{\widehat{S}}Y^A{}_{,\rho}\,g^{\rho\mu}\left(\Omega_{(\widehat{S})\alpha\gamma}\Omega_{(\widehat{S})\mu\beta} - \Omega_{(\widehat{S})\alpha\beta}\Omega_{(\widehat{S})\mu\gamma}\right)
$$
$$
+ \sum_{\widehat{S}=n+1}^{N}\sum_{\widehat{P}=n+1}^{N} \varepsilon_{\widehat{S}}\varepsilon_{\widehat{P}}X_{\widehat{P}}{}^A\left(\Omega_{(\widehat{S})\alpha\beta}\mu_{[\widehat{P}\,\widehat{S}]\gamma} - \Omega_{(\widehat{S})\alpha\gamma}\mu_{[\widehat{P}\,\widehat{S}]\beta}\right) - R^{\rho}{}_{\alpha\beta\gamma}(g)Y^A{}_{,\rho} = 0,
$$
$$\tag{7.88}$$

where $R^A{}_{MNP}(G)$ is the Riemann tensor of U_N. Since $\{Y^A{}_{,\alpha}, X_{\widehat{B}}{}^A\}$ are a basis of the tangent space, Eqs. (7.88) are equivalent to the collection of projections of (7.88) on $\{Y^A{}_{,\alpha}\}$ and $\{X_{\widehat{B}}{}^A\}$. Contracting (7.88) with $G_{AQ}Y^Q{}_{,\delta}$ and with $G_{AQ}X_{\widehat{T}}{}^Q$ and using (7.65) – (7.68) we get, respectively

$$R_{\delta\alpha\beta\gamma}(g) = R_{QMNP}(G)Y^Q{}_{,\delta}\, Y^M{}_{,\alpha}\, Y^N{}_{,\beta}\, Y^P{}_{,\gamma}$$

$$+ \sum_{\widehat{S}=n+1}^{N} \varepsilon_{\widehat{S}} \left(\Omega_{(\widehat{S})\alpha\gamma}\Omega_{(\widehat{S})\delta\beta} - \Omega_{(\widehat{S})\alpha\beta}\Omega_{(\widehat{S})\delta\gamma} \right), \qquad (7.89)$$

$$\Omega_{(\widehat{T})\alpha\beta;\gamma} - \Omega_{(\widehat{T})\alpha\gamma;\beta} = -R_{QMNP}(G)X_{\widehat{T}}{}^Q Y^M{}_{,\alpha}\, Y^N{}_{,\beta}\, Y^P{}_{,\gamma}$$

$$- \sum_{\widehat{S}=n+1}^{N} \varepsilon_{\widehat{S}} \left(\Omega_{(\widehat{S})\alpha\beta}\mu_{[\widehat{T}\widehat{S}]\gamma} - \Omega_{(\widehat{S})\alpha\gamma}\mu_{[\widehat{T}\widehat{S}]\beta} \right). \qquad (7.90)$$

Equations (7.89) – (7.90) are called the **Gauss–Codazzi equations**. When $N = n+1$, this is the full set of necessary and sufficient conditions for V_n to be embeddable in U_N. If $N > n+1$, then (7.89) – (7.90) must be supplemented with the integrability conditions of (7.87), $X_{\widehat{S}}{}^B{}_{;\beta\gamma} - X_{\widehat{S}}{}^B{}_{;\gamma\beta} = 0$. Using (7.87) and (7.90), eliminating the second derivatives of Y^A by (7.71) and remembering that the μ-s are antisymmetric in their Latin indices, we find these conditions to be

$$\sum_{\widehat{P}=n+1}^{N} \varepsilon_{\widehat{P}}X_{\widehat{P}}{}^B \left(\mu_{[\widehat{P}\,\widehat{S}]\beta;\gamma} - \mu_{[\widehat{P}\,\widehat{S}]\gamma;\beta} \right)$$

$$+ \sum_{\widehat{P}=n+1}^{N} \sum_{\widehat{R}=n+1}^{N} \varepsilon_{\widehat{P}}\varepsilon_{\widehat{R}}X_{\widehat{R}}{}^B \left(\mu_{[\widehat{P}\,\widehat{S}]\beta}\mu_{[\widehat{R}\,\widehat{P}]\gamma} - \mu_{[\widehat{P}\,\widehat{S}]\gamma}\mu_{[\widehat{R}\,\widehat{P}]\beta} \right)$$

$$+ g^{\mu\nu} \sum_{\widehat{P}=n+1}^{N} \varepsilon_{\widehat{P}}X_{\widehat{P}}{}^B \left(\Omega_{(\widehat{S})\mu\gamma}\Omega_{(\widehat{P})\nu\beta} - \Omega_{(\widehat{S})\mu\beta}\Omega_{(\widehat{P})\nu\gamma} \right) + R^B{}_{ACD}(G)Y^C{}_{,\beta}\, Y^D{}_{,\gamma}\, X_{\widehat{S}}{}^A$$

$$+ g^{\mu\nu}Y^B{}_{,\nu}\, R_{QMNP}(G)X_{\widehat{S}}{}^Q Y^M{}_{,\mu}\, Y^N{}_{,\beta}\, Y^P{}_{,\gamma} = 0. \qquad (7.91)$$

As before, this is equivalent to the set of projections on $\{Y^A{}_{,\alpha}\}$ and $\{X_{\widehat{B}}{}^A\}$. However, the projection on $\{Y^A{}_{,\alpha}\}$ is zero, so the other one fully represents (7.91). Contracting (7.91) with $G_{BQ}X_{\widehat{T}}{}^Q$ we obtain

$$\mu_{[\widehat{T}\,\widehat{S}]\beta;\gamma} - \mu_{[\widehat{T}\,\widehat{S}]\gamma;\beta} + \sum_{\widehat{P}=n+1}^{N} \varepsilon_{\widehat{P}} \left(\mu_{[\widehat{P}\,\widehat{S}]\beta}\mu_{[\widehat{T}\,\widehat{P}]\gamma} - \mu_{[\widehat{P}\,\widehat{S}]\gamma}\mu_{[\widehat{T}\,\widehat{P}]\beta} \right)$$

$$+ g^{\mu\nu} \left(\Omega_{(\widehat{S})\mu\gamma}\Omega_{(\widehat{T})\nu\beta} - \Omega_{(\widehat{S})\mu\beta}\Omega_{(\widehat{T})\nu\gamma} \right)$$

$$+ R_{QACD}(G)X_{\widehat{T}}{}^Q Y^C{}_{,\beta}\, Y^D{}_{,\gamma}\, X_{\widehat{S}}{}^A = 0. \qquad (7.92)$$

In relativity, (7.89), (7.90) and (7.92) appear almost always in the special case $N = n+1$ (actually, most often with $N = 4$ and $n = 3$, i.e. for hypersurfaces in spacetime). In that case, they simplify. Equations (7.92) are fulfilled identically (because $\mu_{[\widehat{R}\,\widehat{S}]\beta} = 0$ in this case, and the indices with a hat have just one value $N = n+1$, while all terms in (7.92) are antisymmetric in $[\widehat{T}\,\widehat{S}]$). The Gauss–Codazzi equations then become

$$R_{\delta\alpha\beta\gamma}(g) = R_{QMNP}(G)Y^Q{}_{,\delta}\, Y^M{}_{,\alpha}\, Y^N{}_{,\beta}\, Y^P{}_{,\gamma} + \varepsilon\left(\Omega_{\alpha\gamma}\Omega_{\delta\beta} - \Omega_{\alpha\beta}\Omega_{\delta\gamma} \right), \qquad (7.93)$$

$$\Omega_{\alpha\beta;\gamma} - \Omega_{\alpha\gamma;\beta} = -R_{QMNP}(G)X^Q Y^M{}_{,\alpha}\, Y^N{}_{,\beta}\, Y^P{}_{,\gamma}, \qquad (7.94)$$

where X^Q is the single normal vector to V_n and $\varepsilon = G_{AB}X^A X^B$.

The expression (7.76) may be simplified further when the coordinates in U_{n+1} are adapted to V_n as follows. Through every point of V_n we run a curve C in U_{n+1} orthogonal to V_n and choose the arc length s on it as the Y^{n+1} coordinate in U_{n+1} in such a way that $Y^{n+1} = A = $ constant on V_n. The equations $A \neq Y^{n+1} = $ constant then define other hypersurfaces in U_{n+1}. The $\{Y^1, \ldots, Y^n\}$ coordinates in U_{n+1} are chosen so that in V_n they coincide with the intrinsic coordinates of V_n, $Y^\alpha = x^\alpha$, $\alpha = 1, \ldots, n$. In such coordinates $G_{(n+1)\alpha}(V_n) = 0$ and $G_{\alpha\beta}(V_n) = g_{\alpha\beta}$. Then (7.76) may be written as

$$\Omega_{\alpha\beta} = -X_{\alpha;\beta}. \tag{7.95}$$

In some textbooks, (7.95) is used as the definition of the second fundamental form of a hypersurface. Although correct in principle, it is rather misleading, since it has the appearance of a fully covariant definition, which it is not. It holds only in the adapted coordinates, and the semicolon in (7.95) denotes not the covariant derivative in V_n, but the V_n-components of the covariant derivative in U_N.

Another context in which the Gauss–Codazzi equations sometimes appear in relativity is the problem of embedding a given spacetime V_n in a *flat* Riemann space of higher dimension. Then, (7.89), (7.90) and (7.92) should be fulfilled with $R_{ABCD} = 0$. With flat G_{AB}, (7.65) are a set of $n(n+1)/2$ differential equations for N unknown functions Y^A. A simple accounting suggests that if $N = n(n+1)/2$, then the set should have a solution (for $n = 4$, $N = 10$). However, this does not take into account various subtle possibilities. For example, if G_{AB} is positive-definite while $g_{\alpha\beta}$ is not, the set (7.65) will be unsolvable with any N. We do not know what signature G_{AB} should have, so the signs in the canonical form of G_{AB} are additional, discrete unknowns.

This problem of embedding in flat Riemann spaces has not been solved in general, and $N \leq n(n+1)/2$ is only a plausible hint. However, for various special cases the embeddings were demonstrated by explicit calculations and often the dimension of U_N is considerably smaller than $n(n+1)/2$. For the smallest N for which an embedding of a given Riemann space in a flat Riemann space is possible, the number $(N - n)$ is called the **class** of the Riemann space. For example, all conformally flat Riemann spaces can be embedded in flat Riemann spaces of dimension $(n + 2)$ (Plebański, 1967) and thus are of class 2; the 4-dimensional Riemann space corresponding to a spherically symmetric gravitational field in vacuum can be embedded in a 6-dimensional flat space (see Section 14.10), i.e. is also of class 2. Several other 4-dimensional Riemann spaces can be embedded in a 5-dimensional flat Riemann space (Stephani, 1967b).

In the adapted coordinates with $N = n + 1$, Eq. (7.75) reduces to another useful, although noncovariant, form. In these coordinates, X^B has only the $(n+1)$-st component, thus $X^\alpha = 0$. Since $Y^\alpha = x^\alpha$ on V_n, the first term in (7.75) becomes $-G_{\alpha(n+1)}X^{n+1},_\beta = 0$ (because $G_{\alpha(n+1)} = 0$). The second term becomes $\left(-g_{\alpha\rho} \left\{ {}^{\ \rho}_{(n+1)\beta} \right\}_G X^{n+1} \right)$. It is easy to verify that $\left\{ {}^{\ \rho}_{(n+1)\beta} \right\}_G = \frac{1}{2} g^{\rho\sigma} g_{\beta\sigma,(n+1)}$, thus

$$\Omega_{\alpha\beta} = -\frac{1}{2} g_{\alpha\beta,(n+1)} X^{n+1}. \tag{7.96}$$

Thus, in the adapted coordinates, $\Omega_{\alpha\beta}$ is proportional to the directional derivative of the metric in V_n along the normal vector.

7.16 The Petrov classification

One of the unsolved problems of general relativity is that of detecting invariant differences between metrics. Two metrics that seem to be different might be representations of the same metric in two coordinate systems. Finding a transformation between them or proving that no such transformation exists requires solving a set of partial differential equations, for which no general method is known. Therefore, each invariant criterion that allows us to detect coordinate-independent differences between metrics is very useful.

One of such criteria is the Petrov classification of algebraic types of the Weyl tensor, first introduced by Petrov (1954). Later, a few other methods of introducing the same classification were presented by Pirani (1957) and Debever (1959), and the simplest one, based on spinorial techniques, was proposed by Penrose (1960) (see descriptions of all methods by Stephani et al. (2003); the Penrose method is presented in our Chapter 11). We shall use here the method that can be introduced in the briefest way (Ehlers and Kundt, 1962; Barnes, 1984), but it is not simple in practical application. **Note:** this classification applies only in four dimensions and for the Lorentzian signature $(+ - - -)$ (and its equivalents). Analogous classifications may exist for higher dimensions and for other signatures, but they have not been considered in the literature.

Let u^α be an arbitrary timelike vector field with $g_{\mu\nu}u^\mu u^\nu = 1$. We define the tensors

$$E_{\alpha\gamma} \stackrel{\text{def}}{=} C_{\alpha\mu\gamma\nu}u^\mu u^\nu, \tag{7.97}$$

$$H_{\alpha\gamma} \stackrel{\text{def}}{=} \frac{1}{2}\sqrt{-g}\epsilon_{\alpha\sigma\mu\nu}C^{\mu\nu}{}_{\gamma\lambda}u^\sigma u^\lambda, \tag{7.98}$$

where $g \stackrel{\text{def}}{=} \det\|g_{\alpha\beta}\|$ ($g < 0$ because of the signature). The $E_{\alpha\gamma}$ and $H_{\alpha\gamma}$ are called, respectively, the **electric part** and the **magnetic part** of the Weyl tensor. These names have nothing to do with physical interpretation, they refer to the algebraic analogy between (7.97) – (7.98) and the decomposition of the tensor of electromagnetic field into the electric and magnetic fields – see Chapter 13. Both these tensors are symmetric, although the symmetry of $H_{\alpha\gamma}$ is not self-evident (see Exercise 9). Also, $E_{\alpha\gamma}$ and $H_{\alpha\gamma}$ represent the Weyl tensor uniquely, see Exercise 10. We use them to form a new complex tensor

$$Q_{\alpha\gamma} \stackrel{\text{def}}{=} E_{\alpha\gamma} + iH_{\alpha\gamma} = Q_{\gamma\alpha}. \tag{7.99}$$

Since $E_{\alpha\gamma}u^\gamma = H_{\alpha\gamma}u^\gamma \equiv 0$, we have $Q_{\alpha\gamma}u^\gamma \equiv 0$, so $Q_{\alpha\gamma}$ is in fact a tensor in a 3-dimensional space. Since, in consequence of $C_{\alpha[\beta\gamma\delta]} = 0$,

$$g^{\alpha\gamma}Q_{\alpha\gamma} \equiv 0, \tag{7.100}$$

the sum of eigenvalues of Q must be zero. Let us investigate the minimal equation for the matrix $Q = \|Q_{\alpha\gamma}\|$. The following possibilities then arise (I denotes the unit matrix):

THE MINIMAL EQUATION	PETROV TYPE
$(Q - \lambda_1 I)(Q - \lambda_2 I)(Q - \lambda_3 I) = 0$	I
$(Q - \lambda I)^2 (Q + 2\lambda I) = 0$	II
$(Q - \lambda I)(Q + 2\lambda I) = 0$	D
$Q^3 = 0$	III
$Q^2 = 0$	N
$Q = 0$	O

The relations between different Petrov types are shown in Fig. 7.3.

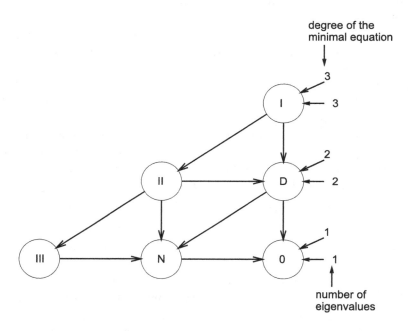

Fig. 7.3 The Petrov classification. Arrows show possible specialisations.

The Petrov classification is important because it is coordinate-independent. Two metrics whose Weyl tensors are of different Petrov types cannot be different coordinate representations of the same metric (but one metric *can* be a limiting case of the other). If the Petrov type is the same for two metrics, then the question is still undecided, and other criteria must be used. In general, this **equivalence problem** is not algorithmic, but attempts to solve it 'in practice' are under way, with some success already (Stephani et al., 2003, Chapter 9; see also Musgrave, https://github.com/grtensor/grtensor).

As seen from the formulae above, in order to determine the Petrov type by this particular method, one must choose a timelike vector field u^α. But the Petrov type does not depend on u^α (see Exercise 11).

7.17 Exercises

1. Find the metric tensors of (i) a cone; (ii) a paraboloid of revolution; (iii) one- and two-sheeted hyperboloids of revolution; (iv) a torus; (v) a cylinder. Calculate the curvature of a cone, a cylinder, a one-sheeted hyperboloid and a torus.

2. Show that the geodesics on a 2-dimensional sphere are great circles.

Hint. First derive the equation of a great circle in spherical coordinates, then solve the geodesic equation in the same coordinates.

3. Solve the geodesic equation on a cylinder. What curves are the geodesics?

4. Show that the Mercator mapping used in cartography is a conformal mapping of a sphere on a cylinder. This mapping is obtained when a cylinder is wrapped around the globe so that they are tangent all along the equator. The images of the points on Earth's surface are obtained by drawing straight lines through the centre of the globe. The image of the point where the straight line intersects the globe's surface is the point where the same line intersects the cylinder. (The poles of the globe get thereby mapped 'to infinity', and the neighbourhoods of the poles are strongly stretched in the meridional direction.)

5. Show that a conformal image of a geodesic line of zero length in V_n is also a geodesic line of zero length in U_n.

6. Verify that the Weyl tensor indeed has the properties mentioned after (7.49), i.e. $C_{\alpha\beta\gamma\delta} = C_{[\alpha\beta]\gamma\delta} = C_{\alpha\beta[\gamma\delta]} = C_{\gamma\delta\alpha\beta}$, $C_{\alpha[\beta\gamma\delta]} = 0$ and $C^{\alpha\rho}{}_{\gamma\rho} = 0$. The other traces are zero in consequence of this one and of the (anti-)symmetries.

7. Verify (7.51)–(7.54) and then verify that the Weyl tensors corresponding to conformally related Riemann spaces are indeed equal or proportional, depending on the positions of indices. Namely, if $\widetilde{g}_{\alpha\beta}$ is the metric tensor in $\widetilde{V}_n$ and $g_{\alpha\beta}$ is the metric tensor in V_n, and $\widetilde{g}_{\alpha\beta} = \phi^{-2} g_{\alpha\beta}$, then $\widetilde{C}^{\alpha}{}_{\beta\gamma\delta} = C^{\alpha}{}_{\beta\gamma\delta}$.

8. Find the second fundamental form for a plane and for a cylinder embedded in a Euclidean E^3.

9. Verify that the $H_{\alpha\gamma}$ of (7.98) is symmetric in α and γ.

Hint. First verify that $\epsilon^{\alpha\gamma\rho\sigma} H_{\alpha\gamma} = 0$ using (3.34), the explicit expression for $\delta^{\gamma\rho\sigma}_{\beta\mu\nu}$ and the results of Exercise 6. Then note that $H_{[\alpha\gamma]} = \frac{1}{2} \delta^{\lambda\kappa}_{\alpha\gamma} H_{\lambda\kappa} \equiv \frac{1}{4} \delta^{\lambda\kappa\rho\sigma}_{\alpha\gamma\rho\sigma} H_{\lambda\kappa}$, so $H_{[\alpha\gamma]} = 0$.

10. Verify that the tensors $E_{\alpha\gamma}$ and $H_{\alpha\gamma}$ and the vector field u^{α} uniquely determine the Weyl tensor $C^{\alpha\beta}{}_{\gamma\delta}$.

Hint. Note that to calculate $E_{\alpha\gamma}$ and $H_{\alpha\gamma}$ you need to know only $C^{\alpha\beta}{}_{\gamma\lambda} u^{\lambda}$, not the whole of $C^{\alpha\beta}{}_{\gamma\lambda}$. So, the inverse mapping will directly determine also only $C^{\alpha\beta}{}_{\gamma\lambda} u^{\lambda}$:

$$C^{\alpha\beta}{}_{\gamma\lambda} u^{\lambda} = \frac{1}{\sqrt{-g}} \epsilon^{\rho\alpha\beta\sigma} u_{\sigma} H_{\rho\gamma} - u^{\alpha} E^{\beta}{}_{\gamma} + u^{\beta} E^{\alpha}{}_{\gamma}. \tag{7.101}$$

In verifying (7.101) you will need the following identities

$$\epsilon_{\alpha\beta\gamma\delta} = g^{-1} g_{\alpha\mu} g_{\beta\nu} g_{\gamma\rho} g_{\delta\sigma} \epsilon^{\mu\nu\rho\sigma}, \tag{7.102}$$

$$\epsilon^{\alpha\beta\gamma\delta} = g g^{\alpha\mu} g^{\beta\nu} g^{\gamma\rho} g^{\delta\sigma} \epsilon_{\mu\nu\rho\sigma}. \tag{7.103}$$

To see that $E_{\alpha\gamma}$, $H_{\alpha\gamma}$ and u^{α} uniquely determine $C^{\alpha\beta}{}_{\gamma\delta}$ begin by choosing an orthonormal basis of vector fields $e_i{}^{\alpha}$ (see Sec. 4.3) such that $e_{\hat{0}}{}^{\alpha} = u^{\alpha}$ (numerical values of the tetrad indices are marked with an overhat, to avoid confusion with coordinate indices). Orthonormal means that the fields $e_i{}^\alpha$ obey $g_{\alpha\beta}e_i{}^\alpha e_j{}^\beta = \eta_{ij} = \text{diag}(+1,-1,-1,-1)$. Then project (7.97) – (7.98) on the $e_i{}^\alpha$ vectors. The new formulae are

$$E_{ij} = C_{i\hat{0}j\hat{0}}, \tag{7.104}$$

$$H_{ij} = \frac{1}{2}\epsilon_{i\hat{0}mn}C^{mn}{}_{j\hat{0}} \tag{7.105}$$

(verify how $\sqrt{-g}\epsilon_{\alpha\rho\mu\nu}$ transforms into ϵ_{ijmn}!). Now calculate all the components of E_{ij} and H_{ij} explicitly, one by one, for example $E_{\hat{1}\hat{3}} = C_{\hat{1}\hat{0}\hat{3}\hat{0}} = C_{\hat{0}\hat{1}\hat{0}\hat{3}} = E_{\hat{3}\hat{1}}$ and $H_{\hat{2}\hat{3}} = -C_{\hat{0}\hat{3}\hat{1}\hat{3}} = C_{\hat{0}\hat{2}\hat{1}\hat{2}} = H_{\hat{3}\hat{2}}$. In this way you will find that E_{ij} and H_{ij} contain all those components of C_{ijkl} for which at least one of $\{ijkl\}$ is zero (you will have to verify that H_{ij} is symmetric in $\{ij\}$; see Exercise 9). The remaining components of C_{ijkl} can be calculated from E_{ij} and H_{ij} using $C^{is}{}_{js} = 0$. In this way, u^i, E_{ij} and H_{ij} contain the whole information about C_{ijkl} and so uniquely determine it.

11. Show that the Petrov type determined by the method of Section 7.16 does not depend on the choice of the vector field u^α.

Hint. Choose an orthonormal basis of vector fields $e_i{}^\alpha$ such that $e_{\hat{0}}{}^\alpha = u^\alpha$ (see Exercise 10) and another orthonormal basis $\tilde{e}_{i'}^\alpha$ such that $\tilde{e}_{\hat{0}}^\alpha = w^\alpha$, where w^α is the new unit timelike field to define the new $\tilde{E}_{\alpha\beta}$ and $\tilde{H}_{\alpha\beta}$. The relation between the bases is

$$\tilde{e}_{i'}^\alpha = A_{i'}{}^i e_i{}^\alpha,$$

where $\det\left(A_{i'}{}^i\right) \neq 0$ and

$$\eta_{ij}A_{i'}{}^i A_{j'}{}^j = \eta_{i'j'}.$$

In the new basis

$$\tilde{E}_{j'}^{i'} = A^{i'}{}_i A_{j'}{}^j E^i{}_j,$$

and similarly for $H^i{}_j$ and $Q^i{}_j$, $A^{i'}{}_i$ is the inverse matrix to $A_{i'}{}^i$.

Now, using (3.38), write the characteristic equation for $\tilde{Q}_{j'}^{i'}$ and verify that $\tilde{Q}_{j'}^{i'}$ has the same set of eigenvalues as $Q^i{}_j$.

8

Symmetries of Riemann spaces, invariance of tensors

8.1 Symmetry transformations

We noted in Section 3.10 that a coordinate transformation on a manifold may be interpreted as a mapping of the manifold into itself. Now we shall interpret the associated transformations of the tensor fields.

Let $F : M_n \to M_n$ be an isomorphism between two subsets of the manifold M_n, $p \in M_n$ and $F(p) = p' \in M_n$. Then, the mappings associated with F carry tensors from p to p'. Thereby, a tensor T that was attached to p before the transformation becomes T' attached to p'. Now consider a subset $U \subset M_n$ and its image $F(U) \subset M_n$. Suppose that F is an element of a continuous group of mappings $\{F_t\}$, with $\{F_{t_0}\}$ being the identity map. If $F = F_{t_1}$ and $|t_1 - t_0|$ is sufficiently small, then $F(U) \cap U \neq \emptyset$. So let $p, p' \in F(U) \cap U$. Then p is an image of another point q, $p = F(q)$, and the tensors that were attached to q before the transformation were sent into p (see Fig. 8.1). Hence, we have two tensors attached to each point p: $T(p)$ that was there before the transformation and $T'(p)$ that was sent to p from q by the transformation. The latter can be calculated from the old $T(q)$ by (3.10). Consequently, we can compare $T'(p)$ with $T(p)$.

If it so happens that $T'(p) = T(p)$ for all points of the manifold, then we call the tensor field T **invariant under the action of** F, and we call F an **invariance transformation of** T. If M_n is a Riemann space, and the metric tensor of M_n is invariant under F, then the mapping F is called a **symmetry** or an **isometry** of M_n.

There is no general theory of all kinds of invariances, so this chapter does not cover, for example, discrete invariances (like reflections or the groups known in crystallography). However, there exists a theory of those mappings that constitute continuous groups, and it will be the subject of this chapter.

8.2 The Killing equations

Let Γ be a one-parameter family of mappings of a manifold M_n into itself such that to every value of the parameter $t \in [t_1, t_2] \overset{\text{def}}{=} B \subset \mathbb{R}^1$ there corresponds a mapping $f_t : M_n \to M_n$:

$$\mathbb{R}^1 \supset B \overset{\text{def}}{=} [t_1, t_2] \ni t \to x'^{\alpha} = f^{\alpha}(t, \{x\}), \tag{8.1}$$

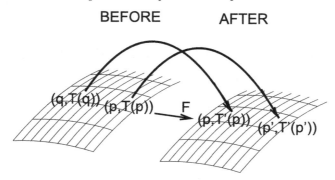

Fig. 8.1 A mapping F of the manifold M_n into itself takes the point q to p, and the associated mapping transforms the tensors $T(q)$ and $T(p)$ into $T'(p)$ and $T'(p')$, respectively. Thus, after the transformation we have two tensors at the same point: $T(p)$ that was there before, and $T'(p)$ that was brought to p by the mapping. If $T'(p) = T(p)$ for all $p \in M_n$, then the tensor field T is invariant under F.

where f_t is the collection of all the functions f^α at a given t and Γ is the collection of all the mappings f_t for every value of $t \in B$. Let us also assume that for $t = t_0 \in [t_1, t_2]$ the mapping f_{t_0} is an identity:

$$f^\alpha(t_0, \{x\}) = x^\alpha. \tag{8.2}$$

Example: let $B = [0, 2\pi]$, $M_n = \mathbb{R}^3$, and let f_t be the rotation of $\mathbb{R}^3$ around a fixed axis A by the angle t. Then Γ is the collection of rotations of $\mathbb{R}^3$ around A by all angles in the range $0 \leq t < 2\pi$ and $t_0 = 0$.

Now let $p \in M_n$, and apply to p the mappings f_t corresponding to all $t \in B$. The collection of all images of p will then be an arc of a curve in M_n passing through $p = f_{t_0}(p)$, and each $p \in M_n$ may be used to generate such an arc. The arc is called the **orbit of p under the action of Γ**, and p is called the initial point of the orbit (although in 'practical' instances the orbits are closed or infinite curves with no endpoints).

We assume that (1) the functions $f^\alpha(t, \{x\})$ are of class C^2 with respect to t; (2) each f_t is invertible and (3) its inverse (denoted f_t^{-1}) is also of class C^2. Assumption (1) implies that along each orbit a field of tangent vectors exists and is continuously differentiable. Assumption (2) implies that the mappings of the family Γ and their inverses form a group G. The group multiplication is the superposition of the mappings:

$$(f_{t_2} \circ f_{t_1})(x) = f_{t_2}(f_{t_1}(x)), \tag{8.3}$$

where f_{t_1} is represented by $f^\alpha(t_1, x)$ and f_{t_2} is represented by $f^\alpha(t_2, f_{t_1}(x))$. For each mapping f_t we may then write (from Taylor's formula):

$$x'^\alpha = x^\alpha + \left.\frac{\partial f^\alpha}{\partial t}\right|_{t=t_0}(t - t_0) + O(\varepsilon^2), \tag{8.4}$$

where $x'^\alpha = f^\alpha(t, \{x\})$, $x^\alpha = f^\alpha(t_0, \{x\})$, $\varepsilon \stackrel{\text{def}}{=} t - t_0$ and $O(\varepsilon^2)$ has the property

$$\lim_{\varepsilon \to 0} \frac{O(\varepsilon^2)}{\varepsilon} = 0. \tag{8.5}$$

Note that we are not making any approximation here; Eq. (8.4) is exact, but the form of $O(\varepsilon^2)$ will in the end turn out to be irrelevant. All such irrelevant terms will be denoted by the same symbol $O(\varepsilon^2)$ even though they may not be identical to each other.

The quantities

$$k^\alpha \stackrel{\text{def}}{=} \left. \frac{\partial f^\alpha}{\partial t} \right|_{t=t_0} \tag{8.6}$$

are components of the vector field tangent to the orbits at their initial points and are called the **generators** of the group G.

Suppose that a tensor $T_{\alpha\beta}$ is invariant under all the transformations in Γ:

$$T'_{\alpha\beta}(p) = T_{\alpha\beta}(p) \qquad \text{for all } p \in M_n \text{ and } t \in B. \tag{8.7}$$

What condition must $T_{\alpha\beta}$ fulfil? Again from Taylor's formula:

$$T'_{\alpha\beta}(p') = T'_{\alpha\beta}(p) + \varepsilon T'_{\alpha\beta,\mu}(p) k^\mu + O(\varepsilon^2), \tag{8.8}$$

where p' has the coordinates $x'^\alpha = f^\alpha(t, \{x\})$ and p has the coordinates $x^\alpha = f^\alpha(t_0, \{x\})$. From (8.4) and (8.6) we have

$$
\begin{aligned}
\frac{\partial x^\mu}{\partial x'^\alpha} &= \frac{\partial}{\partial x'^\alpha} \left[x'^\mu - \varepsilon k^\mu - O\left(\varepsilon^2\right) \right] = \delta^\mu{}_\alpha - \varepsilon k^\mu{}_{,\rho} \frac{\partial x^\rho}{\partial x'^\alpha} - O\left(\varepsilon^2\right) \\
&= \delta^\mu{}_\alpha - \varepsilon k^\mu{}_{,\rho} \left(\delta^\rho{}_\alpha - \varepsilon k^\rho{}_{,\sigma} \frac{\partial x^\sigma}{\partial x'^\alpha} - O(\varepsilon^2) \right) - O(\varepsilon^2) \\
&= \delta^\mu{}_\alpha - \varepsilon k^\mu{}_{,\alpha} + O(\varepsilon^2).
\end{aligned} \tag{8.9}
$$

(We do not differentiate ε because we have advanced along the orbits by the fixed parameter distance $\varepsilon = t - t_0$, which is coordinate-independent.) Hence, from (8.9) and the transformation law for $T_{\alpha\beta}$

$$
\begin{aligned}
T'_{\alpha\beta}(P') &= \frac{\partial x^\mu}{\partial x'^\alpha} \frac{\partial x^\nu}{\partial x'^\beta} T_{\mu\nu}(P) \\
&= T_{\alpha\beta}(P) - \varepsilon k^\mu{}_{,\alpha} T_{\mu\beta}(P) - \varepsilon k^\nu{}_{,\beta} T_{\alpha\nu}(P) + O(\varepsilon^2).
\end{aligned} \tag{8.10}
$$

Comparing (8.8) with (8.10) and using (8.7), we have

$$\varepsilon \left(k^\mu T_{\alpha\beta,\mu} + k^\mu{}_{,\alpha} T_{\mu\beta} + k^\nu{}_{,\beta} T_{\alpha\nu} \right) + O(\varepsilon^2) = 0. \tag{8.11}$$

We now divide (8.11) by ε, let $\varepsilon \to 0$ and recall (8.5). The result is

$$k^\mu T_{\alpha\beta,\mu} + k^\mu{}_{,\alpha} T_{\mu\beta} + k^\mu{}_{,\beta} T_{\alpha\mu} = 0. \tag{8.12}$$

These are the **Killing equations**, and their solutions $k^\mu(\{x\})$ are called **Killing vector fields**. Every field of tangent vectors to orbits of invariance transformations of $T_{\alpha\beta}$ must obey (8.12), and every solution of (8.12) generates an invariance group of $T_{\alpha\beta}$. How to find the invariance transformations given k^α and vice versa will be shown in the next section.

Equations (8.12) can be rewritten in the equivalent form,

$$k^\mu T_{\alpha\beta;\mu} + k^\mu{}_{;\alpha} T_{\mu\beta} + k^\mu{}_{;\beta} T_{\alpha\mu} = 0, \tag{8.13}$$

because the Christoffel symbols cancel out in this combination. Equation (8.13) shows that the Killing equations are covariant. If $T_{\alpha\beta} = g_{\alpha\beta}$ (the metric tensor), then, in view of $g_{\alpha\beta;\gamma} = 0$, Eq. (8.13) may be rewritten as

$$k_{\alpha;\beta} + k_{\beta;\alpha} \equiv 2k_{(\alpha;\beta)} = 0. \tag{8.14}$$

In this form, the Killing equations are most easy to remember, but less convenient to work with (because (8.14) suggests that one must calculate the Christoffel symbols to handle the equations, which is not the case) and apply only to the metric tensor.

Note that (8.12) applies only to a doubly covariant tensor field, and only in this case are the generators of invariances called Killing vector fields. We shall deal with invariances of other tensor fields in Section 8.6 and thereafter.

8.3 The connection between generators and the invariance transformations

If $x'^\alpha = f^\alpha(t, \{x\})$ is a family of invariance transformations of a certain tensor field, then the corresponding generator is given by (8.6), where $t = t_0$ corresponds to the identity transformation. Finding the family of invariances given k^α is less straightforward. Any orbit $B \ni t \to x'^\alpha(t) = f^\alpha(t, \{x\})$ is at its every point tangent to $k^\alpha(p')$ (where $\{x'\}$ are the coordinates of p'). Hence the coordinates $y^\alpha(t)$ of points on the orbits must obey

$$\frac{dy^\alpha}{dt} = k^\alpha(\{y\}) \tag{8.15}$$

with the initial conditions

$$y^\alpha|_{t=t_0} = x^\alpha. \tag{8.16}$$

Equations (8.15) – (8.16) are to be understood as follows. A solution to (8.15) will be a family of curves $y^\alpha = f^\alpha(t, C_1, \ldots, C_k)$, labelled by k parameters $(C_1, \ldots, C_k)$. The condition (8.16) allows one to express the constants C_α in terms of the coordinates x^α of the initial points of the curves. In this way, we obtain the set of functions

$$y^\alpha = f^\alpha(t, \{x\}) \tag{8.17}$$

that satisfies Eqs. (8.15) and the initial conditions (8.16).

8.4 Finding the Killing vector fields

The Killing equations are applied to two kinds of problems:

1. Finding the metric tensor of a Riemann space whose symmetries are assumed – then they are equations determining $g_{\alpha\beta}$, with k^α given.
2. Finding the symmetries of a Riemann space whose metric tensor is given – then they are equations determining k^α with $g_{\alpha\beta}$ given.

An example of the first application will be shown in Section 8.10. The second application requires additional explanation. The Killing equations are linear and homogeneous in k^α, which means that if k^α and l^α are Killing fields, then so is $(Ak^\alpha + Bl^\alpha)$, where A and B are arbitrary constants. A general solution of the Killing equations should thus be a linear combination of basis solutions.

Does there exist a finite basis in the space of solutions of the Killing equations? The answer is: yes, but only for the proper Killing equations, i.e. for the generators of invariances of the metric tensor. The proof given below (borrowed from Stephani (1990)) does not work if K_α generates an invariance group of a tensor field other than the metric tensor, and examples of infinite bases are known (Krasiński, 1983).

For a field K_α generating symmetries of M_n we have from (8.14)

$$K_{\alpha;\beta} = -K_{\beta;\alpha}, \tag{8.18}$$

and from the Ricci formula

$$K_{\alpha;\beta\gamma} - K_{\alpha;\gamma\beta} = R^\rho{}_{\alpha\beta\gamma}K_\rho. \tag{8.19}$$

Because $R^\rho{}_{[\alpha\beta\gamma]} \equiv 0$, we have from the above

$$(K_{\alpha;\beta} - K_{\beta;\alpha})_{;\gamma} + (K_{\gamma;\alpha} - K_{\alpha;\gamma})_{;\beta} + (K_{\beta;\gamma} - K_{\gamma;\beta})_{;\alpha} = 0. \tag{8.20}$$

Using now (8.18), the above reduces to

$$K_{\alpha;\beta\gamma} + K_{\gamma;\alpha\beta} + K_{\beta;\gamma\alpha} = 0, \tag{8.21}$$

and, again from (8.18), this yields

$$K_{\gamma;\alpha\beta} = K_{\beta;\alpha\gamma} - K_{\beta;\gamma\alpha} = R^\rho{}_{\beta\alpha\gamma}K_\rho. \tag{8.22}$$

Thus, in a given Riemannian manifold (where $g_{\alpha\beta}$ and $R^\rho{}_{\alpha\beta\gamma}$ are given as functions of $\{x\}$ on open neighbourhoods of any nonsingular point), (8.22) allows us to calculate $K_{\gamma;\alpha\beta}(p_0)$ *algebraically* if $K_\gamma(p_0)$ is given. If $K_{\gamma;\delta}(p_0)$ is given as well, then from the derivative of (8.22) we can *algebraically* calculate $K_{\gamma;\alpha\beta\epsilon}(p_0)$. By differentiating (8.22) consecutively, we can then calculate all covariant derivatives of K_γ at p_0. Furthermore, having all these derivatives (and, consequently, all partial derivatives of K_γ at p_0), we can calculate $K_\gamma(p)$ where $p \in M_n$ lies in such a neighbourhood of p_0 in which the Taylor series for $K_\alpha(p)$ is convergent. However, after each differentiation a new derivative of the Riemann tensor appears, so, in order that the series is convergent, $R^\rho{}_{\alpha\beta\gamma}$ must be analytic in that neighbourhood.[1] From this argument we see that $K_\gamma(p_0)$ and $K_{\gamma;\delta}(p_0)$ at any chosen $p_0 \in M$ are the data needed to determine $K_\gamma(p)$ uniquely. But $K_{\gamma;\delta}$ obey (8.18), so $K_{\gamma;\delta}(p_0)$ are $\frac{1}{2}n(n-1)$ constants, and $K_\gamma(p_0)$ are n constants in an n-dimensional manifold. Thus, the Taylor series for $K_\gamma(p)$ will contain at most $\frac{1}{2}n(n+1)$ arbitrary constants multiplying various functions of $\{x\}$. The multipliers of those constants will be the basis solutions, so their number cannot exceed $\frac{1}{2}n(n+1)$. $\square$

The prescription for finding the basis of the Killing vector fields for the metric tensor is therefore the following:

---

[1] The curvature tensor does not have to be analytic if another reasoning is followed, see Section 8.12.

1. Solve the Killing equations. The general solution will depend on $N \leq \frac{1}{2}n(n+1)$ arbitrary constants, $k^{\mu} = K^{\mu}(A_1, \ldots, A_N, \{x\})$.

2. Calculate the basis

$$k^{\mu}_{(i)} \overset{\text{def}}{=} \frac{\partial K^{\mu}}{\partial A_i}, \qquad i = 1, \ldots, N. \tag{8.23}$$

Each $k^{\mu}_{(i)}$ generates a one-parameter subgroup of symmetries discussed in Section 8.2.

A possible confusion has to be explained here. For brevity, we say 'Killing vectors', but in truth these are *vector fields*, whose components are functions. Hence, the number of linearly independent Killing vector *fields* can be larger than the dimension of the manifold. For example, in a flat Riemann space the number of linearly independent Killing vector fields is equal to the maximal one, $\frac{1}{2}n(n+1)$.

For tensor fields other than the metric tensor a finite basis may not exist, i.e. the general solution of the invariance equations will contain arbitrary functions rather than arbitrary constants. This is the case, e.g., for the invariance group of $R^{\alpha\beta}{}_{\gamma\delta}$ (note the positions of indices!) in a space of constant curvature: any arbitrary vector field l has the property that $R'^{\alpha\beta}{}_{\gamma\delta}(\{x\}) = R^{\alpha\beta}{}_{\gamma\delta}(\{x\})$ for transformations generated by l (see Krasiński (1983) for less trivial examples). Thus, any coordinate transformation $x'^{\alpha} = f^{\alpha}(\{x\})$ is an invariance transformation of $R^{\alpha\beta}{}_{\gamma\delta}(\{x\})$.

8.5 A Killing vector field along a geodesic is a geodesic deviation field

Consider a Killing vector field K_{γ} along a geodesic line G with the tangent vector field ℓ^{α}, so $\ell^{\rho}\ell^{\alpha}{}_{;\rho} = 0$. Let s be the affine parameter on G. Then contract (8.22) with $\ell^{\alpha}\ell^{\beta}$. Because of $\ell^{\rho}\ell^{\alpha}{}_{;\rho} = 0$, the resulting equation can be written as

$$\ell^{\beta}\left(\ell^{\alpha}K_{\gamma;\alpha}\right)_{;\beta} \equiv \frac{\mathrm{D}^2}{\mathrm{d}s^2}K_{\gamma} = R^{\rho}{}_{\beta\alpha\gamma}\ell^{\alpha}\ell^{\beta}K_{\rho} = R_{\gamma\alpha\beta\rho}\ell^{\alpha}\ell^{\beta}K^{\rho}, \tag{8.24}$$

which is the geodesic deviation equation (6.56) with K_{γ} as the geodesic deviation vector. Hence,

Theorem 8.1 *A Killing vector filed taken along a geodesic G coincides with the geodesic deviation field along G (Serbenta and Korzyński, 2022).*

8.6 Invariance of other tensor fields

We investigated the conditions of invariance of the metric tensor in more detail because they are the most important and are most frequently met. Sometimes, though, we need to know the invariance transformations of other tensor fields. Repeating the reasoning (8.7) – (8.12) for the field of contravariant vectors, that is, assuming the condition $V'^{\alpha}(x) = V^{\alpha}(x)$, we would obtain the following equation:

$$k^{\rho}V^{\alpha}{}_{,\rho} - V^{\rho}k^{\alpha}{}_{,\rho} = 0, \tag{8.25}$$

where k^{α} is the generator of the transformation group.

Invariance conditions for other tensor fields are given in the exercises.

8.7 The Lie derivative

If we trace the derivation of the Killing equations (8.12) and of the other invariance equations, then we will notice that they are obtained in the following steps:

1. Take the tensor field T (arbitrary indices) to be $T(t_0)$, where t_0 is the value of the orbit parameter corresponding to the identity transformation.

2. Using the transformation law for T under coordinate changes, calculate $T(t)$ – the value of T transported to another point along the group orbit. Display the result up to terms linear in $(t - t_0)$ (the remaining terms are not neglected, but left in implicit form).

3. Calculate the quantity

$$- \lim_{t \to t_0} \frac{T(t) - T(t_0)}{t - t_0} \stackrel{\text{def}}{=} \underset{k}{\pounds} T, \qquad (8.26)$$

where $k^\alpha = \left. \mathrm{d}x'^\alpha / \mathrm{d}t \right|_{t=t_0}$, and equate the result to zero.

The quantity $(-\underset{k}{\pounds} T)$ defined in (8.26) is thus the derivative of the tensor field T by the parameter of the transformation. As seen from (8.26), that derivative measures the speed of changes of the field T transported along the orbits tangent to k^α with respect to the values of T defined at the consecutive points. If $\underset{k}{\pounds} T = 0$, then the field transported along the orbit everywhere coincides with the tensor T defined before the transformation. The quantity $\underset{k}{\pounds} T$ is called the **Lie derivative** of the tensor field T along the vector field k and can be calculated also if T is not invariant under the action of Γ.[2] We have thus:

$$(\underset{k}{\pounds} T = 0) \Longleftrightarrow (T'(P) \equiv T(P)). \qquad (8.27)$$

The Lie derivative has all the algebraic properties of ordinary differentiation: it is linear with respect to addition, gives zero when acting on a constant, and when acting on a tensor product it obeys

$$\underset{k}{\pounds}(T_1 \otimes T_2) = (\underset{k}{\pounds} T_1) \otimes T_2 + T_1 \otimes (\underset{k}{\pounds} T_2). \qquad (8.28)$$

These properties allow us to derive the formula for the components of the Lie derivative of any tensor field $T^{\alpha_1 \ldots \alpha_k}_{\beta_1 \ldots \beta_l}$ along any vector field k^α using the results of Section 8.6 and of Exercises 7 and 8:

$$\underset{k}{\pounds} T^{\alpha_1 \ldots \alpha_k}_{\beta_1 \ldots \beta_l} = k^\rho T^{\alpha_1 \ldots \alpha_k}_{\beta_1 \ldots \beta_l, \rho} - \sum_{i=1}^{k} k^{\alpha_i}{}_{,\rho_i} T^{\alpha_1 \ldots \rho_i \ldots \alpha_k}_{\beta_1 \ldots \beta_l} + \sum_{j=1}^{k} k^{\rho_j}{}_{,\beta_j} T^{\alpha_1 \ldots \alpha_k}_{\beta_1 \ldots \rho_j \ldots \beta_l}, \qquad (8.29)$$

where sums over i and j extend over all positions of ρ_i and ρ_j, respectively, in the sequence of indices. Thus, the Lie derivative acts similarly to a directional covariant derivative along k, the factors $(-k^\mu{}_{,\nu})$ playing the role of the Christoffel symbols projected on k, $-k^\mu{}_{,\nu} \longleftrightarrow \left\{ {\mu \atop \nu\rho} \right\} k^\rho$. With the help of (8.29) one can now verify that for any tensor field T and any vector fields k and l we have

$$\underset{[k,l]}{\pounds} T \equiv [\underset{k}{\pounds}, \underset{l}{\pounds}] T, \qquad (8.30)$$

so $[k, l]$ always generates an invariance of T if k and l do.

---

[2] The *notion* of the Lie derivative was introduced by Ślebodziński (1931), the *name* was proposed by van Dantzig and made popular by Schouten (Schouten and van Kampen, 1934; Schouten and Struik, 1935).

8.8 The algebra of Killing vector fields

Since for the Killing fields a finite basis exists, we conclude from the last statement of the previous section that there exist such constants $C^l{}_{ij}$ that, for the basis fields,

$$[\underset{(i)}{k},\ \underset{(j)}{k}]^\alpha = C^l{}_{ij}\underset{(l)}{k}^\alpha \qquad \text{(sum over } l). \qquad (8.31)$$

The constants $C^l{}_{ij}$ are called **structure constants** of the symmetry group. We see thus that the set of Killing fields for a given manifold M_n is a Lie algebra. For generators of invariances of other tensor fields, the coefficients $C^l{}_{ij}$ are not necessarily constant.

8.9 Surface-forming vector fields

Let us consider two linearly independent vector fields k and l defined on M_n. They define two families of their tangent curves by the equations $k^\alpha = \mathrm{d}x^\alpha/\mathrm{d}\lambda$ and $l^\alpha = \mathrm{d}x^\alpha/\mathrm{d}\lambda$, where λ is a parameter on each curve. Take a single curve C of the family tangent to the field l, and then consider all the curves tangent to the field k that intersect C (see Fig. 8.2). They form a surface S out of a single curve tangent to l. It is clear that those vectors of the field l that are not tangent to C need not be tangent to S. However, if they are tangent to every such S, then the vector fields k and l are called **surface-forming**.

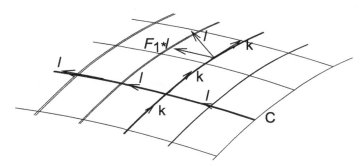

Fig. 8.2 We take a single curve C of the curves tangent to the field l and consider all curves tangent to the field k that intersect C. In this way, we form a surface S, to which other vectors of the field l may or may not be tangent. If they are tangent, then the vector fields k and l are called surface-forming. The field $F_{1*}l$ is l transported along a k-curve by the associated mapping – see text.

What is the condition for two vector fields to be surface-forming? Consider the family of curves $x^\alpha(\lambda)$ tangent to the vectors of the field k, lying in the surface S. They define a family F of mappings of the surface S into itself (the image of a point $S \ni p = x^\alpha(\lambda_0)$ is the point $p' = x^\alpha(\lambda_0 + \Delta\lambda)$, where $\Delta\lambda$ is the same for all $p \in S$). The mapping F_{1*} associated to F maps then vectors tangent to S to other vectors tangent to S. Hence, starting with the vectors l that are tangent to our initial curve C, we can construct the field $(F_{1*}l)$ of vectors tangent to S. The vectors of the field k attached to points of S are tangent to S, too – because S was constructed in this way. Hence, the vectors of the field l that are attached to points of S will be everywhere tangent to S if they are everywhere spanned on the vectors k and $(F_{1*}l)$, thus

$$l^\alpha = \widetilde{a}k^\alpha + \widetilde{b}\,(F_{1*}l)^\alpha\,, \tag{8.32}$$

where $\widetilde{a}(x)$ and $\widetilde{b}(x)$ are arbitrary scalar functions and $\widetilde{b}(x) \neq 0$ (with $\widetilde{b}(x) = 0$, the fields k and l would be linearly dependent, contrary to our assumption). This condition may be rewritten as follows:

$$l^\alpha - (F_{1*}l)^\alpha = \frac{\widetilde{a}}{\widetilde{b}}k^\alpha + \frac{\widetilde{b}-1}{\widetilde{b}}l^\alpha \stackrel{\text{def}}{=} ak^\alpha + bl^\alpha. \tag{8.33}$$

But the rate of change of $[l^\alpha - (F_{1*}l)^\alpha]$ along the curves tangent to k is, by definition, the Lie derivative $(\underset{k}{\pounds}l)^\alpha = [k,l]^\alpha$. Finally then, the necessary and sufficient condition for the vector fields k and l to be surface-forming is

$$[k,l]^\alpha = ak^\alpha + bl^\alpha. \tag{8.34}$$

8.10 Spherically symmetric 4-dimensional Riemann spaces

We call a Riemann space **spherically symmetric** when the group of rotations around a point, $O(3)$, is its isometry group. Its metric tensor must thus obey the Killing equations for each of the three generators of the group $O(3)$. We shall first derive the formulae for these generators.

The orbits of $O(3)$ are 2-dimensional spheres. Each sphere can be embedded in a 3-dimensional Euclidean space E^3. Its equation is then

$$x^2 + y^2 + z^2 = R^2, \tag{8.35}$$

where R is the radius of the sphere, and x, y, z are Cartesian coordinates in E^3. The rotation around the centre of the sphere by the angle α in the plane (x^i, x^j) is then described by the transformation

$$\begin{aligned}
x'^i &= x^i \cos\alpha + x^j \sin\alpha, \qquad x'^j = -x^i \sin\alpha + x^j \cos\alpha,\\
x'^k &= x^k \qquad \text{for } i \neq k \neq j.
\end{aligned} \tag{8.36}$$

The angle α is here the group parameter. From (8.6), the corresponding Killing vector is

$$\underset{[i,j]}{k^\mu} = x^j \delta^\mu{}_i - x^i \delta^\mu{}_j. \tag{8.37}$$

An arbitrary transformation of the sphere into itself can be described as a composition of three consecutive rotations around different axes. Hence, a basis of the space of Killing vectors will be three generators corresponding to rotations around three different axes.

It is often convenient to represent the Killing vectors by the corresponding operators of directional derivatives, also called generators:

$$\underset{(i)}{J} \stackrel{\text{def}}{=} \underset{(i)}{k^\mu} \frac{\partial}{\partial x^\mu}. \tag{8.38}$$

We will choose as our basis of Killing fields the generators of rotations around the three axes of the rectangular Cartesian coordinate system:

$$\underset{[xy]}{J} = x\frac{\partial}{\partial y} - y\frac{\partial}{\partial x}, \qquad \underset{[yz]}{J} = y\frac{\partial}{\partial z} - z\frac{\partial}{\partial y}, \qquad \underset{[xz]}{J} = x\frac{\partial}{\partial z} - z\frac{\partial}{\partial x}. \tag{8.39}$$

Now let us transform the generators to the spherical coordinates

$$x = r\sin\vartheta\cos\varphi, \qquad y = r\sin\vartheta\sin\varphi, \qquad z = r\cos\vartheta. \tag{8.40}$$

In these coordinates the generators become, up to sign,

$$\underset{[xy]}{J} = \frac{\partial}{\partial\varphi}, \qquad \underset{[yz]}{J} = \sin\varphi\frac{\partial}{\partial\vartheta} + \cos\varphi\cot\vartheta\frac{\partial}{\partial\varphi},$$

$$\underset{[xz]}{J} = \cos\varphi\frac{\partial}{\partial\vartheta} - \sin\varphi\cot\vartheta\frac{\partial}{\partial\varphi}. \tag{8.41}$$

Since the coordinates ϑ and φ are defined inside the spheres, we can use them as coordinates in the whole Riemann space. Let us denote the two remaining coordinates t and r. We shall now solve the Killing equations for the metric tensor $g_{\alpha\beta}(t, r, \vartheta, \varphi)$, where $x^0 = t$, $x^1 = r$, $x^2 = \vartheta$, $x^3 = \varphi$, with the Killing vectors given by (8.41), thus

$$\underset{(1)}{k^\alpha} = \delta^\alpha{}_3, \qquad \underset{(2)}{k^\alpha} = \sin\varphi\delta^\alpha{}_2 + \cos\varphi\cot\vartheta\delta^\alpha{}_3,$$

$$\underset{(3)}{k^\alpha} = \cos\varphi\delta^\alpha{}_2 - \sin\varphi\cot\vartheta\delta^\alpha{}_3. \tag{8.42}$$

Note that in assuming the form (8.42) of the Killing vectors for the whole Riemann space we have tacitly made one more assumption. Namely, we assumed that the (ϑ, φ) coordinates on different spheres are correlated in such a way that the rotation of the whole space is described by the same formula as the rotation of a single sphere. One can easily find examples of coordinate systems that do not obey this condition. For example, suppose that we choose the (r, φ) coordinates in the E^2 plane so that the curves $r = $ constant are nonconcentric circles, and the azimuthal angle is measured on each circle independently, beginning from a certain reference direction chosen to be $\varphi = 0$. In such coordinates, the rotation of the plane by the angle α will not be described by $\varphi' = \varphi + \alpha$; the φ' will in general be a nonlinear function of r and φ, which will reduce to $(\varphi + \alpha)$ only on that circle whose centre coincides with the centre of rotation of the plane (and such a circle will not necessarily exist).

As another example of an untypical coordinate system, we may choose the 'generalised spherical coordinates' in the Euclidean space E^3, in which the surfaces $r = $ constant will be concentric spheres, but the poles of the spherical coordinates on different spheres will not lie on a straight line. A rotation by the angle α around an axis will be described by $\varphi' = \varphi + \alpha$ only on that sphere whose pole lies on the axis. On other spheres the rotation will be given by a complicated function of r, ϑ and φ, and the generators will not have the form (8.41).

We should thus expect that the Killing equations for (8.42) will imply a limitation not only on the geometry of the space but also on the coordinates. We shall come back to this question – see the paragraph containing (8.54).

The Killing equations for $\underset{(1)}{k^\alpha}$ reduce to $g_{\alpha\beta,3} = 0$, that is, the whole metric tensor is independent of $x^3 = \varphi$. For $\underset{(2)}{k^\alpha}$ and $\underset{(3)}{k^\alpha}$ the Killing equations are

$$\sin\varphi\frac{\partial}{\partial\vartheta}g_{\alpha\beta} \quad +(\sin\varphi)_{,\alpha}\,g_{2\beta} + (\sin\varphi)_{,\beta}\,g_{\alpha 2}$$

$$+ (\cos\varphi\cot\vartheta)_{,\alpha}\,g_{3\beta} + (\cos\varphi\cot\vartheta)_{,\beta}\,g_{\alpha 3} = 0, \tag{8.43}$$

$$\cos\varphi \frac{\partial}{\partial\vartheta} g_{\alpha\beta} \quad +(\cos\varphi)_{,\alpha}\, g_{2\beta} + (\cos\varphi)_{,\beta}\, g_{\alpha 2}$$

$$-(\sin\varphi\cot\vartheta)_{,\alpha}\, g_{3\beta} - (\sin\varphi\cot\vartheta)_{,\beta}\, g_{\alpha 3} = 0. \tag{8.44}$$

In order to simplify further calculations, we will replace these equations by two combinations thereof. Multiply (8.43) by $\cos\varphi$, (8.44) by $\sin\varphi$ and subtract one result from the other. Using the two identities

$$\sin\varphi(\sin\varphi)_{,\alpha} + \cos\varphi(\cos\varphi)_{,\alpha} \quad \equiv 0,$$

$$\cos\varphi(\sin\varphi)_{,\alpha} - \sin\varphi(\cos\varphi)_{,\alpha} \quad \equiv \varphi_{,\alpha} \tag{8.45}$$

we obtain

$$\varphi_{,\alpha}\, g_{2\beta} + \varphi_{,\beta}\, g_{\alpha 2} + (\cot\vartheta)_{,\alpha}\, g_{3\beta} + (\cot\vartheta)_{,\beta}\, g_{\alpha 3} = 0. \tag{8.46}$$

Now multiply (8.43) by $\sin\varphi$, multiply (8.44) by $\cos\varphi$ and add the results. Using (8.45) again we obtain

$$\frac{\partial}{\partial\vartheta} g_{\alpha\beta} - \varphi_{,\alpha}\cot\vartheta g_{3\beta} - \varphi_{,\beta}\cot\vartheta g_{\alpha 3} = 0. \tag{8.47}$$

Equation (8.46) is algebraic. Taking consecutively the various values of the indices α and β, we obtain the following results:

- For $2 \neq \alpha \neq 3$, $2 \neq \beta \neq 3$ the equation is fulfilled identically.
- For $2 \neq \alpha \neq 3$, $\beta = 2$:

$$g_{\alpha 3} = 0, \qquad \alpha = 0, 1. \tag{8.48}$$

- For $2 \neq \alpha \neq 3$, $\beta = 3$:

$$g_{\alpha 2} = 0, \qquad \alpha = 0, 1. \tag{8.49}$$

- For $\alpha = 2$, $\beta = 2$:

$$g_{23} + g_{32} \equiv 2g_{23} = 0. \tag{8.50}$$

- For $\alpha = 2$, $\beta = 3$:

$$g_{33} = g_{22}\sin^2\vartheta. \tag{8.51}$$

- For $\alpha = \beta = 3$ Eq. (8.50) follows once more.

Now we take the same cases for (8.47) and obtain the following:

- For $2 \neq \alpha \neq 3$, $2 \neq \beta \neq 3$: $(\partial/\partial\vartheta)g_{\alpha\beta} = 0$, $\alpha, \beta = 0, 1$.
- For $2 \neq \alpha \neq 3$, $\beta = 2$ an identity in consequence of (8.49).
- For $2 \neq \alpha \neq 3$, $\beta = 3$ an identity in consequence of (8.48).
- For $\alpha = 2$, $\beta = 2$: $(\partial/\partial\vartheta)g_{22} = 0$.
- For $\alpha = 2$, $\beta = 3$ an identity in consequence of (8.50).
- For $\alpha = \beta = 3$ an identity in consequence of (8.51).

Hence, finally, the solution of (8.46) – (8.47) is

$$ds^2 = \alpha(t,r)dt^2 + 2\beta(t,r)dtdr + \gamma(t,r)dr^2 + \delta(t,r)\left(d\vartheta^2 + \sin^2\vartheta d\varphi^2\right). \qquad (8.52)$$

This is the general 4-dimensional spherically symmetric metric form. Note that, in consequence of the 2-spheres being subspaces of the Riemann space (they are the orbits of the symmetry group), we have obtained a limitation on the signature: the signs at $d\vartheta^2$ and at $d\varphi^2$ must be the same. This is an illustration of the remark made in Section 7.15 on embedding the Riemann spaces in spaces of higher dimension: an inconsistency between the signatures may render the embedding impossible.

Note that we have not assumed anything about the subspaces of the variables (t,r). Hence, arbitrary nonsingular coordinate transformations can be carried out within them:

$$t = f(t',r'), \qquad r = g(t',r'), \qquad (8.53)$$

where f and g are arbitrary functions subject to the condition that $\partial(t,r)/\partial(t',r') \neq 0$. After such a transformation the function $\delta(t,r)$ will preserve its value, while α, β and γ will change to combinations of α, β and γ that will still depend only on t' and r'.

Now we can illustrate the remarks made after (8.42). We can carry out a coordinate transformation on (8.52) that does not obey (8.53), for example

$$r = r' + h(\vartheta,\varphi), \qquad (8.54)$$

where $h(\vartheta,\varphi)$ is an arbitrary function. The result will be

$$ds^2 = \alpha dt^2 + 2\beta dtdr' + 2\beta h_{,\vartheta}\, dtd\vartheta + 2\beta h_{,\varphi}\, dtd\varphi + \gamma dr'^2$$

$$+ 2\gamma h_{,\vartheta}\, dr'd\vartheta + 2\gamma h_{,\varphi}\, dr'd\varphi$$

$$+ \left(\delta + \gamma h_{,\vartheta}{}^2\right)d\vartheta^2 + 2\gamma h_{,\vartheta}\, h_{,\varphi}\, d\vartheta d\varphi + \left(\delta\sin^2\vartheta + \gamma h_{,\varphi}{}^2\right)d\varphi^2 \qquad (8.55)$$

(note that both r and r' are present here; r' is one of the coordinates, while r, the second argument in α, β, γ and δ, is the function defined by (8.54)). The metric (8.55) is still spherically symmetric because it resulted from (8.52) by a coordinate transformation. Nevertheless, (8.55) is not of the form (8.52) because after the transformation (8.54) the spherical coordinates on different spheres ($t = \text{constant}, r = \text{constant}$) are not correlated in the simple way implicit in (8.42).

The orbits of the group $O(3)$ are the subspaces $\{t = \text{constant}, r = \text{constant}\}$ whose metric form, in the coordinates of (8.52), is $ds_2^2 = \delta\left(d\vartheta^2 + \sin^2\vartheta\ d\varphi^2\right)$ (in these subspaces δ is constant). The centre of symmetry is where $\delta(t,r) = 0$ if such a point exists within the manifold. For example, if δ is constant throughout the Riemann space (which is a property invariant under (8.53)), then the centre of symmetry does not exist. This is an analogy to a cylinder or a one-sheeted hyperboloid: these surfaces are rotationally symmetric, but no point on the surface is the centre of rotation.

If the functions α, β, γ and δ in (8.52) are independent of r, then there exists a fourth Killing field $k^{\alpha}_{(4)} = \delta^{\alpha}{}_1$. Then the spacetime has the **Kantowski–Sachs symmetry** (Kantowski and Sachs, 1966). In such spacetimes, those hypersurfaces $t = \text{constant}$ for which $\delta(t) \neq 0$ have no centre. We will come back to them in Chapter 10.

8.11 * Conformal Killing fields and their finite basis

If, after a coordinate transformation, the new expression for the metric tensor is conformally equivalent to the old one, then such a transformation is called a **conformal symmetry** of the Riemann space (and of the metric). The generators of these transformations are called **conformal Killing vectors** and obey the equations

$$k^\rho g_{\alpha\beta,\rho} + k^\rho{}_{,\alpha} g_{\rho\beta} + k^\rho{}_{,\beta} g_{\alpha\rho} \equiv k_{\alpha;\beta} + k_{\beta;\alpha} = \lambda g_{\alpha\beta}, \tag{8.56}$$

where λ is a scalar function.

By a similar method to that in Section 8.4 we will now show that for the Riemann spaces of dimension $n > 2$ a finite basis of generators of conformal symmetries always exists; i.e. that the dimension of the group of conformal symmetries is finite if $n > 2$. (We shall deal with the case $n = 2$ at the end of this section – then, no finite basis exists since all 2-dimensional Riemann spaces are conformally related.)

Contracting (8.56) with $g^{\alpha\beta}$ we obtain

$$\lambda = \frac{2}{n} k^\mu{}_{;\mu}. \tag{8.57}$$

From the Ricci formula we obtain (8.19) and (8.20) again, but instead of (8.18) we have

$$- k_{\beta;\alpha} = k_{\alpha;\beta} - \lambda g_{\alpha\beta}. \tag{8.58}$$

Hence, using (8.56) and $g_{\alpha\beta;\gamma} = 0$ in (8.20), we obtain

$$k_{\alpha;\beta\gamma} + k_{\gamma;\alpha\beta} + k_{\beta;\gamma\alpha} = \frac{1}{2} \left(\lambda_{,\gamma} g_{\alpha\beta} + \lambda_{,\beta} g_{\alpha\gamma} + \lambda_{,\alpha} g_{\beta\gamma} \right). \tag{8.59}$$

From (8.58) and (8.59):

$$k_{\gamma;\alpha\beta} = k_{\beta;\alpha\gamma} - k_{\beta;\gamma\alpha} + \frac{1}{2} \left(-\lambda_{,\gamma} g_{\alpha\beta} + \lambda_{,\beta} g_{\alpha\gamma} + \lambda_{,\alpha} g_{\beta\gamma} \right). \tag{8.60}$$

To the first two terms we now apply the Ricci formula, and obtain

$$k_{\gamma;\alpha\beta} = R^\rho{}_{\beta\alpha\gamma} k_\rho + \frac{1}{2} \left(-\lambda_{,\gamma} g_{\alpha\beta} + \lambda_{,\beta} g_{\alpha\gamma} + \lambda_{,\alpha} g_{\beta\gamma} \right). \tag{8.61}$$

The second derivatives of the conformal Killing fields are thus determined by the fields k_α themselves and by the gradient of λ.

The first derivatives of the generators are connected by Eqs. (8.56) that determine the symmetric part of the matrix $k_{\alpha;\beta}$. Hence, in order to know the first derivatives of k_α, we have to know the antisymmetric part of $k_{\alpha;\beta}$, i.e. $\frac{1}{2}n(n-1)$ function values, plus the value of λ. Together with k_α and $\lambda_{,\alpha}$ this makes

$$\frac{1}{2}n(n-1) + n + n + 1 = \frac{1}{2}(n+1)(n+2) \tag{8.62}$$

constants that must be known in order to determine $k_{\alpha;\beta}$ and $k_{\gamma;\alpha\beta}$.

Now take the covariant derivative of (8.61) with the index δ:

$$k_{\gamma;\alpha\beta\delta} = R^\rho{}_{\beta\alpha\gamma;\delta} k_\rho + R^\rho{}_{\beta\alpha\gamma} k_{\rho;\delta} + \frac{1}{2} \left(-\lambda_{;\gamma\delta} g_{\alpha\beta} + \lambda_{;\beta\delta} g_{\alpha\gamma} + \lambda_{;\alpha\delta} g_{\beta\gamma} \right). \tag{8.63}$$

From this, we subtract the corresponding expression with the indices $(\beta\delta)$ interchanged, and we apply the Ricci formula on the left-hand side:

$$R^\rho{}_{\gamma\beta\delta}k_{\rho;\alpha} + R^\rho{}_{\alpha\beta\delta}k_{\gamma;\rho} = R^\rho{}_{\beta\alpha\gamma;\delta}k_\rho - R^\rho{}_{\delta\alpha\gamma;\beta}k_\rho + R^\rho{}_{\beta\alpha\gamma}k_{\rho;\delta}$$
$$- R^\rho{}_{\delta\alpha\gamma}k_{\rho;\beta} + \frac{1}{2}\left(-\lambda_{;\gamma\delta}\,g_{\alpha\beta} + \lambda_{;\gamma\beta}\,g_{\alpha\delta} + \lambda_{;\alpha\delta}\,g_{\beta\gamma} - \lambda_{;\alpha\beta}\,g_{\gamma\delta}\right). \qquad (8.64)$$

If (8.64) is solvable with respect to $\lambda_{;\beta\delta}$, then the second derivatives of λ will be determined by k_α and $k_{\alpha;\beta}$. In that case, to calculate the third derivatives of k_α from (8.63) we need to know only those constants that were counted in (8.62).

In order to find $\lambda_{;\gamma\delta}$ from (8.64), we contract it with $g^{\alpha\beta}$:

$$R^\rho{}_\gamma{}^\alpha{}_\delta k_{\rho;\alpha} - R^\rho{}_\delta k_{\gamma;\rho} = -R^\rho{}_{\gamma;\delta}k_\rho - R^\rho{}_\delta{}^\alpha{}_{\gamma;\alpha}k_\rho$$
$$- R^\rho{}_\gamma k_{\rho;\delta} - R^\rho{}_\delta{}^\alpha{}_\gamma k_{\rho;\alpha} + \frac{1}{2}\left[-(n-2)\lambda_{;\gamma\delta} - g_{\gamma\delta}\lambda^{;\alpha}{}_{;\alpha}\right]. \qquad (8.65)$$

The first term on the left and the last term containing curvature on the right can be combined as follows:

$$R^\rho{}_\gamma{}^\alpha{}_\delta k_{\rho;\alpha} + R^\rho{}_\delta{}^\alpha{}_\gamma k_{\rho;\alpha} = R^\rho{}_\gamma{}^\alpha{}_\delta\left(k_{\rho;\alpha} + k_{\alpha;\rho}\right) = R^\rho{}_\gamma{}^\alpha{}_\delta\lambda g_{\alpha\rho} = \lambda R_{\gamma\delta}. \qquad (8.66)$$

Now using (8.66), we rewrite (8.65) as follows:

$$(n-2)\lambda_{;\gamma\delta} + g_{\gamma\delta}\lambda^{;\mu}{}_{;\mu} =$$
$$- 2\lambda R_{\gamma\delta} + 2R^\rho{}_\delta k_{\gamma;\rho} - 2R^\rho{}_{\gamma;\delta}k_\rho - 2R^\rho{}_\delta{}^\alpha{}_{\gamma;\alpha}k_\rho - 2R^\rho{}_\gamma k_{\rho;\delta}. \qquad (8.67)$$

We contract this with $g^{\gamma\delta}$ and obtain (remembering that $R^{\rho\gamma}k_{\gamma\rho} = R^{\rho\gamma}k_{(\gamma\rho)} = R^{\rho\gamma}k_{(\rho\gamma)}$)

$$(n-1)\lambda^{;\mu}{}_{;\mu} = -\lambda R - 2R^{\rho\gamma}{}_{;\gamma}k_\rho = -\lambda R - R_{,\gamma}k^\gamma \qquad (8.68)$$

(in the second step we used the contracted Bianchi identities $\left(R^{\mu\nu} - \frac{1}{2}\,g^{\mu\nu}R\right)_{;\nu} \equiv 0$). From here we substitute for $\lambda^{;\mu}{}_{;\mu}$ in (8.67) and obtain

$$(n-2)\lambda_{;\gamma\delta} = \frac{1}{n-1}g_{\gamma\delta}\left(\lambda R + R_{,\gamma}k^\gamma\right)$$
$$- 2\lambda R_{\gamma\delta} + 2R^\rho{}_\delta k_{\gamma;\rho} - 2R^\rho{}_{\gamma;\delta}k_\rho - 2R^\rho{}_\delta{}^\alpha{}_{\gamma;\alpha}k_\rho - 2R^\rho{}_\gamma k_{\rho;\delta}. \qquad (8.69)$$

(It can be verified, with use of (7.14) and (8.56), that the right-hand side of (8.69) is symmetric in γ and δ.) Hence, for $n > 2$, the second derivatives of λ are determined by k_α, $k_{\alpha;\beta}$ and λ itself. Thus, λ, $\lambda_{,\alpha}$, k_α and $k_{(\alpha;\beta)}$ determine $k_{\alpha;\beta\gamma}$, $k_{\alpha;\beta\gamma\delta}$ and $\lambda_{;\gamma\delta}$, and, consequently, determine all higher derivatives of k_α and λ. In this way, the functions $k_\alpha(x)$ and $\lambda(x)$ are determined in that neighbourhood of the point investigated within which these functions are analytic.[3] Consequently, with $n > 2$, the conformal symmetry transformation can have at most $\frac{1}{2}(n+1)(n+2)$ parameters, i.e. a finite basis of generators exists. Note that $\frac{1}{2}(n+1)(n+2)$ is at the same time the maximal dimension of the isometry group of an $(n+1)$-dimensional manifold.

It remains to verify the subcase $n = 2$. Then, from (7.61) we obtain:

$$R^{\alpha\beta}{}_{\gamma\delta} = \frac{1}{2}R\delta^{\alpha\beta}_{\gamma\delta}, \qquad R^\alpha{}_\beta = \frac{1}{2}R\delta^\alpha_\beta,$$

---

[3] The argument presented in the next section applies to conformal symmetries as well, so this limitation on the number of conformal symmetries applies also for nonanalytic functions.

$$R^\rho{}_\delta{}^\alpha{}_\gamma = g_{\delta\mu}g^{\alpha\sigma}R^{\rho\mu}{}_{\sigma\gamma} = \frac{1}{2}R\left(g_{\delta\gamma}g^{\alpha\rho} - \delta_\delta^\alpha\delta_\gamma^\rho\right). \tag{8.70}$$

After substituting these in (8.69) we obtain the identity $0 \equiv 0$. Hence, for 2-dimensional surfaces, the only limitation that follows from (8.67) is (8.68). But (8.67) is just one consequence of (8.64), so we must now consider the other consequences.

In two dimensions, every object that is antisymmetric in two indices must be proportional to the Levi-Civita symbol. Since the last parenthesis in (8.64) is antisymmetric in both $[\beta\delta]$ and $[\alpha\gamma]$, the following must hold:

$$-\lambda_{;\gamma\delta}\,g_{\alpha\beta} + \lambda_{;\beta\gamma}\,g_{\alpha\delta} + \lambda_{;\alpha\delta}\,g_{\beta\gamma} - \lambda_{;\alpha\beta}\,g_{\delta\gamma} = \psi\epsilon_{\alpha\gamma}\epsilon_{\beta\delta}. \tag{8.71}$$

Contracting this with $\epsilon^{\alpha\gamma}\epsilon^{\beta\delta}$ and using $\epsilon^{\beta\delta} = gg^{\beta\mu}g^{\delta\nu}\epsilon_{\mu\nu}$ we obtain

$$\psi = -g\lambda^{;\mu}{}_{;\mu}. \tag{8.72}$$

Now substituting (8.71), (8.72) and (8.70) in (8.64) gives:

$$g_{\gamma\sigma}R\delta^{\rho\sigma}_{\beta\delta}k_{\rho;\alpha} + g_{\alpha\sigma}R\delta^{\rho\sigma}_{\beta\delta}k_{\gamma;\rho} - g_{\beta\sigma}R_{,\delta}\,k_\rho\delta^{\rho\sigma}_{\alpha\gamma} + g_{\delta\sigma}R_{,\beta}\,k_\rho\delta^{\rho\sigma}_{\alpha\gamma}$$
$$- \;\; g_{\beta\sigma}R\delta^{\rho\sigma}_{\alpha\gamma}k_{\rho;\delta} + g_{\delta\sigma}R\delta^{\rho\sigma}_{\alpha\gamma}k_{\rho;\beta} = -g\lambda^{;\mu}{}_\mu\epsilon_{\alpha\gamma}\epsilon_{\beta\delta}. \tag{8.73}$$

The left-hand side above is antisymmetric in $[\beta\delta]$. Using (8.56) one can verify that it is also antisymmetric in $[\alpha\gamma]$. Hence, (8.73) will hold if the contractions of both sides of it with $\epsilon^{\alpha\gamma}\epsilon^{\beta\delta}$ are equal. Doing this contraction and using again $\epsilon^{\beta\delta} = gg^{\beta\mu}g^{\delta\nu}\epsilon_{\mu\nu}$ we obtain

$$R_{,\rho}\,k^\rho + Rk^\mu{}_{;\mu} = -\lambda^{;\mu}{}_{;\mu}. \tag{8.74}$$

Comparing this with (8.57) in the case $n = 2$ we see that (8.74) is equivalent to (8.68).

Consequently, the only integrability condition of (8.63) is (8.68). This equation puts a limit only on $\left(\lambda^{;1}{}_{;1} + \lambda^{;2}{}_{;2}\right)$, leaving the other $\lambda^{;\mu}{}_{;\nu}$ undetermined. Hence, when differentiated, (8.63) will produce still higher derivatives of λ that we will not be able to express through lower derivatives. Let us also note the form of (8.63) in the case $n = 2$:

$$(k_{\gamma;\alpha\beta})_{;\delta} = \frac{1}{2}g_{\beta\sigma}\delta^{\rho\sigma}_{\gamma\delta}\,(Rk_\rho)_{;\delta} + \frac{1}{2}\left(-\lambda_{,\gamma}\,g_{\alpha\beta} + \lambda_{,\beta}\,g_{\alpha\gamma} + \lambda_{,\alpha}\,g_{\beta\gamma}\right)_{;\delta}. \tag{8.75}$$

Both sides of this are covariant derivatives of some expressions, and subsequent differentiations of them will have their integrability conditions identically fulfilled. In order to determine λ and k_α we will thus need an infinite number of constants, i.e. no finite basis of the conformal Killing fields exists in two dimensions. This is a consequence of the fact that every 2-dimensional metric is conformally flat: *any* transformation will keep it conformally flat, thus conformally equivalent to the initial metric.

8.12 * The maximal dimension of an invariance group

In Section 8.4 we proved that the dimension of the symmetry group of an n-dimensional Riemann space cannot be larger than $\frac{1}{2}n(n+1)$. However, in the proof we assumed that the Riemann tensor was analytic in the neighbourhood considered. We will argue now that the result applies generally. It will follow from the sequence of theorems presented below.

Theorem 8.2

Assumptions:

(1) The metric tensor $g_{\alpha\beta}$ of the Riemann space V_n depends on a number of constant parameters $u \overset{\text{def}}{=} \{u_1, \dots, u_k\}$.

(2) The metric $h_{\alpha\beta}$ results from $g_{\alpha\beta}$ when some of the parameters go to certain limits:

$$u_0 \overset{\text{def}}{=} \{u_{10}, \dots, u_{i0}, u_{i+1}, \dots, u_k\}, \qquad h_{\alpha\beta} = \lim_{u \to u_0} g_{\alpha\beta}, \qquad (8.76)$$

where $i \leq k$.

(3) The limit is nonsingular, i.e. no vector field on V_n vanishes identically in the limit.

Thesis:

The dimension of the symmetry group of $h_{\alpha\beta}$ is larger than or equal to the dimension of the symmetry group of $g_{\alpha\beta}$.

Proof:

Let $\underset{(1)}{k}, \dots, \underset{(p)}{k}$ be the generators of symmetries of $g_{\alpha\beta}$, and consider the equations $\underset{k_{(m)}}{\pounds}\, g_{\alpha\beta} = 0$, $m = 1, \dots, p$. Since they are fulfilled at all values of the parameters $\{u_1, \dots, u_k\}$, they must still hold in the limit. Since the limit is nonsingular, none of the generators will go to zero. Hence, $h_{\alpha\beta}$ must inherit all the symmetries of $g_{\alpha\beta}$. $\square$

Comments:

(a) Limits of metric tensors are coordinate-dependent, but the theorem holds for every nonsingular limit.

(b) New symmetries may appear in the limit if there exists a vector field w^α on V_n such that $\underset{w}{\pounds} g_{\alpha\beta} \neq 0$ but $\lim_{u \to u_0} \underset{w}{\pounds} g_{\alpha\beta} = 0$.

(c) The symmetry group H of $h_{\alpha\beta}$ may be nonisomorphic to the symmetry group G of $g_{\alpha\beta}$, even if $\dim H = \dim G$. This may happen because some of the structure constants of G may become zero or take other privileged values in the limit.

(d) The theorem holds for invariance groups of any tensor fields, not just for isometries. Thus, for example, the group of invariance transformations of the Riemann tensor (called **collineations**) may only increase or preserve its dimension in a limit.

(e) The theorem holds for conformal symmetries, too.

Theorem 8.3 *The flat Riemann space is contained as a limit in every nonsingular region of every curved Riemann space.*

Proof:

At any nonsingular point of a curved manifold the flat space of the corresponding signature is its tangent space. Thus we may imagine the flat limit as follows: we decrease the curvature to zero (this may require introducing some free parameters by a coordinate transformation), and in the process open subsets of the tangent spaces become isometric with open subsets of the manifold itself. $\square$

Comments:

(a) This theorem is in fact one of the fundamental postulates of relativity. In the physical language it reads as follows: special relativity is the zero-curvature limit of general relativity, i.e. every curved spacetime has the Minkowski spacetime as a limit.

(b) The construction will fail at singular points, where the curvature is infinite (such as the vertex of a cone). At those points, no tangent space exists.

(c) In general, we do not expect the tangent spaces at different points to coincide in the limit. For example, a cylinder is flat, but its tangent planes do not coincide.

Theorem 8.4 *No Riemann space can have a symmetry group of higher dimension than the symmetry group of the flat Riemann space.*

Proof:
The proof follows from Theorems 8.2 and 8.3. $\square$

The n-dimensional Euclidean space has the symmetry group of dimension $\frac{1}{2}n(n+1)$, the highest possible. One can convince oneself about it by considering generators of the form (8.37) for all possible $\{x^i, x^j\}$ coordinate planes; there are $\frac{1}{2}n(n-1)$ of them. In addition, translations along n linearly independent directions, $x'^i = x^i + a^i$ with constant a^i are also symmetries. In an n-dimensional flat Riemann space of indefinite signature, in the $\{x^k, x^l\}$ planes of signature $(+-)$ the rotations (8.36) are replaced by the Lorentz transformations $x'^k = x^k \cosh u + x^l \sinh u, x'^l = x^k \sinh u + x^l \cosh u$. The dimension of the whole symmetry group is still $\frac{1}{2}n(n+1)$.

There exist Riemann spaces whose symmetry groups are of the same dimension $\frac{1}{2}n(n+1)$ and which are not locally isometric to the flat space. They are the **spaces of constant curvature**, whose Riemann tensor is

$$R^{\alpha\beta}{}_{\gamma\delta} = \frac{1}{n(n-1)} \, R\delta^{\alpha\beta}_{\gamma\delta}, \qquad (8.77)$$

where the scalar curvature R is constant. Their symmetry groups are nonisomorphic in each of the three cases $R > 0$, $R = 0$ and $R < 0$. In relativity, they are the de Sitter spacetimes, see Exercise 13 and Section 14.4. They are the only Riemann spaces that have the symmetry groups of this maximum dimension (Stephani 1990, see Exercise 14).

8.13 Exercises

1. Find the coordinate transformation corresponding to the field of generators $k^\alpha = \delta^\alpha{}_{\alpha_0}$, where α_0 is the label of one of the coordinates.

2. Find the coordinate transformation corresponding to the field of generators $k^\mu = x^i\delta^\mu{}_j - x^j\delta^\mu{}_i$, where i and j are the labels of fixed coordinates. Show that, when (x^i, x^j) are Cartesian coordinates, the transformation is the rotation in the plane of (x^i, x^j).

3. Solve the Killing equations for $k^\alpha = \delta^\alpha{}_{\alpha_0}$.

4. Show that if the parameter λ of the integral lines of a generator $k^\alpha = \mathrm{d}x^\alpha/\mathrm{d}\lambda$ is chosen as a coordinate in the Riemann space, then a tensor invariant under the transformations generated by k^α is independent of λ.

5. Show that if there exist two linearly independent fields k^α and l^α generating invariances of a tensor field $T_{\alpha\beta}$, then the corresponding orbit parameters λ and τ can be chosen as coordinates on M_n if and only if $[k, l]^\alpha = k^\rho l^\alpha,_\rho - l^\rho k^\alpha,_\rho = 0$. In that case $\partial T_{\alpha\beta}/\partial\lambda = \partial T_{\alpha\beta}/\partial\tau = 0$.

Note: This is just a different wording of Exercise 2 in Chapter 6.

6. Prove that if k^α and l^α are Killing fields in a Riemann space, then so is $[k, l]^\alpha$.

Hint. Use (8.14) rather than (8.13), and then use (7.13).

7. Show that the condition of invariance of a scalar field ϕ with respect to the transformation group generated by the vector field k^α is

$$k^\alpha \phi,_\alpha = 0. \tag{8.78}$$

8. Show that the condition of invariance of a covariant vector field ω_α with respect to the transformation group generated by k^α is

$$k^\rho \omega_{\alpha,\rho} + k^\rho,_\alpha \omega_\rho = 0. \tag{8.79}$$

9. Let the spacetime coordinates be $(x^0, x^1, x^2, x^3) = (t, r, \vartheta, \varphi)$. Show that a covariant vector field u_α that is invariant under the $O(3)$ group with generators given by (8.42) must have $u_2 = u_3 = 0$, and u_0 and u_1 depending only on t and r. Show that the same is true for the components of a contravariant u^α that is invariant under (8.42).

10. Prove that the generators of conformal symmetry obey (8.56) and that λ is related to the conformal factor Φ in $g'_{\alpha\beta}(x') = \Phi(x')g_{\alpha\beta}(x')$ by $\lambda = -k^\alpha \Phi,_\alpha$.

Hint. Take the condition $g'_{\alpha\beta}(x') = \Phi(x')g_{\alpha\beta}(x')$ and note that x'^α are functions of the group parameters t_i. Then take the limit of this equation as $t_i \to t_{0i}$, where $x'^\alpha(t_{0i}, x) = x^\alpha$, i.e. is an identity transformation. It follows that Φ must obey $\Phi(x')|_{x'=x} = 1$.

11. Find and interpret all the Killing fields for the Minkowski spacetime in the Cartesian coordinates, in which $\mathrm{d}s^2 = \mathrm{d}t^2 - \mathrm{d}x^2 - \mathrm{d}y^2 - \mathrm{d}z^2$. Find the corresponding isometries. Identify the isometries that should be known to you from a course on special relativity: the special Lorentz transformations along the x-, y- and z-directions, and the rotations in the planes $\{x, y\}$, $\{y, z\}$ and $\{x, z\}$. Verify that they are indeed isometries (i.e. substitute these transformations into the metric form and see what happens). Calculate all the structure constants of the full group.

12. Find all the conformal Killing fields for the Minkowski spacetime in the Cartesian coordinates, $\mathrm{d}s^2 = \mathrm{d}t^2 - \mathrm{d}x^2 - \mathrm{d}y^2 - \mathrm{d}z^2$. Show that those generators that do not correspond to isometries generate the following mappings of the Minkowski spacetime into itself:

(a) Dilatation:

$$x'^\alpha = x^\alpha/C, \tag{8.80}$$

where C is the group parameter.

(b) The so-called acceleration transformations (for which Plebański proposed the name 'Haantjes transformations' (Haantjes, 1937 and 1940))[4]:

$$x'^\alpha = \frac{x^\alpha + C^\alpha x_\rho x^\rho}{1 + 2C_\sigma x^\sigma + C_\nu C^\nu x_\sigma x^\sigma}, \tag{8.81}$$

where C^μ, $\mu = 0, 1, 2, 3$, are group parameters, and the indices are raised and lowered

---

[4] The papers by Haantjes contain only special cases of (8.81), corresponding to (i) $C_\mu C^\mu = 0$ (the 1937 paper) and (ii) Only one of the C^μ being nonzero (the 1940 paper). The oldest source known to us in which (8.81) is given in full generality is the text by Plebański (1967).

by the Minkowski metric. Verify that this is an Abelian group, that the composition of two Haantjes transformations, with parameters C^μ and D^μ, is a Haantjes transformation with parameters $(C^\mu + D^\mu)$ and that the transformation inverse to (8.81) is a Haantjes transformation with parameters $\widetilde{C}^\mu = -C^\mu$. Verify that (8.81) is a composition of the following three transformations:

- the inversion in the pseudosphere of radius L with its centre at $x^\mu = 0$:

$$x'^\mu = \frac{L^2}{x_\rho x^\rho} x^\mu; \tag{8.82}$$

- translation by the vector C^μ:

$$x''^\mu = x'^\mu + L^2 C^\mu; \tag{8.83}$$

- the inversion in the pseudosphere of radius L with its centre at $x''^\mu = 0$:

$$x'''^\mu = \frac{L^2}{x''_\rho x''^\rho} x''^\mu. \tag{8.84}$$

The transformations (8.82) and (8.84) are discrete conformal symmetries of the Minkowski spacetime in Cartesian coordinates, and (8.83) is a conformal symmetry of the Minkowski spacetime in the coordinates x'^μ.

Verify that the Haantjes transformations are conformal symmetries also for a flat space of any dimension and with any signature. Equations (8.81) – (8.84) apply then with the indices running through n values.

Hint. Verify that (8.81) has the following properties:

$$x'_\mu x'^\mu = x_\mu x^\mu / T, \qquad \mathrm{d}x'_\mu \mathrm{d}x'^\mu = \mathrm{d}x_\mu \mathrm{d}x^\mu / T^2, \tag{8.85}$$

where

$$T \overset{\text{def}}{=} 1 + 2C_\sigma x^\sigma + C_\nu C^\nu x_\sigma x^\sigma. \tag{8.86}$$

The second of (8.85) proves that (8.81) is a conformal symmetry of the flat metric, independently of n and the signature.

13. One of the coordinate representations of 4-dimensional spaces of constant curvature with the signature $(+ - - -)$ (the de Sitter (1917) spacetimes) is

$$\mathrm{d}s^2 = \left(1 - \Lambda r^2\right) \mathrm{d}t^2 - \frac{1}{1 - \Lambda r^2} \mathrm{d}r^2 - r^2 \left(\mathrm{d}\vartheta^2 + \sin^2 \vartheta \mathrm{d}\varphi^2\right), \tag{8.87}$$

where Λ is an arbitrary constant. Find all the Killing fields for this metric, in each of the cases $\Lambda > 0$, $\Lambda < 0$ and $\Lambda = 0$ (the last case is just the Minkowski spacetime in spherical coordinates). Find the structure constants of these groups. Take the limit $\Lambda \to 0$ of the first and second cases and see what happens with the structure constants.

Hint. You may prefer to find the Killing fields for the case $\Lambda = 0$ by transforming the results of Exercise 10 to spherical coordinates.

14. Verify that the Riemann spaces of constant curvature (8.77) are the only ones that have symmetry groups of the maximum dimension $\frac{1}{2} n(n + 1)$ (Stephani 1990, Sec. 19.4).

Hint. Differentiate covariantly (8.22) by x^δ, then write the result with β and δ interchanged and subtract the second equation from the first. On the left-hand side of the resulting equation you will have the commutator of second covariant derivatives acting on $k_{\gamma;\alpha}$, to which you can apply (6.9). After reordering you will obtain

$$(R^\rho{}_{\beta\alpha\gamma;\delta} - R^\rho{}_{\delta\alpha\gamma;\beta}) k_\rho + \left(-R^\rho{}_{\gamma\beta\delta}\delta^\sigma_\alpha + R^\rho{}_{\alpha\beta\delta}\delta^\sigma_\gamma\right.$$
$$\left. +R^\rho{}_{\beta\alpha\gamma}\delta^\sigma_\delta - R^\rho{}_{\delta\alpha\gamma}\delta^\sigma_\beta\right) k_{\rho;\sigma} = 0. \tag{8.88}$$

But in a maximally symmetric space k_ρ and $k_{\rho;\sigma}$ are the initial data that determine all symmetry generators, and they cannot be constrained by any equations. This means that the coefficients of k_ρ and $k_{\rho;\sigma}$ in (8.88) must vanish. Equate these coefficients to zero remembering that $k_{\rho;\sigma} = -k_{\sigma;\rho}$ by the Killing equations, so only the antisymmetric part of the coefficient of $k_{\rho;\sigma}$ must be zero. Then contract the resulting equation with δ^α_σ and use the identity (7.17). The result will be

$$(n-1)R^\rho{}_{\gamma\beta\delta} = R_{\gamma\delta}\delta^\rho_\beta - R_{\beta\gamma}\delta^\rho_\delta. \tag{8.89}$$

Contract this with $g^{\beta\gamma}$ obtaining $nR^\rho{}_\sigma = R\delta^\rho_\sigma$. Using this in (8.89) you will obtain (8.77). Then, the coefficient of k_ρ in (8.88) being zero will imply $R_{,\delta} = 0$. $\square$

Note that this result does not depend on the signature.

9

Methods to calculate the curvature quickly: differential forms and algebraic computer programs

The calculus of differential forms is a mighty tool used in many branches of theoretical physics. This chapter is narrowly focussed on the application of differential forms to calculating the Riemann tensor out of a given metric tensor. A rather broad presentation of this calculus can be found in the first volume of Maurin (1973). An extended pedagogical introduction to differential forms is the book by Flanders (1963).

9.1 The basis of differential forms

Let us recall the conclusion of Section 4.3: if a field of bases of contravariant (covariant) vectors is given on a manifold, then this field determines the dual field of bases of covariant (contravariant) vectors via (4.13) or (4.14), and then both bases can be used to uniquely represent arbitrary tensor densities by sets of scalars.

In Chapter 4 we assumed the bases to be given, but they could be arbitrary. We are thus free to choose such a basis that the set of scalars representing a given tensor field is particularly simple. For example, the scalars representing the metric tensor

$$\eta_{ij} = e_i{}^\alpha e_j{}^\beta g_{\alpha\beta} \tag{9.1}$$

(where i and j label different vectors) can all be constant. This is possible for any $g_{\alpha\beta}$: the transition from $g_{\alpha\beta}$ to η_{ij} is, in the language of linear algebra, a transformation of the matrix of a quadratic form induced by a change of basis of the underlying vector space. Equation (9.1) determines the basis up to the transformations that preserve η_{ij}. For example, when $\dim V_n = 3$ and $||\eta_{ij}|| = \mathrm{diag}(1,1,1)$, the $e_i{}^\alpha$ are determined up to the orthogonal transformations $O(3)$, and when $\dim V_n = 4$ and $||\eta_{ij}|| = \mathrm{diag}(+1,-1,-1,-1)$, the $e_i{}^\alpha$ are determined up to the Lorentz transformations $L(1,3)$.

A given basis $e_i{}^\alpha$ and a given scalar image of the metric η_{ij} uniquely determine the metric tensor by a formula inverse to (9.1):

$$g_{\alpha\beta} = e^i{}_\alpha e^j{}_\beta \eta_{ij}. \tag{9.2}$$

In a 4-dimensional manifold, the basis $e_i{}^\alpha$ is called a **tetrad of vector fields**, and η_{ij} is called a **tetrad metric**.

As shown in Sec. 3.2, every coordinate system naturally defines a field of vector bases; they are the vectors orthogonal to the hypersurfaces $f^i = $ constant, defined by

$$e_i{}^\alpha f^j{}_{,\alpha} = \delta^i{}_j, \tag{9.3}$$

where $\{f^j\}$, $j = 1, \ldots, n$ are the coordinates. However, not every field of bases defines a coordinate system because vector fields are in general not orthogonal to families of hypersurfaces. For general fields, (9.3) will not be integrable.

Covariant fields $e^i{}_\alpha$ can be uniquely represented by the differential forms

$$e^i = e^i{}_\alpha \mathrm{d}x^\alpha. \tag{9.4}$$

Multiplying both sides of this equation by $e_i{}^\beta$ and using (4.13), we obtain

$$\mathrm{d}x^\beta = e_i{}^\beta e^i. \tag{9.5}$$

From (9.2) and (9.4) we see that the metric can be represented in the basis $\{e^i\}$ in the following way:

$$\mathrm{d}s^2 = g_{\alpha\beta}\mathrm{d}x^\alpha\mathrm{d}x^\beta = \eta_{ij}e^i e^j. \tag{9.6}$$

9.2 The connection forms

The quantities

$$\Gamma^i{}_{jk} \overset{\mathrm{def}}{=} -e^i{}_{\rho;\sigma}e_j{}^\rho e_k{}^\sigma \tag{9.7}$$

are called the **Ricci rotation coefficients**. Comparing this with (4.19) we find

$$\Gamma^i{}_{jk} = e^i{}_\alpha e_j{}^\beta e_k{}^\gamma \left\{{\alpha \atop \beta\gamma}\right\} - e^i{}_{\beta,\gamma}e_j{}^\beta e_k{}^\gamma, \tag{9.8}$$

$$\left\{{\alpha \atop \beta\gamma}\right\} = e_i{}^\alpha e^j{}_\beta e^k{}_\gamma \Gamma^i{}_{jk} + e_i{}^\alpha e^i{}_{\beta,\gamma}. \tag{9.9}$$

The basis vectors and the Ricci rotation coefficients are thus a unique representation of the Christoffel symbols.

Now let us calculate the exterior derivatives of the forms e^i:

$$\mathrm{d}e^i = -e^i{}_{[\beta,\gamma]}\mathrm{d}x^\beta \wedge \mathrm{d}x^\gamma = -e^i{}_{[\beta;\gamma]}e_j{}^\beta e_k{}^\gamma e^j \wedge e^k = \Gamma^i{}_{[jk]}e^j \wedge e^k \tag{9.10}$$

(along the way, we have used the antisymmetry of the exterior products and the symmetry of the Christoffel symbols in their subscripts that allowed us to replace $e^i{}_{[\beta,\gamma]}$ with $e^i{}_{[\beta;\gamma]}$). Hence, calculating all $\mathrm{d}e^i$ and then decomposing the result in the basis $e^j \wedge e^k$ we find the $\Gamma^i{}_{[jk]}$. Now we will show that $\Gamma_{ijk} \overset{\mathrm{def}}{=} \eta_{is}\Gamma^s{}_{jk}$ are antisymmetric in (ij) by virtue of their definition. Note that $\Gamma^i{}_{jk}$ can be equivalently written as

$$\Gamma^i{}_{jk} = e^i{}_\rho e_j{}^\rho{}_{;\sigma}e_k{}^\sigma. \tag{9.11}$$

Note also that the following is true:

$$e^i{}_\alpha = \eta^{is}g_{\alpha\rho}e_s{}^\rho, \qquad e_i{}^\alpha = \eta_{is}g^{\alpha\rho}e^s{}_\rho. \tag{9.12}$$

To see this, multiply both sides of the first equation by $e_j{}^\alpha$, to obtain $e^i{}_\alpha e_j{}^\alpha = \delta^i{}_j$, which means that the $e^i{}_\alpha$ defined by (9.12) are identical with the inverse matrices to $e_i{}^\alpha$, i.e.

with the $e^i{}_\alpha$ as originally defined by (4.13) and (4.14). Our notation is thus self-consistent: the $e^i{}_\alpha$ result from the $e_i{}^\alpha$ by raising the scalar index with η^{ij} and lowering the tensor index with $g_{\alpha\beta}$ (where η^{ij} is the inverse matrix to η_{ij}). The second equation in (9.12) is verified in a similar way.

Now let us calculate the directional derivatives of (9.1) along $e_k{}^\gamma$:

$$
\begin{aligned}
0 &= (\eta_{ij})_{,\gamma}\, e_k{}^\gamma = \left(e_i{}^\alpha e_j{}^\beta g_{\alpha\beta}\right)_{;\gamma} e_k{}^\gamma \\
&= e_i{}^\alpha{}_{;\gamma}\, e_j{}^\beta g_{\alpha\beta} e_k{}^\gamma + e_i{}^\alpha e_j{}^\beta{}_{;\gamma}\, g_{\alpha\beta} e_k{}^\gamma \\
&= e_i{}^\alpha{}_{;\gamma}\, g^{\beta\rho}\eta_{js} e^s{}_\rho g_{\alpha\beta} e_k{}^\gamma + g^{\alpha\rho}\eta_{is} e^s{}_\rho e_j{}^\beta{}_{;\gamma}\, g_{\alpha\beta} e_k{}^\gamma \\
&= \eta_{js} e_i{}^\rho{}_{;\gamma}\, e^s{}_\rho e_k{}^\gamma + \eta_{is} e_j{}^\rho{}_{;\gamma}\, e^s{}_\rho e_k{}^\gamma = \Gamma_{jik} + \Gamma_{ijk}.
\end{aligned} \tag{9.13}
$$

Along the way we used: the η_{ij} being constant scalars, $g_{\alpha\beta}$ being covariantly constant, Eqs. (9.12), $g^{\alpha\rho}g_{\beta\rho} = \delta^\alpha{}_\beta$ and (9.11). The final result is as announced:

$$
\Gamma_{ijk} = -\Gamma_{jik}. \tag{9.14}
$$

Using this property, one may easily verify that

$$
\Gamma_{ijk} = \Gamma_{i[jk]} - \Gamma_{j[ik]} - \Gamma_{k[ij]}. \tag{9.15}
$$

Thus, having found $\Gamma^i{}_{[jk]}$ from (9.10), we can find the full $\Gamma^i{}_{jk} = \eta^{is}\Gamma_{sjk}$ from the above. We now use the Ricci rotation coefficients to form the **connection forms**:

$$
\Gamma^i{}_j = \Gamma^i{}_{jk} e^k. \tag{9.16}
$$

The tetrad image of the covariant derivative of any tensor field can be expressed fully in terms of the Ricci rotation coefficients in place of the Christoffel symbols and directional derivatives in place of partial derivatives. Thus, for example, for the Riemann tensor

$$
\begin{aligned}
e^a{}_\alpha e_b{}^\beta e_c{}^\gamma e_d{}^\delta e_e{}^\varepsilon R^\alpha{}_{\beta\gamma\delta;\varepsilon} &\equiv e_e{}^\rho R^a{}_{bcd,\rho} \\
&+ \Gamma^a{}_{se} R^s{}_{bcd} - \Gamma^s{}_{be} R^a{}_{scd} - \Gamma^s{}_{ce} R^a{}_{bsd} - \Gamma^s{}_{de} R^a{}_{bcs}.
\end{aligned} \tag{9.17}
$$

9.3 The Riemann tensor

The expression $d\Gamma^i{}_j + \Gamma^i{}_s \wedge \Gamma^s{}_j$ is a 2-form that can be decomposed in the basis $e^k \wedge e^l$:

$$
d\Gamma^i{}_j + \Gamma^i{}_s \wedge \Gamma^s{}_j = \frac{1}{2} R^i{}_{jkl} e^k \wedge e^l. \tag{9.18}
$$

This is a definition of the coefficients $R^i{}_{jkl}$, but they turn out to be the scalar image of the Riemann tensor

$$
R^i{}_{jkl} = e^i{}_\alpha e_j{}^\beta e_k{}^\gamma e_l{}^\delta R^\alpha{}_{\beta\gamma\delta}. \tag{9.19}
$$

This is verified as follows

$$
\begin{aligned}
d\Gamma^i{}_j \ + \Gamma^i{}_s \wedge \Gamma^s{}_j &= d\left(\Gamma^i{}_{jk} e^k\right) + \Gamma^i{}_{sk}\Gamma^s{}_{jl} e^k \wedge e^l \\
&= \left(-\Gamma^i{}_{j[k,|\alpha|} e_l{}^\alpha + \Gamma^i{}_{js}\Gamma^s{}_{[kl]} + \Gamma^i{}_{s[k}\Gamma^s{}_{|j|l]}\right) e^k \wedge e^l,
\end{aligned} \tag{9.20}
$$

where we have used (9.16), (9.10), the antisymmetry of the exterior product and (9.5).
Hence, from (9.18) and (9.20) we obtain:

$$\frac{1}{2}R^i{}_{jkl} = -\Gamma^i{}_{j[k,|\alpha|}e_{l]}{}^\alpha + \Gamma^i{}_{js}\Gamma^s{}_{[kl]} + \Gamma^i{}_{s[k}\Gamma^s{}_{|j|l]}. \tag{9.21}$$

Now, using (9.7) and (8.19) we obtain (9.19).

It is now easy to verify that

$$R_{jl} \stackrel{\text{def}}{=} R^i{}_{jil} \equiv e_j{}^\alpha e_l{}^\beta R_{\alpha\beta}, \tag{9.22}$$

i.e. that the R_{jl} defined above is the scalar image of the Ricci tensor $R_{\alpha\beta}$ given by (7.46).
It is also easy to see that the scalar curvature is

$$R = \eta^{ij}R_{ij} \equiv g^{\alpha\beta}R_{\alpha\beta}. \tag{9.23}$$

The differential forms allow us to carry out many calculations faster and easier than the
traditional tensor calculus. Indeed, some modern courses of relativity avoid tensor calculus
altogether, using the differential forms as a primary device. However, that approach,
although more efficient from the point of view of calculations, makes it more difficult to
connect to the historical roots of relativity.

As examples of the great simplifications to which the differential forms lead, let us
note that the identity $R^\alpha{}_{[\beta\gamma\delta]} \equiv 0$ is an immediate consequence of (9.10). Since the
antisymmetrisation in (9.10) is carried out automatically (it was written out explicitly
only for better clarity), we can make use of (9.16) and write (9.10) in the form

$$\mathrm{d}e^i = -\Gamma^i{}_j \wedge e^j. \tag{9.24}$$

Substituting (9.24) and (9.18) in the identity $\mathrm{d}^2 e^i \equiv 0$ we obtain $\frac{1}{2}R^i{}_{jkl}\, e^j \wedge e^k \wedge e^l \equiv 0$,
which is equivalent to $R^i{}_{[jkl]} \equiv 0$. This identity is thus the integrability condition for
(9.24). Similarly, the Bianchi identities (7.14) are the integrability condition of (9.18).

In relativity, with its 4-dimensional Riemann spaces of signature $(+ - - -)$, the most
frequently used tetrads are the **orthonormal** one (in which $||\eta_{ij}|| = \mathrm{diag}(+1, -1, -1, -1)$),
the **null tetrad** (in which $\eta_{01} = \eta_{10} = 1$, $\eta_{22} = \eta_{33} = -1$, all other components of η are
zero) and the **double-null tetrad** (in which $\eta_{01} = \eta_{10} = -\eta_{23} = -\eta_{32} = 1$, all other
components of η are zero). See Sections 16.6 and 16.8 for an application of the double-null
tetrad; in it the forms e^3 and e^4 are complex and conjugate to each other.

9.4 Using computers to calculate the curvature

Calculating the curvature tensor from a given metric tensor is tedious and time-consuming,
and seemingly innocent errors made along the way cause great chaos in the final results.
In order to obtain a reliable result, every step of the calculation must be carefully verified.
A relatively simple calculation typically takes several hours; in complicated cases it can
extend to months. At the same time, this is routine work that does not really require
intelligence, but just careful application of a set of rules. (Intelligence is required only to
understand the rules.) Thus, it is a typical task for a computer.

This fact was noticed and made use of long ago. Ever since the 1960s, several computer systems have been created that can calculate the Riemann tensor and the associated quantities from a given metric. We mean here symbolic calculations, in which the computer transforms mathematical expressions without expecting them to have numerical values. Some of those systems are parts of big general-purpose algebraic systems (like Maple or Mathematica), some are specialised programs written in generally accessible programming languages. The language of choice for computer algebra is Lisp (abbreviation for List Processing). The market situation is rather dynamic. At the moment of writing this text, the broadest and most up-to-date overview of the programs and systems that existed in the past and of those that are available now is the article by MacCallum (2018).

The modern computer algebra programs are fairly easy to use, and the reduction in time and effort required to do the calculation is dramatic. Instead of doing routine calculations for weeks, one can have the result in less than a minute. Of course, this does not include the time required to write the data, and usually several runs are needed before the result becomes acceptable (for example, additional simplifications might be needed). Still, the gain is obvious, and these calculations are now done 'by hand' essentially only by students for educational purposes. In research work, the computers have taken over the field completely.

9.5 Exercises

1. Verify (9.17).

2. Calculate the exterior derivative of (9.18) and show that it is equivalent to the Bianchi identities (7.14).

3. Take a simple metric, for example the de Sitter metric of (8.87), and calculate its Riemann tensor first by tensor calculus, then by using the differential forms in the orthonormal tetrad. Compare the time and effort required in the two cases.

4. Verify (9.19) by substituting (9.7) into (9.21).

Hint. Note that $\Gamma^i{}_{jk}$ are scalars, so $\Gamma^i{}_{jk,\alpha} = \Gamma^i{}_{jk;\alpha}$. During the calculation you will have to use the Ricci formula (8.19) and the identity $e^s{}_{\rho;\sigma} e_j{}^\rho = -e^s{}_\rho e_j{}^\rho{}_{;\sigma}$.

10

The spatially homogeneous Bianchi-type spacetimes

10.1 The Bianchi classification of 3-dimensional Lie algebras

As seen from (8.31), generators of symmetries of a manifold form a Lie algebra. For reasons to be explained further, the case when this algebra is 3-dimensional is important for relativity. The aim of the Bianchi classification is to sort out those 3-dimensional algebras which are inequivalent in the following sense.

A basis of generators is not defined uniquely. If $k_{(i)}$ is a basis, then

$$\widetilde{k}^{\alpha}_{(i')} \stackrel{\text{def}}{=} A_{i'}{}^{j} k^{\alpha}_{(j)} \qquad \text{(sum over } j) \tag{10.1}$$

is also a basis, provided that the matrix $A_{i'}{}^{j}$ (of constant elements) is nonsingular. Such a change of basis induces the following change of the structure constants

$$\widetilde{C}^{l'}_{i'j'} = \left(A^{-1}\right)^{l'}{}_{l} A_{i'}{}^{i} A_{j'}{}^{j} C^{l}_{ij}. \tag{10.2}$$

Thus, two sets of structure constants that are related by (10.2) correspond to two bases that span isomorphic algebras. How can one recognise whether any matrix A obeying (10.2) exists for two given sets of C^{l}_{ij}? The answer is provided by the Bianchi classification. Note that no reference will be made in this section to the way in which the corresponding group acts on the manifold.

The method of classifying 3-dimensional algebras used here was invented by Schücking in the 1950s, but he presented it only in a seminar talk. It diffused into public knowledge via notes taken by Kundt that were published much later (Kundt, 2003); see Krasiński et al. (2003) for details of this story. The earliest papers that used this approach and introduced it into the literature were by Estabrook, Wahlquist and Behr (1968), Ellis and MacCallum (1969) and by MacCallum and Ellis (1970). The classification was originally introduced by Bianchi (1898), but his presentation is long and complicated, so it is no longer in use.[1] Bianchi sorted out the different types by the properties of the derived algebras.

---

[1] Bianchi considered his classification from the point of view of the theory of Lie algebras. Its importance for relativity was recognised much later. The first paper that explicitly introduced the Bianchi types into relativity was that by Taub (1951), but the inspiration is said to have come from Gödel (1949). See Jantzen (2001) for more on this story.

From (8.31) it follows that $C^l{}_{ij} = -C^l{}_{ji}$, so for an N-dimensional group the number of structure constants cannot exceed $N \cdot \frac{1}{2}N(N-1)$. Through the Jacobi identity

$$[[\underset{(i)}{k}, \underset{(j)}{k}], \underset{(l)}{k}] + [[\underset{(j)}{k}, \underset{(l)}{k}], \underset{(i)}{k}] + [[\underset{(l)}{k}, \underset{(i)}{k}], \underset{(j)}{k}] = 0, \tag{10.3}$$

the definition (8.31) implies a further limitation on $C^l{}_{ij}$:

$$C^m{}_{ij}C^n{}_{ml} + C^m{}_{jl}C^n{}_{mi} + C^m{}_{li}C^n{}_{mj} = 0. \tag{10.4}$$

We shall take (10.4) into account later.

For $N = 3$, $\frac{1}{2}N^2(N-1) = 9$ and equals the number of elements of a general 3×3 matrix. Thus for a 3-dimensional algebra all structure constants can be put into a single 3×3 matrix, let us call it H^{ab}. The one-to-one correspondence between $C^l{}_{ij}$ and the elements of H^{ab} is

$$H^{ab} = \frac{1}{2}\epsilon^{akl}C^b{}_{kl} \Longleftrightarrow C^i{}_{jk} = \epsilon_{sjk}H^{si}, \tag{10.5}$$

where ϵ^{akl} and ϵ_{sjk} are the Levi-Civita symbols. The matrix H^{ab} can now be split into its symmetric part

$$n^{ab} \stackrel{\text{def}}{=} H^{(ab)} = \frac{1}{2}\left(H^{ab} + H^{ba}\right) \tag{10.6}$$

and its antisymmetric part $H^{[ab]}$. But in a 3-dimensional space there is a one-to-one correspondence between antisymmetric matrices and vectors, so $H^{[ab]}$ can be represented by the vector a defined by

$$a_i = -\frac{1}{2}\epsilon_{ijk}H^{[jk]} \Longleftrightarrow H^{[ij]} = -\epsilon^{ijm}a_m. \tag{10.7}$$

From (10.5) – (10.7) we have

$$C^l{}_{jk} = \epsilon_{sjk}n^{sl} - \delta^{lm}_{jk}a_m. \tag{10.8}$$

The identity (10.4) for $N = 3$ may be written as $C^m{}_{ij}C^n{}_{ml}\epsilon^{ijl} = 0$, and with use of (10.8) implies that

$$n^{is}a_s = 0, \tag{10.9}$$

i.e. either $a_i = 0$ or else n^{is} has at least one zero eigenvalue.

From (10.1) and (10.2) it is seen that n^{ab} will transform, under the change of basis (10.1), like a twice contravariant tensor density

$$\tilde{n}^{i'j'} = (\det A)\left(A^{-1}\right)^{i'}{}_i\left(A^{-1}\right)^{j'}{}_j n^{ij}, \tag{10.10}$$

while a_i will transform like a covariant vector, $\tilde{a}_{i'} = A_{i'}{}^i a_i$. The transformation (10.10) can be used to diagonalise n^{ij}. Let us then assume that the basis $\underset{(i)}{k}$ was chosen so that

$$n^{ij} = \begin{pmatrix} n_1 & 0 & 0 \\ 0 & n_2 & 0 \\ 0 & 0 & n_3 \end{pmatrix}. \tag{10.11}$$

This does not yet fix the basis uniquely; transformations that permute (n_1, n_2, n_3) are still allowed. Therefore, if any of the n_i is zero, it can always be moved by transformations of the basis to the upper left corner in (10.11), i.e. we can always assume that

$$n_1 = 0 \qquad \text{if} \qquad a_i \neq 0. \tag{10.12}$$

If $n_1 = 0$ is the single zero eigenvalue, then in view of (10.9) the vector a will assume, after such a choice of basis, the form

$$a_i = [a, 0, 0]. \tag{10.13}$$

If $n_1 = 0$ is a multiple eigenvalue, then (10.12) still allows us to rotate a within the eigenspace of the $n_1 = 0$ eigenvalue. We can then rotate a so that it will assume the form (10.13). Equation (10.13) covers the subcase $a_i = 0$. In such a basis, (10.9) becomes

$$a n_1 = 0. \tag{10.14}$$

Using all the information about $C^l{}_{ij}$, the commutators become:

$$\left[\underset{(1)}{k}, \underset{(2)}{k}\right] = a \underset{(2)}{k} + n_3 \underset{(3)}{k}, \qquad \left[\underset{(2)}{k}, \underset{(3)}{k}\right] = n_1 \underset{(1)}{k}, \tag{10.15}$$

$$\left[\underset{(3)}{k}, \underset{(1)}{k}\right] = n_2 \underset{(2)}{k} - a \underset{(3)}{k}. \tag{10.16}$$

This form of the commutation relations was obtained using rotations of the vector fields $\underset{(i)}{k}$. From now on, no further rotations are allowed, but we may still scale $\underset{(i)}{k}$ without changing their directions, $\underset{(i)}{k} = C_i \underset{(i)}{k}'$ (no sum). After such a scaling, (10.15) – (10.16) change to

$$\left[\underset{(1)}{k}, \underset{(2)}{k}\right] = \frac{a}{C_1} \underset{(2)}{k} + \frac{C_3}{C_1 C_2} n_3 \underset{(3)}{k}, \qquad \left[\underset{(2)}{k}, \underset{(3)}{k}\right] = \frac{C_1}{C_2 C_3} n_1 \underset{(1)}{k}, \tag{10.17}$$

$$\left[\underset{(3)}{k}, \underset{(1)}{k}\right] = \frac{C_2}{C_1 C_3} n_2 \underset{(2)}{k} - \frac{a}{C_1} \underset{(3)}{k} \tag{10.18}$$

(primes were dropped for clarity). We now want to use these scalings to simplify a, n_1, n_2 and n_3. Using C_1, C_2 and C_3 we can scale those of the parameters (a, n_1, n_2, n_3) that are nonzero. A nonzero value can never be made zero by scaling. Hence, the preliminary classification is as shown in Table 10.1, where S stands for 'something' (different from zero). But not all the entries in Table 10.1 are inequivalent. Permutations of the basis vectors that do not violate (10.14) are still allowed. Any permutation of (n_1, n_2, n_3) is allowed when $a = 0$, and with $a \neq 0$ we are still allowed to permute n_2 and n_3. Hence, only the cases indicated in the last line of the table have a chance to be inequivalent.

We will see that the classification of the algebras into inequivalent types does not match the columns of Table 10.1 – the table will serve only to make the presentation orderly. The labels for the types were introduced by Bianchi (1898), who, as already mentioned, used a different method. By tradition, his numbering is still in use, although it does not seem natural in the derivation presented below.

Let us consider the consecutive columns of Table 10.1.

(1) $a = n_1 = n_2 = n_3 = 0$.
This is Bianchi type I where all commutators are zero.

(2) $a = n_2 = n_3 = 0$, $n_1 \neq 0$.

Table 10.1 A preliminary Bianchi classification

a	0	0	0	0	0	0	0	0	S	S	S	S
n_1	0	0	0	0	S	S	S	S	0	0	0	0
n_2	0	0	S	S	0	0	S	S	0	0	S	S
n_3	0	S	0	S	0	S	0	S	0	S	0	S
Case Number	1	/	/	/	2	/	3	4	5	6	/	7

Taking $C_1 = C_2 C_3/n_1$ we obtain $n_1' = 1$. This is Bianchi type II.

(3) $a = n_3 = 0$, $n_1 \neq 0 \neq n_2$.

Taking $C_1 = C_2 C_3/n_1$ we obtain $n_1' = 1$. However, as seen from (10.17) – (10.18), we have then $n_2' = n_1 n_2/C_3{}^2$ and we will not be able to change the sign of n_2' by choice of C_3. Therefore, we must consider two cases:

(3a) $n_1 n_2 > 0$.

Then we take $C_3 = (n_1 n_2)^{1/2}$ and obtain $n_2' = 1$. This is a subcase of Bianchi's type VII. Bianchi himself called it type VII_1, today it is called type VII_0.

(3b) $n_1 n_2 < 0$.

Then we take $C_3 = (-n_1 n_2)^{1/2}$ and obtain $n_2' = -1$. This is a subcase of Bianchi's type VI, called today VI_0. Bianchi noted that this case requires a separate treatment, but the final result fits well within the general type VI, so he did not give it any special name.

(4) $a = 0$, $n_1 n_2 n_3 \neq 0$.

We take $C_1 = C_2 C_3/n_1$ and obtain $n_1' = 1$. But then, as before, $n_2' = n_1 n_2/C_3{}^2$, and the two signs of $n_1 n_2$ have to be considered separately.

(4a) $n_1 n_2 > 0$.

We take $C_3 = (n_1 n_2)^{1/2}$ and obtain $n_2' = 1$. However, $n_3' = n_1 n_3/C_2{}^2$, and two further subcases arise:

(4a$_1$) $n_1 n_3 > 0$.

Then we take $C_2 = (n_1 n_3)^{1/2}$ and obtain $n_3' = 1$. Thus finally, $n_1' = n_2' = n_3' = 1$, $a = 0$. This is Bianchi type IX.

(4a$_2$) $n_1 n_3 < 0$.

Then we take $C_2 = (-n_1 n_3)^{1/2}$ and obtain $n_3' = -1$. Hence finally, $n_1' = n_2' = 1 = -n_3'$. This is Bianchi type VIII.

(4b) $n_1 n_2 < 0$.

We then take $C_3 = (-n_1 n_2)^{1/2}$ and obtain $n_2' = -1$. But then we must again consider the same two subcases as before for C_2:

(4b$_1$) $n_1 n_3 > 0$.

Then we take $C_2 = (n_1 n_3)^{1/2}$ and obtain $n_3' = 1$. Through the basis change (still allowed!)

$$\left(\underset{(2)}{k}, \underset{(3)}{k} \right) = \left(-\underset{(3)}{\tilde{k}}, \underset{(2)}{\tilde{k}} \right)$$

we then obtain the same parameter values as in case (4a$_2$).

(4b$_2$) $n_1 n_3 < 0$.

Then we take $C_2 = (-n_1 n_3)^{1/2}$ and obtain $n_3' = -1$. Again through the basis change

$$\left(\underset{(1)}{k}, \underset{(2)}{k}, \underset{(3)}{k}\right) = \left(\underset{(3)}{\widetilde{k}}, \underset{(1)}{\widetilde{k}}, -\underset{(2)}{\widetilde{k}}\right)$$

we arrive back at the case (4a$_2$).

(5) $a \neq 0$, $n_1 = n_2 = n_3 = 0$.

Taking $C_1 = a$ we obtain $a' = 1$. This is Bianchi type V.

(6) $a \neq 0 \neq n_3$, $n_1 = n_2 = 0$.

We take $a = C_1$, $C_3 = aC_2/n_3$ and obtain $a' = 1 = n_3'$. This is Bianchi type IV.

(7) $a n_2 n_3 \neq 0$, $n_1 = 0$.

Taking $C_2 = C_1 C_3/n_2$ we obtain $n_2' = 1$. We lose again the possibility of changing the sign of $n_3' = n_2 n_3/C_1{}^2$ and must consider two cases:

(7a) $n_2 n_3 > 0$.

Then we take $C_1 = (n_2 n_3)^{1/2}$ and obtain $n_3' = 1$. Since, however, we fix C_1 in this way, we have no possibility left to scale a. Hence, with $n_3' = 1$, the value of a remains arbitrary. Then the algebras corresponding to $n_2 = n_3 = 1$ but different values of a *are not* equivalent. Bianchi called this case type VII$_2$, but, as opposed to previous types, this is a one-parameter family of inequivalent types. With $a = 0$, an algebra equivalent to the one obtained in (3a) results. This is seen if, with $a = 0$, we carry out the basis change

$$\left(\underset{(1)}{k}, \underset{(3)}{k}\right) = \left(\underset{(3)}{\widetilde{k}}, -\underset{(1)}{\widetilde{k}}\right).$$

The general type VII, with $a \neq 0$, is today denoted VII$_h$.

(7b) $n_2 n_3 < 0$.

Then we take $C_1 = (-n_2 n_3)^{1/2}$ and obtain $n_3' = -1$. Just as before, we are left then with no possibility of changing a. Again, this is a one-parameter family of inequivalent types. Bianchi denoted it VI; today it is denoted VI$_h$. With $a = 0$, the algebra obtained in (3b) results, i.e. type VI$_0$. To see this, the same basis change as in (7a) is necessary.

In Bianchi's scheme the subcase of type VI corresponding to $a = 1$ emerged as a separate type which he called type III. In it, $a = n_2 = 1 = -n_3$, $n_1 = 0$.

The final classification is shown in Table 10.2.

Table 10.2   The Bianchi classification

a	0	0	0	0	0	0	1	1	a	a	1
n_1	0	1	1	1	1	1	0	0	0	0	0
n_2	0	0	1	-1	1	1	0	0	1	1	1
n_3	0	0	0	0	1	-1	0	1	1	-1	-1
Bianchi Type	I	II	VII$_0$	VI$_0$	IX	VIII	V	IV	VII$_h$	VI$_h$	III

10.2 The dimension of the group versus the dimension of the orbit

In Section 8.2 we defined the orbits of one-parameter families of mappings. At this point it should be clear that the notion of an orbit can be defined also for a family (or group) of mappings that has several parameters; such orbits can be multidimensional curved spaces.

Let us recall that the Bianchi classification was done without any reference to the way in which the generators act on the manifold M_n. Two situations are possible:

1. The three generators (which are by definition linearly independent as *vector fields*) may also be linearly independent as *vectors* at each single point of M_n. This happens e.g. for generators of translations in $\mathbb{R}^3$.

2. The three generators may be linearly dependent at each single point of M_n. This happens e.g. for the generators of the group $O(3)$, which is a 3-dimensional group, but has 2-dimensional orbits.

A 3-dimensional isometry group cannot have 1-dimensional orbits: with 1-dimensional orbits any two generators would be proportional to each other, and then, from the Killing equations, the proportionality factor would be constant.

If the orbits are 2-dimensional, then their curvature is characterised by one scalar that must be invariant under the action of the group, i.e. constant. The constant curvature may be positive (the orbits are then 2-dimensional spheres), zero (the orbits are then 2-planes) or negative (such a surface has, in polar coordinates, the metric $ds^2 = (1+a^2r^2)^{-1}dr^2 + r^2d\phi^2$, but cannot be embedded in the $\mathbb{R}^3$ with a positive-definite Euclidean metric). Then, the three algebras of generators are of Bianchi types IX, VII_0 and VIII, respectively.

If a 3-dimensional group has 3-dimensional orbits, then the scalar curvature of the orbits must also be constant. However, in three dimensions the scalar curvature does not fully characterise the curvature of space, there exists also the Ricci tensor. Therefore more geometries are possible; they will be briefly discussed in Section 10.6.

10.3 Action of a group on a manifold

If a 3-dimensional group of mappings of M_n into itself has 3-dimensional orbits, then several situations are possible, e.g. the following:

1. The mappings may or may not be symmetries of M_n (they could be for example conformal symmetries of M_n).

2. The orbits may be timelike, spacelike or null hypersurfaces in M_n.

In each situation one can consider the various Bianchi types. Examples of solutions of Einstein's equations are known for which the orbits are timelike (e.g. Krasiński (1974) for Bianchi type I and Krasiński (1998, 2001a) for more general types; in the 1998 and 2001 papers there are examples of null orbits). A systematic investigation of all the spacetimes with 3-dimensional timelike orbits of the Bianchi groups was done by Harness (1982). Examples are also known in which 3 dimensional groups act as groups of symmetries on certain preferred 3-dimensional submanifolds of M_n, but are not symmetries of the whole M_n (these are the so-called spacetimes with intrinsic symmetries (Collins, 1979), for examples see Krasiński (1981) and Wolf (1985)). A brief discussion of general properties of spacetimes for which a 3-dimensional symmetry group has 3-dimensional null orbits is given in Stephani et al. (2003, section 24.2). A systematic investigation was done of the case when the group has 3-dimensional spacelike orbits and the group transformations are conformal symmetries of the manifold with the conformal factor ϕ in $\underset{k}{\pounds} g_{\alpha\beta} = \phi g_{\alpha\beta}$ being constant (these are the so-called **self-similar** spacetimes (Eardley, 1974)).

Most effort went into investigating the case when the 3-dimensional orbits are spacelike and the group in question is a group of symmetries of M_n (but examples are known where the orbits are spacelike in one part of M_n and timelike elsewhere, see Collins and Wainwright (1983) and Collins and Ellis (1979)). Before we consider this case in more detail, a few definitions must be given.

10.4 Groups acting transitively, homogeneous spaces

Let $\text{Orb}(p, G)$ denote the orbit of the point $p \in M_n$ under the action of the group G. We say that G **acts transitively** on a manifold $S \subset M_n$ when, for every $q \in S$, $\text{Orb}(q, G) = S$. *Examples*: the group $O(3)$ acts transitively on the surface of a sphere, the group of arbitrary translations in $\mathbb{R}^n$ acts transitively on $\mathbb{R}^n$. The space S on which G acts transitively is called **homogeneous with respect to** G.

If for every $q \in S$ (assumed homogeneous with respect to G) there exists a subgroup $H \subset G$ such that $Hq = q$, then G is said to act **multiply transitively** on S. A group acting transitively, but not multiply, acts **simply transitively**. *Examples*: the group of arbitrary translations in $\mathbb{R}^n$ acts simply transitively on $\mathbb{R}^n$; the group $O(3)$ acts multiply transitively on a 2-sphere since each point p of the sphere remains unchanged by rotations around the axis which passes through p.

A 4-dimensional manifold M_n with the metric tensor g of signature $(+ - - -)$ is called a **spatially homogeneous spacetime of Bianchi type** if it has the following properties:

1. It has a 3-dimensional symmetry group G.
2. The orbits of G are spacelike and G acts on them simply transitively.

A Bianchi-type spacetime may have a symmetry group of dimension larger than 3; the definition requires then that the full symmetry group has a 3-dimensional subgroup that acts simply transitively on spacelike hypersurfaces.

10.5 Invariant vector fields

As noted in Section 8.6, a vector field X^α is invariant under the group G of transformations generated by the vector field k^α if

$$\underset{k}{\pounds} X^\alpha \equiv [k, X]^\alpha = 0. \tag{10.19}$$

Let S_m be an m-dimensional manifold, let $\overline{X}^\alpha$ be a vector at a point $p \in S_m$ and let $\underset{(i)}{k}$, $i = 1, \dots, m$ be a set of linearly independent vector fields on S_m. Then at every point $p \in S_m$, the set of vectors $\{\underset{(i)}{k}(p)\}, i = 1, \dots, m$ is a basis in the tangent space to S_m at p. Consequently, the matrix K where $K_i{}^\alpha = \underset{(i)}{k^\alpha}(p)$ is nonsingular at every $p \in S_m$, and so defines the inverse matrix κ.

Let us write (10.19) at a $p \in S_m$ for all the fields $\underset{(i)}{k}{}^{\alpha}$:

$$\underset{(i)}{k}{}^{\rho} X^{\alpha}{}_{,\rho} - X^{\rho} \underset{(i)}{k}{}^{\alpha}{}_{,\rho} = 0. \tag{10.20}$$

Let us take (10.20) to be the definition of a vector field X^{α} on S_m, with the initial conditions $X^{\alpha}(p) = \overline{X}^{\alpha}$. Let us multiply the left-hand side of (10.20) by the matrix κ. The result is

$$X^{\alpha}{}_{,\beta} - \kappa^i{}_{\beta} \underset{(i)}{k}{}^{\alpha}{}_{,\rho} X^{\rho} = 0 \qquad \text{(sum over } i\text{)}. \tag{10.21}$$

This formula is similar to the condition of covariant constancy of a vector field, where

$$G^{\alpha}{}_{\beta\gamma} \overset{\text{def}}{=} - \kappa^i{}_{\beta} \underset{(i)}{k}{}^{\alpha}{}_{,\gamma} \qquad \text{(sum over } i\text{)} \tag{10.22}$$

plays the role of the affine connection. Let us follow this analogy and calculate the curvature tensor defined by $G^{\alpha}{}_{\beta\gamma}$. It turns out to be equal to zero. Therefore, (10.21) defines a transport of the vector $\overline{X}^{\alpha}$ along the vector fields $\underset{(i)}{k}{}^{\alpha}$ that is formally analogous to the parallel transport on S_m and has vanishing curvature. Consequently, the transport (10.21) is path-independent. Thus, if $X^{\alpha}(p)$ is defined at any single point $p \in S_m$, then (10.21) will uniquely define X^{α} at any other point of S_m.

We can then define a vector basis $\underset{(i)}{\overline{X}}{}^{\alpha}(p)$ at $p \in S_m, i = 1, \ldots, m$, and use (10.21) (called the **Lie transport**) to transplant the basis to all other points of S_m. The vector fields $\underset{(i)}{X}{}^{\alpha}$ on S_m thus obtained will be automatically invariant with respect to the transformations generated by the fields $\underset{(i)}{k}{}^{\alpha}$. We have thus proven

Theorem 10.1 *If a set of vector fields $\underset{(i)}{k}{}^{\alpha}$ on S_m exists such that $\underset{(i)}{k}{}^{\alpha}(p)$ form a basis of the tangent space to S_m at each $p \in S_m$, then there exists a set of vector fields $\underset{(i)}{X}{}^{\alpha}$ on S_m that are invariant under the group of transformations generated by $\underset{(i)}{k}{}^{\alpha}$ and also form a basis at each $p \in S_m$.*

Now let us assume that the fields $\underset{(i)}{k}{}^{\alpha}$ are Killing fields, and let us calculate the Lie derivatives along $\underset{(i)}{k}{}^{\alpha}$ of the scalars

$$g_{(j)(k)} \overset{\text{def}}{=} \underset{(j)}{X}{}^{\alpha} \underset{(k)}{X}{}^{\beta} g_{\alpha\beta}, \tag{10.23}$$

where $g_{\alpha\beta}$ is the metric tensor on S_m. Since $\underset{(i)}{X}$ are invariant, we have $\underset{\underset{(i)}{k}}{\pounds} \underset{(j)}{X} = 0$, and since $\underset{(i)}{k}$ are now Killing fields, we have $\underset{\underset{(i)}{k}}{\pounds} g_{\alpha\beta} = 0$, and so $\underset{\underset{(i)}{k}}{\pounds} g_{(j)(k)} = 0$. But $g_{(j)(k)}$ are scalars, so from (8.78) it follows that

$$0 = \underset{\underset{(i)}{k}}{\pounds} g_{(j)(k)} = \underset{(i)}{k}{}^{\rho} g_{(j)(k),\rho} \tag{10.24}$$

and since $\underset{(i)}{k}$ form a basis, this means that $g_{(j)(k)}$ are constants.

If $\underset{(i)}{X}$ and $\underset{(j)}{X}$ are invariant, then so is $[\underset{(i)}{X}, \underset{(j)}{X}]$. Since $\underset{(i)}{X}(p)$ form a basis at every $p \in S$, it follows that $[\underset{(i)}{X}, \underset{(j)}{X}]$ can be decomposed in this basis,

$$[\underset{(i)}{X}, \underset{(j)}{X}] = D^l{}_{ij}\underset{(l)}{X}, \tag{10.25}$$

where the scalar coefficients $D^l{}_{ij}$ could be expected to depend on the point p. However, calculating the Lie derivative of (10.25) along $\underset{(i)}{k}$, using the invariance of $[\underset{(i)}{X}, \underset{(j)}{X}]$ and of $\underset{(l)}{X}$, and using the fact that $\underset{(l)}{X}$ form a basis at each $p \in S_m$ we conclude that $D^l{}_{ij}$ are constants.

The basis $\underset{(i)}{X}$ was defined by choosing $\underset{(i)}{X}(p_0)$ arbitrarily at a $p_0 \in S_m$ and transporting it off p_0 by (10.21). We can thus assume that $\underset{(i)}{X}{}^\alpha(p_0) = \underset{(i)}{k}{}^\alpha(p_0)$. In this case

$$D^l{}_{ij} = -C^l{}_{ij}, \tag{10.26}$$

where $C^l{}_{ij}$ are the structure constants defined by the commutators of $\underset{(i)}{k}$ – see Exercise 3.

10.6 The metrics of the Bianchi-type spacetimes

Let $\underset{(i)}{k}$ be the three Killing fields required by the definition of a Bianchi-type spacetime, and let x^I be the coordinates in the homogeneous hypersurfaces ($i, I = 1, 2, 3$). These hypersurfaces are tangent to $\underset{(i)}{k}$ and define the vector field m orthogonal to them,

$$m^\alpha = |g|^{-1/2}\epsilon^{\alpha\beta\gamma\delta}\underset{(1)(2)(3)}{k_\beta k_\gamma k_\delta}, \tag{10.27}$$

where $g = \det||g_{\alpha\beta}||$. Let us choose the parameter on the curves tangent to m as the t-coordinate in M_n: $m^\alpha = \partial x^\alpha / \partial t$. Since $m^\alpha \underset{(i)}{k}{}_\alpha$, $i = 1, 2, 3$, in such coordinates we have

$$g_{0I} = 0. \tag{10.28}$$

From the (00) component of the Killing equations, using (10.28) and $\underset{(i)}{k}{}^0 = 0$, we further obtain $\underset{(i)}{k}{}^I g_{00,I} = 0$, i.e. $g_{00} = g_{00}(t)$. By the next transformation, $t' = \int \sqrt{g_{00}(t)}dt$, we obtain (dropping the prime)

$$ds^2 = dt^2 - g_{IJ}dx^I dx^J. \tag{10.29}$$

From the $(0, I)$ components of the Killing equations we then conclude that $\underset{(i)}{k}{}^I{}_{,t} = 0$ and from the (I, J) components that $\underset{(i)}{k}$ are Killing fields also for the 3-metric g_{IJ}. Since for the homogeneous hypersurfaces with the 3-metric g_{IJ} the fields $\underset{(i)}{k}$ form a basis of the tangent space at each point, we can apply the results of Section 10.5 to each single hypersurface. Here, (10.24) will read $\underset{(i)}{k}{}^R g_{(j)(k),R} = 0$, but $(\partial/\partial t)g_{(j)(k)}$ is not determined. Consequently, the $g_{(j)(k)}$ will be functions of t. Let us denote by $\omega^{(i)}{}_I$ the matrix inverse to $\underset{(i)}{X}{}^I$ (where $\underset{(i)}{X}$

are the vector fields invariant with respect to the transformations generated by $k_{(i)}$); then, from (10.23):

$$g_{IJ} = g_{(j)(k)}(t)\omega^{(j)}{}_I\omega^{(k)}{}_J \tag{10.30}$$

and

$$ds^2 = dt^2 - g_{(j)(k)}(t)\omega^{(j)}{}_I\omega^{(k)}{}_J dx^I dx^J. \tag{10.31}$$

In order to find $\omega^{(i)}{}_I$ one has to construct the Killing fields and their invariant fields for each Bianchi type separately. The resulting formulae can be found e.g. in Stephani et al. (2003, Chapters 13 and 14, in particular Table 13.4 and p. 209).

10.7 The isotropic Bianchi-type (Robertson–Walker) spacetimes

In astrophysics, such Bianchi-type spacetimes are important that are not only spatially homogeneous but also spherically symmetric (= **isotropic**). We shall deal with their (astro-)physical implications in Chapter 17, now we will derive their metric.

The form (8.52) of the metric is preserved by the transformations $t = f(t', r')$, $r = g(t', r')$, which can be used to simplify (8.52) further. In order to transform (8.52) into the form (10.31) we can choose f and g so that the new $\beta = 0$. In the new form, only those transformations will be permissible which preserve the hypersurfaces $t = $ constant. They are $t = f(t')$, $r = g(r')$. In agreement with (10.29), α should then depend only on t and be transformable to 1. So, finally, a spherically symmetric metric can possibly be homogeneous in the Bianchi sense only if it can be put in the form

$$ds^2 = dt^2 + \gamma(t,r)dr^2 + \delta(t,r)\left(d\vartheta^2 + \sin^2\vartheta d\varphi^2\right). \tag{10.32}$$

With such a metric one should now solve the Killing equations.

Some properties of the solution can be guessed in advance. The spacetime with the metric (10.32) has $O(3)$ as a symmetry group. Since $O(3)$ has 2-dimensional orbits, it cannot be the homogeneity group H existing in the Bianchi-type spacetimes. Hence, the full symmetry group must contain $O(3)$ and H as two subgroups. The $O(3)$ cannot be a subgroup of H because it acts multiply transitively. The $O(3)$ and H cannot have any common 2-dimensional subgroup because $O(3)$ has no 2-dimensional subgroups at all. All 1-dimensional subgroups of $O(3)$ are rotations around an axis, which act multiply transitively and so cannot be subgroups of H. Thus, $O(3)$ and H cannot have any common subgroup apart from the identity transformation. Hence, the full symmetry group will have *at least* six parameters, three of them connected with $O(3)$ and three with H. On the other hand, we showed in Section 8.4 that an n-dimensional manifold can have a symmetry group of *at most* $\frac{1}{2}n(n+1)$ parameters, i.e. at most six when $n = 3$. Consequently, the spatially homogeneous and isotropic spacetimes have symmetry groups with exactly six parameters.

We shall now indicate how to solve the Killing equations for the metric (10.32). The calculation is laborious, but uses only routine mathematics. We recall (see Section 10.6) that in a general Bianchi-type spacetime the Killing fields have no time-component ($k^0 = 0$), while the components (00) and (0, I) of the Killing equations have already been solved with the result $k^I{}_{,t} = 0$. (We could proceed without using this information, but then among the solutions there would be those with 4-dimensional orbits of the symmetry group – the Minkowski and de Sitter metrics. They are subcases of the Bianchi spacetimes.)

The remaining Killing equations are

$$k^1 \gamma_{,r} + 2k^1{}_{,r}\,\gamma = 0, \tag{10.33}$$

$$k^2{}_{,r}\,\delta + k^1{}_{,\vartheta}\,\gamma = 0, \tag{10.34}$$

$$k^3{}_{,r}\,\delta \sin^2 \vartheta + k^1{}_{,\varphi}\,\gamma = 0, \tag{10.35}$$

$$k^1 \delta_{,r} + 2k^2{}_{,\vartheta}\,\delta = 0, \tag{10.36}$$

$$k^3{}_{,\vartheta} \sin^2 \vartheta + k^2{}_{,\varphi} = 0, \tag{10.37}$$

$$k^1 \delta_{,r} + 2k^2 \cot \vartheta \delta + 2k^3{}_{,\varphi}\,\delta = 0. \tag{10.38}$$

We seek solutions with the physical signature $(+ - - -)$, so $\gamma < 0$ and $\delta < 0$. We also demand that $k^1 \neq 0$, for otherwise the orbits of the symmetry group would come out 2-dimensional, contrary to a basic property of the Bianchi-type spacetimes. The set (10.33) – (10.38) is overdetermined, so, whichever equation we solve first, the solution will be further limited by the remaining equations. Limitations are thereby imposed not only on the Killing fields but also on the metric components γ and δ. Along the way, some of the equations lead to alternatives of the type $(ab = 0 \implies a = 0$ or $b = 0)$, and this is where the special cases appear.

One of the special solutions that emerge is the metric[2]

$$\mathrm{d}s^2 = \mathrm{d}t^2 - R^2(t)\mathrm{d}r^2 - S^2(t)\left(\mathrm{d}\vartheta^2 + \sin^2 \vartheta \mathrm{d}\varphi^2\right), \tag{10.39}$$

whose generators of symmetries are

$$\underset{(1)}{J} = \frac{\partial}{\partial r}, \qquad \underset{(2)}{J} = \cos\varphi\frac{\partial}{\partial\vartheta} - \sin\varphi\cot\vartheta\frac{\partial}{\partial\varphi},$$

$$\underset{(3)}{J} = \sin\varphi\frac{\partial}{\partial\vartheta} + \cos\varphi\cot\vartheta\frac{\partial}{\partial\varphi}, \qquad \underset{(4)}{J} = \frac{\partial}{\partial\varphi}. \tag{10.40}$$

The generators $\underset{(2)}{J}$, $\underset{(3)}{J}$ and $\underset{(4)}{J}$ generate the $O(3)$ group that was assumed from the beginning. The four-parameter group generated by all of (10.40) has 3-dimensional orbits, and has no three-parameter simply transitive subgroup. This is because $O(3)$ has 2-dimensional orbits and so cannot itself be simply transitive, and has no 2-dimensional subgroups that could be combined with the transformations generated by $\underset{(1)}{J}$ into a three-parameter group. Hence, (10.39) does not belong among the Bianchi-type spacetimes. It is a metric of the Kantowski–Sachs (1966) class.[3]

---

[2] A transformation of the r coordinate is required to achieve the form (10.39).

[3] Metrics with the Kantowski–Sachs symmetry have a longer history than most people suspect. A *generalisation* of such a metric first appeared in the paper by Datt (1938), but the author instantly dismissed it as unphysical. Some physical properties of the metrics (10.39) were investigated by Kompaneets and Chernov (1964). The symmetry was noted and investigated by Kantowski (1965), and became a classical piece of knowledge after the paper by Kantowski and Sachs (1966). The geometric properties of the Datt metric were first investigated by Ruban (1968, 1969). See the reprintings of the Datt, Kantowski and Ruban papers in the *General Relativity and Gravitation* journal for more on this story.

The generic solution of (10.33) – (10.38) is

$$
\underset{(1)}{J} = -V\cos\vartheta\frac{\partial}{\partial r} + W\sin\vartheta\frac{\partial}{\partial\vartheta},
$$

$$
\underset{(2)}{J} = V\sin\vartheta\cos\varphi\frac{\partial}{\partial r} + W\cos\vartheta\cos\varphi\frac{\partial}{\partial\vartheta} - W\frac{\sin\varphi}{\sin\vartheta}\frac{\partial}{\partial\varphi},
$$

$$
\underset{(3)}{J} = V\sin\vartheta\sin\varphi\frac{\partial}{\partial r} + W\cos\vartheta\sin\varphi\frac{\partial}{\partial\vartheta} + W\frac{\cos\varphi}{\sin\vartheta}\frac{\partial}{\partial\varphi},
$$

$$
\underset{(4)}{J} = \cos\varphi\frac{\partial}{\partial\vartheta} - \cot\vartheta\sin\varphi\frac{\partial}{\partial\varphi},
$$

$$
\underset{(5)}{J} = \sin\varphi\frac{\partial}{\partial\vartheta} + \cot\vartheta\cos\varphi\frac{\partial}{\partial\varphi}, \qquad \underset{(6)}{J} = -\frac{\partial}{\partial\varphi}, \tag{10.41}
$$

where

$$
W(r) \overset{\text{def}}{=} \frac{1}{r} - \frac{1}{4}kr, \qquad V(r) \overset{\text{def}}{=} 1 + \frac{1}{4}kr^2. \tag{10.42}
$$

The metric with these symmetries is

$$
\mathrm{d}s^2 = \mathrm{d}t^2 - \frac{R^2(t)}{\left(1 + \frac{1}{4}kr^2\right)^2}\left[\mathrm{d}r^2 + r^2\left(\mathrm{d}\vartheta^2 + \sin^2\vartheta\mathrm{d}\varphi^2\right)\right], \tag{10.43}
$$

where $R(t)$ is an arbitrary function and k is an arbitrary constant.

Special cases of this metric form, corresponding to $k > 0$ and $k < 0$, were first derived, by a rather loose argument, by Friedmann(1922,1924), who thus became, unknowingly, the father of modern cosmology. (He died long before he could witness his success.) A mathematically rigorous derivation, by methods different from ours, and with all signs of k included, was given independently by Robertson (1929, 1933) and Walker (1935). (Paradoxically, Friedmann overlooked the simplest case with $k = 0$.) The metric form (10.43) is thus frequently called the **Robertson–Walker metric**. Other names attached to it in various combinations are Friedmann and Lemaître, but those refer to various special cases of (10.43); we shall come back to this point in Chapter 17.

The last three generators in (10.41) are easily recognised as those of $O(3)$. We shall use the abbreviation $[i,j] \overset{\text{def}}{=} [\underset{(i)}{J}, \underset{(j)}{J}]$. The commutators are

$$
\begin{aligned}
&[1,2] = k\underset{(4)}{J} \\
&[1,3] = k\underset{(5)}{J} && [2,3] = k\underset{(6)}{J} \\
&[1,4] = -\underset{(2)}{J} && [2,4] = \underset{(1)}{J} && [3,4] = 0 \\
&[1,5] = -\underset{(3)}{J} && [2,5] = 0 && [3,5] = \underset{(1)}{J} \\
&[1,6] = 0 && [2,6] = -\underset{(3)}{J} && [3,6] = \underset{(2)}{J} \\
&[4,5] = \underset{(6)}{J} && [4,6] = -\underset{(5)}{J} && [5,6] = \underset{(4)}{J}.
\end{aligned} \tag{10.44}
$$

The last three commutators are those of the algebra of $O(3)$. In (10.44) such relations should be found that correspond to the Bianchi algebras. However, the Bianchi classification introduced standard bases, and (10.41) may contain transformed generators. Indeed, some of the Bianchi bases are linear combinations of (10.41).

In the standard Bianchi bases, the nonzero structure constants were scaled to $+1$ or -1 whenever possible. Thus, for comparing (10.44) with (10.15) – (10.16) and Table 10.2, we have to scale out $k \neq 0$ in (10.44); the scaling is

$$\left(\underset{(1)}{J}, \underset{(2)}{J}, \underset{(3)}{J} \right) = \sqrt{|k|} \left(\underset{(1)}{\tilde{J}}, \underset{(2)}{\tilde{J}}, \underset{(3)}{\tilde{J}} \right),$$

with other generators unchanged. The result is as if $k = -1$ when $k < 0$ and $k = +1$ when $k > 0$. The formulae below show the scaled generators, with tildes omitted.

For $k > 0$, the Bianchi sub-basis is of type IX:

$$\underset{(1)}{L} = \frac{1}{2} \left(\underset{(1)}{J} + \underset{(6)}{J} \right), \qquad \underset{(2)}{L} = \frac{1}{2} \left(\underset{(2)}{J} - \underset{(5)}{J} \right), \qquad \underset{(3)}{L} = \frac{1}{2} \left(\underset{(3)}{J} + \underset{(4)}{J} \right). \tag{10.45}$$

The new generators are determined up to arbitrary orthogonal transformations because the matrix n^{ij} has a triple eigenvalue.

For $k < 0$, two Bianchi algebras are found in (10.41), one of type V:

$$\underset{(1)}{L} = -\underset{(1)}{J}, \qquad \underset{(2)}{L} = \underset{(2)}{J} + \underset{(4)}{J}, \qquad \underset{(3)}{L} = \underset{(3)}{J} + \underset{(5)}{J}, \tag{10.46}$$

and one of type VII_h:

$$\underset{(1)}{L} = -a \underset{(1)}{J} + \underset{(6)}{J}, \qquad \underset{(2)}{L} = \underset{(2)}{J} + \underset{(4)}{J}, \qquad \underset{(3)}{L} = \underset{(3)}{J} + \underset{(5)}{J}. \tag{10.47}$$

For $k = 0$, two standard Bianchi bases are contained in (10.41): $\{ \underset{(1)}{J}, \underset{(2)}{J}, \underset{(3)}{J} \}$ of Bianchi type I and $\{ \underset{(1)}{J}, \underset{(2)}{J}, \underset{(4)}{J} \}$ of Bianchi type VII_0.

It is easy to verify that the examples of bases shown are indeed the standard Bianchi bases of the types indicated. It is, however, more difficult to prove that *only* these Bianchi algebras are contained in (10.41). This was proved by Grishchuk (1967). The relation between the symmetries of (10.43) and the possible Bianchi algebras is discussed in more detail in Ellis and MacCallum (1969).

10.8 Exercises

1. Verify that the curvature tensor calculated using (10.22) as the definition of the connection coefficients is indeed zero.

2. Verify the statement made above (10.25): If $\underset{(i)}{X}$ and $\underset{(j)}{X}$ are invariant, then so is $[\underset{(i)}{X}, \underset{(j)}{X}]$. In other words, if $\pounds_{k} \underset{(i)}{X}^{\alpha} = \pounds_{k} \underset{(j)}{X}^{\alpha} = 0$, then $\pounds_{k} [\underset{(i)}{X}, \underset{(j)}{X}]^{\alpha} = 0$ for any vector field k.

3. Verify (10.26).

Hint. Since both $\underset{(i)}{k}$ and $\underset{(i)}{X}$ are bases at every $p \in S_m$ and $\underset{(i)}{k} = \underset{(i)}{X}$ at p_0, there exists a matrix $M(x)$ (point-dependent!) such that $X^\alpha = M_i{}^j \underset{(j)}{k^\alpha}$ (sum over j) and $M_i{}^j(p_0) = \delta_i{}^j$. Substitute this in (10.25), use $\underset{k}{\pounds} \underset{(i)}{X^\alpha} = \underset{k}{\pounds} \underset{(j)}{X^\alpha} = 0$ and take the result at p_0.

4. Since the invariant fields are defined by $[\underset{(i)}{k}, X] = 0$ and the $\underset{(i)}{k}$ do not depend on t, we can solve this set assuming that X are independent of t. However, a general solution does depend on time. Show that the time-dependent fields are $\underset{(j)}{X^I}(t) = B_j{}^k(t)\underset{(k)}{X^I}(t_0)$, where t_0 is an initial value and $B_j{}^k(t)$ is a nonsingular 3×3 matrix. Hence, the time dependence can be hidden in $\widetilde{g}_{(j)(k)}(t) = g_{(r)(s)}(t)B^{-1^r}{}_j B^{-1^s}{}_k$, and we can use $X^I(t_0)$ as the new invariant fields.

Hint. Show first that if X is an invariant field, then so is dX/dt, and thus can be decomposed in the basis of the invariant fields.

11

* The Petrov classification by the spinor method

This chapter is mostly based on the notes taken by A. K. during J. Plebański's lectures at the Warsaw University in 1968–1971. Plebański (1974) is a largely expanded version.

11.1 What is a spinor?

Spinors are tensor densities in a 2-dimensional vector space over the body of complex numbers. A covariant 1-spinor ψ_A has two complex components. When the basis in the vector space is transformed by $e^{A'} = L^{A'}{}_A e^A$, $A, A' = 1, 2$, the ψ_A transforms by

$$\psi_{A'} = \left(L^{-1}\right)_{A'}{}^A \psi_A. \tag{11.1}$$

In the body of complex numbers, we have to consider the complex conjugation, which reduces to identity when applied to real numbers. Consequently, a new classification of objects, in addition to that into covariant and contravariant, arises: some spinors transform linearly, some transform **antilinearly** under a change of basis. Antilinearly means that $\overline{\left(L^{-1}\right)_{A'}{}^A}$, the complex conjugate of $\left(L^{-1}\right)_{A'}{}^A$, appears in the transformation law. Indices of objects that transform antilinearly are marked by an overdot, thus

$$\overline{\psi_A} = \psi_{\dot{A}}. \tag{11.2}$$

The same spinor may have all four kinds of indices, for example $\chi^{\dot{A}B}{}_{\dot{C}D}$.

The spinor methods are used in this book as an auxiliary tool for just one task – the Petrov classification. However, spinors are a powerful tool that often makes complicated problems miraculously simple (at the cost of hard work required to master the spinor formulae). A useful introduction to spinors in relativity is the paper by Penrose (1960), extended treatments can be found in Penrose and Rindler (1984) and Plebański (1974).

There is no metric in the space of spinors; indices are raised and lowered by the Levi-Civita symbols $\epsilon^{\dot{A}\dot{B}}$, $\epsilon_{\dot{A}\dot{B}}$, ϵ^{AB} and ϵ_{AB}. Since ϵ^{AB} is antisymmetric, a convention on manipulating indices has to be fixed. We adopt the convention that subscripts go before superscripts, and the dummy indices must be made adjacent (Plebański, 1974), so

$$\psi^A = \psi_S \epsilon^{SA}, \qquad \psi_A = \epsilon_{AS} \psi^S, \tag{11.3}$$

and similarly for dotted indices. Because of antisymmetry of the ϵ-s, (11.3) implies

$$\psi_A\phi^A = -\phi_A\psi^A, \qquad \psi_A\psi^A \equiv 0, \qquad \forall\psi_A. \tag{11.4}$$

11.2 Translating spinors to tensors and vice versa

Spinors will be used here to represent tensors defined on a 4-dimensional spacetime of signature $(+ - - -)$. We introduce four Hermitean 2×2 **Pauli matrices** $g^{\alpha\dot{A}B}$ that are a basis of the space of all 2×2 complex Hermitean matrices. With respect to the index α they transform like components of a contravariant vector, with respect to the other two indices they transform as spinors with one contravariant antilinear index and one contravariant linear index. We will find below that we must in fact require that the Pauli matrices are spinor *densities*; their weight will follow. The fact that they are Hermitean means that

$$\overline{g^{\alpha\dot{A}B}} \equiv g^{\alpha A\dot{B}} = g^{\alpha\dot{B}A}. \tag{11.5}$$

The first spinor index labels rows, the other one labels columns of the matrix. The spinor image of a covariant vector v_α is defined as follows:

$$v^{\dot{A}B} = v_\alpha g^{\alpha\dot{A}B}. \tag{11.6}$$

The $v^{\dot{A}B}$ is a Hermitean spinor (density! – see below), and a scalar with respect to coordinate transformations on the manifold. Since the $g^{\alpha\dot{A}B}$ are a basis in the space of Hermitean 2×2 matrices, the coefficients of decomposition of $v^{\dot{A}B}$ in that basis are uniquely determined, so there must be an inverse linear mapping from $v^{\dot{A}B}$ to v_α; we denote it

$$v_\alpha = \frac{1}{2} g_{\alpha\dot{A}B} v^{\dot{A}B}. \tag{11.7}$$

Since this must hold for arbitrary vectors v_α and spinors $v^{\dot{A}B}$, the Pauli matrices and their reciprocal matrices $g_{\alpha\dot{A}B}$ must obey

$$\frac{1}{2} g_{\alpha\dot{A}B} g^{\beta\dot{A}B} = \delta^\beta{}_\alpha \qquad \frac{1}{2} g_{\alpha\dot{C}D} g^{\alpha\dot{A}B} = \delta^{\dot{A}}{}_{\dot{C}} \delta^B{}_D. \tag{11.8}$$

To show that this notation is self-consistent, namely that the reciprocal Pauli matrices are obtained from the proper Pauli matrices by lowering the tensor index with a metric and lowering the spinor indices with the Levi-Civita symbols, we have for a moment to resort to an explicit representation of the Pauli matrices in the Minkowski spacetime.

The Pauli matrices could be defined as just any basis of Hermitean 2×2 matrices. By tradition, for the flat Minkowski manifold, the Pauli matrices are defined as follows:

$$\eta^{i\dot{A}B} = \left[\begin{pmatrix} 1 & 0 \\ 0 & 1 \end{pmatrix}, \begin{pmatrix} 0 & 1 \\ 1 & 0 \end{pmatrix}, \begin{pmatrix} 0 & -i \\ i & 0 \end{pmatrix}, \begin{pmatrix} 1 & 0 \\ 0 & -1 \end{pmatrix} \right]. \tag{11.9}$$

One can always find a set of vectors $e^i{}_\alpha$, $i, \alpha = 0, 1, 2, 3$ such that $g_{\alpha\beta} e_i{}^\alpha e_j{}^\beta = \eta_{ij}$, where η_{ij} is the Minkowski metric (see Chapter 9). Thus, vectors from the Minkowski spacetime can always be transformed into the corresponding vector fields in a curved spacetime by the mapping defined by $e_i{}^\alpha$: the image of the vector v^i of the Minkowski spacetime in the curved spacetime is $v^\alpha = e_i{}^\alpha v^i$. Consequently, the Pauli matrices for a curved spacetime with the metric $g_{\alpha\beta}$ are defined by

$$g^{\alpha \dot{A}B} = e_i{}^\alpha \eta^{i \dot{A}B}. \tag{11.10}$$

Example. For the Schwarzschild metric

$$ds^2 = \left(1 - \frac{2m}{r}\right) dt^2 - \frac{1}{1 - \frac{2m}{r}} dr^2 - r^2 \left(d\vartheta^2 + \sin^2 \vartheta d\varphi^2\right) \tag{11.11}$$

the basis of orthonormal contravariant vectors that maps (11.11) into the Minkowski metric is (indices label vectors, not their components)

$$e_0 = \left(1/\sqrt{1 - \frac{2m}{r}}, 0, 0, 0\right), \qquad e_1 = \left(0, \sqrt{1 - \frac{2m}{r}}, 0, 0\right),$$
$$e_2 = (0, 0, 1/r, 0), \qquad e_3 = (0, 0, 0, 1/(r \sin \vartheta)). \tag{11.12}$$

For the remaining part of the proof that the notation in (11.6) – (11.8) is self-consistent see the hint to Exercise 1.

Let us split the quantity $\epsilon_{\dot{R}\dot{S}} g^{\alpha \dot{R}A} g^{\beta \dot{S}B}$ into the part $A^{\alpha\beta AB}$ symmetric in $\alpha\beta$ and the part $S^{\alpha\beta AB}$ antisymmetric in $\alpha\beta$:

$$A^{\alpha\beta AB} = \frac{1}{2} \epsilon_{\dot{R}\dot{S}} \left(g^{\alpha \dot{R}A} g^{\beta \dot{S}B} + g^{\beta \dot{R}A} g^{\alpha \dot{S}B}\right), \tag{11.13}$$

$$S^{\alpha\beta AB} = \frac{1}{2} \epsilon_{\dot{R}\dot{S}} \left(g^{\alpha \dot{R}A} g^{\beta \dot{S}B} - g^{\beta \dot{R}A} g^{\alpha \dot{S}B}\right). \tag{11.14}$$

It is easy to verify that $A^{\alpha\beta AB}$ is antisymmetric in AB, while $S^{\alpha\beta AB}$ is symmetric in AB. It is also easy to note that in a 2-dimensional vector space every object antisymmetric in AB must be proportional to ϵ^{AB} (the same is true for $_{AB}$ and ϵ_{AB}, and for the dotted indices). Just imagine fixing the values of all other indices; then the components with $A = B = 1$ and with $A = B = 2$ must be zero, while that with $A = 1, B = 2$ is the negative of that with $A = 2, B = 1$. Consequently, $A^{\alpha\beta AB} = \Lambda^{\alpha\beta} \epsilon^{AB}$, so

$$\Lambda^{\alpha\beta} = \frac{1}{2} \epsilon_{AB} A^{\alpha\beta AB} = \frac{1}{2} \epsilon_{AB} \epsilon_{\dot{R}\dot{S}} g^{\alpha \dot{R}A} g^{\beta \dot{S}B}. \tag{11.15}$$

The $\Lambda^{\alpha\beta}$ coincides with the inverse metric tensor (see Exercise 2). We would like the metric to be a scalar with respect to spinor transformations. Because the Levi-Civita symbols in (11.15) are densities of weight $w = 1$ and $\dot{w} = 1$, the Pauli matrices must be taken to be spinor densities of weights $w = \dot{w} = -1/2$.

The object $S^{\alpha\beta AB}$ is called **spin-tensor**. It is a proper tensor (not a density) with respect to $\{\alpha, \beta\}$, a spinor density of weight $w = -1$ and, as we have already noted,

$$S^{\alpha\beta AB} = S^{[\alpha\beta]AB} = S^{\alpha\beta(AB)}. \tag{11.16}$$

It can be used to transform tensors antisymmetric in two indices into spinors symmetric in two indices. Two other properties are:

$$S^{\alpha\beta AB} = \frac{i}{2\sqrt{-g}} \epsilon^{\alpha\beta}{}_{\gamma\delta} S^{\gamma\delta AB}, \qquad S^{\alpha\beta \dot{A}\dot{B}} = -\frac{i}{2\sqrt{-g}} \epsilon^{\alpha\beta}{}_{\gamma\delta} S^{\gamma\delta \dot{A}\dot{B}}, \tag{11.17}$$

where $g = \det \|g_{\alpha\beta}\|$. The dots in $\epsilon^{\cdots}$ show from where the indices were lowered. (Since $\epsilon^{\cdots}$ and $\epsilon_{\cdots}$ are tensor densities of different weights, the covariant one is not equal to the contravariant one with lowered indices.) The second of (11.17) is just a complex conjugate of the first one, for hints on how to verify the first one see Exercise 4. The first property means that the spin-tensor is **self-dual**, the second means that the complex-conjugate spin-tensor is **anti-self-dual**. A further property follows (see Exercise 5):

$$S^{\alpha\beta AB} S_{\alpha\beta \dot{C}\dot{D}} = 0. \tag{11.18}$$

The spinor image of a null vector has a special property. Let k^{α} be a null vector, $k^{\alpha}k_{\alpha} = 0$. From (11.8) this implies that the spinor image has then the property $k_{\dot{A}B} k^{\dot{A}B} = 0$. But because of the Levi-Civita symbols used to manipulate indices, $k_{\dot{A}B} k^{\dot{A}B} = 2 \det \left\| k^{\dot{A}B} \right\|$. A Hermitean matrix whose determinant is zero has the property $k_{\dot{A}B} = k_{\dot{A}} k_B$, i.e. there exists a spinor with one index k_A that obeys this. Hence, the spinor image of a null vector is a spinor with one index. (Note: the k_A is determined only up to the phase, i.e. k_A and $e^{i\phi} k_A$, where ϕ is a real function, determine the same null vector field k^{α}.)

Two more properties of the spin-tensor are useful in calculations (see Exercise 7):

$$S^{\alpha\beta AB} S_{\alpha\beta CD} = 4 \left(\delta^A{}_C \delta^B{}_D + \delta^A{}_D \delta^B{}_C \right), \tag{11.19}$$

$$S_{\alpha\beta}{}^{AB} S^{\gamma\delta}{}_{AB} + S_{\alpha\beta}{}^{\dot{A}\dot{B}} S^{\gamma\delta}{}_{\dot{A}\dot{B}} = 4\delta_{\alpha\beta}^{\gamma\delta}. \tag{11.20}$$

11.3 The spinor image of the Weyl tensor

The spinor image of the Weyl tensor is defined as follows:

$$C_{ABCD} = \frac{1}{64} S^{\alpha\beta}{}_{AB} S^{\gamma\delta}{}_{CD} C_{\alpha\beta\gamma\delta}. \tag{11.21}$$

Because of the properties of $S^{\alpha\beta}{}_{AB}$ and $C_{\alpha\beta\gamma\delta}$, the following is true (see Exercise 8):

$$S^{\alpha\beta}{}_{AB} S^{\gamma\delta}{}_{\dot{C}\dot{D}} C_{\alpha\beta\gamma\delta} \equiv 0. \tag{11.22}$$

In consequence of (11.22), the inverse mapping to (11.21) is (see Exercise 9)

$$C_{\alpha\beta\gamma\delta} = S_{\alpha\beta}{}^{AB} S_{\gamma\delta}{}^{CD} C_{ABCD} + S_{\alpha\beta}{}^{\dot{A}\dot{B}} S_{\gamma\delta}{}^{\dot{C}\dot{D}} C_{\dot{A}\dot{B}\dot{C}\dot{D}}. \tag{11.23}$$

Now one of the miraculous properties of spinors shows up: all the complicated identities obeyed by the Weyl tensor $C_{\alpha\beta\gamma\delta}$ translate into one simple property of C_{ABCD}: it is symmetric in all its indices,

$$C_{ABCD} = C_{(ABCD)}, \tag{11.24}$$

see Exercise 10 for hints on how to verify this.

11.4 The Petrov classification in the spinor representation

Consider the following 4-linear form:

$$\Omega(\zeta) = C_{ABCD} \zeta^A \zeta^B \zeta^C \zeta^D, \tag{11.25}$$

where $\zeta^A = (\zeta^1, \zeta^2)$ is an arbitrary spinor. Assume that $\zeta^1 \neq 0$ and let $z = \zeta^2/\zeta^1$. Then $\Omega(\zeta) = (\zeta^1)^4 P_4(z)$, where $P_4(z)$ is a polynomial of 4-th degree. It has four complex roots, and so can be factorised:

$$
\begin{aligned}
\Omega(\zeta) &= (\zeta^1)^4 \, a \, (z_1 - z)(z_2 - z)(z_3 - z)(z_4 - z) \\
&= (\alpha_A \zeta^A)(\beta_A \zeta^A)(\gamma_A \zeta^A)(\delta_A \zeta^A),
\end{aligned} \tag{11.26}
$$

where $\{z_1, z_2, z_3, z_4\}$ are the roots of P_4, and the spinors $\{\alpha_A, \beta_A, \gamma_A, \delta_A\}$ are defined by the above equation as follows: $\alpha_A = (az_1\zeta_1, -a\zeta_2)$, $\beta_A = (z_2\zeta_1, -\zeta_2)$, etc. Hence, finally:

$$
C_{ABCD}\zeta^A\zeta^B\zeta^C\zeta^D = \alpha_A\beta_B\gamma_C\delta_D\zeta^A\zeta^B\zeta^C\zeta^D = \alpha_{(A}\beta_B\gamma_C\delta_{D)}\zeta^A\zeta^B\zeta^C\zeta^D. \tag{11.27}
$$

Since the spinor ζ^A was arbitrary, this means that

$$
C_{ABCD} = \alpha_{(A}\beta_B\gamma_C\delta_{D)}. \tag{11.28}
$$

Thanks to the total symmetry of C_{ABCD}, no information was lost in (11.25).

The quantities $\{\alpha_A, \beta_B, \gamma_C, \delta_D\}$ are called the **principal spinors** of the Weyl tensor, also **Debever spinors**. In a 2-dimensional space, only two of them can be linearly independent, but we can ask how many pairs of linearly independent spinors exist in this set. There are six different cases: (I) no two Debever spinors are collinear; (II) exactly two Debever spinors are collinear; (III) three Debever spinors are collinear; (D) two Debever spinors are collinear, the two others are collinear, too, but these pairs are distinct; (N) all four Debever spinors are collinear; (0) the Weyl tensor is identically zero. We have denoted the different cases by the same symbols as the Petrov types because these *are* the Petrov types – see Sec. 11.6. This approach was first introduced by Penrose (1960). The resulting diagram can be drawn in a more illustrative way than in Section 7.16, see Fig. 11.1. It remains to prove that this is indeed the same classification. First, however, we shall determine some other algebraic properties of the Weyl spinor.

11.5 The Weyl spinor represented as a 3×3 complex matrix

Now let us interpret the Weyl spinor C_{ABCD} as a 3×3 complex matrix:

$$
C \stackrel{\text{def}}{=} ||C^{AB}{}_{CD}||, \tag{11.29}
$$

whose elements are labelled by the 'superindices' (AB) and (CD). Each 'superindex' takes three values: $(AB) = \{(11), (12), (22)\}$. In the set of such matrices, the unit matrix is

$$
\mathbb{I}^{AB}_{CD} = \delta_C^{(A}\delta_D^{B)}. \tag{11.30}
$$

Since C_{ABCD} is symmetric in all indices, and indices are raised by ϵ^{AB}, the matrix C has all its traces equal to zero. Consider the characteristic equation for C, $\det||C - \lambda\mathbb{I}|| = 0$. As can be verified using (3.38), for a 3×3 matrix, the characteristic equation is

$$
\lambda^3 - (\text{Tr}C)\lambda^2 + \frac{1}{2}\left[(\text{Tr}C)^2 - \text{Tr}\left(C^2\right)\right]\lambda - \det(C)\mathbb{I} = 0. \tag{11.31}
$$

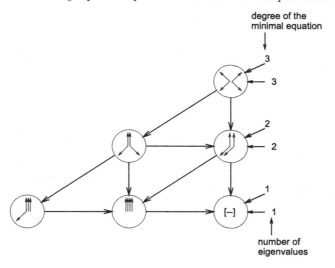

Fig. 11.1 The Petrov classification by the spinor method. Arrows show possible specialisations.

Since for the Weyl 'supermatrix' $\mathrm{Tr}C = 0$, this simplifies to

$$\lambda^3 - \frac{1}{2}\left[\mathrm{Tr}\left(C^2\right)\right]\lambda - \det(C)\mathbb{I} = 0. \tag{11.32}$$

The Hamilton–Cayley equation for C is thus

$$C^3 - \frac{1}{2}\left[\mathrm{Tr}\left(C^2\right)\right]C - \det(C)\mathbb{I} = 0. \tag{11.33}$$

The property $\mathrm{Tr}C = 0$ has one more consequence: taking the trace of (11.33) we obtain

$$\det(C) = \frac{1}{3}\mathrm{Tr}\left(C^3\right). \tag{11.34}$$

In the equations below we will use the following shorthand notation (recall (11.4))

$$(\beta\delta) \stackrel{\mathrm{def}}{=} \beta_A\delta^A \equiv -\delta_A\beta^A = -(\delta\beta). \tag{11.35}$$

Using (11.28) and the above notation, we find for $\mathrm{Tr}\left(C^2\right)$ (see Exercise 11):

$$\mathrm{Tr}\left(C^2\right) = \frac{1}{24}\left\{[(\alpha\beta)(\gamma\delta) + (\alpha\gamma)(\beta\delta) + (\alpha\delta)(\beta\gamma)]^2 - 4(\alpha\beta)(\beta\gamma)(\gamma\delta)(\alpha\delta)\right\}. \tag{11.36}$$

The general formula for the determinant is too unwieldy, but we will not need it. When two Debever spinors, say α^A and β^A, are collinear, $\alpha^A = \mu\beta^A$, the determinant is

$$\det(C) = \frac{1}{3}\mathrm{Tr}\left(C^3\right) \equiv \frac{1}{3}C^{AB}{}_{CD}C^{CD}{}_{EF}C^{EF}{}_{AB} = -\frac{1}{108}\mu^3(\beta\gamma)^3(\beta\delta)^3. \tag{11.37}$$

In the same special case, the formula for $\mathrm{Tr}\left(C^2\right)$ simplifies to

$$\mathrm{Tr}\left(C^2\right) = \frac{1}{6}\mu^2(\beta\gamma)^2(\beta\delta)^2. \tag{11.38}$$

Knowing all this, we can now verify that the following is true:

- In general, (11.32) has three different roots, so (11.33) is the minimal equation.
- When two of the Debever spinors are collinear, say $\alpha^A = \mu\beta^A$, (11.37) and (11.38) apply, and (11.33) becomes

$$\left(C - \frac{a}{6}\mathbb{I}\right)^2 \left(C + \frac{a}{3}\mathbb{I}\right) = 0, \qquad a \stackrel{\text{def}}{=} \mu(\beta\gamma)(\beta\delta). \tag{11.39}$$

  With no further limitations on the Debever spinors, this is the minimal equation, and there are only two distinct eigenvalues.
- When $\alpha^A = \mu\beta^A$ and $\gamma^A = \nu\beta^A$ (three Debever spinors are collinear), we have $\text{Tr}\left(C^2\right) = \det(C) = 0$, and (11.33) becomes

$$C^3 = 0. \tag{11.40}$$

  With no further limitations on the Debever spinors, this is the minimal equation, and the only eigenvalue is 0, triply degenerate.
- When $\alpha^A = \mu\beta^A$ and $\gamma^A = \rho\delta^A$ (i.e. the Debever spinors (α^A, β^A) are collinear and (γ^A, δ^A) are collinear, but the two pairs are distinct), (11.37) and (11.38) simplify to

$$\det(C) = -\frac{1}{108}\mu^3\rho^3(\beta\delta)^6, \qquad \text{Tr}\left(C^2\right) = \frac{1}{6}\mu^2\rho^2(\beta\delta)^4. \tag{11.41}$$

Now (11.33) becomes

$$\left(C - \frac{b}{6}\mathbb{I}\right)^2 \left(C + \frac{b}{3}\mathbb{I}\right) = 0, \qquad b \stackrel{\text{def}}{=} \mu\rho(\beta\delta)^2, \tag{11.42}$$

but the minimal equation is of degree 2 (see Exercise 13):

$$\left(C - \frac{b}{6}\mathbb{I}\right)\left(C + \frac{b}{3}\mathbb{I}\right) = 0. \tag{11.43}$$

- When all four Debever spinors are collinear, $\alpha^A = \mu\delta^A$, $\gamma^A = \rho\delta^A$, $\beta^A = \sigma\delta^A$, so $b = 0$, Eq. (11.40) still holds, but the minimal equation follows from (11.43) and it is $C^2 = 0$.
- When $C = 0$, this is the minimal equation.

In this way, we have verified all the information in Figs. 7.3 and 11.1.

11.6 The equivalence of the Penrose classes to the Petrov classes

It remains to verify that the algebraic types found by the Penrose (1960) method in this chapter coincide with those found by the Ehlers–Kundt (1962) method used in Section 7.16. From (7.97) – (7.98), using (11.23) and (11.17), we find (see Exercise 14):

$$Q_{\alpha\gamma} = E_{\alpha\gamma} + iH_{\alpha\gamma} = 2\left(u^\beta S_{\alpha\beta}{}^{\dot{A}\dot{B}}\right)\left(u^\delta S_{\gamma\delta}{}^{\dot{C}\dot{D}}\right)C_{\dot{A}\dot{B}\dot{C}\dot{D}}. \tag{11.44}$$

The quantity $\left(u^\beta S_{\alpha\beta}{}^{\dot{A}\dot{B}}\right)$ is in effect a 3×3 matrix. It has one symmetric 'superindex' $(\dot{A}\dot{B})$ that takes three values, and it obeys (because of antisymmetry in $(\alpha\beta)$) $u^\alpha\left(u^\beta S_{\alpha\beta}{}^{\dot{A}\dot{B}}\right) \equiv$

0; hence it operates in the 3-dimensional hypersurface element orthogonal to u^α. Consequently, the tensor index α takes values in a 3-dimensional space. Thus, $\left(u^\beta S_{\alpha\beta}{}^{\dot{A}\dot{B}}\right)$ provides an invertible linear mapping from the complex space V of vectors labelled by (AB) to the complex space W of vectors labelled by α (see Exercise 15). Equation (11.44) is the corresponding bilinear mapping of matrices over V to matrices over W. Such mappings preserve the eigenvalues and other invariants of matrices. Hence, the classification by the Penrose method is equivalent to that by the Ehlers–Kundt method. $\square$

11.7 The Petrov classification by the Debever method

The spinor image of the Weyl tensor has the following properties:

- In the most general case, each of the Debever spinors obeys

$$\alpha^A \alpha^B \alpha^C \alpha^D C_{ABCD} \equiv 0, \tag{11.45}$$

$$\alpha_{[A} C_{B]CD[E} \alpha_{F]} \alpha^C \alpha^D \equiv 0. \tag{11.46}$$

Both equations are easy to verify using (11.28).
- For type II, the nondegenerate Debever spinor obeys (11.45) – (11.46), while the preferred (double) Debever spinor α_A obeys

$$\alpha^A \alpha^B \alpha^C C_{ABCD} = 0, \tag{11.47}$$

$$\alpha_{[A} C_{B]CDE} \alpha^C \alpha^D = 0. \tag{11.48}$$

- For type D, each of the (double) Debever spinors α_A and γ_A obeys (11.47) and (11.48).
- For type III, the nondegenerate Debever spinor obeys (11.45) – (11.46), while the preferred (triple) Debever spinor α_A obeys

$$\alpha^B \alpha^C C_{ABCD} = 0. \tag{11.49}$$

- For type N, the only existing (quadruple) Debever spinor α_A obeys

$$\alpha^A C_{ABCD} \equiv 0. \tag{11.50}$$

We now observe that using (11.20) we may uniquely represent any antisymmetric second-rank tensor $T_{\alpha\beta}$ by its (symmetric) spinor image

$$T_{AB} = S_{\alpha\beta AB} T^{\alpha\beta}; \qquad T^{\alpha\beta} = \frac{1}{8} \left(S^{\alpha\beta AB} T_{AB} + S^{\alpha\beta \dot{A}\dot{B}} T_{\dot{A}\dot{B}} \right). \tag{11.51}$$

In particular, $T^{\alpha\beta} = 0$ if and only if $T_{AB} = 0$.

Now let

$$k_\mu \stackrel{\text{def}}{=} \frac{1}{2} g_{\mu\dot{A}B} \alpha^{\dot{A}} \alpha^B \tag{11.52}$$

be the null vector defined by the Debever spinor. A long calculation shows that

$$k_{[\alpha} C_{\beta]\mu\nu[\gamma} k_{\delta]} k^\mu k^\nu = 0. \tag{11.53}$$

The null vector k^μ defined by (11.52) is called **Debever vector**. In verifying (11.53) we make use of the statement $T^{\alpha\beta} = 0 \iff T_{AB} = 0$: Eq. (11.53) will hold if and only if

$$S^{\alpha\beta AB}S^{\gamma\delta CD}k_\alpha C_{\beta\mu\nu\gamma}k_\delta k^\mu k^\nu = 0. \tag{11.54}$$

See Exercise 18 for hints how to verify (11.54).

Equation (11.53) holds even in the most general case (Petrov type I). With more special Petrov types, the vanishing quantity becomes progressively simpler. Specifically:

- For Petrov type II, each of the two nondegenerate Debever vectors obeys (11.53), but the Debever vector that is the tensor image of the double Debever spinor obeys

$$k_{[\alpha}C_{\beta]\mu\nu\gamma}k^\mu k^\nu = 0, \tag{11.55}$$

by virtue of (11.47) and its complex conjugate.
- For Petrov type D, both Debever vectors obey (11.55).
- For Petrov type III, the tensor image of the nondegenerate Debever spinor obeys (11.53), while the tensor image of the triple Debever spinor obeys

$$C_{\beta\mu\nu\gamma}k^\mu k^\nu = 0, \tag{11.56}$$

by virtue of (11.49) and its complex conjugate.
- For Petrov type N, the only existing Debever vector obeys

$$C_{\beta\mu\nu\gamma}k^\mu = 0, \tag{11.57}$$

by virtue of (11.49) and its complex conjugate.

The method of verification is similar in each case. For (11.55), we calculate $S^{\alpha\beta AB}k_\alpha C_{\beta\mu\nu\gamma}k^\mu k^\nu$, whereas for (11.56) and (11.57) we only express $C_{\beta\mu\nu\gamma}$ through its spinor image. In each case, we use the formulae in the hint to Exercise 18 and remember that the contractions with $C_{....}$ enforce symmetrisation.

This approach to the Petrov classification was introduced by Debever (1959, 1964).

11.8 Exercises

1. Show that the reciprocal Pauli matrices $g_{\beta\dot{C}D}$ result from the proper ones $g^{\alpha\dot{A}B}$ by lowering the tensor index with $g_{\alpha\beta}$ and lowering the spinor indices with $\epsilon_{\dot{C}\dot{A}}$ and ϵ_{DB}.

Hint. Lower all three indices of the flat $\eta^{i\dot{A}B}$ given by (11.9) to find

$$\eta_{j\dot{C}D} = \eta_{ji}\epsilon_{\dot{C}\dot{A}}\epsilon_{DB}\eta^{i\dot{A}B} = \left[\begin{pmatrix} 1 & 0 \\ 0 & 1 \end{pmatrix}, \begin{pmatrix} 0 & 1 \\ 1 & 0 \end{pmatrix}, \begin{pmatrix} 0 & i \\ -i & 0 \end{pmatrix}, \begin{pmatrix} 1 & 0 \\ 0 & -1 \end{pmatrix}\right]. \tag{11.58}$$

Next, verify that the objects (11.9) and (11.58) do obey both of (11.8). (Remember: the first spinor index labels rows, the second one labels columns of each matrix.) Thus, the notation is consistent in the Minkowski spacetime. Invert (11.10) to find $\eta^{j\dot{A}B} = e^j{}_\alpha g^{\alpha\dot{A}B}$. Define $h_{\alpha\dot{A}B} = e^i{}_\alpha \eta_{i\dot{A}B} = g_{\alpha\gamma}\epsilon_{\dot{A}\dot{C}}\epsilon_{BD}g^{\gamma\dot{C}D}$, by analogy to (11.10), then calculate $h_{\alpha\dot{A}B}g^{\beta\dot{A}B} = 2\delta_\alpha{}^\beta$ and $h_{\alpha\dot{A}B}g^{\alpha\dot{C}D} = 2\delta_{\dot{A}}{}^{\dot{C}}\delta_B{}^D$. Since the $h_{\alpha\dot{A}B}$ obey (11.8), they must coincide with the reciprocal Pauli matrices $g_{\alpha\dot{A}B}$.

2. Prove that the $\Lambda^{\alpha\beta}$ defined by (11.15) is equal to $g^{\alpha\beta}$.

Hint. Using the knowledge acquired up to (11.15), show that $\Lambda^{\alpha\beta}v_\beta = v^\alpha$ for any v^α.

3. Verify that the spin-tensor is symmetric in its spinorial indices.

4. Verify (11.17).

Hint. Rewrite the first of (11.17) in the tetrad components $S^{ijAB} = e^i{}_\alpha e^j{}_\beta S^{\alpha\beta AB}$, where $e^i{}_\alpha$ is the orthonormal tetrad discussed in Chapter 9. This is achieved as follows:

Using $g^{\alpha\beta} = \eta^{ij}e_i{}^\alpha e_j{}^\beta$, $\det||\eta^{ij}|| = \det||\eta_{ij}|| = -1$ and $\det||e_i{}^\alpha|| = e$ (see (4.17)), verify that

$$g = -1/e^2. \tag{11.59}$$

Now use (3.25) applied to $||e_i{}^\alpha||$:

$$\epsilon^{\alpha\beta\gamma_1\delta_1} = \frac{1}{e}\,\epsilon^{ijkl}e_i{}^\alpha e_j{}^\beta e_k{}^{\gamma_1} e_l{}^{\delta_1}. \tag{11.60}$$

With (11.59) and (11.60), the right-hand side of the first of (11.17) becomes

$$\frac{\mathrm{i}}{2}\,e_i{}^\alpha e_j{}^\beta \epsilon^{ijkl}S_{kl}{}^{AB}. \tag{11.61}$$

Apply now an analogue of (3.25) to $||\eta^{ij}||$ and use $\det||\eta^{ij}|| = -1$. The result is

$$\epsilon^{ijkl} = -\eta^{ip}\eta^{jq}\eta^{kr}\eta^{ls}\epsilon_{pqrs}. \tag{11.62}$$

Using (11.62) in (11.61), the hypothesis to verify becomes

$$S^{ijAB} = -\frac{\mathrm{i}}{2}\,\eta^{ip}\eta^{jq}\epsilon_{pqrs}S^{rsAB}, \tag{11.63}$$

where S^{ijAB} is the spin-tensor in the Minkowski spacetime. From (11.14),

$$S^{ijAB} = \frac{1}{2}\left(\eta^{i\dot1 A}\eta^{j\dot2 B} - \eta^{i\dot2 A}\eta^{j\dot1 B} - \eta^{j\dot1 A}\eta^{i\dot2 B} + \eta^{j\dot2 A}\eta^{i\dot1 B}\right). \tag{11.64}$$

Its independent nonzero components are

$$\begin{array}{ll}
S^{0111} = -S^{0122} = 1, & S^{1212} = \mathrm{i}, \\
S^{0211} = S^{0222} = \mathrm{i}, & S^{1311} = S^{1322} = -1, \\
S^{0312} = -1 & -S^{2311} = S^{2322} = \mathrm{i}.
\end{array} \tag{11.65}$$

The other nonzero components are obtained from (11.65) by $S^{\beta\alpha AB} = -S^{\alpha\beta AB}$ and $S^{\alpha\beta BA} = S^{\alpha\beta AB}$. Using (11.65), Eq. (11.63) can be verified component by component (with the symmetries that $S^{\alpha\beta AB}$ has, this is 18 equations).

5. Verify (11.18).

Hint. Substitute for both S-s from (11.17). They will be more handy in the form:

$$S^{\alpha\beta AB} = \frac{\mathrm{i}}{2\sqrt{-g}}\,g_{\gamma\rho}g_{\delta\sigma}\epsilon^{\alpha\beta\rho\sigma}S^{\gamma\delta AB}, \tag{11.66}$$

$$S_{\alpha\beta\dot C\dot D} = \frac{\mathrm{i}}{2}\,\sqrt{-g}\epsilon_{\alpha\beta\mu\nu}\epsilon_{\dot C\dot E}\epsilon_{\dot D\dot F}S^{\mu\nu\dot E\dot F}. \tag{11.67}$$

Then use (3.32) and (3.37). In the end it will turn out that $S^{\alpha\beta AB}S_{\alpha\beta\dot{C}\dot{D}} = -S^{\alpha\beta AB}S_{\alpha\beta\dot{C}\dot{D}}$.
In verifying (11.66) – (11.67) you will find the following useful:

$$\epsilon_{\alpha\beta\mu\nu} = \frac{1}{g} g_{\alpha\rho}g_{\beta\sigma}g_{\mu\lambda}g_{\nu\tau}\epsilon^{\rho\sigma\lambda\tau} \tag{11.68}$$

$$\epsilon^{\alpha\beta\gamma\delta} = g\, g^{\alpha\alpha_1}g^{\beta\beta_1}g^{\gamma\gamma_1}g^{\delta\delta_1}\epsilon_{\alpha_1\beta_1\gamma_1\delta_1}, \tag{11.69}$$

$$\delta^{\alpha\beta\gamma\delta}_{\rho\sigma\mu\nu} = \delta^{\alpha}_{\rho}\delta^{\beta\gamma\delta}_{\sigma\mu\nu} - \delta^{\alpha}_{\sigma}\delta^{\beta\gamma\delta}_{\rho\mu\nu} + \delta^{\alpha}_{\mu}\delta^{\beta\gamma\delta}_{\rho\sigma\nu} - \delta^{\alpha}_{\nu}\delta^{\beta\gamma\delta}_{\rho\sigma\mu}, \tag{11.70}$$

and similarly for $\delta^{\alpha\beta\gamma}_{\rho\sigma\mu}$. This and (11.70) follow from (3.30) by the usual formula for determinants.

6. Prove that a Hermitean 2-spinor $k_{\dot{A}B}$ whose determinant is zero does indeed define a spinor with one index by $k_{\dot{A}B} = k_{\dot{A}}k_B$.

7. Verify (11.19) and (11.20) by direct substitution from (11.14). In verifying (11.20) be careful to rename the dummy indices wherever appropriate.

8. Verify (11.22).

Hint. Begin by substituting (11.66) and its complex conjugate in (11.22), then apply (11.69) to one of the ϵ-s. Use (11.70) and watch out for terms that disappear because of $g^{\alpha\gamma}C_{\alpha\beta\gamma\delta} = 0$ and the like. Then watch for simplifications resulting from $S^{\alpha\beta}{}_{AB} = -S^{\beta\alpha}{}_{AB}$. In the end you will find that the left-hand side of (11.22) must be equal to its own negative.

9. Verify that (11.23) is the inverse transformation to (11.21).

Hint. Substitute (11.23) in (11.21), then use (11.22) and (11.19). Note that C_{ABCD} is symmetric in AB and in CD in consequence of $S^{\alpha\beta AB} = S^{\alpha\beta BA}$.

10. Verify (11.24).

Hint. Note that $C_{ABCD} = C_{(AB)CD} = C_{AB(CD)} = C_{CDAB}$ are obvious consequences of the definition (11.21). So it is only necessary to prove that $C_{A[BC]D} = 0$. Note first that $C_{A[BC]D}$ is antisymmetric also in AD, so $C_{A[BC]D} = \lambda\epsilon_{AD}\epsilon_{BC}$. By contracting this with $\epsilon^{AD}\epsilon^{BC}$ it follows that $\lambda = \frac{1}{4}C_{ABCD}\epsilon^{AD}\epsilon^{BC}$, so

$$C_{A[BC]D} = \frac{1}{4}\epsilon_{AD}\epsilon_{BC}C_{RS}{}^{RS}. \tag{11.71}$$

Using (11.18), (11.17) and (11.20) show that $C_{RS}{}^{RS} = -C_{\dot{R}\dot{S}}{}^{\dot{R}\dot{S}}$, so

$$C_{A[BC]D} = -\frac{1}{4}\epsilon_{AD}\epsilon_{BC}C_{\dot{R}\dot{S}}{}^{\dot{R}\dot{S}}. \tag{11.72}$$

Now calculate $C_{A[BC]D}$ by a different method. Begin by using (11.21) to find

$$C_{A[BC]D} = \frac{1}{2}\delta^{RS}_{BC}C_{ARSD} = \frac{1}{2\cdot64}\delta^{RS}_{BC}S^{\alpha\beta}{}_{AR}S^{\gamma\delta}{}_{SD}C_{\alpha\beta\gamma\delta}, \tag{11.73}$$

then substitute the definitions of $S^{\cdots}_{\cdots}$ from (11.14) and simplify the result to

$$C_{A[BC]D} = \frac{1}{2\cdot64}\epsilon^{RS}\epsilon_{BC}\epsilon_{\dot{P}\dot{Q}}\epsilon_{\dot{R}\dot{S}}g^{\alpha\dot{P}}{}_A\, g^{\beta\dot{Q}}{}_R\, g^{\gamma\dot{R}}{}_S\, g^{\delta\dot{S}}{}_D\, C_{\alpha\beta\gamma\delta}. \tag{11.74}$$

Consider the sub-expression $\epsilon^{RS} g^{\beta \dot{Q}}{}_R g^{\gamma \dot{R}}{}_S$. Split it into the part symmetric and anti-symmetric in $\beta\gamma$. The symmetric part will be $g^{\beta\gamma} \epsilon^{\dot{Q}\dot{R}}$, just as with $A^{..}_{..}$ in (11.15), so it will give zero contribution to (11.74) because $g^{\beta\gamma} C_{\alpha\beta\gamma\delta} = 0$. The antisymmetric part will be $S^{\beta\gamma\dot{Q}\dot{R}}$ ($g^{\alpha R\dot{A}} = g^{\alpha\dot{A}R}$ because $g^{\alpha\dot{A}R}$ are Hermitean), and will have its spinor indices lowered. So, finally, remembering the rule of manipulating the spinor indices (11.3):

$$C_{A[BC]D} = -\frac{1}{2 \cdot 64} \, \epsilon_{BC} g^{\alpha\dot{P}}{}_A \, g^{\delta\dot{S}}{}_D \, C_{\alpha[\beta\gamma]\delta} S^{\beta\gamma}{}_{\dot{P}\dot{S}}. \tag{11.75}$$

Now use the identity $C_{\alpha[\beta\gamma\delta]} = 0$ (see Exercise 6 to Chapter 7). It can be written as $C_{\alpha\beta\gamma\delta} + C_{\alpha\gamma\delta\beta} + C_{\alpha\delta\beta\gamma} = 0$, so

$$C_{\alpha[\beta\gamma]\delta} = \frac{1}{2} \, (C_{\alpha\beta\gamma\delta} - C_{\alpha\gamma\beta\delta}) = \frac{1}{2} \, (C_{\alpha\beta\gamma\delta} + C_{\alpha\gamma\delta\beta}) = -\frac{1}{2} \, C_{\alpha\delta\beta\gamma}. \tag{11.76}$$

Using (11.76) in (11.75):

$$C_{A[BC]D} = \frac{1}{4 \cdot 64} \, \epsilon_{BC} g^{\alpha\dot{P}}{}_A \, g^{\delta\dot{S}}{}_D \, C_{\alpha\delta\beta\gamma} S^{\beta\gamma}{}_{\dot{P}\dot{S}}. \tag{11.77}$$

Now use (11.18) and (11.23) in this to get

$$C_{\alpha\delta\beta\gamma} S^{\beta\gamma}{}_{\dot{P}\dot{S}} = S_{\alpha\delta}{}^{\dot{A}\dot{B}} C_{\dot{A}\dot{B}\dot{C}\dot{D}} S^{\beta\gamma\dot{C}\dot{D}} S_{\beta\gamma\dot{P}\dot{S}}. \tag{11.78}$$

Using the complex conjugate of (11.19) this simplifies to

$$C_{\alpha\delta\beta\gamma} S^{\beta\gamma}{}_{\dot{P}\dot{S}} = 8 S_{\alpha\delta}{}^{\dot{A}\dot{B}} C_{\dot{A}\dot{B}\dot{P}\dot{S}}. \tag{11.79}$$

Now substitute for $S_{\alpha\delta}{}^{\dot{A}\dot{B}}$ from (11.14) and use (11.8). After this, find from (11.77)

$$C_{A[BC]D} = \frac{1}{8} \, \epsilon_{AD} \epsilon_{BC} C_{\dot{P}\dot{S}}{}^{\dot{P}\dot{S}}. \tag{11.80}$$

Comparing this with (11.72) $C_{A[BC]D} = 0$ follows. □

11. Verify (11.36).

Hint.

$$\mathrm{Tr}(C^2) = C^{AB}{}_{CD} C^{CD}{}_{AB} \equiv C_{ABCD} C^{ABCD}, \tag{11.81}$$

so in consequence of (11.28) the symmetrisation of C^{ABCD} will occur automatically. Therefore, it suffices to explicitly write out the symmetrisation for C_{ABCD} and take from C^{ABCD} only one term, for example $\alpha^A \beta^B \gamma^C \delta^D$. Remember the first of (11.4).

12. Verify (11.37).

Hint. Write (11.37) as $C^{ABCD} C_{CDEF} C^{EF}{}_{AB}$ and note that the symmetrisations $C^{ABCD} C_{(CD)(EF)} C^{EF}{}_{(AB)}$ will occur automatically. Thus, for example, $\gamma_E \delta_F + \gamma_F \delta_E$ in $C_{(CD)(EF)}$ may be replaced by $2\gamma_E \delta_F$. This is a good example of usefulness of algebraic computer programs: my (A.K.'s) program Ortocartan (Krasiński 2001b) did this calculation in 0.47 s, preparing the data took about half an hour.

13. Verify (11.43).

Hint. In verifying this, (11.30) is helpful, and so is the observation (already made in connection with (11.14)): in two dimensions, every spinor antisymmetric in 2 indices must be proportional to the corresponding Levi-Civita symbol, e.g. $\beta_A \delta_B - \beta_B \delta_A = x\epsilon_{AB}$. Contracting this with ϵ^{AB} we find $x = -(\beta\delta)$. In consequence, $\beta^A \delta_B - \beta_B \delta^A = -(\beta\delta)\delta^A{}_B$.

14. Verify (11.44).

Hint. Substitute for $E_{\alpha\gamma}$ and $H_{\alpha\gamma}$ from (7.97) – (7.98), then for $C_{\alpha\beta\gamma\delta}$ from (11.23). Use (11.17), but only in the terms multiplied by i, and only for the $S_{..}^{\cdot\cdot}$ with indices AB and $\dot{A}\dot{B}$. Then use $\epsilon_{\alpha\beta\mu\nu}\epsilon^{\mu\nu\rho\sigma} = 2\delta^{\rho\sigma}_{\alpha\beta}$ and $\delta^{\lambda\kappa}_{\alpha\beta}S_{\lambda\kappa}{}^{AB} = 2S_{[\alpha\beta]}{}^{AB} = 2S_{\alpha\beta}{}^{AB}$.

15. Let the vector w^α be orthogonal to u^α and let

$$w^{AB} = u^\beta S_{\alpha\beta}{}^{AB} w^\alpha. \tag{11.82}$$

Verify that w^{AB} and $w^{\dot{A}\dot{B}}$ uniquely determine w^α.

16. Verify (11.45) and (11.46).

17. Verify (11.51).

18. Verify (11.54).

Hint. Express the Weyl tensor $C_{\beta\mu\nu\gamma}$ through its spinor image by (11.23) and use the following:

$$S_{\beta\mu EF}k^\mu = -\frac{1}{2}g_{\beta\dot{P}E}\alpha^{\dot{P}}\alpha_F - \frac{1}{2}g_{\beta\dot{P}F}\alpha^{\dot{P}}\alpha_E; \tag{11.83}$$

$$S^{\alpha\beta AB}g_{\beta\dot{P}E} = -\epsilon_{\dot{P}\dot{S}}\left(\delta^B{}_E g^{\alpha\dot{S}A} + \delta^A{}_E g^{\alpha\dot{S}B}\right), \tag{11.84}$$

$$g^{\alpha\dot{S}A}k_\alpha = \alpha^{\dot{S}}\alpha^A. \tag{11.85}$$

During the calculation watch out for symmetrisations that will be automatically enforced by the complete symmetry of C_{ABCD}. In the end, the first part of (11.23) will give zero contribution because it will create factors like $\alpha^{\dot{P}}\alpha_{\dot{P}} \equiv 0$, while the second part will be proportional to $C_{\dot{E}\dot{F}\dot{G}\dot{H}}\alpha^{\dot{E}}\alpha^{\dot{F}}\alpha^{\dot{G}}\alpha^{\dot{H}}$, which is zero by virtue of (11.45).

19. Verify, by any of the methods introduced in this chapter, that the Schwarzschild metric (11.11) is of Petrov type D.

Part II
The theory of gravitation

12

The Einstein equations
and the sources of a gravitational field

12.1 Why Riemannian geometry?

As argued in Section 1.4, gravitational forces can be simulated by inertial forces in accelerated motion. Special relativity describes relations between objects in uniform motion with respect to inertial frames, while gravitational interactions are neglected. The metric of the Minkowski spacetime in an inertial reference frame has constant components. If we transform that metric to an accelerated frame, its components become functions. Hence, a gravitational field should have the same effect: the metric of a spacetime with a gravitational field should have nonconstant components. Unlike in the Minkowski spacetime, in general it should not be possible to make the metric components constant by a coordinate transformation. This was, in great abbreviation, the basic observation that led Einstein (1916) to general relativity.

This idea had to be supplemented with equations that would generalise Newton's laws of gravitation and relate the metric to the gravitational field. The derivation of these equations, together with several related matters, is presented in this chapter.

12.2 Local inertial frames

Let us recall the conclusion of Chapter 1: the Universe is permeated by gravitational fields that cannot be screened. Their intensity can be decreased by going far from the sources, but one can never go below the minimum intensity determined by the local mean density of matter in the Universe. For this reason, no body in the Universe moves freely in the sense of Newton's mechanics. Consequently, inertial frames can be realised only with a limited precision. Moreover, there exists no natural standard of a straight line, so the departures of real motions from rectilinearity cannot be measured.

However, let us recall that for a body falling freely in a gravitational field the inertial force caused by the acceleration balances the gravitational force. Assume for a moment that the gravitational field is homogeneous. Then, two bodies falling freely in it will have the same acceleration all the time, so their relative acceleration will be zero and, relative to each other, they will either be at rest or move with a constant velocity. It is known from precise experiments that the gravitational mass is for all material bodies proportional to the inertial mass (Will, 2018), and the factor of proportionality is a universal constant (taken to be just 1). Hence, all bodies will experience the same acceleration in a given gravitational field. Consequently, the frame of reference connected with a body falling freely in a homogeneous gravitational field is inertial.

Such fields do not exist in Nature. However, along the trajectory of a body falling freely in a real gravitational field, the gravitational force and the inertial force cancel each other at every point. Hence, if the gravitational field is continuous (which is the case in all practical instances), then, in a *sufficiently small neighbourhood* of the falling body, the inertial forces will be arbitrarily small. Given the precision ε attainable in measurements, around a falling body B there will exist, at every point of its trajectory, a sphere of radius δ inside which the inertial force will be smaller than ε, i.e. unmeasureable. Inside that sphere the reference system defined by the body B will thus be 'practically' inertial. It is called the **local inertial frame** of B. (In fact, local inertial frames are defined not by individual bodies, but by their trajectories.) It is called 'local' because it differs from the universal inertial frame postulated in Newton's theory: there is an infinity of local inertial frames. It is easy to see that different local inertial frames will in general move with acceleration relative to each other. (For example, consider two bodies falling freely towards the Earth from opposite directions.) Thus, at a large distance from a freely falling body its local inertial frame ceases to be inertial.

12.3 Trajectories of free motion in Einstein's theory

Let us recall one more conclusion of Chapter 1: since no standard for a straight line exists in Nature, it will be simpler to assume that the geometry of our space is non-Euclidean, and in that geometry the trajectories of material bodies are free-motion trajectories. What was called 'gravitational field' in Newton's theory will be a consequence of the non-Euclidean geometry in which the free motions take place. By the argument of Section 12.1, the Riemann geometries should be appropriate. They do contain the Euclidean geometry as a special case and, because of the coordinate-independent formalism they use, are consistent with the postulate of equivalence of all reference systems.

The generalisation of the notion of a straight line that exists in a Riemann space is the geodesic line. If in a *proper* Riemann space (i.e. one with a positive-definite metric) the curvature goes to zero, the geometry goes over into the Euclidean geometry and the geodesic lines go over into straight lines. Hence, geodesic lines are natural candidates for the trajectories of free motion. In addition, they have the following property:

Theorem 12.1 *For a timelike geodesic G, the coordinates in a Riemann space can be chosen so that the Christoffel symbols vanish along G.*

The Christoffel symbols being zero means that near to G the gravitational field will be approximately zero. (It is not exactly zero because the derivatives of the Christoffel symbols will not vanish, and so the curvature will be nonzero.)

Proof:
Latin indices will label vectors; the lower case ones will run through the values $0, 1, 2, 3$, the upper case ones will run through the values $1, 2, 3$. Coordinate indices running through

the values $1, 2, 3$ will also be denoted by Latin letters. Wherever confusion might arise, vector indices will have a hat over them.

At a point $p_0 \in G$, we choose such a basis in the tangent space that $e_{\hat{0}}{}^{\alpha}(p_0)$ is tangent to G, and the other vectors, $e_A{}^{\alpha}(p_0)$, $A = 1, 2, 3$ are orthogonal to $e_{\hat{0}}{}^{\alpha}(p_0)$, i.e. $g_{\alpha\beta}e_{\hat{0}}{}^{\alpha}e_A{}^{\alpha}\big|_{p_0} = 0$. Then we define the bases $e_i{}^{\alpha}(p)$ in a neighbourhood of G as follows:

1. For $p \in G$, we transport $e_i{}^{\alpha}(p_0)$ parallely from p_0 to p along G.

2. For $p \notin G$, we draw through p a geodesic G'_p that intersects G orthogonally (see Fig. 12.1). Let $p' \in G$ be the point of intersection of G'_p and G. Then we transport the basis $e_i{}^{\alpha}(p')$ (already defined in point 1) parallely to p along G'_p.

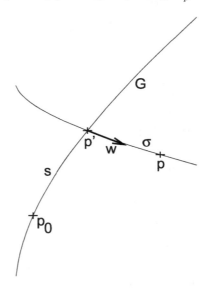

Fig. 12.1 Construction of the Fermi coordinates in which the Christoffel symbols vanish along a given timelike geodesic G. More explanation in the text.

This procedure works when there are no singular points on G'_p. It gives a unique result provided that p is not too distant from G, otherwise the geodesics orthogonal to G might intersect each other.

The following equations hold on G:

$$e_i{}^{\alpha}{}_{;\beta}\big|_G = 0, \qquad e^i{}_{\alpha;\beta}\big|_G = 0. \tag{12.1}$$

The first one follows because (i) $e_i{}^{\alpha}{}_{;\mu}e_{\hat{0}}{}^{\mu}\big|_G = 0$ in consequence of the basis $e_i{}^{\alpha}$ being transported parallely along G and $e_{\hat{0}}{}^{\mu}$ being tangent to G and (ii) $e_i{}^{\alpha}{}_{;\mu}e_S{}^{\mu}\big|_G = 0$ ($S = 1, 2, 3$) in consequence of $e_S{}^{\mu}$ at G being tangent to one of the G'_p used to transplant the bases parallely to other points. The second of (12.1) follows from $e_i{}^{\alpha}e^i{}_{\beta} = \delta^{\alpha}{}_{\beta}$.

Now, using (12.1) in (4.19), we obtain

$$\left\{ \begin{matrix} \alpha \\ \beta\gamma \end{matrix} \right\}\bigg|_G = e_s{}^{\alpha}e^s{}_{\beta,\gamma}\big|_G. \tag{12.2}$$

Since the Christoffel symbols are symmetric, this implies that

$$e^i{}_{\beta,\gamma}|_G = e^i{}_{\gamma,\beta}|_G . \tag{12.3}$$

We have defined the basis (up to rotations of $e_A{}^\alpha(p_0)$, $A = 1, 2, 3$), but we have not so far defined the coordinates. Let w^α be the unit tangent vector to G'_p, and let σ be the affine parameter on G'_p; $\sigma = 0$ for points on G. Let s be the affine parameter on G with $s = 0$ at p_0. For the point p we define the time coordinate as $x^0 = s$ and the space coordinates by

$$x^A = \sigma w^\rho(p') e^{\widehat{A}}{}_\rho(p'), \qquad A = 1, 2, 3. \tag{12.4}$$

In the coordinates (s, x^A) thus defined, called **Fermi coordinates** (Fermi 1922),

$$e_{\widehat{0}}{}^\alpha(p') = \left.\frac{\partial x^\alpha}{\partial s}\right|_G \underset{*}{=} \delta^\alpha{}_0, \tag{12.5}$$

since $e_{\widehat{0}}{}^\alpha$ is tangent to G, and

$$e_{\widehat{A}}{}^0(p') \underset{*}{=} 0, \qquad \widehat{A} = 1, 2, 3, \qquad w^0 \underset{*}{=} 0 \tag{12.6}$$

because $e_{\widehat{A}}{}^\alpha(p')$ and w^α lie in the subspace of the tangent space in which $x^0 = \text{constant}$. By (12.5) and (12.6) the matrix $e_i{}^\alpha(p')$ has a block form, so the inverse matrix $e^i{}_\alpha(p')$ must have a block form, too:

$$e^{\widehat{0}}{}_\alpha(p') \underset{*}{=} \delta^0{}_\alpha, \qquad e^{\widehat{A}}{}_0(p') \underset{*}{=} 0. \tag{12.7}$$

Now let us differentiate (12.4) by σ. Since $w^\rho(p')$ and $e^{\widehat{A}}{}_\rho(p')$ do not depend on σ, we have

$$w^A(\sigma) = \frac{\mathrm{d}x^A}{\mathrm{d}\sigma} = w^\rho(p') e^{\widehat{A}}{}_\rho(p') \underset{*}{=} w^A(p'), \tag{12.8}$$

which means that w^A does not depend on σ. In consequence of (12.6) and (12.7), this implies that $w^A(p') \underset{*}{=} w^K(p') e^{\widehat{A}}{}_K(p')$, $K = 1, 2, 3$, so

$$e^{\widehat{A}}{}_K(p') \underset{*}{=} \delta^{\widehat{A}}{}_K, \tag{12.9}$$

since $w^K(p')$ is an arbitrary unit 3-vector at p'. From (12.7) and (12.9) we then obtain

$$e^i{}_\alpha(p') \underset{*}{=} \delta^i{}_\alpha, \qquad e_i{}^\alpha(p') \underset{*}{=} \delta^\alpha{}_i. \tag{12.10}$$

This implies that $e_i{}^\alpha(p')$ is constant along G, so

$$e^i{}_{\alpha,\beta} e_{\widehat{0}}{}^\beta|_G = \frac{\partial}{\partial s} e^i{}_\alpha(p') \underset{*}{=} e^i{}_{\alpha,0}|_G = 0, \qquad e^i{}_{0,\alpha}|_G = 0. \tag{12.11}$$

(The last equation follows from the previous one by virtue of (12.3).)

Now note that

$$\frac{\mathrm{d}}{\mathrm{d}\sigma}\left(w^\alpha e^i{}_\alpha\right) \equiv \frac{D}{\mathrm{d}\sigma}\left(w^\alpha e^i{}_\alpha\right) = 0, \tag{12.12}$$

because w^α is tangent to a geodesic on which σ is the affine parameter, while $e^i{}_\alpha$ is by definition parallely transported along that geodesic. Moreover, $\mathrm{d}w^\alpha/\mathrm{d}\sigma = 0$ by (12.8), so

$$0 = \frac{\mathrm{d}}{\mathrm{d}\sigma}\left(w^\alpha e^i{}_\alpha\right) = w^\alpha \frac{\mathrm{d}e^i{}_\alpha}{\mathrm{d}\sigma} = w^\alpha w^\beta e^i{}_{\alpha,\beta} = w^A w^B e^i{}_{A,B}, \tag{12.13}$$

which means that $e^i{}_{(A,B)}\big|_G = 0$, because along G w^A is an arbitrary vector orthogonal to G. Then, from (12.3),

$$e^i{}_{A,B}\big|_G = 0. \tag{12.14}$$

Equations (12.11) and (12.14) imply that $e^i{}_{\alpha,\beta}\big|_G = 0$, so, from (12.2):

$$\left\{ \begin{matrix} \alpha \\ \beta\gamma \end{matrix} \right\}\bigg|_G \underset{*}{=} 0. \tag{12.15}$$

□

The physical meaning of (12.15) is that, in a sufficiently small neighbourhood of each $p' \in G$, other geodesics emanating from p' are, up to first-order terms, approximated by straight lines. This neighbourhood is thus a local inertial frame. This is one more suggestion that geodesics should be the trajectories of free motion in relativity.

12.4 Special relativity versus gravitation theory

We have so far dealt with the postulate that general relativity should reduce to Newton's kinematics of free motion in the limit of vanishing gravitational field. In between these extremes there is special relativity that describes the kinematics of free particles in the absence of gravitation, but takes into account velocities comparable to the velocity of light. In the previous sections we have thus discussed a two-stage limiting transition: with the gravitational field to zero and with the velocity of light to infinity. When we switch off the gravitational field, but put no limits on the velocities, the geometric theory of gravitation should reproduce special relativity.

The spacetime of special relativity is a 4-dimensional flat Riemann space of signature $(+ - - -)$. The Riemann space of general relativity should thus be of the same signature. This is because a change of signature means either a discontinuity in some metric components or at least one component of the metric passing through zero value, while we expect that the 'switching off' of the gravitational field can be done in a continuous way and does not lead through any singularities.

12.5 The Newtonian limit of general relativity

We assumed that with the gravitational field switched off, general relativity should reproduce special relativity and, in addition, with the velocity of light becoming infinite it should reproduce the Newtonian kinematics of free motion. A logical consequence of these two postulates is the requirement that when the velocity of light is made infinite while the gravitational field is still there, general relativity should reproduce Newton's theory of gravitation. Hence, the field equations of general relativity should, in the limit $c \to \infty$, reproduce the Poisson equation

$$\Delta\phi = 4\pi G\rho, \tag{12.16}$$

where ϕ is the gravitational potential and ρ is the density of matter generating the gravitational field.

12.6 Sources of the gravitational field

In the theory of gravitation that we are now constructing, the gravitational field should manifest its presence as nonflatness of the metric. Consequently, gravitation should be coded in the metric tensor. In four dimensions, in the most general case, the metric tensor has 10 independent components. Hence, we will need 10 equations to determine it.

The description of the sources of gravitation should thus be correspondingly elaborate. The source in the equations generalising (12.16) should be a quantity that is a generalisation of mass density. According to special relativity, the mass of a body depends on its energy. Hence, the energy of motion of a continuous medium should contribute to the gravitational field and so should the internal energy, e.g. pressure in a fluid and stresses in a solid.

In special relativity, the physical state of a continuous medium is described by the energy-momentum tensor $T_{\alpha\beta}$. In any chosen coordinate system, its component T_{00} is equal to the energy density ρc^2, the components T_{0I}, $I = 1, 2, 3$, form the 3-dimensional vector of energy stream through a unit area of surface orthogonal to the direction of flow, and the components T_{ij} form the stress tensor. The tensor $T_{\alpha\beta}$ is symmetric, and so has in general 10 independent components. Consequently, it is a natural candidate for the source in the field equations of gravitation.

In the Cartesian coordinates, the equations of motion are $T^{\alpha\beta}{}_{,\beta} \underset{*}{=} 0$. In any other coordinates then, these equations take the form

$$T^{\alpha\beta}{}_{;\beta} = 0. \tag{12.17}$$

The field equations of gravitation in which the energy-momentum tensor is the source must be consistent with (12.17).

12.7 The Einstein equations

Since we expect the gravitational field to be a manifestation of a nonflat geometry, it should be connected with the curvature tensor that contains second derivatives of the metric. If the left-hand side of the field equations depends on the curvature, then these equations will be of second order in the metric. This suggests that the components of the metric should be analogues of the Newtonian gravitational potential. If so, then the trajectories of motion (the geodesics) would be determined by the first derivatives of the potential, just like in Newton's theory.

The Riemann tensor is of rank 4, while the energy-momentum tensor is of rank 2. Hence, if the Riemann tensor were equal to a quantity constructed from the energy-momentum tensor, then the source of the gravitational field would be a quadratic function of matter density, which would complicate the transition to Newton's theory. We should thus rather equate a quantity constructed from the Riemann tensor to the energy-momentum tensor. We already know one good candidate, the Ricci tensor (7.46). It is symmetric and linear in the second derivatives of the metric. The field equations might thus read $R_{\alpha\beta} = \kappa T_{\alpha\beta}$, where κ is a constant coefficient. However, such equations are not consistent with (12.17) because in general $R^{\alpha\beta}{}_{;\beta} \neq 0$. But the Ricci tensor obeys an identity similar to (12.17), which is a consequence of the Bianchi identities (7.14). They may be written as

$$R^{\alpha}{}_{\beta\gamma\delta;\epsilon} + R^{\alpha}{}_{\beta\delta\epsilon;\gamma} + R^{\alpha}{}_{\beta\epsilon\gamma;\delta} = 0. \tag{12.18}$$

Contracting this equation with $\delta^{\gamma}{}_{\alpha}g^{\beta\delta}$ we obtain

$$G^{\alpha\beta}{}_{;\beta} = 0, \tag{12.19}$$

where

$$G^{\alpha\beta} \stackrel{\text{def}}{=} R^{\alpha\beta} - \frac{1}{2}g^{\alpha\beta}R \tag{12.20}$$

is called the **Einstein tensor**. It is symmetric, linear in second derivatives of the metric and obeys (12.19), which is identical in form to (12.17). Hence, $G^{\alpha\beta}$ is a better candidate for the left-hand side of the field equations, which should thus read

$$R_{\alpha\beta} - \frac{1}{2}g_{\alpha\beta}R = \kappa T_{\alpha\beta}. \tag{12.21}$$

These are the **Einstein equations**. The coefficient κ will be determined in Section 12.11 from the condition of correspondence of (12.21) to the Poisson equation (12.16).

In vacuum, the Einstein equations become $G^{\alpha\beta} = 0$, which is equivalent to $R^{\alpha\beta} = 0$. Then (7.48) – (7.49) show that outside the sources the Riemann tensor coincides with the Weyl tensor. Thus, the Weyl tensor represents that part of the gravitational field that propagates into vacuum. The Ricci tensor represents that part of curvature that is algebraically determined by matter and vanishes in vacuum.

It may be incredible that such nontrivial equations should be derived almost without calculations and without experimental hints, by a speculation that was in places non-unique. However, this is how it was; Einstein actually guessed his equations by a reasoning described above in great abbreviation (it had taken him about 10 years to arrive at them). Therefore, it is not easy to present a convincing, brief and logical derivation of (12.21). The most convincing route to them is Einstein's own. It was described in more detail by Mehra (1974) and also in papers by Einstein himself (Einstein et al., 1923).

Because of nonuniqueness of the reasoning that led to (12.21), Einstein's general relativity (GR) is not the only geometric theory of gravitation that can be built upon the Newton theory and special relativity. Nearly each one of the intuitive assumptions made on the way can be modified so that another set of equations will result. A few of those alternative theories will be briefly presented in Section 12.16. However, GR has successfully passed all experimental tests (Will, 2018), whereas the other theories were either proved wrong or else found experimentally to imply such small modifications of the GR results that it does not really pay off to abandon GR in their favour.

12.8 Hilbert's derivation of the Einstein equations

The derivation of Einstein's equations given in the previous section was based on the original reasoning of Einstein himself. However, David Hilbert had been working on deriving these equations simultaneously with Einstein by a different method. There was some exchange of information between them, but, taking things formally, Hilbert was the first to publish the correct result (see more details in Mehra (1974)).[1]

Hilbert proposed that all theories in mathematics and physics should be derived by deduction from sets of axioms. This programme contributed greatly to clarifying the logical structure of physical theories. However, the physical justification of some of the postulates is still lacking.

For the gravitation theory, Hilbert postulated to use a variational principle. The reasoning leading to his lagrangian, similarly to Einstein's reasoning, contains a few assumptions that are justified only intuitively. Most of the other geometric theories of gravitation were derived from variational principles, which confirms that this method is not unique, either.

Hilbert proposed the following axioms:

I. The field equations of gravitation should follow from a variational principle. The independent variables in the action integral should be the components of the metric tensor.

II. The action functional should be a scalar.

III. The Euler–Lagrange equations (i.e. the field equations) that will follow should be differential equations of second order in $g_{\mu\nu}$.

From this point on, further reasoning leads to Einstein's equations almost uniquely, but the postulates themselves leave several questions open. For example, some unanswered questions regarding axiom I are the following:

What is the geometrical and physical interpretation of the action functional?

What is the geometrical and physical meaning of the extremum found from the Euler–Lagrange equations?

Why should the field equations of gravitation be equivalent to an extremum of some abstract quantity?

Axiom II is justified by the observation that if an integral should be a tensorial object, then the integrand must be a scalar. Integrals of other tensorial objects are not tensors. Deriving covariant equations from a noncovariant action integral would thus be inconsistent with the underlying philosophy of the theory.

Axiom III makes use of the correspondence principle in the simplest way. Since the Poisson equation, which should follow from the geometric equations as a limit, is of second order, and most other physical theories are based on second-order equations, the field equations of gravitation should be of second order, too. However, this is not a unique argument either: theories of gravitation with field equations of fourth order do exist (see, for example, Capozziello and de Laurentis (2011)). In the limiting transition to Newton's theory, the terms with derivatives of order higher than 2 disappear.

The conclusions of Hilbert's axioms are the following:

---

[1] Hilbert presented the subclass of (12.21), corresponding to $T_{\alpha\beta} = 0$ and to $T_{\alpha\beta}$ being the energy-momentum tensor of the electromagnetic field, at the meeting of the Royal Academy of Sciences in Göttingen on 20 November 1915. Einstein presented Eqs. (12.21) at the meeting of the Prussian Academy of Sciences in Berlin on 25 November 1915. This game with dates is only a historical curiosity. Einstein was the unquestionable spiritual father of relativity, and Hilbert himself made that point repeatedly. Einstein had worked on relativity since about 1907 and had published several papers explaining the basic ideas and preliminary results. Hilbert joined in about 1913 and was influenced by Einstein's ideas from the beginning.

I. In a 4-dimensional spacetime, the action integral should be $\int_{V_4} \mathcal{H} d_4 x$, where V_4 is a certain 4-volume in the spacetime, on whose boundary ∂V_4 the variations $\delta g_{\mu\nu}$ vanish. The functional $\mathcal{H}$ depends on the metric tensor $g_{\mu\nu}$ and its derivatives $g_{\mu\nu,\lambda}$, $g_{\mu\nu,\lambda\sigma}$, etc.

II. If $\int_{V_4} \mathcal{H} d_4 x$ is to be a scalar, then $\mathcal{H}$ should be a scalar density of weight -1 because the volume element $d_4 x$ is a scalar density of weight $+1$. The simplest scalar density of weight -1 is $\sqrt{-g}$. Thus, $\mathcal{H} = \sqrt{-g} H$, where H is a proper scalar.

III. The order of the Euler–Lagrange equations is twice the order of the highest derivative in the action integral. It would thus be best if H were a function only of $g_{\mu\nu}$ and $g_{\mu\nu,\lambda}$. However, one cannot form a nontrivial scalar by combining algebraically $g_{\mu\nu}$ and $g_{\mu\nu,\lambda} = \left\{{\rho \atop \mu\lambda}\right\} g_{\nu\rho} + \left\{{\rho \atop \nu\lambda}\right\} g_{\mu\rho}$, since the Christoffel symbols are not tensors.

The second derivatives of $g_{\mu\nu}$, if present in H, will not give a contribution to the Euler–Lagrange equations if they can be collected into a divergence of an expression that vanishes on the boundary of the integration volume. This will be possible if $g_{\mu\nu,\rho\sigma}$ enter H linearly, i.e. are contracted only with terms that do not contain $g_{\mu\nu,\rho\sigma}$. Expressions linear in the Riemann tensor are of this form.

The only expressions that can be built from the Riemann tensor and are linear in $g_{\mu\nu,\rho\sigma}$ are $R^{\rho}{}_{\beta\rho\delta} = R_{\beta\delta} = -R^{\rho}{}_{\beta\delta\rho}$ and $g^{\alpha\beta} R_{\alpha\beta} = R$. Hence, $H = g^{\alpha\beta} R_{\alpha\beta} = R$ and

$$W_g = \int_{V_4} \sqrt{-g} R d_4 x. \tag{12.22}$$

More exactly, this should be the action integral for the left-hand side of the field equations. The full action integral should be

$$W = W_g + \int_{V_4} L d_4 x, \tag{12.23}$$

where $L = 0$ in vacuum.

Before we calculate $\delta W_g / \delta g_{\alpha\beta}$, let us note two auxiliary equations:

$$\begin{aligned} \delta g &= \frac{1}{3!} \epsilon^{\alpha_1 \ldots \alpha_4} \epsilon^{\beta_1 \ldots \beta_4} g_{\alpha_1\beta_1} g_{\alpha_2\beta_2} g_{\alpha_3\beta_3} \delta g_{\alpha_4\beta_4} \\ &= \frac{1}{3!} g \epsilon^{\alpha_1 \ldots \alpha_4} g^{\beta_1\rho_1} \ldots g^{\beta_4\rho_4} \epsilon_{\rho_1 \ldots \rho_4} g_{\alpha_1\beta_1} g_{\alpha_2\beta_2} g_{\alpha_3\beta_3} \delta g_{\alpha_4\beta_4} \\ &= \frac{1}{3!} g \delta^{\alpha_1 \ldots \alpha_4}_{\rho_1 \ldots \rho_4} \delta^{\rho_1}{}_{\alpha_1} \delta^{\rho_2}{}_{\alpha_2} \delta^{\rho_3}{}_{\alpha_3} g^{\beta_4\rho_4} \delta g_{\alpha_4\beta_4} = g g^{\alpha\beta} \delta g_{\alpha\beta}, \end{aligned} \tag{12.24}$$

$$\delta g^{\alpha\beta} = -g^{\alpha\rho} g^{\beta\sigma} \delta g_{\rho\sigma}. \tag{12.25}$$

In (12.24), we used (3.39), its derivative, (7.103), (3.32) and (3.37). Equation (12.25) follows by differentiating the equation $g^{\alpha\rho} g_{\beta\rho} = \delta^{\alpha}{}_{\beta}$.

Note that variations of the Christoffel symbols are differences between Christoffel symbols calculated from different metric tensors at the same point,

$$\delta \left\{{\alpha \atop \beta\gamma}\right\} = \left\{{\alpha \atop \beta\gamma}\right\} [g_{..} + \delta g_{..}] - \left\{{\alpha \atop \beta\gamma}\right\} [g_{..}]. \tag{12.26}$$

Hence, $\delta \left\{{\alpha \atop \beta\gamma}\right\}$ is a tensor because the nontensorial terms of (4.23) cancel out in the combination (12.26). Writing out the covariant derivative of $\delta \left\{{\alpha \atop \beta\gamma}\right\}$ and comparing it with $\delta R_{\alpha\beta}$, which is calculated from the definition of $R_{\alpha\beta}$, we find that

$$\delta R_{\alpha\beta} = -\left(\delta\left\{{\rho \atop \alpha\rho}\right\}\right)_{;\beta} + \left(\delta\left\{{\rho \atop \alpha\beta}\right\}\right)_{;\rho}. \tag{12.27}$$

Hence

$$
\begin{aligned}
\delta W_g &= \int_{\mathcal{V}_4} \mathrm{d}_4 x \left\{ -\frac{1}{2\sqrt{-g}}\, g g^{\alpha\beta}\delta g_{\alpha\beta} R - \sqrt{-g}\, g^{\alpha\rho} g^{\beta\sigma} R_{\alpha\beta}\delta g_{\rho\sigma} \right. \\
&\quad \left. + \sqrt{-g}\, g^{\alpha\beta}\left[-\left(\delta\left\{{\rho \atop \alpha\rho}\right\}\right)_{;\beta} + \left(\delta\left\{{\rho \atop \alpha\beta}\right\}\right)_{;\rho} \right] \right\}.
\end{aligned} \tag{12.28}
$$

The expression in the second line can be written as

$$-\left(\sqrt{-g}\, g^{\alpha\beta}\delta\left\{{\rho \atop \alpha\rho}\right\}\right)_{;\beta} + \left(\sqrt{-g}\, g^{\alpha\beta}\delta\left\{{\rho \atop \alpha\beta}\right\}\right)_{;\rho} \tag{12.29}$$

because $\sqrt{-g}$ and $g^{\alpha\beta}$ are covariantly constant. The expressions in parentheses are vector densities of weight -1. Thus, as shown in Section 4.1, their covariant divergences equal their ordinary divergences:

$$
\begin{aligned}
&\sqrt{-g}\, g^{\alpha\beta}\left[-\left(\delta\left\{{\rho \atop \alpha\rho}\right\}\right)_{;\beta} + \left(\delta\left\{{\rho \atop \alpha\beta}\right\}\right)_{;\rho}\right] \\
&= \left(-\sqrt{-g}\, g^{\alpha\beta}\delta\left\{{\rho \atop \alpha\rho}\right\} + \sqrt{-g}\, g^{\alpha\rho}\delta\left\{{\beta \atop \alpha\rho}\right\}\right)_{,\beta}
\end{aligned} \tag{12.30}
$$

(in the second term we interchanged ρ and β). Now we see that (12.30) is an ordinary divergence. Hence, from the Stokes theorem,

$$
\begin{aligned}
&\int_{\mathcal{V}_4}\left(-\sqrt{-g}\, g^{\alpha\beta}\delta\left\{{\rho \atop \alpha\rho}\right\} + \sqrt{-g}\, g^{\alpha\rho}\delta\left\{{\beta \atop \alpha\rho}\right\}\right)_{,\beta} \mathrm{d}_4 x \\
&= \int_{\partial\mathcal{V}_4}\left(-\sqrt{-g}\, g^{\alpha\beta}\delta\left\{{\rho \atop \alpha\rho}\right\} + \sqrt{-g}\, g^{\alpha\rho}\delta\left\{{\beta \atop \alpha\rho}\right\}\right) n_\beta \mathrm{d}_3 x,
\end{aligned} \tag{12.31}
$$

where n_β is the normal vector to the boundary $\partial\mathcal{V}_4$ of the region $\mathcal{V}_4$, while $\mathrm{d}_3 x$ is the volume element in $\partial\mathcal{V}_4$. From the assumption $\delta g_{\alpha\beta}(\partial\mathcal{V}_4) = 0$ it now follows that the integral of (12.29) is zero. Hence, in (12.28):

$$\delta W_g = \int_{\mathcal{V}_4} \mathrm{d}_4 x \left[\frac{1}{2}\sqrt{-g}\, g^{\alpha\beta} R - \sqrt{-g}\, R^{\alpha\beta}\right]\delta g_{\alpha\beta} = 0. \tag{12.32}$$

Thus, if we denote $\delta L/\delta g_{\alpha\beta} = -\kappa\sqrt{-g}\, T^{\alpha\beta}$, then the equation $\delta W = 0$ implies (12.21) by virtue of (12.23) and (12.32). Hilbert's method does not say what $T^{\alpha\beta}$ should be; he managed to specify it only for the electromagnetic field in vacuum.

Hilbert's variational principle works in deriving the Einstein equations in full generality. With a less-than-general metric, for example with symmetries, the variational principle can lead to equations with no relation to Einstein's. This is known to happen for a large subset of the Bianchi-type models, because of their spatial homogeneity. When all fiduciary metrics are spatially homogeneous, their variations are spatially homogeneous, too. Hence, they cannot be assumed to vanish on the boundary – they vanish either everywhere or nowhere. In consequence, the boundary terms cannot be neglected (see MacCallum (1979) for more details). No criterion is known for deciding when variational principles lead to correct field equations – see Sec. 13.6 for an example of their mischievous use.

12.9 The Palatini variational principle

Einstein's equations can be derived from a still more general variational principle, called the Palatini (1919) principle.[2] In this approach, the metric tensor and the connection coefficients are treated as two independent sets of variables. The connection is assumed only to be symmetric, $\Gamma^\alpha{}_{\beta\gamma} = \Gamma^\alpha{}_{(\beta\gamma)}$. The vanishing of coefficients of variations of $\Gamma^\alpha{}_{\beta\gamma}$ implies then (12.21), while the vanishing of coefficients of variations of the metric implies $\Gamma^\alpha{}_{\beta\gamma} = \left\{{\alpha \atop \beta\gamma}\right\}$, i.e. the covariant constancy of $g_{\alpha\beta}$.

12.10 The asymptotically Cartesian coordinates and the asymptotically flat spacetime

In the next section we will investigate the limiting transitions from Einstein's theory to Newton's theory and to special relativity. This section is a comment on the interpretation of those operations.

A flat space (spacetime) is a background in both of these theories. When we go far from the sources of the gravitational field, gravitation becomes ever weaker and disappears in the limit of infinite distance. In that limit, the metric tensor of general relativity should tend to the flat metric of special relativity. This is a 'thought experiment' because, as mentioned earlier, in the real Universe the gravitational force can never be smaller than the value determined by the local average matter density in the Universe.

The spacetime whose geometry becomes flat in the limit of large distances from the sources of the gravitational field is called **asymptotically flat**. Only in such spacetimes can we consider the Newtonian limit of general relativity. The coordinates that, in the same limit, go over into the Cartesian coordinates are called **asymptotically Cartesian**.

12.11 The Newtonian limit of Einstein's equations

This section presents the Newtonian limit of Einstein's equations by a simple old-fashioned method. A rigorous treatment of this topic can be found in Ehlers (1981).

From Section 12.5 it follows that in the limit $c \to \infty$ general relativity should reduce to Newton's theory of gravitation. Hence, in the same limit, the equation of a geodesic should reduce to the Newtonian equation of motion of a particle in a gravitational field,

$$m\frac{\mathrm{d}v_I}{\mathrm{d}t} = -m\phi_{,I}\,, \tag{12.33}$$

where v_I is the velocity of the particle and ϕ is the gravitational potential. We will deal with the field equations later in this section. Investigating the limit for the equations

---

[2] This is how this approach is commonly called. Ferraris, Francaviglia and Reina (1982) argued that it was not Palatini (1919) who invented it, but Einstein (1925).

themselves would be difficult. It will be simpler to do it for the lagrangians. Newton's equations of motion follow from the variational principle

$$\delta \int_{t_1}^{t_2} L dt = 0, \qquad L \stackrel{\text{def}}{=} \frac{1}{2} m \delta_{IJ} v^I v^J - m\phi, \qquad (12.34)$$

while the equations of a geodesic follow from

$$\delta \int_{t_1}^{t_2} \mathcal{L} dt = 0, \qquad \mathcal{L} \stackrel{\text{def}}{=} \frac{ds}{dt}. \qquad (12.35)$$

The Euler–Lagrange equations do not change when the Lagrange function is multiplied by a constant or when a constant is added to it. Hence, the condition of correspondence is

$$\lim_{c \to \infty} (C_1 \mathcal{L} + C_2) = L, \qquad (12.36)$$

where C_1 and C_2 are constants, as yet unknown. The Newtonian lagrangian has the dimension of energy. The relativistic lagrangian is

$$\mathcal{L} = \frac{ds}{dt} = c\sqrt{g_{00} + 2g_{0I}\frac{v^I}{c} + g_{IJ}\frac{v^I v^J}{c^2}}, \qquad (12.37)$$

where $I, J = 1, 2, 3$ and $v^I = dx^I/dt$ is the Newtonian velocity of the particle. Since the expression under the square root is dimensionless, the dimension of $\mathcal{L}$ is c. Hence, in order that the dimensions of both sides in (12.36) are the same, we must have $C_1 = \alpha mc$, where α is a dimensionless coefficient.

The Newtonian lagrangian contains the kinetic energy of the particle. In special relativity, the kinetic energy is a part of the total energy $mc^2/\sqrt{1 - (v/c)^2}$. Hence, the constant C_2 in (12.36) must compensate for the rest energy contained in $C_1 \mathcal{L}$ so that $(C_1 \mathcal{L} + C_2)$ contains only the kinetic energy. We do not know yet with what sign the rest energy will be contained in $C_1 \mathcal{L}$, so we look for C_2 in the form $C_2 = \beta mc^2$, $\beta = \pm 1$, and the sign of β will follow later. Finally, (12.36) becomes

$$\lim_{c \to \infty} mc^2 \left(\alpha\sqrt{g_{00} + 2g_{0I}\frac{v^I}{c} + g_{IJ}\frac{v^I v^J}{c^2}} + \beta \right) = \frac{1}{2}\delta_{IJ}mv^I v^J - m\phi. \qquad (12.38)$$

The Taylor formula applied to the square root up to terms of second degree in v^I/c gives

$$\sqrt{g_{00} + 2g_{0I}\frac{v^I}{c} + g_{IJ}\frac{v^I v^J}{c^2}} = \sqrt{g_{00}} + \frac{g_{0I}}{\sqrt{g_{00}}}\frac{v^I}{c}$$
$$+ \frac{1}{2}\left(\frac{g_{IJ}}{\sqrt{g_{00}}} - \frac{g_{0I}g_{0J}}{g_{00}^{3/2}} \right)\frac{v^I v^J}{c^2} + O\left(\frac{v^3}{c^3} \right). \qquad (12.39)$$

There are no terms linear in v^I/c on the right-hand side in (12.38), so

$$\frac{g_{0I}}{\sqrt{g_{00}}} = O\left(\frac{1}{c^2} \right). \qquad (12.40)$$

Then, $(-g_{0I}g_{0J}/g_{00}^{3/2})$ in (12.39) becomes a correction of order $1/c^4$ to $g_{IJ}/\sqrt{g_{00}}$. Hence, using (12.39) and (12.40) we obtain in (12.38)

$$\lim_{c\to\infty} \left[c^2 \left(\alpha\sqrt{g_{00}} + \beta \right) + \alpha \frac{\tilde{g}_{IJ}v^I v^J}{2\sqrt{g_{00}}} + O\left(\frac{1}{c}\right) \right] = \frac{1}{2}\, \delta_{IJ}v^I v^J - \phi, \tag{12.41}$$

where $\tilde{g}_{IJ} = g_{IJ} - O\left(1/c^4\right)$. The equation above should be an identity in v^I. Equating the terms of corresponding orders in v^I we obtain

$$c^2 \left(\alpha\sqrt{g_{00}} + \beta \right) = -\phi + O\left(\frac{1}{c}\right), \tag{12.42}$$

$$\frac{\alpha}{\sqrt{g_{00}}}\, \tilde{g}_{IJ} = \delta_{IJ} + O\left(\frac{1}{c}\right). \tag{12.43}$$

From (12.42) we have

$$\sqrt{g_{00}} = \frac{1}{\alpha} \left[-\beta - \frac{\phi}{c^2} + O\left(\frac{1}{c^3}\right) \right]. \tag{12.44}$$

We will now impose the condition that, in the limit of vanishing gravitation, the lagrangian (12.35) goes over into the lagrangian for geodesic motion in special relativity:

$$\frac{1}{\alpha} \left[-\frac{\phi}{c^2} + O\left(\frac{1}{c^3}\right) \right] \xrightarrow[\phi\to 0]{} 0. \tag{12.45}$$

For $\phi = 0$ we should obtain the Minkowski metric, hence $-\beta/\alpha \xrightarrow[\phi\to 0]{} 1$. But α and β are constants, so $-\beta/\alpha = 1$. In the signature $(+ - - -)$, (12.43) implies $\alpha < 0$, while $\beta = \pm 1$ by definition, so $\alpha = -1$, $\beta = +1$. The final result is thus, from (12.44), (12.40) and (12.43):

$$g_{00} = 1 + \frac{2\phi}{c^2} + O\left(\frac{1}{c^3}\right), \qquad g_{0I} = O\left(\frac{1}{c^2}\right), \qquad g_{IJ} = -\delta_{IJ} + O\left(\frac{1}{c}\right). \tag{12.46}$$

So far, we have investigated the Newtonian and special relativistic limits of the equations of motion. Now we shall investigate the limit $c \to \infty$ for the field equations.

From the equation $g_{\alpha\rho}g^{\beta\rho} = \delta^\beta{}_\alpha$, taking the cases $\{\alpha = 0, \beta = I \neq 0\}$, $\{\alpha = \beta = 0\}$ and $\{\alpha = I \neq 0, \beta = J \neq 0\}$, we conclude that

$$g^{00} = 1 - \frac{2\phi}{c^2} + O\left(\frac{1}{c^3}\right), \qquad g^{0I} = O\left(\frac{1}{c^2}\right), \qquad g^{IJ} = -\delta^{IJ} + O\left(\frac{1}{c}\right). \tag{12.47}$$

Using these in the formulae for the Christoffel symbols we obtain[3]

$$\begin{Bmatrix} 0 \\ 00 \end{Bmatrix} = O\left(\frac{1}{c^3}\right), \tag{12.48}$$

$$\begin{Bmatrix} 0 \\ 0I \end{Bmatrix} = \frac{\phi_{,I}}{c^2} + O\left(\frac{1}{c^3}\right) = \begin{Bmatrix} I \\ 00 \end{Bmatrix}, \tag{12.49}$$

$$\begin{Bmatrix} 0 \\ IJ \end{Bmatrix} = O\left(\frac{1}{c^2}\right) = \begin{Bmatrix} I \\ 0J \end{Bmatrix}, \tag{12.50}$$

$$\begin{Bmatrix} I \\ JK \end{Bmatrix} = O\left(\frac{1}{c}\right). \tag{12.51}$$

---

[3] Note that $x^0 = ct$, so each differentiation by x^0 changes $O\left(1/(c^k)\right)$ to $O\left(1/(c^{k+1})\right)$.

Hence, furthermore,

$$R_{00} = \frac{\phi_{,II}}{c^2} + O\left(\frac{1}{c^3}\right) \qquad \text{(sum over } I\text{)}, \tag{12.52}$$

$$R_{0I} = O\left(\frac{1}{c^2}\right), \qquad R_{IJ} = O\left(\frac{1}{c}\right). \tag{12.53}$$

The Einstein equations (12.21) can be written in the equivalent form

$$R_{\alpha\beta} = \kappa\left(T_{\alpha\beta} - \frac{1}{2}g_{\alpha\beta}T\right), \qquad T \overset{\text{def}}{=} g^{\alpha\beta}T_{\alpha\beta}. \tag{12.54}$$

How should the right-hand side of (12.54) behave in the Newtonian limit? Since T_{00} is the energy density, it contains a contribution from the rest energy. All other contributions to T_{00} must be by at least one order in c smaller than that, so we expect that $T_{00} = \rho c^2 + O(c)$, where ρ is the mass density. The components T_{0I} form the vector of energy stream. In the limit $c \to \infty$, motion of matter should have no influence on the gravitational field it generates.[4] Consequently, T_{0I} should be by at least one order in c smaller than T_{00} so that they become negligible in the limit $c \to \infty$, thus $T_{0I} = O(c)$. The components T_{IJ} describe the stress energy density. Compared to the rest energy, any other kind of energy is by two orders in c smaller, so $T_{IJ} = O(1)$. Hence, finally,

$$T = g^{\alpha\beta}T_{\alpha\beta} = \rho c^2 + O(c). \tag{12.55}$$

From this,

$$R_{00} = \kappa\left(\frac{1}{2}\rho c^2 + O(c)\right), \tag{12.56}$$

$$R_{0I} = \kappa O(c), \qquad R_{IJ} = \kappa O(c^2). \tag{12.57}$$

From (12.56) and (12.52) we have

$$\frac{\phi_{,II}}{c^2} + O\left(\frac{1}{c^3}\right) = \kappa\left(\frac{1}{2}\rho c^2 + O(c)\right). \tag{12.58}$$

In the limit $c \to \infty$ this should be equivalent to the Poisson equation (12.16), so

$$\kappa = \frac{8\pi G}{c^4}. \tag{12.59}$$

The remaining equations, (12.57), impose limitations on the terms $O(1/c^2)$ in g_{0I} and $O(1/c)$ in g_{IJ}. These become meaningful (and can be read out) if we consider the Einstein theory as a small perturbation imposed on the Newton theory in a weak gravitational field. There are various approaches to this application of general relativity, for example the Parametrised Post-Newtonian (PPN) formalism (Misner, Thorne and Wheeler, 1973; Will, 2018). We will briefly describe one such approach (the weak-field approximation) in Section 12.18. Formally, in the limit $c \to \infty$, Eqs. (12.57) are fulfilled identically.

In the equation $R_{00} = \kappa\left(T_{00} - \frac{1}{2}g_{00}T\right)$ we had to include terms of order $1/c^2$ because

---

[4] If the mass distribution does not change with time, then the motion of matter is not detectable via its Newtonian gravitational field. For example, the exterior gravitational field of a mass in axisymmetric rotation is not distinguishable from the field generated by a static mass with the same mass distribution.

the coordinate was $x^0 = ct$, so every differentiation by x^0 introduced the factor $1/c$. In order to liberate R_{00} from the factor $1/c^2$ thus introduced, we had to multiply (12.56) by c^2. To achieve the same in the first equation of (12.57) we would have to multiply it by c, and then we would obtain the identity $0 = 0$ in the limit $c \to \infty$.

12.12 Examples of sources in the Einstein equations: perfect fluid and dust

A fluid whose pressure obeys the Pascal law, while the transport of energy occurs only by means of mass flow is called a **perfect fluid**. It does not conduct heat or electric current, and its viscosity is zero. From this definition, we will now deduce the form of its energy-momentum tensor.

If u^α is the velocity field of an arbitrary continuous medium, while s is the proper time on the flow lines of that medium, then

$$\mathrm{d}s^2 = g_{\alpha\beta}\mathrm{d}x^\alpha \mathrm{d}x^\beta, \qquad u^\alpha = \mathrm{d}x^\alpha/\mathrm{d}s. \tag{12.60}$$

These imply

$$g_{\alpha\beta}u^\alpha u^\beta = 1. \tag{12.61}$$

This holds *for every continuous velocity field* and for any medium.

Now let us choose, for a while, the coordinates adapted to u^α in which $u^\alpha \stackrel{*}{=} \delta^\alpha{}_0$, i.e. $x^0 = s$, and the spatial coordinates of the particles of the fluid $x^I(s)$ obey $\mathrm{d}x^I(s)/\mathrm{d}s \stackrel{*}{=} 0$, $I = 1, 2, 3$. In these coordinates the particles of the fluid do not move with respect to the timelike hypersurfaces of the coordinate system. Such coordinates are called **comoving**, and they are frequently used in relativistic hydrodynamics. Their construction can be visualised as follows. We choose an arbitrary hypersurface S that intersects all flow lines of the fluid, and an arbitrary coordinate system within S. To each particle of the fluid we then assign the spatial coordinate x^I of the point where it crossed S. To assign the time-coordinate to a point p in spacetime we take the flow line C_p that passes through p. The time-coordinate of p is the proper time that elapsed between the event of C_p intersecting S and the event of C_p passing through p.

By definition, $T_{00}(p)$ is, in any coordinate system, the energy density measured at a given point p. In the comoving coordinates, the only energy that a particle can have is its inner energy $\epsilon = $ (rest energy) + (energy of thermal motion of its particles) + (chemical energy). Hence, $T_{00} \stackrel{*}{=} \epsilon$ and, at the same time, $T_{00} \stackrel{*}{=} T_{\alpha\beta}u^\alpha u^\beta$, so

$$T_{\alpha\beta}u^\alpha u^\beta = \epsilon. \tag{12.62}$$

This is now an equality of two scalars, so it holds in any coordinates.

By definition, $T^I{}_0$, $I = 1, 2, 3$ is the vector of the energy stream. But in the comoving coordinates, by the definition of a perfect fluid, there are no energy flows, so consequently $T^I{}_0 \stackrel{*}{=} T^I{}_\beta u^\beta \stackrel{*}{=} 0$. This implies that $T^\alpha{}_\beta u^\beta \stackrel{*}{=} \lambda u^\alpha$, where λ is an unknown coefficient. From (12.61) and (12.62) we find that $\lambda = \epsilon$, so

$$T^\alpha{}_\beta u^\beta = \epsilon u^\alpha, \tag{12.63}$$

which is again a tensor equation.

Now choose a point q in the spacetime and a vector $v^\alpha(q)$ at q orthogonal to $u^\alpha(q)$:

$$u^\alpha(q)v_\alpha(q) = 0. \tag{12.64}$$

The vector $v^\alpha(q)$ points from q toward a neighbouring particle. Since (12.64) says that the projection of the velocity on $v^\alpha(q)$ is zero, it follows that the particle to which $v^\alpha(q)$ points does not move relative to q. The linearly independent vectors having the property (12.64) thus determine a 3-dimensional volume element comoving with the particle that was at q. Consequently, the Pascal law must apply in this volume element: the pressure p exerted on the surface element σ in the fluid creates the force $f = p\sigma$ in the direction n^I, $I = 1, 2, 3$, orthogonal to σ. The pressure p and the force f do not depend on the direction of n^I, i.e. on the orientation of σ within the fluid. Let $-T^I{}_J$ denote the Newtonian (3-dimensional) stress tensor. (The minus sign is a consequence of the signature $(+ - - -)$; in this signature the spatial part of the energy-momentum tensor $T^\alpha{}_\beta$ is not the stress tensor $\tau^I{}_J$ itself, but $-\tau^I{}_J$.) By the definition of the stress tensor, the following must hold:

$$- T^I{}_J \sigma n^J = f n^I \equiv p\sigma n^I, \tag{12.65}$$

which implies

$$T^I{}_J n^J = -p n^I. \tag{12.66}$$

The n^I was an arbitrary vector in the 3-dimensional subspace orthogonal to $u^\alpha(q)$. Equation (12.66) shows that every such vector is an eigenvector of the matrix $T^I{}_J$ connected with the eigenvalue $(-p)$, which implies

$$T^I{}_J = -p\delta^I{}_J. \tag{12.67}$$

The vectors n^I in (12.65) – (12.67) are in general not orthogonal to hypersurfaces, and so cannot define a coordinate system. However, at every point q of the manifold, in the tangent space we can choose a subspace orthogonal to u^α, and choose in that subspace the orthonormal basis $e_{\hat{I}}{}^\alpha$, $I = 1, 2, 3$. In that basis, (12.67) will be fulfilled, where

$$T_{\hat{I}\hat{J}} = e_{\hat{I}}{}^\alpha e_{\hat{J}}{}^\beta T_{\alpha\beta}. \tag{12.68}$$

(In other words, we choose coordinates that are adapted to the basis vectors $e_{\hat{I}}{}^\alpha$ at one point q only, which is always possible.) Now, from (12.62) – (12.63), (12.67) and (12.68) we will deduce the general formula for the energy-momentum tensor of a perfect fluid. Let us choose an orthonormal tetrad $e_i{}^\alpha$, $i = 0, 1, 2, 3$, in spacetime such that $e_{\hat{0}}{}^\alpha = u^\alpha$, while each $e_{\hat{I}}{}^\alpha$, $I = 1, 2, 3$, obeys (12.66). Then

$$g_{\alpha\beta}e_i{}^\alpha e_j{}^\beta = \eta_{ij} = \mathrm{diag}(+1, -1, -1, -1) \tag{12.69}$$

and from (12.62) – (12.63), (12.68) and (12.69) we have for the tetrad components of $T_{\alpha\beta}$:

$$\begin{aligned}
T_{\hat{0}\hat{0}} &= T_{\alpha\beta}u^\alpha u^\beta = \epsilon, & \tag{12.70} \\
T_{\hat{0}\hat{A}} &= T_{\alpha\beta}u^\alpha e_{\hat{A}}{}^\beta = \epsilon u_\beta e_{\hat{A}}{}^\beta = 0, & A = 1, 2, 3, \tag{12.71} \\
T_{\hat{A}\hat{B}} &= -T^{\hat{A}}{}_{\hat{B}} = p\delta^{\hat{A}}{}_{\hat{B}} = -p\eta_{\hat{A}\hat{B}}. & \tag{12.72}
\end{aligned}$$

Applying the inverse projection (4.16) we obtain

$$
\begin{aligned}
T_{\alpha\beta} &= e^i{}_\alpha e^j{}_\beta T_{ij} = u_\alpha u_\beta T_{\hat{0}\,\hat{0}} + e^{\hat{A}}{}_\alpha e^{\hat{B}}{}_\beta T_{\hat{A}\,\hat{B}} \\
&= \epsilon u_\alpha u_\beta - p\eta_{\hat{A}\,\hat{B}} e^{\hat{A}}{}_\alpha e^{\hat{B}}{}_\beta - p u_\alpha u_\beta + p u_\alpha u_\beta \\
&= (\epsilon + p) u_\alpha u_\beta - p g_{\alpha\beta}.
\end{aligned}
\tag{12.73}
$$

A perfect fluid whose pressure is identically zero is called **dust**. It follows that for dust

$$
T_{\alpha\beta} = \epsilon u_\alpha u_\beta.
\tag{12.74}
$$

12.13 Equations of motion of a perfect fluid

The equations of motion of matter, $T^{\alpha\beta}{}_{;\beta} = 0$, for a general perfect fluid are equivalent to

$$
(\epsilon + p)_{,\beta}\, u^\alpha u^\beta + (\epsilon + p) u^\alpha{}_{;\beta}\, u^\beta + (\epsilon + p) u^\alpha u^\beta{}_{;\beta} - p_{,\beta}\, g^{\alpha\beta} = 0.
\tag{12.75}
$$

The identity (12.61) implies

$$
u^\alpha u_{\alpha;\beta} = 0.
\tag{12.76}
$$

Contracting (12.75) with u_α and using (12.61) and (12.76) we obtain

$$
\epsilon_{,\beta}\, u^\beta + (\epsilon + p) u^\beta{}_{;\beta} = 0.
\tag{12.77}
$$

This is the energy conservation equation which says that the volume work $-p u^\beta{}_{;\beta}$ generates the energy stream ϵu^β. Now using (12.77) in (12.75) we obtain

$$
(\epsilon + p) u^\alpha{}_{;\beta}\, u^\beta - p_{,\beta}\, g^{\alpha\beta} + p_{,\beta}\, u^\beta u^\alpha = 0.
\tag{12.78}
$$

These are the general relativistic equations of motion of a perfect fluid. In the Newtonian limit $(c \to \infty)$ and in asymptotically Cartesian coordinates, they go over into the Euler equations of motion $\rho d\mathbf{v}/dt = -\text{grad} p$.

Equations (12.77) and (12.78) simplify for dust; (12.77) then becomes

$$
\left(\epsilon u^\beta\right)_{;\beta} = 0,
\tag{12.79}
$$

which is the relativistic equation of continuity (mass conservation) that in the Newtonian limit goes over into $\partial\rho/\partial t + \text{div}(\rho\mathbf{v}) = 0$. Equation (12.78) becomes $u^\alpha{}_{;\beta}\, u^\beta = 0$. This means that the covariant derivative of u^α along itself is zero, so, by Sec. 5.2, u^α is tangent to affinely parametrised geodesics. Consequently, *dust moves along geodesics*, and *the proper time of the particles of dust is an affine parameter on these geodesics*.

In fact, the necessary and sufficient condition for geodesic motion of a perfect fluid is weaker than $p = 0$. As seen from (12.78), it is $p_{,\beta}\left(g^{\alpha\beta} - u^\alpha u^\beta\right) = 0$. The operator $h^{\alpha\beta} = g^{\alpha\beta} - u^\alpha u^\beta$ projects tensors on the hypersurface elements orthogonal to the vector field u^α. The equation $h^{\alpha\beta} p_{,\alpha} = 0$ means that the gradient of pressure is collinear with the velocity field, i.e. there are no spatial gradients of pressure in the comoving coordinates.

12.14 The cosmological constant

When general relativity was created, it was clear that its predictions would significantly differ from those of Newton's theory in two situations:

(1) in strong gravitational fields, when the extent of the regions where local inertial frames exist becomes small; and

(2) in large sub-volumes of the Universe, where even small departures of the real geometry from Euclidean cumulate over large distances and become visible in observations of distant objects.

Consequently, when Einstein thought about physical applications of his theory, one of the first things he tried was to construct a model of the Universe. At that time everybody was *sure* that the geometry and matter distribution in the Universe are time- and space-independent. Calculations showed, however, that these assumptions are self-contradictory – such a model of the Universe did not exist in relativity.

At that moment, Einstein was on the verge of making another great discovery. The year was 1916, and the first observations proving the expansion of the Universe were published only in 1927. Yet this time Einstein's belief in prejudice turned out to be stronger than his tendency to think boldly. While searching for a reason of the failure of his first attempt, Einstein did not question the *assumption* that the Universe is static, but turned his suspicion against his equations (12.21). He soon found a gap in his reasoning: the left-hand side of the field equations that has all the required properties need not necessarily be the Einstein tensor $R_{\mu\nu} - \frac{1}{2}g_{\mu\nu}R$. One can add to it any symmetric tensor $H_{\mu\nu}$ whose covariant divergence vanishes and that does not depend on the derivatives of the metric. Such a correction will not increase the order of the field equations, while the equation $\left(R^{\mu\nu} - \frac{1}{2}g^{\mu\nu}R + H^{\mu\nu}\right)_{;\nu} = 0$ will still hold. The simplest tensor of this property is $\Lambda g^{\mu\nu}$, where Λ should be a universal constant. The modified Einstein equations

$$R^{\mu\nu} - \frac{1}{2}g^{\mu\nu}R + \Lambda g^{\mu\nu} = \kappa T^{\mu\nu} \tag{12.80}$$

in the limit $c \to \infty$ go over into the modified Poisson equation

$$\Delta\phi - 2\Lambda c^2 = 4\pi G\rho. \tag{12.81}$$

Hence, the **cosmological constant** Λ describes an effect absent from the ordinary Newton theory: a universal attraction (for $\Lambda > 0$) or repulsion (for $\Lambda < 0$) of matter particles.

The modified equations (12.80) allowed for the existence of a static, homogeneous and isotropic solution of the form

$$ds^2 = c^2 dt^2 - R^2 d\chi^2 - R^2 \sin^2\chi \left(d\vartheta^2 + \sin^2\vartheta d\varphi^2\right), \tag{12.82}$$

where

$$R = \frac{1}{\sqrt{-\Lambda}} = \frac{c}{2\sqrt{\pi G\rho}}, \tag{12.83}$$

ρ being the average mass density in the Universe, constant by assumption. The spacetime corresponding to the metric (12.82) is, by tradition, still called the **Einstein Universe**, although it is no longer considered a model of the real Universe. Since $\Lambda < 0$ here, the 'cosmological repulsion' balances the gravitational attraction, which allows the system to be static (but unstable, as we will see in Section 17.7).

This brief history of the introduction of the cosmological constant shows that it appeared in consequence of an error. Had Einstein not insisted on obtaining a static model of the Universe, he would have had a chance to predict that the Universe should expand or collapse, 11 years before the expansion was observed. When he realised later how close he was to making that prediction, he said that the introduction of the constant was 'the biggest blunder of his life' (Misner, Thorne and Wheeler, 1973, pp. 410–411).[5]

Today, the cosmological constant is more popular among particle physicists than among specialists in relativity, and the observations of brightness of distant supernovae are even said to indicate that Λ is strictly negative; see Section 17.13. Its absolute value must be very small (less than 10^{-50} cm^{-2} (Misner, Thorne and Wheeler, 1973, pp. 410–411)), so it can play a role only in the evolution of the Universe. In the Solar System, it has no observable influence on the motion of planets. A great number of solutions of the modified Einstein equations (12.80) is known (Stephani et al., 2003, section 24.2), both inside matter distributions and in vacuum. Examples will be given later in this text (see Section 14.4).

12.15 An example of an exact solution of Einstein's equations: a Bianchi type I spacetime with dust source

The Einstein equations had for many years been rumoured to be very difficult, almost too difficult to find any exact solution. This opinion lingers until today, although it ceased to have a basis long ago. No formula for a general solution of Einstein's equations has been found that would generalise $\phi(\mathbf{x}) = -\int \mathrm{d}_3 x' G\rho(\mathbf{x}')/|\mathbf{x} - \mathbf{x}'|$, which is the formal solution of the Poisson equation. The reason of the difficulty is the nonlinearity of the Einstein equations. In general relativity, the superposition of two solutions is not their sum and a general law of composition of solutions is not known.[6] Consequently, finding a solution for a simple situation does not help much in looking for more general solutions, except that it increases the basis of experience and knowledge. Nevertheless, for special cases (like high symmetry, special properties of the source or of the Weyl tensor), hundreds if not thousands of solutions are known, and any attempt to list and compare them is a major undertaking (Stephani et al., 2003; Krasiński, 1997). In order to prove to the reader that the task of finding a solution is not at all hopeless and to demonstrate a few characteristic methods of calculation, we shall present a derivation of a certain exact solution.[7]

We assume that the spacetime is spatially homogeneous, with the symmetry group of

---

[5] That story has a complicated continuation. Hubble, who is credited with the discovery of the expansion of the Universe (Hubble, 1929), had not believed, until the end of his life, that the Universe is actually expanding. He insisted, even in his last paper that appeared in print after his death (Hubble, 1953), that expressing the observed redshifts in spectra of galaxies through their equivalent velocities of recession is merely a convenient mathematical device (Krasiński and Ellis, 1999). Then, Milne and McCrea (1934) (see Section 17.6) showed that expansion of the Universe could and should have been predicted on the basis of Newton's theory of gravitation in the eighteenth century since all the mathematical knowledge necessary for that purpose had existed already at that time. The prediction was not made because nobody tried – just because everybody was sure that the Universe is static.

[6] But methods of generating new solutions from known ones are worked out for large classes of metrics, for example for the stationary-axisymmetric ones – see Belinskii and Verdaguer (2001).

[7] The derivation is borrowed from Stephani (1990) and a little re-edited.

Bianchi type I, and that the source in the equations is dust. The cosmological constant is assumed zero. We thus begin with the metric (10.29). For a Bianchi type I algebra all generators commute to zero. Hence (see Exercise 5 in Chapter 8) we can choose the spatial coordinates (x, y, z) so that the generators become

$$\underset{(I)}{k^\alpha} = \delta^\alpha{}_I, \qquad x^1 = x, \quad x^2 = y, \quad x^3 = z \tag{12.84}$$

(which means that x^I is a parameter on the curve tangent to $\underset{(I)}{k^\alpha}$ and obeys $\underset{(I)}{k^\alpha} = \partial x^\alpha/\partial x^I$).
For the generators (12.84), the Killing equations are

$$g_{IJ,K} = 0, \qquad I, J, K = 1, 2, 3, \tag{12.85}$$

i.e. the g_{IJ} depend only on t. Knowing this, we find for the Christoffel symbols, the Riemann tensor and the Ricci tensor (the components not listed are zero):

$$\begin{Bmatrix} 0 \\ IJ \end{Bmatrix} = \frac{1}{2}g_{IJ,t}, \qquad \begin{Bmatrix} I \\ 0J \end{Bmatrix} = \frac{1}{2}g^{IS}g_{JS,t}; \tag{12.86}$$

$$R^0{}_{I0J} = \frac{1}{2}g_{IJ,tt} - \frac{1}{4}g^{RS}g_{IR,t}g_{JS,t}, \tag{12.87}$$

$$R^I{}_{JKL} = \frac{1}{4}g^{IR}g_{KR,t}g_{JL,t} - \frac{1}{4}g^{IR}g_{LR,t}g_{JK,t}; \tag{12.88}$$

$$R_{00} = -\frac{1}{2}g^{RS}g_{RS,tt} + \frac{1}{4}g^{LM}g^{RS}g_{LR,t}g_{MS,t}, \tag{12.89}$$

$$R_{JL} = \frac{1}{2}g_{JL,tt} - \frac{1}{4}g^{RS}g_{JR,t}g_{LS,t} + \frac{1}{4}g^{KR}g_{KR,t}g_{JL,t} - \frac{1}{4}g^{KR}g_{LR,t}g_{JK,t}. \tag{12.90}$$

The equation $R_{0I} = 0$ implies $u_0 u_I = 0$. Since u_0 cannot vanish (then u_α would be a spacelike vector, i.e. the dust would move with a superluminal velocity), we must have

$$u_I = 0 = u^I \implies u_0 = u^0 = 1,$$
$$T_{00} = \epsilon, \qquad T_{\alpha I} = 0, \qquad I = 1, 2, 3, \qquad \alpha = 0, 1, 2, 3, \tag{12.91}$$

where ϵ is the energy density of the dust. In the following, we will use the formulae that are obtained by a similar method as (12.24) and (12.25):

$$g^{RS}g_{RS,t} = g_{,t}/g, \tag{12.92}$$
$$g^{MS}{}_{,t} = -g^{LM}g^{RS}g_{LR,t}, \tag{12.93}$$

where $g = \det ||g_{\alpha\beta}|| = -\det ||g_{KL}||$.
From (12.79) we have $(\sqrt{-g}\epsilon u^\alpha)_{,\alpha} = 0$, which in our coordinates becomes $(\sqrt{-g}\epsilon)_{,t} = 0$. Since the metric depends only on t, also g and ϵ depend only on t, so

$$\kappa\epsilon\sqrt{-g} = M = \text{constant.} \tag{12.94}$$

From (12.90) and (12.91), using (12.92) and (12.93) we obtain

$$R = R_{00} - g^{JL}R_{JL} = -g^{RS}g_{RS,tt} - \frac{3}{4}g^{RS}{}_{,t}g_{RS,t} - \frac{1}{4}(g_{,t}/g)^2. \tag{12.95}$$

From $R_{00} - \frac{1}{2}g_{00}R = \kappa\epsilon$, using (12.90) and (12.93) – (12.95) we now obtain

$$\frac{1}{8}g^{RS}{}_{,t}\,g_{RS,t} + \frac{1}{8}\left(g_{,t}/g\right)^2 = M/\sqrt{-g}. \tag{12.96}$$

The Einstein equations imply that for dust

$$R = g^{\alpha\beta}R_{\alpha\beta} = -\kappa\epsilon. \tag{12.97}$$

We substitute (12.95) and (12.94) in the above and obtain

$$g^{RS}g_{RS,tt} + \frac{3}{4}g^{RS}{}_{,t}\,g_{RS,t} + \frac{1}{4}\left(g_{,t}/g\right)^2$$

$$\equiv \left(g^{RS}g_{RS,t}\right)_{,t} - \frac{1}{4}g^{RS}{}_{,t}\,g_{RS,t} + \frac{1}{4}\left(g_{,t}/g\right)^2 = M/\sqrt{-g}. \tag{12.98}$$

Now we use (12.96) to eliminate $g^{RS}{}_{,t}\,g_{RS,t}$, and obtain

$$\left(g^{RS}g_{RS,t}\right)_{,t} + \frac{1}{2}\left(g_{,t}/g\right)^2 = 3M/\sqrt{-g}. \tag{12.99}$$

Using (12.92) we obtain $\left(\sqrt{-g}\right)_{,tt} = 3M/2$, which is integrated to give

$$\sqrt{-g} = \frac{3}{4}Mt^2 + At + B, \tag{12.100}$$

where A and B are arbitrary constants. We still have to integrate the equations $G_{JL} = 0$. Using (12.97), (12.94), (12.92) and (12.93) they are reduced to

$$0 = -G^K{}_L = g^{KJ}R_{JL} - \frac{1}{2}\delta^K{}_L M/\sqrt{-g} = \frac{1}{2}g^{KS}g_{LS,tt}$$

$$+ \frac{1}{2}g^{KS}{}_{,t}\,g_{LS,t} + \frac{1}{4}\frac{g_{,t}}{g}\,g^{KS}g_{LS,t} - \frac{1}{2}\delta^K{}_L M/\sqrt{-g}. \tag{12.101}$$

Hence,

$$\sqrt{-g}\,\left(g^{KS}g_{LS,t}\right)_{,t} - \frac{1}{2}\left(g_{,t}/\sqrt{-g}\right)g^{KS}g_{LS,t} - M\delta^K{}_L = 0, \tag{12.102}$$

which is integrated with the result

$$\sqrt{-g}\,g^{KS}g_{LS,t} = Mt\delta^K{}_L + A^K{}_L = 0, \tag{12.103}$$

where $A^K{}_L$ is a matrix of constant elements. From this, contracting with g_{KM}, we obtain

$$g_{LM,t} = \frac{Mt}{\sqrt{-g}}g_{LM} + \frac{A^K{}_L g_{KM}}{\sqrt{-g}}. \tag{12.104}$$

This shows that $A_{ML} = A_{LM}$. Furthermore, for any fixed instant $t = t_0$ the matrix $g_{KM}(t_0)$ can be transformed into the unit matrix by transformations of the form $x'^I = b^I{}_J x^J$, where the matrix $b^I{}_J$ has constant components. Since A_{KL} is symmetric, we can then use the orthogonal transformations preserving the simplified $g_{KM}(t_0)$ to diagonalise A_{KL}, too. Assume that we have done so. Then, with use of (12.100), Eq. (12.104) becomes

$$g_{LM,t} = \frac{Mt + A^L{}_L}{\frac{3}{4}Mt^2 + At + B}\,g_{LM} \qquad \text{(no sum over L),} \tag{12.105}$$

and the solution of this is

$$g_{LM}(t) = g_{LM}(t_0) \exp \left(\int_{t_0}^{t} \frac{M\tau + A^L{}_L}{\frac{3}{4}M\tau^2 + A\tau + B} \, \mathrm{d}\tau \right). \tag{12.106}$$

Thus, if $g_{LM}(t_0)$ is diagonal, then so is $g_{LM}(t)$ at all times.

(But it is not always possible to diagonalise the metric tensor in a Bianchi-type space-time. In our case, the diagonal form followed from the field equations.)

Equation (12.106) is not the final result yet because the g_{LM} still have to obey (12.100) and (12.96). The constants $A^L{}_L$ are by tradition parametrised as follows:

$$A^L{}_L = 2Ap_L \qquad \text{(no sum over } L\text{)}, \tag{12.107}$$

and then, by virtue of (12.100) and (12.96), the parameters p_L must obey (see Exercise 4):

$$p_1 + p_2 + p_3 = 1, \qquad \frac{1}{2}A^2 \left[1 - \left(p_1{}^2 + p_2{}^2 + p_3{}^2 \right) \right] = MB. \tag{12.108}$$

When $M = 0$, our metric becomes a vacuum solution and can be transformed to

$$\mathrm{d}s^2 = \mathrm{d}t^2 - t^{2p_1}\mathrm{d}x^2 - t^{2p_2}\mathrm{d}y^2 - t^{2p_3}\mathrm{d}z^2, \tag{12.109}$$

where the constants p_1, p_2 and p_3 obey

$$p_1 + p_2 + p_3 = p_1{}^2 + p_2{}^2 + p_3{}^2 = 1. \tag{12.110}$$

This solution was found by Kasner (1921). It was one of the earliest exact solutions of Einstein's equations published in the literature.

12.16 * Other gravitation theories

As mentioned at the end of Sec. 12.7, there exist several gravitation theories different from general relativity (GR). In this section we will present a few of them. The examples are meant to illustrate the various possibilities used by different authors. Except for Rosen's theory, they are generalisations of GR and contain it as a limit. A more complete overview of such theories is given in Thorne, Lee and Lightman (1973), and their comparison with experimental tests can be found in Will (2018). The Kaluza–Klein theory is one of the generalisations, but its presentation must be preceded by a description of the Maxwell theory in curved spacetime, therefore we postpone it to the next chapter.

12.16.1 The Brans–Dicke theory

This theory was first published in 1961 (Brans and Dicke, 1961), and the best source for studying its origins is the book by Dicke (1964). It is conceptually the most developed of all theories of gravitation that once competed with general relativity (GR), with its logical structure and interpretation well clarified. Its field equations are

$$R_{\mu\nu} - \frac{1}{2}g_{\mu\nu}R = \frac{8\pi}{\phi}T_{\mu\nu} + \frac{\omega}{\phi^2}\left(\phi_{,\mu}\,\phi_{,\nu} - \frac{1}{2}g_{\mu\nu}\phi_{,\sigma}\,\phi^{,\sigma} \right) + \frac{1}{\phi}\left(\phi_{;\mu\nu} - g_{\mu\nu}g^{\mu\nu}\phi_{;\mu\nu} \right),$$

$$\tag{12.111}$$

$$g^{\mu\nu}\phi_{;\mu\nu} \;-\; \frac{1}{2\phi}\,\phi_{,\mu}\,\phi^{,\mu} + \frac{\phi}{2\omega}\,R = 0, \qquad \omega = \text{constant}. \tag{12.112}$$

This is a generalisation of GR in which the gravitational constant G is replaced by the scalar field ϕ. The gravitational 'constant' c^4/ϕ thus varies in time and space, and its changes are induced by the scalar curvature R via (12.112). The scalar curvature in turn depends on the distribution of matter via (12.111). In this way, the distribution of matter influences the intensity of gravitation, which is a partial realisation of the Mach principle.

General relativity follows from the BD theory as the limit $\phi = c^4/G = \text{constant}$, $\omega \to \infty$. The value of ω is related to the PPN parameter γ by $\gamma = (1+\omega)/(2+\omega)$ (Will, 2006), so $\gamma = 1$ for GR. With any $0 \le \omega < \infty$, $\gamma < 1$ and discrepancies with GR increase with decreasing ω. Thus, modifications of GR would be the more pronounced the smaller ω. However, observations of light deflection in the gravitational field of the Sun (see Sec. 14.7) imply $|\gamma - 1| \le 4 \times 10^{-4}$ and observations of the Shapiro time delay imply $|\gamma - 1| \le 2.3 \times 10^{-5}$ (both numbers taken from Will, 2006). These translate to $\omega \ge 2502$ and $\omega > 4 \times 10^4$, respectively. With such values, the predictions of the BD theory are observationally indistinguishable from those of GR. For this reason, the BD theory is currently out of favour with theoreticians.

12.16.2 The Bergmann–Wagoner theory

The field equations of this theory are (Bergmann, 1968; Wagoner, 1970)

$$R_{\mu\nu} - \tfrac{1}{2}g_{\mu\nu}R \;-\lambda(\phi)g_{\mu\nu} = \frac{8\pi}{\phi}T_{\mu\nu} + \frac{\omega(\phi)}{\phi^2}\left(\phi_{,\mu}\,\phi_{,\nu} - \frac{1}{2}g_{\mu\nu}\phi_{,\sigma}\,\phi^{,\sigma}\right)$$
$$+\frac{1}{\phi}\left(\phi_{;\mu\nu} - g_{\mu\nu}g^{\rho\sigma}\phi_{;\rho\sigma}\right), \tag{12.113}$$

$$g^{\mu\nu}\phi_{;\mu\nu} + \frac{1}{2}\phi_{,\mu}\,\phi^{,\mu}\frac{\mathrm{d}}{\mathrm{d}\phi}\ln[\omega(\phi)/\phi] + \frac{\phi}{2\omega(\phi)}\left[R + 2\frac{\mathrm{d}}{\mathrm{d}\phi}\left(\phi\lambda(\phi)\right)\right] = 0, \tag{12.114}$$

where $\lambda(\phi)$ and $\omega(\phi)$ are arbitrary functions. This is a generalisation of the Brans–Dicke theory, which results as the limit $\omega = \text{constant}$, $\lambda = 0$. It is presented here just as a curiosity because it is far from being well understood. Its main weakness is the existence of two arbitrary functions that are not determined by any field equations. The function $\lambda(\phi)$ becomes the cosmological constant when $\phi = \text{constant}$.

12.16.3 The Einstein–Cartan theory

In this theory (Trautman, 1972), the connection coefficients are not symmetric. The Ricci tensor is built of the full, nonsymmetric connection coefficients and obeys equations formally identical to the Einstein equations (12.21), but $R_{\mu\nu}$ and $T_{\mu\nu}$ are not symmetric. The torsion tensor $Q^\rho{}_{\mu\nu}$ obeys a separate set of equations:

$$Q^\rho{}_{\mu\nu} + \delta^\rho{}_\mu Q^\sigma{}_{\nu\sigma} - \delta^\rho{}_\nu Q^\sigma{}_{\mu\sigma} = \kappa s^\rho{}_{\mu\nu}, \tag{12.115}$$

where the tensor $s^\rho{}_{\mu\nu}$ describes the density of spin of matter, and it vanishes in vacuum. When $s^\rho{}_{\mu\nu} = 0$, also $Q^\rho{}_{\mu\nu} = 0$. Thus, in this theory torsion does not propagate out into vacuum; it may be nonzero only inside matter. With $s^\rho{}_{\mu\nu} = 0$, the Einstein–Cartan (EC) theory reduces to GR. In vacuum, it is not distinguishable from GR.

The advantage of the EC theory is that it allows a singularity-free universe model (Kopczyński, 1972). (GR predicts that a model of the Universe must have a singularity in the past or in the future – see Section 15.4.) However, it has not been accepted as a replacement for Einstein's theory, mainly because of the impossibility of distinguishing it experimentally from GR – since all experiments testing GR are carried out in vacuum.

12.16.4 The bi-metric Rosen theory

It is not a generalisation, but an alternative to Einstein's theory (Rosen, 1973). There are two metric tensors in it: $g_{\alpha\beta}$, of the same interpretation as in general relativity, and the flat $\gamma_{\alpha\beta}$, for which the curvature tensor $B^\alpha{}_{\beta\gamma\delta}(\gamma) \equiv 0$. Its field equations are

$$N_{\mu\nu} - \frac{1}{2}g_{\mu\nu}N = \kappa T_{\mu\nu}, \tag{12.116}$$

where $N = g^{\mu\nu}N_{\mu\nu}$ and

$$N_{\mu\nu} = \frac{1}{2}\gamma^{\rho\sigma}g_{\mu\nu|\rho\sigma} - \frac{1}{2}\gamma^{\rho\sigma}g^{\lambda\tau}g_{\mu\lambda|\rho}g_{\nu\tau|\sigma}; \tag{12.117}$$

the vertical stroke denotes the covariant derivative with respect to $\gamma_{\mu\nu}$. The author's argument in favour of this theory was its greater simplicity compared to GR. Its experimental predictions are not much different from those of GR (Will, 2006). When the gravitational field vanishes, $g_{\mu\nu} = \gamma_{\mu\nu}$. A majority of relativists consider the flat background metric $\gamma_{\mu\nu}$ to be an unnatural element that spoils the elegance of the relativistic approach (the opinion of the author himself was exactly the opposite). Many papers had been published on this theory, but in the end it has not established itself as a viable alternative to GR.

12.17 Matching solutions of Einstein's equations

Just as in electrodynamics, in gravitation theory we sometimes have to match solutions of Einstein's equations obtained separately for different spacetime regions. Most often, we want to determine whether a given vacuum solution can be interpreted as the exterior gravitational field to a material body for whose interior we have found another solution. In consequence of matching, the arbitrary constants in the vacuum region are related to the parameters of the interior metric.

Sometimes we are interested in describing a surface matter distribution on the boundary between two regions, but we shall exclude this case here. Generally, we assume that there are no singularities of the type of Dirac δ function in components of the curvature tensor and that the hypersurface across which we match the two metrics is nonnull. Together with surface matter distributions we thus exclude shock waves in which the discontinuity in the Riemann tensor is matched to vacuum solutions on both sides. Null matching hypersurfaces pose additional problems, see Manzano and Mars (2021).

With these assumptions, the components of the Riemann tensor can at worst be discontinuous across the matching hypersurface Σ. We have to allow discontinuities because, for example, the mass density on the surface of a perfect fluid body, which equals the component $T^0{}_0$ of the energy-momentum tensor in comoving coordinates, is nonzero at the surface, but zero at adjacent vacuum points.

In order to discuss these conditions in detail, it is most convenient to use the coordinates adapted to Σ, the same that were introduced at the end of Section 7.15. This time, $N = 4$, $n = 3$, $V_n = \Sigma$, and the spacetime metric in the adapted coordinates is

$$\mathrm{d}s^2 = g_{IJ}\mathrm{d}x^I\mathrm{d}x^J + \varepsilon\mathcal{N}^2\left(\mathrm{d}x^4\right)^2, \tag{12.118}$$

where the x^4 coordinate may be timelike (then $\varepsilon = +1$) or spacelike ($\varepsilon = -1$). The unit normal vector to the boundary $x^4 = A =$ constant is $X^\alpha = (0, 0, 0, 1/\mathcal{N})$. For $x^4 > A$ we have one metric $g^+_{\alpha\beta}$ (e.g. vacuum); for $x^4 < A$ we have another $g^-_{\alpha\beta}$ (e.g. the interior of a material body). Since Σ has to be the same, whichever of the two 4-metrics is used to describe it, the components g_{IJ} have to be continuous across Σ. Since on Σ they are functions of the x^I-coordinates, the same functions no matter which 4-metric is used to calculate the g_{IJ}, the continuity of all the derivatives of g_{IJ} along Σ, i.e. of the derivatives by x^K, is automatically guaranteed, and we only need to take care about $g_{IJ,4}$.

We have made sure that the boundary hypersurface Σ has the same intrinsic geometry in both metrics. However, if the spacetime regions on the opposite sides of Σ are to be parts of the same manifold, then Σ must have the same extrinsic geometry with respect to both metrics, i.e. be embeddable in both in the same way. Without that, we might end up trying to identify a cylinder with a plane, for example. Hence, the second fundamental form of Σ must be the same, no matter in which of the two metrics it is calculated.

Thus, finally, the conditions that two metrics, $g^+_{\alpha\beta}$ and $g^-_{\alpha\beta}$, can be considered to describe two parts of the same manifold are as follows:

A hypersurface Σ must exist such that, in coordinates adapted to Σ as in (12.118), the metrics induced on Σ are the same, $g^+_{\alpha\beta}(\Sigma) = g^-_{\alpha\beta}(\Sigma)$, and the second fundamental form of Σ must be the same, whichever metric, $g^+_{\alpha\beta}$ or $g^-_{\alpha\beta}$, is used to calculate it.

Now we will investigate the consequences of these conditions for the Riemann and Einstein tensors of the spacetime and for the Riemann tensor of Σ. We will use the Gauss–Codazzi equations (7.93) – (7.94), but we have to adapt the notation. Now U_{n+1} is the spacetime (here denoted V_4) and the V_n of Section 7.15 is the boundary hypersurface Σ. Thus, the capital Latin indices of Section 7.15 become now the Greek spacetime indices, and the Greek indices of Section 7.15 become now the capital Latin indices for Σ. The covariant derivative in the V_n of Section 7.15, denoted by a semicolon, now becomes the covariant derivative in Σ, denoted by the vertical stroke |. With $X^\alpha = (0, 0, 0, 1/\mathcal{N})$ and Y^α coinciding with x^I in Σ for $\alpha, I = 1, 2, 3$, Eqs. (7.93) – (7.94) become

$$R_{DABC}(\Sigma) = R_{DABC}(V_4) + \varepsilon\left(\Omega_{AC}\Omega_{DB} - \Omega_{AB}\Omega_{DC}\right), \tag{12.119}$$

$$\Omega_{IJ|K} - \Omega_{IK|J} = -R_{4IJK}(V_4)/\mathcal{N}. \tag{12.120}$$

The components of $R_{DABC}(\Sigma)$ are calculated from the 3-metric that was assumed the same, whether Σ was approached from the $+$ side or from the $-$ side, ditto for Ω_{AB}.

Consequently, $R_{DABC}(V_4)$, i.e. the Σ-components of the 4-dimensional Riemann tensor, will be continuous across Σ. The covariant derivatives of Ω_{AB} in (12.120) are taken within Σ. This means that $R_{4IJK}(V_4)/\mathcal{N}$ has to be continuous across Σ, but $\mathcal{N}$ and $R_{4IJK}(V_4)$ individually need not be – they may have a discontinuity that cancels out in the quotient.

For completeness, we have now to consider the components $R_{4I4J}(V_4)$ that are not determined by the Gauss–Codazzi equations. They are (see Exercise 6)

$$
\begin{aligned}
R_{4I4J} &= \varepsilon \mathcal{N}^2 R^4{}_{I4J} \\
&= \varepsilon \mathcal{N}^2 \left[-\frac{1}{4} g^{44}{}_{,J}\, g_{44,I} - \frac{1}{2} g^{44} g_{44,IJ} - \frac{1}{4} g^{44}{}_{,4}\, g_{IJ,4} - \frac{1}{2} g^{44} g_{IJ,44} \right. \\
&\qquad\left. + \frac{1}{2} g^{44} g_{44,R} \left\{ {R \atop IJ} \right\}_{(\Sigma)} + \frac{1}{4} g^{44} g^{RS} g_{RJ,4} g_{IS,4} \right] \\
&= -\varepsilon \mathcal{N} \mathcal{N}_{|IJ} + \mathcal{N}^2 g^{RS} \Omega_{RJ} \Omega_{IS} + \mathcal{N}^2 \Omega_{IJ,\rho} X^\rho .
\end{aligned} \tag{12.121}
$$

These can be discontinuous across Σ because of $g_{IJ,44}$, which are not restricted by the requirement of continuity of Ω_{IJ}.

Now we find for the components of the Einstein tensor:

$$
\begin{aligned}
G^4{}_4 &= -\frac{1}{2} g^{IJ} R^S{}_{ISJ}, \qquad G_{4I} = R^S{}_{4SI}, \\
G_{IJ} &= R^S{}_{ISJ} + R^4{}_{I4J} - \frac{1}{2} g_{IJ} g^{KL} R^S{}_{KSL} - g_{IJ} g^{KL} R^4{}_{K4L}.
\end{aligned} \tag{12.122}
$$

This shows that $G^4{}_4$ is continuous across Σ, while G_{4I} and G_{IJ} can be discontinuous.

Note what this means for matching a perfect fluid solution to vacuum across a timelike hypersurface such that the velocity has no component in the x^4 direction. Then $G^4{}_4 = -\kappa p$, and since $G^4{}_4 = 0$ in vacuum and has to be continuous across Σ, this means that $p = 0$ must hold on Σ, which agrees with the expectation based on physics.

12.18 The weak-field approximation to general relativity

Although considerable progress has been made in the search for exact solutions of Einstein's equations (Stephani et al., 2003), real physical or astrophysical situations are often too complicated to be captured by an exact solution. In those cases, one must resort to approximate calculations, for which several methods have been developed. We shall introduce one such method here, which is useful also for interpreting exact solutions because it allows one to recognise the physical meaning of parameters in the metric. The presentation is based on Stephani (1990), with some modifications.

Suppose that the metric includes the Minkowski limit $\eta_{\mu\nu}$. Write it as

$$
g_{\mu\nu} = \eta_{\mu\nu} + h_{\mu\nu}, \tag{12.123}
$$

and choose coordinates in which the $\eta_{\mu\nu}$ has the form $\eta_{\mu\nu} = \mathrm{diag}(+1, -1, -1, -1)$. Then assume that $h_{\mu\nu}$ is a small correction to $\eta_{\mu\nu}$, thus $|h_{\mu\nu}| \ll 1$ for all indices and $|h_{\mu\nu} h_{\rho\sigma}| \ll |h_{\alpha\beta}|$ for all (μ, ν, ρ, σ) and all $h_{\alpha\beta} \neq 0$. Assume that also the first and second derivatives of $h_{\mu\nu}$ are small, i.e. $|h_{\mu\nu,\alpha}| \ll 1$, $|h_{\mu\nu,\alpha\beta}| \ll 1$. Under these assumptions, all terms in the Christoffel symbols and in the curvature tensor that are nonlinear in $h_{\mu\nu}$ can be neglected.

This set of assumptions is called the **weak-field approximation** to the Einstein theory, and the resulting scheme is sometimes called the 'linearised theory of gravitation'. Within this scheme, it is assumed that the energy-momentum tensor $T_{\mu\nu}$ has its special-relativistic form and obeys the equations of motion in flat space

$$T^{\mu\nu}{}_{,\nu} = 0. \tag{12.124}$$

The matrix $h_{\mu\nu}$ does not appear in (12.124), so, in this approximation, the matter coded in $T^{\mu\nu}$ moves independently of the gravitational field it produces. This is a strong qualitative difference between the exact and the approximate theory. If, however, this assumption of no **back-reaction** holds to a satisfactory degree of precision in a given situation, then the gravitational field generated by a given source can be calculated (approximately) in a much simpler way than in the full Einstein theory.

We adopt the convention that all indices are raised and lowered by the flat metric $\eta_{\mu\nu}$. Since the determinant of $g_{\mu\nu}$ has a form similar to (12.123) ($g = -1 + (\text{small perturbation})$), it follows that also $g^{\mu\nu}$ has the property $g^{\mu\nu} = \eta^{\mu\nu} + f^{\mu\nu}$, $|f^{\mu\nu}| \ll 1$, and that

$$
\begin{aligned}
g^{\mu\nu} &= \eta^{\mu\nu} - h^{\mu\nu} \equiv \eta^{\mu\nu} - \eta^{\mu\rho}\eta^{\nu\sigma}h_{\rho\sigma}, \\
\left\{ {\alpha \atop \beta\gamma} \right\} &= \frac{1}{2}\eta^{\alpha\rho}\left(h_{\beta\rho,\gamma} + h_{\gamma\rho,\beta} - h_{\beta\gamma,\rho} \right), \\
R^{\alpha}{}_{\beta\gamma\delta} &= \frac{1}{2}\eta^{\alpha\rho}\left(h_{\rho\delta,\beta\gamma} - h_{\rho\gamma,\beta\delta} + h_{\beta\gamma,\rho\delta} - h_{\beta\delta,\rho\gamma} \right).
\end{aligned}
\tag{12.125}
$$

The linearised $R^{\alpha}{}_{\beta\gamma\delta}$ of (12.125) is seen to have all the symmetries in the indices of the exact Riemann tensor. The linearised Einstein equations are

$$-\frac{1}{2}\left(h_{\alpha\beta}{}^{,\rho}{}_{,\rho} + h^{\rho}{}_{\rho,\alpha\beta} - h_{\alpha}{}^{\rho}{}_{,\beta\rho} - h_{\beta}{}^{\rho}{}_{,\alpha\rho} \right) + \frac{1}{2}\eta_{\alpha\beta}\left(h^{\rho}{}_{\rho}{}^{,\sigma}{}_{,\sigma} - h^{\rho\sigma}{}_{,\rho\sigma} \right) = \overline{\kappa}T_{\alpha\beta}. \tag{12.126}$$

The κ defined in (12.59) is now replaced by $\overline{\kappa} = \kappa c^2 = 8\pi G/c^2$ because in this section the components of $T_{\alpha\beta}$ will be expressed in units of mass density, not energy density.

Now we define:

$$\widetilde{h}_{\alpha\beta} \stackrel{\text{def}}{=} h_{\alpha\beta} - \frac{1}{2}\eta_{\alpha\beta}h^{\rho}{}_{\rho} \implies \widetilde{h}^{\rho}{}_{\rho} = -h^{\rho}{}_{\rho}, \qquad h_{\alpha\beta} = \widetilde{h}_{\alpha\beta} - \frac{1}{2}\eta_{\alpha\beta}\widetilde{h}^{\rho}{}_{\rho}, \tag{12.127}$$

$$\implies g^{\alpha\beta} = \eta^{\alpha\beta} - \widetilde{h}^{\alpha\beta} + \frac{1}{2}\eta^{\alpha\beta}\widetilde{h}^{\rho}{}_{\rho}, \tag{12.128}$$

and we obtain in (12.126)

$$-\widetilde{h}_{\alpha\beta}{}^{,\rho}{}_{,\rho} + \widetilde{h}_{\alpha}{}^{\rho}{}_{,\beta\rho} + \widetilde{h}_{\beta}{}^{\rho}{}_{,\alpha\rho} - \eta_{\alpha\beta}\widetilde{h}^{\rho\sigma}{}_{,\rho\sigma} = 2\overline{\kappa}T_{\alpha\beta}. \tag{12.129}$$

We have not yet specified the coordinates. We carry out the transformation

$$\overline{x}^{\alpha} = x^{\alpha} + b^{\alpha}(x), \tag{12.130}$$

where, to preserve the property $|h_{\alpha\beta}| \ll 1$, the functions b^{α} are assumed to be of the same order as $h_{\alpha\beta}$. This means that $\left| b^{\alpha}b^{\beta} \right| \ll \left| b^{\gamma} \right|$, $\left| b^{\alpha}h_{\beta\gamma} \right| \ll \left| h_{\delta\epsilon} \right|$, $\left| b^{\alpha}b^{\beta} \right| \ll \left| h^{\gamma\delta} \right|$ and similarly for the derivatives. In the calculations below, it must be remembered that it is the full metric $g_{\alpha\beta}$ which undergoes a linearised tensor transformation, not $\widetilde{h}_{\alpha\beta}$. The transformation law for $\widetilde{h}_{\alpha\beta}$ follows by linearizing the equations $g_{\alpha'\beta'} = x^\rho{}_{,\alpha'} x^\sigma{}_{,\beta'} g_{\rho\sigma}$. After the transformation (12.130) we obtain

$$\widetilde{h}^\mu{}_\mu = \overline{h}^\mu{}_\mu - 2b^\mu{}_{,\mu}, \qquad \widetilde{h}^{\alpha\beta} = \overline{h}^{\alpha\beta} + b^{\alpha,\beta} + b^{\beta,\alpha} - \eta^{\alpha\beta}b^\mu{}_{,\mu}. \tag{12.131}$$

Calculating $\partial/\partial x^\alpha$ of the second equation in (12.131) we find

$$\widetilde{h}^{\alpha\beta}{}_{,\alpha} = \overline{h}^{\alpha\beta}{}_{,\alpha} + b^{\beta,\mu}{}_{,\mu}. \tag{12.132}$$

This simplifies if we impose the following condition on b^β:

$$b^{\beta,\mu}{}_{,\mu} = \widetilde{h}^{\alpha\beta}{}_{,\alpha}, \tag{12.133}$$

which results in

$$\overline{h}^{\alpha\beta}{}_{,\alpha} = 0. \tag{12.134}$$

Equation (12.129) holds with $\widetilde{h}^{\alpha\beta}$ replaced by $\overline{h}^{\alpha\beta}$, and with (12.134) it becomes

$$-2\overline{\kappa}T_{\alpha\beta} = \eta^{\mu\nu}\overline{h}_{\alpha\beta,\mu\nu} \equiv \Box\overline{h}_{\alpha\beta}. \tag{12.135}$$

For each pair of indices $(\alpha\beta)$ the equation for $\overline{h}_{\alpha\beta}$ in (12.135) is independent of the others and $\overline{h}_{\alpha\beta}$ is determined by just one component of $T_{\alpha\beta}$.

Equation (12.135) is the inhomogeneous wave equation, for which a general formal solution is the same as in electrodynamics (Jackson, 1975, Sec. 6.6):

$$\overline{h}_{\alpha\beta}(t,\mathbf{r}) = -\frac{\overline{\kappa}}{2\pi}\int_V \frac{1}{|\mathbf{r}-\mathbf{r}'|}T_{\alpha\beta}(\mathbf{r}',t-|\mathbf{r}-\mathbf{r}'|/c)\mathrm{d}_3r', \tag{12.136}$$

where $\mathbf{r}$ are the spatial coordinates of the observer's location, $\mathbf{r}'$ are the spatial coordinates of a field point and the volume of integration is the set on which $T_{\alpha\beta} \neq 0$.

We assume now that the volume V is finite and that the observer, located at $\mathbf{r}$, is far from V, thus $|\mathbf{r}| \gg |\mathbf{r}'|$ for all points inside V. Then $|\mathbf{r}-\mathbf{r}'|$ and $1/|\mathbf{r}-\mathbf{r}'|$ can be expanded in power series with respect to x'^I, the components of $\mathbf{r}'$:

$$\begin{aligned}
|\mathbf{r}-\mathbf{r}'| &= \sqrt{\mathbf{r}^2 - 2\mathbf{r}\mathbf{r}' + \mathbf{r}'^2} \\
&= r - \frac{x^I x'^I}{r} - \frac{x^I x^J}{2r^3}\left(x'^I x'^J - r'^2\delta^{IJ}\right) + \dots, \tag{12.137}
\end{aligned}$$

$$1/|\mathbf{r}-\mathbf{r}'| = \frac{1}{r} + \frac{x^I x'^I}{r^3} + \frac{x^I x^J}{2r^5}\left(3x'^I x'^J - r'^2\delta^{IJ}\right) + \dots, \tag{12.138}$$

where $I, J, K = 1, 2, 3$. The higher-order terms will not appear explicitly.

Then, we expand $T_{\alpha\beta}$ with respect to $(t-|\mathbf{r}-\mathbf{r}'|/c)$ around $(t-|\mathbf{r}|/c)$. It is justifiable to truncate the expansion because $t-|\mathbf{r}-\mathbf{r}'|/c - (t-|\mathbf{r}|/c) \equiv (|\mathbf{r}|-|\mathbf{r}-\mathbf{r}'|)/c$ is small compared with $|\mathbf{r}|/c$, see Fig. 12.2. Thus,

$$\begin{aligned}
T_{\alpha\beta}(\mathbf{r}',t-|\mathbf{r}-\mathbf{r}'|/c) &= T_{\alpha\beta}(\mathbf{r}',t-|\mathbf{r}|/c) + \frac{1}{c}\dot{T}_{\alpha\beta}(\mathbf{r}',t-|\mathbf{r}|/c)(r-|\mathbf{r}-\mathbf{r}'|) \\
&\quad + \frac{1}{2c^2}\ddot{T}_{\alpha\beta}(\mathbf{r}',t-|\mathbf{r}|/c)(r-|\mathbf{r}-\mathbf{r}'|)^2 + \dots. \tag{12.139}
\end{aligned}$$

Now we substitute (12.137) – (12.139) in (12.136) and truncate the series at terms of order

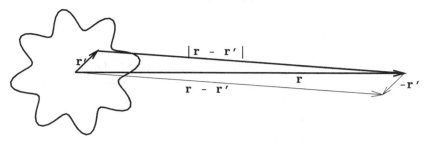

Fig. 12.2 Vectors and distances involved in calculating the linearised metric of a finite body in the weak-field approximation. See the text for explanations.

r^{-3}. We also assume that there are no fast motions inside the body, so $(\mathrm{d}T_{\alpha\beta}/\mathrm{d}t)/c$ is of order no higher than $1/r$. Consequently, the truncation will be done at the *combined* order r^{-3} (i.e. at r^{-3} in $T_{\alpha\beta}$, r^{-2} in $\dot{T}_{\alpha\beta}$ and r^{-1} in $\ddot{T}_{\alpha\beta}$). The result is (see Exercise 8):

$$\frac{1}{|\mathbf{r} - \mathbf{r}'|}T_{\alpha\beta}(\mathbf{r}', t - |\mathbf{r} - \mathbf{r}'|/c)$$

$$= T_{\alpha\beta}\left[\frac{1}{r} + \frac{x^I x'^I}{r^3} + \frac{x^I x^J}{2r^5}\left(3x'^I x'^J - r'^2\delta^{IJ}\right)\right]$$

$$+ \frac{1}{c}\dot{T}_{\alpha\beta}\left[\frac{x^I x'^I}{r^2} + \frac{x^I x^J}{2r^4}\left(3x'^I x'^J - r'^2\delta^{IJ}\right)\right] + \frac{1}{2c^2}\ddot{T}_{\alpha\beta}\frac{x^I x^J}{r^3}x'^I x'^J \quad (12.140)$$

(the arguments of $T_{\alpha\beta}$ on the right-hand side are $\mathbf{r}'$ and $(t - r/c)$).

Now, in calculating the integral in (12.136) it must be noted that the coordinate x^0 is ct and so, since we assume no fast motions in the source, T^{IJ} is of order lower by one in v/c than T^{0J}, and this in turn is of order lower by one than T^{00}. We are working within the first-order post-Newtonian approximation (i.e. in the linearised Einstein theory), and the lowest non-special-relativistic correction to T^{IJ} is of the order $v/c \approx r'/r$. Accordingly, we have to calculate T^{0I} and T^{00} to orders $(v/c)^2 \approx (r'/r)^2$ and $(v/c)^3 \approx (r'/r)^3$, respectively – and the same follows for $\overline{h}^{\alpha\beta}$ in (12.136). The following quantities will appear in the calculations (the arguments of $T^{\alpha\beta}$ are the same as in (12.140)):

$$\begin{array}{llll}
M & \overset{\text{def}}{=} & \int_V T^{00}\mathrm{d}_3 x', & d^I & \overset{\text{def}}{=} & \int_V T^{00}x'^I\mathrm{d}_3 x', \\
d^{IJ} & \overset{\text{def}}{=} & \int_V T^{00}x'^I x'^J\mathrm{d}_3 x', & p^I & \overset{\text{def}}{=} & \int_V T^{0I}\mathrm{d}_3 x', \\
b^{IJ} & \overset{\text{def}}{=} & -\int_V T^{0I}x'^J\mathrm{d}_3 x', & a^{IJ} & \overset{\text{def}}{=} & \int_V T^{IJ}\mathrm{d}_3 x',
\end{array} \quad (12.141)$$

all integrals are over the volume V of the body. From the physical interpretation of the components of $T^{\alpha\beta}$ we recognise (up to the choice of units, i.e. up to a constant factor) M as the mass of the source, d^I as the mass dipole moment, d^{IJ} as the mass quadrupole moment, p^I as the momentum, b^{IJ} as the matrix whose antisymmetric part is the angular momentum and a^{IJ} as the energy of stress.

The equations of motion can be written as

$$T^{\alpha 0}{}_{,0} = -T^{\alpha S}{}_{,S} .$$ (12.142)

Integrating these over a volume W containing V, applying the Gauss law $\int_W v^I{}_{,I} \, d_3 x' = \int_{\partial W} v^I d_2 x'^I$ and taking into account that the surface ∂W lies in vacuum, where $T^{\alpha\beta} = 0$, we obtain the conservation laws $T^{\alpha 0}{}_{,0} = 0$, i.e.

$$\begin{aligned} M &= \text{constant} & \text{for } \alpha = 0, \\ p^I &= \text{constant} & \text{for } \alpha = I. \end{aligned}$$ (12.143)

Using (12.142) in dd^I/dx^0 we obtain

$$\frac{1}{c} \frac{dd^I}{dt} = \int_W T^{00}{}_{,0} \, x'^I d_3 x' = -\int_W T^{0S}{}_{,S} \, x'^I d_3 x'$$ (12.144)

$$= -\int_W \left(T^{0S} x'^I\right)_{,S} d_3 x' + \int_W T^{0I} d_3 x' = -\int_{\partial W} T^{0S} x'^I d_2 x'^S + p^I = p^I.$$

Now we use again the same trick in the time-derivative of

$$B^{IJ} \stackrel{\text{def}}{=} b^{[IJ]} = \frac{1}{2} \left(b^{IJ} - b^{JI}\right),$$ (12.145)

and use the fact that $T^{IJ} = T^{JI}$. The result is the conservation of the angular momentum,

$$B^{IJ} = \text{constant.}$$ (12.146)

The constant momentum p^I can be set to zero by transforming the coordinates to the rest frame of the body. This is a Lorentz transformation that does not change the property (12.123) and will change only the numerical values of the other quantities in (12.141), keeping constant those that were constant. With $p^I = 0$, d^I becomes constant and can be set to zero by moving the origin of coordinates to the centre of mass of the body.

The equation $T^{00}{}_{,0} + T^{0S}{}_{,S} = 0$ implies that

$$\left(T^{00} x'^I x'^J\right)_{,0} + \left(T^{0S} x'^I x'^J\right)_{,S} = T^{0I} x'^J + T^{0J} x'^I.$$ (12.147)

Differentiating this by x^0 and applying (12.142) again, we obtain

$$\left(T^{00} x'^I x'^J\right)_{,00} = \left(T^{RS} x'^I x'^J\right)_{,RS} - 2\left(T^{IS} x'^J + T^{JS} x'^I\right)_{,S} + 2T^{IJ}.$$ (12.148)

By integrating (12.147) and (12.148) over the volume W and applying the Gauss law to eliminate the divergence terms we obtain, respectively

$$-\frac{1}{c} \dot{d}^{IJ} = b^{IJ} + b^{JI}, \qquad \frac{1}{2c^2} \ddot{d}^{IJ} = a^{IJ}.$$ (12.149)

Thus $\dot{d}^{IJ} = -2cb^{(IJ)}$, so

$$b^{IJ} = B^{IJ} - \frac{1}{2c} \dot{d}^{IJ},$$ (12.150)

while the stresses are determined by $\ddot{d}^{IJ}$. So, M, B^{IJ} and d^{IJ} fully characterise the body to our desired order of approximation.

Substituting (12.140) in (12.136), employing the definitions (12.141) and the relations

(12.149) – (12.150), we obtain for the metric perturbations (recall: $p^I = d^I = 0$ and the indices are raised and lowered by $\eta_{\mu\nu}$, see the paragraph containing (12.125))

$$-\frac{2\pi}{\kappa}\overline{h}_{00} = \frac{M}{r} + \frac{x^I x^J}{2r^5}(3d_{IJ} - d_{SS}\delta_{IJ})$$
$$+ \frac{x^I x^J}{2cr^4}\left(3\dot{d}_{IJ} - \dot{d}_{SS}\delta_{IJ}\right) + \frac{x^I x^J}{2c^2 r^3}\ddot{d}_{IJ},$$

$$-\frac{2\pi}{\kappa}\overline{h}_{0I} = \frac{B_{IJ}x^J}{2r^3} - \frac{\dot{d}_{IJ}x^J}{2cr^3} - \frac{\ddot{d}_{IJ}x^J}{2c^2 r^2},$$

$$-\frac{2\pi}{\kappa}\overline{h}_{IJ} = \frac{\ddot{d}_{IJ}}{2c^2 r}. \tag{12.151}$$

Finally, we wish to find the $h_{\alpha\beta}$ defined in (12.123). Reversing the chain of transformations $h_{\alpha\beta} \longrightarrow \widetilde{h}_{\alpha\beta} \longrightarrow \overline{h}_{\alpha\beta}$ given in (12.127) and (12.131) we obtain

$$h_{\alpha\beta} = \overline{h}_{\alpha\beta} + b_{\alpha,\beta} + b_{\beta,\alpha} - \frac{1}{2}\,\eta_{\alpha\beta}\overline{h}^\rho{}_\rho. \tag{12.152}$$

However, we cannot calculate the explicit form of b_α because it was nowhere given – we only showed that b_α *can be chosen* to fulfil (12.133) without knowing what $\widetilde{h}_{\alpha\beta}$ was. With b_α unknown we cannot calculate $\widetilde{h}_{\alpha\beta}$ from (12.131) or $h_{\alpha\beta}$ from (12.152). All that we know from (12.152) is that *there exists* a transformation (12.130) that takes $\overline{h}_{\alpha\beta}$ to $h_{\alpha\beta}$.

From (12.151) we calculate $-(2\pi/\kappa)\overline{h}^\rho{}_\rho = -(2\pi/\kappa)\left(\overline{h}_{00} + \overline{h}^S{}_S\right)$ and then, up to our desired order

$$-\frac{2\pi}{\kappa}\left(\overline{h}_{00} - \frac{1}{2}\,\eta_{00}\overline{h}^\rho{}_\rho\right) = \frac{M}{2r} + \frac{x^I x^J}{4r^5}(3d_{IJ} - d_{SS}\delta_{IJ})$$
$$+ \frac{x^I x^J}{4cr^4}\left(3\dot{d}_{IJ} - \dot{d}_{SS}\delta_{IJ}\right) + \frac{x^I x^J}{4c^2 r^3}\ddot{d}_{IJ} + \frac{\ddot{d}_{SS}}{4c^2 r},$$

$$-\frac{2\pi}{\kappa}\left(\overline{h}_{IJ} - \frac{1}{2}\,\eta_{IJ}\overline{h}^\rho{}_\rho\right) = -\eta_{IJ}\frac{M}{2r}. \tag{12.153}$$

We could now assume that the source of the gravitational field is unchanging in time, so $\dot{d}_{IJ} = \ddot{d}_{IJ} = 0$, and take (12.153) to be the final result, with $b_\alpha = 0$ in (12.152). However, it turns out that the time-derivatives of d_{IJ} in (12.153) can be removed, up to our assumed order in $1/r$ and v/c, by suitably choosing b_α, the choice is

$$\frac{2\pi}{\kappa}b_0 = \frac{x^I x^J}{8r^4}(3d_{IJ} - \delta_{IJ}d_{SS}) + \frac{x^I x^J}{8cr^3}\left(\dot{d}_{IJ} + \delta_{IJ}\dot{d}_{SS}\right),$$

$$\frac{2\pi}{\kappa}b_I = -\frac{d_{IJ}x^J}{2r^3} - \frac{\dot{d}_{IJ}x^J}{2cr^2} + \frac{3d_{JK}x^J x^K x^I}{4r^5} + \frac{\left(\dot{d}_{KL} + \delta_{KL}\dot{d}_{SS}\right)x^K x^L x^I}{8cr^4}. \tag{12.154}$$

The fact that the time-derivatives of d_{IJ} can be removed from $g_{\alpha\beta}$ by a *coordinate transformation* should be a warning for the readers: applying approximate methods to an unknown situation can lead, without anything visibly wrong in the formulae, to fictitious results. The assumptions underlying the approximation must be constantly kept under watch.

The final $h_{\alpha\beta}$ are obtained by substituting (12.153) and (12.154) in (12.152), and the final result for $g_{\alpha\beta}$, up to the desired combined order 3 in v/c and $r'/r \approx v/c$, is[8]

$$g_{00} = 1 - \frac{2m}{r} - \frac{x^I x^J}{r^5}(3D_{IJ} - D_{SS}\delta_{IJ}) + O(1/r^4), \qquad (12.155)$$

$$g_{0I} = -\frac{2\epsilon_{IJK}x^J P^K}{r^3} + O(1/r^3), \qquad (12.156)$$

$$g_{IJ} = \eta_{IJ}\left(1 + \frac{2m}{r} + O(1/r^2)\right), \qquad (12.157)$$

where ϵ_{IJK} is the Levi-Civita symbol and the new constants are related to the old ones by

$$m \stackrel{\text{def}}{=} \frac{\overline{\kappa}M}{8\pi}, \qquad P^I \stackrel{\text{def}}{=} \frac{\overline{\kappa}}{8\pi}\epsilon^{IJK}B_{JK}, \qquad D_{IJ} \stackrel{\text{def}}{=} \frac{\overline{\kappa}}{8\pi}d_{IJ}. \qquad (12.158)$$

For m and D_{IJ} these are just new units, while P^I is the **vector of angular momentum**.

The presence of P^I in the metric marks an important difference between general relativity and Newton's theory of gravitation. Let us take the Minkowski metric in the Lorentzian coordinates, $ds^2 = dt^2 - dx^2 - dy^2 - dz^2$, and let us transform it to the rotating coordinates, in which $x^I = x'^I + \epsilon^{IKL}\omega^K x'^L t$ for one fixed I, where the angular velocity of rotation ω^K is constant. We see that after the transformation g_{0I} becomes similar to (12.156). The contribution from rotation in (12.156) decreases quickly with distance, because of the r^3 in the denominator (missing in the transformed flat metric). The coordinates of (12.155) – (12.157) become inertial at infinity ($r \to \infty$), which means that the local inertial frame of the body discussed here is rotating with respect to the inertial frame of a distant observer. Thus, a rotating body in relativity is **dragging the inertial frames** with it; this phenomenon is also known as the **Lense–Thirring (1918) effect**.

Schiff predicted in 1960 that the dragging of inertial frames should cause precession of the axis of a gyroscope on an orbit around the Earth (Schiff, 1960a, b). The challenge was taken up by a group of experimental physicists at the Stanford University, led by Francis Everitt. The work began in 1963 (Everitt, 1974) and, in consequence of extreme technical requirements, lasted for more than 40 years. The final results, confirming the prediction of general relativity, were first announced at the 18th International Conference on General Relativity and Gravitation in 2007 (Everitt et al., 2008), then officially published (Everitt et al., 2011 and 2015).

In Newton's gravitation theory, in contrast to relativity, there is no direct influence of rotation of the source on the exterior field. There is only indirect influence because the centrifugal force caused by rotation flattens the source, and the changed distribution of mass changes the exterior field by producing a quadrupole moment.

12.19 Exercises

1. Verify that the Newtonian limit of (12.78) are the Euler equations of motion of a perfect fluid $\rho d\mathbf{v}/dt = -\text{grad}p$.

---

[8] The formula for g_{IJ} in Stephani (1990) resulted by taking into account terms of too high an order.

Hint. Use the asymptotically Cartesian coordinates in which the metric has the form (12.46) – (12.47), and observe that in these coordinates the 4-velocity u^α has the form $u^0 = 1$, $u^I = v^I/c$, $I = 1, 2, 3$.

2. Verify that (12.79) becomes the continuity equation $\partial\rho/\partial t + \mathrm{div}(\rho\mathbf{v}) = 0$ in the Newtonian limit.

3. Verify that Eqs. (12.82) – (12.83) do indeed represent a solution of the modified Einstein equations (12.80). The energy-momentum tensor is that of a perfect fluid, the velocity field is $u^\alpha = \delta^\alpha{}_0/c$. Do the calculation both by the tensor method and by using the differential forms. Find the pressure.

4. Prove that Eqs. (12.108) follow from (12.106) via (12.100) and (12.96).

Hint. Recall that the $g_{LM}(t_0)$ in (12.106) is the unit matrix. Consequently, the determinant g equals

$$-\exp\left[\int_{t_0}^t \frac{3M\tau + 2A\left(p_1 + p_2 + p_3\right)}{\frac{3}{4}M\tau^2 + A\tau + B}\, \mathrm{d}\tau\right].$$

5. Find the coordinate transformation from (12.106) to (12.109) when $M = 0$.

6. Verify (12.121).

Hint. Calculate $\left\{\begin{smallmatrix}4\\44\end{smallmatrix}\right\}$, $\left\{\begin{smallmatrix}4\\4I\end{smallmatrix}\right\}$, $\left\{\begin{smallmatrix}4\\IJ\end{smallmatrix}\right\}$, etc. (knowing that $g_{I4} = 0$ from (12.118)), and substitute the results in the definition of $R^4{}_{I4J}$. This will yield the second and third line of (12.121), using $g^{44} = 1/g_{44}$ along the way. Then recall that the coordinates of (12.121) are adapted to Σ and use (7.96) with notation modified as explained in the paragraph above (12.119). In the coordinates of (12.118) $f_{,4} = \mathcal{N}f_{,\rho} X^\rho$ for any function f.

7. Verify that the linearised $R^\alpha{}_{\beta\gamma\delta}$ of (12.125) indeed has the symmetries $R_{\alpha\beta\gamma\delta} = R_{\gamma\delta\alpha\beta} = -R_{\beta\alpha\gamma\delta}$ and $R_{\alpha[\beta\gamma\delta]} = 0$.

8. Derive (12.140) using (12.137) – (12.139).

Hint. Use (12.139) and (12.138) directly, then apply (12.137) to the factor $(r - |\mathbf{r} - \mathbf{r'}|)$. The rest is just discarding too high powers of $1/r$.

9. Derive (12.151).

10. Derive (12.155) – (12.158).

Hint. Use the hints given in the text. You will find that a great number of terms deriving from the b_α given by (12.154) drop out because they are of too high order in $r'/r \approx v/c$ (in fact, the b_I contribute nothing to g_{IJ}). Be careful to discard them before you begin to calculate. Note that $\partial/\partial x^I$ increases the order in $(1/r)$ by one.

13

The Maxwell and Einstein–Maxwell equations and the Kaluza–Klein theory

13.1 The Lorentz-covariant description of electromagnetic field

The separation of electromagnetic field into the electric and magnetic fields is not covariant with the Lorentz transformation; it depends on the velocity of the observer. In relativity, electromagnetic field is described by the antisymmetric **tensor of electromagnetic field** $F_{\mu\nu}$. In any fixed coordinate system, the electric field is

$$F^{0I} = E^I, \qquad I = 1, 2, 3 \tag{13.1}$$

and the magnetic field is

$$H^I = \frac{1}{2}\,\epsilon^{IJK} F_{JK}, \qquad I, J, K = 1, 2, 3. \tag{13.2}$$

The formula inverse to (13.2) is $F_{IJ} = \epsilon_{IJK} H^K$. If E^I and H^I are transformed by the Lorentz transformation, then $F_{\mu\nu}$ transforms like a tensor. The change in $F_{\mu\nu}$ implied by a coordinate transformation is interpreted as the relation between the electric and magnetic fields measured by two observers.

13.2 The covariant form of the Maxwell equations

In special relativity, the Maxwell equations are written as

$$F^{\mu\nu}{}_{,\nu} = \frac{4\pi}{c}\,j^\mu, \tag{13.3}$$

$$F_{[\mu\nu,\lambda]} = 0, \tag{13.4}$$

where j^μ is the 4-vector of current. Equation (13.3) is equivalent to the set $\{\text{div }\mathbf{E} = 4\pi\sigma,\ \text{rot }\mathbf{H} = (4\pi/c)\mathbf{j} + (1/c)\partial\mathbf{E}/\partial t\}$ and (13.4) to the set $\{\text{div }\mathbf{H}=0,\ \text{rot }\mathbf{E} + (1/c)\partial\mathbf{H}/\partial t=0\}$, where σ is the density of electric charge and $\mathbf{j}$ is the ordinary current vector. The covariance of the Maxwell equations with the Lorentz transformation is automatically guaranteed by $F_{\mu\nu}$ being a tensor. We also have

$$j^0 = c\sigma, \qquad \{j^I\} = \mathbf{j}. \tag{13.5}$$

In general relativity, the partial derivatives in (13.3) – (13.4) are replaced with covariant derivatives, thus (13.3) becomes

$$F^{\mu\nu}{}_{;\nu} = \frac{4\pi}{c}\, j^\mu. \tag{13.6}$$

Equation (13.4) need not be changed because the Christoffel symbols cancel out in the combination $F_{[\mu\nu;\lambda]}$.

One can define the differential form $F = F_{\mu\nu}\mathrm{d}x^\mu \wedge \mathrm{d}x^\nu$, and then (13.4) becomes $\mathrm{d}F = 0$. This is the condition of existence of a 1-form A such that $F = \mathrm{d}A$, where $A = A_\mu \mathrm{d}x^\mu$. The consequence (and solution) of (13.4) is thus the existence of the 4-potential A_μ. The continuity equation $j^\mu{}_{;\mu} = 0$ is an identical consequence of (13.6).

13.3 The energy-momentum tensor of electromagnetic field

The electromagnetic field $(\mathbf{E}, \mathbf{H})$ has the energy density

$$u = \frac{1}{8\pi}\left(\mathbf{E}^2 + \mathbf{H}^2\right), \tag{13.7}$$

its energy stream is

$$\mathbf{s} = \frac{1}{4\pi}\left(\mathbf{E} \times \mathbf{H}\right) \tag{13.8}$$

and its stress tensor is (Jackson 1975)

$$T_{IJ} = \frac{1}{4\pi}\left[\mathbf{E}_I\mathbf{E}_J + \mathbf{H}_I\mathbf{H}_J - \frac{1}{2}\left(\mathbf{E}^2 + \mathbf{H}^2\right)\delta_{IJ}\right]. \tag{13.9}$$

Just as we have done with the perfect fluid in Section 12.12, we can now compose these quantities into a 4-dimensional tensor of energy-momentum of electromagnetic field:

$$\overset{\text{e-m}}{T^{\alpha\beta}} = \frac{1}{4\pi}\left(F^{\alpha\mu}F_\mu{}^\beta + \frac{1}{4}g^{\alpha\beta}F_{\mu\nu}F^{\mu\nu}\right). \tag{13.10}$$

It has the property $T^\rho{}_\rho \equiv 0$. This can be written in the equivalent form

$$\overset{\text{e-m}}{T^{\alpha\beta}} = \frac{1}{8\pi}\left(F^{\alpha\mu}F_\mu{}^\beta + {}^*F^{\alpha\mu}\,{}^*F_\mu{}^\beta\right), \tag{13.11}$$

where

$${}^*F_{\mu\nu} \overset{\text{def}}{=} \frac{1}{2}\sqrt{-g}\epsilon_{\mu\nu\rho\sigma}F^{\rho\sigma} \tag{13.12}$$

is called the **dual** tensor to $F^{\mu\nu}$; it has the property ${}^{**}F_{\mu\nu} = -F^{\mu\nu}$. In the form (13.11) it is easy to verify that the energy-momentum tensor does not change when the electromagnetic field is transformed by the following operation, called the **duality rotation**:

$$\widetilde{F}_{\mu\nu} = F_{\mu\nu}\cos\delta + {}^*F_{\mu\nu}\sin\delta, \tag{13.13}$$

where δ is an arbitrary constant. Equation (13.13) implies that

$${}^*\widetilde{F}_{\mu\nu} = -F_{\mu\nu}\sin\delta + {}^*F_{\mu\nu}\cos\delta. \tag{13.14}$$

Consequently, although a given electromagnetic tensor uniquely defines the electromagnetic energy-momentum tensor (13.10), the converse is not true. Given $T^{\alpha\beta}$, the $F^{\mu\nu}$ is defined only up to duality rotations.

From (13.12) it is seen that duality interchanges the electric and magnetic fields: the electric field of $F_{\mu\nu}$ becomes the magnetic field of ${}^*F_{\mu\nu}$ and vice versa. The new electric and magnetic fields defined by $\widetilde{F}_{\mu\nu}$ are linear combinations of the old fields defined by $F_{\mu\nu}$.

The second set of Maxwell's equations, (13.4), is equivalent to

$$ {}^*F^{\mu\nu}{}_{;\nu} = 0. \tag{13.15} $$

This means that no magnetic monopoles (and currents) exist. The 'duality-rotated' ${}^*\widetilde{F}_{\mu\nu}$ will not always have this property. If the 'magnetic current' ${}^*\widetilde{F}^{\mu\nu}{}_{;\nu} = m^\mu \neq 0$ was created by the duality rotation, then of course the inverse duality rotation ((13.13) with δ replaced by $-\delta$) will remove it. But if ${}^*F^{\mu\nu}{}_{;\nu} = m^\mu \neq 0$ for the original $F^{\mu\nu}$, then a duality rotation setting $m^\mu = 0$ may not exist. The condition for it to exist is that the 'magnetic current' ${}^*F^{\mu\nu}{}_{;\nu}$ and the electric current $F^{\mu\nu}{}_{;\nu}$ are linearly dependent.

13.4 The Einstein–Maxwell equations

The energy-momentum tensor of electromagnetic field is a source of a gravitational field on equal footing with the energy-momentum tensor of matter. In the presence of electromagnetic field, the field equations of gravitation are

$$ R_{\alpha\beta} - \frac{1}{2}g_{\alpha\beta}R = \kappa\overset{\text{mat}}{T_{\alpha\beta}} + \overset{\text{e-m}}{T_{\alpha\beta}}, \tag{13.16} $$

where $\overset{\text{mat}}{T_{\alpha\beta}}$ and $\overset{\text{e-m}}{T_{\alpha\beta}}$ are the energy-momentum tensors of matter and of electromagnetic field, (13.10), respectively. The electromagnetic field must obey the Maxwell equations (13.4) and (13.6). The set $\{(13.4), (13.6), (13.10), (13.16)\}$ is called the **Einstein–Maxwell equations**.

The contracted Bianchi identities $\left(R^{\alpha\beta} - \frac{1}{2}g^{\alpha\beta}R\right)_{;\beta} = 0$ imply the equations of motion $T^{\alpha\beta}{}_{;\beta} = 0$ for the *whole* energy-momentum tensor $T_{\alpha\beta}$. In the absence of matter $(\overset{\text{mat}}{T_{\alpha\beta}} = 0)$ and currents $(j^\mu = 0)$, the Maxwell equations $\{(13.4), (13.6)\}$ do imply $\overset{\text{e-m}}{T^{\alpha\beta}}{}_{;\beta} = 0$. But the converse is not really true: the sourceless Maxwell equations (13.4) have to be assumed, then $\overset{\text{e-m}}{T^{\alpha\beta}}{}_{;\beta} = 0$ imply $F^\alpha{}_\rho F^{\rho\beta}{}_{;\beta} = 0$. In order to get (13.3) with $j^\mu = 0$ from this, one must assume that $\det\left(F^{\alpha\beta}\right) \neq 0$ (which is not always the case, for example, $\det\left(F^{\alpha\beta}\right) = 0$ if $F^{\alpha\beta}$ has zero magnetic field in any coordinates).

13.5 * The variational principle for the Maxwell and Einstein–Maxwell equations

The Maxwell equations follow from the variational principle $\delta\overset{\text{e-m}}{W} = 0$, where

$$ \overset{\text{e-m}}{W} \overset{\text{def}}{=} \int_{\mathcal{V}} \mathrm{d}_4 x \sqrt{-g} \left(\frac{1}{16\pi} F_{\mu\nu}F^{\mu\nu} + \frac{2}{c} j^\mu A_\mu\right), \qquad F_{\alpha\beta} \overset{\text{def}}{=} A_{\alpha,\beta} - A_{\beta,\alpha}, \tag{13.17} $$

i.e. the sourceless Maxwell equations (13.4) are pre-assumed and built into the lagrangian. The independent variables are A_α that obey the boundary condition $\delta A_\alpha|_{\partial V} = 0$.

In considering the variational principle for the full set of the Einstein–Maxwell equations we will assume $j^\mu = 0$ for simplicity. This is because one cannot consider the electric current without taking into account the motion of matter that generates the current, and we wish to avoid the associated complications. The action integral for that case is

$$W_g + \overset{\text{e-m}}{W} + \overset{\text{mat}}{W} = \int_V d_4 x \sqrt{-g}\left(R + \frac{1}{16\pi}\,F_{\mu\nu}F^{\mu\nu}\right) + \overset{\text{mat}}{W}. \tag{13.18}$$

To obtain the Einstein equations with electromagnetic field as a source, one must write $F_{\mu\nu}F^{\mu\nu} = g^{\mu\rho}g^{\nu\sigma}F_{\mu\nu}F_{\rho\sigma}$ and calculate the variation of (13.18) taking the inverse metric components $g^{\alpha\beta}$ as the independent variables.

13.6 * The Kaluza–Klein theory

In the 1920s, the only interactions known were electromagnetism and gravitation. Finding a geometric theory that would unite them was thus an attractive goal. The first such attempt was undertaken by Kaluza (1921). Several improvements of his theory were made in later years, but complete success has still not been achieved. The other basic reference is Klein (1926). The exposition given below is based on Appelquist and Chodos (1983).

The basic idea of Kaluza was that our world has five dimensions, that A_α are the components of the metric tensor along the fifth dimension and that the Einstein–Maxwell equations follow from vacuum Einstein equations in five dimensions. In the formulae below, the capital Latin indices A, B, C, ... will run through the values 0, 1, 2, 3, 4, and the Greek indices will run through the values 0, 1, 2, 3. We make the following assumptions (Appelquist and Chodos, 1983):

1. The 5-dimensional metric tensor $\|G_{AB}\|$ is built from a scalar field ϕ, the electromagnetic 4-potential A_μ and the spacetime metric $g_{\mu\nu}$ in the following way:

$$\|G_{AB}\| = \left(\begin{array}{c|c} g_{\mu\nu} + \phi A_\mu A_\nu & \phi A_\mu \\ \hline \phi A_\nu & \phi \end{array}\right). \tag{13.19}$$

2. The vector field $k^\mu = \delta^\mu{}_4$ tangent to the 5-dimensional manifold along the fifth dimension is a Killing field of G_{AB}, i.e. all components of G_{AB} are independent of x^4.

3. The fifth dimension is not observable because the 5-space is closed in the x^4-direction, and the circumference along it is very small.

One disadvantage of the KK theory is visible already here: it is not covariant with respect to 5-dimensional coordinate transformations. The metric (13.19) has the determinant

$$G = \det\|G_{AB}\| = \phi g, \tag{13.20}$$

where $g = \det \|g_{\mu\nu}\|$. Hence,

$$\|G^{AB}\| = \left(\begin{array}{c|c} g^{\mu\nu} & -A^\mu \\ \hline -A^\nu & \phi^{-1} + A_\mu A^\mu \end{array} \right). \tag{13.21}$$

For the various quantities below we will indicate the dimension of the space to which they refer by a subscript in parentheses. The 5-dimensional Christoffel symbols are

$$\left\{ \begin{array}{c} \alpha \\ \beta\gamma \end{array} \right\}_{(5)} = \left\{ \begin{array}{c} \alpha \\ \beta\gamma \end{array} \right\}_{(4)} + \frac{1}{2}\phi \left(A_\beta F^\alpha{}_\gamma + A_\gamma F^\alpha{}_\beta \right) - \frac{1}{2} g^{\alpha\rho} \phi_{,\rho} A_\beta A_\gamma,$$

$$\left\{ \begin{array}{c} \alpha \\ \beta 4 \end{array} \right\}_{(5)} = \frac{1}{2} \phi F^\alpha{}_\beta - \frac{1}{2} g^{\alpha\rho} \phi_{,\rho} A_\beta,$$

$$\left\{ \begin{array}{c} \alpha \\ 44 \end{array} \right\}_{(5)} = -\frac{1}{2} g^{\alpha\rho} \phi_{,\rho},$$

$$\left\{ \begin{array}{c} 4 \\ \beta\gamma \end{array} \right\}_{(5)} = -\frac{1}{2}\phi A^\rho A_\beta F_{\rho\gamma} - \frac{1}{2}\phi A^\rho A_\gamma F_{\rho\beta} + \frac{1}{2}\phi_{,\rho} A^\rho A_\beta A_\gamma$$

$$+ \frac{1}{2\phi} \left(\phi_{,\gamma} A_\beta + \phi_{,\beta} A_\gamma \right) + \frac{1}{2} \left(A_{\beta;\gamma} + A_{\gamma;\beta} \right),$$

$$\left\{ \begin{array}{c} 4 \\ \beta 4 \end{array} \right\}_{(5)} = -\frac{1}{2}\phi A^\rho F_{\rho\beta} + \frac{1}{2} A^\rho \phi_{,\rho} A_\beta + \frac{1}{2\phi} \phi_{,\beta},$$

$$\left\{ \begin{array}{c} 4 \\ 44 \end{array} \right\}_{(5)} = \frac{1}{2} A^\rho \phi_{,\rho}. \tag{13.22}$$

The components of the Ricci tensor are now found to be

$$R_{(5)\beta\delta} = R_{(4)\beta\delta} - \frac{1}{2}\phi F^\rho{}_\beta F_{\rho\delta} + \frac{3}{4} A_\beta \phi_{,\rho} F^\rho{}_\delta + \frac{3}{4} A_\delta \phi_{,\rho} F^\rho{}_\beta$$

$$+ \frac{1}{2}\phi A_\beta F^\rho{}_{\delta;\rho} + \frac{1}{2}\phi A_\delta F^\rho{}_{\beta;\rho} - \frac{1}{2}\phi^{;\rho}{}_{;\rho} A_\beta A_\delta + \frac{1}{4}\phi^2 A_\beta A_\delta F^{\sigma\rho} F_{\sigma\rho}$$

$$+ \frac{1}{4\phi} A_\beta A_\delta g^{\sigma\rho} \phi_{,\sigma} \phi_{,\rho} + \frac{1}{4\phi^2} \phi_{,\beta} \phi_{,\delta} - \frac{1}{2\phi} \phi_{;\beta\delta},$$

$$R_{(5)4\delta} = \frac{1}{2}\phi F^\rho{}_{\delta;\rho} - \frac{1}{2}\phi^{;\rho}{}_{;\rho} A_\delta + \frac{3}{4}\phi_{,\rho} F^\rho{}_\delta + \frac{1}{4}\phi^2 A_\delta F^{\sigma\rho} F_{\sigma\rho}$$

$$+ \frac{1}{4\phi} A_\delta g^{\sigma\rho} \phi_{,\sigma} \phi_{,\rho},$$

$$R_{(5)44} = -\frac{1}{2}\phi^{;\rho}{}_{;\rho} + \frac{1}{4}\phi^2 F^{\sigma\rho} F_{\sigma\rho} + \frac{1}{4\phi} g^{\sigma\rho} \phi_{,\sigma} \phi_{,\rho}, \tag{13.23}$$

and the 5-dimensional scalar curvature is

$$R_{(5)} = R_{(4)} - \frac{1}{4}\phi F^{\sigma\rho} F_{\sigma\rho} - \phi^{-1} \phi^{;\rho}{}_{;\rho} + \frac{1}{2\phi^2} g^{\sigma\rho} \phi_{,\sigma} \phi_{,\rho}. \tag{13.24}$$

The scalar field ϕ is an element of unknown interpretation, no object in Nature is known to correspond to it. This was initially seen as a weakness of the Kaluza–Klein theory, before the scalar field became one of favourite subjects in particle physics.

With $\phi = 1/(4\pi) = \text{constant}$, the 5-dimensional vacuum Einstein equations $R_{(5)AB} = 0$ reproduce the 4-dimensional Einstein equations with electromagnetic field as a source and the Maxwell equations $F^\rho{}_{\delta;\rho} = 0$, but impose the additional condition $F^{\mu\nu} F_{\mu\nu} = 0$, which means that $\mathbf{E}^2 = \mathbf{H}^2$. Thus, $R_{(5)AB} = 0$ do not reproduce the Einstein–Maxwell equations in their full generality, and this is another weakness of the Kaluza–Klein theory that remains uncorrected to this day.

The full Einstein–Maxwell equations may be recovered by the following trick. Take the action integral for the 5-dimensional vacuum Einstein equations:

$$W_{(5)} = \int_{\mathcal{V}_{(5)}} \mathrm{d}_5 x \sqrt{|G|} R_{(5)}. \tag{13.25}$$

Since the circumference of the 5-dimensional space along the fifth dimension, ℓ, is finite and G_{AB} does not depend on x^4, one can integrate with respect to x^4 in (13.25). Using (13.20) and (13.24), we obtain

$$W_{(5)} = \ell \int_{\mathcal{V}_{(4)}} \mathrm{d}_4 x \sqrt{|g\phi|} \left(R_{(4)} - \frac{1}{4} \phi F^{\mu\nu} F_{\mu\nu} - \phi^{-1} \phi^{;\rho}{}_{;\rho} + \frac{1}{2\phi^2} g^{\sigma\rho} \phi_{,\sigma} \phi_{,\rho} \right). \tag{13.26}$$

With $\phi = -1/(4\pi) = $ constant (yes, opposite sign to the one in (13.23)) and $\overset{\text{mat}}{W} = 0$, Eq. (13.26) reduces to (13.18). In this crooked sense, the Kaluza–Klein theory is the unified theory of gravitation and electromagnetism.[1]

The Kaluza–Klein theory became a prototype for other attempts to unify physical theories. The common idea is to take a multidimensional metric and interpret its components in higher dimensions as physical fields in spacetime. Although none of these theories passed experimental verification, the idea still remains popular and inspires new attempts.

13.7 Exercises

1. Verify that in the Minkowski spacetime the tensor (13.10) reduces to (13.7) – (13.9).
   Hint. Verify first, using (13.1) and (13.2), that $F_{\mu\nu} F^{\mu\nu} = 2 \left(\mathbf{H}^2 - \mathbf{E}^2 \right)$. Then recall what was stated in Sec. 12.12: in the signature $(+ - - -)$ the spatial part of the energy-momentum tensor (13.10) is $[-(\text{the stress tensor})]$, so $\overset{\text{e-m}}{T_{IJ}} = - (\text{the } T_{IJ} \text{ of } (13.9))$.
2. Show that the duality operation (13.13) has the property

$$^{**}F_{\mu\nu} = -F_{\mu\nu}. \tag{13.27}$$

Derive (13.14) from (13.13). Verify that the duality rotation does not change $T_{\alpha\beta}$.
3. Derive the Maxwell equations (13.6) from $\delta \overset{\text{e-m}}{W} / \delta A_\alpha = 0$ with $\overset{\text{e-m}}{W}$ given by (13.17).

---

[1] As we showed in Section 12.8, a variational principle with the lagrangian (13.25) leads to the full Einstein equations (12.21) with $T_{\alpha\beta} = 0$. This result is independent of the dimension of the manifold. The fact that taking the symmetry $g_{AB,4} = 0$ into account and transforming (13.25) to (13.26) leads to a *different* set of field equations is an illustration of the remark made at the end of Section 12.8.

14

Spherically symmetric gravitational fields
of isolated objects

In this chapter we will discuss spherically symmetric gravitational fields generated by isolated objects, such as stars and black holes.[1] Just as in Newton's theory, the spherically symmetric case is a foundation for the description of many phenomena taking place in real, more complicated objects.

14.1 The curvature coordinates

In Section 8.10 we showed that the metric form of a spherically symmetric spacetime, in a suitably adapted class of coordinates, has the form (8.52). The coordinates ϑ and φ are spherical coordinates whose values fix a point on a 2-sphere; t and r are arbitrary coordinates whose values are constant on each sphere – the orbit of the symmetry group. The general form of (8.52) and the formulae for the Killing fields (8.42) do not change under transformations that are composites of the following two classes:

- I. The isometries generated by the Killing fields via (8.15) – (8.16). These transformations preserve not just the form (8.52) but also the values of the individual components of the metric. An example is a rotation by the angle α_0 around the axis $\{\vartheta = \pi/2, \varphi = 0\}$ (this axis passes through the centre of the sphere and through a point on the equator). In the (ϑ, φ)-coordinates, such a rotation is given by

$$
\vartheta = \arctan\left[\frac{\sqrt{N}}{\cos\alpha_0 \cos\vartheta' - \sin\alpha_0 \sin\vartheta' \sin\varphi'}\right],
$$

$$
N \stackrel{\text{def}}{=} \sin^2\vartheta' \cos^2\varphi' + (\cos\alpha_0 \sin\vartheta' \sin\varphi' + \sin\alpha_0 \cos\vartheta')^2,
$$

$$
\varphi = \arctan\left(\cos\alpha_0 \tan\varphi' + \sin\alpha_0 \frac{\cot\vartheta'}{\cos\varphi'}\right). \tag{14.1}
$$

- II. The transformations (8.53) that preserve only the general form (8.52), but mix the functions α, β, γ and δ, and are of the form

$$
t = F(t', r'), \qquad r = G(t', r'), \tag{14.2}
$$

---

[1] Spherically symmetric solutions of Einstein's equations are also applied in cosmology, these will be discussed separately in Chapter 18.

where F and G are arbitrary functions of two variables subject to $\partial(F, G) \,/\, \partial(t', r') \neq 0$. After such a transformation, α, β and γ change as follows:

$$\widetilde{\alpha} = \alpha F_{,t'}{}^2 + 2\beta F_{,t'}\, G_{,t'} + \gamma G_{,t'}{}^2, \tag{14.3}$$

$$\widetilde{\beta} = \alpha F_{,t'}\, F_{,r'} + \beta \left(F_{,t'}\, G_{,r'} + F_{,r'}\, G_{,t'}\right) + \gamma G_{,t'}\, G_{,r'}, \tag{14.4}$$

$$\widetilde{\gamma} = \alpha F_{,r'}{}^2 + 2\beta F_{,r'}\, G_{,r'} + \gamma G_{,r'}{}^2, \tag{14.5}$$

while δ transforms like a scalar, $\widetilde{\delta}(t', r') = \delta(F, G)$.

Thus, using the functions F and G one can manipulate α, β and γ, but not δ. Because of the assumed signature $(+ - --)$ the determinant of the metric must be negative

$$\alpha\gamma - \beta^2 < 0. \tag{14.6}$$

Thanks to this inequality, the equation $\widetilde{\alpha} = 0$ can be solved for $F_{,t'}$; if $\alpha \neq 0$, then

$$F_{,t'} = \frac{1}{\alpha}\left(-\beta \pm \sqrt{\beta^2 - \alpha\gamma}\right) G_{,t'}. \tag{14.7}$$

We can thus take an arbitrary G and choose F to obey (14.7). Consequently, we lose no generality when we begin with coordinates in which $\alpha = 0$ already. Then we see from (14.4) that we can carry out such a transformation, after which $\widetilde{\beta} = 0$ (but the new $\widetilde{\alpha}$ will be nonzero). To achieve $\widetilde{\beta} = 0$ while $\alpha = 0$ we need to solve the equation

$$\beta G_{,r'}\, F_{,t'} + \beta G_{,t'}\, F_{,r'} + \gamma G_{,t'}\, G_{,r'} = 0. \tag{14.8}$$

Thus, G can be chosen arbitrarily, and then (14.8) becomes a quasilinear inhomogeneous partial differential equation for F, for which standard methods of solving exist.

We achieved $\beta = 0$ in two steps, and in both G was arbitrary. Hence, we can now choose G so as to achieve some further simplification.

This is where most textbooks make a mistake – they say they choose G so that $\widetilde{\delta} = -r'^2$. But δ is a scalar under the transformations (14.2). Hence, if $\delta = $ constant before the transformation, then $\widetilde{\delta} = \delta = $ constant, and no condition can be imposed on it. Thus, the further simplification of the form (8.52) must be considered more carefully.

Suppose that we wish to transform the coordinates of (8.52) so that

$$\widetilde{\beta} = 0, \qquad \widetilde{\delta} = -r'^2. \tag{14.9}$$

When is this possible? From the second of (14.9) we have

$$r' = \sqrt{-\delta(F(t', r'), G(t', r'))}. \tag{14.10}$$

Differentiating this by t' and by r' we obtain

$$\delta_{,F}\, F_{,t'} + \delta_{,G}\, G_{,t'} = 0, \tag{14.11}$$

$$\delta_{,F}\, F_{,r'} + \delta_{,G}\, G_{,r'} = -2\sqrt{-\delta} = -2r'. \tag{14.12}$$

While solving this set for $(G_{,t'}, G_{,r'})$, two cases must be considered:

- Ia. $\delta_{,G} \neq 0$. Then, substituting for $G_{,t'}$ from (14.11) in (14.3), we obtain

$$\widetilde{\alpha} = \frac{F_{,t'}{}^2}{\delta_{,G}{}^2}\left(\alpha\delta_{,G}{}^2 - 2\beta\delta_{,F}\, \delta_{,G} + \gamma\delta_{,F}{}^2\right). \tag{14.13}$$

We want t' to be the time coordinate, so $\tilde{\alpha} > 0$, which means that

$$\alpha \delta_{,G}{}^2 - 2\beta \delta_{,F}\, \delta_{,G} + \gamma \delta_{,F}{}^2 > 0. \tag{14.14}$$

This is easily verified to be equivalent to

$$g^{\alpha\beta}\delta_{,\alpha}\,\delta_{,\beta} < 0. \tag{14.15}$$

Hence, if $\delta_{,G} \neq 0$, then $\tilde{\delta} = -r'^2$ can be fulfilled only if $\delta_{,\alpha}$ is a spacelike vector.

- Ib. $\delta_{,G} = 0$. Then, from (14.11), either $\delta_{,F} = 0$, which means $\delta = $ constant (this case will be discussed below) or $\delta_{,F} \neq 0$ and $F_{,t'} = 0$, and then (14.3) implies that

$$\tilde{\alpha} = \gamma G_{,t'}{}^2. \tag{14.16}$$

Hence, $\tilde{\alpha} > 0$ only if $\gamma > 0$, which again implies (14.15) via (14.14). (Equation (14.16) shows that in this case r is the time-coordinate before the transformation.)

Hence, in either case the following lemma applies:

Lemma 14.1 (Case I) *The condition $\tilde{\delta} = -r'^2$ can be fulfilled only if $\delta_{,\alpha}$ is a spacelike vector.*

However, there are three other cases in which the set (14.11) – (14.12) cannot be solved at all. We leave it as an exercise for the reader to prove the remaining lemmas.

Lemma 14.2 (Case II) *If $\delta_{,\alpha}$ is a timelike vector, then one can choose the coordinates of (8.52) so that*

$$\tilde{\beta} = 0, \qquad \tilde{\delta} = -t'^2. \tag{14.17}$$

Lemma 14.3 (Case III) *If $\delta_{,\alpha}$ is a nontrivial null vector ($g^{\alpha\beta}\delta_{,\alpha}\,\delta_{,\beta} = 0$ but $\delta_{,\alpha} \neq 0$), then one can choose the coordinates of (8.52) so that $\tilde{\delta} = -r'^2$, but this automatically implies $\tilde{\alpha} = 0$. Hence, it is not possible to simultaneously achieve $\tilde{\beta} = 0$ because of (14.6), but it is possible to have in addition $\tilde{\gamma} = 0$.*

(Case IV) When δ is constant, no condition at all can be imposed on it. The spacetime is then a Cartesian product of a sphere of radius $l = \sqrt{-\delta}$ and a 2-dimensional surface with the metric $\alpha(t,r)\mathrm{d}t^2 + 2\beta(t,r)\mathrm{d}t\mathrm{d}r + \gamma(t,r)\mathrm{d}r^2$.

The possible existence of regions of spacetimes with $\delta_{,\alpha}$ being either spacelike or timelike was first noted by Novikov (1962b), who called them *R-regions* and *T-regions*, respectively.

Case II is ignored by most textbooks, even by those that subsequently describe the extension of the Schwarzschild manifold (see Section 14.9). Cases III and IV are ignored almost universally, with the notable exception of Stephani et al. (2003). The situation of these cases versus the Einstein equations is this:

For the vacuum equations with zero cosmological constant, the cases $g^{\alpha\beta}\delta_{,\alpha}\,\delta_{,\beta} < 0$ and $g^{\alpha\beta}\delta_{,\alpha}\,\delta_{,\beta} > 0$ are complementary to each other: they cover different parts of the same manifold, as we will see in Section 14.9.

However, cases III and IV require separate treatment. Case III has been investigated by Foyster and McIntosh (1972). It leads to a contradictory set of equations in pure vacuum, and the solution of the vacuum Einstein–Maxwell equations with this geometry is given in Stephani et al. (2003, Eq. (15.18)). Case IV leads to a contradiction when $T_{\alpha\beta} = 0 = \Lambda$, but a few solutions of Einstein's equations with more general sources are known. The vacuum solution with $\Lambda \neq 0$ was found by Nariai (1950), and discussed by Krasiński and Plebański (1980). The solutions of the Einstein–Maxwell equations with this geometry and with the source being the vacuum electrostatic field were found by Bertotti (1959) and Robinson (1959); they are compared with the Nariai solution in Krasiński (1999).

In this chapter, we will deal only with the case $g^{\alpha\beta}\delta_{,\alpha}\,\delta_{,\beta} < 0$. We choose the coordinates so as to achieve (14.9) and we denote $\widetilde{\alpha} = \mathrm{e}^{2\nu(t,r)}$, $\widetilde{\gamma} = -\mathrm{e}^{2\mu(t,r)}$, then

$$\mathrm{d}s^2 = \mathrm{e}^{2\nu(t,r)}\mathrm{d}t^2 - \mathrm{e}^{2\mu(t,r)}\mathrm{d}r^2 - r^2\left(\mathrm{d}\vartheta^2 + \sin^2\vartheta\mathrm{d}\varphi^2\right). \qquad (14.18)$$

This form of the metric still allows a subgroup of the transformations (14.2). We have $\delta = -r^2$ before and $\widetilde{\delta} = -r'^2$ after the transformation. Since δ transforms like a scalar, this means $r = r'$, so $G_{,t'} = 0$ and $G_{,r'} = 1$ in (14.3) – (14.5). We also require that $\beta = 0$ before and $\widetilde{\beta} = 0$ after the transformation, so from (14.4)

$$\alpha F_{,t'}\,F_{,r'} = 0. \qquad (14.19)$$

But $\alpha \neq 0$ necessarily, or else the metric would be singular. Also $F_{,t'} \neq 0$ because, with $G_{,t'} = 0$ already required, the Jacobian of the transformation would be zero. Hence, the consequence of (14.19) is $F_{,r'} = 0$, i.e. the transformations still allowed are $t = F(t')$, where F is an arbitrary function of one variable.

The coordinates of (14.18) are called **curvature coordinates** because in them the radial coordinate r is connected with the curvature $\mathcal{R}$ of the symmetry orbits in the same way as in the flat space, $\mathcal{R} = 1/r^2$.

14.2 Symmetry inheritance

Solutions of nonvacuum Einstein equations give rise to the **symmetry inheritance** problem: to what extent do the sources in the Einstein equations, coded in the energy-momentum tensor, inherit the symmetries of the metric of the spacetime?

For a perfect fluid, the matter density, pressure and velocity field are uniquely determined by the metric tensor. The energy-momentum tensor $T_{\alpha\beta}$ is simply a function of the metric, the matter density and pressure are the eigenvalues of $T_{\alpha\beta}$ and the velocity field is the unique timelike eigenvector of $T_{\alpha\beta}$. Hence, if the metric tensor does not change under some transformation, then neither does $T_{\alpha\beta}$, so also the matter density, pressure and velocity field remain unchanged.

However, if the source in the Einstein equations is an electromagnetic field, then, even though the electromagnetic energy-momentum tensor (13.10) does inherit all the symmetries of the metric, the tensor of electromagnetic field $F^{\mu\nu}$ does not necessarily do the same. The equation $\underset{k}{\pounds}T_{\alpha\beta} = 0$ puts some limitation on $F^{\mu\nu}$, but does not imply $\underset{k}{\pounds}F^{\mu\nu} = 0$, and examples of $F^{\mu\nu}$ not being invariant under the isometries of the metric are known (Wainwright and Yaremovicz, 1976; Li and Liang, 1985). However, if $T_{\alpha\beta}$ is invariant under the whole group $O(3)$, then so is $F^{\mu\nu}$, as proven by Wainwright and Yaremovicz (1976). We will accept this result without re-deriving it, and will make use of it in Section 14.4.

14.3 Spherically symmetric electromagnetic field in vacuum

In Section 8.10 we solved the equations $\underset{k_{(i)}}{\pounds}\, g_{\alpha\beta} = 0$ for the generators of the group $O(3)$, given by (8.42). This result can be used in solving $\underset{k_{(i)}}{\pounds}\, F_{\alpha\beta} = 0$ for the same generators. There is one point to be cautious about: since $g_{\alpha\beta} = g_{\beta\alpha}$, Eq. (8.50) implied $g_{23} = 0$. However, $F_{\alpha\beta} = -F_{\beta\alpha}$, so it fulfils (8.50) identically, i.e. F_{23} does not have to be zero. Also, because of antisymmetry, $F_{00} = F_{11} = F_{22} = F_{33} = 0$. Knowing this, and proceeding exactly like in Section 8.10, we conclude that only F_{01} and F_{23} can be nonzero, and then they must be of the form

$$F_{01} = f_{01}(t, r), \qquad F_{23} = f_{23}(t, r)\sin\vartheta, \qquad (14.20)$$

where f_{01} and f_{23} are arbitrary functions of two variables.

Substituting (14.20) in the Maxwell equations in Minkowski spacetime we find that F_{23} is the exterior field of a magnetic monopole. In agreement with classical electrodynamics ('no magnetic monopoles exist') we would thus tend to assume $F_{23} = 0$. However, this follows from experiments, not from spherical symmetry – Maxwell's equations do allow such a solution. We will keep $F_{23} \neq 0$ for a while, and then we will find out that in vacuum the magnetic monopole may be eliminated by the duality rotation (13.13).

The equations $F_{[\mu\nu,\lambda]} = 0$ do not impose any limitation on F_{01}, but for F_{23} they imply

$$f_{23} = \sqrt{8\pi}q = \text{constant} \qquad (14.21)$$

(the coefficient $\sqrt{8\pi}$ was introduced in order to simplify later calculations). The equations $F^{\mu\nu}{}_{;\nu} = 0$ may be written in the form $(\sqrt{-g}F^{\mu\nu})_{,\nu} = 0$; then they become

$$\left(r^2 e^{\mu+\nu}F^{01}\right)_{,t} = 0 = \left(r^2 e^{\mu+\nu}F^{01}\right)_{,r}, \qquad (14.22)$$

while $F^{23} = \sqrt{8\pi}q/\left(r^4\sin\vartheta\right)$ obeys $F^{\mu\nu}{}_{,\nu} = 0$ identically. The solution of (14.22) is

$$F^{01} = \sqrt{8\pi}e\; e^{-\mu-\nu}/r^2, \qquad (14.23)$$

where e is another arbitrary constant (and the coefficient $\sqrt{8\pi}$ was again introduced for later convenience). Now the duality rotation (13.13) with $\delta = -\arctan(q/e)$ leads to the new field $\widetilde{F}^{\mu\nu}$ in which $\widetilde{F}_{23} = 0$, thus $\widetilde{q} = 0$ and $\widetilde{e} = -\text{sign}(e)\sqrt{e^2 + q^2}$, while $\overset{\text{e-m}}{T}_{\alpha\beta}$ does not change.[2] Hence, with no loss of generality we may now assume $q = 0 = F_{23}$.

14.4 The Schwarzschild and Reissner–Nordström solutions

As stated in the last sentence above, we assume $q = 0 = F_{23}$. For calculating the energy-momentum tensor of the $F_{\alpha\beta}$ given by (14.23), the following formulae are useful:

$$\begin{aligned}
F_{0\rho}F^{\rho}{}_0 &= 8\pi e^2 e^{2\nu}/r^4, & F_{1\rho}F^{\rho}{}_1 &= -8\pi e^2 e^{2\mu}/r^4, \\
F_{\mu\nu}F^{\mu\nu} &= -16\pi e^2/r^4; & & \qquad (14.24)
\end{aligned}$$

---

[2] Several solutions of the Einstein–Maxwell equations include the electric and magnetic charge through the combination $e^2 + q^2$. The generalisation with respect to pure electric charge is thus rather illusory, since the magnetic charge may then be generated/removed/changed by duality rotations and does not independently influence the geometry of spacetime.

the components of $F_{\alpha\rho}F^\rho{}_\beta$ that are not listed are identically zero. For the metric (14.18), the most convenient tetrad of differential forms is the orthonormal one

$$e^0 = e^\nu dt, \qquad e^1 = e^\mu dr, \qquad e^2 = rd\vartheta, \qquad e^3 = r\sin\vartheta d\varphi. \qquad (14.25)$$

The nonzero tetrad components of the Ricci tensor for (14.25) are

$$\begin{aligned}
R_{\widehat{0}\widehat{0}} &= e^{-2\mu}\nu'' - e^{-2\mu}\nu'\mu' + e^{-2\mu}\nu'^2 + 2e^{-2\mu}\nu'/r \\
&\quad -e^{-2\nu}\ddot\mu + e^{-2\nu}\dot\nu\dot\mu - e^{-2\nu}\dot\mu^2, &(14.26)
\end{aligned}$$

$$R_{\widehat{0}\widehat{1}} = 2e^{-\nu-\mu}\dot\mu/r, \qquad (14.27)$$

$$\begin{aligned}
R_{\widehat{1}\widehat{1}} &= -e^{-2\mu}\nu'' + e^{-2\mu}\nu'\mu' - e^{-2\mu}\nu'^2 + 2e^{-2\mu}\mu'/r \\
&\quad +e^{-2\nu}\ddot\mu - e^{-2\nu}\dot\nu\dot\mu + e^{-2\nu}\dot\mu^2, &(14.28)
\end{aligned}$$

$$R_{\widehat{2}\widehat{2}} = R_{\widehat{3}\widehat{3}} = e^{-2\mu}(\mu' - \nu')/r + \left(1 - e^{-2\mu}\right)/r^2, \qquad (14.29)$$

where $\dot{} = \partial/\partial t$ and $' = \partial/\partial r$. The nonzero tetrad components of $F_{\alpha\rho}F^\rho{}_\beta$ are now found from (14.24) and (14.25):

$$F_{\widehat{0}s}F^s{}_{\widehat{0}} = -F_{\widehat{1}s}F^s{}_{\widehat{1}} = 8\pi e^2/r^4. \qquad (14.30)$$

We will solve the Einstein–Maxwell equations with the cosmological constant and with the energy-momentum tensor (13.10), so, in tetrad components

$$R_{ij} - \frac{1}{2}\eta_{ij}R + \Lambda\eta_{ij} = \frac{1}{4\pi}\left(F_{is}F^s{}_j + \frac{1}{4}\eta_{ij}F_{\mu\nu}F^{\mu\nu}\right). \qquad (14.31)$$

The trace of (14.31) implies $R = 4\Lambda$. Substituting this in (14.31) we obtain:

$$R_{ij} = \Lambda\eta_{ij} + \frac{1}{4\pi}\left(F_{is}F^s{}_j + \frac{1}{4}\eta_{ij}F_{\mu\nu}F^{\mu\nu}\right). \qquad (14.32)$$

Since $||\eta_{ij}|| = \mathrm{diag}(+1, -1, -1, -1)$, Eqs. (14.30) and (14.32) imply

$$R_{\widehat{0}\widehat{0}} + R_{\widehat{1}\widehat{1}} = 0. \qquad (14.33)$$

Substituting (14.26) and (14.28) in this we obtain $2e^{-2\mu}(\mu' + \nu')/r = 0$, which solves as

$$\nu = -\mu + f(t), \qquad (14.34)$$

where $f(t)$ is an arbitrary function. Then, from $R_{\widehat{0}\widehat{1}} = 0$ we obtain $\mu = \mu(r)$. Now recall that the metric (14.18) does not change its form under the transformations $t' = g(t)$, where $g(t)$ is an arbitrary function. Consequently, we transform t in (14.18) by

$$t' = \int e^{f(t)}dt, \qquad (14.35)$$

where $f(t)$ is the function from (14.34). The result of (14.35) is as if $f(t) \equiv 0$. Hence, with no loss of generality, $\nu = -\mu$, and the whole metric tensor becomes independent of the time coordinate. Moreover, in (14.23) we obtain $F^{01} = \sqrt{8\pi}e/r^2$, which means that the electromagnetic tensor is also independent of time.

The statement thus obtained goes by the name of **Birkhoff theorem**: the spherically symmetric gravitational field in vacuum is static.(Here we have in fact derived a generalisation–our source is a vacuum electromagnetic field.) However, calling it a theorem is an exaggeration. It is a side-result of straightforward calculations and finds no other application. Moreover, it is, in this formulation, false: it holds only for those spacetime regions where $\delta_{,\alpha}$ is spacelike.

From $R_{\widehat{22}} = -\Lambda - [1/(16\pi)]F_{\mu\nu}F^{\mu\nu}$ we now obtain with the help of (14.29) and (14.34):

$$1 - e^{-2\mu} + 2re^{-2\mu}\mu' = -\Lambda r^2 + e^2/r^2, \tag{14.36}$$

which is easily integrated with the result

$$e^{-2\mu} = 1 + \frac{C}{r} + \frac{1}{3}\Lambda r^2 + \frac{e^2}{r^2} = e^{2\nu}, \tag{14.37}$$

where C is an arbitrary constant. The components $i = j = 0$ and $i = j = 1$ of (14.32) are now fulfilled identically. For the strictly vacuum field ($\Lambda = 0 = e$) we obtain from here

$$g_{00} = e^{2\nu} = 1 + C/r, \tag{14.38}$$

while for weak gravitational fields (12.46) should hold. In Newton's theory, for a spherically symmetric gravitational field in vacuum we have

$$\phi = -GM/r, \tag{14.39}$$

where G is the gravitational constant and M is the mass of the source of the field. Equations (12.46) and (14.39) are consistent with (14.38) when $C = -2GM/c^2 \overset{\text{def}}{=} -2m$.

Finally then, we found the following solution:

$$ds^2 = e^{2\nu}dt^2 - e^{-2\nu}dr^2 - r^2\left(d\vartheta^2 + \sin^2\vartheta d\varphi^2\right), \tag{14.40}$$

where

$$e^{2\nu} = 1 - \frac{2m}{r} + \frac{e^2}{r^2} + \frac{1}{3}\Lambda r^2. \tag{14.41}$$

Special cases of this solution were derived by different authors during the first few years after general relativity was published in 1915 and today they are known under different names. The case $\Lambda = e = 0$ was found by Schwarzschild (1916a) and was historically the first exact solution of the Einstein equations ever found.[3] A large part of relativistic astrophysics and several experimental tests of general relativity are based on it. The subcase $\Lambda = 0$ was found by Reissner (1916) and Nordström (1918). The subcase $m = e = 0$ was found by de Sitter (1916). The subcase $e = 0$ was found by Kottler (1918). In full generality, the metric (14.40) – (14.41) appeared as a subcase of still more general solutions found by Cahen and Defrise (1968) and by Kinnersley (1969).

We will mostly deal with the Schwarzschild solution for which

$$e^{2\nu} = 1 - \frac{2m}{r}. \tag{14.42}$$

Note that for $r \to 2m$ this metric apparently has a singularity, since $g_{00} \to 0$ and $g_{11} \to -\infty$. However, if we substitute in (14.40) and (14.42) a value $r < 2m$, then the

---

[3] A year later, Droste (1917) presented another, mathematically more elegant derivation of the same solution and discussed its geometrical properties. His paper was surprisingly insightful and ahead of its time. Consequently, it would be quite appropriate to refer to this metric as the Schwarzschild–Droste solution (Rothman, 2002). An English translation of the Schwarzschild paper was published in *Gen. Relativ. Gravit.* **35**, 951 (2003), but the editorial note to it contains incorrect claims about its interpretation – see Senovilla (2007) for rectifications.

metric is still nonsingular, but $g_{00} < 0$ and $g_{11} > 0$, i.e. the r-coordinate becomes time and the t-coordinate becomes a measure of the radial distance. The region $r < 2m$ is the complementary part to $r > 2m$ that was mentioned after Lemma 14.3. In that region $g^{\alpha\beta}\delta_{,\alpha}\,\delta_{,\beta} > 0$, and the 'Birkhoff theorem' does not hold.[4] We will discuss the geometrical and physical interpretation of the spurious singularity at $r = 2m$ in Section 14.11.

14.5 Orbits of planets in the gravitational field of the Sun

In general relativity, just as in Newton's theory, it is assumed that the planets in their motion on the orbits behave like point masses, while their own gravitational fields are weak and can be neglected in comparison with the gravitational field of the Sun. These two assumptions in fact contradict each other. The gravitational field of a point mass is singular at the mass' position, and so stronger than any exterior field. In Newton's theory this difficulty is solved by the observation that the centre of mass of any extended body follows the same trajectory as a point body with no self-gravitation would follow. At the centre of mass, the body's own gravitational field vanishes.

In relativity, so far there is not even a generally accepted definition of the centre of mass, although work on this problem is being done (see e.g. Dixon, 2015). Thus, when we consider the orbits of point bodies in relativity, we are in fact extending the theory into the domain in which it has not been worked out yet. Nevertheless, these results agree with observational tests.

We assume that the gravitational field of the Sun is spherically symmetric, that the space around the Sun does not contain electromagnetic fields and that the cosmological constant is zero. This means that the spacetime will be described by the metric form (14.40) – (14.42). The orbits of planets should be timelike geodesics in this spacetime, and hence should be solutions of (5.14), where the derivative D/ds is calculated along the geodesic. Contracting (5.14) with $g_{\alpha\beta}\mathrm{d}x^{\beta}/\mathrm{d}s$ and using (5.14) in the result, we obtain

$$0 = \frac{\mathrm{D}}{\mathrm{d}s}\left(g_{\alpha\beta}\frac{\mathrm{d}x^{\alpha}}{\mathrm{d}s}\frac{\mathrm{d}x^{\beta}}{\mathrm{d}s}\right) = \frac{\mathrm{d}}{\mathrm{d}s}\left(g_{\alpha\beta}\frac{\mathrm{d}x^{\alpha}}{\mathrm{d}s}\frac{\mathrm{d}x^{\beta}}{\mathrm{d}s}\right). \tag{14.43}$$

Hence, the quantity $\ell \stackrel{\mathrm{def}}{=} g_{\alpha\beta}\left(\mathrm{d}x^{\alpha}/\mathrm{d}s\right)\left(\mathrm{d}x^{\beta}/\mathrm{d}s\right)$ is constant along every geodesic. This means in particular that its sign cannot change during motion; a geodesic is thus timelike $(\ell > 0)$, null $(\ell = 0)$ or spacelike $(\ell < 0)$ along its whole length.

The affine parameter s is determined up to inhomogeneous linear transformations, so for timelike geodesics it can be scaled so that

$$g_{\alpha\beta}\frac{\mathrm{d}x^{\alpha}}{\mathrm{d}s}\frac{\mathrm{d}x^{\beta}}{\mathrm{d}s} = 1. \tag{14.44}$$

This affine parameter is $s = c\times$(the physical time), where c is the velocity of light. On the orbit we wish to use the physical units, so in the following our time coordinate will be ct.

---

[4] This region is an example of a spacetime with the Kantowski–Sachs symmetry, see the footnote after (10.40). In fact, it is the unique vacuum solution of Einstein's equations with this symmetry.

With this, the Christoffel symbols for the metric (14.40) – (14.42) are:

$$\begin{Bmatrix} 0 \\ 01 \end{Bmatrix} = \frac{m}{r^2} \frac{1}{1 - 2m/r} = -\begin{Bmatrix} 1 \\ 11 \end{Bmatrix},$$

$$\begin{Bmatrix} 1 \\ 00 \end{Bmatrix} = \frac{mc^2}{r^2} - \frac{2m^2c^2}{r^3},$$

$$\begin{Bmatrix} 1 \\ 22 \end{Bmatrix} = 2m - r = \begin{Bmatrix} 1 \\ 33 \end{Bmatrix} / \sin^2 \vartheta,$$

$$\begin{Bmatrix} 2 \\ 12 \end{Bmatrix} = \frac{1}{r} = \begin{Bmatrix} 3 \\ 13 \end{Bmatrix},$$

$$\begin{Bmatrix} 2 \\ 33 \end{Bmatrix} = - \cos \vartheta \sin \vartheta, \qquad \begin{Bmatrix} 3 \\ 23 \end{Bmatrix} = \cot \vartheta. \tag{14.45}$$

Hence, the equations of a geodesic are:

$$c\frac{\mathrm{d}^2 t}{\mathrm{d}s^2} + \frac{2m}{r^2} \frac{1}{1 - 2m/r} c\frac{\mathrm{d}t}{\mathrm{d}s} \frac{\mathrm{d}r}{\mathrm{d}s} = 0, \tag{14.46}$$

$$\frac{\mathrm{d}^2 r}{\mathrm{d}s^2} + \left(\frac{mc^2}{r^2} - \frac{2m^2c^2}{r^3} \right) \left(\frac{\mathrm{d}t}{\mathrm{d}s} \right)^2 - \frac{m}{r^2} \frac{1}{1 - 2m/r} \left(\frac{\mathrm{d}r}{\mathrm{d}s} \right)^2$$

$$+ (2m - r) \left(\frac{\mathrm{d}\vartheta}{\mathrm{d}s} \right)^2 + (2m - r) \sin^2 \vartheta \left(\frac{\mathrm{d}\varphi}{\mathrm{d}s} \right)^2 = 0, \tag{14.47}$$

$$\frac{\mathrm{d}^2 \vartheta}{\mathrm{d}s^2} + \frac{2}{r} \frac{\mathrm{d}r}{\mathrm{d}s} \frac{\mathrm{d}\vartheta}{\mathrm{d}s} - \cos \vartheta \sin \vartheta \left(\frac{\mathrm{d}\varphi}{\mathrm{d}s} \right)^2 = 0, \tag{14.48}$$

$$\frac{\mathrm{d}^2 \varphi}{\mathrm{d}s^2} + \frac{2}{r} \frac{\mathrm{d}r}{\mathrm{d}s} \frac{\mathrm{d}\varphi}{\mathrm{d}s} + 2 \cot \vartheta \frac{\mathrm{d}\vartheta}{\mathrm{d}s} \frac{\mathrm{d}\varphi}{\mathrm{d}s} = 0. \tag{14.49}$$

The last equation may be written as $(\mathrm{d}/\mathrm{d}s)\left[r^2 \sin^2 \vartheta (\mathrm{d}\varphi/\mathrm{d}s) \right] = 0$, so

$$\frac{\mathrm{d}\varphi}{\mathrm{d}s} = \frac{J_0}{r^2 \sin^2 \vartheta}, \tag{14.50}$$

where J_0 is an arbitrary constant. Substituting this in (14.48) we obtain

$$\frac{\mathrm{d}^2 \vartheta}{\mathrm{d}s^2} + \frac{2}{r} \frac{\mathrm{d}r}{\mathrm{d}s} \frac{\mathrm{d}\vartheta}{\mathrm{d}s} - \frac{J_0^2 \cos \vartheta}{r^4 \sin^3 \vartheta} = 0. \tag{14.51}$$

A first integral of this equation is

$$r^4 \left(\frac{\mathrm{d}\vartheta}{\mathrm{d}s} \right)^2 + J_0^2 \cot^2 \vartheta = A = \text{constant}. \tag{14.52}$$

We still have the freedom to rotate the axes of the spherical coordinates – the metric is invariant under such rotations, and we have not applied any of them so far. Suppose that at the initial instant $\vartheta(s_0) = \vartheta_0$. Rotate the axis of the spherical coordinates so that at s_0 the planet is on the equator, i.e. $\vartheta_0 = \pi/2$. Then we still have the freedom to rotate the spherical coordinates around the single axis passing through $x^\alpha(s_0)$. Apply this rotation by such an angle that the initial velocity of the planet will lie in the plane of the equator, i.e. $\mathrm{d}\vartheta/\mathrm{d}s|_{s=s_0} = 0$. Substituting the initial conditions $\vartheta_0 = \pi/2$ and $\mathrm{d}\vartheta/\mathrm{d}s|_{s=s_0} = 0$ in (14.52) we obtain $A = 0$. But A is a sum of two nonnegative terms, hence both of them must be zero at all times. Disregarding the uninteresting case $J_0 = 0$, which is a straight radial trajectory, $A = 0$ means that

$$\vartheta = \pi/2 \tag{14.53}$$

on the whole orbit. Substituting (14.53) in (14.50) we obtain

$$\frac{d\varphi}{ds} = \frac{J_0}{r^2}. \tag{14.54}$$

A first integral of (14.46) is

$$\left(1 - \frac{2m}{r}\right) c\frac{dt}{ds} = E, \tag{14.55}$$

where E is an arbitrary constant. The only equation of the set (14.46) – (14.49) not yet used is (14.47), but we can replace it by (14.44), which, using (14.53) – (14.55), becomes

$$\frac{E^2}{1 - 2m/r} - \frac{1}{1 - 2m/r} \left(\frac{dr}{ds}\right)^2 - \frac{J_0{}^2}{r^2} = 1, \tag{14.56}$$

which can be written as[5]

$$\frac{dr}{ds} = \left[E^2 - \left(1 + \frac{J_0{}^2}{r^2}\right)\left(1 - \frac{2m}{r}\right)\right]^{1/2}. \tag{14.57}$$

Now we have three first-order equations to solve, (14.54), (14.55) and (14.57). Usually, we are interested only in the shape of the orbit given by (14.54) and (14.57). The third equation gives information on the dependence of the position of the planet on time.

If $J_0 = 0$, then, from (14.54), $\varphi = $ constant, which means that the 'planet' falls directly towards the centre of the Sun. We omit this case, and assume $J_0 \neq 0$. Then (14.54) and (14.57) can be replaced by just one equation

$$\left(\frac{J_0}{r^2} \frac{dr}{d\varphi}\right)^2 = E^2 - \left(1 + \frac{J_0{}^2}{r^2}\right)\left(1 - \frac{2m}{r}\right). \tag{14.58}$$

Now we introduce the new variable $\sigma = 1/r$ and obtain in (14.58)

$$\left(J_0 \frac{d\sigma}{d\varphi}\right)^2 = E^2 - \left(1 + J_0{}^2\sigma^2\right)(1 - 2m\sigma). \tag{14.59}$$

An exact solution of this equation would lead to elliptic integrals. We prefer to solve it approximately, up to first order in the small parameter defined below. For that calculation, it is more convenient to go one step back and differentiate (14.59) by φ to obtain

$$\frac{d\sigma}{d\varphi} \cdot 2J_0{}^2 \frac{d^2\sigma}{d\varphi^2} = \frac{d\sigma}{d\varphi} \left[-2J_0{}^2\sigma + 2m\left(1 + 3J_0{}^2\sigma^2\right)\right]. \tag{14.60}$$

---

[5] We omit the minus sign in front of the right-hand side in (14.57) because the resulting solutions describe motions along the same orbits in the opposite direction. In a spherically symmetric spacetime, the reversal of the sense of orbital motion is equivalent to a coordinate transformation (reflection). This is not so when the source of the gravitational field is rotating, where the orbital motion can be **direct**, i.e. with angular velocity parallel to that of the source, or **retrograde**, i.e. with the two angular velocities antiparallel. We shall meet this problem in the Kerr spacetime, see Section 21.8.

One solution of this is $d\sigma/d\varphi = 0$, i.e. motion on a circular orbit with $r = $ constant. When $d\sigma/d\varphi \neq 0$, (14.60) becomes

$$\frac{d^2\sigma}{d\varphi^2} + \sigma = \frac{m}{J_0{}^2} + 3m\sigma^2. \tag{14.61}$$

This differs from the corresponding Newtonian equation by the last term. It is seen from (14.54) that $J_0 \propto 1/c$. In order that the first term on the right takes the Newtonian form $GM\mu^2/J^2$ at $c \to \infty$ we assume $J_0 = J/(\mu c)$ (M is the mass of the source, μ is the mass of the planet and J is the orbital angular momentum of the planet). We denote

$$\frac{mc^2\mu^2}{J^2} \equiv \frac{GM\mu^2}{J^2} \stackrel{\text{def}}{=} \frac{1}{p}, \qquad 3m \equiv \frac{3GM}{c^2} \stackrel{\text{def}}{=} \alpha. \tag{14.62}$$

In the Newtonian limit, α becomes zero, while p is the **parameter** of the orbit: the distance from the Sun to the planet at that instant at which the Sun–planet direction is perpendicular to the Sun–perihelion direction. In this notation, (14.61) becomes

$$\frac{d^2\sigma}{d\varphi^2} + \sigma = \frac{1}{p} + \alpha\sigma^2. \tag{14.63}$$

We will solve this equation up to linear terms in α, taking α to be the small parameter. More precisely, since α has the dimension of distance, the small parameter will be the dimensionless number $\alpha\sigma$. For the Sun, $\alpha \approx 4.5$ km, while the radius of the Sun is approximately 7×10^5 km, so the planetary orbits of necessity have still larger radii, and $\alpha\sigma < 0.64 \times 10^{-5}$. Since σ and $1/p$ are comparable, $\alpha\sigma^2 \ll \sigma$ and $\alpha\sigma^2 \ll 1/p$, and the last term in (14.63) is small compared to all other terms. However, for strongly concentrated objects such as neutron stars or black holes, the orbits may come much closer to $r = 0$, and the approximation used here becomes invalid. In those cases, one has to resort either to numerical calculations or to calculations with elliptic functions.[6]

In the zeroth order of approximation, the solution of (14.63) is the elliptic orbit

$$\sigma_0 = \frac{1}{p} + \frac{\varepsilon}{p} \cos(\varphi - \varphi_0), \tag{14.64}$$

where ε is the **eccentricity** of the orbit, which is also a small quantity for all solar planets (for Pluto, $\varepsilon = 0.2444$, for Mercury $\varepsilon = 0.205628$; for other planets it is still smaller).

In the first-order approximation, we assume $\sigma = \sigma_{[1]} = \sigma_0 + \sigma_1$, where σ_1 is of the order of α. We substitute $\sigma_{[1]}$ for σ in (14.63), use (14.64), neglect terms of order α^2, and obtain

$$\frac{d^2\sigma_1}{d\varphi^2} + \sigma_1 = \frac{\alpha}{p^2} \left[1 + 2\varepsilon\cos(\varphi - \varphi_0) + \varepsilon^2\cos^2(\varphi - \varphi_0)\right]. \tag{14.65}$$

The solution of the homogeneous part of (14.65) is $(\varepsilon/p)\cos(\varphi - \varphi_0)$. To find a special solution of the full inhomogeneous equation is a textbook problem; it is

$$\sigma_1 = \frac{\alpha}{p^2}\left(1 + \frac{\varepsilon^2}{3}\right) + \frac{\alpha\varepsilon}{p^2}(\varphi - \varphi_0)\sin(\varphi - \varphi_0) + \frac{\alpha\varepsilon^2}{3p^2}\sin^2(\varphi - \varphi_0). \tag{14.66}$$

---

[6] In real astrophysical situations, orbits are perturbed in so many ways that numerical treatment becomes necessary anyway. Examples: gravitation of other planets, rotation of the star or black hole, lack of spherical symmetry caused by rotation of the central body, loss of orbital angular momentum because of friction with interplanetary matter.

However, the $\sigma_0 + \sigma_1$ implied by such σ_1 is in violent disagreement both with Newton's theory and with observations. At $\varphi - \varphi_0 = n\pi$, $n = 1, 2, 3, \ldots$, the curve given by (14.66) passes through always the same two points, but in the sectors where $\sin(\varphi - \varphi_0) > 0$ σ_1 increases systematically with every revolution, so $r \to 0$ as $\varphi \to \infty$. In the sectors where $\sin(\varphi - \varphi_0) < 0$, σ_1 decreases systematically with every revolution and will at a certain instant cause that $\sigma_{[1]} = 0$, i.e. $r \to \infty$. The term responsible for this unusual behaviour is $(\alpha\varepsilon/p^2)(\varphi - \varphi_0)\sin(\varphi - \varphi_0)$. On the other hand, it is easy to note that with $E^2 < 1$ the solution of (14.59) must be bounded (see Exercise 4). We conclude that this trouble-causing term must be the first term of a power series representing a bounded function $f(\varphi)$ and that in the first order solution to (14.63) we must include $f(\varphi)$ in full. Going to higher orders of approximation to identify $f(\varphi)$ is a laborious way of solving the problem, since, with every added order, many new terms appear. We want the solution to reduce to (14.66) only to first order, and any bounded function with this property will do. After some guesswork, the function

$$f(\varphi) = \frac{\varepsilon}{p}\,\cos\left[\left(1 - \frac{\alpha}{p}\right)(\varphi - \varphi_0)\right] \tag{14.67}$$

is found to be the right one; it includes the cos-term of (14.64). This function is bounded for all values of α, but replacing it by the 'approximation' linear in α caused the nonsensical behaviour of (14.66). We will come back to this at the end of this section.

Hence, the correct approximation to the solution of (14.63) up to terms linear in α is:

$$\sigma = \frac{1}{p} + \frac{\alpha}{p^2}\left(1 + \frac{\varepsilon^2}{3}\right) + \frac{\varepsilon}{p}\,\cos\left[\left(1 - \frac{\alpha}{p}\right)(\varphi - \varphi_0)\right] + \frac{\alpha\varepsilon^2}{3p^2}\,\sin^2(\varphi - \varphi_0). \tag{14.68}$$

Since $\varepsilon\alpha/p \ll 1$, the third term on the right dominates over the last one, which can thus be neglected. After the planet has made one full revolution, $\varphi \to \varphi + 2\pi$, σ does not return to its initial value. Instead, it returns (approximately) to its initial value after the planet has revolved by $2\pi/(1 - \alpha/p) > 2\pi$. Consequently, the aphelion and the perihelion revolve around the Sun in the same direction as the planet. Two consecutive perihelia are separated by the angle $\Delta\varphi = 2\pi/(1 - \alpha/p) - 2\pi = 2\pi\alpha/p + O((\alpha/p)^2)$, and the formula for this perihelion shift can be written in three equivalent ways:

$$\Delta\varphi \approx \frac{2\pi\alpha}{p} = \frac{6\pi GM}{c^2 p} = \frac{6\pi GM}{c^2 b(1 + \varepsilon)} = \frac{6\pi GM}{c^2 a\,(1 - \varepsilon^2)}, \tag{14.69}$$

where b is the minimal distance of the planet from the Sun and a is the semimajor axis of the orbit. The shape of the orbit is shown in Fig. 1.1, with $\Delta\varphi$ greatly exaggerated.

Note that $\Delta\varphi$ is larger, the smaller b is. Hence, Mercury offers the best conditions for measuring this effect. Since $\Delta\varphi$ is very small for a single revolution, but cumulates with time, the most convenient unit to measure $\Delta\varphi$ in astronomical practice is not radians per revolution, as in (14.69), but arc seconds per century. In these units, we have

$$\Delta\Phi = \frac{C}{T} \times \frac{360 \times 60 \times 60}{2\pi} \times \frac{6\pi GM}{c^2 a\,(1 - \varepsilon^2)}, \tag{14.70}$$

where $C = 100$ years, T is the orbital period of Mercury measured in terrestrial years (t. y.) and the second factor converts radians to arc seconds. Table 14.1 shows the values of

Table 14.1   Quantities determining the orbit of Mercury

quantity	name	value
T	orbital period	0.24085 t.y.
G	gravitational constant	$6.670 \times 10^{-8} \text{ cm}^3\text{g}^{-1}\text{s}^{-2}$
M	mass of the Sun	$1.989 \times 10^{33}\text{g}$
c^2	velocity of light squared	$8.987554 \times 10^{20}(\text{cm/s})^2$
a	semimajor axis	$57.9 \times 10^{11}\text{cm}$
ε	eccentricity	0.205628

these parameters, after Allen (1973). Substituting all these values in (14.70) we obtain[7]
$\Delta\Phi = 43.03$ arc seconds per century.

Equation (14.69) is the famous result (the **relativistic perihelion shift**) that predicted the orbital motion of Mercury in agreement with astronomical observations. As described in Chapter 1, when this correction is added to the perturbations of Mercury's orbit by other planets, the sum is in good agreement with the observed value.

This effect is measured by registering the positions of planets at different instants, reconstructing their orbits in space and determining the positions of perihelia. As mentioned in Chapter 1, the perihelion shift of Mercury had been observed already in the first half of the nineteenth century, and its Newtonian components were identified by LeVerrier in 1859 (Dicke, 1964). He was also the person who noted the discrepancy between the observed value and the prediction of Newton's theory. There is one more component in the observed effect: the astronomical observations are carried out from the Earth, so their raw results are expressed in the geocentric reference system. This gives the largest component of the perihelion shift. According to modern observations (Will, 2018; Lang, 1974), the full observed perihelion shift of Mercury is $5599.74 \pm 0.41''$ per century, of which $\approx 5000''$ is caused by the geocentric reference system, $\approx 280''$ is caused by the gravitational perturbations from Venus, $\approx 150''$ by perturbations from Jupiter and $\approx 100''$ by perturbations from the remaining planets (Will, 2018). The residue is $43.11 \pm 0.45''$ per century (Lang, 1974), which agrees very well with the value calculated from relativity.

For other planets, the relativistic perihelion shift is not larger than a few arc seconds per century (Will, 2018). Moreover, their orbits have small eccentricities, so the positions of their perihelia are known with precision unsuitable for testing relativity.

Now a few comments on the procedure we used to obtain the approximate orbit equation:

1. We were able to recognise that (14.66) is an incorrect solution because we knew from elsewhere that bound orbits exist. However, perturbative methods are often applied in relativistic astrophysics in situations where nothing is known about the expected result. So, one must be careful with drawing conclusions from approximate calculations.

---

[7] Each pocket calculator will show a slightly different result, because of inconsistent roundoff errors. The value quoted here is the official result obtained by carefully tuning the precision of all factors.

2. While going to higher orders of approximation, an avalanche of new terms appears. This is typical of all perturbative schemes. Thus, the hope that something useful can be inferred about the exact result by considering consecutive approximations may be illusory. There are some results, for example, in higher post-Newtonian orders of the relativistic two-body problem (see, e.g., Damour, Jaranowski and Schäfer, 2016), but the formulae are so unwieldy that they can be handled only by algebraic computer programs.

3. We emphasise again that the results (14.69) – (14.70) were obtained under the assumption that the radius of the orbit, at every point, is much larger than the gravitational radius of the central body, GM/c^2. They *do not* apply to those orbits around neutron stars and black holes that come close to the central body.

14.6 Deflection of light rays in the Schwarzschild field

We will now investigate the orbit of a light ray in the spherically symmetric gravitational field. Hence, we have to solve the equations of null geodesics in the Schwarzschild geometry.

The conclusion that coordinates may be chosen so that $\vartheta = \pi/2$ on the whole orbit is valid also for null orbits. But instead of (14.44) we now have

$$g_{\alpha\beta} \frac{\mathrm{d}x^\alpha}{\mathrm{d}s} \frac{\mathrm{d}x^\beta}{\mathrm{d}s} = 0. \tag{14.71}$$

Moreover, now s is not time, but any parameter on the curve.

The partly integrated equations of a geodesic are (14.54), (14.55) and

$$E^2 - \left(\frac{\mathrm{d}r}{\mathrm{d}s}\right)^2 - \frac{J_0{}^2}{r^2} \left(1 - \frac{2m}{r}\right) = 0. \tag{14.72}$$

Radial motion ($J_0 = 0$) is again possible. With $J_0 \neq 0$, we obtain from (14.54) and (14.72):

$$\frac{\mathrm{d}r}{\mathrm{d}\varphi} = \frac{r^2}{J_0} \sqrt{E^2 - \frac{J_0{}^2}{r^2} \left(1 - \frac{2m}{r}\right)}. \tag{14.73}$$

(We take only the $+$ sign for the square root for the same reason as before.) We will handle (14.73) as in the previous section. Introducing $\sigma = 1/r$, we obtain

$$\left(J_0 \frac{\mathrm{d}\sigma}{\mathrm{d}\varphi}\right)^2 = E^2 - J_0{}^2 \sigma^2 \left(1 - 2m\sigma\right). \tag{14.74}$$

From here, differentiating by φ,

$$\frac{\mathrm{d}\sigma}{\mathrm{d}\varphi} \cdot \frac{\mathrm{d}^2\sigma}{\mathrm{d}\varphi^2} = \frac{\mathrm{d}\sigma}{\mathrm{d}\varphi} \left(-\sigma + 3m\sigma^2\right). \tag{14.75}$$

As before, we have two cases: either $\mathrm{d}\sigma/\mathrm{d}\varphi = 0$ (a circular orbit) or $\mathrm{d}\sigma/\mathrm{d}\varphi \neq 0$, then

$$\frac{\mathrm{d}^2\sigma}{\mathrm{d}\varphi^2} + \sigma = \alpha\sigma^2, \tag{14.76}$$

where $\alpha = 3m$, as before. We solve this equation by the same method as (14.63). In the zeroth (Newtonian) approximation, the solution is

$$\frac{1}{r} = \sigma_0 = \frac{1}{R} \cos(\varphi - \varphi_0),$$ (14.77)

where $R = $ constant. In the Newtonian limit, (r, ϑ, φ) are the spherical coordinates in the Euclidean space. Then (14.77) is the equation of a straight line passing at the distance R from the point $r = 0$.

In the first approximation we look for a solution in the form $\sigma_{[1]} = \sigma_0 + \sigma_1$, where $\sigma_1/\sigma_0 \propto \alpha$. The function σ_1 must then obey

$$\frac{\mathrm{d}^2\sigma_1}{\mathrm{d}\varphi^2} + \sigma_1 = \frac{\alpha}{R^2} \cos^2(\varphi - \varphi_0).$$ (14.78)

A particular integral of (14.78) is

$$\sigma_1 = \frac{\alpha}{3R^2} \left[1 + \sin^2(\varphi - \varphi_0)\right].$$ (14.79)

The full solution of (14.76) up to terms of order α is thus

$$\frac{1}{r} = \frac{1}{R} \cos(\varphi - \varphi_0) + \frac{\alpha}{3R^2} \left[1 + \sin^2(\varphi - \varphi_0)\right].$$ (14.80)

Note that (14.80) has only one solution for $\cos(\varphi - \varphi_0)$ (the other one would have an absolute value greater than 1). The solution for φ is

$$\varphi_\pm = \varphi_0 \pm \arccos\left[\frac{3R}{2\alpha}\left(1 - \sqrt{1 + \frac{8\alpha^2}{9R^2} - \frac{4\alpha}{3r}}\right)\right].$$ (14.81)

This orbit has two asymptotes whose directions are calculated from here in the limit $r \to \infty$. The angle between them is (see Fig. 14.1)

$$\Delta\varphi = \lim_{r\to\infty} (\varphi_+ - \varphi_-) - \pi = 2\arccos\left[\frac{3R}{2\alpha}\left(1 - \sqrt{1 + \frac{8\alpha^2}{9R^2}}\right)\right] - \pi.$$ (14.82)

Equation (14.80) is an approximate solution to (14.76) up to terms linear in α. Hence, it suffices to take (14.82) with the same precision, and then it becomes

$$\Delta\varphi = \frac{4\alpha}{3R} = \frac{4GM}{c^2R}.$$ (14.83)

This is another famous result. Before we confront it with observations, we will first rectify a misunderstanding that still lingers in the literature.

Some authors wish to calculate the angle of gravitational deflection of a light ray in a quick and easy way using a combination of special relativity and Newton's theory. They assume that a photon has mass (whose value is irrelevant because it cancels out in the final formula) and moves like a massive body. We will show below that the result obtained in this way is one half of the correct (verified observationally!) Eq. (14.83).

The Newtonian orbit in polar coordinates is given by (14.64). From there

$$\varphi_\pm = \varphi_0 \pm \arccos\left[\frac{1}{\varepsilon}\left(\frac{p}{r} - 1\right)\right].$$ (14.84)

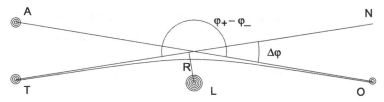

Fig. 14.1 Gravitational deflection of light rays in a spherically symmetric gravitational field. T is the true position of the source of light, A is the apparent position seen by the observer O in consequence of the deflection. According to Newton's theory, the light ray emitted from T in the direction TN would move along the straight line TN. In reality, and according to relativity, it follows the curve TO. The straight line TN is the asymptote of the initial part of the ray's path; AO is the asymptote of the final part. L is the deflecting body, R is the distance between the centre of the body L and the Newtonian path of the light ray, $\varphi_+ - \varphi_-$ is the angle between the asymptotes calculated from (14.81), $\Delta\varphi$ is the angle of deflection. The origin of the polar coordinates is at the centre of L; φ increases in the clockwise direction.

The deflection angle is thus

$$\Delta\varphi = \lim_{r\to\infty} (\varphi_+ - \varphi_-) - \pi = 2\arccos\left(-\frac{1}{\varepsilon}\right) - \pi = 2\arcsin\left(\frac{1}{\varepsilon}\right). \tag{14.85}$$

For a Newtonian orbit the eccentricity ε is

$$\varepsilon = \sqrt{1 + \frac{2E_0 J^2}{G^2 M^2 \mu^3}} \tag{14.86}$$

($\varepsilon > 1$ because the orbit is hyperbolic), where E_0 is the total energy of the particle on the orbit, J is its angular momentum and μ is its mass. In our case the 'particle' is a photon moving at the velocity c, so

$$E_0 = \frac{1}{2}\mu c^2, \qquad J = \mu c R. \tag{14.87}$$

Hence, in (14.86):

$$\varepsilon = \sqrt{1 + \frac{R^2 c^4}{G^2 M^2}} \gg 1. \tag{14.88}$$

Up to linear terms in $GM/(c^2 R)$ we thus have from (14.85)

$$\Delta\varphi = \frac{2GM}{c^2 R}, \tag{14.89}$$

i.e. half of the prediction of general relativity. Note that we arranged for the discrepancy to be small: had we taken the special relativistic equation $E = \mu c^2$ instead of (14.87), the final result would have been $\Delta\varphi = \sqrt{2}GM/(c^2 R)$, differing from (14.83) by still more. The lesson from this is: picking out elements of different theories and doing simple arithmetics on them are not guaranteed to lead to correct results. Each physical theory has its inner logic that has to be applied consistently.

The same result (14.89) can be derived from the pure Newton theory, without mixing in special relativity; see Exercise 6.

14.7 Measuring the deflection of light rays

It is seen from (14.83) that $\Delta\varphi \propto 1/R$, thus the angle of deflection is greater for those rays that approach the central star to within a smaller distance. (But recall: this formula was derived, and applies, under the assumption of weak gravitation. It does not apply to rays getting near to black holes or neutron stars.) For a light ray grazing the surface of the Sun, $R = 6.9599 \times 10^{10}$ cm. Taking this number, and the other quantities in (14.83) as in Section 14.5, we obtain $\Delta\varphi = 1.75''$. This is the maximum value available in observations.[8] For stars other than the Sun M and R cannot be measured with sufficient precision, and the deflection angles would be much too small to be measured (see later in this section). Planets cause unmeasurably small deflection angles.

We discussed the deflection of light rays, but of course the result applies to all kinds of radiation, e.g. X rays, γ rays and microwaves. But during the first decades of the twentieth century the whole of observational astronomy relied on optical observations; other kinds of radiation from space had been impossible to detect with the technology of those times. Consequently, in the first attempt to measure the deflection of light by the Sun, undertaken by Eddington in 1919 (Dyson, Eddington and Davidson, 1920), optical observations were applied. Observing light rays that graze Sun's surface is possible only during a total eclipse. The idea was this:

1. Find two stars that will be visible at the very edge of the Sun during the eclipse; best of all on opposite ends of Sun's diameter.

2. Take a photograph of these stars during the eclipse.

3. Take a photograph of the same two stars several months later, when the Sun is on the opposite side of the Earth.

4. Measure the difference between the positions of the stars in the two photographs and calculate the angle of deflection.

The geometry of this measurement is shown in Fig. 14.2. With the Sun being away, the observer O sees the two stars at their true positions T_1 and T_2, at angular separation α_1. With the Sun between them, the stars are seen at the apparent positions A_1 and A_2, at angular separation $\alpha_2 > \alpha_1$. The drawing is not to scale. The distance from the Sun to the stars is much greater than to the Earth. So much greater that the paths of the deflected and the nondeflected ray are 'practically' parallel to each other before they reach the Sun. Under this assumption, the angle of deflection is $\Delta\varphi = (\alpha_2 - \alpha_1)/2$.

With the technology of 1919, this measurement posed severe technical difficulties. The expected effect was so small that mechanical deformations of photographic plates stored for months could visibly disturb the result. Total eclipses of the Sun typically occur in the tropical zone – in the oceans or in jungles and deserts, far away from well-equipped observatories. Eddington's expedition consisted of two groups, carrying out the observations in Sobral in Brazil and on Principe Island in the Guinea Bay off Africa.

During an eclipse the temperature of the atmosphere drops, which causes some turbulence of the air and cooling of the telescope. The precision achievable under such conditions was limited, but Eq. (14.83) was confirmed. The deflection angle measured at Sobral was

---

[8] Today, gravitational deflection of light is observed also for distant galaxies, in **gravitational lenses**, see the next section. These objects are not used for quantitative testing of relativity because their mass distributions and radii are not known with sufficient precision.

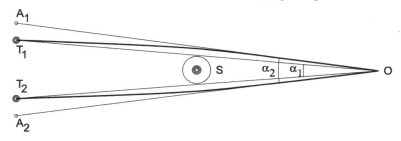

Fig. 14.2 Measuring the deflection of light rays by the Sun by Eddington's method. When the Sun is far away, the observer O sees the stars at their true positions T_1 and T_2, at angular separation α_1. Then, the light from the stars reaches the observer along the straight lines T_1O and T_2O. When the Sun S is seen between the stars, their light follows the curved arcs T_1O and T_2O, and the observer sees the images of the stars at the apparent positions A_1 and A_2, at angular separation $\alpha_2 > \alpha_1$. The straight lines A_1O and A_2O are tangent to the arcs T_1O and T_2O at the point O. The prediction of relativity is $\Delta\varphi = (\alpha_2 - \alpha_1)/2 = 1.75''$.

$1.98 \pm 0.16''$, that on Principe Island was $1.61 \pm 0.40''$ (Will, 2018). Unlike the anomalous perihelion shift of Mercury that had been known for many years before 1915, the deflection of light rays was *predicted* by relativity. The calculations by Cavendish and Soldner (Will, 1988; Soldner, 1804; Schneider, Ehlers and Falco, 1992) were long-forgotten by then. Eddington's positive result catapulted relativity and Einstein personally to the fame and public fascination that still last today.

Later, a different method of measuring the deflection was invented. The electromagnetic waves used in the observation were microwaves, and the observing devices were radio telescopes. Three radio sources, 0119+11, 0116+08 and 0111+02, lie nearly on a straight line, inclined at a large angle to the Sun's yearly trajectory through the sky. (The symbols of the radio sources are their coordinates on the sky, the right ascension in hours and minutes and the declination in degrees.) Each year, for about three weeks in March/April, the Sun passes in front of the middle radio source, see Fig. 14.3. The other two are then so far from the Sun's disc on the sky that their radiation is not measurably deflected. Thus, the real position of the middle source relative to the two others is measured when the Sun is elsewhere in the sky and compared with the observed position while the Sun's edge approaches the line of sight. The Sun is a weak radio source and its own microwave radiation does not disturb the measurement; the object 0116+08 can be observed right up to the moment when it vanishes behind the Sun. The only complication is that the plasma in the solar corona also deflects the microwaves. However, the plasma deflection depends, in a known way, on the wavelength of the radiation, while the gravitational deflection is the same for all wavelengths. Thus, carrying out the measurement at two wavelengths suffices to disentangle the two effects.

This measurement was carried out for the first time in 1974 by Fomalont and Sramek (1975, 1976, 1977) at the National Radio Astronomy Observatory in Green Bank (West Virginia). With this method, the observation is done under laboratory conditions and so is more precise. It has become customary to quote the observational results as the value

Fig. 14.3 Measuring the deflection of microwaves by the Sun. When the Sun is away, the three radio sources P_1, P_2 and P_3 are seen nearly on a straight line (left). The true position of the middle source P_2 can then be measured relative to the other two. With the disc of the Sun approaching (right), the image of P_2 moves away from the Sun. Its true position (small circle) can then be calculated from the positions of P_1 and P_3 (which do not change measurably) and compared with the observed apparent position to calculate the angle of deflection. The shift of the image of P_2 is exaggerated for clarity.

of γ – the ratio of the measured quantity to the value predicted by relativity; $\gamma = 1$ means perfect agreement. The Fomalont–Sramek result was 1.007 ± 0.009 (standard error).

Later measurements by the same method confirmed the general relativity value with much improved precision, the best of which was, as of 2004, $\gamma = 1 \pm 2 \times 10^{-4}$ (Will, 2006).

14.8 Gravitational lenses

A **gravitational lens** is a body deflecting light rays so that they intersect at the observer's position. The theory and observations of gravitational lenses have become a science in itself (Schneider, Ehlers and Falco, 1992), and we will touch on this subject here only very briefly.

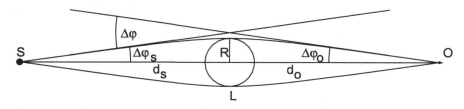

Fig. 14.4 A spherical star as a gravitational lens. In the approximation used here, the radius of the star, R, equals the distance from the star's centre to the point where the two straight lines intersect. Everything else is elementary geometry – see the text.

A spherical star acting as a gravitational lens is schematically shown in Fig. 14.4. S is the source of light, L is the deflecting star (the lens) and O is the observer; R is the radius of the star, d_S and d_O are the distances from the lens to the source and to the observer, respectively; $\Delta\varphi_S$ and $\Delta\varphi_O$ are the angles filled by the radius R at the position of the source and of the observer; $\Delta\varphi$ is the deflection angle of the ray that grazes the surface of the lens. Neglecting the curvature of space, we obtain, approximately for small angles,

$$R = d_S \Delta\varphi_S = d_O \Delta\varphi_O. \tag{14.90}$$

On the other hand, using (14.83) and Fig. 14.4 we have

$$\Delta\varphi = \Delta\varphi_S + \Delta\varphi_O = \frac{4GM}{c^2 R}. \tag{14.91}$$

From (14.90) and (14.91) we obtain the 'equation of a gravitational lens':

$$\frac{1}{d_S} + \frac{1}{d_O} = \frac{4GM}{c^2 R^2}. \tag{14.92}$$

Gravitational lenses, unlike the optical ones, do not focus the light rays: the rays flying farther from the optical axis are deflected by smaller angles. Hence, it is not possible to 'view' anything through a gravitational lens as if it were a magnifying glass – the image is very distorted. Nevertheless, there is some intensification of light: the rays that would disperse in the absence of a lens intersect again. Gravitational lenses thus increase the range of optical observations (Schneider, Ehlers and Falco, 1992).

Would it be possible for an observer on the Earth to use the Sun as a gravitational lens? Equation (14.92) allows us to answer this question decisively: no. The first rays (those that graze the surface of the Sun) intersect at the distance d_O that is smaller the greater d_S is. Hence, the minimal d_O calculated from (14.92) in the limit $d_S \to \infty$ is $d_{\min} = (cR)^2/(4GM) = 8.2 \times 10^{10}$ km, while the radius of the Earth's orbit is 1.49597892×10^8 km. (The $d_{\min}$ is more than 13 times the radius of the orbit of Pluto.) Likewise, there is no chance to observe the lensing by other stars. Even for the closest star, which is 4.5 light years away from the Earth, assuming it has the same radius as the Sun, the angle $\Delta\varphi_O$ would be $3.4 \times 10^{-3}{''}$ – too little to be measured.[9]

By extending the formulae (14.83) and (14.92) to intergalactic distances (which is not correct – see below) we can conclude that galaxies do have a chance to be gravitational lenses. Substituting in (14.92) the mass of our Galaxy,[10] $M = 1.4 \times 10^{11} M_\odot$ and its smallest diameter (thickness of the galactic disc) $R = 5$ kpc, with 1 kpc $= 3.0857 \times 10^{21}$ cm, we get $d_{\min} = 9.33 \times 10^2$ Mpc. By the Hubble formula the luminosity distance is $D_L = zc/H_0$, where $H_0 \approx 67.11$ km/(s×Mpc) (Planck 2014), this $d_{\min}$ corresponds to $z \approx 0.2$. Taking the largest diameter of our Galaxy, $R = 30$ kpc we get $z \approx 7.2$. The range of redshift for quasars is not much different (Bisogni, Risaliti and Lusso, 2018), and indeed, most observed gravitational lenses are quasars. Thus, in spite of the crude approximations, our estimate gave a realistic result. For a galaxy of diameter 30 kpc at the distance 1.34×10^4 Mpc (which corresponds to $z \approx 3$) the angle $\Delta\varphi_O \approx 0.2{''}$, which is measureable.

Equations (14.83) and (14.92) do not apply to quasars. They apply (approximately) only in the gravitational field of a single spherical star. The distances from the Earth to quasars are large in the cosmological scale. For calculating the deflection of light over such distances one should consider null geodesics in a model of the Universe. In astronomical practice, gravitational lenses are described by a sort of geometric optics based on the Newtonian description of propagation of light (Schneider, Ehlers and Falco, 1992). Nevertheless, it gives testable results that are in approximate agreement with observations.

---

[9] Such 'microlensing' has successfully been observed by measuring the changes in intensity of light from more distant stars when they are eclipsed by the lenses (Wambsganss 2006).

[10] The numbers for the sizes and the mass of our Galaxy are different in different sources. The numbers given here, except for the Hubble parameter, are rough approximations guessed by this author (A. K.).

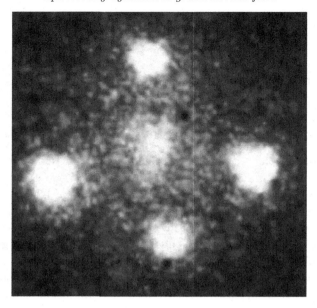

Fig. 14.5 The most famous gravitational lens – the 'Einstein cross'. See explanation in the text.

Figure 14.5 shows the most famous gravitational lens, called the Einstein cross[11] or the Huchra lens, after its discoverer John Huchra. The four peripheral bright spots are images of the quasar QSO 2237+0305 created by the nearer galaxy ZW 2237+030. The galaxy is seen as the haze surrounding all the spots; the bright spot at the centre is its nucleus. The complicated image is a consequence of the lens having no symmetry.

An even more interesting example of a gravitational lens is the galaxy LRG 3-757 shown in Fig. 14.6.[12] The lens lies nearly on the straight line connecting the Earth with the farther galaxy, which is not seen directly. In consequence of the nearly axially symmetric arrangement, the light of the far galaxy forms a nearly full ring around the nearer one.

14.9 The spurious singularity of the Schwarzschild solution at $r = 2m$

The Schwarzschild metric given by (14.40) and (14.42) has a singularity at $r = 2m = 2GM/c^2$: as $r \to 2m$, the component g_{00} goes to zero, while g_{11} becomes infinite. However, for the determinant of the metric, $g = -r^4 \sin^4 \vartheta$, the value $r = 2m$ is not in any way special. Also, the components of the Riemann tensor, R_{ijkl} in the tetrad defined by (14.25), are all of the form $\alpha m/r^3$, where $\alpha = 0, \pm 1, \pm 2$, i.e. are regular at $r = 2m$. These facts suggest that the singularity at $r = 2m$ is spurious – created by the coordinates used

---

[11] Credit: NASA, ESA, and STScI, CC BY-SA 3.0. This image was copied from https://upload.wikimedia .org/wikipedia/commons/c/c8/Einstein_cross.jpg

[12] Credit: ESA/Hubble & NASA. This image was copied from https://en.wikipedia.org/wiki/Gravitational_lens#/media/File:A_Horseshoe_Einstein_Ring_from_ Hubble.jpg

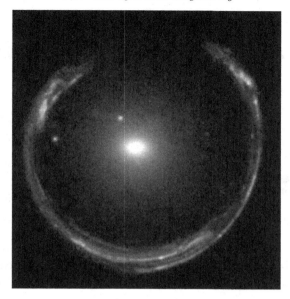

Fig. 14.6 Another conspicuous gravitational lens – LRG 3-757. See explanation in the text.

there, and not by the geometry of the manifold.[13] To see that this is possible, let us recall the transformation law of tensors:

$$g_{\alpha'\beta'} = \frac{\partial x^\alpha}{\partial y^{\alpha'}} \frac{\partial x^\beta}{\partial y^{\beta'}} g_{\alpha\beta}. \tag{14.93}$$

If $g_{\alpha\beta}$ is regular in the old coordinates at $x^\alpha = x_0^\alpha$, but some of the functions $\partial x^\alpha / \partial y^{\alpha'}$ have a singularity at $x^\alpha = x_0^\alpha$, then some components of $g_{\alpha'\beta'}$ will be singular at $y^\alpha = y^\alpha(x_0)$. Such a singularity clearly can be removed by the inverse transformation $y^\alpha \to x^\alpha$.

There is no general criterion to distinguish a 'true' singularity of the Riemannian geometry from one introduced by the coordinates.[14] However, if a coordinate transformation that removes the singularity is found, then this proves that the singularity was spurious. For the Schwarzschild solution several such transformations are known. The most general of them is the one found (independently and almost simultaneously) by Kruskal (1960) and by G. Szekeres (1960).[15] It leads to coordinates that reveal the global structure of the Schwarzschild manifold.

---

[13] But the singularity at $r = 0$ is real, as will be seen from the following.

[14] If scalars connected with the geometry of the manifold, such as tetrad components of the curvature, become infinite at certain points, then this is an indication that the singularity is genuine.

[15] The Kruskal–Szekeres transformation was a crowning of a series of less general results. Probably the oldest proof that the singularity at $r = 2m$ is spurious was given by Lemaître (1933a), who noticed that in coordinates connected with freely falling observers the surface $r = 2m$ is no obstacle to their motion. (This interpretation of the Lemaître coordinates was provided later by Novikov (1964b).) Other transformations removing the spurious singularity at $r = 2m$ were found by Raychaudhuri (1953) and by Finkelstein (1958).

Let us write the Schwarzschild solution in the form[16]

$$ds^2 = (1 - 2m/r) \left[dt^2 - \frac{dr^2}{(1 - 2m/r)^2} \right] - r^2 \left(d\vartheta^2 + \sin^2 \vartheta d\varphi^2 \right) \tag{14.94}$$

and transform the coordinates as follows:

$$\widetilde{p} = t + \xi, \qquad \widetilde{q} = t - \xi, \tag{14.95}$$

where

$$\xi \stackrel{\text{def}}{=} \int \frac{dr}{1 - 2m/r} = r + 2m \ln \left(\frac{r}{2m} - 1 \right). \tag{14.96}$$

Then the metric becomes

$$ds^2 = \left(1 - \frac{2m}{r} \right) d\widetilde{p} d\widetilde{q} - r^2 \left(d\vartheta^2 + \sin^2 \vartheta d\varphi^2 \right), \tag{14.97}$$

where $r = r(\xi)$ is now the function inverse to the $\xi(r)$ of (14.96). Since $\xi_{,r} > 0$ for $r > 2m$, the transformation (14.96) is invertible in the region in which the Schwarzschild solution was originally defined. The function (14.96) can be formally extended to the values $r < 2m$, by writing $\xi = r + 2m \ln |r/(2m) - 1|$, Fig. 14.7 shows the graph of the extension. The inverse functions exist in both domains, $r > 2m$ and $r < 2m$, but there exists no single inverse function in the domain $0 < r < \infty$.

Now let us carry out the next coordinate transformation:

$$(p, q) = \left(e^{\widetilde{p}/a}, e^{-\widetilde{q}/a} \right) \iff (\widetilde{p}, \widetilde{q}) = a(\ln p, -\ln q), \tag{14.98}$$

where a is a constant whose value will be chosen later. In these coordinates we have

$$ds^2 = -\left(1 - \frac{2m}{r} \right) \frac{a^2}{pq} dp dq - r^2 \left(d\vartheta^2 + \sin^2 \vartheta d\varphi^2 \right), \tag{14.99}$$

where this time $r = r(p, q)$; the transformation $(\widetilde{p}, \widetilde{q}) \longrightarrow (p, q)$ is invertible for all positive values of p and q. Let us substitute (14.98) and (14.95) − (14.96) in (14.99). The result is:

$$ds^2 = -\left(1 - \frac{2m}{r} \right) a^2 e^{-2r/a} \left(\frac{r}{2m} - 1 \right)^{-4m/a} dp dq - r^2 \left(d\vartheta^2 + \sin^2 \vartheta d\varphi^2 \right). \tag{14.100}$$

If $a = 4m$, then the factors that cause the singularity cancel out and the result is

$$ds^2 = -\frac{32m^3}{r} e^{-r/(2m)} dp dq - r^2 \left(d\vartheta^2 + \sin^2 \vartheta d\varphi^2 \right). \tag{14.101}$$

The hypersurfaces of constant p and of constant q are null, which is not always convenient. Therefore, we now introduce the further new coordinates v and u by:

$$v = \frac{1}{2}(p - q) \equiv e^{r/(4m)} \sqrt{\frac{r}{2m} - 1} \sinh \left(\frac{t}{4m} \right),$$

---

[16] Even though the set $r = 2m$ is the boundary of the region covered by the curvature coordinates, (14.94) still makes sense when $0 < r < 2m$. In that region, t becomes a spacelike coordinate and r becomes time. The resulting metric is the unique vacuum Kantowski–Sachs (K–S) spacetime (Kantowski and Sachs, 1966) (see also Sections 8.10, 10.7 and 19.5 for brief descriptions of the K–S metrics). This is the case that is commonly overlooked in textbooks and that was mentioned in Section 14.1, Eq. (14.17).

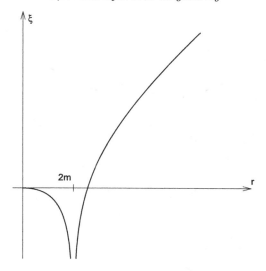

Fig. 14.7 A graph of the function $\xi(r) = r + 2m \ln |r/(2m) - 1|$.

$$u = \frac{1}{2}(p + q) \equiv e^{r/(4m)} \sqrt{\frac{r}{2m} - 1} \cosh\left(\frac{t}{4m}\right). \tag{14.102}$$

In the coordinates $(v, u, \vartheta, \varphi)$, the metric (14.101) becomes

$$ds^2 = \frac{32m^3}{r} e^{-r/(2m)} \left(dv^2 - du^2\right) - r^2 \left(d\vartheta^2 + \sin^2 \vartheta d\varphi^2\right). \tag{14.103}$$

By following all the intermediate steps, it is seen that the transformation $(t, r) \to (v, u)$ is single-valued and nonsingular for all $r \geq 2m$, so it is invertible. At $r = 2m$, it is also well defined, but $\partial v/\partial r$ and $\partial u/\partial r$ go to infinity as $r \to 2m$. This is the reason why (14.94) has a singularity at $r = 2m$. In (14.103), r can go through $r = 2m$ with no problem, and can approach $r = 0$ arbitrarily near. But the tetrad components of the Riemann tensor go to infinity as $r \to 0$. This shows that the set $r = 0$ is a singularity of the geometry.

For $r < 2m$, the transformation (14.102) cannot be carried out, but it can be formally extended to all $r > 0$ by writing it in the form

$$u^2 - v^2 = e^{r/(2m)} \left(\frac{r}{2m} - 1\right), \tag{14.104}$$

$$v/u = \tanh\left(\frac{t}{4m}\right). \tag{14.105}$$

Hence, in the surface of the (v, u) variables, the lines $r = $ constant are hyperbolae, the set $r = 2m$ is the pair of straight lines $u = \pm v$ and the lines $t = $ constant are straight lines passing through the point $u = v = 0$ with different slopes; for $t \to \pm\infty$, $v/u = \pm 1$. The singularity $r = 0$ is the pair of hyperbolae $u^2 - v^2 = -1$. The (u, v) surface thus looks as

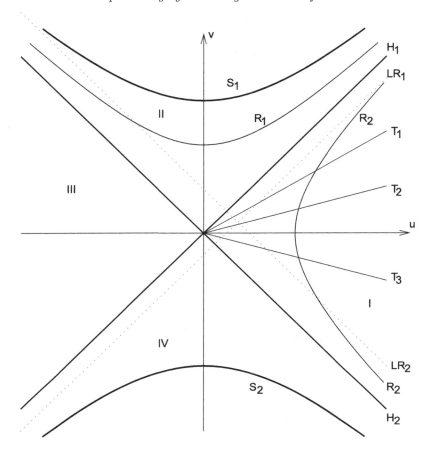

Fig. 14.8 The Kruskal diagram of the maximally extended Schwarzschild spacetime. S_1 and S_2 are the singularities at $r = 0$; the regions above and below them are not parts of the spacetime. The straight lines H_1 and H_2 are the event horizons; they divide the plane into four sectors denoted I, ..., IV. The curvature coordinates cover only sector I. Curvature coordinates applied inside the horizons cover sectors II and IV, although one does not see then that the two sectors do not coincide. The dotted lines LR_1 and LR_2 are exemplary paths of radial light rays; LR_1 passes out from the past singularity S_2 and from under the horizon and escapes to infinity, LR_2 falls from infinity and is trapped inside the horizon, eventually hitting the future singularity S_1. The hyperbolae R_1 and R_2 are exemplary lines of constant r, one inside the horizon (with $r < 2m$) and the other outside (with $r > 2m$). On the straight lines T_1, T_2 and T_3 t is constant. Note that on the horizons $r = 2m$ and $|t| = \infty$, with $t = +\infty$ on H_1 and $t = -\infty$ on H_2. This misled some early researchers into believing that the surface $r = 2m$ can never be reached by a material object. But t is not the physical time; it is just a badly behaving parameter used to measure time.

in Fig. 14.8. The radial null geodesics, which are $u \pm v =$ constant in (14.103), are in this figure straight lines inclined at 45° to the axes and they divide the figure into four sectors. The original curvature coordinates cover only sector I or sector III, but not both at once. The region $r < 2m$ in the curvature coordinates corresponds to sector II or sector IV, but again not to both of them at once.

Figure 14.8 is called the **Kruskal diagram**. It not only provides new useful coordinates for the Schwarzschild solution but is at the same time the **maximal extension** of the original Schwarzschild spacetime. The incompleteness of the Schwarzschild spacetime as represented in the curvature coordinates was recognised when it turned out that there exist timelike and null geodesics that escape the range of the Schwarzschild map without hitting any singularity. The Kruskal–Szekeres representation of the Schwarzschild solution does not have this defect – any timelike or null (or spacelike, too) geodesic either can be continued to infinite values of the affine parameter, or hits the singularity $r = 0$. Note that, contrary to what the original Schwarzschild representation would have us expect, the radial null geodesics reaching the observer in sector I come from a different part of the manifold than the one to which the observer can send her light signals, and the singularity set consists of two separate parts. An observer in sector I can never get an answer to a signal that she/he would send through the hypersurface $r = 2m$, because any such answer will hit the future singularity. However, observers in sector IV do have such a possibility: after moving into sector II, they can get an answer from sectors I and III to the signals they had sent in sector IV, but they will have no chance to reply.

A similar consideration was applied to the Reissner–Nordström metric by Graves and Brill (1960), and the result was even more strange (see Section 14.15). When the electric charge obeys $e^2 < m^2$, there are two hypersurfaces analogous to Schwarzschild's $r = 2m$, and the maximally extended manifold is an infinite chain of copies of a set similar to the one in Fig. 14.8. The same strange behaviour is found in the Kerr metric that generalises Schwarzschild's for rotation of the source; we will deal with it in Chapter 21.

14.10 * Embedding the Schwarzschild spacetime in a flat Riemannian space

We will now represent the Schwarzschild manifold as a subspace of a 6-dimensional flat space. We first take the 2-dimensional metric defined by $t = $ constant, $\vartheta = \pi/2$:

$$ds^2 = \frac{dr^2}{1 - 2m/r} + r^2 d\varphi^2. \tag{14.106}$$

Since the Schwarzschild metric in curvature coordinates does not depend on t, the Schwarzschild manifold is in these coordinates a family of identical copies of the hypersurface $t = $ constant. Then, $\vartheta = \pi/2$ is the subspace in which motion on timelike and null geodesics occurs (see the text after (14.52)). The surface (14.106) is thus physically important.

Equation (14.106) is the metric of the surface in Euclidean 3-space given by

$$r = r, \qquad \varphi = \varphi, \qquad z^2 = 8mr - 16m^2, \tag{14.107}$$

where r, φ and z are cylindrical coordinates in the Euclidean space. The surface defined by (14.107) is shown in Fig. 14.9. It is generated by rotating the parabola $r = z^2/(8m) + 2m$ around the z-axis. For large r its geometry becomes approximately flat.[17]

This picture does not leave any place for the region $r < 2m$ – on the other side of the ring $r = 2m$ (which is an intersection of the paraboloid with the plane $z = 0$) there is another copy of the same sheet. The surface shown in Fig. 14.9 coincides with the surface $\{v = 0, \vartheta = \pi/2\}$ in the Kruskal–Szekeres coordinates because $v = 0$ at $t = 0$. From Fig.

---

[17] Nonetheless, it is not correct to draw this surface as if it were actually becoming a plane far away from the 'throat' at $z = 0$. It is a simple exercise to verify that no asymptotic plane exists for this surface.

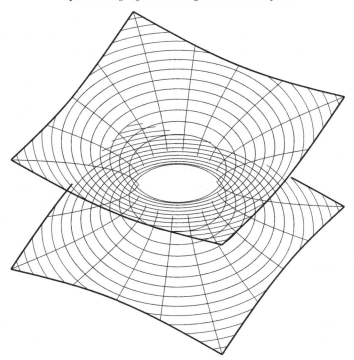

Fig. 14.9 The surface $\{t = \text{constant}, \vartheta = \pi/2\}$ in the Schwarzschild spacetime. The z-axis runs vertically through the middle of the 'throat', the r-coordinate is the distance from that axis.

14.8 it is seen that the smallest value of r on the surface $\{v = 0, \vartheta = \pi/2\}$, which is attained at $u = 0$, is $r = 2m$. It is also seen that on surfaces with $\{v = \text{constant}, \vartheta = \pi/2\}$, where $0 < v < 1$, the smallest value of r (the radius of the smallest ring in the throat) becomes ever smaller than $2m$, going down to zero at $v = \pm 1$ (see (14.104)). For $|v| = \text{constant} > 1$, the bridge connecting the two sheets disappears; each of those surfaces consists of two disjoint subsets.

The full 4-dimensional Schwarzschild spacetime can be embedded in a flat 6-dimensional Riemannian space (Fronsdal, 1959) with the metric

$$ds_6{}^2 = dZ_1{}^2 - dZ_2{}^2 - dZ_3{}^2 - dZ_4{}^2 - dZ_5{}^2 - dZ_6{}^2, \tag{14.108}$$

where

$$Z_1 = 4m\sqrt{1 - \frac{2m}{r}}\,\sinh\left(\frac{t}{4m}\right), \qquad Z_2 = 4m\sqrt{1 - \frac{2m}{r}}\,\cosh\left(\frac{t}{4m}\right),$$

$$Z_3 = \int_{2m}^{r}\sqrt{\frac{2m'}{r} + \left(\frac{2m'}{r}\right)^2 + \left(\frac{2m'}{r}\right)^3}\,dr'$$

$$Z_4 = r\sin\vartheta\cos\varphi, \qquad Z_5 = r\sin\vartheta\sin\varphi, \qquad Z_6 = r\cos\vartheta. \tag{14.109}$$

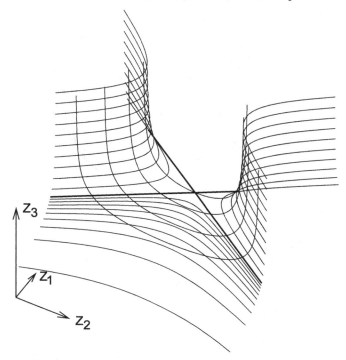

Fig. 14.10 Embedding of the Schwarzschild spacetime in six dimensions projected on the space of the variables (Z_1, Z_2, Z_3). The thick straight lines are the horizons $r = 2m$; they cross at the saddle where $Z_1 = Z_2 = Z_3 = 0$. The variable r increases from bottom to top, at $r \to 0$ $Z_3 \to -\infty$ and $Z_1 \to \pm\infty$, see the text for more details. The curves running horizontally are the $Z_3 = $ constant curves, the U-shaped curves are given by $Z_1 = $ constant, with $Z_1 = 0$ running through the saddle. The surface is mirror-symmetric with respect to the $Z_1 = 0$ plane, but the back side below the saddle is not shown for the sake of clarity. It is also mirror-symmetric with respect to the $Z_2 = 0$ plane.

The first two equations in (14.109) make sense only in the region $r \geq 2m$, but they can be replaced by one in which going into $r < 2m$ poses no problem:

$$Z_2{}^2 - Z_1{}^2 = 16m^2 \left(1 - \frac{2m}{r} \right). \tag{14.110}$$

In this form, the embedding covers both sides of the $r = 2m$ hypersurface, so it is at the same time another way of extending the Schwarzschild spacetime through it.

Figures 14.10 and 14.11 show the projections of the 6-dimensional space of variables $(Z_1, \ldots, Z_6)$ on the 3-dimensional subspaces of the variables (Z_1, Z_2, Z_3) (Fig. 14.10) and (Z_3, Z_4, Z_5) (Fig. 14.11).[18] In order to draw Fig. 14.10, one must observe that $Z_3 = $ constant and $Z_2{}^2 - Z_1{}^2 = $ constant when $r = $ constant. The saddle-shaped surface is the surface of the Schwarzschild variables (t, r). The pair of thicker straight lines is the image

---

[18] Reconstructing a 3-dimensional object in a Euclidean space from projections on different 2-planes is what engineers routinely do. Are you good enough to be an engineer in a 6-dimensional space?

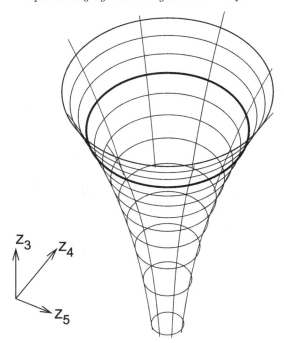

Fig. 14.11 Embedding of the Schwarzschild spacetime in six dimensions projected on the space of the variables (Z_4, Z_5, Z_3). r increases from bottom to top, with $r \to 0$ at the tip of the funnel (where $Z_3 \to -\infty$, $Z_4 \to 0$ and $Z_5 \to 0$). The thick circle is the horizon $r = 2m$.

of $r = 2m$. Further properties of this surface that help in reading the diagram are

$$\left| \frac{dZ_3}{dZ_2} \right| \underset{\substack{r \to \infty \\ Z_1 = \text{const}}}{\longrightarrow} +\infty, \qquad\qquad \left| \frac{dZ_3}{dZ_1} \right| \underset{\substack{r \to 0 \\ Z_2 = 0}}{\longrightarrow} 1, \qquad\qquad (14.111)$$

$$|Z_1| \underset{\substack{r \to 0 \\ Z_2 = \text{const}}}{\longrightarrow} +\infty, \qquad\qquad Z_3 \underset{r \to 0}{\longrightarrow} -\infty. \qquad\qquad (14.112)$$

In consequence of (14.110), on each surface $Z_1 = \text{constant}$ r must obey

$$r \geq r_{\min} = \frac{32m^3}{16m^2 + Z_1{}^2}. \qquad\qquad (14.113)$$

At $r = r_{\min}$, $Z_2 = 0$. Likewise, on a $Z_2 = \text{constant}$ surface with $|Z_2| < 4m$, r must obey

$$r \leq r_{\max} = \frac{32m^3}{16m^2 - Z_2{}^2}, \qquad\qquad (14.114)$$

and at $r = r_{\max}$, $Z_1 = 0$. When $Z_2 \to \pm 4m$, $r_{\max} \to \infty$, and for $|Z_2| \geq 4m$ there is no upper limit for r on a $Z_2 = \text{constant}$ surface.

The surface of the Schwarzschild variables (r, φ) in the subspace (Z_3, Z_4, Z_5), shown in Fig. 14.11, can be drawn after investigating the properties of dZ_3/dr, where $r^2 = Z_4{}^2 + Z_5{}^2$. It is seen that $dZ_3/dr > 0$, $dZ_3/dr \underset{r \to 0}{\longrightarrow} +\infty$, $dZ_3/dr \underset{r \to \infty}{\longrightarrow} 0$, $Z_3 \underset{r \to \infty}{\longrightarrow} +\infty$ and $Z_3 \underset{r \to 0}{\longrightarrow} -\infty$.

14.11 Interpretation of the spurious singularity at $r = 2m$; black holes

We verified that the Schwarzschild spacetime has no singularity at $r = 2m$. Nevertheless, this set does have some special properties. Equation (14.57), rewritten in the form

$$\left(\frac{dr}{ds}\right)^2 = E^2 - \left(1 + \frac{J_0{}^2}{r^2}\right)\left(1 - \frac{2m}{r}\right), \qquad (14.115)$$

shows that for $r \leq 2m$ the derivative dr/ds cannot be zero. Hence, if any body enters the region $r \leq 2m$ with $dr/ds < 0$, then dr/ds will not be able to change its sign and the motion will be continued into the singularity at $r = 0$. If, however, $dr/ds > 0$ at an initial instant in the region $r \leq 2m$, then motion will be continued until values $r > 2m$ are attained, and only then can the recession possibly be reversed to become a fall. Equation (14.72) shows that light rays have the same property. Consequently, massive bodies and light rays, once they cross the hypersurface $r = 2m$ in the direction of decreasing r, will never be able to turn back and will continue their motion until $r = 0$, also when $J_0 \neq 0$.

For ordinary astronomical objects, such as the Sun or the Earth, the quantity $2m = 2GM/c^2$, called the **gravitational radius**, is very small compared with the physical radius. For example, for the Sun $2m \approx 2.95$ km, for the Earth $2m = 0.89$ cm (recall the physical radii: Sun 696, 000 km and Earth 6378 km on the equator). The point at the distance $2m$ from the centre of such an object is hidden deep inside it, where the spacetime metric is different from Schwarzschild's. For such objects, the set $r = 2m$ has no physical meaning. This is why the strange properties of this set had been looked upon as mathematical curiosities well into the 1960s. If the Sun had collapsed to a size $r_0 \leq 2m$, then its mean mass density would go up to $\rho_0 \geq 1.85 \times 10^{16}$ g/cm^3, which is 80 times the density inside a uranium nucleus. Note, however, that $2m$ is proportional to the mass M of the object, while the physical radius r is approximately proportional to $M^{1/3}$ (because $M = (4/3)\pi\bar{\rho}r^3$, where $\bar{\rho}$ is the mean mass density in the object). Hence, for M sufficiently large, $2m(M)$ increases faster than $r(M)$.[19] For every finite density such a mass exists, at which the whole object will be inside the radius $r = 2m$. For $\rho = 1$ g/cm^3 (the density of water) this mass is equal to 2.7×10^{38} kg $\approx 1.36 \times 10^8$ solar masses ≈ 0.0001 of the mass of our Galaxy. The corresponding radius is 2.68 astronomical units, which is more than the radius of the orbit of Mars and less than the radius of the orbit of Jupiter.

Those objects whose physical radii are smaller than $2m$ are called **black holes** because everything that falls into them can never come back out. Note, however, that the theory predicts also the inverse behaviour: objects of radius smaller than $2m$ may emit matter as long as there is any supply inside. These (so far only hypothetical) objects are called **white holes**. They can be imagined as the initial cosmological singularity (the **Big Bang**, see Sections 17.4 and 18.3) still going off at an isolated location.

Astronomers have several candidates for black holes among observed objects, and the number of such objects is increasing. It is commonly believed today that every galaxy contains a very massive black hole at its centre. The accretion disk around the central black hole of the M87 galaxy in the Virgo cluster was actually photographed[20] in 2019 (EHTC

---

[19] To see this more clearly, imagine the graphs of the functions $f(x) = ax^{1/3}$ and $g(x) = bx$ with $a > 0$ and $b > 0$. No matter what values a and b have, g will be greater than f beginning with a certain x.

[20] More precisely, the Earth-wide array of radiotelescopes known as the Event Horizon Telescope registered

Collaboration, 2019). The one at the centre of our Galaxy has also been photographed in 2022, although the image is of a somewhat worse quality (Bower et al., 2022). See Frolov and Novikov (1998) for an exhaustive description of the theory and observations of black holes. We will come back to the subject of black holes in Section 18.9, where we will discuss the formation of spherically symmetric black holes in the Universe, and in Chapter 21, where we will briefly discuss rotating black holes.

So far, the only sure identification of a white hole is the whole Universe. As we will see in Chapters 16 and 17, in the currently accepted cosmological models the evolution of the Universe begins with an explosion called the Big Bang (BB). In the most commonly used models, those of Robertson–Walker class, the BB is a single event in the spacetime. The process of expansion away from the BB is the time-reverse of the collapse to a singularity, i.e. it is precisely what a white hole should do. In fact, as we will see in Chapter 17, the R–W models are exceptional in almost every possible respect. In more general models (see Chapters 18 and 20), the BB is not a single event, but a process extended in time. In such models, 'lagging cores' of expansion may exist that would be visible to distant observers as white holes. They were once proposed as the explanation of the energy source in quasars (Novikov, 1964a; Neeman and Tauber, 1967), but this explanation was later abandoned in favour of black holes with orbiting discs of matter.

In Newton's theory of gravitation, the escape velocity v_e from the surface of a spherical object of mass M and radius r is $v_e = \sqrt{2GM/r}$. Note that $r_g = 2GM/c^2$ is the radius of the object of mass M at whose surface $v_e = c$. This observation was first made by Laplace in the eighteenth century (Laplace, 1795). Hence, the notion of a black hole had in fact been introduced already then, although the name was coined only in the 1960s.[21]

The Kruskal diagram allows us to follow an object falling into a black hole. The lines $u \pm v = $ constant are intersections of the diagram with light cones. Any timelike line (not necessarily geodesic) in that diagram must thus have its tangent inclined to the v-axis at an angle smaller than 45° at every point. Imagine an object proceeding towards the black hole and emitting light signals at regular time intervals. Suppose that they are picked up by an observer resting at $r = r_O \gg 2m$. When the emitter approaches the surface $r = 2m$, the observer receives the signals at increasing intervals. These intervals tend to infinity as $r_{\text{emitter}} \to 2m$. The signal sent at $r = 2m$ will stay within the set $r = 2m$ for ever. For this reason, this set is called the **event horizon**. All signals sent at $r < 2m$ will hit the singularity at $r = 0$. Thus, from the point of view of the distant observer, the process of falling into a black hole lasts infinitely long. The same is true for the surface of an object that collapses to form a black hole. Hence, one should not imagine a black hole as a 'finished' object. A distant observer has a chance to see it only during formation, as an object whose light is becoming darker and redder until it disappears from sight. Objects falling into a black hole will disappear from sight *before* they hit the event horizon.[22]

---

the intensity of microwave rays emanating from various points in the accretion disk. The second black hole was 'photographed' in the same sense in 2022.

[21] The analogy between a black hole and Laplace's 'dark star' should not be taken literally. From under the surface of a black hole there is no return. In Newton's theory, no object can escape from the 'dark star' *to infinity*, but the surface $r = 2GM/c^2$ is freely traversable in both directions.

[22] Because of the high symmetry of the Schwarzschild solution, the event horizon in it coincides with two other entities that are in general distinct. One of them is the *apparent horizon*. The future apparent

For an observer who decided to fall into a black hole, the proper time needed to reach the event horizon is finite. This is seen from (14.115), which, for radial motion($J_0 = 0$) towards the centre, becomes

$$\frac{dr}{ds} = -\left(E^2 - 1 + \frac{2m}{r}\right)^{1/2}, \tag{14.116}$$

and then the time of flight from $r = r_0$ to $r = 2m$ is, for every $r_0 < \infty$,

$$s_h = -\int_{r_0}^{2m} \left(E^2 - 1 + \frac{2m}{r}\right)^{-1/2} dr < \infty. \tag{14.117}$$

14.12 The Schwarzschild metric in other coordinate systems

The curvature coordinates were the first used for the Schwarzschild metric. However, for investigating some advanced topics, in particular the relation of the Schwarzschild metric to other ones, other coordinates are more useful. Two such sets are used most often.

The first are the **isotropic coordinates**, in which the subspace $t = $ constant is explicitly conformally flat; in them a large number of nonvacuum generalisations of the Schwarzschild metric was found (Krasiński, 1997). Take the Schwarzschild metric in curvature coordinates, (14.40) and (14.42), and transform r as follows:

$$r = r'\left(1 + \frac{m}{2r'}\right)^2. \tag{14.118}$$

The resulting metric is:

$$ds^2 = \frac{[1 - m/(2r')]^2}{[1 + m/(2r')]^2} dt^2 - \left(1 + \frac{m}{2r'}\right)^4 \left[dr'^2 + r'^2\left(d\vartheta^2 + \sin^2 \vartheta d\varphi^2\right)\right]. \tag{14.119}$$

The spurious singularity now appears at $r' = m/2$.

The other important coordinates are called **geodesic-**, **Lemaître-** or **Novikov coordinates**. They are comoving coordinates of a congruence of radially freely falling observers. We shall consider the generalisation of the classic Lemaître–Novikov coordinates (that apply in the $\Lambda = e = 0$ subcase of (14.41)) to the general metric (14.40) – (14.41)) because this generalisation is nearly trivial. The metric in those coordinates is:

$$ds^2 = dt^2 - \frac{R_{,r}^2 dr^2}{1 + 2E(r)} - R^2(t,r)\left(d\vartheta^2 + \sin^2 \vartheta d\varphi^2\right), \tag{14.120}$$

where the function $R(t, r)$ obeys the equation

$$R_{,t}^2 = 2E(r) + \frac{2m}{R} - \frac{e^2}{R^2} - \frac{1}{3}\Lambda R^2 \tag{14.121}$$

horizon is a hypersurface in spacetime within which all null geodesics can proceed only towards the singularity, never away from it. As we will see in Chapter 18, in nonstatic spacetimes the apparent horizon and the event horizon in general do not coincide. Then, in gravitational fields of rotating bodies, the *infinite redshift hypersurface* (IRH) does not in general coincide with the event horizon, as we will see in Chapter 21. From within this hypersurface, light signals arrive at infinity being infinitely redshifted, yet the IRH is freely traversable both ways for material particles and light rays.

(the signs of the m, Λ and e^2 terms are opposite to those in (14.41)!), $E(r)$ being an arbitrary function. See Exercise 10 for a hint on how to derive (14.120) – (14.121).

For the subcase $E = \Lambda = e = 0$ these coordinates were first introduced by Lemaître (1933a).[23] The cases $E \neq 0$ (but still with $\Lambda = e = 0$) were investigated by Novikov (1964b), who provided the physical interpretation of these coordinates. Equation (14.121) is the equation of radial free fall in the spherically symmetric gravitational field generated by the mass m, with $E(r)$ being the total conserved energy of the motion. An observer at a fixed r (note that transformations of the form $r = f(r')$ do not change this metric) proceeds with time to other values of R, i.e. is either receding from or approaching the centre of symmetry, in accordance with the equation of free fall. When $E > 0$, the observers can recede to infinity and still have nonzero kinetic energy there. With $E = 0$, they can still recede to infinity, but their kinetic energy decreases to zero. With $E < 0$, the observers can fly away from the centre of symmetry only out to a finite distance, and then fall back. For each sign of E, (14.121) may be explicitly solved for $t(R)$, but for $E \neq 0$ the solutions cannot be inverted to define an explicit elementary function $R(t, r)$.

With $E \geq 0$ and $R_{,r} \neq 0$ the metric (14.120) has no singularity anywhere apart from $R = 0$. As mentioned under (14.93), this is how Lemaître first noticed that the singularity of the Schwarzschild solution at $r = 2m$ is only an artefact of the coordinates used.

The Schwarzschild solution in the form (14.120) – (14.121) emerges as the vacuum limit of the **Lemaître–Tolman** cosmological model that will be discussed in Chapter 18. Equation (14.121) with $e = 0$ frequently appears in cosmology – it governs the Lemaître–Tolman model and the Friedmann model that is a spatially homogeneous subcase of the former.

14.13 The equation of hydrostatic equilibrium

We will now investigate the Einstein equations *inside* a spherically symmetric body of perfect fluid, assuming that matter in the body is at rest in the coordinates of Section 14.1. Then the velocity field is

$$u^\alpha = \mathrm{e}^{-\nu} \delta^\alpha{}_0, \qquad u_\alpha = \mathrm{e}^\nu \delta^0{}_\alpha, \tag{14.122}$$

where $\nu(t, r)$ is an unknown function, while the pressure and density of the perfect fluid are constant along the lines of flow, i.e. depend only on r. Choosing an orthonormal tetrad in which $e_0{}^\alpha = u^\alpha$ (so $u^i = \delta^i{}_0$) and using (14.26) – (14.29), we obtain:[24]

$$G_{00} = \mathrm{e}^{-2\mu} \left(\frac{2}{r} \mu' - \frac{1}{r^2} \right) + \frac{1}{r^2} = \frac{8\pi G}{c^4} \epsilon, \tag{14.123}$$

$$G_{01} = \frac{2}{r} \mathrm{e}^{-\mu-\nu} \dot{\mu} = 0, \tag{14.124}$$

$$G_{11} = \mathrm{e}^{-2\mu} \left(\frac{2}{r} \nu' + \frac{1}{r^2} \right) - \frac{1}{r^2} = \frac{8\pi G}{c^4} p, \tag{14.125}$$

---

[23] Lemaître's form was somewhat unreadable because he parametrised his metric in such a way that the limit $\Lambda \to 0$ could not be directly taken.

[24] By choosing the metric in the form (14.18) we left aside the cases when the $\delta_{,\alpha}$ in (8.52) is a timelike or null vector, or is constant. Those cases do not contain static sources for the Schwarzschild metric, but cannot be neglected when considering nonstatic ones.

$$G_{22} = G_{33} = e^{-2\mu} \left(\nu'' - \mu'\nu' + \nu'^2 + \frac{\nu' - \mu'}{r} \right)$$

$$- e^{-2\nu} \left(\ddot{\mu} - \dot{\mu}\dot{\nu} + \dot{\mu}^2 \right) = \frac{8\pi G}{c^4} p. \tag{14.126}$$

Equation (14.124) implies that $\dot{\mu} = 0$, like in the Schwarzschild solution. Since $\dot{p} = \dot{\epsilon} = 0$, differentiating (14.125) by t we obtain $\dot{\nu}' = 0$, i.e. $\nu = f(t) + g(r)$. Then, the coordinate transformation $t' = \int e^{f(t)} dt$ gives $f = 0$ in the new coordinates, and $\nu = \nu(r)$.

Equation (14.123) can now be integrated. Assuming that $\epsilon(0)$ is finite, we obtain

$$e^{-2\mu} = 1 - \frac{8\pi G}{c^4 r} \int_0^r \epsilon(r') r'^2 dr'. \tag{14.127}$$

We denote

$$M(r) \stackrel{\text{def}}{=} \frac{4\pi}{c^2} \int_0^r \epsilon(r') r'^2 dr'. \tag{14.128}$$

This quantity has the dimension of mass. If $\epsilon(r) \equiv 0$ for $r \geq R$ (i.e. if the surface of the matter distribution is at $r = R$), then $M(R)$ is equal to the mass parameter of the Schwarzschild metric. Consequently, we interpret $M(r)$ as the mass within the sphere of coordinate radius r. Note that this mass is smaller than the sum of rest masses of all the particles inside the body. The density of rest mass is $\rho(r) = \epsilon(r)/c^2$, so the sum of rest masses within the sphere of radius r is, using (14.18),

$$M_{\text{rest}} = \int_{Vol_3(r)} \rho(r') \sqrt{-g_3} d_3 x = \frac{4\pi}{c^2} \int_0^r \epsilon(r') r'^2 e^{\mu(r')} dr'. \tag{14.129}$$

Since $e^{\mu} > 1$ (from (14.127)), we have $M_{\text{rest}} > M(r)$. This is the large-scale (astrophysical) analogue of the 'mass defect' known from nuclear/elementary particle physics: the mass of a bound object is smaller than the sum of masses of its components. The mass defect multiplied by c^2 is equal to the energy that would have to be supplied in order to break the object up into separate particles.

Using the notation (14.128), we obtain in (14.127)

$$e^{-2\mu} = 1 - \frac{2GM(r)}{c^2 r}. \tag{14.130}$$

The equations of motion $T^{\alpha\beta}{}_{;\beta} = 0$ must hold in consequence of the field equations, and in the present case they will be first integrals of the field equations. For $\alpha = 0, 2, 3$ they are fulfilled identically, while $T^{1\beta}{}_{;\beta} = 0$ implies $e^{-2\mu} [p' + \nu'(\epsilon + p)] = 0$, which means that

$$\nu' = -\frac{p'}{\epsilon + p}. \tag{14.131}$$

To integrate this, an equation of state must be assumed (necessarily of the form $\epsilon = \epsilon(p)$).

Using (14.130) and (14.131), Eq. (14.125) can now be written as

$$\frac{dp}{dr} = -\frac{G \left(\rho + p/c^2 \right) \left[M(r) + 4\pi r^3 p/c^2 \right]}{r^2 \left[1 - 2GM(r)/(c^2 r) \right]}. \tag{14.132}$$

Equation (14.126) is now fulfilled by virtue of those already solved, see Exercise 11.

Equation (14.132) is called the **equation of hydrostatic equilibrium**. In the Newtonian limit $c \to \infty$ it becomes

$$\frac{\mathrm{d}p}{\mathrm{d}r} = -\frac{G\rho M(r)}{r^2}. \tag{14.133}$$

Comparing (14.133) with (14.132) we see that in relativity the pressure increases the gravitational attraction because it appears in (14.132) as a positive contribution to the mass and mass density. Given the same mass, density and pressure the *gradient* of pressure, opposing the gravitational attraction, is greater in (14.132) than in (14.133). The greater pressure gradient implies faster growth of pressure towards the centre of the object, and this increased pressure then requires a still greater gradient. One can thus imagine a situation in which the equilibrium maintained over some time by a certain process (such as energy production in a star) is perturbed, and then the growth of pressure will lead to a loss of stability: no pressure gradient will be able to keep the object static again. A collapse to a size smaller than the gravitational radius $r_g = 2GM/c^2$ will then occur, and a black hole will be created. The Newtonian equilibrium equation (14.133) does not allow such a situation: the gradient of pressure needed to maintain equilibrium is determined by the mass density at the distance r from the centre of the object and does not depend on the value of pressure, so the solution exists for every density and mass.

14.14 The 'interior Schwarzschild solution'

In this section we will provide an example of a material source for the Schwarzschild metric. Equation (14.131) can be integrated after an equation of state has been given. Various equations of state and mass distributions intended to imitate real conditions inside various stellar objects are discussed in relativistic astrophysics. Typically, such general equations of state lead to complicated forms of (14.131) that can be integrated only numerically. We will deal here with an example that is unrealistic and of academic interest only, but it illustrates the problems encountered while matching matter solutions with vacuum solutions. We assume that $\epsilon = $ constant. From (14.126) – (14.128) we then obtain:

$$\mathrm{e}^{-2\mu} = 1 - \frac{8\pi G}{3c^4}\epsilon r^2 \stackrel{\text{def}}{=} 1 - Dr^2, \tag{14.134}$$

$$M(r) = \frac{4\pi}{3c^2}\epsilon r^3 = \frac{c^2}{2G}Dr^3, \tag{14.135}$$

where we used the abbreviation $D \stackrel{\text{def}}{=} 8\pi G\epsilon/(3c^4) = $ constant > 0. Subtracting (14.125) from (14.126) and using (14.134) we obtain

$$\left(1 - Dr^2\right)\left(\nu'' + \nu'^2\right) - Dr\nu' - \frac{1}{r}\left(1 - Dr^2\right)\nu' = 0, \tag{14.136}$$

which is integrated with the result $\mathrm{e}^\nu = C - B\sqrt{1 - Dr^2}$, where B and C are constants. The pressure can now be calculated from (14.125):

$$p = \frac{1}{3}\,\epsilon\,\frac{3B\sqrt{1 - Dr^2} - C}{C - B\sqrt{1 - Dr^2}}. \tag{14.137}$$

This pressure obeys (14.132), so we have a complete solution of Eqs. (14.123) – (14.126):

$$ds^2 = \left(C - B\sqrt{1 - Dr^2}\right)^2 dt^2 - \frac{dr^2}{1 - Dr^2} - r^2\left(d\vartheta^2 + \sin^2\vartheta d\varphi^2\right). \qquad (14.138)$$

This is the **interior Schwarzschild solution** (Schwarzschild, 1916b). It is conformally flat (which is not evident in the coordinates of (14.138) – but its Weyl tensor is zero).

Now let us verify the matching conditions between (14.138) and the vacuum Schwarzschild solution, (14.40) and (14.42), at the hypersurface Σ given by $r = R = $ constant. The coordinates on both sides of Σ are already adapted, so we can use the formalism of Section 12.17. Here, $x^4 = r$. The continuity of the metric components g_{IJ}, $I \neq 4 \neq J$, requires

$$\left(C - B\sqrt{1 - DR^2}\right)^2 = 1 - \frac{2GM}{c^2R}. \qquad (14.139)$$

The continuity of the derivatives of the metric of Σ in directions tangent to Σ is guaranteed, as stated in Section 12.17. In the present case, this is fulfilled trivially: the derivatives by t and by φ are all zero, the derivative by ϑ is obviously the same. The single nonzero component of the unit normal vector to $r = R$ is $1/\mathcal{N} = \sqrt{1 - 2GM/(c^2R)}$ in the exterior metric and $1/\mathcal{N} = \sqrt{1 - DR^2}$ in the interior metric. Following the recipe of Section 12.17, with the second fundamental form of the $r = R$ hypersurface given by (7.96), we require the continuity of $g_{IJ,r}/\mathcal{N}$ at $r = R$. This imposes two more equations,

$$\sqrt{1 - DR^2} = \sqrt{1 - \frac{2GM}{c^2R}} \qquad (14.140)$$

from the continuity of $g_{22,r}/\mathcal{N}$ and $g_{33,r}/\mathcal{N}$ at $r = R$, and

$$2BDR\left(C - B\sqrt{1 - DR^2}\right) = \frac{2GM}{c^2R^2}\sqrt{1 - \frac{2GM}{c^2R}} \qquad (14.141)$$

from the continuity of $g_{00,r}/\mathcal{N}$. The solution of (14.139) – (14.141) is

$$D = \frac{2GM}{c^2R^3}, \qquad B = \pm\frac{1}{2}, \qquad C = \pm\frac{3}{2}\sqrt{1 - \frac{2GM}{c^2R}}. \qquad (14.142)$$

This guarantees that $p(R) = 0$; see (14.137).

14.15 * The maximal analytic extension of the Reissner–Nordström metric

The Reissner–Nordström (R–N) metric, given by (14.40) – (14.41) with $\Lambda = 0$, can have its spurious singularities where $e^{2\nu} = 0$. There are three cases to consider separately:

- 1. When $m^2 - e^2 < 0$, $e^{2\nu}$ does not vanish at any value of r, and no spurious singularity exists. This case has no Schwarzschild limit.
- 2. When $m^2 - e^2 > 0$, $e^{2\nu}$ vanishes at two different values of r

$$r_- = m - \sqrt{m^2 - e^2}, \qquad r_+ = m + \sqrt{m^2 - e^2}. \qquad (14.143)$$

These are spurious singularities – the tetrad components of the Riemann tensor have well-defined values at those points. In the Schwarzschild limit $e \to 0$, the inner spurious singularity at $r = r_-$ collapses onto the genuine singularity at $r = 0$, while the outer one goes over into the event horizon at $r = 2m$.

- 3. When $m^2 - e^2 = 0$, $e^{2\nu}$ vanishes at just one value of r, $r = m$. This case has no Schwarzschild limit, either: with $e = 0$ it reduces to the Minkowski metric.

Similarly to what was done with the Schwarzschild solution in Section 14.9 the spurious singularities of the R–N solution can be removed by a coordinate transformation. Following Graves and Brill (1960), we will show how to do it in a more general static metric,

$$\mathrm{d}s^2 = \phi \mathrm{d}t^2 - \frac{1}{\phi}\mathrm{d}r^2 - r^2 \left(\mathrm{d}\vartheta^2 + \sin^2\vartheta \mathrm{d}\varphi^2\right), \tag{14.144}$$

where $\phi(r)$ is any function. We introduce such coordinates $u(t, r)$ and $v(t, r)$ in which

$$\mathrm{d}s^2 = f^2(u, v) \left(\mathrm{d}v^2 - \mathrm{d}u^2\right) - r^2(u, v)\left(\mathrm{d}\vartheta^2 + \sin^2\vartheta \mathrm{d}\varphi^2\right). \tag{14.145}$$

The functions f, u and v must obey

$$f^2 \left(v_{,t}^2 - u_{,t}^2\right) = \phi(r), \qquad f^2 \left(v_{,r}^2 - u_{,r}^2\right) = -\frac{1}{\phi(r)},$$

$$v_{,t} v_{,r} - u_{,t} u_{,r} = 0. \tag{14.146}$$

The last equation says that $v_{,t}/u_{,t} = u_{,r}/v_{,r}$. Dividing the first equation by the second and using this we obtain $u_{,t}^2/v_{,r}^2 = \phi^2(r)$. From here and from the last equation in (14.146) we then obtain the set

$$u_{,t} = \phi(r)v_{,r}, \qquad v_{,t} = \phi(r)u_{,r}. \tag{14.147}$$

This is easily solved if we introduce the new variable $r^*(r)$ defined by $\mathrm{d}r^*/\mathrm{d}r = 1/\phi$:

$$u = h(r^* + t) + g(r^* - t), \qquad v = h(r^* + t) - g(r^* - t), \tag{14.148}$$

where h and g are arbitrary functions. A prime will denote derivatives of h and g by their arguments. Using (14.148), we find from (14.146)

$$f^2 = \frac{\phi(r)}{4h'(r^* + t)g'(r^* - t)}. \tag{14.149}$$

Any singularity or zero of $\phi(r)$ must now be cancelled by the product in the denominator, and the resulting f must be time-independent (success is not guaranteed, but such choices of g' and h' were demonstrated to exist for the Schwarzschild metric and for the R–N metric). The product $h'(r^* + t)g'(r^* - t)$ will be independent of t only if

$$h = Ae^{\gamma(r^* + t)} + C, \qquad g = Be^{\gamma(r^* - t)} + D, \tag{14.150}$$

where A, B, C, D and γ are arbitrary constants; we shall take $A = B$ and $C = D = 0$. The formula for f^2 then becomes

$$f^2 = \frac{\phi(r)}{4A^2\gamma^2 e^{2\gamma r^*}}. \tag{14.151}$$

Now, the constant γ must be suitably chosen. Substituting (14.151) in (14.148) we obtain the formulae for the transformation $(t, r) \to (v, u)$:

$$u = Ae^{\gamma r^*}\left(e^{\gamma t} + e^{-\gamma t}\right) \equiv 2Ae^{\gamma r^*}\cosh(\gamma t),$$

$$v = Ae^{\gamma r^*} \left(e^{\gamma t} - e^{-\gamma t} \right) \equiv 2Ae^{\gamma r^*} \sinh(\gamma t). \tag{14.152}$$

The inverse transformation $(v, u) \to (t, r)$ is implicitly given by

$$F(r) \stackrel{\text{def}}{=} 4A^2 e^{2\gamma r^*} = u^2 - v^2, \qquad t = \frac{1}{\gamma} \text{artanh}(v/u), \tag{14.153}$$

where artanh is the inverse function to tanh. Thus, in the (v, u) coordinates, lines of constant t are straight lines through the origin, $v/u = \text{constant}$, and lines of constant r are the hyperbolae $u^2 - v^2 = \text{constant}$.

For the R–N metric in the case $e^2 < m^2$ we have

$$\phi = 1 - \frac{2m}{r} + \frac{e^2}{r^2} = \frac{(r - r_+)(r - r_-)}{r^2}, \tag{14.154}$$

and the expression for $r^*(r)$ is

$$r^* = r + \frac{r_+^2}{r_+ - r_-} \ln |r - r_+| - \frac{r_-^2}{r_+ - r_-} \ln |r - r_-|. \tag{14.155}$$

With the ϕ of (14.154) we have in (14.151)

$$f^2 = \frac{(r - r_+)^{1 - 2\gamma r_+^2/(r_+ - r_-)}(r - r_-)^{1 + 2\gamma r_-^2/(r_+ - r_-)}}{4A^2 \gamma^2 r^2 e^{2\gamma r}}. \tag{14.156}$$

Now γ can be chosen to cancel *one* of the spurious singularities, but not both at once. Let the index $i = 1$ refer to r_+ and $i = 2$ to r_-. To cancel the singularity at $r = r_i$ γ must be

$$\gamma_i = \frac{r_i - r_j}{2r_i^2}, \tag{14.157}$$

where $i \neq j$.

If we choose γ so as to cancel the spurious singularity at $r = r_+$, then, in the (v, u) coordinates, we can proceed from large r towards smaller r across the set $r = r_+$. In order to continue across $r = r_-$ we then have to go back to the original (t, r) coordinates and transform them to such (v, u) that cancel the second spurious singularity. In order to visualise the extended manifold, we first have to locate the spurious singularities and the true singularity in the (v, u) plane.

We see from (14.155), (14.157) and (14.153) that $F(r_i) = 0$ when $\gamma = \gamma_i$. Thus, the set $r = r_i$, in the coordinate patch that makes it nonsingular, has the equation $u = \pm v$. By virtue of the definition (14.153) $F(r)$ is a positive constant at $r = 0$. Hence, in the (v, u) coordinates, the equation of the singular set is $u^2 - v^2 = \text{constant} > 0$, i.e. it is a pair of hyperbolae that intersect the horizontal u-axis. (Note: these hyperbolae are *timelike*, i.e. they are rotated by 90° with respect to those of the Kruskal diagram.)

The (v, u) plane is infinite, so in order to visualise its properties it is useful to map it into a finite patch of the plane in such a way that null geodesics go over into null geodesics. The picture of the spacetime in the new coordinates is called the **conformal diagram** and the employed mapping is called the **Penrose transformation**. The radial null geodesics in the metric (14.145) are $du/dv = \pm 1$. Therefore, we first go over to the null coordinates

$$p = u + v, \qquad q = u - v, \tag{14.158}$$

and then employ the following Penrose transformation:

$$P = \tanh p, \qquad Q = \tanh q. \tag{14.159}$$

In the (Q, P) coordinates, the equation of the spurious singularity at $u = v$ is $Q = 0$, and of the one at $u = -v$ it is $P = 0$. The (v, u) plane now fits in the square $\{P, Q\} = \{[-1, 1] \times [-1, 1]\}$, and the **null infinities** $p = \pm\infty$, $q = \pm\infty$ are mapped into the sets $P = \pm 1$, $Q = \pm 1$. We can introduce the usual time–space coordinates in this square by

$$U = (P + Q)/2, \qquad V = (P - Q)/2. \tag{14.160}$$

The removed spurious singularity has the equation $U = \pm V$, while the infinities of p and q become the four straight line segments $P = U + V = \pm 1$, $Q = U - V = \pm 1$.

The equation of the true singularity, $u^2 - v^2 = 4A^2 e^{2\gamma r^*(0)} \overset{\text{def}}{=} \mathcal{A} = $ constant, in the (p, q) coordinates has the form $pq = \mathcal{A}$, so the singular set contains the points $\{p = \pm\infty, q = 0\}$ and $\{p = 0, q = \pm\infty\}$, whose (P, Q) coordinates are $\{P = \pm 1, Q = 0\}$ and $\{P = 0, Q = \pm 1\}$. These four points thus lie in the images of null infinities and in the spurious singularities $P = 0$ or $Q = 0$. Consequently, the spurious singularities $r = r_+$, the true singularities and the images of infinities do have common points, as will be seen in the picture.

Just as in the Schwarzschild spacetime, when we proceed from a point A in the $r > r_+$ region back in time along a $q = $ constant null geodesic and cross the spurious singularity $r = r_+$ at $p = 0$, we land in a different region of spacetime from that which would be reached by proceeding from A to the future along a $p = $ constant null geodesic. By extending these two kinds of null geodesics, we recover the analogues of sectors I, II and IV of Fig. 14.8. By sending null geodesics back in time from sector II and to the future from sector IV, we can also recover the analogue of sector III. Now, when we are in one of the $r < r_+$ regions, we change back to the (t, r) coordinates, transform them so as to cancel the $r = r_-$ spurious singularity and then carry out the Penrose transformation once more. We draw the new conformal diagrams in such a way that their images of the $r = r_+$ spurious singularities coincide with the $r = r_+$ singularities of the original Penrose diagram (this in fact involves some deformation). The $r = r_-$ singularities are now again straight lines and, as before, we find that their conformal images do have common points with the endpoints of the true singularities. In this way, by patching together conformal diagrams of different parts of the original manifold, we arrive at the manifold shown within the rectangle in Fig. 14.12.

The thin straight segments in Fig. 14.12 (two of them are marked ∞) are the conformal images of the null infinities, where $r \to \infty$. Timelike and spacelike infinities are the endpoints of the null infinity segments. The thin hyperbolae segments are the $r = $ constant lines; they are timelike for $r > r_+$ and $r < r_-$ and spacelike for $r_- < r < r_+$. The thick straight segments are the spurious singularities at $r = r_+$ and $r = r_-$. The hatched hyperbolae segments are the true singularities at $r = 0$. Roman numbers label sectors analogous to those of Fig. 14.8. Radial null geodesics (none are shown in Fig. 14.12) would be straight lines parallel to the spurious singularities. This shows that the spurious singularities are in fact event horizons: no future-directed null geodesic can cross from sector II to sector I or sector III. Unlike in the Schwarzschild case, the true singularity here is timelike and it leaves open tunnels to the future and to the past.

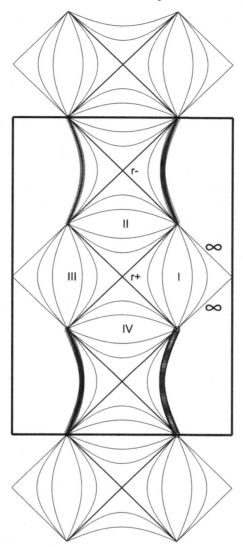

Fig. 14.12 The conformal diagram of the maximally extended Reissner–Nordström spacetime. Explanation is given in the text.

Now we have to decide how to interpret the image within the rectangle in Fig. 14.12. We note that the upper tunnel between the true singularities is a copy of the lower tunnel. We can identify the two tunnels, thus making the extended manifold finite and cyclic in the timelike direction. Alternatively, we can continue to send null geodesics to the future from the upper tunnel and to the past from the lower tunnel, thus constructing more and more copies of the sectors already constructed, and extend the picture indefinitely in both directions, obtaining an infinite chain. It had been said (Carter, 1973) that the identification would result in an acausal spacetime, in which a signal sent to the future would reach the sender from the past before it was emitted. This would lead to paradoxes such as the order to kill the sender's father before the sender was conceived. However, if the signal were a radial null geodesic, then it would have to be reflected at a point within the tunnel (otherwise, the ray would hit the singularity), so this is not a realistic example. In technical terms, the problem is whether there exists a pair of events E_1 and E_2 in sector I that has the following properties: (I) E_1 and E_2 can be connected by a timelike curve C (geodesic or not); (II) there exists a timelike or null geodesic G that connects E_2 to an event E_3 on $\widetilde{C}$ (the copy of C in the future copy of sector I) such that E_3 lies on $\widetilde{C}$ earlier than the copy of E_2. (G must be a geodesic or else the particle or ray going along it would need pre-designed exterior steering.) If such pairs exist, then causality is violated. No solution of this problem is known to this author (A.K.).

The extension of the R–N metric composed of two coordinate patches was first constructed by Graves and Brill (1960). The infinite mosaic of conformal diagrams shown in Fig. 14.12 first appeared in a paper by Carter (1966a) and was described in more detail in another article by Carter (1973).

Just like we did for the Schwarzschild solution, we can consider the embedding of the $\{t = \text{constant}, \vartheta = \pi/2\}$ 2-surface in the Euclidean space. This surface has the metric

$$ds^2 = \frac{1}{1 - 2m/r + e^2/r^2}\,dr^2 + r^2 d\varphi^2, \tag{14.161}$$

and, if this is going to be the metric of the surface $z(r)$ in the Euclidean space with $ds^2 = dz^2 + dr^2 + r^2 d\varphi^2$, then

$$z(r) = \int_{r_+}^{r} \sqrt{\frac{1}{1 - 2m/r' + e^2/r'^2} - 1}\,dr' \equiv \int_{r_+}^{r} \sqrt{\frac{2mr' - e^2}{r'^2 - 2mr' + e^2}}\,dr'. \tag{14.162}$$

This reproduces the z of (14.107) when $e = 0$. Figure 14.13 shows the comparison of the surface (14.161) with the one in Fig. 14.9: the presence of charge makes the throat thinner.

Like in the Schwarzschild spacetime, the surface $\{t = \text{constant}, \vartheta = \pi/2\}$ coincides with the surface $v = 0$ in those (u, v) coordinates that cancel the $r = r_+$ singularity, i.e. with the horizontal section of the manifold shown in Fig. 14.12 that goes through the crossing of the $r = r_+$ horizons. Imagine this section being moved parallely upwards from $v = 0$. The minimal r in the surface first decreases from r_+ to r_- and then increases to r_+ again – i.e. the throat never shrinks to a point and the two sheets of the surface never split. Some authors like to say that the flux of the electric field through the throat prevents it from collapsing to a point, instead it pulsates periodically between the radii r_- and r_+.

The surface $v = \text{constant}$ changes its geometry completely as it enters the tunnel between the true singularities – then it does not recede to infinite distances as in Fig. 14.13, but remains finite in extent. Then, with $r \leq r_-$, embedding the surface with the metric (14.161) in a Euclidean space by the same method as before requires solving the equation

$$z_{,r}{}^2 = \frac{2mr - e^2}{r^2 - 2mr + e^2}. \tag{14.163}$$

A solution will exist only if $e^2/(2m) \leq r \leq r_-$, because for $r < e^2/(2m)$ the right-hand

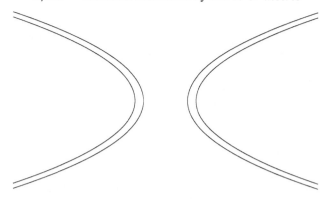

Fig. 14.13 A comparison of the 'throat' in the Schwarzschild spacetime (larger diameter) and in the R–N spacetime with $e^2 < m^2$ (smaller diameter). The curves are axial cross-sections of surfaces like the one in Fig. 14.9. The mass parameters for both surfaces are the same. The presence of charge makes the throat longer and thinner; it becomes infinitely long when $e^2 \to m^2$; see the comment after (14.166).

side of (14.163) is negative. The other part of the $\{t = \text{constant}, \vartheta = \pi/2\}$ surface can be embedded in a flat 3-space with the indefinite-signature metric $(\mathrm{d}r^2 + r^2\mathrm{d}\varphi^2 - \mathrm{d}u^2)$. The solution of (14.163) with $e^2/(2m) \le r \le r_-$ is

$$z(r) = \pm \int_r^{r_-} \sqrt{\frac{2mr' - e^2}{r'^2 - 2mr' + e^2}} \, \mathrm{d}r'. \tag{14.164}$$

With $r < e^2/(2m)$, the solution for $u(r)$ is

$$u(r) = \pm \left[z\left(\frac{e^2}{2m}\right) + \int_r^{e^2/(2m)} \sqrt{\frac{e^2 - 2mr'}{r'^2 - 2mr' + e^2}} \, \mathrm{d}r' \right]. \tag{14.165}$$

The surfaces $v = 0$ corresponding to $z(r)$ defined by (14.164) and to $u(r)$ defined by (14.165) are shown side by side in Fig. 14.14. The same graphs, put together and adjusted to the scale of Fig. 14.13, are shown in Fig. 14.15.

For the extreme case $e^2 = m^2$ the metric can be written as

$$\mathrm{d}s^2 = \left(\frac{r-m}{r}\right)^2 \mathrm{d}t^2 - \left(\frac{r}{r-m}\right)^2 \mathrm{d}r^2 - r^2 \left(\mathrm{d}\vartheta^2 + \sin^2\vartheta \mathrm{d}\varphi^2\right). \tag{14.166}$$

This case is essentially different from $e^2 < m^2$. In the latter, the radial distance from a point with coordinate r to the nearest spurious singularity is finite (for $r > r_+$ it is $\ell(r) = \int_{r_+}^r \left(r'/\sqrt{r'^2 - 2mr' + e^2}\right) \mathrm{d}r'$). With $e^2 = m^2$, the radial distance from a point $r \ne m$ to the spurious singularity $r = m$ is infinite, as may be verified using (14.166).

In order to find the extension, we consider the surface $\{\vartheta = \text{constant}, \varphi = \text{constant}\}$, and write the metric as

$$\mathrm{d}s^2 = \left(\frac{r-m}{r}\right)^2 \left[\mathrm{d}t - \left(\frac{r}{r-m}\right)^2 \mathrm{d}r\right]\left[\mathrm{d}t + \left(\frac{r}{r-m}\right)^2 \mathrm{d}r\right]. \tag{14.167}$$

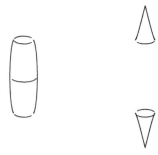

Fig. 14.14 Embeddings of the $v = 0$ surface that passes through $r = r_-$. The figure on the left corresponds to the function $z(r)$ defined by (14.164) and an embedding in an ordinary Euclidean 3-space, the figure on the right corresponds to the function $u(r)$ defined by (14.165) and an embedding in a flat 3-space with the indefinite metric of signature $(+ + -)$. The 'equator' of the left surface is at $r = r_-$, the upper and lower ends of the left surface are at $r = e^2/(2m)$, where the embedding in the Euclidean space breaks down. The tips of the right figure are at the singularity $r = 0$, the gap between the two parts of the figure has its edges at $r = e^2/(2m)$, where the embedding in this space also breaks down.

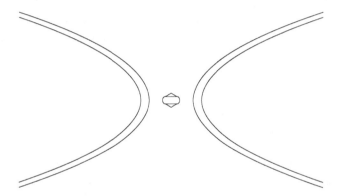

Fig. 14.15 Cross-sections of the surfaces of Fig. 14.14 put together and transformed to the scale of Fig. 14.13. The horizontal lines separate the part embedded in a Euclidean space (between them) from the part embedded in the space with indefinite metric. The placement of the small object relative to the large one is for comparing the size only – the two objects do not exist at the same time.

We introduce the null coordinates p and q by

$$p = t + \zeta, \qquad q = t - \zeta,$$

$$\zeta \overset{\text{def}}{=} \int \frac{r^2}{(r-m)^2} \mathrm{d}r = r - m - \frac{m^2}{r-m} + 2m \ln |r - m|. \tag{14.168}$$

The metric then becomes

$$\mathrm{d}s^2 = \left[\left(\frac{r-m}{r} \right)^2 \mathrm{d}p\mathrm{d}q - r^2 \left(\mathrm{d}\vartheta^2 + \sin^2 \vartheta \mathrm{d}\varphi^2 \right) \right]_{r=r(p,q)}. \tag{14.169}$$

The sets $\{r = m, \vartheta = \text{constant}, \varphi = \text{constant}\}$ are seen to be composed of radial null geodesics, so in the (p, q) coordinates they are represented by lines parallel to $p = \text{constant}$ or $q = \text{constant}$.

The metric is regular for all real values of p and q, and the spurious singularity at $r = m$ lies at $p = -q = -\infty$ when approached from $r > m$, and at $p = -q = +\infty$ when approached from $r < m$, but these two regions are covered by two different infinite coordinate patches. The infinity $r = \infty$ is at $p = -q = +\infty$ in the first patch. The singularity $r = 0$ lies on the straight line $p - q = 4m \ln m$ in the second patch. We bring the infinite values of p and q to finite distances by the transformation

$$p = \tan P, \qquad q = \tan Q. \tag{14.170}$$

In this way, the images of $r = m$ in the two patches can be laid side by side; they are now at $P = -Q = -\pi/2$ in the $r > m$ patch, and at $P = -Q = \pi/2$ in the $r < m$ patch. The infinity is at $P = -Q = \pi/2$ in the $r > m$ patch. The singularity is on the line $\tan P - \tan Q = 4m \ln m$ in the $r < m$ patch. Note that, just as in the case $e^2 < m^2$, the image of the singularity includes points at which simultaneously $P = Q = \pi/2$ and those where $P = Q = -\pi/2$, i.e. the singularity has common points with the images of the spurious singularity. However, this time the singularity will have no common points with the infinities. Putting all those bits of information together, we obtain the infinite chain of conformal diagrams shown in Fig. 14.16.

This extension and Fig. 14.16 were first presented by Carter (1966a, 1973).

Just like we did for the Schwarzschild metric and for the R–N metric with $e^2 < m^2$, we can embed the surface $\{t = \text{constant}, \vartheta = \pi/2\}$ in a flat 3-dimensional space. As before, the embedding is different on each side of the spurious singularity $r = m$, and again the region with $r < m$ cannot all be embedded in a Euclidean space. From $\mathrm{d}(z(r))^2 + \mathrm{d}r^2 + r^2\mathrm{d}\varphi^2 = r^2\mathrm{d}r^2/(r - m)^2 + r^2\mathrm{d}\varphi^2$ we obtain for the regions $r > m$ and $m/2 \leq r < m$:

$$z(r) = \sqrt{2m}\left(2\sqrt{r - m/2} + \sqrt{\frac{m}{2}}\ \ln\left|\frac{\sqrt{r - m/2} - \sqrt{m/2}}{\sqrt{r - m/2} + \sqrt{m/2}}\right|\right). \tag{14.171}$$

This tends to $-\infty$ at $r \to m$, to $+\infty$ at $r \to \infty$ and to 0 at $r \to m/2$. For $r \leq m/2$, we can embed the surface $\{t = \text{constant}, \vartheta = \pi/2\}$ in a flat 3-space with the indefinite metric $-\mathrm{d}(u(r))^2 + \mathrm{d}r^2 + r^2\mathrm{d}\varphi^2$; the equation of embedding is

$$u(r) = -2\sqrt{2m}\left(\sqrt{m/2 - r} - \sqrt{m/2}\arctan\sqrt{1 - 2r/m}\right). \tag{14.172}$$

This tends to 0 at $r \to m/2$ and to the finite value $m(\pi - 2)$ at $r = 0$.

The embeddings are shown in Fig. 14.17.

14.16 Motion of particles in the Reissner–Nordström spacetime with $e^2 < m^2$

This section is mostly borrowed from Graves and Brill (1960).

A free particle in a gravitational field moves on a geodesic $x^\gamma(s)$, where s is an affine parameter. The geodesic equation $\mathrm{D}^2 x^\gamma/\mathrm{d}s^2 = 0$ is a generalisation of the Newtonian

Fig. 14.16 The conformal diagram for the maximal extension of the extreme $(e^2 = m^2)$ R–N metric. The thin straight segments are the images of the null infinities, where $r \to \infty$. Thin hyperbolae segments are the $r = $ constant lines; they are timelike except the $r = m$ lines that are null. The thick straight segments are the spurious singularities at $r = m$, they are event horizons. The hatched curve segments are the true singularities at $r = 0$; they are timelike, too. Just as in the $e^2 < m^2$ case, we can choose to identify the square at the bottom with the second one up.

statement that the acceleration along the trajectory of free motion is zero.[25] A charged particle moving in a combined gravitational–electromagnetic field will experience a force acting on it. In the particle's rest frame, the force is $q\mathbf{E}$, where q is the particle's charge and $\mathbf{E}$ is the intensity of the electric field. In agreement with (13.1), the electric field consists of the components F^{0I} of the electromagnetic tensor, and these components in the rest frame of the particle are $F^{\mu}{}_{\nu} dx^{\nu}/ds$, where dx^{ν}/ds is the particle's velocity. The acceleration of such a particle will equal the force divided by the particle's mass. Putting all this together,

[25] See Section 15.2 for the definition of acceleration in relativity.

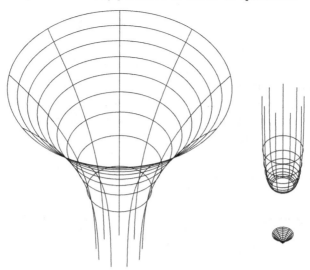

Fig. 14.17 Embeddings of the $\{t = \text{constant}, \vartheta = \pi/2\}$ surface of the extreme $(e^2 = m^2)$ R–N metric. Left: the embedding of the part $r > m$. The funnel is infinitely long downwards, and it approaches asymptotically the cylinder of radius m from the outside as $r \to m$. It goes infinitely far outwards and upwards. Right: the embedding of the part $0 \le r < m$ in a flat 3-space. The upper surface goes infinitely far upwards and approaches asymptotically the same cylinder from the inside as $r \to m$. The surface is cut off at $r = m/2$ (where $z = 0$) because the remaining part cannot be embedded in a Euclidean space. The lower cone is the embedding of the part $r \le m/2$ in the pseudo-Euclidean 3-space with the metric $-du^2 + dr^2 + r^2 d\varphi^2$. It is cut off at $r = m/2$, and its vertex touches the singularity at $r = 0$.

the equation of motion of a charged particle in a gravitational–electromagnetic field is

$$\frac{d^2 x^\gamma}{ds^2} + \left\{ \begin{matrix} \gamma \\ \alpha\beta \end{matrix} \right\} \frac{dx^\alpha}{ds} \frac{dx^\beta}{ds} = \frac{q}{\sqrt{8\pi\mu}} F^\gamma{}_\nu \frac{dx^\nu}{ds}, \tag{14.173}$$

where μ is the mass of the particle and the factor $1/\sqrt{8\pi}$ was introduced in order to simplify the subsequent formulae (it merely redefines the unit of charge). By (14.23), the electromagnetic tensor in the R–N metric has only two nonzero components, $F^{01} = -F^{10} = \sqrt{8\pi}e/r^2$. Knowing this, we find that two of the equations in the set (14.173) (those corresponding to $\gamma = 2$ and $\gamma = 3$) coincide with (14.48) and (14.49), so Eqs. (14.53) − (14.54) remain in force. Using this result, the remaining two equations are

$$\frac{d^2 t}{ds^2} + \frac{1}{\phi}\phi_{,r} \frac{dt}{ds}\frac{dr}{ds} = -\frac{qe}{\mu r^2 \phi}\frac{dr}{ds}, \tag{14.174}$$

$$\frac{d^2 r}{ds^2} + \frac{1}{2}\phi_{,r}\left[\phi\left(\frac{dt}{ds}\right)^2 - \frac{1}{\phi}\left(\frac{dr}{ds}\right)^2\right] - r\phi\left(\frac{d\varphi}{ds}\right)^2 = -\frac{qe\phi}{\mu r^2}\frac{dt}{ds}, \tag{14.175}$$

where we denoted

$$\phi = 1 - \frac{2m}{r} + \frac{e^2}{r^2}. \tag{14.176}$$

Multiplying (14.174) by $2\phi dt/ds$, (14.175) by $-2(dr/ds)/\phi$ and adding the results, we obtain an equation that is easily integrated with the result

$$\phi \left(\frac{dt}{ds} \right)^2 - \frac{1}{\phi} \left(\frac{dr}{ds} \right)^2 - \frac{J_0{}^2}{r^2} = \varepsilon, \tag{14.177}$$

where ε is a constant of integration. This is the same integral that exists for geodesics; it says that the tangent vector to the trajectory has a constant length. By adjusting the parameter, the constant can be made equal to 1. But we shall keep a general ε because later we will use (14.177) for null geodesics with $\varepsilon = 0$.

Equation (14.174) can be integrated with the result

$$\phi \frac{dt}{ds} = \frac{qe}{\mu r} + \Gamma, \tag{14.178}$$

where Γ is a constant of integration. Using this to eliminate dt/ds from (14.177), we obtain

$$\frac{1}{\phi} \left(\frac{qe}{\mu r} + \Gamma \right)^2 - \frac{1}{\phi} \left(\frac{dr}{ds} \right)^2 = \varepsilon + \frac{J_0{}^2}{r^2}. \tag{14.179}$$

Since $\varepsilon = +1$, this shows that the turning points of the trajectory ($dr/ds = 0$) can exist only outside the outer horizon ($r > r_+$) or inside the inner horizon ($r < r_-$), where $\phi > 0$.

We now rewrite (14.179) in yet another form

$$\left(\frac{dr}{ds} \right)^2 = \left(\frac{qe}{\mu r} + \Gamma \right)^2 - \left(\varepsilon + \frac{J_0{}^2}{r^2} \right) \left(1 - \frac{2m}{r} + \frac{e^2}{r^2} \right). \tag{14.180}$$

The motion can take place only in those regions where the right-hand side of (14.180) is positive. With $J_0 \neq 0$, the term $-J_0{}^2 e^2/r^4$ will dominate over the other terms when $r \to 0$, and will render the right-hand side negative. Thus, the neighbourhood of $r = 0$ is inaccessible for motion, which means that no charged particle can ever hit the central singularity. This conclusion still holds with $J_0 = 0$, provided $q^2 < \mu^2$ – then the dominating term is $(q^2/\mu^2 - 1)e^2/r^2$. This means that even a radially moving charged particle will be repelled by the singularity, provided its charge is *small* enough compared to the mass, irrespective of the signs of the charges of the particle (q) and of the R–N black hole (e). Strangely, this conclusion continues to hold for neutral particles for which $q = 0$, even with $J_0 = 0$. Then the dominating terms are $-J_0{}^2 e^2/r^4$ and $-e^2/r^2$, respectively. Thus, a charge on matter creates antigravitation – repulsion that, in spite of its electromagnetic origin, acts also on electrically neutral particles. We will come back to this in Sec. 19.4.3, where we will discuss the gravitational and electric fields inside a charged dust sphere.

Equation (14.180) can be used to calculate the proper time it takes to reach the horizon $r = m$ along a timelike radial geodesic in the special case $e^2 = m^2$. In spite of the infinite distance to this horizon within a $t = $ constant space, the proper time along a timelike geodesic turns out to be finite. The same is true for the value of the affine parameter along a null geodesic; see Exercise 13 for both results. Thus, the manifold of the extreme R–N metric in the original R–N coordinates is indeed incomplete.

With $q^2 < \mu^2$ (14.180) has an even stronger implication than the remark under (14.179): the turning point of a radial trajectory of a charged or uncharged particle must be within the inner horizon, i.e. at $r < r_-$; see Exercise 14.

14.17 Exercises

1. Prove that (14.1) is indeed a rotation around the axis $\{\vartheta = \pi/2, \varphi = 0\}$ (e.g. find the corresponding Killing field and recognise that it is the $\underset{(2)}{k^\alpha}$ of (8.42)). Verify that the components of the metric (8.52) do not change after the transformation (14.1). Note: it is advisable to use a computer-algebra program for this calculation. Doing it all by hand might not be enjoyable. See Sec. 24.1 for further comments.

2. Prove Lemmas 14.2 and 14.3.

3. Prove that the duality rotation (13.13) with the parameter $\delta = -\arctan(q/e)$ does indeed transform the electromagnetic field (14.23), (14.21) and (14.20), with the metric (14.18), into a new field for which the magnetic charge $\tilde{q} = 0$ and the new electric charge is $\tilde{e} = -\text{sign}(e)\sqrt{e^2 + q^2}$.

4. Prove that with $E^2 < 1$ the σ determined by (14.59) cannot go down to 0, so $r < \infty$. This means that bounded solutions of (14.59) exist, contrary to what (14.66) implies.

Hint. Solutions of (14.59) exist only when the right-hand side is nonnegative.

5. Show, using (14.74), that a circular photon orbit can exist only when $|J_0/E| = 3\sqrt{3}m$, and then its coordinate radius is $r = 3m$.

Hint. For a light ray on a circular orbit $d\sigma/d\varphi \equiv 0$ in (14.74), so the radius of this orbit must obey $E^2 r^3 - J_0{}^2(r - 2m) = 0$. Show that this equation has at most two positive roots. With two distinct roots, $d\sigma/d\varphi = 0$ indicates a passage through a turning point, not a circular orbit. On a circular orbit, the right-hand side of (14.74) must have a local minimum and a zero at the same r.

6. Prove that the same (incorrect) result (14.89) for the deflection of light rays can be obtained by purely Newtonian methods.

Hint. Calculate the deflection angle as in (14.85) (that far, no special relativity was used) and use (14.86). Note that $J = \mu R v_R$, where R is the smallest distance between the particle and the central star, and v_R is the velocity at that smallest distance. Calculate v_R by equating the total energy (kinetic + potential) at that point to the energy at infinity, $\frac{1}{2}\mu v_\infty{}^2$. Assume that the 'particle' is a photon and that $v_\infty = c$. Finally, calculate the result up to linear terms in GM/c^2.

Remark: This is how the angle of deflection of a light ray was calculated by Cavendish in the eighteenth century, in unpublished notes (Will, 1988), and by von Soldner in 1804 (Soldner, 1804; Schneider, Ehlers and Falco, 1992). Einstein himself found at first this incorrect result, before he formulated his field equations (12.21).

7. Calculate the tetrad components R_{ijkl} of the Riemann tensor for the Schwarzschild solution (14.40) – (14.42) and verify that they are all regular at $r = 2m$.

8. Verify that the integral in (14.117) is finite for every $r_0 < \infty$.

9. Assume geodesic coordinates in (8.52), so that $ds^2 = dt^2 - S^2(t, r) dr^2 - R^2(t, r) \left(d\vartheta^2 + \sin^2\vartheta d\varphi^2\right)$ (make sure first that the transformations (8.53) do really allow such a choice!). Then solve the vacuum Einstein equations $G_{\mu\nu} = 0$ for this metric and verify that (14.120) – (14.121) is the general solution that results when $R_{,r} \neq 0$.

10. Prove by direct coordinate transformation that (14.120) – (14.121) is indeed a representation of the general metric (14.40) – (14.41).

Hint. Let

$$F \stackrel{\text{def}}{=} e^{2\nu} = 1 - \frac{2m}{r} + \frac{1}{3}\Lambda r^2 + \frac{e^2}{r^2}. \tag{14.181}$$

Take the metric (14.40) – (14.41) and transform the coordinates by $t = f(\tau, u)$, $r = R(\tau, u)$, where f and R are functions as yet unknown. Then demand that in the new coordinates $g_{\tau\tau} = 1$ and $g_{\tau u} = 0$. Solve the resulting equations algebraically for the derivatives of f. Impose the integrability condition $f_{,\tau u} = f_{,u\tau}$. (The τ and u are the (t, r) coordinates of (14.120) – (14.121).) Do not substitute for F up to this point, but use the fact that F depends only on R, so $F_{,\tau} = F_{,R} R_{,\tau}$ and $F_{,u} = F_{,R} R_{,u}$. This substitution will reduce the equation $f_{,\tau u} - f_{,u\tau} = 0$ to

$$-\frac{1}{2} F^2 R_{,u} (F_{,R} + 2R_{,\tau\tau}) = 0. \tag{14.182}$$

The equation $F_{,R} + 2R_{,\tau\tau} = 0$ has the integral

$$F + R_{,\tau}^2 = 1 + 2E(u) \tag{14.183}$$

(with $E(u)$ being arbitrary), which is (14.121). The rest is easy.

Note that in this whole calculation we have nowhere used the explicit form of F, we only used the property $F = F(r) \implies F_{,t} = 0$. Hence, the coordinate transformation leading to the form (14.120) exists for every metric (14.40) with $F = e^{2\nu}$ depending only on r. The function R must then obey $R_{,t}^2 = -F + 1 + 2E(r)$.

11. Verify that (14.126) is fulfilled by virtue of (14.123) – (14.125).

Hint. Differentiate (14.125) by r, then manipulate (14.123) – (14.126) and (14.131) until you reproduce (14.126).

12. Verify that the Weyl tensor of (14.138) is zero.

13. Show that the proper time s needed to reach the horizon $r = m$ of the extreme R–N metric from any point along a radial timelike geodesic is finite. Show that, for a radial null geodesic starting at a finite r, the value of the affine parameter s at $r = m$ is also finite.

Hint. The curve defined by (14.180) becomes geodesic when $q = 0$.

14. Prove that the inner turning point of a radial trajectory of a charged or uncharged particle in the R–N spacetime with $e^2 < m^2$ is within the inner horizon if $q^2/\mu^2 < 1$.

Hint. As stated under (14.179), r at a turning point r_{tp} must obey $r_{\text{tp}} < r_-$ or $r_{\text{tp}} > r_+$. Thus, it suffices to verify that $r_{\text{tp}} > r_+$ leads to a contradiction. This is true in all three cases that must be considered separately: $\Gamma^2 > 1$, $\Gamma^2 < 1$ and $\Gamma^2 = 1$. See Sec. 24.2 for more detailed hints.

15

Relativistic hydrodynamics and thermodynamics

15.1 Motion of a continuous medium in Newtonian hydrodynamics

Let x_i, $i = 1, 2, 3$, be the rectangular Cartesian coordinates in the Euclidean 3-space. We assume that one line of flow of a fluid passes through every point of a certain region in space. Let the velocity field of the fluid, $v_i(t, x_j)$, be differentiable at every point x_i and at every instant t. Then

$$\frac{\mathrm{d}x_j(t)}{\mathrm{d}t} = v_j(t, y_i)|_{x_j=y_j}. \tag{15.1}$$

The $x_j(t)$ on the left are coordinates of the flowing fluid particle, while the y_i on the right are the coordinates of a point of space.

Let us follow the motion of the fluid particle P, which at the instant t occupies the position $\{x_i\}$ and moves with the velocity $v_j(t, x_i)$, and of an adjacent particle Q that, *at the same instant t*, occupies the position $\{x_i + \delta x_i\}$ and moves with the velocity $v_j(t, x_i + \delta x_i)$. Up to terms linear in δx_i, the velocity of the particle Q relative to P is

$$(v_{QP})_j(t) \equiv v_j(t, x_i + \delta x_i) - v_j(t, x_i) = v_{j,k}(t, x_i)\delta x_k + O\left(\delta x^2\right) \tag{15.2}$$

(sums over all repeated indices are implied in all the equations in this section), where $O\left(\delta x^2\right)$ denotes terms of order 2 and higher in δx_k. Hence, at the instant $(t + \Delta t)$ the position of the particle Q relative to P will be given by the vector

$$\delta x'_j = \delta x_j + (v_{QP})_j \Delta t + O\left(\Delta t^2\right) = \delta x_j + v_{j,k}\delta x_k \Delta t + O\left(\Delta t^2, \delta x^2\right). \tag{15.3}$$

Thus, the matrix $v_{j,k}$ determines the relative velocity of two neighbouring particles of the fluid. Under transformations between Cartesian coordinates $v_{j,k}$ transforms like a tensor. It can be decomposed into three parts that transform independently:

$$v_{j,k} = \sigma_{jk} + \omega_{jk} + \frac{1}{3}\delta_{jk}\theta, \tag{15.4}$$

where

$$\theta = v_{j,j}, \tag{15.5}$$

$$\sigma_{jk} = v_{(j,k)} - \frac{1}{3}\delta_{jk}\theta, \tag{15.6}$$

$$\omega_{jk} = v_{[j,k]}. \tag{15.7}$$

This decomposition can be done for every tensor of rank 2. However, for $v_{j,k}$ each part has a physical interpretation. The easiest way to read it out is to consider three types of motion, each with just one part being nonzero.

- I. Let $\sigma_{jk} = \omega_{jk} = 0 \neq \theta$. Then, from (15.2) and (15.3):

$$\delta x'_j = \left(1 + \frac{1}{3}\theta\Delta t\right)\delta x_j + O\left(\Delta t^2, \delta x^2\right). \tag{15.8}$$

The new vector that connects P and Q at the instant $(t + \Delta t)$ has in this case the same direction as the old vector from P to Q, but a different length, and $\theta = 3\mathrm{d}(\ln|\delta\mathbf{x}|)/\mathrm{d}t$. The particles P and Q either recede from each other (when $\theta > 0$) or approach each other (when $\theta < 0$) along the straight line PQ. Such motion is called **isotropic expansion**, and the quantity θ is called the **scalar of expansion**.

- II. Let $\theta = 0$ and $\sigma_{jk} = 0 \neq \omega_{jk}$. Then, from (15.3) and (15.4):

$$\delta x'_i = (\delta_{ik} + \omega_{ik}\Delta t)\,\delta x_k + O\left(\Delta t^2, \delta x^2\right). \tag{15.9}$$

Let us calculate the length $\delta\ell$ of the vector $\delta x'_i$. Since ω_{ik} is antisymmetric, we obtain

$$\begin{aligned}\delta\ell' & = (\delta x'_i \delta x'_i)^{1/2} = \left[\delta x_k \delta x_k + O\left(\Delta t^2, \delta x^3\right)\right]^{1/2} \\ & = (\delta x_k \delta x_k)^{1/2} + O\left(\Delta t^2, \delta x^3\right) = \delta\ell + O\left(\Delta t^2, \delta x^3\right). \end{aligned} \tag{15.10}$$

Hence, the derivative of $\delta\ell$ along v_i is zero. What happens with the direction of δx_i?

$$(\delta x'_i - \delta x_i)\,\delta x_i = \omega_{ik}\delta x_i \delta x_k \Delta t + O\left(\Delta t^2, \delta x^3\right) = O\left(\Delta t^2, \delta x^3\right). \tag{15.11}$$

So, the rate of change of δx_i projected orthogonally on δx_i is zero. The properties (15.10) and (15.11) are characteristic for rotational motion. Thus, in this type of motion, Q revolves around P.

Now let us calculate the vector of angular velocity of this motion, $\vec{\omega}$. By definition:

$$\mathbf{v}_{QP} = \vec{\omega} \times \boldsymbol{\delta}\mathbf{x} \implies (v_{QP})_i = \epsilon_{ikl}\omega_k \delta x_l. \tag{15.12}$$

At the same time, from (15.2) in the present case we have $(v_{QP})_i = \omega_{il}\delta x_l$. These two equations must hold for every δx_l, hence

$$\omega_{il} = -\epsilon_{ilk}\omega_k. \tag{15.13}$$

Inverting this equation and using (15.7) we obtain

$$\omega_j = \frac{1}{2}\epsilon_{jil}\omega_{li} = \frac{1}{2}\epsilon_{jil}v_{l,i} \implies \vec{\omega} = \frac{1}{2}\mathrm{rot}\ \mathbf{v}. \tag{15.14}$$

The ω_{ik} is called the **rotation tensor**. The quantity $\omega^2 \overset{\text{def}}{=} \omega_i\omega_i = \omega_{kl}\omega_{kl}/2$ is always nonnegative and vanishes only when $\omega_{kl} = 0$. Hence, also the **scalar of rotation** ω can be used to distinguish rotational and irrotational motions.

- III. Let $\theta = 0$ and $\omega_{ij} = 0 \neq \sigma_{ij}$. Consider three particles Q_1, Q_2 and Q_3 neighbouring P and occupying, at the instant t, the positions relative to P given by the vectors $\delta\mathbf{x}$,

$\delta\mathbf{y}$ and $\delta\mathbf{z}$ attached to P. The volume of the parallelepiped spanned on these vectors at the instant t is

$$\delta V = \delta\mathbf{x} \cdot (\delta\mathbf{y} \times \delta\mathbf{z}) = \epsilon_{ijk}\delta x_i \delta y_j \delta z_k. \tag{15.15}$$

At the instant $(t + \Delta t)$, the corresponding volume will be equal to

$$\delta V' = \delta V + D\Delta t + O\left(\Delta t^2, \delta x^4\right), \tag{15.16}$$

where

$$D \stackrel{\text{def}}{=} (\epsilon_{lmk}\sigma_{kn} + \epsilon_{ljn}\sigma_{jm} + \epsilon_{imn}\sigma_{il})\,\delta x_l \delta y_m \delta z_n. \tag{15.17}$$

Using $\sigma_{ij} = \sigma_{ji}$ and $\sigma_{ii} = 0$ one can now verify that $D = 0$ (see Exercise 1), i.e. that $\delta V' = \delta V + O\left(\Delta t^2, \delta x^4\right)$. Hence, in this type of motion, the rate of change of volume of the parallelepiped spanned on the vectors $\delta\mathbf{x}$, $\delta\mathbf{y}$ and $\delta\mathbf{z}$ is zero. However, the shape of the parallelepiped is changing because the vector $\sigma_{ik}\delta x_k$ has in general a different direction and different length from δx_i.[1]

This kind of motion is called **shearing motion** and σ_{ij} is called the **shear tensor**. Similarly to rotation, $\sigma_{ij} = 0$ if and only if the scalar σ defined by $\sigma^2 = \sigma_{ij}\sigma_{ij}/2$, called simply **shear**, is zero.

15.2 Motion of a continuous medium in relativistic hydrodynamics

In relativity, the motion of a fluid is described in a similar way to that in Newtonian hydrodynamics. We assume that one line of flow of the fluid passes through every point $\{x^\alpha\}$ of a certain region in spacetime, and that the velocity field of the fluid $u^\alpha(x)$ is differentiable in this whole region. Then $dx^\alpha/ds = u^\alpha(x^\beta)$, and, just as in (15.1), the $x^\alpha(s)$ on the left-hand side are coordinates of the flowing fluid element, while the x^β on the right-hand side are the coordinates of that point of spacetime in which the $x^\alpha(s)$ equals x^α. The parameter s is the proper time on the worldlines of the fluid, hence

$$u_\alpha(x)u^\alpha(x) = 1. \tag{15.18}$$

The tensor

$$h^\alpha{}_\beta \stackrel{\text{def}}{=} \delta^\alpha{}_\beta - u^\alpha u_\beta \tag{15.19}$$

projects vectors on the hypersurface orthogonal to u^α at a given point x. This is because $h^\alpha{}_\beta u^\beta = 0$ and, for any arbitrary vector B^α, $u_\alpha \cdot \left(h^\alpha{}_\beta B^\beta\right) = 0$. The quantity $B^\alpha_\perp \stackrel{\text{def}}{=} h^\alpha{}_\beta B^\beta$ is the component of the vector B^α along the direction perpendicular to u_α. Note also that

$$g_{\alpha\beta}B^\alpha_\perp B^\beta_\perp = h_{\alpha\beta}B^\alpha_\perp B^\beta_\perp \tag{15.20}$$

---

[1] If $\delta\mathbf{x}$, $\delta\mathbf{y}$ and $\delta\mathbf{z}$ are collinear with the eigenvectors of the matrix σ, then their directions do not change during the motion, but, in consequence of $\sigma_{ii} \equiv \text{Tr}(\sigma) = 0$, the sum of changes of their lengths must be zero. Hence, as long as $\sigma \neq 0$, if two of the vectors become longer, the third one must become shorter. Consequently, the shape of the parallelepiped will be changed also in this case.

for any arbitrary vector B^α. Thus, the tensor $h_{\alpha\beta}$ plays the role of the metric tensor in the hypersurfaces orthogonal to u^α.[2]

Let us choose s as the time-coordinate x^0 in spacetime. Then

$$\underset{*}{u'^\alpha} = \delta^\alpha{}_0. \tag{15.21}$$

Let us next choose the spatial coordinates x^I, $I = 1, 2, 3$, so that their parametric lines are contained in the hypersurfaces $S_P(s)$ orthogonal to one fixed worldline P. In such coordinates $g_{0I}|_P \underset{*}{=} 0$, so, for an arbitrary vector field B^α orthogonal to $u^\alpha_P(s)$, we have $0 = \left(g_{\alpha\beta}u^\alpha B^\beta\right)_P \underset{*}{=} g_{00}|_P B^0_P$, i.e. $B^0_P = 0$ (since $g_{00} \neq 0$ in consequence of (15.21) and (15.18)). Consider the curves $x^\alpha(\lambda)$ tangent to the vector field B^α, given by $B^\alpha = \mathrm{d}x^\alpha/\mathrm{d}\lambda$. In our chosen coordinates we have $0 \underset{*}{=} B^0_P = \mathrm{d}x^0/\mathrm{d}\lambda|_P$, i.e. at points of the curve P these lines are tangent to the hypersurface $x^0 = $ constant. Hence, the hypersurfaces $x^0 = $ constant and the hypersurfaces orthogonal to P are tangent to each other along P. Consequently, the hypersurface $S_P(s_0)$ is called the **hypersurface of events simultaneous with $P(s_0)$** or the **hypersurface of constant time $s = s_0$** for the observer P.

Let the particle moving along the curve P occupy at the instant s the point P_0 in spacetime. Let Q be an adjacent worldline, and let δx^α be a vector joining P_0 to an arbitrary point on Q. The event Q_0 simultaneous with P_0 is then at the position relative to P_0 given by the vector $\delta_\perp x^\alpha = h^\alpha{}_\beta (P_0) \delta x^\beta$. The velocity of the fluid at P_0 is $u^\alpha(x^\beta)$, and the velocity at Q_0 is $u^\alpha(x^\beta + \delta_\perp x^\beta)$. After the time Δs, the particle that at s occupied the point P_0 will be at P_1 of coordinates $x'^\alpha = x^\alpha + u^\alpha \Delta s$. Where will then be the particle that occupied the point Q_0 at s? Its position relative to P_0 will be determined by the vector $\delta_\perp x^\alpha + v^\alpha \Delta s$, attached to P_0, where v^α is determined by $u^\alpha(x^\beta + \delta_\perp x^\beta)$. However, v^α cannot be equal to $u^\alpha(x^\beta + \delta_\perp x^\beta)$, because the latter is attached at a different point than $\delta_\perp x^\alpha$. In order to add the vectors, we must first transport one of them parallely to the point of attachment of the other. Thus, v^α must be the vector $u^\alpha(x^\beta + \delta_\perp x^\beta)$ parallely transported from $(x^\beta + \delta_\perp x^\beta)$ to x^β.

Let us apply (5.6) to our present case. The $v^\alpha_\parallel(\tau_2)$ of (5.6) is our v^α, and the $v^\alpha(\tau_1)$ of (5.6) is our $u^\alpha(x^\beta + \delta_\perp x^\beta)$. Hence we have

$$v^\alpha = u^\alpha(x^\beta + \delta_\perp x^\beta) - \int_{x^\beta + \delta_\perp x^\beta}^{x^\beta} \Gamma^\alpha{}_{\sigma\rho}(x)u^\sigma(x)\mathrm{d}x^\rho. \tag{15.22}$$

We now apply the mean value theorem to the integral, and also develop the vector $u^\alpha(\cdot)$ by the Taylor formula up to terms linear in $\delta_\perp x^\beta$. The result is

$$v^\alpha = u^\alpha(x^\beta) + u^\alpha{}_{,\rho}(x^\beta)\delta_\perp x^\rho + \Gamma^\alpha{}_{\sigma\rho}(\overline{x})u^\sigma(\overline{x})\delta_\perp x^\rho + O\left(\delta_\perp x^2\right). \tag{15.23}$$

The point of coordinates $\overline{x}$ is intermediate between x^β and $(x^\beta + \delta_\perp x^\beta)$. When we replace $\overline{x}$ by x^β, the difference will be of the order of $\delta_\perp x^\beta$, and, since the whole expression is

---

[2] Each of the hypersurfaces meant here is orthogonal to a single flow line of the fluid. For a vector *field*, being orthogonal to a family of hypersurfaces is a rather special property which does not hold in general. As an exercise, readers may wish to verify that such a family of hypersurfaces exists if and only if the rotation tensor, defined later in this section, is zero.

multiplied by $\delta_\perp x^\beta$, the difference in the equation will be of order $O\left(\delta_\perp x^2\right)$ and can be neglected. Then we obtain

$$v^\alpha = u^\alpha(x^\beta) + u^\alpha{}_{,\rho}(x^\beta)\delta_\perp x^\rho + \Gamma^\alpha{}_{\sigma\rho}(x)u^\sigma(x)\delta_\perp x^\rho + O\left(\delta_\perp x^2\right). \tag{15.24}$$

This is equivalent to

$$v^\alpha = u^\alpha(x^\beta) + u^\alpha{}_{;\rho}(x^\beta)\delta_\perp x^\rho + O\left(\delta_\perp x^2\right). \tag{15.25}$$

So, the particle that occupied the point Q_0 at the instant s will at the later instant $(s+\Delta s)$ occupy the point of coordinates

$$x''^\alpha = x^\alpha + \delta_\perp x^\alpha + u^\alpha(x^\beta)\Delta s + u^\alpha{}_{;\rho}(x^\beta)\delta_\perp x^\rho \Delta s + O\left(\delta_\perp x^2, \Delta s^2\right). \tag{15.26}$$

Hence, the new position of the particle Q relative to P will be given by the vector

$$\delta_\perp x'^\alpha = x''^\alpha - x'^\alpha = \delta_\perp x^\alpha + u^\alpha{}_{;\rho}\,\delta_\perp x^\rho \Delta s + O\left(\delta_\perp x^2, \Delta s^2\right). \tag{15.27}$$

Thus, $u^\alpha{}_{;\rho}$ determines the rate of change of the position of the particle Q with respect to the particle P. But not the whole matrix $u^\alpha{}_{;\rho}$ gives a contribution to (15.27). We have

$$u^\alpha{}_{;\rho} \equiv u^\alpha{}_{;\sigma}\,\delta^\sigma{}_\rho \equiv u^\alpha{}_{;\sigma}\left(h^\sigma{}_\rho + u^\sigma u_\rho\right). \tag{15.28}$$

The second term of (15.28) gives a zero contribution in (15.27). Hence, finally

$$\delta_\perp x'^\alpha = \delta_\perp x^\alpha + u^\alpha{}_{;\sigma}\,h^\sigma{}_\rho\delta_\perp x^\rho \Delta s + O\left(\delta_\perp x^2, \Delta s^2\right). \tag{15.29}$$

The following identities hold:

$$u^\alpha{}_{;\sigma}\,u_\alpha \equiv 0, \qquad u^\alpha{}_{;\sigma} \equiv h^\alpha{}_\mu u^\mu{}_{;\sigma}. \tag{15.30}$$

So, $u^\alpha{}_{;\sigma}\,h^\sigma{}_\rho$ is an operator acting in the 3-dimensional hypersurface orthogonal to u^α. Comparing (15.29) with (15.3) we see that $u^\alpha{}_{;\sigma}\,h^\sigma{}_\beta$ plays in relativistic hydrodynamics the same role as $v_{j,k}$ played in Newtonian hydrodynamics. This time, $u^\alpha{}_{;\sigma}\,h^\sigma{}_\beta$ is a genuine tensor, and we can decompose it into the three independent parts in the same way:

$$u_{\alpha;\sigma}h^\sigma{}_\beta = \sigma_{\alpha\beta} + \omega_{\alpha\beta} + \frac{1}{3}\theta h_{\alpha\beta}, \tag{15.31}$$

where

$$\theta = u^\alpha{}_{;\sigma}\,h^\sigma{}_\alpha \equiv u^\alpha{}_{;\alpha}, \tag{15.32}$$

$$\omega_{\alpha\beta} = u_{[\alpha;|\sigma|}h^\sigma{}_{\beta]} \equiv u_{[\alpha;\beta]} - \dot{u}_{[\alpha}u_{\beta]}, \tag{15.33}$$

$$\dot{u}_\alpha = u_{\alpha;\beta}u^\beta, \tag{15.34}$$

$$\sigma_{\alpha\beta} = u_{(\alpha;|\sigma|}h^\sigma{}_{\beta)} - \frac{1}{3}\theta h_{\alpha\beta} \equiv u_{(\alpha;\beta)} - \dot{u}_{(\alpha}u_{\beta)} - \frac{1}{3}\theta h_{\alpha\beta}. \tag{15.35}$$

These quantities are called, respectively, the **scalar of expansion**, the **rotation tensor**, the **acceleration vector** and the **shear tensor**. The vector $\dot{u}^\alpha$ is called acceleration because $\dot{u}^\alpha = 0$ is the necessary and sufficient condition for the field u^α to be tangent to geodesic lines, i.e. for the fluid to move under the influence of gravitation only, which means free motion in the language of relativity.

The tensor of rotation can be uniquely represented by the **rotation vector**

$$w^\alpha \overset{\text{def}}{=} \frac{1}{2\sqrt{-g}}\, \epsilon^{\alpha\beta\gamma\delta} u_\beta \omega_{\gamma\delta}, \tag{15.36}$$

and the rotational motion may be characterised by the scalar w:

$$w^2 \overset{\text{def}}{=} -w_\alpha w^\alpha = \frac{1}{2}\omega_{\alpha\beta}\omega^{\alpha\beta} \geq 0, \tag{15.37}$$

called the **rotation scalar**. Similarly, the shearing motion may be characterised by the **shear scalar** σ defined by

$$\sigma^2 \overset{\text{def}}{=} \frac{1}{2}\sigma_{\alpha\beta}\sigma^{\alpha\beta} \geq 0. \tag{15.38}$$

The definitions (15.32) – (15.35) were introduced by Ehlers (1961).

It can be verified that the quantities given by (15.32), (15.33) and (15.35) are, in suitably chosen coordinates, proportional to their Newtonian counterparts given by (15.5) – (15.7). Namely, at a fixed point p_0 of the spacetime we can choose coordinates in which $\left\{ {\alpha \atop \beta\gamma} \right\}(p_0) = 0$, i.e. $D/\partial x^\alpha|_{p_0} = \partial/\partial x^\alpha$. Then, after carrying out all the operations in (15.32) – (15.35), we substitute $v/c = 0$. Marking the relativistic quantities with the subscript R and their Newtonian counterparts with the subscript N, we obtain

$$\dot{u}^\alpha \underset{v/c\to 0}{\longrightarrow} 0, \qquad \underset{R}{\omega_{0i}} \underset{v/c\to 0}{\longrightarrow} 0, \qquad \underset{R}{\sigma_{0i}} \underset{v/c\to 0}{\longrightarrow} 0;$$

$$c\theta_R = \theta_N, \qquad \underset{R}{c\omega_{ij}} = \underset{N}{\omega_{ij}}, \qquad \underset{R}{c\sigma_{ij}} = \underset{N}{\sigma_{ij}}. \tag{15.39}$$

15.3 The equations of evolution of θ, $\sigma_{\alpha\beta}$, $\omega_{\alpha\beta}$ and $\dot{u}^\alpha$; the Raychaudhuri equation

The equations derived in this section will be consistency conditions imposed by the Ricci identities on the hydrodynamical quantities introduced in the previous section and on the curvature of spacetime. Although they may thus seem to be of secondary importance, they are surprisingly powerful in their applications.[3]

From (6.7) applied to u^α, we have

$$u_{\gamma;\delta\sigma} - u_{\gamma;\sigma\delta} = -R_{\gamma\rho\delta\sigma}u^\rho. \tag{15.40}$$

Let us contract both sides of this equation with $u^\sigma h^\gamma{}_\alpha h^\delta{}_\beta$. In the second term on the left we then transfer the derivative with the index δ from $u_{\gamma;\sigma}$ to u^σ, while on the right we will use the antisymmetries of the Riemann tensor to eliminate some terms. The result is

$$h^\gamma{}_\alpha h^\delta{}_\beta (u_{\gamma;\delta})^\cdot - h^\gamma{}_\alpha h^\delta{}_\beta \dot{u}_{\gamma;\delta} + h^\gamma{}_\alpha h^\delta{}_\beta u^\sigma{}_{;\delta}\, u_{\gamma;\sigma} = -R_{\alpha\rho\beta\sigma}u^\rho u^\sigma, \tag{15.41}$$

where the overdot denotes the directional covariant derivative along the velocity field, $\cdot \overset{\text{def}}{=} u^\mu \nabla_\mu$. Now we contract (15.41) with $g^{\alpha\beta}$ and obtain

$$h^{\gamma\delta}(u_{\gamma;\delta})^\cdot - h^{\gamma\delta}\dot{u}_{\gamma;\delta} + h^{\gamma\delta}u^\sigma{}_{;\delta}\, u_{\gamma;\sigma} + R_{\rho\sigma}u^\rho u^\sigma = 0. \tag{15.42}$$

---

[3] In the paper in which these equations were first derived (Ellis, 1971), and in probably all papers in which they were applied, the signature $(-+++)$ was used, as opposed to $(+---)$ used here. This is why the equations of this section will differ from those in other sources – but they are equivalent.

Up to this point, these were general equations of differential geometry. Now we use the Einstein equations with a perfect fluid source to replace

$$R_{\alpha\beta} = \kappa \left(T_{\alpha\beta} - \frac{1}{2} g_{\alpha\beta} T \right) = \kappa \left[(\epsilon + p) u_\alpha u_\beta + \frac{1}{2} (p - \epsilon) g_{\alpha\beta} \right], \tag{15.43}$$

and then we apply (15.31) to $h^{\gamma\delta} u^\sigma{}_{;\delta}$. In the other two terms of (15.42) we substitute (15.19) and we transfer the differentiation from the derivatives of u_γ to u^γ. The result is

$$\left(u^\gamma{}_{;\gamma} \right){}^{\cdot} - \left(u^\gamma u_{\gamma;\delta} \right){}^{\cdot} u^\delta + \dot{u}^\gamma u^\delta u_{\gamma;\delta} - \dot{u}^\gamma{}_{;\gamma} + \left(u^\gamma \dot{u}_\gamma \right){}_{;\delta} u^\delta$$

$$- u^\gamma{}_{;\delta} u^\delta \dot{u}_\gamma + u_{\gamma;\sigma} \left(\sigma^{\sigma\gamma} + \omega^{\sigma\gamma} + \frac{1}{3} \theta h^{\sigma\gamma} \right) + \frac{1}{2} \kappa(\epsilon + 3p) = 0. \tag{15.44}$$

In this we use (15.34) and (15.30); the latter implies $\dot{u}^\alpha u_\alpha = 0$. We also use the definitions of θ, $\sigma^{\alpha\beta}$ and $\omega^{\alpha\beta}$ and the equations

$$\sigma_{\alpha\beta} u^\beta = \omega_{\alpha\beta} u^\beta = 0 \tag{15.45}$$

that follow from (15.33), (15.35) and (15.30). We obtain then in (15.44):

$$0 = \dot{\theta} + \frac{1}{3} \theta^2 - \dot{u}^\gamma{}_{;\gamma} + 2 \left(\sigma^2 - \omega^2 \right) + \frac{1}{2} \kappa(\epsilon + 3p). \tag{15.46}$$

This equation, in the form quoted above, was derived by Ehlers (1961) and is called the **Raychaudhuri equation** (the name seems to have been introduced by Ellis (1971)). The idea, and a subcase of (15.46) corresponding to $p = 0$, without the definitions of shear and rotation, were first introduced by Raychaudhuri (1955), see also Raychaudhuri (1957).

Taking the antisymmetric part of (15.41), then its symmetric part and using in the second one the Raychaudhuri equation to eliminate $\dot{\theta}$, we obtain two other equations:

- the vorticity propagation equation

$$h^\gamma{}_\alpha h^\delta{}_\beta \dot{\omega}_{\gamma\delta} - h^\gamma{}_\alpha h^\delta{}_\beta \dot{u}_{[\gamma;\delta]} + 2\sigma_{\delta[\alpha} \omega^\delta{}_{\beta]} + \frac{2}{3} \theta \omega_{\alpha\beta} = 0; \tag{15.47}$$

- and the shear propagation equation

$$h^\gamma{}_\alpha h^\delta{}_\beta \dot{\sigma}_{\gamma\delta} - h^\gamma{}_\alpha h^\delta{}_\beta \dot{u}_{(\gamma;\delta)} + \dot{u}_\alpha \dot{u}_\beta + \omega_{\alpha\gamma} \omega^\gamma{}_\beta + \sigma_{\alpha\gamma} \sigma^\gamma{}_\beta$$

$$+ \frac{2}{3} \theta \sigma_{\alpha\beta} + \frac{1}{3} h_{\alpha\beta} \left[2 \left(\omega^2 - \sigma^2 \right) + \dot{u}^\gamma{}_{;\gamma} \right] + E_{\alpha\beta} = 0, \tag{15.48}$$

where $E_{\alpha\beta}$ is the 'electric part' of the Weyl tensor, which was defined in (7.97).

In addition, the following three other equations hold:

$$\omega_{[\alpha\beta;\gamma]} + \dot{u}_{[\alpha;\gamma} u_{\beta]} + \dot{u}_{[\alpha} \omega_{\beta\gamma]} = 0, \tag{15.49}$$

$$h^\alpha{}_\beta \left(\omega^{\beta\gamma}{}_{;\gamma} - \sigma^{\beta\gamma}{}_{;\gamma} + \frac{2}{3} \theta^{;\beta} \right) - \left(\omega^\alpha{}_\beta + \sigma^\alpha{}_\beta \right) \dot{u}^\beta = 0, \tag{15.50}$$

$$2 \dot{u}_{(\alpha} w_{\beta)} - \sqrt{-g} h^\gamma{}_\alpha h^\delta{}_\beta \left(\omega_{(\gamma}{}^{\mu;\nu} + \sigma_{(\gamma}{}^{\mu;\nu)} \right) \epsilon_{\delta)\rho\mu\nu} u^\rho = H_{\alpha\beta}, \tag{15.51}$$

where $H_{\alpha\beta}$ is the 'magnetic part' of the Weyl tensor defined in (7.98), and w_α is the rotation vector defined in (15.36). Equation (15.49) is obtained by antisymmetrising (15.40) in the indices γ, δ and σ, then using $-R_{[\gamma|\rho|\delta\sigma]}u^\rho = u^\rho R_{\rho[\gamma\delta\sigma]} = 0$ and (15.31). Equation (15.50) is obtained by contracting (15.40) with $g^{\gamma\delta}h^{\alpha\sigma}$, then using (15.43) and (15.31). In order to obtain (15.51), one has to rewrite (15.40) in the form

$$- u_{\delta}{}^{;\mu\nu} + u_{\delta}{}^{;\nu\mu} = R^{\mu\nu}{}_{\delta\sigma}u^\sigma, \tag{15.52}$$

then contract both sides of (15.52) with $\frac{1}{2}\sqrt{-g}\epsilon_{\gamma\rho\mu\nu}u^\rho h^\gamma{}_\alpha h^\delta{}_\beta$ and then symmetrise the result with respect to α and β. Along the way, (11.68) and (11.69) will probably be useful. Equations (15.46) – (15.51) are algebraically independent components of (15.40).

A conclusion from (15.46) – (15.51) is that assumptions made about the hydrodynamical tensors can lead to restrictive results. For example, assume that $\sigma = \omega = 0 = \dot{u}^\alpha$. Then (15.47) and (15.49) are fulfilled identically, (15.50) says that the expansion scalar may change only along the flow lines of the perfect fluid, while (15.48) and (15.51) imply that the Weyl tensor is zero (see Exercise 10 to chapter 7). The family of perfect fluid solutions of the Einstein equations for which the Weyl tensor vanishes was found by Stephani (1967a) (see also Stephani et al. (2003)). They have the properties $\sigma = \omega = 0$, but in general $\dot{u}^\alpha \neq 0$. In the limit $\dot{u}^\alpha = 0$ they reduce to the Robertson–Walker metrics of Section 10.7 – which is a rather strong simplification. The full set of solutions of Einstein's equations with a perfect fluid source for which $\sigma = \omega = 0$ was found by Barnes (1973).

15.4 Singularities and singularity theorems

Let us define the function $\ell(x^\mu)$ through the equation

$$\frac{1}{\ell}\frac{d\ell}{ds} = \frac{1}{3}\theta, \tag{15.53}$$

where $d/ds \overset{\text{def}}{=} u^\rho\partial/\partial x^\rho$. This function is a generalisation of the scale factor $R(t)$ of the Robertson–Walker models, see Section 10.7. It can be seen from (15.29), (15.19) and (15.31) that, with $\omega = \sigma = 0 = \dot{u}^\alpha$, the distance between the simultaneous positions of two particles obeys (15.53). In the general case, $\ell(x)$ has no direct physical interpretation and is just a convenient representation of the expansion scalar. We will assume that $\omega = 0 = \dot{u}^\alpha$; then the Raychaudhuri equation becomes

$$3\frac{\ddot{\ell}}{\ell} + 2\sigma^2 + \frac{1}{2}\kappa(\epsilon + 3p) = 0. \tag{15.54}$$

Since $\epsilon + 3p > 0$ for all kinds of matter known from laboratory, (15.54) shows that $d^2\ell/ds^2 < 0$. This means that the function $\ell(s)$ is concave in all its range – if at any point p of the curve $\ell(s)$ we draw a straight line tangent to $\ell(s)$, then the *whole* curve $\ell(s)$ will lie below that straight line. There are two possibilities. If at present $(s = s_0)$ $(d\ell/ds)(s_0) > 0$ (i.e. the fluid expands, curve I in Fig. 15.1), then at a certain instant s_p *in the past*, $s_1 < s_p < s_0$, ℓ *was* zero. If, however, $(d\ell/ds)(s_0) < 0$ at present (i.e. the fluid contracts, curve II in Fig. 15.1), then at a certain instant s_F *in the future*, $s_2 > s_F > s_0$, ℓ *will be* zero. Consequently, in every matter model in which $\dot{u}^\alpha = 0 = \omega$ there exists such an instant, in the past or in the future, at which $\ell \to 0$. This implies, via (15.54), that

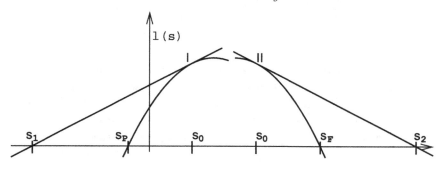

Fig. 15.1 A function that is everywhere concave must go to zero either in the past (at $s_P > s_1$) or in the future (at $s_F < s_2$).

$\epsilon + 3p \to \infty$ or $\sigma \to \infty$. Hence, every such portion of matter must have a singularity either in its future or in its past.

With the help of a similar analysis, based on various less restrictive assumptions, Penrose, Hawking and Ellis proved several **singularity theorems** which imply that quite general fluid configurations must contain singularities. A summary and overview of these theorems is presented in the book by Hawking and Ellis (1973) that caused a certain revision in the understanding of the relativity theory. The theorems were said to show that general relativity cannot be the ultimate theory of space and time. In order to avoid singularities, one would have to resort to a more general theory that would be capable of describing the quantum effects taking place at great densities of matter.

However, the singularity theorems are not as general as it was initially claimed. Several interesting solutions of the Einstein equations that *do not* contain any singularities were found by Senovilla and coworkers (for an extended review see Senovilla (1998)). They have not, so far, been shown to describe any actual astrophysical situation, but their very existence proves that singularities are an inevitable part not of relativity as such, but of the collection of models of matter defined by the assumptions of the singularity theorems.

15.5 Relativistic thermodynamics

The considerations of this section are applied in relativistic astrophysics to interiors of stars or to the Universe as a whole (see Section 16.1). For simplicity, we shall assume that the medium is a one-component perfect fluid. More general media, like viscous, heat-conducting or anisotropic fluids are considered in the literature, but they require more advanced thermodynamics for their description. The equations of motion of a perfect fluid are (12.17), with $T^{\alpha\beta}$ given by (12.73).

Let n denote the particle number density. We shall consider only such processes in which particles are neither created nor annihilated, so the total number of particles contained in a volume at time t_2 will be either the same as at any $t_1 < t_2$, or equal to the sum of the number of particles at t_1 and the number of those that entered/left the volume between

t_1 and t_2. In addition to (12.17) we thus postulate the equation of continuity for n:

$$(nu^\alpha)_{;\alpha} = 0. \tag{15.55}$$

In phenomenological thermodynamics, if the volume V of a given system is determined, then the **enthalpy** of the system is defined by $H = U + pV$, where U is the internal energy of the medium. In cosmology or in considering the interior of stars, the only well-defined volume for local considerations is the volume per particle of the fluid $V_p = 1/n$, and $(\epsilon + p)$ is the enthalpy density. We can thus define the enthalpy per particle

$$\mathcal{H} = (\epsilon + p)/n. \tag{15.56}$$

Now we postulate that for a single 'particle' of the fluid the phenomenological thermodynamics still applies (in the next chapter we will see that in cosmology that 'particle' will be a galaxy cluster or a still larger object). The enthalpy obeys the **Gibbs identity**:

$$dH = V dp + T dS. \tag{15.57}$$

Since ϵ and p are provided by the Einstein equations, and n and $V = 1/n$ have a clear physical interpretation, one may treat H and $\mathcal{H}$ as given. Phenomenological thermodynamics says that at most two state functions are sufficient for a full thermodynamical description of a one-component substance; the other functions can be calculated from the equation of state. In keeping with this, one would say that at most two of the three functions p, V and $\mathcal{H}$ are independent. In that case, the differential 1-form $(d\mathcal{H} - V dp)$ is a form in only two variables, and thus must have an integrating factor. Denoting this factor by $1/T$, we conclude that the form $(1/T)(d\mathcal{H} - V dp)$ is a perfect differential of a function S, so

$$d\mathcal{H} = dp/n + T dS. \tag{15.58}$$

In this way, we have apparently defined the temperature T and the entropy S. The integrating factor $1/T$ and the function S are not determined uniquely, but, still following the rules of phenomenological thermodynamics, one may conclude that T is determined up to linear transformations, i.e. up to the choice of scale (Werle, 1957). Given T, S is determined up to an additive constant.

However, there is a problem with this reasoning. It remained unnoticed for a long time because the solutions of Einstein's equations used in astrophysics are almost exclusively of high symmetry: they are spherically symmetric, or stationary and axisymmetric, or homogeneous of Bianchi type (Robertson–Walker spacetimes being a subcase of the latter). In the first two cases all the metric components, and thus all the thermodynamical quantities, depend only on two variables, so (15.58) may indeed be considered a definition of T and S. In the third case, all thermodynamical quantities depend on just one variable (the comoving time), so an even simpler equation of state of the type $\epsilon = \epsilon(p)$ is imposed on the matter by the assumed symmetry. However, if the metric has a 1-dimensional symmetry group, or no symmetry at all, then the functions ϵ, p and n depend on three or four variables, respectively, and then the existence of an integrating factor for the differential form $(d\mathcal{H} - V dp)$ is an additional postulate. Spacetimes in which the form $(d\mathcal{H} - V dp)$ does have an integrating factor, for which thus temperature and entropy can be defined by the reasoning presented above, are said to admit a **thermodynamical scheme**. The fact that the thermodynamical scheme might not exist in some spacetimes was first noted by Bona and Coll (1985, 1988) and Coll and Ferrando (1989). They also showed that the Stephani universe (Stephani, 1967a), which has in general no symmetry, acquires a 3-dimensional symmetry group acting on 2-dimensional orbits when the thermodynamical scheme is imposed (Bona and Coll, 1988). The problem of existence of the thermodynamical scheme was further discussed by Quevedo and Sussman (1995) and applied to all currently known cosmological solutions with no symmetry by Krasiński, Quevedo and Sussman (1997). Apart from the Stephani universe, the problem of existence of the thermodynamical scheme shows up in the two classes of metrics found by Szafron (1977) (see our Section 20.1) – all other solutions of Einstein's equations known at present are vacuum, have dust source or have a high symmetry. It turns out that in the $\beta' \neq 0$ family of the Szafron metrics the thermodynamical scheme imposes a 3-dimensional symmetry group. For the $\beta' = 0$ family, a subcase with no symmetry survives the imposition of the scheme.

The conclusion from the results briefly reported above is that, in considering spacetimes of no symmetry, one must allow for a more complicated thermodynamical scheme than a single-component perfect fluid.

Having said this, we will now discuss the thermodynamics of single-component perfect fluids in some more detail.

Using (15.55) – (15.56) and (15.58) one can now write (12.17) for a perfect fluid as

$$0 = nu^\beta \left(\mathcal{H}u_\alpha\right)_{;\beta} - p_{,\alpha} = n\left[u^\beta \left(\mathcal{H}u_\alpha\right)_{;\beta} - \mathcal{H}_{,\alpha} + TS_{,\alpha}\right]. \tag{15.59}$$

By virtue of $u_\alpha u^\alpha = 1$ and $u^\beta u_{\beta;\alpha} = 0$ this becomes

$$0 = u^\beta \left[(\mathcal{H}u_\alpha)_{;\beta} - (\mathcal{H}u_\beta)_{;\alpha}\right] + TS_{,\alpha} = u^\beta \left[(\mathcal{H}u_\alpha)_{,\beta} - (\mathcal{H}u_\beta)_{,\alpha}\right] + TS_{,\alpha}. \tag{15.60}$$

From here we see easily that $S_{,\alpha} u^\alpha = 0$, i.e. the entropy is constant along the flow lines of the perfect fluid. A more special kind of motion is often considered, in which

$$S_{,\alpha} = 0, \tag{15.61}$$

i.e. the entropy is constant in the whole volume under consideration. Such motion is called **isentropic**. Then, from (15.56) and (15.58):

$$\frac{\mathrm{d}p}{n} = \mathrm{d}\left(\frac{\epsilon + p}{n}\right) = \frac{\mathrm{d}\epsilon + \mathrm{d}p}{n} - \frac{(\epsilon + p)\mathrm{d}n}{n^2}. \tag{15.62}$$

Hence, $\mathrm{d}\epsilon = (\epsilon + p)\mathrm{d}n/n$, which means that $\epsilon = \epsilon(n)$, and consequently $p = p(n)$ and $\epsilon = \epsilon(p)$. This type of equation of state is called **barotropic**. It is almost exclusively used in cosmology (where it is necessitated by the symmetry of the Robertson–Walker models). However, it is a simplifying assumption when used in less symmetric models.

A reverse theorem also holds: for a perfect fluid, if $\epsilon = \epsilon(p)$ and $(nu^\beta)_{;\beta} = 0$, then either $p_{,\alpha} u^\alpha = 0$ or $S_{,\alpha} = 0$. Proof: for a one-component perfect fluid, we have an equation of state $F(\epsilon, p, n) = 0$. If $\epsilon = \epsilon(p)$, then the equation of state implies immediately that $n = n(p)$, so $\epsilon = \epsilon(n)$. Then, $u_\alpha T^{\alpha\beta}_{;\beta} = 0$ implies that

$$\epsilon_{,\beta} u^\beta + (\epsilon + p)u^\beta_{;\beta} = 0 \tag{15.63}$$

and the Gibbs identity (15.58) becomes

$$\frac{d\epsilon}{n} - \frac{(\epsilon + p)dn}{n^2} = TdS, \tag{15.64}$$

which can be written as

$$\epsilon_{,\alpha} - (\epsilon + p)n_{,\alpha}/n = nTS_{,\alpha}. \tag{15.65}$$

Substituting $u^\beta{}_{;\beta} = -n_{,\beta} u^\beta/n$ (from $(nu^\beta){}_{;\beta} = 0$) in (15.63) we obtain $[d\epsilon/dn - (\epsilon + p)/n] n_{,\beta} u^\beta = 0$. Hence, either $n_{,\beta} u^\beta = 0$ (the particle number density does not change along the flow lines) or $d\epsilon/dn = (\epsilon + p)/n$, so $\epsilon_{,\alpha} = (\epsilon + p)n_{,\alpha}/n$. Substitution of the last equation in (15.65) gives $S_{,\alpha} = 0$. $\square$

15.6 Exercises

1. Show that the quantity D defined in (15.17) is indeed zero.
Hint. Decompose the vectors $\delta\mathbf{x}$, $\delta\mathbf{y}$ and $\delta\mathbf{z}$ in the basis of the eigenvectors of the matrix σ. Use the property $\text{Tr}(\sigma) = 0$.
2. Verify (15.39).

16

Relativistic cosmology I: general geometry

16.1 A continuous medium as a model of the Universe

When describing the Universe as a whole, one assumes that it is filled with a continuous medium (fluid or gas), whose state can be described by scalar fields such as mass density and pressure, vector fields such as the velocity of flow or tensor fields, e.g. an electromagnetic field. This is a rather crude approximation, since our real Universe has a 'granular' structure. Its basic units are stars, and the relevant information from the point of view of observational cosmology is, for example, the number of stars in a given volume rather than the average mass density in that volume. The less-than-perfect adequacy of the fluid approximation is also demonstrated by the fact that the view on which objects should be considered the 'elementary cells' of the cosmic fluid has been changing with time. In the times of Hubble (1920s and 1930s), these were the galaxies. In later times, when galaxy clusters and proper motions of galaxies in clusters were observed, the galaxy clusters took over. In still later years, it was found that galaxies and galaxy clusters tend to occupy edges of large volumes of space (called voids) with few galaxies inside. According to current beliefs, the elementary units of the Universe should be groups of voids. These changes in the definition of the elementary unit of the Universe were, characteristically, adopted in order to save the assumption of homogeneity and isotropy of the Universe 'in the large'.

This assumption deserves a separate comment. The astronomical observations provide reliable quantitative information about a relatively small neighbourhood of the Solar System. With increasing distance from the Earth, the precision of this information is quickly degraded. If we want to describe the Universe as a whole, we have to extrapolate the results of local observations to large volumes, and then test the conclusions from the extrapolation. The extrapolations, however, always contain a large amount of arbitrariness. Hence, if observations tell us that a given extrapolation leads to a correct prediction, this does not mean that the extrapolation was the only one possible.

The most fundamental extrapolation is contained in the so-called **cosmological principle**. It stems from the idea of Copernicus, and so is sometimes called the *Copernican principle*. Copernicus was the first astronomer who noted that the Earth is not at the centre of the Universe, but occupies a very ordinary position in the Solar System. Afterwards, the Earth had been 'degraded' a few more times when it was established that even the Sun is not at the centre of the Universe, but is one of a great number of stars that are

223

similar to each other, and that our Galaxy is also one of many, not the greatest and not placed at any preferred position. The cosmological principle is a summary of this line of thinking. In its weaker form, it says: we (the inhabitants of the Earth) occupy a position in the Universe that is not in any way preferred. Quite often, however, this principle is expressed in the extreme form: all positions in the Universe are equivalent; the geometrical and physical properties of the Universe do not depend on the point from which the Universe is observed. No matter whether one believes in the cosmological principle, and in which version of it, it must be remembered that this principle is not a summary of observational results, but an *assumption*, upon which the theory of the structure of the Universe is built. This assumption was a good working hypothesis when theoreticians constructed the first-ever models of the Universe in the 1920s (Friedmann, 1922; Lemaître, 1927) because at that time there were no observational data to contradict it. Today, the cosmological principle still has no direct observational verification,[1] while models that generalise those of Friedmann–Lemaître and do not obey this principle are known; see Chapters 18 and 20. The fact that virtually the whole of observational cosmology is based on the Friedmann–Lemaître models is a consequence of inertia in thinking and of emotional attachment to the doctrine of equivalence of all positions in the Universe. However, natural sciences, physics and astronomy among them, are said to use the criterion of consistency of theory with observations/experiments. At the very least, in order to verify the cosmological principle, alternatives to it have to be considered and compared with observations. Working always from within the same theory, we make it more difficult to verify its basic assumptions. There are first signs that this kind of critical approach to the cosmological principle is catching on in the astronomical community: in 2021, a large conference took place in Pohang, South Korea, its main purpose being a debate on the assumption that the Universe is isotropic and homogeneous (Aluri et al., 2023).

The cosmological principle is translated into assumptions about the geometry of spacetime in the following way. The observations show that the space is *approximately* isotropic around us.[2] According to the cosmological principle, the space should thus be isotropic around every other point. A space that is isotropic around every point is homogeneous. This argument points to the Robertson–Walker (R–W) spacetimes of Section 10.7. Before we come to them, we shall discuss the fluid model of the Universe in the background of a general geometry. We assume that each point in the Universe can be assigned energy density, pressure and the 4-vector of velocity of the fluid particle that passes through the point. We also assume that the matter of the Universe treated in this way obeys the equations of hydrodynamics known from laboratory – which is another bold assumption.

16.2 The geometric optics approximation

The approach presented below is based on the papers by Ellis (1971, 1973) and Kristian and Sachs (1966). We will apply the geometric optics approximation to the Maxwell equations.

---

[1] Because of difficulties in determining the distances to other galaxies. The isotropy of the cosmic microwave background radiation imposes only weak limitations on the anisotropy of matter distribution; see Chapter 18.

[2] At present, the main argument supporting this statement is the measured isotropy of the microwave background radiation, but the statement first appeared long before the CMB radiation was discovered.

We assume that the electromagnetic field propagates into vacuum, with no charges or currents, and that it is weak enough not to influence the geometry of the spacetime (i.e. it is a *test field* in a given geometry). Then, the Maxwell equations are (13.4) and (13.6) with $j^\mu = 0$. We will seek solutions of this set in the form of waves:

$$F_{\alpha\beta} = G_{\alpha\beta} \sin(S + \phi), \tag{16.1}$$

where the amplitude $G_{\alpha\beta}$ is assumed to vary slowly compared to the phase S; ϕ is a constant. Thus, we assume that each differentiation of $G_{\alpha\beta}$ introduces a small factor ε. If $G_{\alpha\beta}$ can be developed in a power series with respect to ε, then we can write

$$G_{\alpha\beta} = B^0_{\alpha\beta} + \sum_{i=1}^{\infty} \varepsilon^i B^i_{\alpha\beta}. \tag{16.2}$$

We now assume that (13.4) and (13.6) are fulfilled at each order in ε, and substitute (16.2) in them. Denoting[3]

$$S_{,\alpha} = k_\alpha \tag{16.3}$$

we obtain at the zeroth order:

$$(B^0)^{\alpha\beta} k_\beta = 0, \qquad k_{[\alpha} B^0_{\beta\gamma]} = 0. \tag{16.4}$$

and at the first order

$$(B^0)^{\alpha\beta}{}_{;\beta} \sin(S + \phi) = -\varepsilon (B^1)^{\alpha\beta} k_\beta \cos(S + \phi), \tag{16.5}$$

$$B^0_{[\alpha\beta;\gamma]} \sin(S + \phi) = -\varepsilon B^1_{[\alpha\beta} k_{\gamma]} \cos(S + \phi). \tag{16.6}$$

Since the $(B^i)^{\alpha\beta}$, $i = 0, 1, \ldots$, are antisymmetric in $[\alpha\beta]$, the covariant derivative in (16.5) can be written as $(1/\sqrt{-g}) \left[\sqrt{-g} (B^0)^{\alpha\beta} \right]_{,\beta}$, so this expression is small by virtue of our assumptions. The same applies to (16.6) – in this combination, the Christoffel symbols cancel out and the covariant derivative reduces to a partial derivative. The last two equations show that the first-order terms act as sources in the Maxwell equations for the zeroth order terms. Thus, comparing this scheme with the Maxwell equations in empty space, we see that the electromagnetic wave does not in fact propagate into vacuum – the higher-order terms act as a medium with currents and charges, on which the zeroth order wave may be dispersed. Similarly, the first-order terms will be influenced by second-order terms, and so on. This is the influence of curvature on the propagation of the electromagnetic wave. (In a flat space in Cartesian coordinates, a constant $B^0_{\alpha\beta}$, with all $B^i_{\alpha\beta} \equiv 0$ for $i \geq 1$ is a solution of (16.5) – (16.6).) Assuming that the scheme is self-consistent (no formal proof of this assumption seems to be available) and that the 'tail' terms remain small, we will now consider the consequences of (16.4). The second equation of (16.4) can be written as

$$k_\alpha B^0_{\beta\gamma} + k_\beta B^0_{\gamma\alpha} + k_\gamma B^0_{\alpha\beta} = 0. \tag{16.7}$$

---

[3] With (16.3) fulfilled, the *rotation* of the vector field k_α, defined later in this section, is zero. Thus, the geometric optics approach turns out to be more general in this respect: the wave description does not allow rotating congruences of rays.

Contracting this with k^α and making use of (16.4) we immediately obtain

$$k^\alpha k_\alpha = 0, \tag{16.8}$$

i.e. the wave vector of the ray is a null vector. From this it follows that

$$k_{\alpha;\beta} k^\alpha = 0. \tag{16.9}$$

But, since $k_\alpha = S_{,\alpha}$, we have $k_{\alpha;\beta} = k_{\beta;\alpha}$, and then from (16.9)

$$k_{\beta;\alpha} k^\alpha = 0, \tag{16.10}$$

which means that k^α is geodesic, and its parametrisation is affine.

Furthermore, contracting (16.7) with $(B^0)^{\beta\gamma}$ and using (16.4) we obtain

$$B^0_{\alpha\beta}(B^0)^{\alpha\beta} = 0. \tag{16.11}$$

Contracting (16.7) with $(B^0)^{\delta\gamma}$ and using (16.4) once more, we obtain

$$k_\alpha B^0_{\beta\gamma}(B^0)^{\delta\gamma} - k_\beta B^0_{\alpha\gamma}(B^0)^{\delta\gamma} = 0. \tag{16.12}$$

This equation has the form $k_\alpha V_\beta - k_\beta V_\alpha = 0$ (with V having the upper index δ). Such an equation means that the vectors V_α and k_α are proportional (collinear), thus

$$B^0_{\alpha\gamma}(B^0)^{\delta\gamma} = U^\delta k_\alpha, \tag{16.13}$$

where U^δ is the (vectorial) proportionality factor. With the index δ lowered, the left-hand side of (16.13) is symmetric in $(\alpha\delta)$, so $U_\delta k_\alpha - U_\alpha k_\delta = 0$, and by the same argument $U_\alpha = \mu k_\alpha$. Thus, finally

$$B^0_{\alpha\gamma}(B^0)_\delta{}^\gamma = \mu k_\alpha k_\delta. \tag{16.14}$$

From (16.11) and (16.14) we see that the electromagnetic energy-momentum tensor (13.10) of the field (16.1) – (16.2) is

$$T_{\alpha\beta} = \frac{1}{4\pi} \left(\mu k_\alpha k_\beta + O(\varepsilon) \right), \tag{16.15}$$

which is, up to terms linear in ε, a perfect-fluid-type medium with the '4-velocity' k_α. With k_α being null, the velocity of flow equals the velocity of light, so (16.15) corresponds to a stream of photons. From $T^{\alpha\beta}{}_{;\beta} = 0$, and making use of (16.10), we obtain

$$(\mu k^\alpha)_{;\alpha} = 0, \tag{16.16}$$

which means that the stream of photons is conserved.

16.3 The redshift

An observer moving with the 4-velocity u^α will measure the rate of change of phase of the light wave $v_p = S_{,\alpha} u^\alpha = k_\alpha u^\alpha$. Within a short time-interval Δs, the phase will thus change by $\Delta S = k_\alpha u^\alpha \Delta s$. For another observer, moving with the velocity u_1^α and measuring the change of phase at another spacetime point, where $k^\alpha = k_1^\alpha$, the same change of phase ΔS will in general take a different time-interval, Δs_1: $\Delta S = k_{1\alpha}u_1^\alpha \Delta s_1$. Hence,

$$\frac{\Delta s_1}{\Delta s_2} = \frac{(k_\alpha u^\alpha)_2}{(k_\alpha u^\alpha)_1}.$$ (16.17)

If the electromagnetic wave is periodic, then ΔS is connected with the frequency ν by $\Delta S = 2\pi\nu\Delta s$ (this formula applies for an arbitrary time-interval Δs). Hence, for the same change of phase measured by two different observers we have $\nu_1\Delta s_1 = \nu_2\Delta s_2$, so

$$\frac{\nu_2}{\nu_1} = \frac{\Delta s_1}{\Delta s_2} = \frac{(k_\alpha u^\alpha)_2}{(k_\alpha u^\alpha)_1}.$$ (16.18)

This is the formula for the cosmological **redshift**, derived without invoking any definite cosmological model. It can be written in a more familiar form. Let the subscripts e and o denote quantities calculated at the point of emission of the light ray and at the point of detection, respectively. Then, from the definition of the redshift:

$$z = \frac{\lambda_o - \lambda_e}{\lambda_e} = \frac{\lambda_o}{\lambda_e} - 1.$$ (16.19)

But $\lambda_o/\lambda_e = \nu_e/\nu_o$ (because the locally measured velocity of light $c = \lambda\nu$ is constant), so

$$1 + z = \frac{\nu_e}{\nu_o} = \frac{(k_\alpha u^\alpha)_e}{(k_\alpha u^\alpha)_o}.$$ (16.20)

The light ray with the wave vector k^α is received by the observer from the direction determined by the unit spacelike vector n^α:

$$n_\alpha n^\alpha = -1,$$ (16.21)

which is collinear with the projection of k^α on the hypersurface of constant time. Hence, if the 4-velocity of the observer is u^α, then

$$n^\alpha = \rho\left(\delta^\alpha{}_\beta - u^\alpha u_\beta\right) k^\beta.$$ (16.22)

Substituting (16.22) in (16.21) we obtain $\rho^2 = (k_\rho u^\rho)^{-2}$, hence

$$n^\alpha = -\frac{1}{k_\rho u^\rho} k^\alpha + u^\alpha \qquad (\Longrightarrow n^\alpha u_\alpha = 0).$$ (16.23)

We have chosen $\rho = -(k_\rho u^\rho)^{-1}$ because n^α denotes the direction *towards* the source of light, opposite to the direction of the wave vector.

The considerations up to this point applied for an arbitrary distance; the only approximation was that connected with geometric optics. However, in applying (16.20) to results of observations, one has to integrate the equations of a null geodesic, which is almost always difficult. In that case, an approximate version of (16.20) is useful, which applies for small redshifts, $z \ll 1$. Then (16.19) becomes $z = \mathrm{d}\lambda/\lambda$, and in (16.20) we have

$$z = \frac{(k_\alpha u^\alpha)_e - (k_\alpha u^\alpha)_o}{(k_\alpha u^\alpha)_o} = \frac{\mathrm{d}(k_\alpha u^\alpha)}{(k_\alpha u^\alpha)_o}.$$ (16.24)

The change $d(\cdot)$ in (16.24) should be calculated along the light ray connecting the light source with the observer. Denoting the affine parameter on the ray by v, we have

$$d\left(k_\alpha u^\alpha\right) = D\left(k_\alpha u^\alpha\right) = \left(k_{\alpha;\beta}k^\beta u^\alpha\right)_o dv + \left(k_\alpha u^\alpha{}_{;\beta}\,k^\beta\right)_o dv. \tag{16.25}$$

The first term vanishes in consequence of (16.10), and in the second term we substitute the decomposition (15.31) and use the antisymmetry of $\omega_{\alpha\beta}$. The result is

$$d\left(k_\alpha u^\alpha\right) = \left[\left(\sigma_{\alpha\beta}k^\alpha k^\beta\right)_o - \frac{1}{3}\theta(k_\alpha u^\alpha)_o{}^2 + (k_\alpha \dot{u}^\alpha)_o\,(k_\alpha u^\alpha)_o\right] dv. \tag{16.26}$$

Using (16.23) and (15.45) we obtain

$$d\left(k_\alpha u^\alpha\right) = (k_\rho u^\rho)_o{}^2 \left(\sigma_{\alpha\beta}n^\alpha n^\beta - \frac{1}{3}\theta - n_\alpha \dot{u}^\alpha\right)_o dv. \tag{16.27}$$

Equations (16.23) and (16.8) imply

$$k^\rho n_\rho = k^\rho u_\rho, \tag{16.28}$$

and $\left[-\left(k^\rho n_\rho\right)_o dv\right]$ is the distance in the rest space of the observer travelled by the front of the light wave corresponding to the change dv in the affine parameter, i.e. it is the distance between the light source and the observer, which we shall denote by $\delta\ell$. In consequence of this, and using (16.27) in (16.24), we obtain:

$$z = \left(-\sigma_{\alpha\beta}n^\alpha n^\beta + \frac{1}{3}\theta + n_\alpha \dot{u}^\alpha\right)_o \delta\ell. \tag{16.29}$$

This shows that for $z \ll 1$ rotation has no influence on z. It is seen that $\sigma_{\alpha\beta}$ and $\dot{u}^\alpha$ introduce anisotropy in z, whereas with $\sigma_{\alpha\beta} = 0 = \dot{u}^\alpha$ the redshift should be isotropic. Equation (16.29) still applies in every cosmological model, but only for light sources that are near to the observer. Its advantage is that all the quantities in it can be calculated without integrating any differential equations.

16.4 The optical tensors

We shall now apply a reasoning similar to that in Section 15.2 to families of null curves. The tensor of projection on a locally orthogonal space now has to be defined in a different way, because a hypersurface orthogonal to a null vector k contains k. Therefore, given k^α, we first define a second null vector ℓ^α that is tangent to the same light cone at the same point and obeys

$$\ell^\alpha k_\alpha = 1, \qquad \ell^\alpha \ell_\alpha = 0. \tag{16.30}$$

Note that ℓ^α is not defined uniquely. If m^α is an arbitrary spacelike vector of unit length ($m^\alpha m_\alpha = -1$) orthogonal both to k^α and to ℓ^α, then $\ell'^\alpha = \ell^\alpha + \frac{1}{2}b^2 k^\alpha + bm^\alpha$ obeys (16.30) as well, where b is an arbitrary parameter. We will come back to this in Section 16.6.

Then we define the projection tensor on the surface (this time 2-dimensional) that is orthogonal to both ℓ^α and k^α:

$$p_{\alpha\beta} = g_{\alpha\beta} - \ell_\alpha k_\beta - k_\alpha \ell_\beta \implies p_{\alpha\beta}k^\beta = p_{\alpha\beta}\ell^\beta = 0. \tag{16.31}$$

This surface is not tangent to ℓ^α or k^α – it is an exercise in special relativity to prove that a vector orthogonal to two linearly independent null vectors must be spacelike.

Now let k^α be a null vector *field*. It is usually assumed at this point that k^α is geodesic and affinely parametrised so that $k^\nu k_{\mu;\nu} = 0$. We will not assume this; instead we define

$$\dot{k}^\mu \stackrel{\text{def}}{=} k^\mu{}_{;\nu} k^\nu, \tag{16.32}$$

which we shall call the **acceleration** of a light ray. The specialisation to the geodesic affinely parametrised case will then follow immediately by $\dot{k}^\mu = 0$. We will need the more general formulae in Section 16.6.

Being null, the field k^μ obeys

$$k^\mu k_\mu = 0 = k^\mu k_{\mu;\alpha} \implies k^\mu \dot{k}_\mu = 0. \tag{16.33}$$

We define

$$A_{\alpha\beta} \stackrel{\text{def}}{=} k_{\rho;\sigma} p^\rho{}_\alpha p^\sigma{}_\beta. \tag{16.34}$$

By writing out (16.34) we obtain

$$k_{\alpha;\beta} = A_{\alpha\beta} + a_\alpha k_\beta + k_\alpha b_\beta + \dot{k}_\alpha \ell_\beta, \tag{16.35}$$

where

$$\begin{aligned}
a_\alpha &\stackrel{\text{def}}{=} \ell^\rho k_{\alpha;\rho} - \frac{1}{2} k_{\rho;\sigma} \ell^\rho \ell^\sigma k_\alpha, \\
b_\alpha &\stackrel{\text{def}}{=} \ell^\rho k_{\rho;\alpha} - \frac{1}{2} k_{\rho;\sigma} \ell^\rho \ell^\sigma k_\alpha - \dot{k}^\rho \ell_\rho \ell_\alpha
\end{aligned} \tag{16.36}$$

(the term $k_{\rho;\sigma} \ell^\rho \ell^\sigma k_\alpha$ was evenly split between a_α and b_α for later convenience; this splitting is not implied by (16.34)). It follows that $a_\alpha k^\alpha = b_\alpha k^\alpha = 0$. Now we apply to $A_{\alpha\beta}$ the decomposition into the trace, the trace-free symmetric part and the antisymmetric part, just like we did in Section 15.2 with the covariant derivative of a unit timelike field:

$$A_{\alpha\beta} = \omega_{\alpha\beta} + \sigma_{\alpha\beta} + p_{\alpha\beta}\theta \tag{16.37}$$

(by tradition, the last term does not include the coefficient $1/2$), where

$$\omega_{\alpha\beta} \stackrel{\text{def}}{=} A_{[\alpha\beta]} \tag{16.38}$$

is called the **rotation** of the family of null curves,

$$\theta \stackrel{\text{def}}{=} \frac{1}{2} g^{\alpha\beta} A_{\alpha\beta} \tag{16.39}$$

is called the **expansion** of the family, and

$$\sigma_{\alpha\beta} \stackrel{\text{def}}{=} A_{(\alpha\beta)} - p_{\alpha\beta}\theta \tag{16.40}$$

is called the **shear** of the family. The geometric interpretation of rotation, expansion and shear is similar to that in hydrodynamics; this time the changes apply to images of an object projected by the family of light rays on orthogonal 2-surfaces. Like in hydrodynamics, rotation and shear do or do not vanish simultaneously with the scalars defined below. Also, the following equations are useful:

$$
\begin{aligned}
p^\alpha{}_\rho p^\rho{}_\beta &= p^\alpha{}_\beta, & g^{\alpha\beta} p_{\alpha\beta} = p^{\alpha\beta} p_{\alpha\beta} = 2, \\
p^{\alpha\beta} \sigma_{\alpha\beta} &= g^{\alpha\beta} \sigma_{\alpha\beta} = 0, \\
k^\beta \omega_{\alpha\beta} &= k^\beta \sigma_{\alpha\beta} = \ell^\beta \omega_{\alpha\beta} = \ell^\beta \sigma_{\alpha\beta} = 0.
\end{aligned}
\tag{16.41}
$$

The scalars of rotation, expansion and shear are then

$$
\omega^2 \overset{\text{def}}{=} \frac{1}{2} \omega_{\alpha\beta} \omega^{\alpha\beta} = \frac{1}{2} k_{[\alpha;\beta]} \left(k^{\alpha;\beta} - 2\dot{k}^\alpha \ell^\beta \right) - \frac{1}{4} \left(\dot{k}_\rho \ell^\rho \right)^2,
\tag{16.42}
$$

$$
\theta = \frac{1}{2} k^\mu{}_{;\mu} - \frac{1}{2} \dot{k}_\rho \ell^\rho,
\tag{16.43}
$$

$$
\sigma^2 \overset{\text{def}}{=} \frac{1}{2} \sigma_{\alpha\beta} \sigma^{\alpha\beta} = \frac{1}{2} k_{(\alpha;\beta)} \left(k^{\alpha;\beta} - 2\dot{k}^\alpha \ell^\beta \right) + \frac{1}{4} \left(\dot{k}_\rho \ell^\rho \right)^2 - \theta^2.
\tag{16.44}
$$

For affinely parametrised geodesics ($\dot{k}^\mu = 0$) these scalars are independent of ℓ^α.

16.5 The apparent horizon

We have already defined the event horizon for the Schwarzschild metric in Section 14.11. There is one more variety of horizon that is important for studying dynamical black holes, i.e. such that keep swallowing up new matter and increase their masses. The notion of the event horizon has the disadvantage that it can be defined only when we know the whole future evolution of the spacetime. Hence, it is practically useless in observational cosmology. It is more realistic to use only such notions as can be identified by local observations of short duration. The apparent horizon is such a notion.

In a flat spacetime, a flash of light sent from both sides of a closed surface has the property that the light rays are converging inside the surface and diverging outside. The convergence/divergence is measured by the scalar of expansion for the family of (geodesic) rays, $\theta = \frac{1}{2} k^\mu{}_{;\mu}$, defined in the previous section. However, in the vicinity of a singularity the *outward-directed* bundle of rays is convergent. This is visible in the Kruskal diagram (Fig. 14.8): under the horizon H_1 all future-directed rays intersect lines of still-decreasing r (recall that each point in the Kruskal diagram represents a sphere of constant t and r). Hence, the area of the light front is decreasing for both bundles. A closed surface from which it is impossible to send a diverging bundle of rays is called a **closed trapped surface**. Then, an **apparent horizon** is the outer envelope of the region in which closed trapped surfaces exist.

There are two kinds of trapped surfaces and apparent horizons, as seen in Fig. 14.8: the **future-trapped surfaces** and **future apparent horizons** in region II, and the **past-trapped surfaces** and **past apparent horizons** in region IV. For a past-trapped surface, the light rays converge towards the past. A more physical way to formulate this definition is to say that, for a past-trapped surface S_p, both *ingoing* bundles of rays (i.e. those that simultaneously reach S_p both from inside and from outside) are *diverging*, whereas for a future-trapped surface S_f, both *outgoing* bundles of rays are converging (where 'outgoing' means starting their journey simultaneously at S_f, both inward and outward).

In the Schwarzschild spacetime, the apparent horizons coincide with the event horizons, but in nonstatic spacetimes these horizons are different, see examples in Sec. 18.8.

16.6 * The double-null tetrad

We set up a field of null vector bases over the spacetime that includes the k^α and ℓ^α introduced in Section 16.4. The other basis vectors, m^α and $\overline{m}^\alpha$, are complex conjugate to each other, orthogonal both to k^α and to ℓ^α, and obey relations similar to (16.30):

$$g_{\alpha\beta}m^\alpha m^\beta = g_{\alpha\beta}\overline{m}^\alpha\overline{m}^\beta = 0, \qquad g_{\alpha\beta}m^\alpha\overline{m}^\beta = -1,$$
$$g_{\alpha\beta}m^\alpha k^\beta = g_{\alpha\beta}m^\alpha\ell^\beta = 0, \qquad g_{\alpha\beta}\overline{m}^\alpha k^\beta = g_{\alpha\beta}\overline{m}^\alpha\ell^\beta = 0. \tag{16.45}$$

Thus we have [4]

$$e_{\widehat{0}}{}^\alpha = k^\alpha, \qquad e_{\widehat{1}}{}^\alpha = \ell^\alpha, \qquad e_{\widehat{2}}{}^\alpha = m^\alpha, \qquad e_{\widehat{3}}{}^\alpha = \overline{m}^\alpha, \tag{16.46}$$

and, from (9.1), the tetrad metric is

$$[\eta_{ij}] = \begin{bmatrix} 0 & 1 & 0 & 0 \\ 1 & 0 & 0 & 0 \\ 0 & 0 & 0 & -1 \\ 0 & 0 & -1 & 0 \end{bmatrix}. \tag{16.47}$$

The upper and lower tetrad indices are related to each other as follows:

$$v_{\widehat{0}} = v^{\widehat{1}}, \qquad v_{\widehat{1}} = v^{\widehat{0}}, \qquad v_{\widehat{2}} = -v^{\widehat{3}}, \qquad v_{\widehat{3}} = -v^{\widehat{2}} = \overline{v_{\widehat{2}}}. \tag{16.48}$$

From this and (16.46) it follows that

$$e^{\widehat{0}}{}_\alpha = \ell_\alpha, \qquad e^{\widehat{1}}{}_\alpha = k_\alpha, \qquad e^{\widehat{2}}{}_\alpha = -\overline{m}_\alpha, \qquad e^{\widehat{3}}{}_\alpha = -m_\alpha. \tag{16.49}$$

The tetrad components of the basis vectors are thus

$$k^i = \delta^i{}_{\widehat{0}}, \qquad \ell^i = \delta^i{}_{\widehat{1}}, \qquad m^i = -\delta^i{}_{\widehat{2}}, \qquad \overline{m}^i = -\delta^i{}_{\widehat{3}},$$
$$k_i = \delta_i{}^{\widehat{1}}, \qquad \ell_i = \delta_i{}^{\widehat{0}}, \qquad m_i = \delta_i{}^{\widehat{3}}, \qquad \overline{m}_i = \delta_i{}^{\widehat{2}}. \tag{16.50}$$

Some of the Ricci rotation coefficients in this basis, defined by (9.7), vanish in consequence of (9.14), thus

$$\Gamma^0{}_{1a} \equiv \Gamma_{11a} = 0 = \Gamma_{00a} \equiv \Gamma^1{}_{0a},$$
$$\Gamma^2{}_{3a} \equiv -\Gamma_{33a} = 0 = -\Gamma_{22a} \equiv \Gamma^3{}_{2a}. \tag{16.51}$$

Some other Ricci rotation coefficients have physical interpretation. For example, the nonzero tetrad components of the shear tensor (16.40) are:

$$\sigma_{\widehat{22}} = k_{\alpha;\beta}m^\alpha m^\beta = -\Gamma^1{}_{22} \equiv -\Gamma_{022} \equiv \Gamma_{202} \equiv -\Gamma^3{}_{02},$$
$$\sigma_{\widehat{33}} = k_{\alpha;\beta}\overline{m}^\alpha\overline{m}^\beta = -\Gamma^1{}_{33} \equiv -\Gamma_{033} \equiv \Gamma_{303} \equiv -\Gamma^2{}_{03}. \tag{16.52}$$

---

[4] Hats mark tetrad indices. Where no confusion may arise, they will be omitted.

In order to calculate the expansion, we do the following operations:

$$k^{\mu}{}_{;\mu} = g_{\mu\nu} k^{\mu;\nu} = \eta_{ab} e^{a}{}_{\mu} e^{b}{}_{\nu} k^{\mu;\nu},$$

and similarly for the term $\dot{k}_{\rho} \ell^{\rho}$, and then we explicitly run through all the values of a and b. Most of the terms are zero or cancel out, and what remains is

$$\theta = -\frac{1}{2} k^{\mu;\nu} \left(m_{\mu} \overline{m}_{\nu} + \overline{m}_{\mu} m_{\nu} \right) = \frac{1}{2} \left(\Gamma^{1}{}_{23} + \Gamma^{1}{}_{32} \right) \equiv \frac{1}{2} \left(\Gamma^{1}{}_{23} + \overline{\Gamma}^{1}{}_{23} \right). \tag{16.53}$$

For the tetrad components of the rotation tensor we find

$$\omega_{23} = -\omega_{32} = -\frac{1}{2} \left(\Gamma^{1}{}_{23} - \Gamma^{1}{}_{32} \right) = \overline{\omega}_{32} = -\overline{\omega}_{23}, \tag{16.54}$$

all other components being zero. So, the only nonvanishing component of the rotation tensor is pure imaginary, and, as follows from (16.42) and (16.45), it is connected to the scalar of rotation ω by $\omega_{23}{}^{2} = -\omega^{2}$; thus $\omega_{23} = -\mathrm{i}\omega$. Consequently, the expansion and the rotation are the real and the imaginary part of the same complex quantity

$$\theta + \mathrm{i}\omega = \Gamma^{1}{}_{23} \equiv \Gamma_{023} \equiv -\Gamma_{203}. \tag{16.55}$$

Several identities among the Ricci rotation coefficients are not self-evident; examples were given in (16.52) and (16.55).

The double-null tetrad is a basis of the **Newman–Penrose formalism**. Although shuffling indices in it is somewhat confusing (because of the scalar metric being nondiagonal), this tetrad has proven powerful in finding exact solutions of Einstein's equations. For more on this formalism, see Stephani et al. (2003).

The various identities obeyed by the Riemann and Weyl tensors were deduced under the assumption that the Riemann tensor arises from commutators of second covariant derivatives. However, if the basic objects in the theory are the Ricci rotation coefficients, like in the Newman–Penrose formalism, then the curvature tensors are present in first-order equations, and not all the 'identities' will really be fulfilled identically (in fact, only $R_{ijkl} = -R_{ijlk} = -R_{jikl}$ and $C^{i}{}_{jil} = 0$). In Sections 16.8 and 21.2, we will see some of those other 'identities' being imposed as equations to fulfil.

The tetrad $(k, \ell, m, \overline{m})$ is not uniquely defined. One usually begins with a given family of null curves, so the *direction* of k^{α} is fixed, but k^{α} itself may be rescaled by

$$k'^{\alpha} = A k^{\alpha}, \tag{16.56}$$

where A is an arbitrary real function. This change corresponds to changing the parametrisation of the curves tangent to k^{α}. With k^{α} so changed, ℓ^{α} will remain null and will preserve its scalar product with k^{α} when a fixed multiple of k^{α} and a fixed multiple of a fixed vector in the $(m^{\alpha}, \overline{m}^{\alpha})$ plane are added to it, thus

$$\ell'^{\alpha} = \frac{1}{A} \left(\ell^{\alpha} + \overline{B} e^{\mathrm{i}\phi} m^{\alpha} + B e^{-\mathrm{i}\phi} \overline{m}^{\alpha} + B \overline{B} k^{\alpha} \right). \tag{16.57}$$

The vectors m^{α} and $\overline{m}^{\alpha}$ can be rotated in their plane by an arbitrary angle ϕ and will thereafter remain orthogonal to k^{α} and ℓ^{α}. So, they are defined up to the transformations

$$m'^{\alpha} = e^{\mathrm{i}\phi} m^{\alpha}, \tag{16.58}$$

where ϕ is a real function. The scalar products of m^α and $\overline{m}^\alpha$ with k^α and between themselves do not change when a multiple of k^α is added to any of them.

As an example of usefulness of this tetrad, let us note how simply it expresses the criterion for the Weyl tensor being algebraically special, i.e. obeying (11.55). Projecting that equation on the double-null tetrad we obtain the equivalent equation

$$C_{200d} = C_{300d} = 0. \tag{16.59}$$

16.7 * The equations of propagation of the optical scalars

There are analogies between the equations of evolution of hydrodynamical tensors of Sec. 15.3 and the equations of propagation of the optical scalars defined in Sec. 16.4. We will now deal with the latter. Take the Ricci formula for the null field k^α:

$$k_{\alpha;\beta\gamma} - k_{\alpha;\gamma\beta} = R_{\rho\alpha\beta\gamma}k^\rho. \tag{16.60}$$

We will project this on different combinations of the tetrad vectors. The following equations will be useful in this process:

$$g^{\alpha\beta}k_{\alpha;\beta\gamma} \equiv \left(k^\beta{}_{;\beta}\right)_{;\gamma} = 2\theta_{,\gamma} + \left(\dot{k}_\rho\ell^\rho\right)_{,\gamma}, \tag{16.61}$$

$$k^{\gamma;\beta}k_{\beta;\gamma} = 2\left(\sigma^2 - \omega^2 + \theta^2\right) + 2b_\beta\dot{k}^\beta + \left(\ell_\beta\dot{k}^\beta\right)^2, \tag{16.62}$$

$$k^\gamma g^{\alpha\beta}k_{\alpha;\gamma\beta} \equiv k^\gamma k^\beta{}_{;\gamma\beta} = \dot{k}^\beta{}_{;\beta} - k^{\gamma;\beta}k_{\beta;\gamma}, \tag{16.63}$$

$$k^\gamma m^\alpha \overline{m}^\beta k_{\alpha;\beta\gamma} = k^\gamma\left(m^\alpha \overline{m}^\beta k_{\alpha;\beta}\right)_{;\gamma} - k^\gamma m^\alpha{}_{;\gamma}\overline{m}^\beta k_{\alpha;\beta} - k^\gamma m^\alpha \overline{m}^\beta{}_{;\gamma}k_{\alpha;\beta}, \tag{16.64}$$

$$k^\gamma m^\alpha \overline{m}^\beta k_{\alpha;\gamma\beta} = \overline{m}^\beta\left(m^\alpha \dot{k}_\alpha\right)_{;\beta} - \overline{m}^\beta k^\gamma{}_{;\beta}m^\alpha k_{\alpha;\gamma} - \overline{m}^\beta m^\alpha{}_{;\beta}\dot{k}_\alpha, \tag{16.65}$$

$$
\begin{aligned}
m^\alpha{}_{;\gamma} &= g^{\alpha\mu}m_{\mu;\gamma} = \eta^{ab}e_a{}^\alpha e_b{}^\mu m_{\mu;\gamma} \\
&= \left(k^\alpha\ell^\mu + \ell^\alpha k^\mu - m^\alpha \overline{m}^\mu - \overline{m}^\alpha m^\mu\right)m_{\mu;\gamma}.
\end{aligned} \tag{16.66}
$$

Equations (16.61) and (16.62) follow from (16.43) and (16.35) – (16.37). The other ones in the above set are just identities. Equations (16.64) and (16.65) apply together with their complex conjugates. In simplifying the right-hand sides of (16.64) and (16.65) we will use (9.7), the definition of $\Gamma^i{}_{jk}$. The last line of (16.66) is a consequence of (16.46) – (16.47). Its last term is zero in this example in consequence of $m^\mu m_\mu = 0$, but in other applications all the terms between the parentheses have to be included.

Contract (16.60) with $k^\gamma g^{\alpha\beta}$; the result is the equation of propagation of θ, analogous to the Raychaudhuri equation (15.46):

$$k^\gamma\theta_{,\gamma} + \sigma^2 - \omega^2 + \theta^2 + \frac{1}{2}A_{(1)} = -\frac{1}{2}R_{\rho\gamma}k^\rho k^\gamma, \tag{16.67}$$

where $A_{(1)}$ is a collection of terms containing acceleration:

$$A_{(1)} \stackrel{\text{def}}{=} 2b_\beta\dot{k}^\beta + \left(\ell_\beta\dot{k}^\beta\right)^2 - \dot{k}^\beta{}_{;\beta} + k^\gamma\left(\ell_\beta\dot{k}^\beta\right)_{,\gamma} \tag{16.68}$$

and vanishes when $\dot{k}^\alpha = 0$ (the index in parentheses is just a label, it does not refer to the coordinate or tetrad components).

Now contract (16.60) with $k^\gamma \left(m^\alpha \overline{m}^\beta - m^\beta \overline{m}^\alpha \right)$ and use $R_{\rho[\alpha\beta]\gamma} k^\rho k^\gamma \equiv 0$. This is a complicated calculation. In addition to (16.61) – (16.66) one has to use the identities resulting from (16.45), such as $\overline{m}^\mu m_{\mu;\gamma} + \overline{m}^\mu{}_{;\gamma} m_\mu = 0$ and $k^\gamma m_{\mu;\gamma} k^\mu + m_\mu \dot{k}^\mu = 0$, and also (9.7) – the definition of $\Gamma^i{}_{jk}$. The final result is

$$k^\gamma \left(\Gamma^1{}_{[23]} \right)_{,\gamma} + \frac{1}{2} \left[\left(\Gamma^1{}_{23} \right)^2 - \left(\Gamma^1{}_{32} \right)^2 \right] - \frac{1}{2} A_{(2)} = 0, \tag{16.69}$$

where $A_{(2)}$ is another collection of terms containing acceleration:

$$A_{(2)} \overset{\text{def}}{=} -\overline{m}^\beta \left(m^\alpha \dot{k}_\alpha \right)_{,\beta} + m^\beta \left(\overline{m}^\alpha \dot{k}_\alpha \right)_{,\beta} + \overline{m}^\beta m^\alpha{}_{;\beta} \dot{k}_\alpha - m^\beta \overline{m}^\alpha{}_{;\beta} \dot{k}_\alpha$$
$$- m_\rho \dot{k}^\rho \left(2\Gamma^1{}_{13} + \Gamma^2{}_{10} - \Gamma^1{}_{31} \right) + \overline{m}_\rho \dot{k}^\rho \left(2\Gamma^1{}_{12} + \Gamma^3{}_{10} - \Gamma^1{}_{21} \right), \tag{16.70}$$

and vanishes when $\dot{k}^\alpha = 0$ (in accord with the rest of (16.69), $A_{(2)}$ is imaginary). In consequence of (16.54), Eq. (16.69) may be written as

$$i k^\gamma \omega_{,\gamma} + 2i\theta\omega - \frac{1}{2} A_{(2)} = 0. \tag{16.71}$$

Equations (16.67) and (16.71) can be written as one complex equation:

$$k^\gamma Z_{,\gamma} + Z^2 + \sigma^2 + \frac{1}{2} A_{(1)} - \frac{1}{2} A_{(2)} = -\frac{1}{2} R_{\rho\gamma} k^\rho k^\gamma \equiv -\frac{1}{2} R_{\widehat{0}\widehat{0}}, \tag{16.72}$$

where

$$Z \overset{\text{def}}{=} \theta + i\omega \equiv \Gamma^1{}_{23}. \tag{16.73}$$

Next, contract (16.60) with $k^\gamma m^\alpha m^\beta$ and use (16.64) – (16.66). The result is

$$k^\gamma \Gamma^1{}_{22,\gamma} + 2\Gamma^1{}_{22} \left(\theta + \Gamma^3{}_{30} \right) + A_{(3)} = R_{\widehat{0}\widehat{2}\widehat{0}\widehat{2}}, \tag{16.74}$$

where $A_{(3)}$ is yet another collection of terms with acceleration:

$$A_{(3)} \overset{\text{def}}{=} m^\beta m^\alpha \dot{k}_{\alpha;\beta} + m_\rho \dot{k}^\rho \left(2\Gamma^1{}_{12} + \Gamma^1{}_{21} + \Gamma^3{}_{10} \right). \tag{16.75}$$

The result of contracting (16.60) with $k^\gamma \overline{m}^\alpha \overline{m}^\beta$ can be written at once, as the complex conjugate of (16.74):

$$k^\gamma \Gamma^1{}_{33,\gamma} + 2\Gamma^1{}_{33} \left(\theta + \Gamma^2{}_{20} \right) + \overline{A_{(3)}} = R_{\widehat{0}\widehat{3}\widehat{0}\widehat{3}}. \tag{16.76}$$

Finally, contract (16.60) with $k^\gamma m^\alpha \overline{m}^\beta$, use (16.64) – (16.66) and also use (16.52) for $\Gamma^1{}_{22}\Gamma^1{}_{33} = \sigma_{\widehat{2}\widehat{2}}\sigma_{\widehat{3}\widehat{3}} \equiv \frac{1}{2}\sigma_{ij}\sigma^{ij} = \sigma^2$, to obtain

$$k^\gamma \Gamma^1{}_{23,\gamma} + \left(\Gamma^1{}_{23} \right)^2 + \sigma^2 + A_{(4)} = R_{\widehat{0}\widehat{2}\widehat{0}\widehat{3}}, \tag{16.77}$$

where

$$A_{(4)} \overset{\text{def}}{=} \overline{m}^\beta m^\alpha \dot{k}_{\alpha;\beta} + m_\rho \dot{k}^\rho \left(2\Gamma^1{}_{13} + \Gamma^2{}_{10} \right) + \overline{m}_\rho \dot{k}^\rho \Gamma^1{}_{21}. \tag{16.78}$$

Since $\Gamma^1{}_{23} = \theta + i\omega = Z$, we use (16.72) to eliminate $k^\gamma \Gamma^1{}_{23,\gamma}$ from (16.77) and obtain

$$\frac{1}{2} \left(A_{(2)} - A_{(1)} \right) + A_{(4)} = R_{\widehat{0}\widehat{2}\widehat{0}\widehat{3}} + \frac{1}{2} R_{\widehat{0}\widehat{0}}. \tag{16.79}$$

16.8 * The Goldberg–Sachs theorem

The Goldberg–Sachs theorem does not find any direct application in cosmology; it is applied mainly to vacuum solutions, and we will use it in Chapter 21. However, it is so closely related to the previous two sections that it belongs naturally in this place.

There is a connection between properties of the optical tensors of Section 16.4 and the Petrov type of the Weyl tensor – similar to the connection for the hydrodynamical tensors in Section 15.3, where $\sigma_{\alpha\beta} = w_{\alpha\beta} = 0 = \dot{u}_\alpha$ forced the metric to be conformally flat.

In order to prove an analogous property of the optical tensors, we must first list the full set of properties of an algebraically special Weyl tensor in the tetrad $(k, \ell, m, \overline{m})$. With (16.59) assumed, the equations $C^s{}_{isj} = 0$ imply

$$
\begin{aligned}
-C_{0101} &= C_{0213} + C_{0312} = -C_{2323}, \\
C_{0112} &= C_{1223}, \qquad C_{0113} = -C_{1323}, \\
C_{0212} &= C_{0223} = C_{0313} = C_{0323} = C_{1213} = 0.
\end{aligned}
\tag{16.80}
$$

The theorem we are going to prove consists of three parts that we list as separate theorems.

Theorem 16.1

Assumptions:

1. The Weyl tensor is algebraically special, so that (11.55) and (16.59) are fulfilled. We adapt the tetrad of (16.45) – (16.50) to the degenerate Debever field k^α that obeys (11.55).
2. The vacuum Einstein equations are fulfilled, $R_{\alpha\beta} = 0$.

Thesis:

The degenerate Debever vector field k^α is geodesic and shearfree, so, from (16.52)

$$
\dot{k}^\alpha = \phi k^\alpha, \qquad \sigma_{\alpha\beta} = 0 \Longrightarrow \Gamma_{202} = \Gamma_{303} = 0.
\tag{16.81}
$$

Proof:

Take the Bianchi identities $R_{\alpha\beta[\gamma\delta;\epsilon]} = 0$, contract them with $g^{\beta\gamma}$ and use the assumption $R_{\alpha\beta} = 0$. The result is equivalent to $C^\rho{}_{\alpha\beta\gamma;\rho} = 0$. Now use (9.17) contracted with δ^e_a:

$$
e_a{}^\mu C^a{}_{bcd,\mu} + \Gamma^s{}_{rs}C^r{}_{bcd} - \Gamma^s{}_{br}C^r{}_{scd} - \Gamma^s{}_{cr}C^r{}_{bsd} - \Gamma^s{}_{dr}C^r{}_{bcs} = 0.
\tag{16.82}
$$

Take the component $(b, c, d) = (0, 0, 2)$ of this equation, then use (16.59) and (16.80) together with (16.51). The result is

$$
\Gamma_{200}\left(C_{0123} + C_{0213} - C_{0101}\right) = 0.
\tag{16.83}
$$

Then take the component $(0, 0, 3)$ and make similar simplifications. The result is[5]

$$
\Gamma_{300}\left(-C_{0123} + C_{0312} - C_{0101}\right) = 0.
\tag{16.84}
$$

Thus, either $\Gamma_{200} = 0 \Longrightarrow \Gamma_{300} = 0$ or else the expressions in parentheses must vanish. Consider first the second case. Then:

$$
\begin{aligned}
C_{0123} + C_{0213} \phantom{+C_{0312}} - C_{0101} &= 0, \\
-C_{0123} \phantom{+ C_{0213}} + C_{0312} - C_{0101} &= 0,
\end{aligned}
\tag{16.85}
$$

[5] Note that (16.84) is the complex conjugate of (16.83) – as expected, since the interchanges $2 \leftrightarrow 3$ always correspond to complex conjugation in this tetrad. Thus, one could write down (16.84) on the basis of (16.83) without repeating the calculations. This is one of the advantages of the double-null tetrad.

and also the first of (16.80) and $C_{0[bcd]} = 0$ must be obeyed:

$$\begin{aligned} C_{0213} \ + \ C_{0312} \ + \ C_{0101} &= \ 0, \\ C_{0123} \ - \ C_{0213} \ + \ C_{0312} &= \ 0. \end{aligned} \tag{16.86}$$

The solution of the set (16.85) – (16.86) is, using (16.80),

$$C_{0123} = C_{0213} = C_{0312} = C_{0101} = C_{2323} = 0. \tag{16.87}$$

Now take the components $(0,0,1)$ and $(0,2,3)$ of (16.82), with (16.80) and (16.87) assumed. They are

$$\Gamma_{300}C_{0112} + \Gamma_{200}C_{0113} = 0 = \Gamma_{300}C_{1223} + \Gamma_{200}C_{1323}. \tag{16.88}$$

Taking into account the middle part of (16.80), we see that with $\Gamma_{200} \neq 0 \neq \Gamma_{300}$ the solution of (16.88) is

$$C_{0112} = C_{0113} = C_{1223} = C_{1323} = 0. \tag{16.89}$$

Take the components $(0,1,2)$ and $(0,1,3)$ of (16.82), with (16.80), (16.87) and (16.89) assumed. They say that $\Gamma_{300}C_{1212} = 0 = \Gamma_{200}C_{1313}$, i.e. $C_{1212} = C_{1313} = 0$. Together with (16.59), (16.80), (16.87) and (16.89) this means $C_{abcd} = 0$. With $R_{ab} = 0$, this is the Minkowski spacetime, in which a congruence of shearfree null geodesics does exist, as can be verified by calculation in the Cartesian coordinates. The theorem is then true in a modified form: a shearfree null geodesic congruence exists and obviously obeys (11.55).

We return to (16.83) – (16.84) and take the other solution

$$\Gamma_{200} = \Gamma_{300} = 0. \tag{16.90}$$

This means, from the definition of the Ricci rotation coefficients,

$$\begin{aligned} 0 &= \ \Gamma^3{}_{00} = m_{\rho;\sigma}k^\rho k^\sigma = -k_{\rho;\sigma}m^\rho k^\sigma = -m^\rho \dot{k}_\rho, \\ 0 &= \ \Gamma^2{}_{00} = \overline{m}_{\rho;\sigma}k^\rho k^\sigma = -k_{\rho;\sigma}\overline{m}^\rho k^\sigma = -\overline{m}^\rho \dot{k}_\rho. \end{aligned} \tag{16.91}$$

Thus, $\dot{k}^\alpha$ is orthogonal to both m^α and $\overline{m}^\alpha$, so lies in the plane of k^α and ℓ^α. Since $\dot{k}^\alpha k_\alpha = 0$ and $\ell^\alpha k_\alpha = 1$, $\dot{k}^\alpha$ must be proportional to k^α, which is the first of (16.81). As shown in Sec. 5.2, we can then change the parametrisation of k^α so that $\dot{k}^\alpha = 0$.

Now take (16.82) with $(b,c,d) = (2,0,2)$ and $(3,0,3)$, then use (16.59), (16.80) and (16.90). The result is

$$\Gamma_{022}\left(C_{0123} + C_{0213} - C_{2323}\right) = 0 = \Gamma_{033}\left(-C_{0123} + C_{0312} - C_{2323}\right) \tag{16.92}$$

(again, the second equation is just a complex conjugate of the first one). Suppose that $\Gamma_{022} \neq 0 \Longleftrightarrow \Gamma_{033} \neq 0$. Since $C_{2323} = C_{0101}$ (from (16.80)), we obtain a copy of the set (16.85) and (16.86), whose solution is (16.87) once more.

The components $(1,0,2)$ and $(1,0,3)$ of (16.82), using (16.80), now say that $\Gamma_{202}C_{0113} = \Gamma_{303}C_{0112} = 0$. With the assumed $\Gamma_{202} \neq 0 \neq \Gamma_{303}$, and with (16.80), this implies (16.89) again. Then, the components $(1,0,1)$ and $(1,2,3)$ of (16.82) become $\Gamma_{303}C_{1212} + \Gamma_{202}C_{1313} = 0 = -\Gamma_{303}C_{1212} + \Gamma_{202}C_{1313}$. With $\Gamma_{202} \neq 0 \neq \Gamma_{303}$, the solution of this is $C_{1212} = C_{1313} = 0$, i.e. again $C_{abcd} = 0$.

Thus, (16.92) implies $\Gamma_{022} = \Gamma_{033} = 0$, which, as seen from (16.52), means that the shear of the k^α congruence is zero. $\square$

Now comes

Theorem 16.2

Assumptions:
1. *There exists a shearfree geodesic null vector field k^α on a spacetime.*
2. *The vacuum Einstein equations are obeyed, $R_{\alpha\beta} = 0$.*
3. *The Weyl tensor is nonzero.*

Thesis:
The field k^α is collinear with one of the Debever vector fields.

Proof:
In the double-null tetrad, the condition for the field k^α to be collinear with a Debever vector field, i.e. to obey (11.53), is

$$C_{0202} = C_{0303} = C_{0203} = 0. \qquad (16.93)$$

Choose the parametrisation on the curves tangent to k^α so that $\dot{k}^\alpha = 0$. Then, with $\dot{k}^\alpha = 0 = \sigma_{ab}$, in consequence of $R_{\alpha\beta} = 0$ and of (16.52), Eqs. (16.74), (16.76) and (16.79) imply (16.93). $\square$

Now contract (16.60) with $m^\alpha \ell^\beta k^\gamma$, then use the analogues of (16.64) and (16.65) and of the tricks described above (16.69) to obtain

$$C_{0210} = -k^\gamma \Gamma^1{}_{21,\gamma} - \Gamma^1{}_{21}\left(\Gamma^3{}_{30} + \Gamma^1{}_{23}\right) + \Gamma^0{}_{20}\Gamma^1{}_{23}$$
$$+ \Gamma^1{}_{22}\left(\Gamma^0{}_{30} - \Gamma^1{}_{31}\right) + A_{(5)}, \qquad (16.94)$$

where $A_{(5)}$ is another collection of acceleration terms:

$$A_{(5)} \overset{\text{def}}{=} -2m_\rho \dot{k}^\rho \Gamma^1{}_{11} - \ell^\beta m^\alpha \dot{k}_{\alpha;\beta} - \ell_\rho \dot{k}^\rho \Gamma^1{}_{21}. \qquad (16.95)$$

In order to prove the next theorem, we need the following lemma:

Lemma 16.1 *If $\dot{k}^\alpha = 0 = \sigma$ and $\theta + \mathrm{i}\omega \neq 0$, then there exists such a transformation of the basis vectors (16.56) – (16.57) after which $\Gamma^1{}_{21} = 0$. The direction of the new ℓ^α is uniquely determined.*

Proof:
Apply the transformation (16.56) – (16.57) to $\Gamma^1{}_{21}$; with $\dot{k}^\alpha = 0$ the result is

$$\widetilde{\Gamma}^1{}_{21} = \frac{1}{A^2}\left(\mathrm{e}^{\mathrm{i}\phi}\Gamma^1{}_{21} + B\Gamma^1{}_{23} + \mathrm{e}^{2\mathrm{i}\phi}\overline{B}\Gamma^1{}_{22}\right). \qquad (16.96)$$

In view of (16.52) and (16.55) it follows that $\widetilde{\Gamma}^1{}_{21} = 0$ when $\sigma_{\widehat{2}\widehat{2}} = -\Gamma^1{}_{22} = 0$ and $B = -\mathrm{e}^{\mathrm{i}\phi}\Gamma^1{}_{21}/\Gamma^1{}_{23}$. The property $\Gamma^1{}_{21} = 0$ is then preserved by the transformations (16.57) with $B = 0$, so the direction of the new ℓ^α is indeed unique. $\square$

Comment. Note that $\dot{k}^\alpha = 0$ implies $\Gamma^2{}_{00} = 0$ because $\Gamma^2{}_{00} = \overline{m}_{\mu;\nu}k^\mu k^\nu = -\overline{m}_\mu \dot{k}^\mu$; likewise for $\Gamma^3{}_{00}$.

We can now prove the last part of the main theorem:

Theorem 16.3 *If there exists a shearfree geodesic null vector field k^α on the spacetime and $R_{\alpha\beta} = 0$, then the Weyl tensor is algebraically special (i.e. obeys (11.55)).*

Proof:

In view of Theorem 16.2 and of Eq. (16.59), we need only show that $C_{0102} = 0$ in the double-null tetrad. As Lemma 16.1 demonstrates, the case $\theta + i\omega = 0$ has to be considered separately. Thus, we split the proof into two parts. Hats over tetrad indices are omitted because explicit values of the coordinate indices will not appear in this proof.

Part I: $\theta + i\omega = 0$

Calculate $R^1{}_{223}$ as in (9.21), then take into account that $\Gamma^1{}_{22} = \Gamma^1{}_{33} = 0$ because of the assumed shear $= 0$ (from (16.52)) and $\Gamma^1{}_{23} = \Gamma^1{}_{32} = 0$ because of $\theta + i\omega = 0$ (from (16.73)). All that remains from $R^1{}_{223}$ is

$$R^1{}_{223} = \Gamma^1{}_{20}\left(\Gamma^0{}_{23} - \Gamma^0{}_{32}\right). \tag{16.97}$$

Then use the definition of $\Gamma^1{}_{20}$ (Eq. (9.7)) and (16.49) to find

$$\Gamma^1{}_{20} = -k_{\rho;\sigma}m^\rho k^\sigma = -m^\rho \dot{k}_\rho = 0 \tag{16.98}$$

because k^ρ was assumed geodesic ($\Gamma^1{}_{20} = 0$ even if k^ρ is not affinely parametrised – then $\dot{k}^\rho \propto k^\rho$ and $m^\rho k_\rho = 0$ from (16.45)). Consequently,

$$R^1{}_{223} = R_{0223} = 0. \tag{16.99}$$

Now observe that

$$0 = -R_{02} = -R^0{}_{002} - R^3{}_{032} = R_{0102} + R_{2032} \tag{16.100}$$

because of $R_{\alpha\beta} = 0$. Then $R_{2032} = R_{0223} = 0$ by (16.99) and $C_{0102} = R_{0102} = 0$. $\square$

Part II: $\theta + i\omega \neq 0$

Then we choose, as the lemma allows, a direction of ℓ^α for which $\Gamma^1{}_{21} = 0$. As follows from (16.94), we then need only show that $\sigma = 0 = \dot{k}^\alpha$ implies $\Gamma^0{}_{20} \equiv \Gamma^3{}_{10} = 0$. Since $\Gamma^1{}_{21} = 0$ was achieved under the assumptions $\sigma = 0 = \dot{k}^\alpha$, we shall not display the shear and acceleration terms in what follows.

Take the equation $0 = R_{22} \equiv 2R_{0212} = 2R_{\rho\alpha\beta\gamma}k^\rho m^\alpha \ell^\beta m^\gamma$; it implies that the contraction of (16.60) with $m^\alpha \ell^\beta m^\gamma$ is zero. After using the assumptions $\Gamma^1{}_{21} = \Gamma^1{}_{22} = \Gamma^2{}_{00} = 0$ + their complex conjugates we obtain the equation $\Gamma^1{}_{23}\Gamma^0{}_{22} = 0$. Since we are considering the case $\Gamma^1{}_{23} \neq 0$, this means that

$$\Gamma^0{}_{22} \equiv \Gamma^3{}_{12} = 0 \implies \Gamma^0{}_{33} \equiv \Gamma^2{}_{13} = 0. \tag{16.101}$$

Now contract (16.60) with $k^\gamma\left(m^\alpha \ell^\beta - m^\beta \ell^\alpha\right)$; the result simplifies to

$$k^\gamma \Gamma^1{}_{12,\gamma} + \Gamma^3{}_{30}\Gamma^1{}_{12} + \Gamma^1{}_{23}\Gamma^0{}_{20} + \Gamma^1{}_{32}\left(\Gamma^1{}_{12} - \Gamma^0{}_{20}\right) = 0. \tag{16.102}$$

Then calculate $-R_{02} = R_{0102} + R_{0223}$. Since it was assumed that $R_{\alpha\beta} = 0$, it follows that

$$k^\gamma \Gamma^1{}_{12,\gamma} = -\Gamma^1{}_{12}\left(\Gamma^1{}_{32} + \Gamma^1{}_{23} + \Gamma^3{}_{30}\right) + \Gamma^1{}_{32}\Gamma^0{}_{20} - m^\gamma \Gamma^1{}_{23,\gamma}. \tag{16.103}$$

Substituting (16.103) in (16.102) we find

$$m^\gamma \Gamma^1{}_{23,\gamma} = \Gamma^1{}_{23}\left(\Gamma^0{}_{20} - \Gamma^1{}_{12}\right). \tag{16.104}$$

and, with all the current assumptions, (16.77) becomes

$$k^\gamma \Gamma^1{}_{23,\gamma} = - \left(\Gamma^1{}_{23}\right)^2. \tag{16.105}$$

Equations (16.104) and (16.105) require the integrability condition, which, for noncommuting directional derivatives, has the form

$$e_a{}^\mu \left(e_b{}^\nu \phi_{,\nu}\right)_{;\mu} - e_b{}^\mu \left(e_a{}^\nu \phi_{,\nu}\right)_{;\mu} = -2\Gamma^c{}_{[ab]} e_c{}^\nu \phi_{,\nu} \tag{16.106}$$

for an arbitrary function ϕ. Applying this rule to (16.104) and (16.105) we obtain

$$k^\gamma \left(\Gamma^0{}_{20} - \Gamma^1{}_{12}\right)_{,\gamma} = -2\Gamma^1{}_{23}\Gamma^0{}_{20} + \left(\Gamma^2{}_{20} - \Gamma^2{}_{02}\right)\left(\Gamma^0{}_{20} - \Gamma^1{}_{12}\right), \tag{16.107}$$

see Exercise 2 for a hint on how to derive this.

Using (16.103) and (16.104) in this, we obtain

$$k^\gamma \Gamma^0{}_{20,\gamma} = \Gamma^0{}_{20}\left(-3\Gamma^1{}_{23} + \Gamma^2{}_{20}\right). \tag{16.108}$$

Now, calculating $0 = R_{22} = R^0{}_{202} + R^1{}_{212}$ (in fact, $R^1{}_{212} \equiv 0$ because of (16.102) and the other current assumptions) and using (9.21) to calculate $R^i{}_{jkl}$, we obtain

$$m^\gamma \Gamma^0{}_{20,\gamma} = \Gamma^0{}_{20}\Gamma^2{}_{22} - \left(\Gamma^0{}_{20}\right)^2. \tag{16.109}$$

Applying the integrability condition (16.106) to (16.108) and (16.109) we obtain

$$\Gamma^0{}_{20}\left(k^\gamma \Gamma^2{}_{22,\gamma} - m^\gamma \Gamma^2{}_{20,\gamma}\right) = \Gamma^0{}_{20}\left[-\Gamma^0{}_{02}\Gamma^2{}_{20} - 9\Gamma^0{}_{20}\Gamma^1{}_{23}\right.$$
$$\left. - \Gamma^2{}_{02}\left(\Gamma^2{}_{22} - \Gamma^0{}_{20}\right) + \Gamma^2{}_{20}\left(\Gamma^2{}_{22} + \Gamma^0{}_{20}\right)\right]. \tag{16.110}$$

One solution of this is $\Gamma^0{}_{20} = 0$, which is the desired result. In order to verify the implications of the other alternative, we calculate the expression in parentheses on the left-hand side above from the equation $0 = R_{20} \equiv R^1{}_{210} + R^2{}_{220}$, in which we use (9.21) to calculate the tetrad components of the Riemann tensor. The result is, in consequence of all the assumptions made,

$$k^\gamma \Gamma^2{}_{22,\gamma} - m^\gamma \Gamma^2{}_{20,\gamma} = \Gamma^0{}_{20}\left(\Gamma^1{}_{23} + \Gamma^2{}_{20} + \Gamma^2{}_{02}\right) - \Gamma^0{}_{02}\Gamma^2{}_{20}$$
$$+ \Gamma^2{}_{22}\left(\Gamma^2{}_{20} - \Gamma^2{}_{02}\right). \tag{16.111}$$

Substituting (16.111) in (16.110) we obtain $\Gamma^0{}_{20}\Gamma^1{}_{23} = 0$, which, in the case $\Gamma^1{}_{23} \neq 0$ now considered, implies $\Gamma^0{}_{20} = 0$. Thus, $\Gamma^0{}_{20} = 0$ in both branches of (16.110). $\square$

Theorems 16.1–16.3 are summarised in the following

Theorem 16.4 (The Goldberg–Sachs theorem)

A vacuum spacetime is algebraically special if and only if there exists on it a geodesic and shearfree null vector field k^α. If the Weyl tensor is nonzero, then the field k^α is collinear with the degenerate Debever field.

This theorem was first formulated by Goldberg and Sachs (1962).[6] Parts of this result, such as Theorem 16.1 and the other theorems under additional assumptions, had been

---

[6] A few formulae in the later part of the proof in Goldberg and Sachs (1962) have misprints, which were corrected in the 2009 reprinting. The presentation given here has several details filled in because the original text was rather condensed.

proven earlier; references to those results are given in the original paper. The theorem is useful in investigating vacuum solutions of Einstein's equations; see Section 21.1.

A few generalisations of Theorems 16.1–16.4 are known (Stephani et al., 2003), but their theses cannot be stated without more elaborate explanations. One generalisation of Theorems 16.2 and 16.3 can be verified rather easily. The field equations we used in proving these two theorems were not really $R_{\alpha\beta} = 0$, but $R_{\alpha\beta}k^\alpha k^\beta = 0 = R_{\alpha\beta}k^\alpha m^\beta = R_{\alpha\beta}m^\alpha m^\beta$ and their complex conjugates; the other equations used in the proof were properties of the curvature tensor. All the equations that we made use of remain valid when $R_{\alpha\beta} = \Phi k_\alpha k_\beta$. However, such a generalisation does not hold for Theorem 16.1 (Stephani et al., 2003).

16.9 * The area distance

In curved spacetime, it is a problem to define the distance between two objects. Abstract geometric definitions, such as the integral of ds along a uniquely defined spacelike path connecting two worldlines, have no relation to astronomical practice. We need a definition that would yield a quantity measurable by means of astronomical observations. The reasoning used here is adapted from Ellis (1971),[7] and the definitions of distance go back to Etherington (1933). The subscript 'e' will denote quantities calculated at the emitter, 'O' will denote quantities at the observer.

An observer moving with a 4-velocity u_O^α would measure the flux of radiation reaching her (i.e. the energy flowing through a unit surface area in a unit of time) to be equal to $\mathcal{F}_O = cT_{\alpha\beta}u_O^\alpha u_O^\beta$, where $T_{\alpha\beta}$ is the energy-momentum tensor of radiation, given by (16.15), calculated at the observer's position. Thus, neglecting terms of order ε:

$$\mathcal{F}_O = \frac{c\mu_O}{4\pi}\,(k_\alpha u_O^\alpha)^2. \tag{16.112}$$

On the other hand, combining the conservation equation (16.16) with the surface-area-propagation equation $k^\mu{}_{;\mu} = d(\ln \delta S_n)/ds$ (see Exercise 3) we see that $\mu\delta S$ is constant along null geodesics, thus

$$(\mu\delta S)|_{s=s_1} = (\mu\delta S)|_{s=s_2} \tag{16.113}$$

(δS is the area of an orthogonal cross-section of the bundle of rays). Combining (16.112) and (16.113) we obtain $\mathcal{F}_O = \text{constant}\,(k_\alpha u^\alpha)_O^2/(\delta S)_O$. Remembering that $(k_\alpha u^\alpha)_e$ is constant along each null geodesic and using (16.20), we obtain

$$\mathcal{F}_O = \frac{C_{\mathcal{F}}}{\delta S_O}\left[\frac{(k_\alpha u^\alpha)_O}{(k_\alpha u^\alpha)_e}\right]^2 = \frac{C_{\mathcal{F}}}{\delta S_O\,(1+z_O)^2}, \tag{16.114}$$

where $C_{\mathcal{F}}$ is constant in each bundle (it depends on the solid angle occupied by the bundle). Thus, the flux $\mathcal{F}_O$ decreases with increasing redshift. However, δS_O does not necessarily increase with distance, as can be seen from the 'null Raychaudhuri' equation (16.67). For geodesics, $A_{(1)} = 0$. Assume that $\omega = 0$ and $R_{\mu\nu} = 0$ (vacuum) for simplicity. Then (16.67) shows that $\theta_{,\mu}k^\mu < 0$, that is, θ decreases and may become negative. This observation is expressed by saying that *curvature causes null geodesics to converge*, even in vacuum. This

---

[7] Ellis (1971) used units in which the velocity of light $c = 1$. This had the undesirable consequence that the energy density of radiation and its flux became equal.

conclusion remains true in a perfect fluid or dust, where $R_{\alpha\beta}k^\alpha k^\beta = \kappa(\epsilon + p)(k_\alpha u^\alpha)^2 > 0$. Consequently, the situation shown in Fig. 16.1 is possible: an observer at Q will see the light source at G to have the same brightness as seen by the observer at P. For Q, the source will appear anomalously bright and anomalously large.[8]

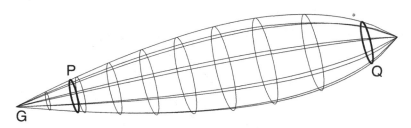

Fig. 16.1 A nonrotating null congruence in vacuum or in a perfect fluid necessarily has its expansion scalar decreasing along the rays. This may (if the distance is sufficiently large) eventually cause the expansion to become negative, which results in refocussing. The observer at Q would see the source G to be anomalously bright and large if she/he were able to cancel the redshift. Actually, when shear is nonzero, the geodesics will be refocussed not to a point but to a caustic surface.

In order to make (16.114) a workable formula, we have to calculate the constant $C_{\mathcal{F}}$. Imagine then that we surround G with a sphere S centred at G and of radius r_S.[9] Equation (16.114) shows that, for a given bundle of null geodesics, $(1+z)^2 \mathcal{F}\delta S$ is constant along the bundle. On the surface of a sphere, δS is proportional to the solid angle $\delta\Omega_G$ subtended by δS, thus $\delta S_G = r_S{}^2 \delta\Omega_G$, where r_S is the radius of the sphere S. Assuming that the source G radiates isotropically, and denoting by L its total luminosity (i.e. energy emitted per unit of time), on S we have $\mathcal{F}_S = L/(4\pi r_S{}^2)$. Also, $z = 0$ on S (because S is close to the source), so, taking (16.114) on S, we have

$$C_{\mathcal{F}} = \frac{L}{4\pi} \delta\Omega_G. \tag{16.115}$$

Now, following Etherington (1933) and Ellis (1971), we define the **source area distance** r_G (from the source to the observer, see Fig. 16.2) by

$$\delta S_O \stackrel{\text{def}}{=} r_G{}^2 \delta\Omega_G. \tag{16.116}$$

Finally, combining (16.114), (16.115) and (16.116) we obtain

$$\mathcal{F}_O = \frac{L}{4\pi} \frac{1}{(1+z_O)^2 r_G{}^2}. \tag{16.117}$$

The quantity $\mathcal{F}_O$ can be measured (this is the flux of energy from a given source through

---

[8] In actual observations, the redshift will decrease the flux of photons. The effect of focussing by curvature would be clearly visible only if the observer moved so as to cancel the redshift and if there were no absorption of light between the emitter and the observer.

[9] We assume the radius of the sphere to be sufficiently small that the space inside it can be approximated as a subset of the flat tangent space to the manifold. Otherwise, it might be impossible to construct such a sphere – for example, when the bundle has nonzero rotation.

a given surface sheet at the observer's position) and so can z_O. However, r_G cannot be measured because we cannot get close enough to the source to measure $\delta\Omega_G$. With L being essentially unknown, (16.117) cannot be observationally tested.

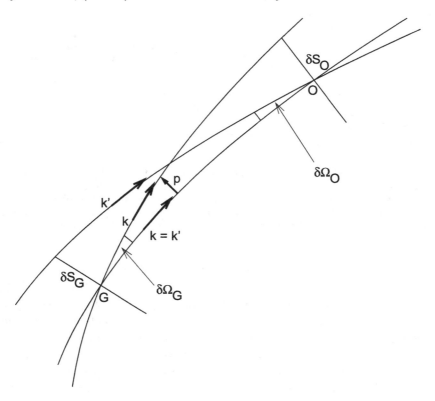

Fig. 16.2 An illustration of the reciprocity theorem. Explanation is given in the text.

But we can define distance in another way. Imagine a surface sheet of area δS_G placed at the source G and a bundle of light rays sent from this sheet in such a way that it converges to a single point at the observer's position O (Fig. 16.2). Assume that the centres of δS_G and of G coincide and that the sheet is orthogonal to the central ray of the bundle. Let the solid angle filled by this bundle as it reaches O be $\delta\Omega_O$. We can then define the **observer area distance** r_O from G to O by

$$\delta S_G \stackrel{\text{def}}{=} r_O{}^2 \delta\Omega_O. \tag{16.118}$$

We can measure $\delta\Omega_O$, and we can *in principle* calculate δS_G (for example, assuming that the bundle subtends the whole source, whose geometric size we know). In this way, we can *in principle* measure r_O. It turns out that r_G is determined by r_O, as we show below.

16.10 * The reciprocity theorem

Figure 16.2 will help in comparing r_O and r_G. The source of light G (suppose, a galaxy) sends a light ray from its middle point that hits the observer O. It also sends a diverging bundle of rays that surrounds the central ray GO and, at G, fills the solid angle $\delta\Omega_G$. The bundle has the projected area δS_O at the observer's position. Another bundle of rays is sent from a surface sheet of area δS_G that is orthogonal to GO at the central point G. The second bundle converges at O's centre, and there it fills the solid angle $\delta\Omega_O$. (Figure 16.2 shows a cross-section through one half of each bundle.) Let the affine parameter and the field of null vectors tangent to the first bundle be v and k^α, respectively, and let the same quantities for the second bundle be v' and k'^α. Along the ray GO, $k^\alpha = k'^\alpha$. Let the field of geodesic deviation defined by the field k^α be p^α, and the deviation for the other bundle be p'^α. By construction, $p^\alpha = 0$ at G and $p'^\alpha = 0$ at O. Then the following is true:

Theorem 16.5 (the reciprocity theorem)
Assumption:
The bundle k^α completely surrounds O, i.e. there are rays of the bundle in every direction from O within δS_O.
Thesis:
The area distances r_O and r_G are related by

$$r_G{}^2 = r_O{}^2 \left(1 + z_O\right)^2 . \tag{16.119}$$

Comment: The assumption would not be fulfilled if O were placed on a reflecting or absorbing surface, e.g. on a boundary between vacuum and opaque matter.

Proof:
Along the central ray we have $v = v'$ and $k^\alpha = k'^\alpha$ (but not $p^\alpha = p'^\alpha$). Thus, the geodesic deviation equation (6.56) and the symmetries of the Riemann tensor in its indices imply that, along this ray,

$$p'^\alpha \frac{D^2 p_\alpha}{dv^2} - p^\alpha \frac{D^2 p'_\alpha}{dv^2} = 0. \tag{16.120}$$

Consequently, along GO

$$p'^\alpha \frac{D p_\alpha}{dv} - p^\alpha \frac{D p'_\alpha}{dv} = \text{constant}. \tag{16.121}$$

Note that (6.54), with zero torsion, implies that $k^\beta p^\alpha{}_{;\beta} = p^\beta k^\alpha{}_{;\beta}$. Using this in (16.121) (where $D p_\alpha / dv = p_{\alpha;\beta} k^\beta$), we see that the connection terms cancel out. From now on we will thus replace the covariant derivatives by partial derivatives. Taking (16.121) at G (where $p^\alpha = 0$) and at O (where $p'^\alpha = 0$), we obtain

$$\left[p'^\alpha \frac{d p_\alpha}{dv} \right]_G = - \left[p^\alpha \frac{d p'_\alpha}{dv} \right]_O . \tag{16.122}$$

The geodesic deviation equation (6.56) determines only the component of δx^α transversal to k^μ, the component along k^μ obeys (6.56) identically and is not determined, also when k^μ is null. We may thus assume that $k_\alpha p^\alpha = k_\alpha p'^\alpha = 0$ all along the central ray GO. Of all these deviation vectors we now choose for further investigation those p^α that are orthogonal to the 4-velocity of the observer at O, and those p'^α that are orthogonal to the 4-velocity of the light source at G. The chosen vectors span 2-dimensional planes tangent to the spacetime at O and G. In the tangent plane at G there exist pairs that obey

$$\left[\frac{dp_1{}^\alpha}{dv}\frac{dp_{2\alpha}}{dv}\right]_G = 0. \tag{16.123}$$

They exist because of $p^\alpha|_G = 0$, so a given $(dp^\alpha/dv)_G$ selects one neighbouring geodesic from the k bundle, and the initial data $p^\alpha|_G = 0$, $(dp^\alpha/dv)_G$ uniquely determine k^α along the whole geodesic. By the assumption of the theorem, neighbouring geodesics exist in all directions around GO, so for any $(dp_1^\alpha/dv)_G$ there exists a $(dp_2^\alpha/dv)_G$ orthogonal to it that selects a neighbouring geodesic.

The vectors p_1^α and p_2^α whose initial data at G obey (16.123) will not in general be orthogonal at O, so let $\pi/2 + \psi$ be the angle between them at O. But the pairs obeying (16.123) can be rotated by any angle, by the assumption of the theorem. Suppose the pair is rotated by $\pi/2$. Then, arbitrarily near G, p_{1G}^α will become p_{2G}^α, and p_{2G}^α will become $-p_{1G}^\alpha$. Now note that if any p^α obeys (6.56), then so does Ap^α with any constant A. In particular, this is true for $A = -1$. Thus, at O, among all the vectors transported from G by (6.56), there will be $-p_{1O}^\alpha$, the image of $-p_{1G}^\alpha$. Since the solutions of the geodesic deviation equation with given initial data are unique, the following will happen at O: p_{1O} will become p_{2O}, and p_{2O} will become $-p_{1O}$. Whatever the angle $\pi/2 + \psi$ was before the rotation, it will become $\pi/2 - \psi$ after the rotation. Consequently, by continuity, if the pair obeying (16.123) is rotated by all angles between 0 and $\pi/2$, then the corresponding pair at O will form all angles between $\pi/2 + \psi$ and $\pi/2 - \psi$, and one of those angles will be $\pi/2$. Choosing the pair for which it happens, we have

$$(p_1^\alpha p_{2\alpha})|_O = 0 \tag{16.124}$$

while (16.123) still holds.

From the field p'^α we now choose the vectors that at G obey

$$\left[p'_1{}^\alpha \frac{dp_{2\alpha}}{dv}\right]_G = 0 = \left[p'_2{}^\alpha \frac{dp_{1\alpha}}{dv}\right]_G. \tag{16.125}$$

Through (16.122) this implies that

$$\left[p_1^\alpha \frac{dp'_{2\alpha}}{dv}\right]_O = 0 = \left[p_2^\alpha \frac{dp'_{1\alpha}}{dv}\right]_O. \tag{16.126}$$

Since all the vectors lie in 2-planes at G and at O, (16.123) and the first equation of (16.125) imply that $p'_1{}^\alpha|_G$ is collinear with $dp_1^\alpha/dv|_G$; denote this by $p'_1|_G \propto dp_1/dv|_G$. Then the second of (16.125) says that

$$(p'_1{}^\alpha p'_{2\alpha})_G = 0. \tag{16.127}$$

By a similar consideration, (16.126) and (16.124) imply that

$$\left[\frac{dp'_1{}^\alpha}{dv}\frac{dp'_{2\alpha}}{dv}\right]_O = 0. \tag{16.128}$$

In consequence of (16.123), (16.124), (16.127) and (16.128), the norms of the vector products $\vec{p_1} \times \vec{p_2}$, $\vec{p_1'} \times \vec{p_2'}$, $dp_1/dv \times dp_2/dv$ and $dp_1'/dv \times dp_2'/dv$ reduce to products of the lengths of the vectors, denoted $|p_i|$, etc. Thus,

$$\delta S_O = [|p_1|\,|p_2|]_O, \qquad\qquad \delta S_G = [|p_1'|\,|p_2'|]_G,$$
$$\delta \Omega_G = \left[\left|\frac{dp_1}{d\ell}\right| \cdot \left|\frac{dp_2}{d\ell}\right|\right]_G, \qquad \delta \Omega_O = \left[\left|\frac{dp_1'}{d\ell}\right| \cdot \left|\frac{dp_2'}{d\ell}\right|\right]_O, \qquad (16.129)$$

where $d\ell$ is the radius at which the solid angles $\delta\Omega_G$ and $\delta\Omega_O$ subtend the surfaces $[|dp_1|\,|dp_2|]$ and $[|dp_1'|\,|dp_2'|]$, respectively. It is related to dv by $d\ell = -k^\rho u_\rho dv$ (see the text after (16.28)). Thus we have from (16.129):

$$\delta S_O \delta\Omega_O = \left[|p_1|\left|\frac{dp_1'}{dv}\right| |p_2|\left|\frac{dp_2'}{dv}\right|\right]_O \frac{1}{(k^\rho u_\rho)_O{}^2}. \qquad (16.130)$$

We already know that $p_1'|_G \propto dp_1/dv|_G$. By the same method we find from (16.123) – (16.128) that $p_2'|_G \propto dp_2/dv|_G$, $p_1|_O \propto dp_1'/dv|_O$ and $p_2|_O \propto dp_2'/dv|_O$. Thus, the scalar products in all these pairs reduce to the products of their lengths, so (16.122) becomes

$$\left[|p_1'|\left|\frac{dp_1}{dv}\right|\right]_G = -\left[|p_1|\left|\frac{dp_1'}{dv}\right|\right]_O, \qquad (16.131)$$

and the same is true for the subscript 2. Using this in (16.130) we obtain

$$\delta S_O \delta\Omega_O = \left[|p_1'|\left|\frac{dp_1}{dv}\right| |p_2'|\left|\frac{dp_2}{dv}\right|\right]_G \frac{1}{(k^\rho u_\rho)_O{}^2} = \frac{(k^\rho u_\rho)_G{}^2}{(k^\rho u_\rho)_O{}^2}\, \delta S_G \delta\Omega_G. \qquad (16.132)$$

This is equivalent to (16.119) by (16.20), (16.116) and (16.118). □

The reciprocity theorem implies that

If the observer moves so as to cancel the redshift, then equal projected surface areas at the source G and at the observer O fill equal solid angles at O and G, respectively.

In Fig. 16.1, both r_G and r_O between the endpoints are zero, so (16.119) still holds, but the above conclusion does not.

Recall that an observer at Q would see the source at G to be not only anomalously bright but also to have an anomalously large angular size. Imagine the observer being placed on the symmetry axis and gradually removed farther from G. The angular size of G would then at first decrease. As the observer moves past the first refocussing point, the angular size would start increasing. We will come back to the topic of refocussing in Secs. 17.5 and 18.8 in the context of the Friedmann and Lemaître–Tolman models.

Substituting (16.119) in (16.117) we obtain an expression for r_O:

$$r_O = \sqrt{\frac{L}{4\pi \mathcal{F}_O}} \frac{1}{(1+z)^2}. \qquad (16.133)$$

This allows us to calculate r_O if we know L from theory and $\mathcal{F}_O$ from measurement. The r_O calculated in this way is called the **corrected luminosity distance**, while

$$D_L \stackrel{\text{def}}{=} \sqrt{\frac{L}{4\pi \mathcal{F}_o}} \qquad (16.134)$$

is the ordinary (uncorrected) **luminosity distance**. It is the distance from which the radiating body, if motionless in a Euclidean space, would produce an energy flux equal to the $\mathcal{F}_O$ measured by the observer.

16.11 Other observable quantities

Having introduced the two definitions of distance, the redshift, the flux of radiation and the reciprocity theorem, one can define further observationally meaningful quantities. One of them is the number of radiation sources in a given volume. Consider again the bundle k'^α in Fig. 16.2. Imagine that we move from the surface sheet δS_G (at the observer area distance r_O from O) to a farther sheet $\delta S'$ at the distance $(r_O + \mathrm{d}\ell)$. The $\mathrm{d}\ell$ is connected to the affine parameter v on the bundle by $\mathrm{d}\ell = -k^\rho u_\rho \mathrm{d}v$; see the text after (16.28). Let $n(x)$ be the number of radiation sources per unit proper volume. The number of sources within the bundle between δS_G and $\delta S'$ will then be

$$\mathrm{d}N = -r_O{}^2 \delta\Omega_O \left(nk^\rho u_\rho\right)_G \mathrm{d}v. \tag{16.135}$$

The integral of this over the volume of the bundle from O to δS_G will be the total number of sources of radiation within this volume. This quantity is considered in observational cosmology, and the procedure of determining it is called **number counts**.

Other observationally meaningful quantities are:

1. The specific flux – the flux per unit frequency interval;
2. The intensity of radiation – the flux per unit solid angle;
3. The specific intensity – the intensity per unit frequency interval;
4. The emission and absorption of light in the intergalactic space, and the optical depth.

While being useful for observations, these quantities do not really involve relativity in their definitions and properties, and are handled by routine mathematics. A good overview source for them is Ellis (1971); we will not discuss them here.

16.12 The Fermi–Walker transport

Vectors can be transported along curves by different prescriptions. We have already learned the parallel transport in Sec. 5.1, its name is self-explanatory, and the Lie transport in Sec. 10.5. A vector field X^α is Lie-transported along the vector field k^α when X^α is invariant under the mapping of the manifold into itself defined by the integral curves of k^α. Now we will get to know one more kind of transport called **Fermi–Walker (FW) transport** (Fermi, 1922; Walker, 1932; Bini and Jantzen, 2002). The idea behind it is to transport a vector V^α along a timelike curve C in such a way that it does not rotate around C.

If C is geodesic, then it does not define any structures on itself, just like a straight line in Euclidean geometry. But if it is not geodesic, then the first thing it defines is the direction of the principal normal. Let T^α be the unit tangent vector to C, $T_\alpha T^\alpha = 1$, then the principal normal has the direction of $\dot{T}^\alpha \stackrel{\text{def}}{=} T^\rho T^\alpha{}_{;\rho}$. (If T^α is the velocity field of a fluid, then $\dot{T}^\alpha$ is the acceleration defined in Sec. 15.2.)

If a vector V^α lies initially in the plane of T^α and $\dot{T}^\alpha$ (call it $\mathcal{P}$), then during the FW transport it is supposed to remain in $\mathcal{P}$ all along C. If it is not in $\mathcal{P}$, then we split it into the component $V_{\mathcal{P}}^\alpha$ in $\mathcal{P}$ and the component $V_{\perp \mathcal{P}}^\alpha$ orthogonal to $\mathcal{P}$. The first component must remain in $\mathcal{P}$, the second one will be parallely transported along C. We will derive the formula for the FW transport by the axiomatic method inspired by Rindler (1980).

We define:

- T^α – the unit tangent vector field to C, $T_\alpha T^\alpha = 1$.
- N^α – the unit vector field along the principal normal to C,

$$N^\alpha \propto T^\rho T^\alpha{}_{;\rho}, \qquad g_{\alpha\beta} N^\alpha N^\beta = -1, \qquad g_{\alpha\beta} T^\alpha N^\beta = 0. \qquad (16.136)$$

When C is geodesic, $T^\rho T^\alpha{}_{;\rho} = 0$, the direction of $\dot{T}^\alpha = 0$ is not determined. Then we require N^α to be any vector obeying the second and third of (16.136).

- N_2^α, N_3^α – arbitrary unit spacelike vectors that form an orthonormal tetrad together with T^α and N^α.

We denote the covariant change of an arbitrary vector V^α along C during the not-yet-defined FW transport by $\overset{*}{V}{}^\alpha$, so

$$\left.\frac{DV^\alpha}{ds}\right|_{\text{FW}} = \left.\frac{DV^\alpha}{\partial x^\rho} T^\rho\right|_{\text{FW}} \overset{\text{def}}{=} \overset{*}{V}{}^\alpha. \qquad (16.137)$$

Since the operation * involves differentiation, it will have the property

$$(V_\alpha W^\alpha)^* = \overset{*}{V}_\alpha W^\alpha + V_\alpha \overset{*}{W}{}^\alpha \qquad (16.138)$$

for arbitrary V_α and W^α. We postulate the following further properties for *:

- **(I)** A vector that initially lies in the plane spanned by T^α and N^α remains in this plane all along C. (Note: this is in general a different plane at each point of C.)
- **(II)** Vectors orthogonal to T^α and N^α are parallely transported along C, so

$$\overset{*}{N}{}_2^\alpha = \overset{*}{N}{}_3^\alpha = 0. \qquad (16.139)$$

- **(III)** Scalar products of vectors are preserved, so, for V_α and W^α

$$(V_\alpha W^\alpha)^* = 0. \qquad (16.140)$$

We take an arbitrary vector V^α and decompose it in the basis $\{T^\alpha, N^\alpha, N_2^\alpha, N_3^\alpha\}$:

$$V^\alpha = \alpha T^\alpha + \beta N^\alpha + \gamma N_2^\alpha + \delta N_3^\alpha. \qquad (16.141)$$

But $\alpha = V_\alpha T^\alpha$, $\beta = -V_\alpha N^\alpha$, $\gamma = -V_\alpha N_2^\alpha$, $\delta = -V_\alpha N_3^\alpha$, so by Postulate **III**:

$$\overset{*}{\alpha} = \overset{*}{\beta} = \overset{*}{\gamma} = \overset{*}{\delta} = 0, \qquad (16.142)$$

which means that $\alpha, \beta, \gamma, \delta$ are constant along C. Using (16.139) and (16.142) we obtain from (16.141):

$$\overset{*}{V}{}^\alpha = \alpha \overset{*}{T}{}^\alpha + \beta \overset{*}{N}{}^\alpha. \qquad (16.143)$$

From Postulate **I** we have

$$\overset{*}{T}{}^\alpha = aT^\alpha + bN^\alpha, \qquad (16.144)$$

$$\overset{*}{N}{}^\alpha \;=\; cT^\alpha + dN^\alpha, \tag{16.145}$$

the functions a, b, c, d to be determined. Since $T_\alpha T^\alpha = 1 = -N_\alpha N^\alpha$, Postulate **III** implies:

$$0 \;=\; (T_\alpha T^\alpha)^* = 2T_\alpha \overset{*}{T}{}^\alpha, \tag{16.146}$$

$$0 \;=\; (N_\alpha N^\alpha)^* = 2N_\alpha \overset{*}{N}{}^\alpha. \tag{16.147}$$

Contracting (16.144) with T_α and (16.145) with N_α, then using (16.146), (16.136) and $T_\alpha T^\alpha = 1$ we get

$$a = d = 0 \;\Longrightarrow\; \overset{*}{T}{}^\alpha = bN^\alpha, \qquad \overset{*}{N}{}^\alpha = cT^\alpha. \tag{16.148}$$

(If C is geodesic, then $\overset{*}{T}{}^\alpha = 0$ from (16.137) and $b = 0$.) From (16.143) and (16.148)

$$\overset{*}{V}{}^\alpha = \alpha bN^\alpha + \beta cT^\alpha = b\,(V_\rho T^\rho)\,N^\alpha - c\,(V_\rho N^\rho)\,T^\alpha. \tag{16.149}$$

But from Postulate **III** $0 = (V_\rho V^\rho)^* = 2V_\rho \overset{*}{V}{}^\rho$, so from (16.141) and (16.149)

$$(V_\mu N^\mu)\,(V_\nu T^\nu)\,(b - c) = 0. \tag{16.150}$$

Since V^α is arbitrary, this implies $b = c$ and (16.149) becomes

$$\overset{*}{V}{}^\alpha = c\,[(V_\rho T^\rho)\,N^\alpha - (V_\rho N^\rho)\,T^\alpha]. \tag{16.151}$$

Now we recall (16.137): $\overset{*}{V}{}^\alpha = DV^\alpha/ds$ for an arbitrary V^α. Applying this to T^α and using (16.148) we obtain $DT^\alpha/ds = cN^\alpha$, so, denoting $DT^\alpha/ds = \dot{T}^\alpha$, we conclude that the vector V^α is FW-transported along the curve C with unit tangent vector T^α when the covariant derivative of V^α along C obeys

$$\overset{*}{V}{}^\alpha = \frac{DV^\alpha}{ds} = (V_\rho T^\rho)\,\dot{T}^\alpha - \left(V_\rho \dot{T}^\rho\right) T^\alpha. \tag{16.152}$$

The **FW transport equation** (16.152) defines the **Fermi–Walker derivative** of V^α along C as follows:

$$\frac{\mathcal{D}_{\mathrm{FW}} V^\alpha}{ds} \overset{\text{def}}{=} V^\alpha{}_{;\rho}\,T^\rho - \left[(V_\rho T^\rho)\,\dot{T}^\alpha - \left(V_\rho \dot{T}^\rho\right) T^\alpha\right], \tag{16.153}$$

so that $\mathcal{D}_{\mathrm{FW}} V^\alpha/ds = 0$ when V^α is FW-transported along C. Note that

$$\frac{\mathcal{D}_{\mathrm{FW}} T^\alpha}{ds} = 0. \tag{16.154}$$

The definition of $\mathcal{D}_{\mathrm{FW}} V^\alpha/ds$ can be extended to arbitrary tensor fields and to spacelike curves (Bini and Jantzen 2002), but such extensions will not be used in what follows.

16.13 Position drift of light sources

We will now deal with the following problem: suppose an observer $\mathcal{O}$ keeps receiving light rays from a distant emitter $\mathcal{E}$ for a period that is long on the cosmological scale. Every light ray on its way from $\mathcal{E}$ to $\mathcal{O}$ passes through evolving cosmic structures such as voids and galaxy clusters. The direction of the ray thereby changes. The fact that moving matter will sweep light rays passing through it seems to have been first noted by Bondi (1947). Following Korzyński and Kopiński (KK, 2018) we call the changes **position drift**.

How can the observer detect those changes? Hasse and Perlick (HP, 1988) proposed a criterion for a cosmological model to be **drift-free** (in their terminology **parallax-free**): an arbitrary observer should see the angle between the rays coming from two sources $\mathcal{E}_1$ and $\mathcal{E}_2$ to be constant in time. Equivalently, if $\mathcal{O}$ sees $\mathcal{E}_1$ and $\mathcal{E}_2$ in the same spatial direction at one instant, then she/he will see these sources in the same direction for all time. HP investigated the consequences of this definition without reference to explicit solutions of Einstein's equations, and presented a few equivalent formulations – one of them was the velocity field of the cosmic fluid being collinear with a conformal Killing vector field.

Later, Krasiński and Bolejko (KB, 2011) investigated the position drift defined in a way equivalent to HP's, but under the still different name of repeatability of light paths (RLP). KB imposed the RLP condition on the cosmological models of Szekeres (1975, see our Chapter 20 for a detailed description of these) and calculated the rate of the drift for rays passing through an evolving cosmic void with fictitious parameters (see Sec. 18.23).

Recently, Korzyński and Kopiński (2018) proposed an elegant definition of the position drift in a general spacetime. The HP criterion is a necessary condition for zero position drift in the KK sense; see Krasiński (2023) for a comparison of the three criteria. We will now present the basic ideas of the KK approach in abbreviation.

We denote by k^μ the tangent vector field to the central ray γ_0 of a bundle received by the observer, and by ξ^μ the geodesic deviation vector field in this bundle. In this section we will use the symbol λ for the affine parameter on the rays. We also introduce the **geodesic deviation operator** $\mathcal{G}[\bullet]$ defined as follows:

$$\mathcal{G}[\xi]^\mu \overset{\text{def}}{=} \frac{D^2\xi^\mu}{d\lambda^2} - R^\mu{}_{\rho\sigma\nu}k^\rho k^\sigma \xi^\nu, \tag{16.155}$$

using which we rewrite the geodesic deviation equation (6.56) as

$$\mathcal{G}[\xi]^\mu = 0. \tag{16.156}$$

Contracting (16.155) with k_μ, using $Dk^\mu/d\lambda = 0$ (because k_μ is geodesic) and noting that for the scalar $k_\mu \xi^\mu$ the covariant derivative reduces to the ordinary one we obtain

$$\frac{d^2}{d\lambda^2}\left(k_\mu \xi^\mu\right) = 0 \implies k_\mu \xi^\mu = A + B\lambda, \tag{16.157}$$

where A and B are constant along γ_0. We also note that if ξ^μ obeys (16.156), then so does

$$\widetilde{\xi}^\mu = \xi^\mu + (C + D\lambda)k^\mu, \tag{16.158}$$

where C and D are constants. In what follows we will often consider a family of null geodesics and geodesic deviation vectors that obey

$$k_\mu \xi^\mu = 0. \tag{16.159}$$

Now let us consider an observer $\mathcal{O}$ with four-velocity u^μ whose worldline intersects the central ray of the bundle, γ_0, at the point P. We construct the orthonormal tetrad

$\left(u^{\mu}, e_{\hat{1}}{}^{\mu}, e_{\hat{2}}{}^{\mu}, n^{\mu}\right)$, where n^{μ} is the projection of k^{μ} on the space $u_{\perp}$ orthogonal to u^{μ}, normalised so that $n_{\mu}n^{\mu} = -1$, see (16.21) – (16.23). The remaining spacelike vectors, $e_A{}^{\mu}$, $A = 1, 2$, are orthogonal to u^{μ} and k^{μ} (and to each other), and are determined up to rotations around n^{μ}. The plane spanned by $\{e_{\hat{1}}{}^{\mu}, e_{\hat{2}}{}^{\mu}\}$ is called **screen space**. The tetrad is attached at P, and let the coordinates of P be $x^{\mu} = 0$.

Let us next consider the ray γ_{ϵ}, $\epsilon \ll 1$, neighbouring to γ_0. The vectors

$$\delta x_{\epsilon}^{\mu} = \epsilon \left(\xi^{\mu} + C k^{\mu}\right), \qquad C = \text{constant} \tag{16.160}$$

point from P to the points on γ_{ϵ}. One of them lies in the screen space of P and is collinear with one of the deviation vectors of the class (16.158); we will denote it $\xi_{\perp}^{\mu}$. It obeys

$$\xi_{\perp}^{\mu} u_{\mu} = 0 = \xi_{\perp}^{\mu} k_{\mu}, \tag{16.161}$$

and this, via (16.23), implies

$$\xi_{\perp}^{\mu} n_{\mu} = 0. \tag{16.162}$$

Imagine now the tetrad $\left(u^{\mu}, e_{\hat{1}}{}^{\mu}, e_{\hat{2}}{}^{\mu}, n^{\mu}\right)$ being parallely transported along γ_0, and denote by $\left(\hat{u}^{\mu}, \hat{e}_{\hat{1}}^{\mu}, \hat{e}_{\hat{2}}^{\mu}, \hat{n}^{\mu}\right)$ the transported tetrad at running points along γ_0. Let us consider a past-directed bundle of null geodesics that emanate from the common observation point $\mathcal{O}$. They are a subset of $\mathcal{O}$'s past light cone. At $\mathcal{O}$, $\xi^{\mu}(\lambda_{\mathcal{O}}) = 0$, and ξ^A along the rays will be determined by

$$\nabla_k \xi^A (\lambda_{\mathcal{O}}) \equiv k^{\mu} \xi^A{}_{,\mu}(\lambda_{\mathcal{O}}). \tag{16.163}$$

The mapping from $\nabla_k \xi^A (\lambda_{\mathcal{O}})$ to $\xi^A(\lambda)$ is represented by the **Jacobi matrix** $\mathcal{D}^A{}_B(\lambda)$:

$$\xi^A(\lambda) = \mathcal{D}^A{}_B(\lambda) \nabla_k \xi^B (\lambda_{\mathcal{O}}). \tag{16.164}$$

The Jacobi matrix maps the differences in initial directions of the rays to orthogonal displacements of points on the rays across the bundle. It would become singular where the rays intersect, but we investigate such regions of spacetime where this does not occur. Consequently, $\mathcal{D}^{-1A}{}_B(\lambda)$ exists. The Jacobi matrix depends on the choice of the affine parameter along γ_0: if $\lambda' = C\lambda + D$ with constant C and D, then

$$k'^{\mu} = \frac{1}{C} k^{\mu}, \qquad \mathcal{D}'^A{}_B = C\mathcal{D}^A{}_B. \tag{16.165}$$

Now let the observer $\mathcal{O}$ move along her worldline $\chi_{\mathcal{O}}$ and keep receiving light rays from the emitter $\mathcal{E}$ that follows its worldline $\chi_{\mathcal{E}}$, as in Fig. 16.3. Let $\chi_{\mathcal{O}}$ be parametrised by the observer's proper time τ, $\chi_{\mathcal{E}}$ by the emitter's proper time σ and the family γ_{τ} of rays by the time τ at which the ray intersects $\chi_{\mathcal{O}}$. The affine parameters on the rays are chosen so that the intersection with $\chi_{\mathcal{O}}$ occurs always at the same value $\lambda_{\mathcal{O}}$, and the intersection with $\chi_{\mathcal{E}}$ always at the same $\lambda_{\mathcal{E}}$. So,

$$\gamma_{\tau}(\lambda_{\mathcal{O}}) = \chi_{\mathcal{O}}(\tau), \qquad \gamma_{\tau}(\lambda_{\mathcal{E}}) = \chi_{\mathcal{E}}(\sigma(\tau)). \tag{16.166}$$

Let us consider the ray γ_0 received by $\mathcal{O}$ at $\tau = \tau_0$ and the later ray received at $\tau_0 + d\tau$.

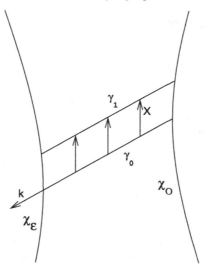

Fig. 16.3 The emitter $\mathcal{E}$ following its worldline $\chi_{\mathcal{E}}$ keeps sending light rays to the observer $\mathcal{O}$ that follows her worldline $\chi_{\mathcal{O}}$; two such rays, γ_0 and γ_1, are shown. X is the geodesic deviation vector field along γ_0 and k is the past-directed tangent vector field to γ_0. (This is simplified Fig. 8 from Korzyński and Kopiński (2018).)

These two rays are connected by the geodesic deviation vector field X^μ, which will be called the **observation time vector**, see Fig. 16.3. It obeys the geodesic deviation equation

$$\frac{\mathrm{D}^2 X^\mu}{\mathrm{d}\lambda^2} = R^\mu{}_{\rho\sigma\nu} k^\rho k^\sigma X^\nu, \tag{16.167}$$

and we assume the following initial conditions:

$$X^\mu(\lambda_{\mathcal{O}}) = u^\mu_{\mathcal{O}}, \qquad X^\mu(\lambda_{\mathcal{E}}) = Cu^\mu_{\mathcal{E}}, \qquad C = \text{constant} > 0, \tag{16.168}$$

i.e. at $\mathcal{O}$ the vector X^μ coincides with the observer's 4-velocity, and at $\mathcal{E}$ it is collinear with the emitter's 4-velocity. That X^μ must be collinear with $u^\mu_{\mathcal{O}}$ and $u^\mu_{\mathcal{E}}$ is evident, but the first of (16.168) fixes the scale of the parameter ϵ labelling the rays in the family. Also, being a geodesic deviation vector, it must obey (6.54):

$$X^\rho k^\mu{}_{;\rho} = k^\rho X^\mu{}_{;\rho} \implies (k^\rho X^\mu{}_{;\rho}) k_\mu = 0. \tag{16.169}$$

The last equation means that $k_\mu X^\mu$ is constant along every γ_τ, so it must be the same at both ends, $\mathcal{O}$ and $\mathcal{E}$. By (16.168) this means $(k_\mu u^\mu)_{\mathcal{O}} = C(k_\mu u^\mu)_{\mathcal{E}}$, so, from (16.20), $C = 1/(1+z)$ and the second of (16.168) becomes

$$X^\mu(\lambda_{\mathcal{E}}) = \frac{1}{1+z} u^\mu_{\mathcal{E}}. \tag{16.170}$$

In order to be able to use the Jacobi matrix formalism for X^μ we have to extract from it a vector that is zero at $\mathcal{O}$ and is orthogonal to γ_0 elsewhere. This is done as follows:

Let $\widehat{u}_{\mathcal{O}}^{\mu}$ be the vector $u_{\mathcal{O}}^{\mu}$ parallely transported along γ_0, so

$$\frac{\mathrm{D}\widehat{u}_{\mathcal{O}}^{\mu}}{\mathrm{d}\lambda} = 0, \qquad \widehat{u}_{\mathcal{O}}^{\mu}(\lambda_{\mathcal{O}}) = u_{\mathcal{O}}^{\mu}. \tag{16.171}$$

Now define b^{μ} by

$$X^{\mu} = \widehat{u}_{\mathcal{O}}^{\mu} + b^{\mu}. \tag{16.172}$$

Consequently, $b^{\mu}(\lambda_{\mathcal{O}}) = 0$, and, in consequence of (16.167), (16.171) and (16.172):

$$\mathcal{G}[b]^{\mu} \equiv \frac{\mathrm{D}^2 b^{\mu}}{\mathrm{d}\lambda^2} - R^{\mu}{}_{\rho\sigma\nu}k^{\rho}k^{\sigma}b^{\nu} = R^{\mu}{}_{\rho\sigma\nu}k^{\rho}k^{\sigma}\widehat{u}_{\mathcal{O}}^{\nu}, \tag{16.173}$$

$$b^{\mu}(\lambda_{\mathcal{E}}) = \frac{1}{1+z} u_{\mathcal{E}}^{\mu} - \widehat{u}_{\mathcal{O}}^{\nu}(\lambda_{\mathcal{E}}). \tag{16.174}$$

This b^{μ} is orthogonal to k^{μ}, see Exercise 5. Together with $b^{\mu}(\lambda_{\mathcal{O}}) = 0$ it would make b^{μ} suitable for applying the Jacobi matrix to it, except that it does not obey the geodesic deviation equation, see (16.173). So now we split b^{μ} as follows:

$$b^{\mu} = \phi^{\mu} + m^{\mu}, \tag{16.175}$$

where we assume for m^{μ}:

$$\mathcal{G}[m]^{\mu} = R^{\mu}{}_{\rho\sigma\nu}k^{\rho}k^{\sigma}\widehat{u}_{\mathcal{O}}^{\nu}, \qquad m^{\mu}(\lambda_{\mathcal{O}}) = 0, \qquad \nabla_k m^{\mu}(\lambda_{\mathcal{O}}) = 0, \tag{16.176}$$

so for ϕ^{μ} it follows that

$$\mathcal{G}[\phi]^{\mu} = 0, \qquad \phi^{\mu}(\lambda_{\mathcal{O}}) = 0, \tag{16.177}$$

$$\phi^{\mu}(\lambda_{\mathcal{E}}) = \frac{1}{1+z} u_{\mathcal{E}}^{\mu} - \widehat{u}_{\mathcal{O}}^{\nu}(\lambda_{\mathcal{E}}) - m^{\mu}(\lambda_{\mathcal{E}}). \tag{16.178}$$

The right-hand side of (16.178) is orthogonal to k^{μ}, see Exercise 6. For further use we only need to know X^{μ} projected on the screen space of an arbitrary observer[10] and k^{μ}, so we will use (16.173) – (16.178) only with $\mu = A = 1, 2$.

Observe that

$$\begin{aligned} \nabla_k \phi^{\mu}(\lambda_{\mathcal{O}}) &= \nabla_k b^{\mu}(\lambda_{\mathcal{O}}) \,[\text{from (16.175) - (16.176)}] \\ &= \nabla_k (X^{\mu} - \widehat{u}_{\mathcal{O}}^{\mu}) \,[\text{from (16.172)}] = \nabla_k X^{\mu}(\lambda_{\mathcal{O}}) \,[\text{from (16.171)}] \end{aligned}$$

and

$$\phi^{A}(\lambda_{\mathcal{E}}) = \mathcal{D}^{A}{}_{B}(\lambda_{\mathcal{E}}) \nabla_k \phi^{B}(\lambda_{\mathcal{O}}) \tag{16.179}$$

because ϕ^{A} obeys the same equation and the same initial conditions as ξ^{A} did for (16.164). But so far $\nabla_k \phi^{B}(\lambda_{\mathcal{O}})$ is not determined, while $\phi^{A}(\lambda_{\mathcal{E}})$ is – by (16.178). So, using (16.179) and (16.178) we calculate

$$\begin{aligned} \nabla_k X^{B}(\lambda_{\mathcal{O}}) &= \nabla_k \phi^{B}(\lambda_{\mathcal{O}}) = \mathcal{D}^{-1B}{}_{C}\phi^{C}(\lambda_{\mathcal{E}}) \\ &= \mathcal{D}^{-1B}{}_{C}\left(\frac{1}{1+z} u_{\mathcal{E}}^{C} - \widehat{u}_{\mathcal{O}}^{C}(\lambda_{\mathcal{E}}) - m^{C}(\lambda_{\mathcal{E}})\right). \end{aligned} \tag{16.180}$$

[10] If the observer has 4-velocity coincident with $u_{\mathcal{E}}^{\mu}$, then $u_{\mathcal{E}}^{A} = 0$ in (16.178).

With $\nabla_k X^B (\lambda_\mathcal{O})$ now determined, we use (16.164) to obtain

$$\phi^A(\lambda) = \mathcal{D}^A{}_B(\lambda)\, \mathcal{D}^{-1B}{}_C(\lambda_\mathcal{E}) \left(\frac{1}{1+z}\, u_\mathcal{E}^C - \widehat{u}_\mathcal{O}^C - m^C\right)\Big|_{\lambda_\mathcal{E}}, \tag{16.181}$$

and we can calculate the whole $X^A = \widehat{u}_\mathcal{O}^A + m^A + \phi^A$ using (16.176).

We denote by $\delta_\mathcal{O}$ the FW derivative along the observer's 4-velocity, and define the position drift of an emitter as $\delta_\mathcal{O} n^\mu$, where n^μ is the unit direction vector towards the light source defined in (16.23). We have $n_\mu \delta_\mathcal{O} n^\mu = 0$ because $n_\mu n^\mu = -1$ and $u_{\mathcal{O}\mu} \delta_\mathcal{O} n^\mu = 0$ from $n_{\mathcal{O}\mu} u_\mathcal{O}^\mu = 0$ and (16.154). This means that $\delta_\mathcal{O} n^\mu$ is orthogonal to both $u_\mathcal{O}^\mu$ and $k_\mathcal{O}^\mu \propto u_\mathcal{O}^\mu - n_\mathcal{O}^\mu$ (see (16.23) again), so must lie in $\mathcal{O}$'s screen space. From (16.153) applied to $V^\alpha = n^\alpha$ and $T^\alpha = u_\mathcal{O}^\alpha = X^\alpha$ we obtain[11]

$$\delta_\mathcal{O} n^\mu = n^\mu{}_{;\rho}\, X^\rho\big|_\mathcal{O} + n_\rho \dot{u}_\mathcal{O}^\rho u_\mathcal{O}^\mu. \tag{16.182}$$

The second term is orthogonal to the screen space and thus irrelevant for calculating the drift. The first term, using (16.23), is

$$n^\mu{}_{;\rho}\, X^\rho\big|_\mathcal{O} = \left(\frac{Q_{,\rho} X^\rho}{Q^2}\, k^\mu - \frac{1}{Q}\, k^\rho X^\mu{}_{;\rho} + X^\rho X^\mu{}_{;\rho}\right)_\mathcal{O}, \qquad Q \overset{\text{def}}{=} k_\nu X^\nu, \tag{16.183}$$

where we have used $X^\rho k^\mu{}_{;\rho} = k^\rho X^\mu{}_{;\rho}$ that follows from (6.54). The first term above is orthogonal to the screen space while the last one is the observer's acceleration (because $X_\mathcal{O}^\mu = u_\mathcal{O}^\mu$). In the second term we use (16.180) to obtain the formula for the position drift:

$$\delta_\mathcal{O} n^A = -\frac{1}{k_\sigma u_\mathcal{O}^\sigma}\, \mathcal{D}^{-1A}{}_B \left(\frac{1}{1+z}\, u_\mathcal{E}^B - \widehat{u}_\mathcal{O}^B - m^B\right)_{\lambda_\mathcal{E}} + \dot{u}_\mathcal{O}^A. \tag{16.184}$$

The papers by Korzyński and Kopiński (2018) and by Grasso, Korzyński and Serbenta (GKS, 2019) contain several other physically motivated formulae, like redshift drift, Jacobi matrix drift and area- and luminosity distance drifts, all in a general geometry. The GKS paper contains an illuminating reformulation of some parts of the KK approach. We refer the interested reader there for a follow-up reading.

16.14 Exercises

1. Prove that any vector orthogonal to two linearly independent null vectors must be spacelike.

Hint. Consider the vertex of the null cone, P_0, and choose the locally Minkowskian coordinates at P_0.

2. Verify that the integrability condition (16.106) applied to (16.104) – (16.105) leads to (16.107). Verify the same for (16.108) and (16.109).

Hint for (16.104) – (16.105). Applying $k^\gamma \partial/\partial x^\gamma$ to (16.104) and $m^\gamma \partial/\partial x^\gamma$ to (16.105) is straightforward. Then the right-hand side of (16.106) becomes

$$-\left[\left(\Gamma^0{}_{02} - \Gamma^0{}_{20}\right) k^\nu + \left(\Gamma^1{}_{02} - \Gamma^1{}_{20}\right) \ell^\nu\right]$$

---

[11] Differences with the KK paper in some of the signs are a consequence of the different signatures: $(-+++)$ in KK and $(+---)$ here.

$$+ \left(\Gamma^2{}_{02} - \Gamma^2{}_{20}\right) m^\nu + \left(\Gamma^3{}_{02} - \Gamma^3{}_{20}\right) \overline{m}^\nu \Big] \Gamma^1{}_{23,\nu}.$$

Now several terms drop out because of the assumptions made in the theorem:

$\Gamma^1{}_{20} = -k_{\mu;\nu} m^\mu k^\nu = -m^\mu \dot{k}_\mu = 0$ because k^μ was assumed geodesic,

$\Gamma^3{}_{02} = -\Gamma_{202} = \Gamma_{022} = \Gamma^1{}_{22} = 0$ because of $\sigma = 0$,

$\Gamma^1{}_{02} = -\Gamma_{002} \equiv 0,$ $\qquad \Gamma^3{}_{20} = -\Gamma_{220} \equiv 0$ because $\Gamma_{ijk} = -\Gamma_{jik}$.

So, the terms containing ℓ^ν and $\overline{m}^\nu$ drop out completely, and what remains leads to (16.107), after factoring out $\Gamma^1{}_{23} \neq 0$. The same method works in deriving (16.110).

3. Using the methods of Section 15.1 prove that the expansion and rotation of a family of null geodesics have the same interpretation as the corresponding quantities in hydrodynamics. Show that shear changes the shape of a parallelepiped in a plane locally orthogonal to a ray, but preserves its surface area. Verify that the expansion scalar obeys

$$2\theta = \frac{\mathrm{d}}{\mathrm{d}v} \ln(\delta S), \tag{16.185}$$

where v is the affine parameter on the null geodesics and δS is the surface area of the propagating light-front of a bundle of rays at the parameter value v.

4. Consider a ray proceeding from a spacetime point P_1 to P_2 and then from P_2 to P_3. Denote the redshifts acquired in the intervals $[P_1, P_2]$, $[P_2, P_3]$ and $[P_1, P_3]$ by z_{12}, z_{23} and z_{13}, respectively. Prove that (16.20) implies

$$1 + z_{13} = (1 + z_{12}) (1 + z_{23}). \tag{16.186}$$

5. Show that the b^μ defined in (16.172) is orthogonal to k^μ all along γ_0.

Hint. From (16.169), $k_\mu X^\mu$ is constant along γ_0. So is $k_\mu \hat{u}^\mu_{\mathcal{O}}$ because k^μ is geodesic and $\hat{u}^\mu_{\mathcal{O}}$ is parallelly transported along $\gamma_0 \Longrightarrow k_\mu b^\mu = k_\mu (X^\mu - \hat{u}^\mu_{\mathcal{O}}) = $ constant on γ_0. But $k_\mu (X^\mu - \hat{u}^\mu_{\mathcal{O}})|_{\mathcal{O}} = 0$. $\square$

6. Show that the right-hand side of (16.178) is orthogonal to k^μ.

Hint. Contract (16.176) with k_μ and use $\mathrm{D}k_\mu/\mathrm{d}\lambda = 0$. It follows that $\mathrm{D}^2 (m_\mu k^\mu)/\mathrm{d}\lambda^2 = 0$. But $m_\mu k^\mu$ and $(\mathrm{D}/\mathrm{d}\lambda) (m_\mu k^\mu)$ are scalars, so $\mathrm{d}^2 (m_\mu k^\mu)/\mathrm{d}\lambda^2 = 0 \Longrightarrow m_\mu k^\mu = C_1 \lambda + C_2$, $C_1, C_2 = $ constant on γ_0. Then contract the last of (16.176) with k_μ and use $\mathrm{D}k_\mu/\mathrm{d}\lambda = 0 \Longrightarrow$ $\mathrm{d}(k_\mu m^\mu)/\mathrm{d}\lambda|_{\mathcal{O}} = 0 \Longrightarrow C_1 = 0$. But the second of (16.176) implies that $(k_\mu m^\mu) (\lambda_{\mathcal{O}}) = 0$ $\Longrightarrow C_2 = 0 \Longrightarrow k_\mu m^\mu = 0$ everywhere on γ_0. $\square$

17

Relativistic cosmology II:
the Robertson–Walker geometry

17.1 The Robertson–Walker metrics as models of the Universe

The Robertson–Walker (R–W) metrics were derived in Section 10.7 as being spatially homogeneous in the Bianchi sense and isotropic:

$$ds^2 = dt^2 - \frac{R^2(t)}{\left(1 + \frac{1}{4}kr^2\right)^2} \left[dr^2 + r^2 \left(d\vartheta^2 + \sin^2 \vartheta d\varphi^2\right)\right], \tag{17.1}$$

where $R(t)$ (called the **scale factor**) is a function to be determined from the Einstein equations. The constant k, called the **curvature index**, when nonzero, can be scaled to $+1$ or -1 by the transformations $r = r'/\sqrt{|k|}$, $R(t) = \sqrt{|k|}\, \widetilde{R}(t)$. After this, the R–W metrics are no longer one continuous family; they become three different metrics and the limiting transition $k \to 0$ becomes impossible. This can cause difficulties when comparing the three classes with each other; see Section 17.4.

Other representations of the R–W metrics are also used. Let

$$r = \frac{2r'}{1 + \sqrt{1 - kr'^2}} \qquad \Longleftrightarrow \qquad r' = \frac{r}{1 + \frac{1}{4}kr^2}. \tag{17.2}$$

This changes (17.1) to (primes dropped for better readability):

$$ds^2 = dt^2 - R^2(t) \left[\frac{dr^2}{1 - kr^2} + r^2 \left(d\vartheta^2 + \sin^2 \vartheta d\varphi^2\right)\right]. \tag{17.3}$$

When $k = +1$, the transformation

$$r = \sin \psi \tag{17.4}$$

changes (17.3) to

$$ds^2 = dt^2 - R^2(t) \left[d\psi^2 + \sin^2 \psi \left(d\vartheta^2 + \sin^2 \vartheta d\varphi^2\right)\right], \tag{17.5}$$

where the hypersurfaces $t = $ constant are 3-dimensional spheres.[1]
With $k = -1$, the transformation

$$r = \sinh \psi \tag{17.6}$$

---

[1] In (17.4) we assumed that $r \leq 1$. With $k = +1$, the range $0 \leq r \leq 1$ covers half of the 3-sphere. To cover the other half, one has to employ a second copy of the coordinate chart in (17.2) – as r goes from 0 to ∞, r' first increases from 0 to $1/\sqrt{k}$, achieved at $r = 2/\sqrt{k}$, and then decreases to 0 as $r \to \infty$.

changes (17.3) to

$$ds^2 = dt^2 - R^2(t) \left[d\psi^2 + \sinh^2 \psi \left(d\vartheta^2 + \sin^2 \vartheta d\varphi^2 \right) \right]. \tag{17.7}$$

A hypersurface $t = $ constant in (17.5) and (17.7) has constant curvature:

$$R^{IJ}{}_{KL} = \pm \frac{1}{R^2(t)} \delta^{IJ}_{KL}, \qquad I, J, K, L = 1, 2, 3;$$

with $+$ for (17.5) and $-$ for (17.7). When $k = 0$, it is seen from (17.1) and (17.3) that the hypersurfaces $t = $ constant are flat. Because of these properties, the R–W spacetimes with $k = +1$ are colloquially called the 'closed Universe', those with $k = -1$ are called the 'open Universe' and the one with $k = 0$ is called the 'flat Universe'.

The first author to investigate these spacetimes was Alexandr Alexandrovich Friedmann[2] in 1922 (the case $k = +1$, in the coordinates of (17.5)) and in 1924 (Friedmann, 1924) (the case $k = -1$; in coordinates different from all those used here, and with partly incorrect conclusions about the evolution (Krasiński and Ellis, 1999)). He solved the Einstein equations for these metrics, with dust source and with the cosmological constant allowed. He himself treated his result as only a mathematical curiosity, since the expansion of the Universe had not been discovered yet at that time, and no application for those solutions in astronomy could be seen. Both papers were quickly forgotten – most astronomers and physicists had not been ready to accept their implications, while Friedmann was not a sufficiently prominent figure to win attention. In 1925, Friedmann died prematurely and had no chance to claim credit for his discovery when the expansion of the Universe was detected by Hubble (1929).[3] The simplest case $k = 0$ was found by Robertson (1929) in a systematic survey of spacetimes with the, so-called today, Robertson–Walker geometries.

The case $k > 0$ was discussed by Georges Lemaître (1927). He generalised Friedmann's solution to nonzero pressure, and was aware that it reflects some properties of the real Universe. Unfortunately, he used to publish his papers in a local Belgian journal, in French, and, being an unknown person at that time, did not win attention for his results, either. In 1930, when the expansion of the Universe was already generally known, Arthur Eddington asked whether exact models of the expanding Universe might be derived from the relativity theory (Royal Astronomical Society Discussion, 1930). It was only then that Lemaître brought his paper from 1927 to Eddington's attention; and the English translation (Lemaître 1931) followed.[4] The recognition of Friedmann's role came gradually later (Krasiński and Ellis, 1999).

[2] The transliteration of his name from Russian is 'Fridman', but, following the original publications in German, the form 'Friedmann' became standard.

[3] As already mentioned at the end of Sec. 12.14, Hubble did not believe that the Universe is actually expanding. He insisted that recalculating the redshifts into velocities of recession is merely a convenient mathematical trick (Hubble, 1929, 1936, 1953).

[4] The 1931 translation omitted some details of the 1927 paper. One of the omitted pieces showed that Lemaître was aware of the expansion of the Universe before Hubble announced his law of redshifts. This gave rise to a public discussion in 2011 on who was the translator of the original paper and who (and why) edited out the pieces in question. Suspicions were cast on Eddington and on Hubble. A scrutiny of documents by M. Livio (2011) showed that 'it is now certain that Lemaître himself translated his article, and that he chose to delete several paragraphs and notes without any external pressure'. Lemaître omitted the few pieces because he found them to be no longer sufficiently important in 1931. This story is described by J. P. Luminet (2013).

The general metric (17.1) was first derived from geometrical considerations by Robertson (1933) and by Walker (1935). Friedmann and Lemaître derived the solutions of Einstein's equations (with slightly different sources) for that metric. In view of this history, the spacetimes with this geometry are often called the Friedmann–Lemaître–Robertson–Walker (FLRW) models, but various subsets of this collection of four names are also in use. It seems appropriate to use the name Robertson–Walker when referring to the general metric (17.1) and the names Friedmann or Lemaître when referring to the explicit solutions.

In each of the coordinate representations, (17.1), (17.3), (17.5) and (17.7), the Einstein tensor for the R–W metric is diagonal. This shows that, if the source in the Einstein equations should be a perfect fluid, then its velocity field can have only the zeroth component, and, because of $u_\alpha u^\alpha = 1$ and $g_{00} = 1$, the velocity must be $u^\alpha = \delta^\alpha{}_0$. This, in turn, means that all those coordinate systems are comoving: each matter particle has fixed spatial coordinates, and its proper time is the time-coordinate in spacetime.

The velocity field $u^\alpha = \delta^\alpha{}_0$ has the following properties: $\dot{u}^\alpha = 0$, $w_{\alpha\beta} = \sigma_{\alpha\beta} = 0$ and $\theta = 3\dot{R}/R$. The velocity of expansion depends on time, but at each instant it is the same at all points of the space of constant t.

The R–W spacetimes are the only perfect fluid solutions of the Einstein equations for which $w_{\alpha\beta} = \sigma_{\alpha\beta} = 0 = \dot{u}^\alpha$ and $\theta \neq 0$. The proof follows from the paper by Stephani (1967a); see also Stephani et al. (2003). As we stated in Section 15.3, with $w_{\alpha\beta} = \sigma_{\alpha\beta} = 0 = \dot{u}^\alpha$ and a perfect fluid source, the metric must be conformally flat. Theorem 37.17, p. 601 in Stephani et al. (2003) says that all conformally flat perfect fluid solutions are those found by Stephani (1967a). Those of them for which $\theta \neq 0$ have $w_{\alpha\beta} = \sigma_{\alpha\beta} = 0$, but in general $\dot{u}^\alpha \neq 0$. In the limit $\dot{u}^\alpha = 0$ they reduce to the R–W metrics.

17.2 Optical observations in an R–W universe

17.2.1 The redshift

In order to apply (16.20) to the R–W spacetimes, we have to know the field of vectors tangent to light rays, k^α. In consequence of the spatial homogeneity of any R–W spacetime, all points within the same space $t = $ constant are equivalent, so the result of any calculation will be independent of the spatial position of the observer. Let us then assume that the observer is at the origin, $r = 0$. A null geodesic sent off from there radially, with $\dot{\vartheta}_0 = \dot{\varphi}_0 = 0$, will preserve the radial direction $\dot{\vartheta} = \dot{\varphi} = 0$ at all other points. Hence, such a geodesic lies in the surface of constant ϑ and φ and obeys the equation

$$0 = dt^2 - \frac{R^2(t)}{\left(1 + \frac{1}{4}kr^2\right)^2}dr^2. \tag{17.8}$$

Let the subscript e denote quantities calculated at the emission point of the ray, and the subscript o – those calculated at the observation point. For a ray proceeding *towards* the observer sitting at $r = 0$ we thus have

$$\int_{t_e}^{t_o} \frac{dt}{R(t)} = -\int_{r_e}^0 \frac{dr}{1 + \frac{1}{4}kr^2}. \tag{17.9}$$

Let us define the new parameter v on the geodesic (17.8) by

$$\frac{dt}{dv} = \frac{1}{R(t)}. \tag{17.10}$$

In this parametrisation, the tangent vector to that geodesic is

$$k^\alpha = \left[\frac{1}{R} , -\frac{1}{R^2} \left(1 + \frac{1}{4}kr^2\right), 0, 0 \right]. \tag{17.11}$$

Now it can be verified that the parameter v is affine, i.e. that $Dk^\mu/dv + \left\{{\mu \atop \rho\sigma}\right\}k^\rho k^\sigma = 0$. The velocity field is $u^\alpha = \delta^\alpha{}_0$ everywhere, so

$$k^\alpha u_\alpha = 1/R, \tag{17.12}$$

and so in (16.20) we have

$$1 + z = R(t_o)/R(t_e). \tag{17.13}$$

When $|R(t_o) - R(t_e)|/R(t_e)$ is small, developing $R(t_o)$ by the Taylor formula around t_e and neglecting terms nonlinear in $(t_o - t_e)$ gives

$$z \approx \left[\dot{R}(t_e)/R(t_e)\right](t_o - t_e) \approx \left.\frac{\dot{R}}{R}\right|_e \delta\ell, \quad \dot{R} \stackrel{\text{def}}{=} dR/dt \tag{17.14}$$

because $t_o - t_e = \delta\ell$ is the distance from the light source to the observer. The same result follows from (16.29) when we recall that $\sigma_{\alpha\beta} = 0 = \dot{u}^\alpha$ and $\theta = 3\dot{R}/R$ in the R–W models. This is the Hubble law, with the Hubble parameter $H = (c/3)\theta = c\dot{R}/R$ (the factor c appeared because the physical time τ is related to the time coordinate t by $\tau = t/c$).

The Hubble law in fact holds exactly in the R–W models. Note that

$$\ell(t) = R(t) \int_0^{r_e} \frac{dr}{1 + \frac{1}{4}kr^2} \tag{17.15}$$

is the distance of the light source at $r = r_e$ from the observer at $r = 0$ calculated within the space of constant t. Since the coordinates are comoving, for each definite source of light we have $r_e = $ constant, and then from (17.15) we find

$$\frac{d\ell}{dt} = \frac{\dot{R}}{R}\ell = \frac{H\ell}{c}. \tag{17.16}$$

17.2.2 The redshift–distance relation

Assume that, this time, the coordinate origin is at the source G in Fig. 16.2. The element of area in the surface of constant $\{t = t_v, r = r_v\}$ is, in the coordinates of (17.1),

$$dS = \left[\frac{R(t_v)r_v}{1 + \frac{1}{4}kr_v^2}\right]^2 \sin\vartheta d\vartheta d\varphi,$$

where $d\Omega = \sin\vartheta d\vartheta d\varphi$ is the element of solid angle. Using (16.116) we see that

$$r_G = \frac{R_o\bar{r}}{1 + \frac{1}{4}k\bar{r}^2}, \tag{17.17}$$

where $\bar{r}$ is the r-coordinate of the observer and R_O is the value of R at the time $t = t_v$, when the ray passes the observer. By the reciprocity theorem (16.119), we then have

$$r_O = \frac{R_O \bar{r}}{(1 + z_O)\left(1 + \frac{1}{4}k\bar{r}^2\right)}. \tag{17.18}$$

This provides the relation between $\bar{r}$ and the observable quantity r_O, and (17.13) then provides a relation between R and the observable quantity z. The integral of (17.8) along a radial *outgoing* null geodesic is

$$\int_{t_e}^{t} \frac{dt'}{R(t')} = \int_0^r \frac{d\tilde{r}}{1 + \frac{1}{4}k\tilde{r}^2}. \tag{17.19}$$

We can calculate this integral when $R(t)$ is given explicitly. We will do it in Section 17.4.

The quantities that are used in astronomy to characterise $R(t)$ are the **Hubble parameter** of (17.16)[5] and the **deceleration parameter** q_0:

$$q_0 \stackrel{\text{def}}{=} -\frac{R R_{,tt}}{R_{,t}^2}\bigg|_{\text{now}} \equiv -1 - \frac{c}{H_0^2}\frac{dH_0}{dt}. \tag{17.20}$$

The sign of q_0 tells us whether the expansion of the Universe accelerates ($R_{,tt} > 0$) or decelerates, and so is relevant for estimating the value of the cosmological constant.

17.2.3 Number counts

This time, the observer will be at $r = 0$, and r will refer to the radial coordinate of the source G. If we assume that the number of radiation sources is conserved $((nu^\alpha)_{;\alpha} = 0)$, then nR^3 must be constant, so

$$n = n_O R_O{}^3/R^3, \tag{17.21}$$

where the index 'O' refers to any fixed instant, e.g. the instant of observation. Taking the full solid angle $\delta\Omega_O = 4\pi$ and substituting for the other quantities in (16.135), namely for r_O from (17.18), for $R_O = R(1 + z_O)$ from (17.13), for $(k^\rho u_\rho)_G = 1/R$ from (17.12) and for $dv = -R^2 dr/\left(1 + \frac{1}{4}kr^2\right)$ from (17.9) – (17.10), we obtain for the total number of sources out to the distance corresponding to $\bar{r}$

$$N(\bar{r}) = 4\pi n_O R_O{}^3 \int_0^{\bar{r}} \frac{r^2 dr}{\left(1 + \frac{1}{4}kr^2\right)^3} \stackrel{\text{def}}{=} 4\pi n_O R_O{}^3 F(\bar{r}). \tag{17.22}$$

The integral is equal to

$$F(\bar{r}) = \begin{cases} \dfrac{1}{2k^{3/2}}(\psi - \sin\psi\cos\psi), & \sin\psi = \dfrac{\sqrt{k}\bar{r}}{1 + \frac{1}{4}k\bar{r}^2} & \text{for } k > 0 \\[2ex] \bar{r}^3/3 & & \text{for } k = 0 \\[2ex] \dfrac{1}{2(-k)^{3/2}}(\sinh\psi\cosh\psi - \psi), & \sinh\psi = -\dfrac{\sqrt{-k}\bar{r}}{1 + \frac{1}{4}k\bar{r}^2} & \text{for } k < 0 \end{cases} \tag{17.23}$$

---

[5] It is most often denoted H_0, to stress that its current value is meant.

(this is the same ψ as in $(17.4) – (17.7)$). Equations $(17.22) – (17.23)$ provide (in principle!) a method to determine the sign of k observationally. However, determinations of distance in cosmology are insufficiently precise to distinguish among the three forms of F.

17.3 The Friedmann equation

With a perfect fluid source, in a general geometry, the following unknown quantities must be determined from the field equations: six out of ten components of the metric tensor $g_{\alpha\beta}$ (the remaining four can be fixed by choice of the coordinates), the energy density, the pressure and three components of the velocity field u^α (the fourth component is determined by $u_\alpha u^\alpha = 1$). In total, this is 11 quantities to be found from 10 equations $G_{\alpha\beta} = \kappa T_{\alpha\beta}$, and the set is underdetermined. To make it determinate, we have to add one more equation, and usually this is the equation of state.[6]

This indeterminacy survives in the R–W metrics, in spite of their high symmetry. They automatically define a perfect fluid energy-momentum tensor. The energy density and pressure of the fluid are determined as functions of the scale factor $R(t)$ and of its derivatives. In order to obtain a definite solution, we have to add an equation of state relating the pressure to the energy density. With all functions depending only on t, this must be a barotropic equation of state $\epsilon = \epsilon(p)$ (see Section 15.5). This is a differential equation that determines $R(t)$.

What equation of state is most appropriate for describing the Universe as a whole? At present, the mass density in the Universe is so small ($< 10^{-28}$ g/cm^3) that pressure does not influence the large-scale dynamics of matter and $p = 0$ is an acceptable hypothesis. This is the assumption that Friedmann made in obtaining his solutions (given in Secs. 17.4 and 17.7). However, if we want to describe the evolution of the Universe with the early period of high density included, then we have to take pressure into account – and the equation of state cannot be the same at all times. A very brief description of the 'history of the Universe' is given in Section 17.14. We shall consider here the later periods and, following Friedmann (1922), we assume $p = 0$.

In the orthonormal tetrad for (17.1) or (17.3)[7] the Einstein equations are

$$G_{00} = \frac{3k}{R^2} + 3\frac{\dot{R}^2}{R^2} = \kappa\epsilon - \Lambda, \tag{17.24}$$

$$G_{11} = G_{22} = G_{33} = -\left(\frac{k}{R^2} + \frac{\dot{R}^2}{R^2} + 2\frac{\ddot{R}}{R}\right) = \kappa p + \Lambda. \tag{17.25}$$

The equations of motion $T^{\alpha\beta}{}_{;\beta} = 0$ reduce here to the single equation

$$\left(\epsilon R^3\right)^{\cdot} + 3R^2\dot{R}p = 0 \tag{17.26}$$

---

[6] In vacuum, because of $G^{\alpha\beta}{}_{;\beta} = 0$, only six of the Einstein equations are independent (Stephani, 1990, pp. 160–168). In matter, the equations of motion $T^{\alpha\beta}{}_{;\beta} = 0$ are in most cases not obeyed before all the field equations are solved. They are usually simpler than the Einstein equations and are solved first.

[7] Since the tetrad components of the Einstein tensor are scalars, and since in the R–W geometry they have to be spatially homogeneous, i.e. independent of the radial coordinate, they come out the same no matter which of the two coordinate representations is used. However, they would be different for (17.5) and (17.7) because there the constant k is scaled to $+1$ or -1, respectively.

(others are fulfilled identically). With $p = 0$, this becomes the mass conservation equation

$$\rho R^3 = \text{constant} \overset{\text{def}}{=} 3\mathcal{M}/(4\pi), \tag{17.27}$$

where $\rho = \epsilon/c^2$ is the mass density. If $\dot{R} \neq 0$, then (17.25) follows from (17.24) and (17.26). If $\dot{R} = 0$, then the solution of (17.24) – (17.25) is a generalisation of the static 'Einstein Universe' (12.82) – (12.83) to arbitrary k and $p \neq 0$ (the 'Einstein Universe' itself follows when $p = 0$ and $k = +1$). We shall not consider this case. With $p = 0$ and $\dot{R} \neq 0$ we thus have only (17.24) to solve, which becomes, in consequence of (17.27),

$$R_{,t}^2 = -k + \frac{2G\mathcal{M}}{c^2 R} - \frac{1}{3}\Lambda R^2. \tag{17.28}$$

With $c\dot{R}/R = H$ – the Hubble 'constant', we obtain from (17.27) and (17.28)

$$\rho = \frac{3H^2}{8\pi G} + \frac{c^2}{8\pi G}\left(\frac{3k}{R^2} + \Lambda\right). \tag{17.29}$$

The quantities ρ and H are *in principle* measurable.[8] The value of Λ may be deduced from observations.[9] Hence, (17.29) gives us a possibility to determine the sign of k for the real Universe – assuming that it really has an R–W geometry on large scale. In more general cosmological models (see Chapters 18–20) the curvature index k is replaced by a function of position in space and the local value of matter density is not connected with the type of evolution – unlike in Friedmann models. We will come to that connection in Sec. 17.4.

Taking $k = 0 = \Lambda$ in (17.29) we obtain the formula for the **critical density**

$$\rho_{cr} \overset{\text{def}}{=} \frac{3H^2}{8\pi G}. \tag{17.30}$$

If we take into account only that matter which is visible in telescopes, then the observed mass density $\rho_o \ll \rho_{cr}$. Taking the smallest value of H_0 consistent with current observations (Planck, 2014),[10] $H_0 = 67.11$ km s^{-1} Mpc^{-1}, we obtain $\rho_{cr} = 8.46 \times 10^{-30}$ g/cm^3, while $\rho_o \leq 10^{-31}$ g/cm^3 (Lang, 1974, p. 558). Greater values of H_0 increase the discrepancy. Orbital velocities of stars in galaxies indicate that galaxies contain large amounts of nonluminous *dark matter*. Estimates of density of that matter imply $\rho_o \leq 0.2\rho_{cr}$ (Coles and Ellis, 1997). Observations of galaxy clusters show that also the intergalactic space contains nonluminous matter, but including it does not allow one to make ρ_o greater than $0.3\rho_{cr}$. The current majority view is that the actual density is such as to ensure $k = 0$ in (17.29), and the invisible part is hidden in the form of dark matter (Padmanabhan, 1993;

---

[8] The difficult element is measuring the distances. All measurements of intergalactic distances rely on assumptions that are difficult to verify, and they can be carried out for relatively near galaxies. For very distant galaxies, in astronomical practice the Hubble law is *assumed* to hold exactly, and then used as a measure of distance.

[9] The currently established value (Planck, 2014) is $-\Lambda \approx 10^{-46}$ km^{-2}. A sphere of this curvature would have the radius of ≈ 109 Mpc.

[10] There is a disagreement (courteously called *tension*) between the value of H_0 measured via the cosmic microwave background radiation (Planck, 2014) and the value obtained from local observations; see de la Macorra, Garrido and Almaraz (2022) for a recent discussion. The values of $G = 6.67430 \times 10^{-11}$m^3/(kg s^2) and 1 Mpc = 3.085678×10^{22} m were taken from https://https://physics.nist.gov/cgi-bin/cuu/Value?bg—search_for=universal_in! and https://www.astronomycenter.net/data.html#len, respectively.

Peebles, 1993) – that can be anything except matter known from laboratory – or in the form of 'dark energy', i.e. the cosmological constant or its imitations (see Section 17.13).

The general solutions of (17.28) can be expressed in terms of elliptic functions. We will discuss them qualitatively in Section 17.7.

17.4 The Friedmann solutions with $\Lambda = 0$

With $\Lambda = 0$ and $dR/dt > 0$, Eq. (17.28) implies

$$\frac{dR}{dt} = \left(\frac{2GM}{c^2 R} - k\right)^{1/2} \overset{\text{def}}{=} \left(\frac{2\alpha}{R} - k\right)^{1/2}. \tag{17.31}$$

The solutions of (17.31) are best represented in a parametric way. For $k < 0$ we obtain:

$$R = -\frac{\alpha}{k}(\cosh \omega - 1), \qquad t - t_B = \frac{\alpha}{(-k)^{3/2}}(\sinh \omega - \omega), \tag{17.32}$$

where ω is the parameter and t_B is an arbitrary constant. For $k = 0$ the solution is

$$R = \left[\frac{9}{2}\alpha (t - t_B)^2\right]^{1/3}. \tag{17.33}$$

In this case the constant α can be scaled to any arbitrary value by transformations of r, so its actual value merely defines the unit of distance and has no physical meaning. (Unlike in the previous case and in the next case, where the value of α is related to the curvature of space at a given time.) Finally, for $k > 0$, the solution of (17.31) is

$$R = \frac{\alpha}{k}(1 - \cos \omega), \qquad t - t_B = \frac{\alpha}{k^{3/2}}(\omega - \sin \omega). \tag{17.34}$$

In all three cases we have taken into account the observed fact that at present $\dot{R} > 0$. As we predicted in Section 15.4, each of these solutions has a singularity at $t \to t_B$ called **Big Bang**, at which $R \to 0$ and $\rho \to \infty$. At that instant (in the past), all matter of the model was condensed in one point. The last model has a second such singularity, in the future, called **Big Crunch**, at

$$t = t_{FS} = t_B + \frac{2\alpha\pi}{k^{3/2}} = t_B + \frac{2\pi GM}{k^{3/2}c^2} = t_B + \frac{8\pi^2 G\rho_o R_o^3}{3k^{3/2}c^2}. \tag{17.35}$$

The models (17.32) and (17.33), if they are expanding at the initial instant, will continue to expand for an infinite time. The model (17.34) has a finite lifetime; at $t = t_{FS}$ its existence is terminated. The graph of $R(t)$ corresponding to different values of k is shown in Fig. 17.1, which again explains the dangers connected with rescaling of $|k| \neq 0$ to 1.

Formally, Eqs. (17.34) describe a flattened or compressed cycloid (when $k < 1$ or $k > 1$, respectively) on which t runs through an infinite range, while R oscillates between 0 and $R_{\max} = 2\alpha/k = 2GM/(kc^2)$. However, $dR/dt \to -\infty$ as $t \to t_{FS}$, and the integration of (17.31) through this point is not possible. This is why the $k > 0$ Friedmann models have only a finite time of existence equal to $(t_{FS} - t_B)$ (note that this quantity depends on $k!$).

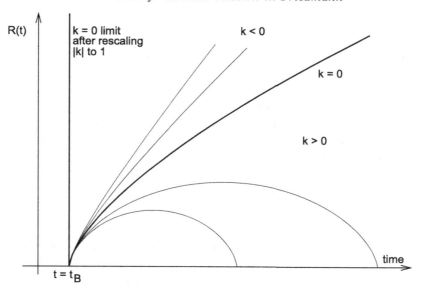

Fig. 17.1 The functions $R(t)$ corresponding to Friedmann models with different values of k. When $k \to 0$, the rescaling of R required to achieve $|k| = 1$ maps the $k = 0$ graph into the vertical straight line at $t = t_B$, and no place is left for the $k < 0$ models. Note that, for models with $k > 0$, the value of k determines the lifetime of the model, so *it has physical meaning*. Similarly, for models with $k < 0$, the value of $-k$ determines the asymptotic expansion velocity, $\dot{R}$, as $t \to \infty$. Thus, rescaling k involves tampering with physical parameters.

17.5 The redshift–distance relation in the $\Lambda = 0$ Friedmann models

Let $\mathcal{H}_0 \overset{\text{def}}{=} H_0/c$. Having found the explicit forms of $R(t)$, we can now integrate (17.19), to find the observer area distance as a function of the redshift, the Hubble parameter and the deceleration parameter. From (17.31) $R_{,tt} = -\alpha/R^2$, and from (17.20) $R_{,tt} = -q_0 R_{,t}^2/R \equiv -q_0 R \mathcal{H}_0^2$. Thus $R = \left[\alpha/\left(q_0 \mathcal{H}_0^2\right)\right]^{1/3}$. Using this, (17.31) can be rewritten as

$$\mathcal{H}_0^2 = 2 q_0 \mathcal{H}_0^2 - \frac{k}{\alpha^{2/3}} \, q_0^{2/3} \mathcal{H}_0^{4/3}. \tag{17.36}$$

Note that the sign of k is the same as the sign of $(2q_0 - 1)$, and $q_0 = 1/2$ when $k = 0$. From here, when $k > 0$,

$$\alpha = \left(\frac{k}{2q_0 - 1}\right)^{3/2} \frac{q_0}{\mathcal{H}_0} \implies R = \frac{1}{\mathcal{H}_0} \sqrt{\frac{k}{2q_0 - 1}}. \tag{17.37}$$

The integrals in (17.19) are different for each k. For $k > 0$ the result is

$$\arctan\left(\frac{\sqrt{k}r}{2}\right) = \frac{1}{2} \left[\omega(t) - \omega\left(t_e\right)\right]. \tag{17.38}$$

But, from (17.34),

$$\omega(t) = \arccos\left(1 - \frac{kR}{\alpha}\right), \qquad \omega\left(t_e\right) = \arccos\left[1 - \frac{kR}{(1+z)\alpha}\right]. \tag{17.39}$$

Substituting for α and R from (17.37) we obtain in (17.18)

$$r_O = \frac{q_0 z + (q_0 - 1)\left(\sqrt{2q_0 z + 1} - 1\right)}{\mathcal{H}_0 q_0^2 (1 + z)^2}.$$

(17.40)

This calculation is lengthy and tricky. See Exercise 7.

Going back to (17.37) and repeating the whole calculation for $k < 0$ we obtain again (17.40), even though the intermediate formulae are different. In particular, k and $(2q_0 - 1)$ are replaced by $(-k)$ and $(1 - 2q_0)$, respectively. See Exercise 7 also for this hint.

With $k = 0$, we substitute $q_0 = 1/2$ in (17.40) and obtain

$$r_O = 2\frac{\sqrt{1 + z} - 1}{\mathcal{H}_0 (1 + z)^{3/2}}.$$

(17.41)

The limit of (17.40) at $q_0 \to 0$ is $r_O = \frac{1}{2\mathcal{H}_0}\left[1 - \frac{1}{(1+z)^2}\right]$. This implies $\alpha = 0$, i.e. $\mathcal{M} = 0$, which is the Minkowski spacetime in coordinates connected with an expanding family of timelike straight lines. Equation (17.40) is called the **Mattig formula** (Mattig, 1958).

Note that $\mathrm{d}r_O/\mathrm{d}z = 1/\mathcal{H}_0 > 0$ at $z = 0$ and becomes negative as $z \to \infty$. Thus, $r_O(z)$ is increasing at small z and decreasing at large z, achieving a maximum in between. This means that there is refocussing in the Friedmann models with $\Lambda = 0$. This is true for every k, including $k = 0$, for which the maximal r_O occurs at $z = 5/4$. This phenomenon was mentioned in Sec. 16.10 and in Fig. 16.1; we will come back to it in Sec. 18.8.

17.6 The Newtonian cosmology

We will now describe, in Newton's theory, the motion of a spherical cloud of particles that interact only by gravitation. Let r_{out} be the radius of the cloud. We will make the same assumptions that led us to the R–W models in relativity: a homogeneous and isotropic mass distribution, $\rho = \rho(t)$, and a spherically symmetric initial distribution of velocities:

$$v_i(t_0) = v(t_0, r)\frac{x_i}{r}.$$

(17.42)

Let us consider the motion of a particle that lies on the sphere of radius $r_p(t) < r_{\mathrm{out}}$. Since the cloud is spherically symmetric, the force exerted on this particle by matter that lies in the region $r > r_p$ is zero (see Exercise 8). The particle thus moves in the gravitational field created by matter inside that sphere, i.e. in the potential

$$V(r) = -\frac{GM}{r} = -\frac{4}{3}\pi G\rho(t)r^2(t).$$

(17.43)

Consequently, the equation of motion of the particle is

$$\frac{\mathrm{d}v_i}{\mathrm{d}t} \equiv \frac{\partial v_i}{\partial t} + v_j\frac{\partial v_i}{\partial x_j} = -\frac{GM}{r^3}x_i = -\frac{4}{3}\pi G\rho(t)x_i(t).$$

(17.44)

Using (17.44) and (17.42) one can easily verify that

$$\frac{\mathrm{d}}{\mathrm{d}t}\left(\frac{v_i}{v}\right) = 0,$$

(17.45)

where $v = \sqrt{v_i v_i}$, i.e. the direction of the velocity vector at every point remains the same for all time. Therefore, using (17.42) we can rewrite (17.44) as follows:

$$\frac{dv}{dt} \equiv \frac{\partial v}{\partial t} + v \frac{\partial v}{\partial r} = -\frac{GM}{r^2} = -\frac{4}{3}\pi G\rho(t)r. \tag{17.46}$$

This is the equation of motion. The equation of continuity $\partial\rho/\partial t + (\rho v_i)_{,i} = 0$, in consequence of $\rho = \rho(t)$, (17.42) and (17.45), takes the form

$$\frac{1}{\rho}\frac{d\rho}{dt} + v_{,r} + \frac{2}{r}v = 0. \tag{17.47}$$

We denote

$$\frac{1}{\rho}\frac{d\rho}{dt} \stackrel{\text{def}}{=} -3F(t). \tag{17.48}$$

Then we have in (17.47) $(r^2 v)_{,r} = 3F(t)r^2$, which is integrated to give

$$v(t,r) = F(t)r + \frac{\mathcal{K}(t)}{r^2}. \tag{17.49}$$

We substitute this in (17.46) and obtain

$$\dot F r + \frac{\dot{\mathcal{K}}}{r^2} + \left(Fr + \frac{\mathcal{K}}{r^2}\right)\left(F - \frac{2\mathcal{K}}{r^3}\right) = -\frac{4}{3}\pi G\rho(t)r. \tag{17.50}$$

This is an algebraic equation in r whose coefficients are functions of t. The coefficients of different powers of r have to vanish separately, hence

$$\mathcal{K}(t) = 0, \qquad \dot F + F^2 = -\frac{4}{3}\pi G\rho(t). \tag{17.51}$$

Consequently, (17.49) becomes

$$v(t,r) = F(t)r, \tag{17.52}$$

and from (17.42) we have $v_i = F(t)x_i$, i.e. the Hubble law.

Equation (17.52) can be integrated again since $v = dr/dt$. We substitute this and (17.48) in (17.52) and obtain $(1/r)dr/dt = -[1/(3\rho)]d\rho/dt$, which is integrated with the result $r(t) = (A/\rho)^{1/3}$, where A is an arbitrary constant. Hence, for each particle separately

$$\rho r^3 = A = \text{constant} \stackrel{\text{def}}{=} \frac{3M}{4\pi}. \tag{17.53}$$

This is the analogue of (17.27) in the Friedmann solution. Substituting (17.53) in (17.48) and then substituting the result in (17.51) gives $\ddot r/r = -GM/r^3$. This implies

$$\dot r^2 = \frac{2GM}{r} - 2K, \tag{17.54}$$

where $K = $ constant. This equation has the same form as (17.31), so it has the same three types of solutions. In relativity, (17.31) defined the scale factor $R(t)$ that determined the changes of distance (see (17.15)). Here, (17.54) determines the distance of an arbitrary particle from the centre of the cloud. In the R–W models, the constant k determined the sign of the spatial curvature. Here, multiplying (17.54) by m, the mass of a fluid particle, we see that the quantity $(-mK)$ is the total energy of the particle. If $K > 0$, then the energy is negative; then at $r \to GM/K$ the velocity of the particle decreases to zero; the particle turns back towards the origin and falls onto it in a finite time. When $K = 0$, the particle can escape infinitely far from the origin, but its velocity tends to zero as $r \to \infty$. If $K < 0$, then $\dot{r} \xrightarrow[r \to \infty]{} \dot{r}_0 > 0$.

Equation (17.54) was first derived and interpreted by Milne (1934) and McCrea, and the authors were surprised that this result had not been known earlier. It is so elementary that it could have been found still in the eighteenth century, if only anybody had allowed for the *possibility* that the Universe is not static (psychology and prejudice have always played a role in science, and continue to do so). However, until the discovery of the expansion of the Universe, everybody believed that the Universe was unchanging in time.

Although the Milne–McCrea equation is identical with the Friedmann equation in its algebraic form, its physical interpretation is different. The solutions of (17.54) contain Newtonian notions such as absolute space and absolute velocity of matter (relative to space points). In contrast to relativity, they do not imply any law of propagation of light and are in conflict with special relativity. Removing that conflict requires an extension of Newton's theory. Such a theory (called 'kinematical relativity') was devised by Milne in later years (Milne, 1948), but it did not gain acceptance by physicists because, unlike relativity, it was well suited only to this particular cosmological model.

17.7 The Friedmann solutions with the cosmological constant

Now we will discuss the full Friedmann equation (17.28). Following Friedmann we denote $\Lambda = -\lambda$. Equation (17.28) becomes then

$$\dot{R}^2 = \frac{2G\mathcal{M}}{c^2 R} - k + \frac{1}{3}\lambda R^2. \tag{17.55}$$

The $\lambda(R)$ graphs of the equation $\dot{R}^2 = 0$ are shown in Fig. 17.2. For $k \leq 0$ the function $\lambda(R)$ is monotonic and negative for all $R < \infty$. With $k > 0$, $\lambda(R)$ increases from $-\infty$ at $R = 0$ through 0 (attained at $R = 2G\mathcal{M}/(c^2 k) \stackrel{\text{def}}{=} R_1$) to the maximum $\lambda_E = c^4 k^3/(3G\mathcal{M})^2$, attained at $R = 3G\mathcal{M}/(c^2 k) \stackrel{\text{def}}{=} R_E$, and then monotonically decreases to zero as $R \to \infty$.

Since λ is a universal constant, R can vary only along straight lines parallel to the R-axis. The λ calculated from (17.55) is never smaller than the λ determined by the $\dot{R} = 0$ curves, so the allowed area for R-values is above the corresponding $\dot{R} = 0$ curve, and the extrema of R lie on $\dot{R} = 0$. By definition, $R(t) > 0$. This implies the following:

- (1) For $\lambda < 0$, only models oscillating between $R = 0$ and $R = R_{\text{max}}$ exist. The R_{max} is greater for $k < 0$ and smaller for $k > 0$. The 'cosmological attraction' implied by $\lambda < 0$ will always halt and reverse the expansion of the Universe.
- (2) For $\lambda = 0$ and $k \leq 0$, matter expands monotonically from infinite density at $R = 0$ to zero density at $R \to \infty$ or collapses from $R = \infty$ to $R = 0$; these models were discussed in Section 17.4.
- (3) For $\lambda = 0$ and $k > 0$, R oscillates between $R = 0$ and $R_1 = 2G\mathcal{M}/(c^2 k)$.
- (4) For $0 < \lambda < \lambda_E$, the following cases exist:

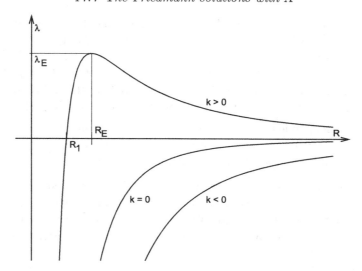

Fig. 17.2 The curves $\dot{R} = 0$ in the (R, λ) plane. The area accessible for evolution is above each curve. $R_1 = 2GM/(c^2 k)$ is the size of the Universe at maximum expansion when $\lambda = 0$ and $k > 0$. $\lambda_E = c^4 k^3/(3GM)^2$ is the minimum value of λ at which R changes between 0 and ∞; it corresponds to the maximum of the $k > 0$ curve that occurs at $R = R_E = 3GM/(c^2 k)$.

- (4a) If $k \leq 0$, the model constantly expands from a singular state at $R = 0$ to zero density at $R \to \infty$, or constantly contracts from $R = \infty$ to $R = 0$. In this case, $\dot{R}$ in (17.55) cannot change sign. Moreover, if $\dot{R} > 0$, then $\dot{R}$ becomes an increasing function of R at large R and the expansion then proceeds with acceleration – this is the effect of the 'cosmological repulsion' implied by $\lambda > 0$.
- (4b) If $k > 0$, then we have two cases:
  - (4b$_A$) If $R < R_E = 3GM/(c^2 k)$, then R oscillates between $R = 0$ and $R_{\max} < R_E$.
  - (4b$_B$) If $R > R_E$, then the model contracts from zero density at $R = \infty$ to a finite maximal density at $R = R_{\min}$ (with $\dot{R}(R_{\min}) = 0$), and then expands again to zero density as $R \to \infty$.

- (5) For $\lambda = \lambda_E$, there are several possibilities:

  - (5a) If $k \leq 0$, then the model constantly expands or contracts, like in case (4a).
  - (5b) If $k > 0$, then with $\lambda = \lambda_E$ the right-hand side of (17.55) has a double root at $R = R_E = 3GM/(c^2 k)$, and then the time required to reach R_E from either side:

$$t = \pm \int_{R(t_0)}^{R_E} \left(\frac{2GM}{c^2 R} - k + \frac{1}{3}\lambda_E R^2 \right)^{-1/2} dR \qquad (17.56)$$

is infinite ($+$ when $R(t_0) < R_E$ and $-$ otherwise). Consequently, when the initial $R < R_E$, the model either expands from a singular state at $R = 0$ asymptotically to $R = R_E$ at $t \to \infty$ or contracts from the asymptotic state $R \to R_E$ at $t \to -\infty$ to a singularity at $R = 0$.

– (5c) If $k > 0$ and $R > R_E$, then the model either expands from an asymptotic state $R \to R_E$ at $t \to -\infty$ to $R \to \infty$ at $t \to \infty$ or contracts from $R \to \infty$ at $t \to -\infty$ to $R \to R_E$ at $t \to \infty$.

– (5d) For $k > 0$, there also exists the static solution $R \equiv R_E$, which is unstable. This is the 'Einstein Universe' (12.82) – (12.83). It is unstable because even a small perturbation of R will cause it to expand as in case (5c) or contract as in case (5b).

• (6) For $\lambda > \lambda_E$, independently of the sign of k, there exist only models that monotonically expand from $R = 0$ to $R \to \infty$ or monotonically contract. Just as in cases (4a) and (5c), for large R the expansion proceeds with acceleration.

These facts are summarised in Figs. 17.3–17.5. Figure 17.3 shows all the possible oscillating models. With λ negative, the maximum R of the oscillating model with $k \le 0$ can be arbitrarily large, and the lifetime of the model can be arbitrarily long. Figure 17.4 shows all the models for which $\lambda = \lambda_E$, and Fig. 17.5 shows the remaining models.

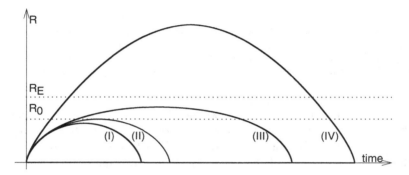

Fig. 17.3 The scale factor as a function of t for recollapsing Friedmann models, for different values of k and λ. Curve (I): $k > 0$, $\lambda < 0$; curve (II): $k > 0$, $\lambda = 0$ (this is one of the models considered in Section 17.4); curve (III): $k > 0$, $0 < \lambda < \lambda_E$; curve IV: $k \le 0$, $\lambda < 0$. The horizontal line R_E is at $R = R_E = 3GM/c^2$ in the static Einstein Universe. The horizontal line R_0 is at the maximum R in the recollapsing $\lambda = 0$ model. As seen from Fig. 17.2, $\lambda < 0$ will always force recollapse, irrespective of the sign of k. However, with $k \le 0$, an arbitrarily large maximum R can occur with a sufficiently small $|\lambda|$. On all curves, $GM/c^2 = 1$. On curve (I) $k = +1$, $\lambda = -0.1$, on curve (II) $k = +1$, $\lambda = 0$, on curve (III) $k = +1$, $\lambda = +0.1$, on curve (IV) $k = -1$, $\lambda = -0.1$. With $\lambda < 0$ of sufficiently large absolute value, the lifetime for a $k \le 0$ model may be shorter than in any given $k > 0$ model.

With $\lambda \ne 0$, the connection between the sign of spatial curvature and the type of motion no longer holds. There are models with $k < 0$ that recollapse after a finite time (case (1)) and models with $k > 0$ that expand to infinite extent (cases (4b_B), (5c) and (6)). The latter do not always attain a singular state.

The fact that, with $\Lambda < 0$, models expanding with acceleration may exist will be important in the discussion of horizons in Sec. 17.11.

The discussion by the method used above was first presented by Friedmann (1922 and 1924); it can also be found in the book by Robertson and Noonan (1968) and, in a more extended form, in Rindler (1980). Friedmann's original discussion was incomplete because he considered only the case $k = +1$, and did not know about the case $k = 0$.

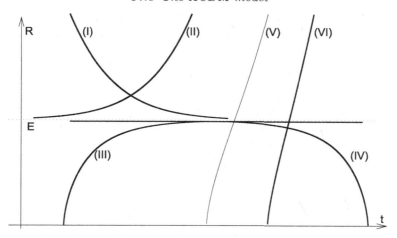

Fig. 17.4 The graphs of $R(t)$ for the Friedmann models with $\lambda = \lambda_E$. The dotted horizontal line marked E is the static Einstein Universe in which $R = R_E = 3GM/(c^2k)$. Curves I–IV represent models that approach the Einstein Universe asymptotically towards the future or towards the past; for all of them $k > 0$. On curve I, $R > R_E$ and $R_{,t} < 0$. On curve II, $R > R_E$ and $R_{,t} > 0$. On curve III, $R < R_E$ and $R_{,t} > 0$. On curve IV, $R < R_E$ and $R_{,t} < 0$. Curve V represents the model with $k = 0$, curve VI the model with $k < 0$. Curves V and VI have inflection points at $R = R_E$. Time-reverses of curves V and VI are also solutions of $\dot{R} = 0$ in (17.55), but are omitted for transparency. Curves III–VI represent models that have singularities in the past or in the future. Curves I and II represent models with no singularity. Note that the Einstein Universe is unstable: any perturbation that sets R off the value R_E will cause the model to expand or contract away from the initial state. The values of parameters in the figure are $GM/c^2 = 1$ on all curves, $k = +1$ for curves I–IV and $k = -1$ for curve VI.

17.8 The ΛCDM model

The notion of critical density mentioned at the end of Sec. 17.3 played an important role in cosmology in those years (approx. 1929–1999) when the majority view was that $\Lambda = 0$. The challenge was then to decide whether the mean mass density in the Universe is smaller than ρ_{cr}, larger than ρ_{cr} or equal to ρ_{cr} (which would imply, via (17.29), $k < 0$, $k > 0$ or $k = 0$, respectively). The current obligatory view in astronomy is that the Universe is expanding at an accelerated rate, see Sec 17.13. This is possible only when Λ, or something that imitates its action, is nonzero, as follows from (17.28) differentiated by t:

$$R_{,tt} = -\frac{GM}{(cR)^2} - \frac{1}{3}\Lambda R. \qquad (17.57)$$

All those 'somethings' that imitate Λ are collectively called 'dark energy'. If $\Lambda \geq 0$, then $R_{,tt} < 0$, i.e. the expansion of the Universe is decelerated all the time. With $\Lambda < 0$, the accelerating expansion sets in when $R = R_i = \left[3GM/\left(-\Lambda c^2\right)\right]^{1/3}$. The value of t at R_i can be calculated once we know the solution of (17.28).

The current conventional wisdom in astronomy says still more: the values of H and Λ are such that they imply $k = 0$ in (17.29), so the Universe is not only expanding with acceleration but also remains spatially flat. Such a model is called ΛCDM, the CDM meaning 'cold dark matter'. The solution of (17.28) with $k = 0$, $\Lambda < 0$ and $R_{,t} > 0$ is

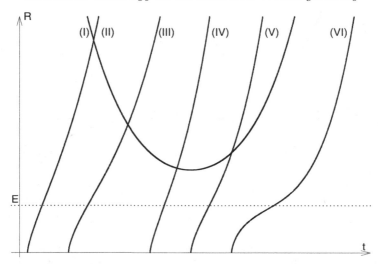

Fig. 17.5 The graphs of $R(t)$ for the remaining Friedmann models, for different values of k and λ. The horizontal line E is at $R = R_E = 3GM/(c^2k)$ in the static Einstein Universe. On curves I – III $0 < \lambda < \lambda_E$; on curves IV – VI $\lambda > \lambda_E$. On curve I $k > 0$ and $R > R_E$, on curve II $k < 0$, on curve III $k = 0$. On curves IV, V and VI $k < 0$, $k = 0$ and $k > 0$, respectively. The inflection points are at $R > R_E$ on Curves II and III and at $R < R_E$ on curves IV to VI. Positive λ implies cosmic repulsion that opposes gravitational attraction. Beyond the inflection point the repulsion prevails and sets the Universe into accelerating expansion. With large initial R, the repulsion prevents collapse even for $k > 0$.

$$R(t) = \left[\frac{6M_0}{(-\Lambda)}\right]^{1/3} \sinh^{2/3}\left[\frac{\sqrt{-3\Lambda}}{2}(t - t_{B\Lambda})\right], \qquad (17.58)$$

where $t_{B\Lambda}$ is the Big Bang (BB) time and $M_0 \overset{\text{def}}{=} GM/c^2$. The inflection $R_{,tt} = 0$ is at

$$t_i - t_{B\Lambda} = \frac{1}{\sqrt{-3\Lambda}}\ln\left(\frac{\sqrt{3}+1}{\sqrt{3}-1}\right). \qquad (17.59)$$

Taking $-\Lambda = 10^{-46}$ km^{-2} (Planck, 2014) and remembering that the physical time is t/c we obtain from (17.59) $(t_i - t_{B\Lambda})/c \approx 8 \times 10^9$ years.[11] The current estimate of the age of the Universe is $\approx 13.67 \times 10^9$ years (Planck, 2014). So, the acceleration began at somewhat earlier than $2/3$ of the present age of the Universe.

The conclusion about the accelerating expansion followed from observations of type Ia supernovae (Riess et al., 1998; Perlmutter et al., 1999), see Sec. 17.13 for more on that. But the two teams of observers did not allow any non-R–W models in their analysis. As we will see in Sec. 18.22, inhomogeneities in mass distribution and expansion rate can simulate accelerating expansion even when $\Lambda = 0$.

---

[11] For this calculation c had to be expressed in km/year, which is $\approx 3 \times 10^5$ km/s $\times$ (the number of seconds in a year ≈ 365 days $\times 24$ hours $\times 3600$ s $= 31,536,000$ s).

17.9 The redshift–distance relation in the $\Lambda \neq 0$ Friedmann models

Let us define the following three dimensionless parameters:

$$(\Omega_m, \Omega_k, \Omega_\Lambda) \stackrel{\text{def}}{=} \frac{1}{3{\mathcal{H}_0}^2} \left(\frac{8\pi G \rho_0}{c^2}, -\frac{3k}{{R_0}^2}, -\Lambda \right) \Bigg|_{t=t_0}, \tag{17.60}$$

where t_0 is the current instant, $\rho_0 = \rho(t_0)$ is the current mean mass density in the Universe, $c\mathcal{H}_0 = H_0$ is the current value of the Hubble parameter and $R_0 = R(t_0)$. They are called, respectively, the density parameter, the curvature parameter and the cosmological constant parameter. Using them, (17.29) becomes

$$\Omega_m + \Omega_k + \Omega_\Lambda = 1. \tag{17.61}$$

We assume that the observer is located at the centre of symmetry of the spacetime, so every light ray received or emitted by her will be radial. In an R–W spacetime this is no limitation of generality: because of homogeneity, every point in a space of constant t is the centre of symmetry. Then, from the formula for the observer area distance r_O (17.18) transformed to (17.3) we see that

$$r_O = (rR)|_{\text{ray}}, \tag{17.62}$$

where $(rR)|_{\text{ray}}$ is calculated along the light ray connecting the source to the observer. In order to calculate r_O we have to solve the equation of radial null geodesics in the metric (17.3). We take the initial point of the geodesic at the observer position, and follow it to the past towards the source. Then

$$\frac{1}{R} \frac{dt}{dr} = -\frac{1}{\sqrt{1 - kr^2}}. \tag{17.63}$$

Integrating this from $r = 0$ to r_e, the position of the light source where the ray was emitted at $t = t_e$, and taking $k < 0$ for the beginning, we obtain

$$\int_{t_0}^{t_e} \frac{dt}{R(t)} = -\int_0^{r_e} \frac{dr}{\sqrt{1 - kr^2}} = -\frac{1}{\sqrt{-k}} \operatorname{arsinh}\left(\sqrt{-k} r_e \right). \tag{17.64}$$

To calculate $\int_{t_0}^{t_e} dt/R(t)$ we first transform the variable by $dt/R(t) = dR/(RR_{,t})$, then substitute for $R_{,t} > 0$ from (17.28) and for $\mathcal{M}$ at $t = t_0$ from (17.27). The result is

$$\int_{t_0}^{t_e} \frac{dt}{R(t)} = \int_{R(t_0)}^{R(t_e)} \frac{dR}{R\sqrt{\dfrac{8\pi G \rho_0 {R_0}^3}{3c^2 R} - k - \frac{1}{3}\Lambda R^2}}. \tag{17.65}$$

Now we use $1 + z = R_0/R_e$, from (17.13). In (17.65), the running R is at the position of the emitter, and R_0 is constant. So, $R = R_0/(1 + z)$ and we now change the integration variable from R to z. At the observer position $z = 0$, so (17.65) becomes

$$\int_{t_0}^{t_e} \frac{dt}{R(t)} = -\int_0^z \frac{dz'}{\sqrt{\dfrac{8\pi G \rho_0 {R_0}^2}{3c^2}(1 + z')^3 - k(1 + z')^2 - \frac{1}{3}\Lambda {R_0}^2}}. \tag{17.66}$$

This we substitute in (17.64) and use (17.60). Solving for r_e we obtain

$$r_e = \frac{1}{\sqrt{-k}} \sinh \left[\int_0^z \frac{\sqrt{\Omega_k} dz'}{\sqrt{\Omega_m(1+z')^3 + \Omega_k(1+z')^2 + \Omega_\Lambda}} \right]. \tag{17.67}$$

We now use (17.62), (16.133), the definition of D_L – (16.134), we again replace the running $R = R|_{ray}$ with $R_0/(1+z)$ and use (17.60). The final result is

$$D_L(z) = \frac{1+z}{\mathcal{H}_0 \sqrt{\Omega_k}} \sinh \left[\int_0^z \frac{\sqrt{\Omega_k} dz'}{\sqrt{\Omega_m(1+z')^3 + \Omega_k(1+z')^2 + \Omega_\Lambda}} \right]. \tag{17.68}$$

This is the 'canonical' form of the (uncorrected) luminosity distance vs. redshift relation.

Equation (17.68) holds also for $k > 0$ if sinh and $\sqrt{\Omega_k}$ are replaced by sin and $\sqrt{|\Omega_k|}$. (This is consistent with the identity $\sinh(ix) \equiv i \sin x$.)

The now-standard ΛCDM model has $k = 0 = \Omega_k$. The limit of (17.68) at $\Omega_k \to 0$ is

$$D_L(z) = \frac{1+z}{\mathcal{H}_0} \int_0^z \frac{dz'}{\sqrt{\Omega_m(1+z')^3 + \Omega_\Lambda}}. \tag{17.69}$$

17.10 The redshift drift: a test for accelerating expansion

The Hubble parameter H depends on the cosmic time coordinate. The first author to consider the possibility of measuring the rate of change of H, called **redshift drift**, was Allan Sandage (1962). He estimated that, with the technology of then, $\approx 10^7$ years of monitoring the value of H would be required to detect its changes (Lazkoz, Leanizbarrutia and Salzano, 2018). By a more recent estimate, a few decades would be sufficient (Loeb, 1998). We will show how the hypothesis of accelerating expansion of the Universe can be confirmed or refuted by detecting just the sign of dH/dt.

If dt_e and dt_o denote time intervals at the emitter and at the observer that correspond to each other (for example, are periods of the same electromagnetic wave), then from (16.18) and (16.20) $dt_o/dt_e = 1 + z$. Using this, (17.13) and also $dR(t_e)/dt_o = dR(t_e)/dt_e \times dt_e/dt_o = (dR(t_e)/dt_e)/(1+z)$, we obtain

$$\frac{dz}{dt_o} = \frac{1}{R(t_e)} \left[\frac{dR(t_o)}{dt_o} - \frac{dR(t_e)}{dt_e} \right]. \tag{17.70}$$

So, $dz/dt > 0$ when dR/dt has increased between t_e and t_o, and $dz/dt < 0$ when it decreased. Hence, if $dz/dt_o > 0$ for some observed objects, then the expansion of the Universe had been accelerating for some time, but if $dz/dt_o < 0$ for all observed objects, then no accelerating expansion has ever occurred.

The conclusion stated above follows when we assume that the real Universe is correctly modelled by the R–W metrics. With more general models, the relation between dz/dt and the acceleration/deceleration may be more complicated. But then, the thesis that the Universe expands with acceleration was deduced by adhering exclusively to the R–W models (see Sec. 17.13), so – let us be self-consistent.

17.11 Horizons in the Robertson–Walker models

This section is a brief presentation of the results of Rindler (1956).

Every observer in the Universe receives information about distant objects via null geodesics. It turns out that in most of the Robertson–Walker models there exist, for every observer, such objects, from which he/she has not yet received any signal. Then, in some of the R–W models (those that expand with acceleration) objects exist from which a given observer has not received and will never receive any signal. The boundaries separating objects already observed from those not yet observed and objects observable from unobservable are called *horizons*. This definition will be made precise below.

The **event horizon** for observer A is the hypersurface in the spacetime that divides the collection of all events into two nonempty classes: those events that have been, are being or will be observed by A, and those that A has never observed and will never be able to observe. Not every R–W spacetime has event horizons.

The **particle horizon** for observer A at $t = t_0$ is a 2-dimensional surface in the space $t = t_0$ that divides all particles (i.e. worldlines of matter) into two nonempty classes: those that had been observed by A up to $t = t_0$, and those that A has not yet observed.

There exist models that do not possess any of these horizons (for example, the Newtonian model of Sec. 17.6). All the Friedmann solutions with $\Lambda = 0$ have particle horizons.

We will use the R–W metric (17.1). Let us define

$$\sigma(r) \stackrel{\text{def}}{=} \int_0^r \frac{\mathrm{d}r'}{1 + \frac{1}{4}kr'^2} = \begin{cases} \dfrac{2}{\sqrt{k}} \arctan\left(\dfrac{\sqrt{k}\, r}{2}\right) & \text{for} \quad k > 0, \\[2mm] r & \text{for} \quad k = 0, \\[2mm] \dfrac{1}{\sqrt{-k}} \ln\left(\dfrac{1 + \frac{1}{2}\sqrt{-k}\, r}{1 - \frac{1}{2}\sqrt{-k}\, r}\right) & \text{for} \quad k < 0. \end{cases} \qquad (17.71)$$

For an observer at $r = 0$, the equation of motion of a particle at $r = r_1$ is

$$\ell(t, r_1) = R(t)\sigma(r_1). \qquad (17.72)$$

Thus, $\sigma(r)$ is proportional to ℓ, the length of the arc of the r-curve (on which t, ϑ, φ are all constant) between the points $r = 0$ and $r = r_1$.

The equation of motion of a photon emitted at $r = r_1$ and proceeding *towards* the observer at $r = 0$ is, from (17.9) and (17.71),

$$\sigma(r) = \sigma(r_1) - \int_{t_1}^t \frac{\mathrm{d}\tau}{R(\tau)}. \qquad (17.73)$$

With $k < 0$ we have $\sigma(r) \to \infty$ for $r \to 2/\sqrt{-k}$, so the range $r \in [0, 2/\sqrt{-k})$ covers the whole space $t = $ constant. For $k > 0$ we have $\sigma \to \pi/\sqrt{k}$ for $r \to \infty$. Hence, $r = \infty$ corresponds to the point that is antipodal to $r = 0$ on each sphere $t = $ constant. An r-curve can be continued beyond this point, although the coordinate r does not cover the extra stretch. The r-line may wind several times around the sphere $t = $ constant. On such multiply wound curves we define $\sigma(r) = n\pi/\sqrt{k} + \widetilde{\sigma}(r)$, where n is the number of passages of the curve through the poles $r = 0$ and $r = \infty$, and $\widetilde{\sigma}(r)$ is the quantity calculated by

(17.71) between the last pole passed and the final r. Hence, also in the model with $k > 0$, a point in the space $t = $ constant can be assigned to every value of σ.

The reasoning below will be mostly directed at models with infinite time of existence. The modifications for recollapsing models will be mentioned in digressions. For them, the limit $t \to \infty$ has to be replaced by $t \to t_{FS}$, where t_{FS} is the instant of the final singularity, and 'finite limit' has to be replaced by 'value smaller than $\pi/\sqrt{k}$'.

With $k \le 0$, the necessary and sufficient condition for the existence of an event horizon is the convergence of the integral $\int_{t_1}^{t} \mathrm{d}\tau/R(\tau)$ to a finite limit at $t \to \infty$. Then, there exist particles at $r = r_H$ (with $\sigma(r_H) < \infty$) such that the photon emitted from there at $t = t_H$ will not reach the observer at $r = 0$ (where $\sigma(0) = 0$) even after an infinite time. From (17.73) we obtain for r_h (the minimal value of r_H):

$$\sigma(r_h) = \int_{t_h}^{\infty} \frac{\mathrm{d}\tau}{R(\tau)} < \infty. \tag{17.74}$$

With $k > 0$, the event horizon exists if $\sigma(r_h)$ defined by (17.74) obeys $\sigma(r_h) \le \pi/\sqrt{k}$. With $\sigma(r_h) > \pi/\sqrt{k}$, each photon sent from $r = r_h$ will manage, in a finite time, to travel farther than half of the circumference of the Universe, and, in consequence of this, the light signal from the emission event will reach the observer at $r = 0$ on one route or another.

For the recollapsing models with $k > 0$, the event horizon may exist even if $\sigma(r_h) \overset{\text{def}}{=} \int_{t_h}^{t_{FS}} \mathrm{d}\tau/R(\tau) > \pi/\sqrt{k}$ (see the example below).

In consequence of isotropy of the R–W models, if there exists r_h obeying (17.74) (or the corresponding condition for $k > 0$), then the event horizon exists for every direction of observation. In consequence of their homogeneity, the origin $r = 0$ may be placed at any point of the space $t = $ constant. Hence, if there exists an event horizon for one given observer, then event horizons exist for all observers.

When the event horizon exists, the events whose coordinate r obeys $r > r_h$ (when $k \le 0$) or $r_h < r \le \pi/\sqrt{k}$ (when $k > 0$) will never be observed by the observer at $r = 0$. How can this be explained intuitively? If $R(t)$ increases sufficiently rapidly, then the spatial distance between the observer at $r = 0$ and the light source at $r = r_1 > r_h$, given by (17.72), increases so fast that the light cannot overcome this 'swelling of space', even though it keeps going towards the observer. Eddington once explained this as follows: 'light is then like a runner on an expanding track, with the winning post receding faster than he can run'.[12]

In the Friedmann models with $\Lambda = 0$, $\dot{R}$ is a decreasing function, so, with $k \le 0$, every event should eventually become visible for every observer. Indeed, from (17.32) – (17.33), it follows that $\int_{t_1}^{t_2} \mathrm{d}\tau/R(\tau) \underset{t_2 \to \infty}{\longrightarrow} \infty$, i.e. the event horizon does not exist. But with $k > 0$, the final singularity is at $\omega = 2\pi$, and then, from (17.73) and (17.34)

$$\sigma(r_1) = \int_{t_1}^{t(2\pi)} \frac{\mathrm{d}\tau}{R(\tau)} = \int_{\omega(t_1)}^{2\pi} \frac{1}{\sqrt{k}} \, \mathrm{d}\omega = \frac{2\pi - \omega(t_1)}{\sqrt{k}}. \tag{17.75}$$

---

[12] Quoted after Rindler (1956). This author (A. K.) was not able to locate the original source.

Hence, if $\omega(t_1) > \pi$ (i.e. $t_1 - t_B > \pi\alpha/k^{3/2}$), then events at $t > t_1$ and $r > r_1$ will not become visible for the observer at $r = 0$ before the final singularity occurs. The event horizon will thus exist (even though $\sigma(r_1) > \pi/\sqrt{k}$ in the period $t_1 - t_B < \pi\alpha/k^{3/2}$).

From (17.73) it is also seen that if a particle was initially in the field of view (i.e. inside the event horizon) for observer A, then it will remain visible to her for ever. This is because if the equation

$$0 = \sigma(r_1) - \int_{t_1}^{t_0} \frac{d\tau}{R(\tau)} \tag{17.76}$$

has a solution for t_1 at a given r_1 and t_0, then it will have a solution for all times $t > t_0$.

Proof: Since $\sigma(r_1)$ does not depend on t, the problem consists in finding an integration interval in which the integral in (17.76) has a given value. From Fig. 17.6 one sees that if one such interval $[t_1, t_0]$ exists, then for every $t_0' > t_0$ there will exist a $t_1' > t_1$ such that

$$\int_{t_1'}^{t_0'} \frac{d\tau}{R(\tau)} = \int_{t_1}^{t_0} \frac{d\tau}{R(\tau)} \tag{17.77}$$

(if $1/R(t)$ is continuous, but this we are assuming all the time). $\square$

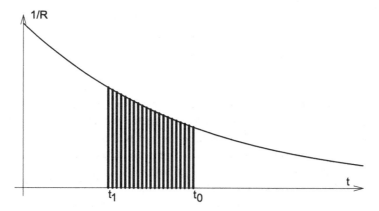

Fig. 17.6 An illustration to (17.76). If $\int_{t_1}^{t_0}(d\tau/R(\tau)) = B$ for a continuous function $1/R$, then for every $t_0' > t_0$ there exists a $t_1' > t_1$ such that the integral of $1/R$ over $[t_1', t_0']$ has the same value B.

If the event horizon exists (i.e. $\sigma(r_1) < \infty$ for $t_0 \to \infty$ in (17.76)), then t_1 goes to a finite limit t_{out} as $t_0 \to \infty$. This means that observer A will see the particle at $r = r_1$ only up to the instant $t = t_{\text{out}}$ on the particle's clock. The signal sent out from $r = r_1$ at $t = t_{\text{out}}$ will reach A after an infinite time ($t_0 \to \infty$), i.e. never. This, in turn, means that the worldline of the particle at $r = r_1$ will intersect the event horizon of A at the instant $t = t_{\text{out}}$. Moreover, a $t_{\text{out}} < \infty$ exists for every $r_1 < r_h$. Consequently, when the event horizon exists, each particle different from A will be visible to A only for a finite period of its history and will eventually escape from A's field of view through the horizon. The crossing of the horizon occurs in the infinite future of A. The farther away the particle, the shorter the period of its history that will be visible to A.

The necessary and sufficient condition for the existence of a particle horizon is the convergence of the integral in (17.76) to a finite limit at $t_1 \to t_B$. (In those models that have no initial singularity, the convergence to a finite limit at $t \to -\infty$. The analysis of this second case is similar and will be omitted here.) When the integral is finite at $t_1 \to t_B$,

$$\sigma(r_1) = \int_{t_B}^{t_0} \frac{\mathrm{d}\tau}{R(\tau)} \stackrel{\text{def}}{=} \phi(t_0) \qquad (17.78)$$

determines the farthest particles from which the observer at $r = 0$ could have received a light signal up to t_0. If $\sigma(r_1) \underset{t_1 \to t_B}{\longrightarrow} \infty$ in (17.76), then the observer at $r = 0$ could receive signals from all other particles for any t_0. At fixed t_0, (17.78) is an equation of a 2-dimensional sphere. Since $1/R > 0$, the function $\phi(t_0)$ is increasing. This means that, if a particle horizon exists, then with time still more particles come within it. Whether every particle will eventually enter the particle horizon depends on whether the integral in (17.78) has a finite limit at $t_0 \to \infty$. If it has, then, according to the previous definition, the event horizon exists and some particles will never enter A's field of view.

The first signal received by an observer A from each particle is the signal sent out at the initial singularity at $t = t_B$. However, by (17.13), this initial signal (for which $R(t_e) = 0$) is received with an infinite redshift.[13] This means that *in the Robertson–Walker models the initial singularity is not observable.*

In the models that contract from $R = \infty$ at $t = -\infty$, (17.13) gives $1 + z = 0$ at $t_e = -\infty$. Since $1 + z = \lambda_o/\lambda_e$, in these models the first signal received from each particle has zero wavelength at the observation point, i.e. is infinitely blueshifted.

In the Friedmann models with $\Lambda = 0$ the particle horizons do exist.

The two kinds of horizon were defined above for observers comoving with matter. One can also consider observers moving independently of the matter background. In some cosmological models there exist horizons also for such observers: irrespective of how fast the observer moves towards a given light source, he/she will not see any event on a worldline that is sufficiently far from the observer's starting point (Rindler, 1956).

Rindler's paper contains several examples of horizons in various R–W models; readers are advised to consult that paper for further information.

There exists one more type of horizon: the *apparent horizon* (see Secs. 14.11 and 16.5). In a Universe expanding from a singularity, it is a closed hypersurface at which both the inward- and outward-going light rays are diverging, i.e. at which the expansion scalar of any bundle of emitted light rays is positive. This notion plays no useful role in R–W cosmologies; it applies mainly to black hole models. Because of the peculiar properties of the R–W models, it is actually easier to interpret an apparent horizon in the more general Lemaître–Tolman models. However, we will calculate its position in the Friedmann models for completeness, as an exercise, in the coordinates of (17.3). The affinely parametrised tangent vector field to a bundle of null geodesics emanating radially from a sphere is found by transforming (17.11) with use of (17.2). Dropping primes, the result is

$$k^\alpha = \left[\frac{1}{R}, \frac{\varepsilon}{R^2} \sqrt{1 - kr^2}, 0, 0\right], \qquad (17.79)$$

---

[13] This is not a universal property of cosmological models; in Sec. 18.12 we will see that in most Lemaître–Tolman models radial rays emitted at the BB are received with infinite blueshift by all observers.

where $\varepsilon = +1$ for outward- and $\varepsilon = -1$ for inward-going rays. The expansion θ of this bundle is

$$2\theta = k^{\alpha}{}_{;\alpha} = \frac{1}{\sqrt{-g}} \left(\sqrt{-g}k^{\alpha}\right)_{,\alpha}. \tag{17.80}$$

Calculating this and substituting for $R_{,t}$ from (17.31) with the $+$ sign (we consider an expanding Universe) we find that $\theta > 0$ is equivalent to

$$\frac{r\sqrt{2\alpha/R - k}}{\sqrt{1 - kr^2}} + \varepsilon > 0. \tag{17.81}$$

This is fulfilled for an outward-going bundle ($\varepsilon = +1$), and it will be fulfilled also for an inward-going bundle ($\varepsilon = -1$) when

$$R < 2\alpha r^2 = \frac{2G\mathcal{M}}{c^2} r^2. \tag{17.82}$$

This always holds in a vicinity of the BB singularity; this is a region of past-trapped surfaces, see Sec. 16.5 for a definition. Its envelope, where $R = 2\alpha r^2$, is the past apparent horizon. Note from (17.17) and (17.2) that, in the coordinates we are now using, rR is the source area distance from the singularity at $R = 0$. Then, by analogy between (17.28) and the Newtonian equation (17.54), note from (17.27) that $\mathcal{M}r^3$ is equal to the mass contained within the source area distance rR from the singularity. Thus, (17.82) can be written in the equivalent form $r_G < 2m$, where $m \stackrel{\text{def}}{=} G\mathcal{M}r^3/c^2$. This is, not accidentally, similar to the condition defining the interior of the event horizon in the Schwarzschild metric. The meaning of this will become clearer in the Lemaître–Tolman models, see Section 18.8.

17.12 The inflationary models and the 'problems' they solved

(The views expressed in this section are A. K.'s. J. P. bears no responsibility for them.)

The R–W metrics are oversimplified models of the Universe. When taken literally, they can lead to puzzles or problems, most of which do not exist in the L–T and Szekeres models of Chapters 18 and 20. Two of such 'problems' gave rise to a booming field of activity in cosmology – the inflationary models. They are of marginal interest for basic relativity, but gained such prominence that we have to mention them.

The two problems are the 'flatness problem' and the 'horizon problem' (Guth, 1981). In brief, the 'flatness problem' is this. With $k = 0$, the density is found from (17.27) and (17.28)[14] to be $c^2\kappa\rho_{\text{cr}}^{\Lambda} = 3\dot{R}^2/R^2 + \Lambda$. Use (17.27) – (17.28) and consider the quantity

$$\widetilde{\Omega} \stackrel{\text{def}}{=} \frac{\rho_{\text{cr}}^{\Lambda}}{\rho} - 1 = -\frac{kR}{2\alpha}, \tag{17.83}$$

where α is defined in (17.31). Since in the ever-expanding models $R \to \infty$ as $t \to \infty$, the parameter $\widetilde{\Omega}$ will tend to infinity, unless $k = 0$. To estimate the value of $\widetilde{\Omega}$ by today, we take the $\Lambda = 0$ Friedmann model,[15] so that (17.27) holds. We will calculate how $\widetilde{\Omega}$

---

[14] We include Λ for later reference, although the original paper (Guth, 1981) assumed $\Lambda = 0$.
[15] In the original paper (Guth, 1981), a radiation distribution with $\Lambda = 0$ and $\epsilon = 3p$ was considered, but the conclusion was similar.

has changed since the instant when the average density in the Universe was equal to that inside a proton,[16] $\rho_N = 6.7 \times 10^{14}$ g/cm^3; the present density is assumed to be $\rho_0 \approx 10^{-31}$ g/cm^3. Using (17.27) we obtain from (17.83):

$$\widetilde{\Omega}_0/\widetilde{\Omega}_N = (\rho_N/\rho_0)^{1/3} \approx 1.9 \times 10^{15}. \qquad (17.84)$$

At present $\rho_{cr}/\rho_0 \approx 84.6$ (see after (17.30)), which means $\widetilde{\Omega}_0 \approx 83.6$. This discrepancy between ρ_{cr} and ρ_0 seems large. However, in order that $\widetilde{\Omega}_0$ can be so small, at the epoch of nuclear density $\widetilde{\Omega}$ had to be equal to $\widetilde{\Omega}_N = (1/1.9) \times 10^{-15}\widetilde{\Omega}_0 \approx 4.4 \times 10^{-16}$. Hence, during the first moments of the existence of the Universe the density had to be very close to the critical density, or else the present discrepancy between these two quantities would be huge.[17] Why had the initial conditions of the evolution of the universe been set up in such a way that it was described by the flat Friedmann universe with such a high precision?

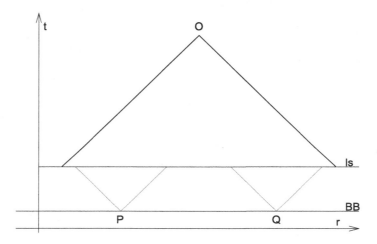

Fig. 17.7 The 'horizon problem' in a Robertson–Walker universe in comoving coordinates. The line BB is the Big Bang (BB) time, the line ls is the *last scattering hypersurface* – the time by which the CMB radiation was emitted. O is the observer today. The past light cone of O encompasses a much larger volume at $t = t_{ls}$ than the volume inside any light cone emitted at the BB (two of which are shown). Hence, several regions we see today had had no chance to come into causal contact prior to last scattering, yet the temperature of the CMB radiation coming from them is nearly the same. Inflationary models avoid this problem by replacing a thin slice of spacetime just after the BB with a model that expands much faster than any matter-filled R–W model. In effect, inflation makes the future light cones of P and Q much wider – encompassing the whole spatial cross-section of O's past light cone at $t = t_{ls}$.

The 'horizon problem' needs to be first explained qualitatively; see Fig. 17.7. Calculations show that the emission process of the cosmic microwave background radiation (CMB) took place during a brief period of **last scattering** (also called the **recombination** epoch) around $t_{ls} \approx 3.8 \times 10^5$ years after the BB (BB) (Maoz, 2016, p. 252). A light cone with its vertex at the BB encompasses a certain volume at $t = t_{ls}$; Fig. 17.7 shows two such

---

[16] https://physics.nist.gov/cgi-bin/cuu/Value?rp—search_for=atomnuc!
[17] The inflationary models consider still earlier periods, where the fine-tuning is by several orders of magnitude finer. We want to be conservative and do not go too far from laboratory physics.

cones. Particles that are still outside P's light cone at t_{ls} could not have interacted with P because the interaction would have had to propagate faster than light. The past light cone of a present observer, with the vertex at O, encompasses at t_{ls} a much larger volume than does any BB cone. Yet the radiation received by O from points in the $t = t_{ls}$ space has the same temperature, as if the various sources had achieved equilibrium beforehand. How is this possible? This is the 'horizon problem'.

To formulate the problem in numbers, we have to assume a model of matter. After t_{ls} we assume dust. It is usually assumed that the Universe prior to t_{ls} was **radiation-dominated**, i.e. that the energy density of radiation was larger than the energy density contained in the massive particles. Since radiation is supposed to obey the Maxwell equations, and the energy-momentum tensor of a Maxwell field is traceless (see (13.10)), it is assumed that radiation-dominated matter is a perfect fluid with the property $T^\rho{}_\rho = 0$. This implies the equation of state $\epsilon = 3p$, and then the equation of motion (17.26) becomes $\kappa \epsilon R^4 = \epsilon_0 = $ constant. Substituting this in (17.24) with $\Lambda = 0$ we obtain an equation for R that is easily integrated (with the initial condition $R(t_B) = 0$), yielding the result

$$R = \begin{cases} \sqrt{\epsilon_0/(3k) - k\left(t - t_B - \frac{1}{k}\sqrt{\epsilon_0/3}\right)^2} & \text{when} \quad k \neq 0, \\ \sqrt{2\sqrt{\epsilon_0/3}\ (t - t_B)} & \text{when} \quad k = 0. \end{cases} \tag{17.85}$$

Following Guth (1981), we take $k = 0$. We also take $r = 0$ for the observer's position, and $t_B = 0$. The radius of the visible region of the Universe at $t = t_{ls}$ for an observer at t_0 is calculated from (17.72) using (17.76) with t_{ls} instead of t_1. Using (17.33) it is

$$\ell_{ls}|_o = R\left(t_{ls}\right) \int_{t_{ls}}^{t_0} \frac{dt}{R(t)} = 3\left(t_{ls}{}^{2/3} t_0{}^{1/3} - t_{ls}\right). \tag{17.86}$$

Taking $t_0 = c \times (13.67 \times 10^9 \text{ years})$ for the present age of the Universe (Planck, 2014) and $t_{ls} = t_{ls}|_o = c \times (4.77 \times 10^5 \text{ years})$ we obtain[18] $\ell_{ls}|_o /c = 4.23 \times 10^7$ years.

The geometric radius of the intersection of the light cone originating at $t = t_B = 0$ with the space $t = t_{ls}$ is calculated in a similar way, but with R given by (17.85) and $t_{ls} = t_{ls}|_{BB} = c \times (3.8 \times 10^5 \text{ years})$

$$\ell_{ls}|_{BB} = R\left(t_{ls}\right) \int_0^{t_{ls}} \frac{dt}{R(t)} = 2t_{ls} = c \times (7.6 \times 10^5 \text{ years}). \tag{17.87}$$

The observed region thus has the radius 55.66 times larger than this, and so contains $(55.66)^3 > 170,000$ times as many particles.[19]

---

[18] An explanation is needed here. Once Λ is specified for the ΛCDM model (17.58), the ratios between $R(t)$ at different times are uniquely determined, and then, given $z = 1090$ for the CMB (Planck, 2014), the time of last scattering $\tau_{ls} = t_{ls}/c$ can be computed. It comes out to be $\tau_{ls}|_o = 4.77 \times 10^5$ y, see Krasiński (2016a) for details. On the other hand, the most-often cited value is $\tau_{ls}|_{BB} = 3.8 \times 10^5$ y (Maoz, 2016). The solution of this seeming contradiction is this: the ΛCDM model of (17.58) does not apply before last scattering because at those times the cosmic matter is not dust. For that epoch, the pressure of radiation and matter cannot be neglected, and a more general model must be used, which, with $z_{ls} = 1090$, implies $\tau_{ls} = \tau_{ls}|_{BB}$ given above. The z_{ls} is calculated by methods of particle physics (Peebles, 1968; Zeldovich et al., 1968). This is why we use $\tau_{ls}|_o$ in (17.86) and $\tau_{ls}|_{BB}$ in (17.87).

[19] Guth (1981) compared the two radii at an earlier time than t_{ls} and obtained a much more impressive ratio of particle numbers, 10^{83}.

Before we discuss other aspects of the 'horizon problem' let us note that it was ill-posed from the beginning. The CMB radiation is emitted when the primordial plasma cools below the ionisation temperature. With a given chemical composition of the plasma (see Sec. 17.14), the ionisation temperature is the same everywhere and no long-distance interaction is needed to equalize it. The puzzle is elsewhere: the fact that the temperature of the CMB radiation is observed today to be *nearly* the same (but not exactly equal!) in all directions means that the time between the emission of the CMB and the present day is nearly the same for all directions. (The 'horizon problem' is thus rather a 'uniform age problem'.) But marvelling over its uniformity is not constructive: the angular fluctuations of the CMB temperature ΔT with respect to the average temperature T are *measured* to be $\Delta T/T \approx 5 \times 10^{-6}$ (Smoot et al., 1992). These can be recalculated into local differences of the age of cosmic matter, i.e. into nonuniformities in the local BB instant, $\Delta t_B/c$ (because it is reasonable to assume that the time interval between the BB and the emission of the CMB radiation was everywhere the same).[20] A calculation based on the ΛCDM metric (17.58) is given in Sec. 18.6.2, it shows that $\Delta T/T \approx 5 \times 10^{-6}$ translates to $\Delta t_B/c \approx 7.39 \times 10^4$ y.

The solution of the two 'problems' proposed by Guth (1981) was that the equation of state of matter during the early phase of evolution (between 10^{-34} and 10^{-32} s after the BB) was such that the Universe expanded exponentially ('inflated'), so that the radius of a light cone originating at the BB was, at $t_{ls}|_{BB}$, much larger than the radius $\ell_{ls}|_o$ of the visible region of the Universe. In Guth's original paper, the de Sitter metric (8.87) was proposed as the model of the inflation process. This metric can formally be treated as a perfect fluid solution of Einstein's equations with the equation of state $\epsilon = -p$.[21] In later papers, the source was assumed to be a scalar field, i.e. an entity $\phi(x)$ whose field equation is $g^{\alpha\beta}\phi_{;\alpha\beta} - \partial V/\partial\phi = 0$, where the potential $V(\phi)$ is chosen so as to give ϕ the desired behaviour. To cause inflation, it suffices that the energy-momentum tensor

$$
\begin{aligned}
T_{\alpha\beta} &= (\epsilon + p)u_\alpha u_\beta - pg_{\alpha\beta}, & u_\alpha &= \phi_{,\alpha}/\sqrt{\phi_{,\rho}\,\phi^{;\rho}}, \\
\epsilon &= \phi_{,\rho}\,\phi^{;\rho}/2 - V, & p &= \phi_{,\rho}\,\phi^{;\rho}/2 + V
\end{aligned}
\tag{17.88}
$$

has the property $p < -\epsilon/3$ (Peebles, 1993).

Thus, the two 'problems' were solved within the class of R–W models by postulating a new equation of state. Nevertheless, the inflationary idea has become a huge success with cosmologists and gave rise to hundreds if not thousands of papers, so the subject cannot be avoided by anyone interested in cosmology. This is so in spite of two circumstances:

1. The two 'problems' that the inflationary models were designed to solve had not been perceived as much of a problem before. So much so that Guth (1981) felt compelled to argue at length to convince readers that they were problems indeed.

2. Inflationary models created new problems besides supposedly solving some.

---

[20] We repeat to stress: the 5×10^{-6} is not an error of measurement but an actually measured value, so nonuniformities in the BB time corresponding to such $\Delta T/T$ should really exist.

[21] Note, however, that the corresponding energy-momentum tensor $T_{\alpha\beta} = -pg_{\alpha\beta}$ does not define any velocity field. This gave rise to the following question: how are the local directions of flow defined after inflation stops? This was listed by Guth (1981) as one of the problems – *created by the idea of inflation* – that needed to be solved.

The 'flatness problem' is completely transformed if we consider the Lemaître–Tolman (L–T) and Szekeres models (see Chapters 18 and 20). In those models, the curvature index is not a constant, but a function of a spatial variable. If we find that it is unusually close to zero, then this means that we live in a privileged position in space. The 'horizon problem' finds a simple solution in the L–T model without recourse to inflation; see Section 18.19. Besides, inflation would have solved this problem if it were able to provide a model in which the Universe is initially inhomogeneous and acquires the uniform temperature of the CMB radiation in the course of its evolution. No such model has been presented. All the inflationary models are homogeneous all the time, in the best case they are anisotropic, of various Bianchi types. A great majority of them has a R–W geometry all the way. So inflation solved the 'horizon problem' by proclaiming: these models expand so fast that they *might* homogenise the CMB radiation!

We will not go into details; overviews of the inflationary scenario may be found in textbooks on cosmology (Peebles, 1993; Padmanabhan, 1993). The main virtue of the inflationary idea is the collection of speculations and hypotheses that it inspired (e.g. concerning the formation of structures in the Universe).

Among the problems that the idea of inflation created are these:

1. The so-called 'graceful exit' problem: once inflation sets in (its onset being a problem in itself), it cannot end without additional assumptions put into the model 'by hand'.

2. The cosmological constant problem: Λ is determined by the value of the scalar field at the end of the inflationary period, and it could be arbitrary. Why is it so close to zero?

3. Inflation overdid the solution of the 'flatness problem' – it implies that k must be very close to zero, i.e. that the density of matter in the Universe must be very near the critical one. Since observations of luminous matter cannot honestly account for more than 20% of the critical density (Coles and Ellis, 1997), inflation now relies on the postulate of existence of the 'dark energy' that makes up for the discrepancy. In simplest inflationary models the 'dark energy' is the energy density connected with the cosmological constant (see Sec. 17.13), and the numbers now favoured by astronomers are: $\approx 30\%$ of the total density is ordinary (but not necessarily luminous) matter and $\approx 70\%$ is generated by the 'dark energy' (Perlmutter et al., 1999). This creates another problem: the cosmological constant does not change with time, while the mass density falls with time by the law $\rho \propto R^{-3}$. They cannot be so close to each other at all times, which means that we live in a special epoch. This is a problem very much like the 'flatness problem'.

Finally, let us mention a problem that does not seem to bother cosmologists. Inflation is supposed to have taken place between 10^{-34} and 10^{-32} s after the BB. Note that, both in the dust era ($p = 0$ for $t \geq 4 \times 10^5$ years) and in the radiation era ($p = \epsilon/3$ for $t < 4 \times 10^5$ years), matter density obeys $\rho (t - t_B)^2 = $ constant. Thus, taking $\rho_0 = 10^{-31}$ g/cm^3 for the present density, $t_0 = 13.67 \times 10^9$ years for the present age of the Universe and 1 year $= 3.1536 \times 10^7$ s, we conclude that during the inflation period the density must have been greater than 10^{68} g/cm^3. This is far outside the range of laboratory physics and of any astronomical observations. Thus, if we wish to stick with the tradition that physics is an empirical science, inflation cannot be called a physical theory.

Some more problems created by inflation were discussed by Rothman and Ellis (1987).

17.13 The value of the cosmological constant

Riess et al. (1998) and Perlmutter et al. (1999) used a collection of supernovae in distant galaxies to find the best-fit cosmological model in the set that included various values of the present density and of Λ. The underlying idea was that supernovae in binary systems with white dwarfs, called type Ia, are 'standard candles', i.e. that their absolute luminosity at maximum is the same for all of them. Thus, knowing L and measuring $\mathcal{F}_O$ and z_O, one can use (16.117) to calculate the source area distances r_G. (In astronomy, a measure of $\mathcal{F}_O$ is used, called *apparent magnitude*, m, and defined by $m = -2.5 \log_{10} \mathcal{F}_O +$ constant.) Then, (16.119) and (17.40) provide a relation between r_G and the (H_0, q_0) parameters. By plotting magnitude against redshift, while knowing the value of H_0, one can determine q_0 and thus discern between different evolution laws. This gives, in principle, the possibility to determine q_0 *if we know all the other parameters with sufficient precision*.

The procedure used by the Riess and Perlmutter teams requires advanced knowledge of astronomy, and we cannot go into any of its details in this book. The authors took several disturbances into account, such as extragalactic extinction of light from the supernovae and the possibility of gravitational lenses in the line of sight or evolutionary changes in the supernovae-to-be (called *progenitors*). The result they obtained is that $\Lambda = 0$ is strongly inconsistent with the measurements. With the Hubble parameter $H_0 = 63$ km/(s × Mpc) (this value was then accepted as the right one), the best fit implied Λ accounting for more than 50% of the observed density. With $k = 0$, this gives $\Omega_M = 0.28^{+0.05}_{-0.04}$ as the percentage of mass in the form of matter, the remaining 0.72 being the cosmological constant. An additional bonus was the determination of the age of the Universe, $t_0 = (14.5 \pm 1.0) \times 10^9$ years (later modified to 13.67×10^9 y by the Planck (2014) satellite team).

This result should be taken with caution (the Nobel Prize of 2011 for the discoverers notwithstanding). Supernova explosions are rare and distant events. It is rather bold to assume that we know enough about the mechanism of the explosion, and to declare that a particular class of the supernovae are 'standard candles'. However, the claim that the Universe expands with acceleration has been welcomed by the astronomical community, and the now-standard model of the Universe is the ΛCDM model with its $k = 0$ and $-\Lambda/(\kappa c^2) = 0.72\rho_{cr}$.

Meanwhile, Célérier (2000) interpreted the same observations in terms of the Lemaître–Tolman model. She found that they can be reproduced by an inhomogeneous mass distribution, with no implications for the value of Λ, and $\Lambda = 0$ not being excluded (see Sec. 18.19 for more on this).

17.14 The 'history of the Universe'

The realisation that the Universe might have a history was coming only gradually to the minds of astronomers and physicists. The Hubble (1929, 1936) discovery of the expansion of the Universe and the realisation that the Friedmann (1922)–Lemaître (1927) solutions of Enstein's equations account for it set this line of thinking in motion. Since the Universe had been denser in the past, it must have been hotter as well. Some time ago then, it should have been sufficiently hot that all atoms were ionised. When the temperature dropped below the ionisation temperature, radiation had to be emitted. Expansion of the Universe should have cooled the radiation. Assuming that it had the black-body spectrum all the time, the evolution of temperature can be calculated. The intensity I of the black-body radiation as a function of frequency ν is given by the Planck formula:

$$I(\nu) = \frac{2h\nu^3}{c^2 \left\{\exp\left[h\nu/(kT)\right] - 1\right\}}, \tag{17.89}$$

where h and k are the Planck and Boltzmann constants, respectively, and T is the temperature of the radiation. The received frequencies of radiation obey (16.20), so $\nu(1+z) = \nu_e = $ constant along each ray. Consequently, to keep the form of the function $I(\nu)$ in (17.89) unchanged, we must have $T(1+z) = $ constant. That radiation should still be present in the Universe today. This idea first occurred to Gamow (1948) and collaborators (Alpher and Herman, 1948). The discovery came in 1965: the radiation exists and has the temperature 2.73 K (Dicke et al., 1965; Penzias and Wilson, 1965).

The speculation went on. Atomic nuclei are built of protons and neutrons. Of these two, only protons are stable. All the known atomic nuclei could, in principle, be built by adding protons and electrons consecutively one by one to the nucleus of hydrogen (the proton). Still further in the past, the Universe must have been hot enough to crash all heavier nuclei; only protons, neutrons and loose electrons could survive. Could it have been that matter originally consisted only of protons and electrons, and heavier nuclei came into existence through collisions between these? This idea again occurred to Gamow (1948) and Alpher and Herman (1948). Computer simulations indicated that something like this was going on, but only the first few nuclei of the Mendeleev table could be created in this way: about 25% of the mass of the Universe would be converted into helium ^{4}He. Tiny traces of deuterium, helium ^{3}He, lithium, beryllium and boron could be created, but the falling temperature of the Universe would stop any further synthesis (Wagoner, Fowler and Hoyle, 1967). (Heavier elements are synthesised later, in stars (Fowler, 1967), but this process is not within the domain of relativistic cosmology.) These calculated proportions of nuclides were confirmed by observations (Boesgaard and Steigman, 1985).

Thinking along these lines, cosmologists reconstructed the possible sequence of events in the evolution of the Universe. Some of the conclusions have been verified observationally (like the ones mentioned above), some others remain speculations. We will not go into details since they are not within the domain of relativity; they can be found in books on astrophysical cosmology (e.g. Padmanabhan, 1993). Here we give only a short list of the most important events.

The leading motive in theoretical cosmology was this: as we proceed backwards in time, approaching the singularity, the density and temperature of matter become arbitrarily large. Thus, whatever processes we know or can think of that should take place at high temperatures must have actually taken place in the early Universe. Why the BB explosion occurred and what preceded it are questions that cannot be answered by means of the currently existing physics or mathematics. Thus, they are usually not asked and we take the BB as a given thing.

The natural and interesting question is, at precisely what times the consecutive stages of evolution took place. The numbers given in different sources are different. The selection of numbers and references given below is random and we make no claim to being precise or up to date.

Assuming the values currently favoured by astronomers, $k = 0$, $\rho/\rho_{total} = 0.27$ and $-\Lambda/(\kappa c^2 \rho_{total}) = 0.73$, the BB took place 13.67×10^9 years ago (Planck, 2014).

During the first 10^{-34} s, the Universe had a temperature of about 10^{27} K and was supposedly described by a 'Grand Unified Theory' (GUT) that unites the strong nuclear, weak nuclear and electromagnetic interactions (Lang, 1999, Vol. II). The theory is still in the making. The elementary particles we know today had not necessarily existed then, and matter might have been composed of loose quarks and gluons.

Between 10^{-34} s and 10^{-32} s after the BB, inflation took place (see Section 17.12). No set of observations necessarily requires inflation for its explanation, but inflation is widely believed to be a proven theory. At the end of the inflation period, the elementary particles we know today should have come into existence, together with seeds for structure formation. The reasons, mechanism and exact instant of creation of the seeds are still a matter of debate (Lang, 1999, Vol. II). The BB did not have to be spatially homogeneous (see Chapter 18), and the inhomogeneities may have existed in the Universe from the very beginning, but this idea is not popular in the astronomical community.

About 1 s after the BB, neutrinos decoupled and should have thereafter propagated freely through space. During the next few seconds, protons, neutrons, electrons, positrons and photons existed in thermal equilibrium, at temperatures $T \geq 10^{10}$ K (Misner, Thorne and Wheeler, 1973).

The light atomic nuclei formed between 2 and 1000 s after the BB (Misner, Thorne and Wheeler, 1973). At the end of this period, the temperature dropped to about 10^9 K.

Later, the Universe continued to be a radiation-dominated plasma, but the radiation energy density ρ_r (which obeyed $\epsilon_r R^4 = $ constant, with $R \propto 1/T$, as follows from (17.89) and (17.13)) was decreasing faster than the energy density of massive particles, ϵ_m (obeying $\epsilon_m R^3 = $ constant). About 3.8×10^5 years after the BB (Maoz, 2016), and at temperatures of the order 10^3 K, ϵ_r became smaller than ϵ_m and radiation decoupled from matter, having too little energy to ionise the atoms that had already captured their electrons. It evolved into the CMB radiation of temperature 2.73 K that is observed today.

Then, structures like galaxies, galaxy clusters, superclusters and still larger condensations, and voids, must have formed. Even though structure formation should have proceeded by gravitation and should be well within the domain of classical gravitation theory, it is still poorly understood. There exists no quantitative account of the emergence of structures, only a general firm belief that structures came about by gravitational magnification of fluctuations created very early. Advances have been achieved in numerical simulations of this process (for example, Springel, Frenk and White, 2006). The Lemaître–Tolman model offers some interesting possibilities in this respect, see Secs. 18.5 and 18.6.

As this very brief account shows, the only input from relativity to this subject was the idea that the Universe began hot and dense, and then cooled as it expanded. Everything else is thermodynamics, hydrodynamics and particle physics. The fact that observations confirm some of the results does not uniquely point to the R–W models. Even in the R–W class, observations are not precise enough to define the evolution uniquely, so there is certainly enough room to consider more general models. It is often claimed that the high isotropy of the CMB radiation (with maximal temperature anisotropies of $\Delta T/T \approx 5 \times 10^{-6}$, Smoot et al., 1992) *proves* that our Universe has a R–W geometry. Such statements are, however, meaningless without a quantitative account of interaction between the CMB radiation and inhomogeneities in matter distribution. Existing estimates (see Section 18.14) show that the interaction is weak, and no temperature anisotropies larger than 10^{-5} should ever have been expected.

The above remarks were meant to convince the reader that, by considering more general (L–T and Szekeres) cosmological models, one does not deny the confirmed successes of the R–W cosmology. The more general models should fill in the finer details that cannot be captured in the R–W geometries, like the structure formation. The R–W models still remain valid as a rough first approximation to a more detailed description. Very unfortunately though, there is a tendency in the astronomical community to treat any consideration departing from the R–W class of geometries as a blasphemy that calls for instant punishment.

17.15 Invariant definitions of the Robertson–Walker models

It is often difficult to recognise a known metric when it is represented in unfamiliar co-ordinates. Metrics of high symmetry can be disguised in particularly elaborate ways; an example is the Schwarzschild solution represented in the Lemaître–Novikov coordinates, see Sec. 14.12. The same applies to the R–W metrics, so we shall list here some invariant criteria by which to recognise them. One invariant criterion, presented in Sec. 10.7, is by the symmetry group, but it requires solving the Killing equations and is therefore rather laborious.

The following set of properties is a necessary and sufficient condition for a spacetime to be in the R–W class:

1. The metric obeys Einstein's equations with a perfect fluid source.[22]
2. The velocity field of the perfect fluid has zero rotation, shear and acceleration. $\square$

The necessity of these properties follows by calculation. The proof of their sufficiency follows from (15.46) – (15.51). Those equations imply that a perfect fluid solution with $\sigma = \omega = 0 = \dot{u}^\alpha$ must be conformally flat. The most general conformally flat perfect fluid solution was found by Stephani (1967a) (Stephani et al., 2003, Theorem 37.17, p. 601). It has $\sigma = \omega = 0$, but in general $\dot{u}^\alpha \neq 0$. Its limit $\dot{u}^\alpha = 0$ is the R–W metric.

In this text, we are concerned only with expanding and nonempty R–W models for which $\theta \neq 0 \neq \epsilon + p$, and these inequalities are always tacitly assumed. The R–W model with $\theta = 0$ is the Einstein static Universe; the ones with $\epsilon + p = 0$ are the de Sitter vacuum Universes (the Minkowski spacetime is included among them as the limit $\Lambda = 0$); neither is appropriate for describing the observed Universe.

---

[22] Some authors considered the R–W metrics with nonperfect fluid and/or shearing/rotating/accelerating contributions to the energy-momentum tensor. Those contributions cancel each other and leave no trace in the metric tensor – which means that, from the point of view of the 4-dimensional geometry, they are undetectable. Hence, considering them is in contradiction to the Ockham razor principle.

Condition 1, that is, the perfect fluid source, is essential. Examples are known (Krasiński, 1997) of solutions for which $w = \sigma = 0 = \dot{u}^\alpha$, but which are not R–W because the source is not a perfect fluid.

There is another invariant definition of the R–W spacetimes, which makes no use of the field equations. The following set of properties is a necessary and sufficient condition for a spacetime to be R–W:

3. The spacetime admits a foliation into spacelike hypersurfaces of constant curvature.
4. The lines orthogonal to the leaves of the foliation are shearfree geodesics.
5. The expansion scalar of the geodesic congruence has its gradient tangent to those geodesics. $\Box$

17.16 Different representations of the R–W metrics

Several representations of the R–W metrics exist in the literature. The forms that appear most often are (17.1), (17.3), (17.5) and (17.7). Another frequently used representation is:

$$ds^2 = dt^2 - R^2(t) \left[dr^2 + f^2(r) \left(d\vartheta^2 + \sin^2 \vartheta d\varphi^2 \right) \right], \tag{17.90}$$

where

$$f(r) = \begin{cases} \sin r & \text{for } k > 0, \\ r & \text{for } k = 0, \\ \sinh r & \text{for } k < 0. \end{cases} \tag{17.91}$$

The three cases of (17.91)) can be written in one formula as

$$ds^2 = dt^2 - R^2(t) \left[dr^2 + \frac{1}{k} \sin^2(\sqrt{k}r) \left(d\vartheta^2 + \sin^2 \vartheta d\varphi^2 \right) \right]. \tag{17.92}$$

The range of r is finite or infinite, depending on the sign of k and on the coordinates used, but this is easy to recognise in each case. Note that in the case $k > 0$ the coordinates of (17.3) cover only half of the 3-sphere $t = $ constant ($0 \le r < 1/\sqrt{k}$) and are unsuitable for considering the geometry in the vicinity of the equator $r = 1/\sqrt{k}$.

Rather unfamiliar is the form of the R–W metric that results from solutions with two commuting Killing vector fields. The easiest way to find it is to demand that the metric be independent of y and z, while the spaces $t = $ constant have constant curvature. Then

$$ds^2 = dt^2 - R^2(t) \left[dx^2 + f_{,x}{}^2 dy^2 + f^2(x)dz^2 \right], \tag{17.93}$$

where

$$f(x) = \begin{cases} \sin(\sqrt{k}x) & \text{for } k > 0, \\ x & \text{for } k = 0, \\ \sinh(\sqrt{-k}x) & \text{for } k < 0. \end{cases} \tag{17.94}$$

Yet another form of the R–W metrics follows from plane symmetric metrics. Then

$$ds^2 = dt^2 - R^2(t) \left[dx^2 + e^{2Cx} \left(dy^2 + dz^2 \right) \right], \tag{17.95}$$

where C is a constant. When $C = 0$, this is evidently the flat R–W metric; when $C \neq 0$, it is the $k < 0$ R–W metric. The $k > 0$ R–W metric is incompatible with plane symmetry.

The R–W metrics result in quite unfamiliar form from the Goode–Wainwright (G–W) representation of the Szekeres models, discussed in Sec. 20.4. One of the G–W forms is:

$$ds^2 = dt^2 - S^2 \left[W^2 f^2 \nu_{,z}{}^2 dz^2 + e^{2\nu} \left(dx^2 + dy^2 \right) \right], \tag{17.96}$$

where

$$W^2 = \frac{1}{\varepsilon - kf^2},$$

$$e^\nu = \frac{f(z)}{a(z)\left(x^2 + y^2 \right) + 2b(z)x + 2c(z)y + d(z)}, \tag{17.97}$$

ε and k are arbitrary constants, and $S(t)$, $f(z)$, $a(z)$, $b(z)$, $c(z)$ and $d(z)$ are arbitrary functions subject to

$$ad - b^2 - c^2 = \varepsilon/4. \tag{17.98}$$

The slices $t = $ constant of the metric (17.96) – (17.98) are spaces of constant curvature equal to k/S^2. The t-coordinate lines are shearfree geodesics with the expansion scalar depending only on t. This is a characteristic property of the R–W spacetimes (see points 3, 4 and 5 in the previous section).

The other G–W form is

$$ds^2 = dt^2 - S^2 \left[A^2 dz^2 + e^{2\nu} \left(dx^2 + dy^2 \right) \right], \tag{17.99}$$

where

$$e^\nu = \frac{1}{1 + \frac{1}{4}k \left(x^2 + y^2 \right)},$$

$$A = e^\nu \left\{ a(z) \left[1 - \frac{1}{4}k \left(x^2 + y^2 \right) \right] + b(z)x + c(z)y \right\}, \tag{17.100}$$

$S(t)$, $a(z)$, $b(z)$ and $c(z)$ are arbitrary functions and k is an arbitrary constant. This metric has exactly the same properties as (17.96) – (17.98).

17.17 Exercises

1. Verify that the Einstein tensor for the R–W metric is diagonal in all the coordinate representations (17.1), (17.3), (17.5) and (17.7).

2. Prove that a null geodesic sent off radially from any point in a R–W spacetime, i.e. with $\dot{\vartheta}_0 = \dot{\varphi}_0 = 0$ at the initial point, will remain radial along its whole length.

3. Verify that the parameter v defined by (17.10) is indeed affine on radial null geodesics, i.e. that $Dk^\mu/dv + \left\{ {}^{\ \mu}_{\rho\sigma} \right\} k^\rho k^\sigma = 0$ when $k^\mu = dx^\mu/dv$.

4. Verify (17.24) – (17.25).

5. Verify that the equations of motion of a perfect fluid $T^{\alpha\beta}{}_{;\beta} = 0$ with the metric (17.3) indeed have only one nontrivial component given by (17.26) (it corresponds to $\alpha = 0$); the other three are fulfilled identically.

6. Verify that with $R_{,t} \neq 0$, Eq. (17.25) follows from (17.24) and (17.26).

7. Derive (17.40) from (17.38).

Hint. From (17.38)

$$\frac{\sqrt{k}r}{2} = \frac{\sin\left\{\frac{1}{2}\left[\omega(t) - \omega(t_e)\right]\right\}}{\cos\left\{\frac{1}{2}\left[\omega(t) - \omega(t_e)\right]\right\}} = \frac{\sqrt{1 - \cos\left[\omega(t) - \omega(t_e)\right]}}{\sqrt{1 + \cos\left[\omega(t) - \omega(t_e)\right]}} \qquad (17.101)$$

$$= \frac{1 - \cos\left[\omega(t) - \omega(t_e)\right]}{\sin\left[\omega(t) - \omega(t_e)\right]} = \frac{1 - \cos\omega(t)\cos\omega(t_e) - \sin\omega(t)\sin\omega(t_e)}{\sin\omega(t)\cos\omega(t_e) - \cos\omega(t)\sin\omega(t_e)}.$$

Multiply the numerator and the denominator by $[\sin\omega(t)\cos\omega(t_e) + \cos\omega(t)\sin\omega(t_e)]$. The result, after simplifying, is

$$\frac{\sqrt{k}r}{2} = \frac{\sin\omega(t) - \sin\omega(t_e)}{\cos\omega(t) + \cos\omega(t_e)}. \qquad (17.102)$$

Now use (17.37) and (17.39) to find

$$\frac{R}{\alpha} = \frac{2q_0 - 1}{kq_0}, \qquad \cos\omega(t) = \frac{1 - q_0}{q_0}, \qquad \cos\omega(t_e) = \frac{q_0(z - 1)}{(1 + z)q_0}, \qquad (17.103)$$

and the corresponding formulae for $\sin(\cdot)$. The final result is

$$\frac{\sqrt{k}r}{2} = \frac{z\sqrt{2q_0 - 1}}{1 + z + \sqrt{2q_0 z + 1}}. \qquad (17.104)$$

Substituting this and the second of (17.37) in (17.18) you will obtain (17.40).

For $k < 0$ the solution of (17.19) using (17.32) is

$$\frac{\sqrt{-k}r}{2} = \tanh\left\{\frac{1}{2}\left[\omega(t) - \omega(t_e)\right]\right\}. \qquad (17.105)$$

In this case, the calculation leading from (17.105) to (17.40) goes through the hyperbolic analogues of (17.101) – (17.102) and (17.39) (the latter are obtained from (17.32)). Here, k is replaced by $(-k)$ and $(2q_0 - 1)$ is replaced by $(1 - 2q_0)$.

8. Let $\rho(t, r)$ be any spherically symmetric (possibly inhomogeneous) finite distribution of matter (i.e. $\rho(t, r) = 0$ for $r > r_0$, $r_0 < \infty$). Let A be a point inside this distribution located at $r = r_1 < r_0$. Show that the total (Newtonian) gravitational force exerted on the point A by matter outside the sphere $r = r_1$ is zero.

9. Verify that with $k > 0$ the function $f(R) \overset{\text{def}}{=} \left(\frac{2GM}{c^2 R} - k + \frac{1}{3}\lambda_E R^2\right)$ in (17.56) has a double root at $R = R_E$, so

$$f = \frac{1}{R}\,(R - R_E)^2\,(\alpha R + \beta),$$

where α and β are independent of R. Show that the integral in (17.56) is then infinite, from whichever side R approaches R_E.

10. Verify that the various forms of the R–W metrics given in Section 17.16 do fulfil the invariant criteria of Sec. 17.15.

Note. It is not advisable to try to prove the equivalence of the G–W forms to R–W by explicit coordinate transformations.

18

Relativistic cosmology III:
the Lemaître–Tolman geometry

18.1 The comoving-synchronous coordinates

The curvature coordinates introduced in Sec. 14.1 were convenient for investigating the Schwarzschild solution. For other purposes the **comoving coordinates** are more useful. They can be introduced whenever there exists a timelike vector field u^α in the spacetime, for example, the 4-velocity of matter. In these coordinates the contravariant u^α has only the time-coordinate, $u^{\alpha'} \propto \delta^{\alpha'}{}_0$. They exist always – one has to solve the set of three equations $u^\alpha x^{I'}{}_{,\alpha} \overset{*}{=} 0$ for $I' = 1, 2, 3$. The transformations preserving the comoving coordinates are $x^I = f^I(x^{J'})$, $I, J' = 1, 2, 3$, $x^0 = f^0(x^{\alpha'})$.

If the vector field u^α has zero rotation, then the comoving coordinates can be chosen so that, in addition, they are **synchronous**, that is, the metric tensor has no time–space components. Here is the proof that vanishing rotation is a necessary condition:

Let comoving-synchronous coordinates $\{x^{\alpha'}\}$ exist. Then $u_{\alpha'} = \lambda \delta^0{}_{\alpha'}$ and, on changing from $\{x^{\alpha'}\}$ to other coordinates $\{x^\alpha\}$, we obtain $u_\alpha = f_{,\alpha} \lambda$, where $f = x^{0'}$. Because of $u_\alpha u^\alpha = 1$ we have $f_{,\rho} u^\rho = 1/\lambda$, and then $\dot{u}_\alpha = f_{,\alpha} \lambda_{,\rho} u^\rho + \lambda f_{,\alpha\rho} u^\rho - \frac{1}{2} u^\mu u^\rho g_{\rho\mu,\alpha}$. But $u^\mu u^\rho g_{\rho\mu,\alpha} \equiv 2 u^\rho u_{\rho,\alpha} = 2 u^\rho (\lambda f_{,\rho\alpha} + \lambda_{,\alpha} f_{,\rho})$. From all this, we find $\omega_{\alpha\beta} = 0$. □

The proof that $\omega_{\alpha\beta} = 0$ is also a sufficient condition for the existence of synchronous coordinates is somewhat complicated. If $\omega_{\alpha\beta} = 0$, then also $2\sqrt{-g} w^\alpha = \epsilon^{\alpha\beta\gamma\delta} u_\beta u_{\gamma,\delta} = 0$. This can be written as $u \wedge du = 0$, where $u \overset{\text{def}}{=} u_\alpha dx^\alpha$. This means that there exists a 1-form τ such that $du = \tau \wedge u$. Then there exist such functions f and λ that $u = \lambda df$, i.e. $u_\alpha = \lambda f_{,\alpha}$ (Flanders, 1963, Section 7.2). The time-coordinate $x^{0'} = f$ is then synchronous.

18.2 The spherically symmetric inhomogeneous models

For a spherically symmetric metric obeying the Einstein equations with a perfect fluid source, rotation of the fluid's velocity is zero (see Exercise 1). Hence, using (8.52), the comoving–synchronous coordinates can be introduced in which

$$ds^2 = e^{C(t,r)}dt^2 - e^{A(t,r)}dr^2 - R^2(t,r)\left(d\vartheta^2 + \sin^2 \vartheta d\varphi^2\right), \tag{18.1}$$

and the velocity field is

$$u^\alpha = e^{-C/2}\delta^\alpha{}_0. \tag{18.2}$$

The function R is connected with the area S of the surface of constant t and r by the Euclidean relation $S = 4\pi R^2$. Hence, R is called the **areal radius**. (N.B. this is the source area distance, defined in (16.116), between an observer at an arbitrary location and the centre at $R = 0$.)

The Einstein equations for this metric are, in coordinate components,

$$G^0{}_0 = e^{-C}\left(\frac{R_{,t}{}^2}{R^2} + \frac{A_{,t}\,R_{,t}}{R}\right) - e^{-A}\left(2\frac{R_{,rr}}{R} + \frac{R_{,r}{}^2}{R^2} - \frac{A_{,r}\,R_{,r}}{R}\right) + \frac{1}{R^2}$$

$$= \kappa\epsilon - \Lambda, \tag{18.3}$$

$$G^1{}_0 = e^{-A}\left(2\frac{R_{,tr}}{R} - \frac{A_{,t}\,R_{,r}}{R} - \frac{R_{,t}\,C_{,r}}{R}\right) = 0, \tag{18.4}$$

$$G^1{}_1 = e^{-C}\left(2\frac{R_{,tt}}{R} + \frac{R_{,t}{}^2}{R^2} - \frac{C_{,t}\,R_{,t}}{R}\right) - e^{-A}\left(\frac{R_{,r}{}^2}{R^2} + \frac{C_{,r}\,R_{,r}}{R}\right) + \frac{1}{R^2}$$

$$= -\kappa p - \Lambda, \tag{18.5}$$

$$G^2{}_2 = G^3{}_3 =$$

$$\frac{1}{4}e^{-C}\left(4\frac{R_{,tt}}{R} - 2\frac{C_{,t}\,R_{,t}}{R} + 2\frac{A_{,t}\,R_{,t}}{R} + 2A_{,tt} + A_{,t}{}^2 - C_{,t}\,A_{,t}\right)$$

$$- \frac{1}{4}e^{-A}\left(4\frac{R_{,rr}}{R} + 2\frac{C_{,r}\,R_{,r}}{R} - 2\frac{A_{,r}\,R_{,r}}{R} + 2C_{,rr} + C_{,r}{}^2 - C_{,r}\,A_{,r}\right)$$

$$= -\kappa p - \Lambda. \tag{18.6}$$

In the calculation that follows we assume $R_{,r} \neq 0 \neq R_{,t}$. The case $R_{,t} = 0$ is not interesting for cosmology: (18.4) implies for it that either $R_{,r} = 0$, which leads to the Nariai (1950) solution, or $A_{,t} = 0$, which, by virtue of (18.3), has density independent of t. But with $R_{,r} = 0$, Eqs. (18.3) – (18.6) do admit an interesting solution that will be dealt with in Sec. 19.5.

Equation (18.3), multiplied by $R^2 R_{,r}$, may be rewritten as

$$\left(R + e^{-C}RR_{,t}{}^2 - e^{-A}RR_{,r}{}^2 + \frac{1}{3}\Lambda R^3\right)_{,r}$$

$$- R\left(e^{-C}R_{,t}{}^2\right)_{,r} + e^{-C}A_{,t}\,RR_{,t}\,R_{,r} = \kappa\epsilon R^2 R_{,r}. \tag{18.7}$$

In consequence of (18.4), the last two terms on the left-hand side sum up to zero, so

$$\left(R + e^{-C}RR_{,t}{}^2 - e^{-A}RR_{,r}{}^2 + \frac{1}{3}\Lambda R^3\right)_{,r} = \kappa\epsilon R^2 R_{,r}. \tag{18.8}$$

Multiplying by $R^2 R_{,t}$ and using (18.4) we transform (18.5) to

$$\left(R + e^{-C}RR_{,t}{}^2 - e^{-A}RR_{,r}{}^2 + \frac{1}{3}\Lambda R^3\right)_{,t} = -\kappa p R^2 R_{,t}. \tag{18.9}$$

From (18.8) we can now recognise that the quantity

$$m \overset{\text{def}}{=} \frac{c^2}{2G}\left(R + e^{-C}RR_{,t}{}^2 - e^{-A}RR_{,r}{}^2 + \frac{1}{3}\Lambda R^3\right) \tag{18.10}$$

has all the properties we expect from mass: assuming $R = 0$ at $r = r_0$ and integrating (18.8) from r_0 to a current r in the hypersurface of constant t, we obtain $m(r) =$

$\int_{r_0}^{r} 4\pi\epsilon c^2 R^2 R_{,r'}\, dr' \equiv \int_{0}^{R(r)} 4\pi\epsilon c^2 u^2 du$. Then, (18.9) can be read as the energy-conservation equation: the time-derivative of mass equals volume-work.

At the surface of a spherical star, where $p = 0$, Eq. (18.9) reduces to $m_{,t} = 0$, i.e. the total mass of a star immersed in vacuum is constant.

The definition of mass given above (in the case $\Lambda = 0$) was first introduced by Lemaître (1933a), and then re-derived by Podurets (1964), but in the literature it is most often credited to Misner and Sharp (1964).

18.3 The Lemaître–Tolman model

In order to solve (18.3) – (18.6), we have to assume an equation of state. A relation of the type $\epsilon = f(p)$ limits the generality of an inhomogeneous model since it implies that the entropy per particle is a universal constant (see the end of Sec. 15.5). Lacking any other workable idea, the most natural equation of state is $p = 0$, i.e. evolution by gravitation only. We shall assume this now.

From (12.78), with $p = 0$ the fluid will move along timelike geodesics. Acceleration being zero then implies $C_{,r} = 0$ in (18.1) (see Exercise 2), and the transformation $t' = \int e^{C/2} dt$ leads to $C = 0$ in the new coordinates. Then (18.4) becomes

$$\left(e^{-A/2} R_{,r}\right)_{,t} = 0. \tag{18.11}$$

The case $R_{,r} = 0$ has already been set aside for separate consideration; see Section 19.5. With $R_{,r} \neq 0$, Eq. (18.11) is integrated with the result

$$e^A = \frac{R_{,r}^2}{1 + 2E(r)}, \tag{18.12}$$

where $E(r)$ is an arbitrary function. This form of the 'integration constant' of (18.11) will make the other equations below more clear. For the metric (18.1) to have the right signature it is necessary that E obeys $E \geq -1/2$ for all r.[1]

Using $C = p = 0$ and then (18.12) to eliminate $R_{,r}^2$ from (18.5) we obtain

$$2\frac{R_{,tt}}{R} + \frac{R_{,t}^2}{R^2} - \frac{2E(r)}{R^2} + \Lambda = 0. \tag{18.13}$$

Since we assumed that $R_{,t} \neq 0$, we multiply (18.13) by $R^2 R_{,t}$ and integrate to obtain

$$R_{,t}^2 = 2E(r) + \frac{2M(r)}{R} - \frac{1}{3}\Lambda R^2, \tag{18.14}$$

$M(r)$ being one more arbitrary function. With $\Lambda = 0$, this equation is formally identical to the Newtonian equation of radial motion in a Coulomb potential. In this Newtonian analogy, $c^2 M(r)/G$ plays the role of the **active gravitational mass** within an $r = $ constant shell, and $c^2 E(r)/G$ plays the role of the total energy within the same shell.

A solution of (18.14) will contain one more arbitrary function, $t_B(r)$, that will appear in the combination $(t - t_B(r))$. It is called the **bang-time function**, and in the case

---

[1] The value $E = -1/2$ is admissible provided that $R_{,r} = 0$ at the same location. This is a *neck* or *wormhole* – see Hellaby and Lake (1984) and Sec. 18.10.

$\Lambda = 0$ we will see below that it does indeed define the t-coordinate of the Big Bang (BB) singularity, which is in general position-dependent. With an arbitrary value of Λ, the BB will not always occur, just as in the Friedmann models.

With $C = 0$ and (18.11) – (18.13) fulfilled, (18.6) becomes an identity, while (18.3) provides the definition of the mass density:

$$\frac{8\pi G}{c^4}\epsilon = \frac{2M_{,r}}{R^2 R_{,r}}. \tag{18.15}$$

The mass density ϵ/c^2 becomes infinite where $R = 0 \neq M_{,r}$ and where $R_{,r} = 0 \neq M_{,r}$. The first of these is the BB, which occurs necessarily whenever $\Lambda = 0$, see the remark below (18.22). The second is a shell crossing singularity, where ϵ goes to infinity and changes sign to negative. At shell crossings, the radial geodesic distance between the point $\{t_0, r_0, \vartheta_0, \varphi_0\}$ and the point $\{t_0, r_0 + dr, \vartheta_0, \varphi_0\}$, equal to $\sqrt{|g_{rr}|}dr$, becomes zero, which means that shells of different values of the r-coordinate coincide. This singularity is avoided when $M(r)$, $E(r)$ and $t_B(r)$ are suitably chosen; see Section 18.10. In the Friedmann limit, defined below, the shell crossing coincides with the BB.

The final solution of (18.3) – (18.6) with $p = 0$ is thus

$$ds^2 = dt^2 - \frac{R_{,r}{}^2}{1 + 2E(r)}dr^2 - R^2(t, r)\left(d\vartheta^2 + \sin^2\vartheta d\varphi^2\right), \tag{18.16}$$

where $R(t, r)$ is determined by (18.14). It was first found and interpreted by Lemaître (1933a). Then, more of its properties were discussed by Tolman (1934) and by Bondi (1947). It will be called here the **Lemaître–Tolman (L–T) model**.[2] Many papers have been published in which various properties of the L–T model were discussed. The book by Krasiński (1997) contains an overview of these, complete until 1994.

The active gravitational mass c^2M/G in (18.14) and (18.15) generates the gravitational field. It is not equal to the sum of masses of particles that formed the gravitating body (compare (14.128) – (14.130)); here we will denote it by c^2N/G. Suppose that matter fills a sphere V with the centre at the centre of symmetry of the space, $r = r_0$, and the surface at $r = r_S$. Then c^2N/G is, from (18.16),

$$c^2N/G = \int_V (\epsilon/c^2) \sqrt{-g_3}d_3V \equiv 4\pi \int_{r_0}^{r_S} \frac{(\epsilon/c^2) R^2 R_{,r}}{\sqrt{1 + 2E(r)}}\, dr, \tag{18.17}$$

while the active gravitational mass is, from (18.15),

$$c^2M/G = 4\pi \int_{r_0}^{r_S} (\epsilon/c^2) R^2 R_{,r}\, dr. \tag{18.18}$$

Depending on the sign of E, M may be larger than N (when $E > 0$), smaller than N (when $E < 0$) or equal to it (when $E = 0$). If $M < N$, then $(N - M)$ is called the **relativistic mass defect** – the general-relativistic analogue of the mass defect known from nuclear and elementary particle physics. In a bound system ($E < 0$), part of the energy contained

---

[2] Even though Tolman made it clear that he was discussing Lemaître's solution, and Bondi cited Tolman, over many years this solution had been labelled the 'Tolman' or 'Tolman–Bondi' model by most authors. It is called here 'Lemaître–Tolman' only to avoid confusion with the Friedmann–Lemaître models, otherwise there would be no reason to add anybody's name to Lemaître's.

in the component particles had to be shed, and this lost energy is responsible for the mass defect. In the opposite case $(E > 0)$, the system is unbound and its excess energy sums up with the energy equivalent to the sum of masses of components. When $E = 0$, the system is 'marginally bound' – no energy has been shed to form it, and there is no excess energy. This interpretation of E was first given by Bondi (1947).

With $\Lambda \neq 0$, the explicit solutions of (18.14) involve elliptic functions. They were discussed by Lemaître (1933a) and Omer (1965). When $\Lambda = 0$, they are as follows:

when $E(r) < 0$:

$$R(t,r) = -\frac{M}{2E}(1 - \cos \eta),$$

$$\eta - \sin \eta = \frac{(-2E)^{3/2}}{M}[t - t_B(r)] ; \qquad (18.19)$$

when $E(r) = 0$:

$$R(t,r) = \left\{ \frac{9}{2}M(r)[t - t_B(r)]^2 \right\}^{1/3} ; \qquad (18.20)$$

when $E(r) > 0$:

$$R(t,r) = \frac{M}{2E}(\cosh \eta - 1),$$

$$\sinh \eta - \eta = \frac{(2E)^{3/2}}{M}[t - t_B(r)] . \qquad (18.21)$$

Actually, the conditions for the occurrence of each case are $E/M^{2/3} <, =$ or > 0, since the conditions of regularity of the spacetime at the centre of symmetry require that $E = 0$ at $M = 0$, see Section 18.4.

The solutions (18.19) – (18.21) have the same algebraic form as the Friedmann solutions (17.32) – (17.34) which follow when

$$M = M_0 r^3, \qquad E = -\frac{1}{2}kr^2, \qquad t_B = \text{constant}. \qquad (18.22)$$

Consequently, in all three cases the Big Bang singularity at $t \to t_B$, where $R \to 0$ and $\epsilon \to \infty$, is present. In the case (18.19) there is a second, Big Crunch, singularity at $\eta = 2\pi$.

The limit (18.22) is coordinate-dependent. We will show that the invariant condition is $\epsilon_{,r} = 0$. In each interval of r in which $M_{,r}$ does not change sign, $M(r)$ can be used as the independent variable instead of r, since (18.15) and (18.16) are covariant under the transformations $r = f(r')$. Using M as the radial coordinate, we rewrite (18.15):

$$\kappa \epsilon = \frac{6}{(R^3)_{,M}} \iff R^3 - R^3(M_0) = \int_{M_0}^M \frac{6}{\kappa \epsilon(\widetilde{M})} \, d\widetilde{M}. \qquad (18.23)$$

Then, the condition of spatial homogeneity is $\epsilon_{,M} = 0$, i.e. $(R^3)_{,MM} = 0$. Working it out in detail, one finds that this is equivalent to

$$E/M^{2/3} = \text{constant}, \qquad t_B = \text{constant}, \qquad (18.24)$$

and these two equations define the Friedmann limit invariantly (see Exercise 3). The functions $(E/M^{2/3})_{,M}$ and $t_{B,M}$ generate, respectively, the increasing and the decreasing perturbation of the mass density in the background Friedmann model (see Silk (1977) and Section 18.21).

The function $E(r)$ has one more interpretation. In the subspace $t = $ constant of (18.16), R depends only on r, so it can be used as the radial coordinate. Taking the orthonormal tetrad defined by the 3-metric of that subspace ($e^1 = dR/\sqrt{1 + 2E}$, $e^2 = Rd\vartheta$, $e^3 = R\sin\vartheta d\varphi$), we find the tetrad components of the 3-dimensional Riemann tensor to be

$$R_{1212} = -\frac{E_{,R}}{R} = R_{1313}, \qquad R_{2323} = -\frac{2E}{R^2}. \tag{18.25}$$

Thus, with $E = 0$, every space $t = $ constant is flat. If $E/R^2 = $ constant, then those spaces have constant curvature proportional to $-E$. A nonconstant E is a measure of their curvature which, unlike in the Friedmann models, is *local* – it depends on r and may be positive in one neighbourhood of the space, but positive elsewhere. This shows that the distinction between the Robertson–Walker models of different values of k is a peculiarity of the R–W class of metrics and not a property of the physical Universe – the same spacetime can be approximately like the $k > 0$ Friedmann model in one neighbourhood and like the $k < 0$ Friedmann model in another. Thus, (18.19) – (18.21) do not always describe different cosmological models – they can hold in different regions of the same spacetime.

Equation (18.15) shows that with $M_{,r} = 0$ the L–T model becomes vacuum. Being spherically symmetric, it must coincide in this limit with the Schwarzschild solution or its extension through the event horizon. Indeed it does, as shown in Section 14.12 (Eqs. (14.120) – (14.121)) and Exercise 10 in Chapter 14. See there for a discussion.

Other coordinate representations of the L–T model show up rarely. The curvature coordinates are rather cumbersome (see Exercise 4) – in them, the simple statements $M_{,t} = E_{,t} = 0$ become differential equations with implicitly defined coefficients. Nevertheless, Gautreau (1984) used the curvature coordinates to arrive at an interesting conclusion about the influence of cosmic expansion on planetary orbits; see Section 18.7.

Stoeger, Ellis and Nel (1992) tested their *observational cosmology programme* on the L–T model. The basis of the programme are coordinates connected with the observer's past light cone. Transforming the L–T solution to such coordinates requires the integration of the equations of null geodesics, which proved to be an impossible task. Therefore, the authors re-derived the model from the Einstein equations using the observational coordinates from the beginning. The L–T metric in these coordinates is (in the original notation):

$$ds^2 = A^2(\eta)\left(-dw^2 + 2dwdy\right) + C^2(w, y)\left(d\vartheta^2 + \sin^2\vartheta d\varphi^2\right), \tag{18.26}$$

where $\eta \overset{\text{def}}{=} w - y$, $A(\eta)$ is a function related to the observed redshift $z(\eta)$ by $A(\eta) = A_0/[1 + z(\eta)]$, $A_0 = $ constant and C is determined by

$$\frac{\partial C}{\partial y} = A(\eta)W(y), \qquad \frac{\partial C}{\partial \eta} = A\sqrt{\frac{2\omega_0(y)}{C} + W^2(y) - 1}, \tag{18.27}$$

$W(y)$ and $\omega_0(y)$ being arbitrary functions; W^2 is the analogue of $(1 + 2E)$. The matter density is $\kappa\epsilon = \epsilon_0(y)/\left(AC^2\right)$, and $\epsilon_0(y)$ is one more arbitrary function. The authors explained how the functions are related to observable quantities, but the connection is not simple, and readers are advised to refer to the original paper. The function $z(y)|_{w=\text{const}}$ can in principle be read out from observations, but in practice the procedure would require cosmological parallax distances to be measured and this is unrealistic with the present-day technology. For comparison, the R–W metric in the observational coordinates is

$$ds^2 = R^2(\eta)\left[dw^2 - 2dwdy - k^{-2}\sin^2(ky)\left(d\vartheta^2 + \sin^2\vartheta d\varphi^2\right)\right]. \tag{18.28}$$

18.4 Conditions of regularity at the centre

We found that the mass density ϵ/c^2 becomes infinite at those locations where $R = 0 \neq M_{,r}$. Thus, not the whole set $R = 0$ is a singularity. Part of this set is the centre of symmetry, and we will now formulate the conditions that the arbitrary functions in the L–T model have to obey in order that the centre of symmetry remains nonsingular.[3]

Let $r = r_c$ be the radial coordinate of the centre of symmetry, where $R(t, r_c) = 0$ for all $t > t_B(r_c)$, and let $\epsilon(t, r_c)/c^2 = \rho(t, r_c) \stackrel{\text{def}}{=} \alpha(t) < \infty$. Assuming that $\rho(t, r)$ is not only finite at $r = r_c$ but also continuous in a neighbourhood of $r = r_c$, $\alpha(t) < \infty$ implies that

$$M(r_c) = 0. \tag{18.29}$$

Applying the de l'Hôpital rule and using (18.23), we then find

$$\lim_{r \to r_c} \frac{R}{M^{1/3}} = \left[\lim_{r \to r_c} \frac{R^3}{M}\right]^{1/3} = \left[\lim_{r \to r_c} (R^3)_{,M}\right]^{1/3} = \left[\frac{6}{\kappa\epsilon(t, r_c)}\right]^{1/3}. \tag{18.30}$$

Thus, R must behave in the neighbourhood of $r = r_c$ as

$$R = \beta(t)M^{1/3} + O_{1/3}(M), \qquad \beta = \left(\frac{6}{\kappa\alpha c^2}\right)^{1/3}. \tag{18.31}$$

where the symbols $O_a(M)$ will denote quantities with the property

$$\lim_{M \to 0} \frac{O_a(M)}{M^a} = 0. \tag{18.32}$$

Thus, if $\epsilon(t, r_c) = 0$, then $\lim_{r \to r_c}(R/M^{1/3}) = \infty$. Note from (18.20) that (18.31) is always fulfilled when $E = 0$. For the other two models, defining

$$\gamma = \frac{1}{2}\beta_{,t}{}^2 - \frac{1}{\beta} + \frac{1}{6}\Lambda\beta^2 \tag{18.33}$$

we get from (18.14) $\gamma = $ constant and

$$E = \gamma M^{2/3} + O_{2/3}(M). \tag{18.34}$$

Finally, from (18.19) – (18.21) and (18.31), we see that $t_B(r_c)$ must have a finite value, which can be formally written as

$$t_B = \tau + O_0(M). \tag{18.35}$$

Equations (18.29) – (18.35) will be assumed from now on for models that have a centre.

---

[3] One could consider models with a permanently existing singularity at the centre of symmetry, but this author (A.K.) does not know about any. Then, there exist L–T models that do not have any centre of symmetry (think of a 2-dimensional analogue, for example a one-sheeted hyperboloid). Examples will be given later in this chapter.

18.5 Formation of voids in the Universe

Voids are large (typically 15 to just over 100 Mpc in diameter, see e.g. Fig. 7 in Sutter et al., 2012) volumes in the intergalactic space with an average matter density ≤ 0.2 of the large-scale average (Sutter et al., 2012). Their observational discovery (Gregory and Thompson, 1978) was a surprise because it contradicted the then-universal belief that galaxies are distributed uniformly in space. But in fact, the first papers indicating that voids should be ubiquitous were published in the 1930s and had not been understood.

The first such indication was given by Tolman (1934) and Sen (1934). Tolman's main result was the proof that the Einstein and Friedmann models are unstable against the growth of inhomogeneities. This is derived as follows.

Let the initial conditions at $t = t_1$ be chosen so that $R(t_1, r)$ in the L–T model is the same function as $rR_F(t_1)$ in a Friedmann model, and $R_{,t}(t_1, r) = rR_{F,t}(t_1)$. The first is just a choice of the coordinate r: whatever r was initially, we choose the new r by $r' = R(t_1, r)/R_F(t_1)$. The quantity $R_{,t}$ is a measure of the velocity of expansion, so with $R_{,t}(t_1, r) = rR_{F,t}(t_1)$ the assumed perturbation of the Friedmann model leaves the initial velocity distribution unperturbed.

With $R(t_1, r)$ and $R_{,t}(t_1, r)$ defined as above, (18.14) imposes a relation between $M(r)$ and $E(r)$, so we still have one free function at our disposal. This means that the mass density $\rho = \epsilon/c^2$ is not yet determined and $\rho(t_1, r)$ may be assumed any.

The assumed initial conditions imply $(R_{,tr}/R_{,r})(t_1) = (R_{F,t}/R_F)(t_1)$, and then

$$\frac{\partial}{\partial t}\left(\ln \rho_{LT} - \ln \rho_F\right)\big|_{t=t_1} = 0, \tag{18.36}$$

where ρ_{LT} and ρ_F are given by (18.15) and (17.27). From (18.15) we find further

$$\left[\frac{\partial^2}{\partial t^2}\ln\epsilon\right]_{LT}(t_1) = \left[-2\frac{R_{,tt}}{R} + 2\frac{R_{,t}^2}{R^2} - \frac{R_{,ttr}}{R_{,r}} + \frac{R_{,tr}^2}{R_{,r}^2}\right]_{LT}(t_1). \tag{18.37}$$

Using (18.14) to find $R_{,tt}$ and $R_{,ttr}$ and then using (18.15), we simplify (18.37) to

$$\left[\frac{\partial^2}{\partial t^2}\ln\epsilon\right]_{LT}(t_1) = \left[\frac{1}{2}\kappa\epsilon + \Lambda + 2\frac{R_{,t}^2}{R^2} + \frac{R_{,tr}^2}{R_{,r}^2}\right]_{LT}(t_1). \tag{18.38}$$

A similar equation follows for the Friedmann models from (17.27) – (17.28):

$$\left[\frac{\partial^2}{\partial t^2}\ln\epsilon\right]_F(t_1) = \left[\frac{1}{2}\kappa\epsilon + \Lambda + 3\frac{R_{,t}^2}{R^2}\right]_F(t_1). \tag{18.39}$$

Subtracting (18.39) from (18.38) and using the assumptions, we find:

$$\frac{\partial^2}{\partial t^2}\left(\ln \epsilon_{LT} - \ln \epsilon_F\right) = \frac{1}{2}\kappa\left(\epsilon_{LT} - \epsilon_F\right). \tag{18.40}$$

Together with (18.36) this means: wherever $\rho_{LT} - \rho_F \neq 0$, the difference in densities will be increasing in time and the sign of $\rho_{LT} - \rho_F$ will be preserved.[4] This means that under the assumptions made by Tolman an L–T model with initial condensations *or voids* will

---

[4] This conclusion does not say that a condensation or a void will remain, respectively, a condensation or a void for ever. This is so under the assumption of equal initial velocity distributions. In general, an initial condensation can evolve into a void or vice versa (Mustapha and Hellaby, 2001).

be evolving away from the background Friedmann model. This was as close as possible at that time to predicting that the Friedmann models are unstable against the formation of condensations and voids. This is how Tolman himself formulated the prediction:

'... at those values of r where the density in the distorted model is different from that in the Friedmann model, there is at least an initial tendency for the differences to be emphasised ... in cases where condensation is taking place ... the discrepancies will continue until we reach a singular state involving infinite density or reach a breakdown in the simplified equations.'

Sen (1934) carried out a complementary study: he assumed the initial density to be unperturbed, with the velocity distribution being non-Friedmannian at the initial time. By a similar method to Tolman's, he concluded explicitly that 'the models are unstable for initial rarefaction'.

The book by Krasiński (1997) contains a complete overview of studies of void formation done on the basis of the L–T model until 1994. They were mostly motivated by astronomical observations and used various shapes of the arbitrary functions to investigate observable effects. The most complete study (Maeda, Sasaki and Sato, 1983; Maeda and Sato, 1983a ; Sato and Maeda, 1983), extensively summarised by Sato (1984), considered the long-term evolution of voids.

A new approach, presented below, was proposed by Krasiński and Hellaby (2002, 2004a).

18.6 Formation of other structures in the Universe

Formation of galaxies (then called 'nebulae') was first considered by Lemaître (1933b). He showed that the initial mass distribution may be set up so that a region of comoving size $r = r_0$ around the centre will recollapse while the region $r > r_0$ will keep expanding forever. In such a configuration, the curvature of the space is positive everywhere and the expansion of the outer region is caused by the cosmological repulsion induced by Λ.

Bonnor's (1956) model was a Friedmann dust tube around the centre of symmetry, surrounded by a L–T transition zone, and that in turn surrounded by another Friedmann dust region so that, at any fixed t, the density in the outer region is different from that in the inner region. The boundaries of the Friedmann regions were assumed to be comoving. If both Friedmann regions have positive spatial curvature and the density in the inner region is higher than that in the outer region, then the inner region will start to recollapse earlier than the background and will form a condensation. Bonnor assumed that the condensation has the mass of a typical galaxy, that is, it contains $N \cong 3 \times 10^{67}$ nucleons, and that it formed at $t_i \cong 1000$ years after the BB. Then the following problem arose: if such a condensation formed as a statistical fluctuation in a homogeneous background, then the initial density contrast is $\delta\epsilon/\epsilon = |\epsilon_c - \epsilon_b|/\epsilon_b \cong N^{-1/2} \cong 10^{-34}$, where ϵ_c is the density in the condensation and ϵ_b is the background density. However, in order to develop into a galaxy of typical density, the initial perturbation at t_i would have to be of the order $\delta\epsilon/\epsilon \cong 10^{-5}$. On the other hand, if a perturbation of the order of 10^{-5} is to arise as a statistical fluctuation, then it can involve only 10^{10} particles.

If the curvature is negative in the outer region and positive in the inner region, then the initial perturbation has to be about 10 times larger than in the preceding case. If both

Friedmann regions have negative curvature, a galaxy cannot form at all. The cosmological constant does not help in a model that begins from a BB. Hence, two possibilities are left: either $\Lambda \neq 0$ and the Universe begins as an instability in the Einstein Universe in the asymptotic past (then there is an arbitrary amount of time available for the statistical fluctuations to grow) or there exists a mechanism for producing large perturbations.

The current thinking is that initial fluctuations of density were generated by quantum fluctuations of the scalar field that drives the inflation and indeed are of the order 10^{-5} (Padmanabhan, 1996). However, one of the results of Krasiński and Hellaby (2004a) is that density fluctuations alone cannot be responsible for the generation of structures: the velocity distribution at the initial time must be taken into account. In particular, an initial condensation can evolve into a void (Mustapha and Hellaby, 2001).

The book by Krasiński (1997) contains an overview of other discussions of structure formation in the L–T model. We shall present here a more recent approach (Krasiński and Hellaby, 2002; 2004a; 2006). It was later implemented also in the quasi-spherical Szekeres geometry by Walters and Hellaby (2012), see Sec. 20.6.

18.6.1 Density to density evolution

The mass M will be used as the radial coordinate. We specify the density distributions at the instants $t = t_i$, $i = 1, 2$ and assume $t_2 > t_1$:

$$\rho_i(M) = \rho(t_i, M) \equiv \epsilon(t_i, M)/c^2. \tag{18.41}$$

Then we calculate $R(t_i, M)$ from (18.23). We also assume $R_2 > R_1$, i.e. that matter expanded between t_1 and t_2. This assumption is dictated by the intended application of the results (structure formation in the Universe), but a similar investigation could be done for collapsing matter. The cases $E > 0$ and $E < 0$ have to be considered separately.[5]

For $E > 0$ we write out the evolution equations (18.21) at t_1 and t_2:

$$R_i(M) = R(t_i, M) = \frac{M}{2E}(\cosh \eta_i - 1),$$

$$\sinh \eta_i - \eta_i = \frac{(2E)^{3/2}}{M}[t_i - t_B(M)]. \tag{18.42}$$

Solving for t_B at t_1 we obtain

$$t_B = t_1 - \frac{M}{(2E)^{3/2}}\left[\sqrt{(1 + 2ER_1/M)^2 - 1} - \operatorname{arcosh}(1 + 2ER_1/M)\right]. \tag{18.43}$$

We substitute this in (18.42) at t_2 and obtain

$$\sqrt{(1 + 2ER_2/M)^2 - 1} - \operatorname{arcosh}(1 + 2ER_2/M)$$

$$-\sqrt{(1 + 2ER_1/M)^2 - 1} + \operatorname{arcosh}(1 + 2ER_1/M) = \frac{(2E)^{3/2}}{M}(t_2 - t_1). \tag{18.44}$$

This equation defines $E(M)$, and then (18.43) defines $t_B(M)$. These two functions then

---

[5] The case $E = 0$ is exceptional because then just one of the distributions $\rho_i(t, M)$ uniquely determines an L–T model. Choosing two $\rho_i(t, M)$, it is practically impossible to hit upon such a pair that is connected by the $E = 0$ evolution. The $E = 0$ model will appear only as an intermediate limiting case.

define the L–T evolution from $\rho(t_1, M)$ to $\rho(t_2, M)$. In fact, this is already a solution of our problem, but we have to answer the following question: does (18.44) have any solution, and if so, is the solution unique? For ease of calculation, let us denote:

$$x \stackrel{\text{def}}{=} 2E/M^{2/3}, \qquad a_i \stackrel{\text{def}}{=} R_i/M^{1/3}, \quad i = 1, 2;$$

$$\begin{aligned}
\psi_H(x) \stackrel{\text{def}}{=} \; & \sqrt{(1 + a_2 x)^2 - 1} - \text{arcosh}(1 + a_2 x) \\
& - \sqrt{(1 + a_1 x)^2 - 1} + \text{arcosh}(1 + a_1 x) - (t_2 - t_1)x^{3/2}.
\end{aligned} \tag{18.45}$$

The problem is now: for what values of $a_2 > a_1$ and $t_2 > t_1$ does the equation $\psi_H(x) = 0$ have a solution $x \neq 0$?[6] From the properties of the function $\psi_H(x)$ (Krasiński and Hellaby, 2002, and Exercise 5) it follows that (18.45) has a solution if and only if

$$t_2 - t_1 < \frac{\sqrt{2}}{3}\left(a_2{}^{3/2} - a_1{}^{3/2}\right), \tag{18.46}$$

and then the solution is unique. This inequality says that the expansion between t_1 and t_2 must have been faster than in the $E = 0$ model.

Equation (18.46) guarantees only the existence of a solution for a given M. Some initial conditions may lead to shell crossings, and these are not excluded by (18.46) – one must check for their presence separately. But the criteria for the occurrence of shell crossings are known, see Section 18.10, so faulty initial conditions can be recognised and eliminated.

For $E < 0$, the inverse of the function cos in the range $[0, \pi]$ is different from that in the range $[\pi, 2\pi]$. Consequently, the L–T model evolving between two given states must be considered separately for the case when the final state is still expanding ($\eta \in [0, \pi]$ in (18.19)) and for the case when the final state is already recollapsing ($\eta \in [\pi, 2\pi]$).

For the still-expanding final state, the variables and the function whose zero has to be found are the a_i from (18.45) and

$$x \stackrel{\text{def}}{=} - 2E/M^{2/3},$$

$$\begin{aligned}
\psi_X(x) \stackrel{\text{def}}{=} \; & \arccos(1 - a_2 x) - \sqrt{1 - (1 - a_2 x)^2} \\
& - \arccos(1 - a_1 x) + \sqrt{1 - (1 - a_1 x)^2} - (t_2 - t_1)x^{3/2}.
\end{aligned} \tag{18.47}$$

The reasoning is analogous to that for (18.45), but this time the arguments of arccos must have absolute values not greater than 1. This implies $x \leq 2/a_i$ for both i, so, since $a_2 > a_1$,

$$0 \leq x \leq 2/a_2. \tag{18.48}$$

The two square roots in (18.47) will then also exist. Equation (18.48) is equivalent to the requirement that $(R_{,t})^2$ (in (18.14) with $\Lambda = 0$) is nonnegative at both t_1 and t_2.

Again, elementary reasoning (Krasiński and Hellaby, 2002) shows that $\psi_X(x) = 0$ has a nonzero solution if and only if

$$\frac{\sqrt{2}}{3}\left(a_2{}^{3/2} - a_1{}^{3/2}\right) < t_2 - t_1 \leq$$
$$(a_2/2)^{3/2}\left[\pi - \arccos(1 - 2a_1/a_2) + 2\sqrt{a_1/a_2 - (a_1/a_2)^2}\right]. \tag{18.49}$$

---

[6] $x = 0$ implies $E = 0$. In fact, we are looking for a zero of the function $\psi_H(x)/x^{3/2}$.

With (18.49) fulfilled, the solution is unique. The first inequality means that the model must have expanded between t_1 and t_2 slower than the $E = 0$ model; the second one means that the final state is still earlier than the instant of maximal expansion. The condition $E \geq -1/2$ does not follow from (18.49) and has to be verified separately.

For the recollapsing final state, the variables are defined as in (18.49), but the function whose zeros are sought is different:

$$\psi_C(x) \overset{\text{def}}{=} \pi + \arccos(-1 + a_2 x) + \sqrt{1 - (1 - a_2 x)^2}$$
$$- \arccos(1 - a_1 x) + \sqrt{1 - (1 - a_1 x)^2} - (t_2 - t_1) x^{3/2}. \tag{18.50}$$

The solution of $\psi_C(x) = 0$ exists if and only if

$$t_2 - t_1 \geq (a_2/2)^{3/2} \left[\pi - \arccos(1 - 2a_1/a_2) + 2\sqrt{a_1/a_2 - (a_1/a_2)^2} \right], \tag{18.51}$$

and then it is unique (Krasiński and Hellaby, 2002).

The results obtained above can be summarised in the following

Theorem 18.1 *Given any two instants t_1 and $t_2 > t_1$, and any two spherically symmetric density profiles $\rho_1(M)$ and $\rho_2(M)$ defined over the same range of M with $R_1(M) < R_2(M)$, an L–T model can be found that evolves from ρ_1 to ρ_2 in time $(t_2 - t_1)$. The inequalities (18.46), (18.49) and (18.51) will tell which L–T evolution applies at each M. The shell crossings and $E < -1/2$ are not excluded, and must be separately checked for.*

18.6.2 Velocity to density evolution

A measure of expansion velocity more convenient than $R_{,t}(t, M)$ is $R_{,t}(t, M)/M^{1/3}$. It is independent of M in the Friedmann limit, and so is also a measure of inhomogeneity of the spacetime. Suppose that the initial state of the Universe is specified by

$$b_1(M) = R_{,t}(t_1, M)/M^{1/3}, \tag{18.52}$$

while the final state is specified, as before, by a density distribution $\rho(t_2, M)$. Again, the two signs of E must be considered separately.

For $E > 0$, with the variables defined as in (18.45) and (18.52), the equation whose solutions are sought is $\Phi_H(x) = 0$, where

$$\Phi_H(x) \overset{\text{def}}{=} \sqrt{(1 + a_2 x)^2 - 1} - \sqrt{\left(\frac{b_1^2 + x}{b_1^2 - x} \right)^2 - 1}$$
$$- \operatorname{arcosh}(1 + a_2 x) + \operatorname{arcosh}\left(\frac{b_1^2 + x}{b_1^2 - x} \right) - x^{3/2}(t_2 - t_1). \tag{18.53}$$

Now the necessary and sufficient condition for the existence of solutions of (18.53) consists of two inequalities (Krasiński and Hellaby, 2004a):

$$t_2 - t_1 < \frac{\sqrt{2}}{3} a_2^{3/2} - \frac{4}{3 b_1^3}, \qquad 2/a_2 < b_1^2. \tag{18.54}$$

The second one is a necessary condition for the existence of a $t_2 > t_1$ obeying the first, and is equivalent to $R_2 > R_1$. With both inequalities fulfilled, the solution of $\Phi_H(x) = 0$ is unique. The first of (18.54) is equivalent to (18.46).

For $E < 0$, as before, the cases when the final state is still expanding and recollapsing have to be considered separately. When the final state is still expanding, the function whose zeros are to be found is

$$\Phi_X(x) \overset{\text{def}}{=} \sqrt{1 - \left(\frac{b_1{}^2 - x}{b_1{}^2 + x}\right)^2} - \sqrt{1 - (1 - a_2 x)^2}$$

$$+ \arccos\left(1 - a_2 x\right) - \arccos\left(\frac{b_1{}^2 - x}{b_1{}^2 + x}\right) - x^{3/2}\left(t_2 - t_1\right), \tag{18.55}$$

and the equation $\Phi_X(x) = 0$ has a solution if and only if

$$\frac{\sqrt{2}}{3} a_2{}^{3/2} - \frac{4}{3 b_1{}^3} < t_2 - t_1 \leq$$

$$(a_2/2)^{3/2}\left[\pi + \frac{b_1\sqrt{2a_2}}{a_2 b_1{}^2/2 + 1} - \arccos\frac{a_2 b_1{}^2/2 - 1}{a_2 b_1{}^2/2 + 1}\right]. \tag{18.56}$$

This is equivalent to (18.49). With the inequalities fulfilled, the solution of the corresponding equation is unique (Krasiński and Hellaby, 2004a).

When the final state is recollapsing, the appropriate function is

$$\Phi_C(x) \overset{\text{def}}{=} \sqrt{1 - \left(\frac{b_1{}^2 - x}{b_1{}^2 + x}\right)^2} + \sqrt{1 - (1 - a_2 x)^2} + \pi$$

$$- \arccos\left(\frac{b_1{}^2 - x}{b_1{}^2 + x}\right) + \arccos\left(-1 + a_2 x\right) - x^{3/2}\left(t_2 - t_1\right), \tag{18.57}$$

and the condition for the existence of the appropriate evolution is

$$t_2 - t_1 \geq (a_2/2)^{3/2}\left[\pi + \frac{b_1\sqrt{2a_2}}{a_2 b_1{}^2/2 + 1} - \arccos\left(\frac{a_2 b_1{}^2/2 - 1}{a_2 b_1{}^2/2 + 1}\right)\right]. \tag{18.58}$$

All derivations can be found in Krasiński and Hellaby (2004a).

Using the methods of this section, examples of (numerical) evolution of various initial configurations, defined at the epoch of last scattering, to models of present galaxy clusters, voids and galaxies with central black holes were found (Krasiński and Hellaby, 2002; 2004; 2006; Bolejko et al., 2010). The diameters of the initial regions of perturbed velocity or density were smaller than the angular resolution of the observed anisotropies of the CMB temperature. Here are some interesting conclusions:

- Two models with the same initial density profile at one time can have totally different evolutions – because of different initial velocity profiles (Krasiński and Hellaby, 2004a).
- An explicit numerical example (not just a proof of existence) was provided of a void evolving into a condensation (Krasiński and Hellaby, 2004a).
- The differences of the BB time at different locations needed to produce a model galaxy cluster at present are of the order of 300 years (Krasiński and Hellaby, 2006). This is to be compared with the differences implied by the measured angular anisotropies of the CMB

temperature, $\Delta T/T = 5 \times 10^{-6}$ (Smoot et al., 1992). Here is the calculation based on the ΛCDM metric (17.58). As stated below (17.89), the CMB temperature obeys $T(1+z) =$ constant, so $\Delta T/T = -\Delta z/(1+z)$. Then, from (17.13), $R(t_e)(1+z) = R(t_o) =$ constant, so $-\Delta z/(1+z) = \Delta R/R$. Now we find from (17.58):

$$\frac{\Delta R}{R} = \sqrt{\frac{-\Lambda}{3}} \coth \left[\frac{\sqrt{-3\Lambda}}{2} (t - t_{B\Lambda}) \right] \Delta t_B. \qquad (18.59)$$

Taking the numerical data given below Eq. (17.59), $-\Lambda = 10^{-46}$ km^{-2}, $t - t_{B\Lambda} = c \times 13.67 \times 10^9$ y, $c = 3 \times 10^5 \times 3.1536 \times 10^7$ km/y we obtain

$$\Delta t_B/c = 7.39 \times 10^4 \text{ y}. \qquad (18.60)$$

This easily accommodates the 300 years mentioned above.

- Quotation from Bolejko et al. (2010):

'For the evolution from a homogeneous initial density to a galaxy cluster ... the velocity amplitude at t_1, $(\Delta b/b)(t_1)|_{\text{max}} = 5.2 \times 10^{-4}$, ... is on the border of the observationally implied range. This means that a pure velocity perturbation can very nearly produce a galaxy cluster. For the evolution from a homogeneous initial velocity to a galaxy cluster (same final profile as above) ... the [needed] density amplitude at t_1, $(\Delta \rho/\rho)(t_1)|_{\text{max}} = 12 \times 10^{-3}$, ... differs from the observationally allowed value by 3 orders of magnitude. Hence, velocity perturbations generate structures much more efficiently than density perturbations. ... Thus the initial density distribution does not determine the final structure that will emerge from it; the velocity distribution can obliterate the initial setup.'

In the R–W models the velocity distribution is rigidly connected with the density distribution by the Hubble law and has no separate influence on anything.

The approach presented in this section has not been entirely successful so far. If the distributions of velocity at t_1 and of density at t_2 are assumed, then the density distribution at t_1 is determined, and has to obey the observational constraints. It is sensitive to the shape of the velocity profile at t_1, and the optimal shape has not been identified yet.

18.6.3 Velocity to velocity evolution

The given quantities are now the velocity distributions at t_1 and t_2:

$$b_i(M) = R_{,t}(t_i, M)/M^{1/3}, \qquad i = 1, 2; \qquad (18.61)$$

the definition of $x = \pm 2E/M^{2/3}$ being the same as in the previous cases. We assume that $t_2 > t_1$ and that the model is expanding at t_1, thus $b_1 > 0$. This way of specifying the data is probably not useful for astrophysics, but is included here for completeness.

For $E > 0$, the function whose zeros are to be found is

$$\chi_H(x) \overset{\text{def}}{=} \sqrt{\left(\frac{b_2{}^2 + x}{b_2{}^2 - x}\right)^2 - 1} - \sqrt{\left(\frac{b_1{}^2 + x}{b_1{}^2 - x}\right)^2 - 1}$$

$$- \operatorname{arcosh}\left(\frac{b_2{}^2 + x}{b_2{}^2 - x}\right) + \operatorname{arcosh}\left(\frac{b_1{}^2 + x}{b_1{}^2 - x}\right) - x^{3/2}(t_2 - t_1). \qquad (18.62)$$

The necessary and sufficient condition for the existence of an $E > 0$ evolution between the two states is the set of two inequalities

$$0 < b_2 < b_1, \qquad t_2 - t_1 > \frac{4}{3}\left(\frac{1}{b_2{}^3} - \frac{1}{b_1{}^3}\right). \tag{18.63}$$

The second inequality becomes more intelligible when it is rewritten as

$$b_2{}^3 > \frac{b_1{}^3}{1 + \frac{3}{4}b_1{}^3\,(t_2 - t_1)}, \tag{18.64}$$

which means that the expansion at t_2 is faster than it would be in the $E = 0$ model (for which $b^3 = 4/[3(t - t_B)]$, giving an equality in (18.64)).

For an $E < 0$ evolution with the final state still expanding, the function χ is

$$\chi_X(x) \stackrel{\text{def}}{=} -\sqrt{1 - \left(\frac{b_2{}^2 - x}{b_2{}^2 + x}\right)^2} + \sqrt{1 - \left(\frac{b_1{}^2 - x}{b_1{}^2 + x}\right)^2}$$

$$+ \arccos\left(\frac{b_2{}^2 - x}{b_2{}^2 + x}\right) - \arccos\left(\frac{b_1{}^2 - x}{b_1{}^2 + x}\right) - x^{3/2}\,(t_2 - t_1). \tag{18.65}$$

A solution of $\chi_X(x) = 0$ exists if and only if

$$0 < b_2 < b_1, \qquad t_2 - t_1 < \frac{4}{3}\left(\frac{1}{b_2{}^3} - \frac{1}{b_1{}^3}\right), \tag{18.66}$$

which now implies that the expansion between t_1 and t_2 must have been slower than it would be in an $E = 0$ model.

These two cases together exhaust all possibilities for $b_2 > 0$. When the final state is already recollapsing, we have $b_2 < 0$. Then the evolution exists for any values of $b_1 > 0$ and $b_2 < 0$. The function χ is here

$$\chi_C(x) \stackrel{\text{def}}{=} \sqrt{1 - \left(\frac{b_1{}^2 - x}{b_1{}^2 + x}\right)^2} - \sqrt{1 - \left(\frac{x - b_2{}^2}{x + b_2{}^2}\right)^2} + \pi$$

$$+ \arccos\left(\frac{x - b_2{}^2}{x + b_2{}^2}\right) - \arccos\left(\frac{b_1{}^2 - x}{b_1{}^2 + x}\right) - x^{3/2}\,(t_2 - t_1), \tag{18.67}$$

and with $t_2 > t_1$ the equation $\chi_C(x) = 0$ always has an $x > 0$ solution.

18.7 The influence of cosmic expansion on planetary orbits

The first formally correct study of the problem of expansion of planetary orbits induced by the expansion of the Universe was carried out by Einstein and Straus (1945, 1946).[7] They showed that the Schwarzschild solution can be matched to any Friedmann model.

---

[7] Some papers on this topic were published earlier, but they were based on questionable assumptions; see Krasiński (1997) for an overview of them. To be able to decide whether an orbit expands or not one must first choose a standard ruler that does not change its length. This problem does not exist in the static Schwarzschild spacetime, but it does exist in one where there is expanding matter. The authors of the old papers did not take proper care about this point.

This implies that the planetary orbits are *in this configuration* not influenced by the expansion of the Universe. In the coordinates of (17.3) the Schwarzschild mass m is related to the Friedmann mass integral $\mathcal{M}$ from (17.27) by

$$m = G\mathcal{M}r_b{}^3/c^2 \stackrel{\text{def}}{=} \mu(r_b), \tag{18.68}$$

where r_b is the radius of the Schwarzschild vacuole. The geodesic radius of the vacuole, $R(t)\int_0^{r_b} \left(1/\sqrt{1-kr^2}\right)dr$, expands together with the Universe.

Equation equivalent to (18.68) was derived by Einstein and Straus in a rather outlandish notation. It can be derived from the Lemaître–Novikov representation of the Schwarzschild solution, (14.120) – (14.121) matched to the Friedmann limit of (18.16) and (18.14) (with $\Lambda = 0$). In the Friedmann limit $R = rR_F(t)$, where R_F is the Friedmann scale factor. The coordinates on both sides of the hypersurface $r = r_b$ are the same, so the matching conditions of Sec. 12.17 imply that $R(t,r_b)$ in (14.120) must be the same *as a function of t* as $R(t,r_b) = r_bR_F(t)$ in (18.16). Consequently, $R_{,t}(t,r_b)$ must be the same in both metrics, and then (18.14) implies that $E(r_b)$ is the same on both sides, and $m = M(r_b)$, where $M(r_b)$ is the L–T mass contained within the $r = r_b$ hypersurface of (18.16). Now we take the Friedmann limit (18.22) of (18.14), (18.16) and (18.18), remembering that in this limit ϵ depends only on t. Finally, we find $M(r_b)$ from (18.18) and use (17.27) to eliminate $(\epsilon/c^2)R_F^3$. The result is (18.68). It says that the Schwarzschild mass at the centre of symmetry must be equal to the Friedmann mass removed from within the sphere $r = r_b$.

Einstein and Straus' result was for many years taken as the general implication of relativity. However, (18.68) need not be fulfilled if the Einstein–Straus configuration is taken only at a single moment $t = t_0$ as an initial condition for an L–T model. Then, the results of other papers (Sato, 1984 and papers cited therein, Lake and Pim, 1985) imply that if $m < \mu(r_b)$, then the boundary of the vacuole will expand faster than the Friedmann background, whereas if $m > \mu(r_b)$, then initial conditions may be set up so that the vacuole will start to collapse. This indicates that the Einstein–Straus configuration is unstable against perturbations of the condition (18.68), that is, it is an exceptional situation.

The same problem was studied by a different method by Gautreau (1984). He based his study on an $E = 0$ L–T model represented in the curvature coordinates of (18.221). In these coordinates, R is the curvature radius of the orbits of the symmetry group. These orbits do not participate in the cosmic expansion and therefore R of any single orbit can be used as a standard of length. In Gautreau's configuration, there is a central mass of finite extent embedded in an expanding background that goes up to the surface of the central object. By investigating the equations of timelike geodesics Gautreau showed that in this model circular orbits do not exist. This is in fact a Newtonian phenomenon: in Gautreau's model the smoothed-out cosmic matter density extends throughout the planetary system, and, as a result of cosmic expansion, matter streams out of every sphere of constant R. Hence, each planet moves under the influence of a gravitational force that is decreasing with time, so the orbit must spiral out. Gautreau derived the Newtonian formula for the rate of change of an orbital radius in such a setup, $dR/dt = 8\pi R^4 H\bar{\rho}/(2\mu)$, where R is the orbital radius, H is the Hubble parameter and $\bar{\rho}$ is the mean cosmic density of matter. The effect is thus greater for larger orbits; for Saturn it is $(dR/dt)_S = 6 \times 10^{-18}$ m per year. This is obviously unmeasurable (one proton diameter per 1000 years!). For a star at the edge of the Andromeda galaxy the effect would be $(dR/dt)_{gal} = 1100$ km per year. The effect is thus extremely small but nonzero. In the Einstein–Straus approach, it was exactly zero. As explained above, the model of Einstein and Straus is unstable against the perturbations of (18.68), and hence is less realistic than Gautreau's.

18.8 * The apparent horizons for a central observer in L–T models

From now on we will investigate the L–T models with $\Lambda = 0$.

As defined in Section 16.5, an apparent horizon (AH) is the outer envelope of a region of closed trapped surfaces, while a closed trapped surface S_t is one from which it is impossible to send a diverging bundle of light rays – because both the outward-directed and the inward-directed bundles immediately converge: $k^\mu{}_{;\mu} \leq 0$ at S_t.

Since the L–T model is spherically symmetric, the AH surrounding the centre of symmetry must be spherically symmetric, too. For the most part of this book the term 'apparent horizon' will be synonymous with this case (however, see Sec. 18.13 – the trapped surfaces and horizons that do not contain the centre are radically different). Hence, to determine and investigate it, it suffices to consider families of null geodesics sent orthogonally from a surface $r = $ constant. We must identify the surface at which $k^\mu{}_{;\mu}$ becomes zero for all future-directed radial null geodesics. For this, we shall use the method of Szekeres (1975b). From (18.16), the tangent vectors to radial null curves obey $k^0 - (\varepsilon R_{,r}/\sqrt{1 + 2E}) k^1 = 0$, $k^2 = k^3 = 0$, where $\varepsilon = +1$ for outward-directed and $\varepsilon = -1$ for inward-directed curves. Because of spherical symmetry, these curves must be geodesics. Now consider a bundle of null geodesics originating at a surface $S_{t,r}$ given by $\{t = t_s, r = r_s\}$. At these constant values of t and r, the affine parameter on the null geodesics may be chosen so that

$$k^0 = \frac{R_{,r}}{\sqrt{1 + 2E}}, \qquad k^1 = \varepsilon \qquad \text{on } S_{t,r}. \tag{18.69}$$

The divergence of this field on $S_{t,r}$ is then

$$\theta \stackrel{\text{def}}{=} k^\mu{}_{;\mu} = k^\mu{}_{,\mu} + \left\{ {\mu \atop \rho\nu} \right\} k^\rho k^\nu \tag{18.70}$$

$$= k^0{}_{,t} + k^1{}_{,r} + \frac{R_{,r}}{\sqrt{1 + 2E}} \left(\frac{R_{,tr}}{R_{,r}} + 2\frac{R_{,t}}{R} \right) + \varepsilon \left(-\frac{E_{,r}}{1 + 2E} + \frac{R_{,rr}}{R_{,r}} + 2\frac{R_{,r}}{R} \right).$$

The k^μ is null, geodesic and affinely parametrised, so $k_\mu k^\mu = 0$ and $k^\mu{}_{;\nu} k^\nu = 0$. Taking $\partial/\partial t$ of the first equation and $\mu = 1$ in the second one, we obtain on $S_{t,r}$

$$k^0{}_{,t} - \frac{R_{,tr}}{\sqrt{1 + 2E}} - \varepsilon\frac{R_{,r}}{\sqrt{1 + 2E}} k^1{}_{,t} = 0, \tag{18.71}$$

$$\theta = \frac{R_{,r}}{\sqrt{1 + 2E}} k^1{}_{,t} + \varepsilon k^1{}_{,r} + 2\varepsilon\frac{R_{,tr}}{\sqrt{1 + 2E}} + \frac{R_{,rr}}{R_{,r}} - \frac{E_{,r}}{1 + 2E}. \tag{18.72}$$

We add (18.72) multiplied by ε to (18.71), and thereby eliminate $k^1{}_{,t}$. The resulting equation is used to eliminate $k^0{}_{,t} + k^1{}_{,r}$ from (18.70), and the result is

$$\theta = 2\frac{R_{,r}}{R} \left(\frac{R_{,t}}{\sqrt{1 + 2E}} + \varepsilon \right). \tag{18.73}$$

One solution of $\theta = 0$ is $R_{,r} = 0$, but this is either a shell crossing or a neck. What happens there will be investigated in Section 18.10. The generic solution of (18.73) is

$$\frac{R_{,t}}{\sqrt{1+2E}} = -\varepsilon. \tag{18.74}$$

For outward-directed geodesics $\varepsilon = +1$, so the solution of (18.74) will exist only in collapsing models, in which $R_{,t} < 0$. For inward-directed geodesics $\varepsilon = -1$, and the solution of (18.74) will exist only in expanding models. In each case, $R_{,t}$ has the sign of $-\varepsilon$. Using (18.14), we have in (18.74)

$$\sqrt{2E + \frac{2M}{R} - \frac{1}{3}\Lambda R^2} = \sqrt{1+2E}. \tag{18.75}$$

With $\Lambda = 0$, the solution of this is

$$R = 2M. \tag{18.76}$$

In the Schwarzschild limit, $M =$ constant, the AH becomes identical to the event horizon. In a general L–T spacetime, the $r =$ constant shell obeying (18.76) is just falling into its own Schwarzschild horizon.

Using the result of Exercise 2 in Chapter 16, the formula for the AH can be derived in an equivalent way. A bundle of rays sent outwards from a spherical surface forms a spherical light front. The AH is where the surface area of this front starts decreasing. Since this surface area is proportional to R^2, it suffices to identify the location where $R(t,r)$ starts decreasing as we go along the null geodesics.

Equation (18.14) with $\Lambda = 0$ can be written as

$$\dot{R} = \ell\sqrt{\frac{2M}{R} + 2E}, \tag{18.77}$$

where $\ell = +1$ for expanding models and $\ell = -1$ for collapsing models. The radial null geodesics, from (18.16), are given by

$$\left.\frac{dt}{dr}\right|_n = \frac{jR_{,r}}{\sqrt{1+2E}}, \tag{18.78}$$

where $j = +1$ for outgoing rays and $j = -1$ for incoming rays. We write the solution of (18.78) as $t = t_n(r)$. Along this ray we have

$$R_n \overset{\text{def}}{=} R(t_n, r), \qquad (R_n)_{,r} = \dot{R}\left.\frac{dt}{dr}\right|_n + R_{,r} = \left(\ell j\frac{\sqrt{2M/R + 2E}}{\sqrt{1+2E}} + 1\right)R_{,r}. \tag{18.79}$$

The AH is the hypersurface in spacetime where R stops increasing along the rays:

$$(R_n)_{,r} = 0 \quad \Rightarrow \quad \sqrt{\frac{2M}{R} + 2E} = -\ell j\sqrt{1+2E} \quad \Rightarrow \quad \ell j = -1 \text{ and } R = 2M. \tag{18.80}$$

Inside the future AH, all light rays proceed towards the final singularity because $(R_n)_{,r} < 0$. The existence of such a region was predicted by Bondi (1947) and by Barnes (1970). Note that, in every L–T model that collapses, the dust must enter the future AH before it hits the final singularity at $R = 0$, and in every L–T model that expands the dust remains inside the past AH for a while after leaving the BB.

The maximum of R at the AH and the subsequent decrease of the light front area is a manifestation of refocussing, mentioned in Fig. 16.1, in Sec. 16.10 and in Sec. 17.5. The zero of $(R_n)_{,r}$ at the AH, if not properly handled in numerical computations, may lead to some quantities becoming spuriously infinite (in truth, they are regular 0/0 limits). A misreading of this phenomenon resulted in a few papers published in 2005–2006 whose authors called the maximum of R a 'critical point' and interpreted its presence as a serious defect of the L–T models. To avoid spreading a false message we do not quote those papers, but instead refer our readers to the paper by Krasiński et al. 2010, in which this misunderstanding (and a few other ones) were pointed out and corrected.

We now wish to establish whether the AH is timelike, spacelike or null. For this purpose, we find dt/dr along the AH by differentiating $R = 2M$:

$$\dot{R}dt + R_{,r}\, dr = 2M_{,r}\, dr. \tag{18.81}$$

The result is

$$t'_{AH} = \left.\frac{dt}{dr}\right|_{AH} = \frac{2M_{,r} - R_{,r}}{\dot{R}} = \frac{2M_{,r} - R_{,r}}{\ell\sqrt{2M/R + 2E}}, \tag{18.82}$$

and, since $R = 2M$ on the AH,

$$\left.\frac{dt}{dr}\right|_{AH} = \frac{\ell\,(2M_{,r} - R_{,r})}{\sqrt{1 + 2E}}. \tag{18.83}$$

Where $M_{,r} = 0$, we have $dt/dr|_{AH} = dt/dr|_n$ since $\ell j = -1$, i.e. the AH is null. Note that $M_{,r} = 0$ can be local, then AH is null only in that region.

For hyperbolic regions, with $E \geq 0$, there is either only expansion or only collapse, i.e. only one AH can exist, the future AH$^+$ or the past AH$^-$. Two AHs can be present only in an elliptic $E < 0$ region. At the moment of maximum expansion, where $R_{,t} = 0$, the maximum R is $R_{\max} = -M/E$. In elliptic regions, where $-1 < 2E < 0$, $R_{\max} > 2M$, so the dust always escapes from the past AH before it falls into the future AH. However, at those locations where $2E = -1$, the maximum R equals $2M$, which means that the past AH touches the future AH. Such a location is a **neck**, the nonvacuum analogue of the Kruskal wormhole. We shall come back to it in Sec. 18.10.

To establish whether the AH is timelike, null or spacelike, we compare the slope of the AH$^+$ with the outgoing light ray (or the slope of the AH$^-$ with the incoming light ray):

$$B \stackrel{\text{def}}{=} \left.\frac{dt}{dr}\right|_{AH} \bigg/ \left.\frac{dt}{dr}\right|_n = -\ell j\left(1 - \frac{2M_{,r}}{R_{,r}}\right) = 1 - \frac{2M_{,r}}{R_{,r}} \tag{18.84}$$

because $\ell j = -1$ from(18.80). Now, since the conditions for no shell crossings (Section 18.10) require $M_{,r} \geq 0$ where $R_{,r} > 0$ and vice versa, we have

$$
\begin{array}{llll}
B = B_{max} = 1, & \to \text{AH}^+ \text{ outgoing null} & (M_{,r} = 0,\ \rho = 0), \\
1 > B > -1, & \to \text{AH}^+ \text{ spacelike} & (\text{for most } M_{,r}), \\
B = -1, & \to \text{AH}^+ \text{ ingoing null} & (M_{,r}/R_{,r} = 1), \\
-1 > B > -\infty, & \to \text{AH}^+ \text{ ingoing timelike} & (M_{,r}/R_{,r} > 1),
\end{array}
\tag{18.85}
$$

so an outgoing timelike AH^+ is not possible. Thus outgoing light rays that reach the AH^+ fall inside the AH^+, except where $M_{,r} = 0$, in which case they move along it. Thus, in a typical situation when both $M_{,r}$, $R_{,r} > 0$, the AH^+ is timelike if

$$\ell \left. \frac{dt}{dr} \right|_{AH} > \frac{R_{,r}}{\sqrt{1 + 2E}}. \tag{18.86}$$

The argument is similar for light rays at the AH^-, except that 'ingoing' should be swopped with 'outgoing'. If ingoing light rays reach the AH^-, they pass out of it or run along it.

The AHs in the Friedmann models with $\Lambda = 0$ are always timelike (see Exercise 8).

In elliptic regions two AHs are present in the same spacetime. We first consider the expansion phase of an elliptic model, where $0 \leq \eta \leq \pi$ and $E < 0$, so $\ell = +1$ and only the AH^- is present. Since $R = 2M$ on an AH, we have from (18.19):

$$\cos \eta_{AH} = 1 + 4E, \tag{18.87}$$

and thus, along a given worldline, the time t of passing through the AH^-, counted from the Bang time t_B, can be calculated from (18.19) with $R = 2M$ to be

$$t_{AH^-} - t_B = M \frac{\arccos(1 + 4E) - 2\sqrt{-2E(1 + 2E)}}{(-2E)^{3/2}}. \tag{18.88}$$

The function $F = (t_{AH^-} - t_B)/M$ of the argument $f = 2E$ has the following properties

$$F(-1) = \pi, \qquad F(0) = 4/3, \qquad dF/df < 0 \quad \text{for} \quad -1 < f < 0, \tag{18.89}$$

i.e. it is decreasing. These properties mean that the AH^- does not touch the BB anywhere except possibly at $M = 0$,[8] this is true also at the locus of $E = 0$. Along all worldlines with $E > -1/2$ the dust particles emerge from the AH^- a finite time after the BB (which occurs at $\eta = 0$) and a finite time before maximum expansion (which occurs at $\eta = \pi$).

Although (18.87) shows that wordlines with larger $|E|$ exit the AH^- at a later stage of evolution (i.e. with larger η_{AH}), this need not correspond to a later time t, or even to a greater $(t - t_B)$. It is not at all necessary that E is a monotonic function of r in an elliptic region; in general it can increase and decrease again any number of times.

A shell of $E = 0$ worldlines occurs at the boundary between elliptic and hyperbolic regions, where $E \to 0$, but $E_{,r} \neq 0$ and $M > 0$. From (18.87) and (18.88) with $\ell = +1$

$$\eta \xrightarrow[E \to 0]{} 0, \qquad t_{AH^-} - t_B \xrightarrow[E \to 0]{} \frac{4M}{3}, \qquad \frac{dt_{AH^-}}{dM} \xrightarrow[E \to 0]{} \frac{4}{3} - \frac{4M}{5} \frac{dE}{dM} + \frac{dt_B}{dM}, \tag{18.90}$$

so the AH^- never touches the bang here, despite η being zero.

The time of the final singularity, where $\eta = 2\pi$, is found from (18.19) to be

$$t_C(r) = t_B(r) + \frac{2\pi M}{(-2E)^{3/2}}, \tag{18.91}$$

---

[8] But whether the coincidence of the AH^- and the BB at $M = 0$ is real or only a coordinate effect depends on the shapes of the functions $E(r)$ and $t_B(r)$. A null or timelike segment of the Big Bang/Big Crunch set can stick out at the centre, and it is invisible in the comoving coordinates because they, too, have a singularity there that squeezes this segment into a point. A signature of such a singularity is a family of different light cones with vertices at the same coordinate point. See Section 18.16.

so it diverges wherever $E \to 0$ at $M \neq 0$. This shows that either the bang time or the crunch time recedes to infinity, or both do. This third possibility will be illustrated in the next section. At these limits (18.90) does not determine the slope of the AH$^-$.

In the collapse phase, keeping t_B as our arbitrary function, we have $\pi \leq \eta \leq 2\pi$. Equation (18.87) still applies, but instead of (18.88) we now obtain

$$t_{\text{AH+}} - t_B = M \, \frac{\pi + \arccos(-1 - 4E) + 2\sqrt{-2E(1 + 2E)}}{(-2E)^{3/2}}. \tag{18.92}$$

The corresponding results in an expanding $E \equiv 0$, L–T model follow from (18.20) with $R = 2M$. Then, (18.90) is replaced by

$$\frac{\mathrm{d}t_{\text{AH}-}}{\mathrm{d}M} = \frac{4}{3} + \frac{\mathrm{d}t_B}{\mathrm{d}M}, \tag{18.93}$$

which is the $E \to 0$ limit of (18.90). The AH$^-$ may still exhibit all possible behaviours of (18.85) with 'outgoing' and 'ingoing' interchanged. In a collapsing parabolic model, time-reversed results apply.

In expanding hyperbolic regions, using the same methods as for expanding elliptic regions, we find that the behaviour of the AH is qualitatively the same. There is of course only one AH, no maximum expansion, and loci where $E = -1/2$ are not possible, but the results for origins and for the parabolic limit both carry over. Collapsing hyperbolic regions are essentially like collapsing elliptic regions.

18.9 * Black holes in the evolving Universe

The process of forming a black hole was first described by Oppenheimer and Snyder (1939) (but the term 'black hole' was coined more than 20 years later). They discussed the collapse of a Friedmann dust cloud matched to the Schwarzschild solution. They found that the collapse to the Schwarzschild horizon $r = 2m$ takes a finite proper time for each infalling particle, but an infinite time for a static distant observer in the Schwarzschild region who would see the star gradually redden. For a comoving observer on the surface of an object of the mass of the Sun, but rarefied so that the initial density equals that of water, the time needed to reach the horizon would be of the order of one day. As the horizon is approached, light can escape outwards within a cone around the radial direction that becomes progressively narrower and closes completely at the horizon.

The theory of black holes developed from the 1960s onwards is based on stationary vacuum solutions, that is, it describes black holes that have always existed and are observed from afar. The pioneering idea in the Oppenheimer–Snyder paper was to study the black hole in the process of its formation. The L–T model allows for an even more sophisticated approach (see below).

Bondi (1947) observed that if matter in the L–T model is collapsing with a great velocity so that $[1 + 2E(r)]^{1/2} + \partial R/\partial t < 0$, then along a light ray emitted in the outward direction with the tangent vector k^μ the quantity $k^\mu R_{,\mu}$ is negative, that is, the ray is forced to move inwards. A necessary condition for this is $R < 2M$, i.e. a sufficiently large mass in a region of a given radius. With hindsight, this was a prediction that black holes would form under certain conditions.

Several properties of apparent horizons and black holes in a L–T model can be explained using the simple example discussed below. Its parameter values are unrealistic, having been chosen in such a way that all the figures are easily readable.

We take an $E < 0$ L–T model whose Big Bang (BB) function is

$$t_B(M) = -bM^2 + t_{B0}, \qquad (18.94)$$

and whose Big Crunch (BC) function is

$$t_C(M) = aM^3 + T_0 + t_{B0}, \qquad (18.95)$$

where T_0 is the lifetime of the central worldline $(M = 0)$. The values of the parameters used in the figures are $a = 2 \times 10^4$, $b = 200$, $t_{B0} = 5$ and $T_0 = 0.05$. They were chosen such as to make the figures readable and illustrative, and are unrelated to any astrophysical quantities. Since $t = t_C$ at $\eta = 2\pi$, we find from (18.19):

$$E(M) = -\frac{1}{2}\left(\frac{2\pi M}{t_C - t_B}\right)^{2/3} = -\left(\frac{\pi^2}{2}\right)^{1/3}\frac{M^{2/3}}{(aM^3 + bM^2 + T_0)^{2/3}}. \qquad (18.96)$$

As $M \to \infty$, we have $t_B \to -\infty$, $t_C \to +\infty$ and $E \to 0$. Hence, the space contains infinite mass and has infinite volume. Unlike in the Friedmann models, positive space curvature does not imply finite volume; this has been known since long ago (Bonnor, 1985; Hellaby and Lake, 1985).

The main features of this model are shown in Fig. 18.1. Note that the AH$^+$ first appears at a finite distance from the centre, where $t_{AH+}(M)$ has its minimum, at $t = t_{hs} < t_C(0)$.

At all times after the crunch first forms, $t > t_C(0)$, the mass M_s already swallowed up by the singularity is necessarily smaller than the mass M_{bh} that had disappeared into the AH$^+$. The mass M_s cannot even be estimated by astronomical methods. The situation is reversed in time for the BB singularity and the AH$^-$.

The 3-dimensional diagram in Fig. 18.2 shows the value of R at each t and each M. Figure 18.3 shows the 'topographic map' of the surface from Fig. 18.2. It contains contours of constant R (the thinner curves) inscribed into Fig. 18.1. It also shows several outgoing radial null geodesics. Each geodesic has a vertical tangent at the centre. This is a consequence of using M as the radial coordinate. Since $dt/dM = (\partial R/\partial M)/\sqrt{1 + 2E}$ on each geodesic and $R \propto M^{1/3}$ close to the centre, $dt/dM \propto M^{-2/3}$ and $dt/dM \to \infty$ as $M \to 0$. Each geodesic proceeds to higher values of R before it meets the apparent horizon AH$^+$, where it is tangent to an $R =$ constant contour, and then proceeds towards smaller R values. The future event horizon consists of those radial null geodesics that approach the AH$^+$ asymptotically. In Fig. 18.2, it lies between geodesics #5 and #6 (counted from the lower right); we shall discuss its location in more detail below.

Geodesic #5 emanates from the centre $M = 0$, where the BB function has a maximum, its tangent is horizontal there. Geodesics to the lower right of this one all begin with a vertical tangent. All the geodesics meet the BC with their tangents being vertical.

By the time the crunch has formed at $t = t_C(0)$, the AH$^+$ already exists (see Figs. 18.1 and 18.3). The shells of constant M first go through the AH, and then hit the singularity at $t = t_C(M)$. At $t = t_2 =$ now the singularity has accumulated the mass M_s, while the

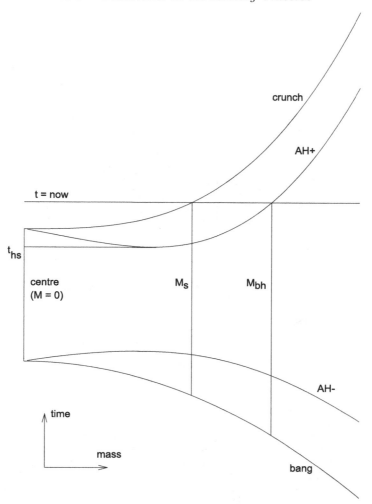

Fig. 18.1 Evolution leading to a black hole in the $E < 0$ L–T model of Eqs. (18.94) − (18.95). The current state is assumed at the instant $t = t_2 =$ now. Worldlines of dust particles are vertical straight lines, each has a constant mass-coordinate. Intersections of the line $t = t_2$ with the lines representing the Big Crunch and the future apparent horizon determine the masses M_s and M_{bh}, respectively.

mass hidden inside the AH at the same time is $M_{bh} > M_s$. Both of them grow with time. From the definitions of M_s and M_{bh} it follows that

$$t_2 = t_C(M_s) = t_{AH+}(M_{bh}). \qquad (18.97)$$

Even though the model has a rather simple geometry, locating the event horizon (EH) is quite a complicated task that requires complete knowledge of the whole spacetime, including the null infinity. Hence, in the real Universe, where our knowledge (mostly

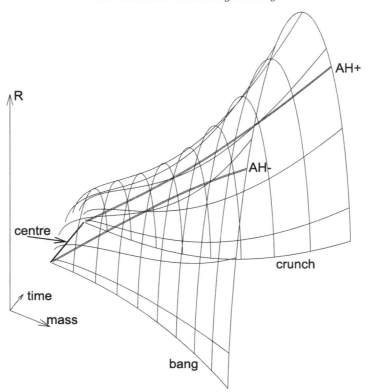

Fig. 18.2 A 3-dimensional graph of the black hole formation process from Fig. 18.1: the areal radius as a function of M and t. Each shell of constant mass evolves in a plane given by $M = $ constant. It starts at $R = 0$, then escapes from the past apparent horizon AH$^-$, then reaches maximum R, then falls into the future apparent horizon AH$^+$, and finally hits the Big Crunch. Note that the surface intersects the $R = 0$ plane perpendicularly all along the $R = 0$ contour. The apparent horizons are intersections of the $R(M, t)$ surface with the plane $R = 2M$. (In the $k > 0$ Friedmann model, the bang and the crunch would be parallel straight lines, and all the $R(t)|_{M=\text{constant}}$ curves would be identical.)

incomplete and imprecise) is limited to a relatively small neighbourhood of our past light cone and our past worldline, the EH simply cannot be located by astronomical observations.

The future event horizon (EH$^+$) is formed by those null geodesics that fall into the AH$^+$ 'as late as possible', i.e. approach it asymptotically. Hence, in order to locate the EH, we must issue null geodesics backwards in time from the 'future endpoint' of the AH$^+$. This cannot be done in the (M, t) coordinates used so far because the spacetime and the AH$^+$ are infinite. Hence, we must first compactify the spacetime. The most convenient compactification for considering null geodesics at null infinity is, theoretically, the Penrose (1964) transform that spreads the null infinities into finite sets at a finite distance. However, in order to find a Penrose transformation, one must first choose null coordinates, and in the L–T model this is an impossible task (Stoeger, Ellis and Nel, 1992; Hellaby, 1996a). Hence, we will use a simpler compactification that will squeeze the null infinities into single points in the 2-dimensional (time–radius) spacetime diagram. It is

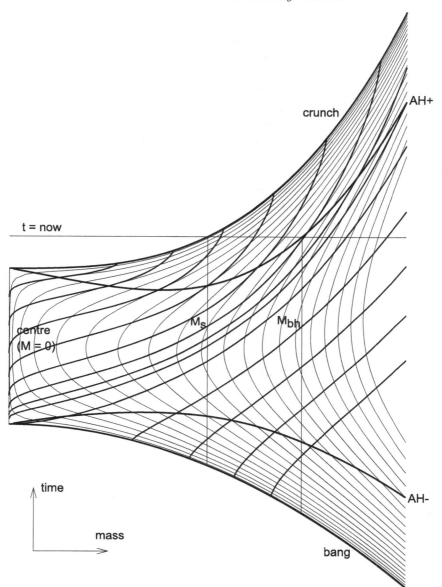

Fig. 18.3 Contours of constant R (the thinner lines) and outgoing radial null geodesics inscribed into the spacetime diagram of Fig. 18.1. The R-values on consecutive contours differ by always the same amount.

provided by the transformation

$$M = \tan(\mu), \qquad t = \tan(\tau). \qquad (18.98)$$

In these coordinates, the $\mathbb{R}^1 \times \mathbb{R}^1_+$ space of Figs. 18.1–18.3 becomes the finite $[-\pi/2, \pi/2] \times [0, \pi/2]$ rectangle; see Fig. 18.4. The upper curve in Fig. 18.4 is the BC singularity; the

AH$^+$ runs so close to it that it seems to coincide with it.[9] The horizontal line is the $\tau =$ now line. The lower curve is the BB and the AH$^-$, again running one on top of the other. The point on the τ-axis where the three lines meet is the image of the $M = 0$ line of Fig. 18.1, squeezed here into a point because of the scale of this figure.

The theoretical method to locate the future EH would now be to run a radial null geodesic backwards in time from the point $(\mu, \tau) = (\pi/2, \pi/2)$, i.e. from the image of the future end of the AH$^+$. However, for the most part, the AH$^+$ runs so close to the BC singularity, and the geodesics intersecting the AH$^+$ are so nearly tangent to the AH$^+$, that numerical instabilities crash any such geodesic into the singularity instantly. This happens all the way down to $\mu = 1.1$ at double precision. The calculation succeeded, with double precision, only at $\mu = 1.0$, and a null geodesic could be traced from there to the centre at $\mu = 0$. At the scale of Fig. 18.4, this whole geodesic seems to coincide with the BC and the AH$^+$. However, it is well visible if one closes in on the image of the area shown in Fig. 18.1; the closeup is shown in the inset.

The EH is transformed back to the (M, t) coordinates and written into the frame of Figs. 18.1–18.3 in Fig. 18.5. As stated earlier, the EH is located between geodesics #5 and #6 from the lower right in Fig. 18.3. By accident (caused by our choice of numerical values in this example), the EH hits the centre very close to the central point of the BB, but does not coincide with it.

This whole construction should make it evident that there is no chance to locate the EH by astronomical observations, even approximately. It only makes sense, in the observational context, to speak about an upper limit on the mass inside the apparent horizon.

18.10 * Shell crossings and necks/wormholes

The mass density in the L–T model becomes infinite where $R_{,r} = 0 \neq M_{,r}$. This singularity is called **shell crossing** (SC) because at those locations the radial distance between two adjacent shells that have different values of r becomes zero. If $R_{,r}$ changes sign there, then the mass density on the other side of the shell crossing becomes negative.

The tetrad components of the Riemann tensor corresponding to the metric (18.16) with (18.14), in the tetrad $e^0 = \mathrm{d}t$, $e^1 = \left(R_{,r}/\sqrt{1 + 2E}\right)\mathrm{d}r$, $e^2 = R\mathrm{d}\vartheta$, $e^3 = R\sin\vartheta\mathrm{d}\varphi$, are:

$$R_{0101} = \frac{2M}{R^3} - \frac{M_{,r}}{R^2 R_{,r}}, \qquad R_{0202} = R_{0303} = -\frac{M}{R^3} = \frac{1}{2}R_{2323},$$

$$R_{1212} = R_{1313} = \frac{M}{R^3} - \frac{M_{,r}}{R^2 R_{,r}}. \tag{18.99}$$

Thus, a shell crossing is a curvature singularity (the R_{ijkl} are scalars and some become infinite at a SC). It is considered less 'dangerous' than the Big Bang for two reasons:

1. In real astrophysical objects pressure gradients are present, and these should be able to prevent the occurrence of SCs. The L–T model is not general enough to describe such

---

[9] From (18.91) – (18.92), the time difference between the BC and the AH$^+$ goes to infinity when $M \to \infty$. However, the *ratio* of this time difference to the crunch time goes to zero, which explains why the two curves in Fig. 18.4 meet at the image of the infinity. The same is true for the Big Bang and the AH$^-$.

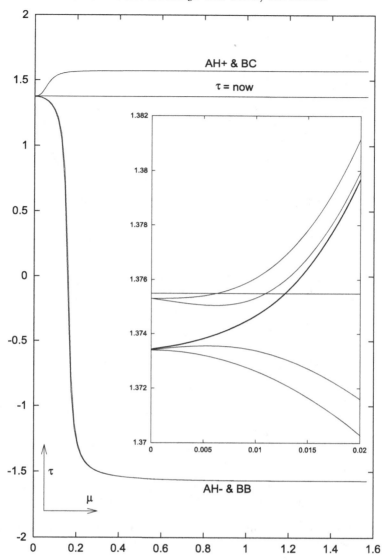

Fig. 18.4 The spacetime diagram of Fig. 18.1 compactified according to Eq. (18.98). The region shown in Fig. 18.1 is squeezed into the point where the three lines meet at the τ-axis. The worldlines of dust are still vertical straight lines here. The upper curve is the future apparent horizon (AH$^+$) and the Big Crunch (BC) singularity, they seem to coincide at the scale of this picture. The lower curve is the past apparent horizon (AH$^-$) and the Big Bang (BB) singularity, again coinciding only spuriously. The horizontal straight line is the $\tau = $ now time. Inset: a closeup view of the image (in the coordinates (μ, τ)) of the region shown in Fig. 18.1. The thicker line is the event horizon. It does not really hit the central point of the Big Bang; the apparent coincidence is just an artefact of the scale.

a situation. It is believed that a SC is a zero-pressure limit of an acoustic wave – of high, but finite, density.

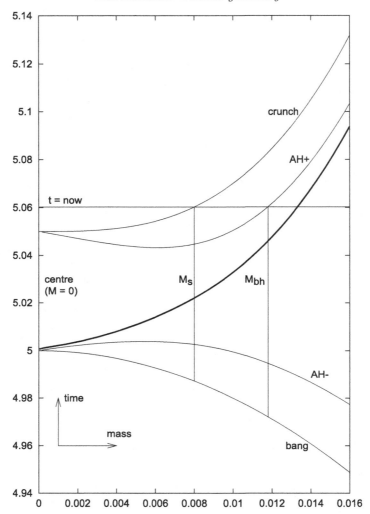

Fig. 18.5 The event horizon (thicker line) written into Fig. 18.1. Its intersection with the $M = 0$ axis does not coincide with the central point of the Big Bang; this is an illusion created by the scale.

2. A bundle of geodesics sent into a SC does not become focussed into a surface or a line, unlike at the Big Bang. This means that material objects hitting a SC would not be crushed (see Joshi (1993) for details). For this reason a SC is called a *weak* singularity.

This singularity can be avoided if the shapes of the arbitrary functions are properly chosen. There are two ways to avoid it:
(1) by choosing the functions so that $R_{,r} \neq 0$ everywhere in the model; and
(2) by choosing the functions so that $R_{,r} = 0$ only at those locations $r = r_w$ where

$$M_{,r}(r_w) = 0 \qquad (18.100)$$

and $\lim_{r \to r_w} |M_{,r}/R_{,r}| < \infty$.

We shall first deal with the second situation, by the method of Hellaby and Lake (1985). We assume that $t \geq t_B$, i.e. we consider an expanding model. For collapse towards the Big Crunch ($t \leq t_C$), it suffices to replace $(t - t_B)$ with $(t_C - t)$ and $t_{B,r}$ with $-t_{C,r}$. Since M does not depend on t, the density can be finite only if the locus of $R_{,r} = 0$ does not depend on time, either. This means that $R_{,r} = 0$ holds along a worldline of dust. Equation (18.100) is only a necessary condition for finite mass density. The sufficient condition is that the limit of $(M_{,r}/R^2 R_{,r})$ at $r = r_w$ is finite. It becomes particularly simple if we choose M as the radial coordinate, then it is

$$\lim_{M \to M(r_w)} (R^3)_{,M} > 0. \tag{18.101}$$

Note that

$$\lim_{r \to r_w} \frac{M}{R^3} = \lim_{r \to r_w} \frac{M_{,r}}{3R^2 R_{,r}}, \tag{18.102}$$

so $\lim_{r \to r_w} M_{,r}/(R^2 R_{,r}) < \infty$ ensures that all the R_{ijkl} in (18.99) are finite at $r \to r_w$.

From (18.20) we see that for the $E = 0$ model, if $R_{,r} = 0 = M_{,r}$ at $r = r_w$, then automatically $t_{B,r}(r_w) = 0$. For the other two models, we find from (18.19) and (18.21)

$$R_{,r} = \left(\frac{M_{,r}}{M} - \frac{E_{,r}}{E}\right) R + \left[\left(\frac{3}{2} \frac{E_{,r}}{E} - \frac{M_{,r}}{M}\right)(t - t_B) - t_{B,r}\right] R_{,t}. \tag{18.103}$$

The R, $R_{,t}$ and $(t - t_B) R_{,t}$ are linearly independent as functions of t, so $R_{,r}(r_w) = 0$ can hold at all t only if the coefficients of these functions in (18.103) vanish identically:

$$M_{,r}(r_w) = E_{,r}(r_w) = t_{B,r}(r_w) = 0. \tag{18.104}$$

These are the necessary conditions for the absence of a shell crossing at such $r = r_w$ at which $R_{,r} = 0$.[10] If these conditions hold and $1 + 2E(r_w) \neq 0$, then (18.16) shows that $g_{11} \to 0$. However, since we have already assumed that $M_{,r}/R_{,r} < \infty$ at $r = r_w$, Eqs. (18.99) and (18.102) show that the curvature is nonsingular there. Hence, with (18.101) and (18.104) fulfilled while $1 + 2E(r_w) \neq 0$, the locus $r = r_w$ is only a coordinate singularity. It can be removed by a coordinate transformation that also removes the property $R_{,r} = 0$, i.e. makes at least one of the functions $M_{,r}$, $E_{,r}$ and $t_{B,r}$ nonzero at r_w.

The situation is different when $E(r_w) = -1/2$. Equations (18.88) and (18.92) show then that the past and future apparent horizons meet at $r = r_w$, $t = t_w = t_{AH}(r_w)$, which is the instant of maximum expansion. An outgoing radial null geodesic that would leave the past AH through the hypersurface $r = r_w$ would immediately fall inside the future AH. An outgoing radial null geodesic that is already out of the past AH with $r < r_w$ must enter the future AH with $t > t_w$, $r < r_w$. Thus, no light ray can be sent from the region $\{t_{AH-} \leq t \leq t_{AH+}, r < r_w\}$ to the region $\{t_{AH-} \leq t \leq t_{AH+}, r > r_w\}$. In this case, the locus $r = r_w$ is called a **neck** or a **wormhole**, and it is a generalisation of the Kruskal–Szekeres 'throat' at $r = 2m$ of the Schwarzschild solution. In the vacuum limit ($M_{,r} = 0$ over an extended region), the neck goes over into the Kruskal–Szekeres throat. In the nonvacuum case, the neck need not be mirror-symmetric with respect to $(r - r_w)$. The possibility of existence of a neck was first noted by Barnes (1970). A generalisation

---

[10] Sufficient conditions more specific than (18.101) still await formulation.

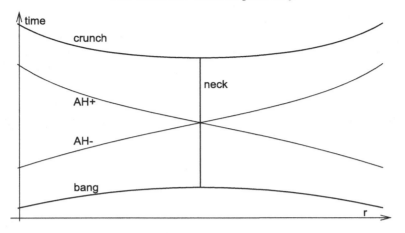

Fig. 18.6 An example of a spacetime region around a neck. The neck is the vertical straight line, which represents a world-tube with spherical cross-sections $r = r_w$. The two apparent horizons touch each other at the neck, at the instant of maximum expansion of the model. Every future-directed radial null geodesic that touches the AH^- is forced to leave the AH^-; every such geodesic that touches the AH^+ is forced to fall into the AH^+. This is why there is no communication between the region outside both AHs to the left of the neck and the corresponding region to the right of the neck. The functions in the figure are mirror-symmetric with respect to the neck, but in general they would not be.

of it appeared in numerical studies of a model with nonzero pressure gradients by Suto et al. (1984). For illustrative examples of necks see Hellaby (1987), Hellaby and Krasiński (2002), Krasiński and Hellaby (2004b) and Fig. 18.6.[11]

Now we come to the conditions for avoiding shell crossings at those points where $M_{,r} \neq 0$. We wish to translate the condition $R_{,r} \neq 0$ into properties of the functions $M(r)$, $E(r)$ and $t_B(r)$. The cases $R_{,r} > 0$ and $R_{,r} < 0$ have to be considered separately. We will write out the conditions only for $R_{,r} > 0$. For $R_{,r} < 0$, the inequalities (18.105), (18.109), (18.110), (18.116) and (18.117) must have their senses inverted. In all cases, in those regions where $R_{,r} > 0$, in order that the mass density is positive, we must have

$$M_{,r} > 0. \tag{18.105}$$

From here on, the three types of models have to be considered separately.

18.10.1 $E < 0$

Making use of (18.14) for expansion ($R_{,t} > 0$) and of (18.19), we rewrite (18.103) as

$$\frac{R_{,r}}{R} = \left(\frac{M_{,r}}{M} - \frac{E_{,r}}{E}\right) + \left(\frac{3}{2}\frac{E_{,r}}{E} - \frac{M_{,r}}{M}\right)\Phi_1(\eta) - \frac{(-2E)^{3/2}}{M}t_{B,r}\Phi_2(\eta) \overset{\text{def}}{=} f(\eta), \tag{18.106}$$

---

[11] The functions used in drawing Fig. 18.6 are: $M(r) = M_w + d(r - r_w)^2$, $t_B(r) = t_{B0} - bd^2(r - r_w)^2$, $E(r) = -(1/2)M_w/\left[M_w + d(r - r_w)^2\right] + ad(r - r_w)^2$ and $t_C(r) = 2\pi M(r)/(-2E(r))^{3/2} + t_B(r)$ – the crunch time; the parameters are $M_w = 20.0$, $d = 0.1$, $r_w = 2.0$, $t_{B0} = 10.0$, $b = 2000.0$ and $a = 1.0$.

where

$$\Phi_1(\eta) \overset{\text{def}}{=} \frac{\sin\eta(\eta - \sin\eta)}{(1 - \cos\eta)^2}, \qquad \Phi_2(\eta) \overset{\text{def}}{=} \frac{\sin\eta}{(1 - \cos\eta)^2}. \tag{18.107}$$

The function $f(\eta)$ should be strictly positive in the whole range $\eta \in [0, 2\pi]$. Note that

$$\lim_{\eta\to 0} \Phi_1(\eta) = 2/3, \qquad \lim_{\eta\to 2\pi} \Phi_1(\eta) = -\infty,$$

$$\lim_{\eta\to 0} \Phi_2(\eta) = \infty, \qquad \lim_{\eta\to 2\pi} \Phi_2(\eta) = -\infty, \qquad \lim_{\eta\to 2\pi} \frac{\Phi_1(\eta)}{\Phi_2(\eta)} = 2\pi. \tag{18.108}$$

These imply that the last term in (18.106) is unbounded when $\eta \to 0$, and positive only if

$$t_{B,r} < 0. \tag{18.109}$$

In the limit $\eta \to 2\pi$ the last two terms of (18.106) become unbounded. Factoring out Φ_2, which goes to $-\infty$, and demanding that the coefficient is negative we obtain:

$$2\pi \left(\frac{3}{2} \frac{E_{,r}}{E} - \frac{M_{,r}}{M} \right) - \frac{(-2E)^{3/2}}{M} t_{B,r} < 0. \tag{18.110}$$

Equations (18.109) and (18.110) imply

$$\frac{3}{2} \frac{E_{,r}}{E} - \frac{M_{,r}}{M} < 0. \tag{18.111}$$

Equations (18.105), (18.109) and (18.110) are the necessary conditions for the absence of shell crossings in the case $R_{,r} > 0$. They are also sufficient (see Exercise 10). The meaning of (18.110) is that the crunch time must be an increasing function of r.

Figure 18.3 can now be used to describe how shell crossings $R_{,r} = 0$ would appear in it. Recall that mass is used as the radial coordinate there, so $M_{,r} = 1 > 0$ is assured. Since the worldlines of the dust are constant-M lines (i.e. vertical straight lines), they can never intersect. A shell crossing would show in this picture as a point where a constant-R contour has a horizontal tangent. (Apart from the central points of the two singularities, $M = 0$ at $R = 0$, no such points exist in this example, because the functions $E(M)$ and $t_B(M)$ were chosen appropriately.) The centre of symmetry $M = 0$, the Big Crunch and the Big Bang together form the $R = 0$ contour. Adjacent contours of small constant R must have a similar shape. Hence, if $R_{,r} \neq 0$ were not guaranteed, segments of some contours would be nonmonotonic functions, whose derivative by M would change sign somewhere. At the changeover points, the tangents to the contours would be horizontal, and these would be the shell crossings. Figure 18.13 illustrates this comment even more graphically.

18.10.2 $E = 0$

We calculate from (18.20):

$$\frac{R_{,r}}{R} = \frac{M_{,r}}{3M} - \frac{\sqrt{2M} t_{B,r}}{R^{3/2}}. \tag{18.112}$$

As $R \to 0$ the second term dominates, so $t_{B,r} < 0$ is necessary for $R_{,r} > 0$ everywhere. Together with (18.105) this is then the necessary and sufficient condition.

18.10.3 $E > 0$

The analogue of (18.106) – (18.107) is here

$$\frac{R_{,r}}{R} = \left(\frac{M_{,r}}{M} - \frac{E_{,r}}{E} \right) + \left(\frac{3}{2} \frac{E_{,r}}{E} - \frac{M_{,r}}{M} \right) \Phi_3(\eta) - \frac{(2E)^{3/2}}{M} t_{B,r} \Phi_4(\eta), \qquad (18.113)$$

where

$$\Phi_3(\eta) \overset{\text{def}}{=} \frac{\sinh \eta (\sinh \eta - \eta)}{(\cosh \eta - 1)^2}, \qquad \Phi_4(\eta) \overset{\text{def}}{=} \frac{\sinh \eta}{(\cosh \eta - 1)^2}. \qquad (18.114)$$

The following properties of $\Phi_3(\eta)$ and $\Phi_4(\eta)$ are useful in calculations:

$$\lim_{\eta \to 0} \Phi_3(\eta) = 2/3, \qquad \lim_{\eta \to \infty} \Phi_3(\eta) = 1,$$

$$\lim_{\eta \to 0} \Phi_4(\eta) = \infty, \qquad \lim_{\eta \to \infty} \Phi_4(\eta) = 0,$$

$$\frac{\mathrm{d}\Phi_3}{\mathrm{d}\eta} > 0 \text{ for } \eta > 0, \qquad \frac{\mathrm{d}\Phi_4}{\mathrm{d}\eta} < 0 \text{ for } \eta > 0. \qquad (18.115)$$

Now take (18.113) at $\eta \to 0$. The last term dominates, and it will be positive if

$$t_{B,r} < 0. \qquad (18.116)$$

Taking (18.113) at $\eta \to \infty$ we easily obtain

$$E_{,r} > 0. \qquad (18.117)$$

Equations (18.105), (18.116) and (18.117) are necessary conditions for $R_{,r} > 0$. To see that they are also sufficient, rewrite (18.113) in the form

$$\frac{R_{,r}}{R} = \frac{M_{,r}}{M} (1 - \Phi_3) + \frac{E_{,r}}{E} \left(\frac{3}{2}\Phi_3 - 1 \right) - \frac{(2E)^{3/2}}{M} t_{B,r} \Phi_4(\eta) \qquad (18.118)$$

and take note of (18.115).

18.10.4 Final comment

For every set of the functions $M(r)$, $E(r)$ and $t_B(r)$ we can construct either an L–T model that explodes at $t = t_B$ or one that collapses towards $t = t_B$. Note that the condition $t_{B,r} < 0$ for avoiding shell crossings in the expanding model *guarantees* the appearance of shell crossings in the model that collapses towards the same $t_B(r)$. Thus the conditions for $R_{,r} \neq 0$ result in not so much removing the shell crossings, but rather in pushing them over to 'the other side of the Big Bang'. This remark will appear in new light in Sec. 19.4.6 where we will discuss the avoidance of the Big Bang in electrically charged dust.

18.11 The redshift along radial rays

In several applications of a cosmological model, we need to calculate the redshift of light emitted by a source at a given time and location and received by an observer located down a null geodesic from the source. We have a ready-to-use formula (16.20), but the field of tangent vectors to the geodesic, k^α, must be affinely parametrised in it. Transforming to such a parametrisation is usually not easy, so, for numerical calculations, it is often more convenient to use other methods.

One of them is the following method (copied from Bondi, 1947). From (18.16), for a radial null geodesic proceeding towards the observer, we find

$$\frac{dt}{dr} = -\frac{R_{,r}(t,r)}{\sqrt{1+2E(r)}}. \tag{18.119}$$

Let two light rays be emitted in the same direction, the second one later by a small time interval τ. Let the equations of the two rays be

$$t = T(r) \quad \text{and} \quad t = T(r) + \tau(r). \tag{18.120}$$

Both rays must obey (18.119), so

$$\frac{dT}{dr} = -\frac{R_{,r}(T(r),r)}{\sqrt{1+2E(r)}}, \qquad \frac{d(T+\tau)}{dr} = -\frac{R_{,r}(T(r)+\tau(r),r)}{\sqrt{1+2E(r)}}. \tag{18.121}$$

Since $\tau(r)$ was assumed small, we have, to first order in τ,

$$R_{,r}(T(r)+\tau(r),r) = R_{,r}(T(r),r) + \tau(r)R_{,tr}(T(r),r). \tag{18.122}$$

Using (18.122) and the first of (18.121) in the second of (18.121) we obtain

$$\frac{d\tau}{dr} = -\tau(r)\frac{R_{,tr}(T(r),r)}{\sqrt{1+2E(r)}}. \tag{18.123}$$

If τ at emission is the period of a wave, then comparing it with the corresponding period at the point of observation we obtain

$$\frac{\tau(r_{\text{obs}})}{\tau(r_{\text{em}})} = 1 + z(r_{\text{em}}). \tag{18.124}$$

Keeping the observer at a fixed position and considering the sources at two distances, r_{em} and $(r_{\text{em}} + dr)$, we find by differentiating the above that $(d\tau/dr)/\tau = -(dz/dr)/(1+z)$. Using this in (18.123):

$$\frac{1}{1+z}\frac{dz}{dr} = \frac{R_{,tr}(T(r),r)}{\sqrt{1+2E(r)}}. \tag{18.125}$$

Hence, the redshift may be calculated numerically from

$$\ln(1+z(r)) = \int_{r_{\text{obs}}}^{r_{\text{em}}} \frac{R_{,tr}(T(r),r)}{\sqrt{1+2E(r)}}dr. \tag{18.126}$$

This formula is equivalent to (16.20) (see Exercise 11), but here the (nonaffine) parameter on the null geodesic is just the coordinate r.

Exercise 11 also shows how to introduce the affine parametrisation for the field k^α. It can be introduced by the general method presented in Section 5.2, and we shall do this now. Let us begin with $\tau = -r$ as the parameter, then, using (18.119) we obtain

$$k^1 = -1, \qquad k^0 = \frac{R_{,r}}{\sqrt{1+2E}}. \tag{18.127}$$

Since this parametrisation is not affine, k^α obeys $k^\alpha{}_{;\beta} k^\beta = -(\mathrm{d}\ln\lambda/\mathrm{d}\tau)\,k^\alpha$ according to (5.11). Substituting here the k^α from (18.127) we find the coefficient to be

$$-\frac{1}{\lambda}\frac{\mathrm{d}\lambda}{\mathrm{d}\tau} = 2\frac{R_{,tr}}{\sqrt{1+2E}} - \frac{R_{,rr}}{R_{,r}} + \frac{E_{,r}}{1+2E}. \tag{18.128}$$

Now the meaning of all the symbols has to be precisely understood. The differential equation (18.128) holds along a single null geodesic, so all quantities in it have to be taken along that geodesic. In particular, wherever any function depends on t, the coordinate t in it must be replaced by the $T(r)$ of (18.120). The derivatives of R in (18.128) had been taken before the quantities were specialised to the null geodesic, so they are partial derivatives. Hence, $R_{,rr}$ is not the total derivative of $R_{,r}$ by r, but just the partial derivative by the second argument of $R_{,r}$. It is related to the total derivative, denoted $\mathrm{D}/\mathrm{d}r$, by

$$\frac{\mathrm{D}R_{,r}\left(T(r),r\right)}{\mathrm{d}r} = R_{,tr}\left(T(r),r\right)\frac{\mathrm{d}T}{\mathrm{d}r} + R_{,rr}\left(T(r),r\right). \tag{18.129}$$

Substituting from (18.121) for $\mathrm{d}T/\mathrm{d}r$ we obtain from the above

$$R_{,rr}\left(T(r),r\right) = \frac{\mathrm{D}R_{,r}\left(T(r),r\right)}{\mathrm{d}r} + R_{,tr}\left(T(r),r\right)\frac{R_{,r}\left(T(r),r\right)}{\sqrt{1+2E(r)}}. \tag{18.130}$$

Substituting this in (18.128) and recalling that $\tau = -r$ we obtain

$$\frac{1}{\lambda}\frac{\mathrm{d}\lambda}{\mathrm{d}r} = \frac{R_{,tr}\left(T(r),r\right)}{\sqrt{1+2E(r)}} - \frac{1}{R_{,r}\left(T(r),r\right)}\frac{\mathrm{D}R_{,r}\left(T(r),r\right)}{\mathrm{d}r} + \frac{E_{,r}}{1+2E(r)}. \tag{18.131}$$

This can be integrated with the result

$$\lambda = C\frac{\sqrt{1+2E}}{R_{,r}\left(T(r),r\right)}\,\exp\left(\int\frac{R_{,tr}\left(T(r),r\right)}{\sqrt{1+2E(r)}}\mathrm{d}r\right), \tag{18.132}$$

where C is an arbitrary constant. From (5.12), the parameter τ of (18.127) is related to the affine parameter s by $\mathrm{d}\tau/\mathrm{d}s = \lambda/C$, so the field k^α of (18.127) is related to the same field in the affine parametrisation, denoted $\widetilde{k}^\alpha$, by $\widetilde{k}^\alpha = (\lambda/C)k^\alpha$. Consequently, the components of the affinely parametrised field are

$$\widetilde{k}^0 = \exp\left(\int\frac{R_{,tr}\left(T(r),r\right)}{\sqrt{1+2E(r)}}\mathrm{d}r\right),$$

$$\widetilde{k}^1 = -\frac{\sqrt{1+2E}}{R_{,r}\left(T(r),r\right)}\exp\left(\int\frac{R_{,tr}\left(T(r),r\right)}{\sqrt{1+2E(r)}}\mathrm{d}r\right). \tag{18.133}$$

18.12 The blueshift

Blueshift occurs when $\nu_o > \nu_e$ in (16.20), so $z < 0$. We call the blueshift infinite when $\nu_o/\nu_e \to \infty$; then $z \to -1$.

The possible existence of blueshifts in L–T models was mentioned without proof by Szekeres (1980) in a conference report. His note seems to imply that infinite blueshift will appear along every light ray emitted from a nonconstant segment of the BB, where

$dt_B/dr \neq 0$. But there is another necessary condition: the ray must be directed radially (Szekeres considered only radial rays!). This second condition was found by Hellaby and Lake (1984), but in their paper it is hidden as two humble numbers in tables and a one-line comment and seems to have been overlooked by all later authors. We will verify it now.

The equations defining the tangent vectors $k^\alpha = dx^\alpha/d\lambda$ to geodesics of the metric (18.16), with λ being the affine parameter, are

$$\frac{dk^t}{d\lambda} + \frac{R_{,r} R_{,tr}}{1 + 2E} (k^r)^2 + RR_{,t} \left[(k^\vartheta)^2 + \sin^2 \vartheta (k^\varphi)^2\right] = 0, \tag{18.134}$$

$$\frac{dk^r}{d\lambda} + 2\frac{R_{,tr}}{R_{,r}} k^t k^r + \left(\frac{R_{,rr}}{R_{,r}} - \frac{E_{,r}}{1 + 2E}\right) (k^r)^2$$
$$- \frac{(1 + 2E)R}{R_{,r}} \left[(k^\vartheta)^2 + \sin^2 \vartheta (k^\varphi)^2\right] = 0, \tag{18.135}$$

$$\frac{dk^\vartheta}{d\lambda} + 2\frac{R_{,t}}{R} k^t k^\vartheta + 2\frac{R_{,r}}{R} k^r k^\vartheta - \cos \vartheta \sin \vartheta (k^\varphi)^2 = 0, \tag{18.136}$$

$$\frac{dk^\varphi}{d\lambda} + 2\frac{R_{,t}}{R} k^t k^\varphi + 2\frac{R_{,r}}{R} k^r k^\varphi + 2\frac{\cos \vartheta}{\sin \vartheta} k^\vartheta k^\varphi = 0. \tag{18.137}$$

The geodesics determined by (18.134) – (18.137) are null when

$$(k^t)^2 - \frac{R_{,r}^2 (k^r)^2}{1 + 2E} - R^2 \left[(k^\vartheta)^2 + \sin^2 \vartheta (k^\varphi)^2\right] = 0. \tag{18.138}$$

Using $R_{,t} k^t + R_{,r} k^r = dR/d\lambda$ and $k^\vartheta = d\vartheta/d\lambda$, the general solution of (18.137) is

$$R^2 \sin^2 \vartheta k^\varphi = J_0, \tag{18.139}$$

where J_0 is constant along the geodesic. Using this, the general solution of (18.136) is

$$R^4 (k^\vartheta)^2 \sin^2 \vartheta + J_0{}^2 = C^2 \sin^2 \vartheta, \tag{18.140}$$

where C^2 is another constant along the geodesic. When $C = 0$, the geodesic is radial. Then $J_0 = 0$ and either (a) $\vartheta = 0$ or π, with φ being undetermined or (b) ϑ is constant and φ is constant in consequence of (18.139).

From (18.139) and (18.140) $(k^\vartheta)^2 + \sin^2 \vartheta (k^\varphi)^2 = C^2/R^4$, and then (18.138) becomes

$$(k^t)^2 = \frac{R_{,r}^2 (k^r)^2}{1 + 2E} + \frac{C^2}{R^2}. \tag{18.141}$$

One can rescale the affine parameter λ to obtain on past-directed rays

$$k^t(t_o) = -1. \tag{18.142}$$

In comoving coordinates $u^\alpha = \delta^\alpha{}_0$, so, using (18.142), we obtain from (16.20)

$$1 + z = -k^t(t_e). \tag{18.143}$$

For nonradial rays, on which $C \neq 0$, the last term in (18.141) will go to infinity when $R \to 0$. Thus, at the BB, independently of whether the $R_{,r}^2$ term is finite or not,

$$\lim_{R \to 0} |k^t| \equiv \lim_{R \to 0} z = \infty. \tag{18.144}$$

Consequently, a necessary condition for infinite blueshift from the BB ($\lim_{R\to 0} z = -1$) is that the ray be radial. Whether this is also a sufficient condition has not so far been proved analytically, there exist only numerical confirmations (Krasiński, 2014c; 2016a) and a proof based on a perturbative calculation with the small parameter being R near the BB (Hellaby and Lake, 1984).

In the real Universe, no observer can receive light directly from the BB because for the first $t_{ls} \approx 380,000$ years after the BB matter is an intransparent plasma. The earliest light that can reach any observer is the CMB radiation emitted at the last-scattering (LS) epoch at $t = t_{ls}$. No dust model can faithfully describe the earlier times. The infinite blueshift from the BB should thus not be used as an argument that the L–T models are acceptable only with constant t_B. Finite blueshifts ($-1 < z < 0$) can in principle be observed on rays emitted at the LS and chances are that they are hiding among the observations already made, as argued by this author in a series of papers (Krasiński 2016a,b; 2018a,b; 2020).

In inhomogeneous models the redshift is not always a monotonic function of the affine parameter along the rays. When shifting the light source along a ray away from an observer towards the past, z can increase at first, then decrease (possibly below $z = 0$), then increase again and go to $+\infty$ at the BB. Examples of such behaviour of z in the Szekeres models are known (Krasiński, 2016b); these models will be discussed in Chapter 20. Therefore, z being a measure of distance is one more peculiarity of the R–W class of models, in general z cannot be used for this purpose.

In the Friedmann models, rays from the BB reach all observers with infinite redshift, as we already know from Sec. 17.11. In general L–T models with nonconstant BB, radial rays from the BB get infinitely blueshifted. In the transition from the L–T models to the Friedmann limit the infinite blueshifts disappear abruptly. This is one more instability of the Friedmann models (see Sec. 18.5 for the other one).

A similar discontinuity occurs within the L–T family of models: the infinite blueshifts disappear when going from radial to nonradial rays, so radial rays are unstable. They are unstable in yet another way: at the point where a radial ray reaches the Big Crunch, its projection on a $t = $ constant hypersurface is orthogonal to the $M = $ constant surface. Nonradial rays reach the BC *tangentially* to the $M = $ constant surfaces, as attested by Fig. 18.7 (copied from Krasiński, 2021): the radial ray #0 proceeds orthogonally to all the circles of constant M, while nonradial rays become tangential to those circles when approaching the BC. This figure will be mentioned again in Sec. 18.13.

It is seen from (18.119) that in the (t, r) variables the tangent to a radial ray becomes horizontal wherever $R_{,r} = 0$. Using (18.103) for the $E \neq 0$ models and (18.20) for the $E = 0$ model we find that the terms not involving $t_{B,r}$ go to zero at any point of the singular set $t = t_B$, independently of the path of approach.[12] If $t_{B,r} \neq 0$ at $r = r_1$, then the term $t_{B,r} R_{,t}$ in (18.103) and the corresponding term calculated from (18.20) will become infinite as $t \to t_B$, $r \to r_1$ because $R_{,t} \xrightarrow[t\to t_B]{} \infty$. Hence, a necessary condition for the light ray to have a horizontal tangent at a point P_1 located in the BB/BC singularity is $t_{B,r}(P_1) = 0$ (the same is true for the $E = 0$ model). Whether it is also sufficient cannot

---

[12] To see this, note that $R_{,t} = \ell\sqrt{2M + 2ER}/\sqrt{R}$, $\ell = \pm 1$ and $\lim_{t\to t_B}(t - t_B)/\sqrt{R} = 0$ (use the de l'Hôpital rule).

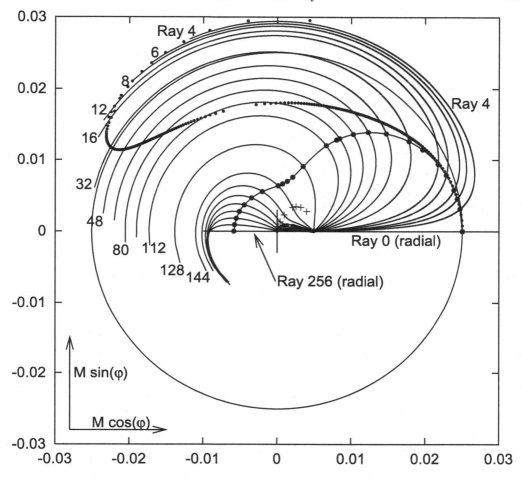

Fig. 18.7 Projections of selected nonradial rays on a surface of constant t along the flow lines of the cosmic dust in the spacetime of Fig. 18.1. The rays approach the Big Crunch tangentially to the surfaces of constant M. The centre of symmetry is at $M = 0$, the origin of the rays is at $(t, M) = (5.05, 0.005), \varphi = 0$. The dots on the continuous curve are the loci of maximum R along the rays. The crosses are the loci of minima of R. The dotted curve is the locus of zero expansion scalar of the ray bundle. The large circle has the radius $M = 0.025058$, this is where the outward radial ray #0 reaches $R = 2M$.

be said in general because the limit of $t_{B,r}R_{,t}$ as $P \to P_1$ may depend on the path of approach to P_1. If we approach it along the $r = r_1$ line, then the limit is zero.

The tangent to a radial ray from the BB being vertical or horizontal in the (t, r) graph has physical consequences. If the tangent is horizontal at the emission point, then the redshift at any other point is infinite. If the tangent is vertical, then elsewhere the ray will be infinitely blueshifted. This was first noted without a proof by Szekeres (1980), and then calculated by Hellaby and Lake (1984) by developing the quantities present in (16.20) into truncated Taylor series in R around $R = 0$. That calculation is rather complicated, so we shall explain this result by intuitive arguments only.

Suppose that the tangent to the ray at the emission point is horizontal. This means that the ray proceeds along the surface of constant time, so there will be an infinite number of cycles of an electromagnetic wave in a unit of time. The frequency at the emission point is then infinite, which, by (16.20), implies infinite redshift.

Now let the tangent at the emission point be vertical. In the comoving coordinates, the velocity of the dust is vertical everywhere, i.e. it is tangent to the light cone at emission. Thus, the dust is leaving the BB with the velocity of light. Consequently, an observer sitting on a dust particle would be riding on the crest of a light wave and would see zero frequency. By (16.20), this means that the ratio of the later-observed frequency to the frequency at emission is infinite, so at the observation point there would be an infinite blueshift. However, this infinity is a consequence of the frequency of the emitter being zero – the frequency at the observation point is finite.

18.13 * Apparent horizons for noncentral observers

This section is a brief account of Krasiński (2021, see online version for colour figures).

So far we have considered the AHs in the L–T models for bundles of light rays emitted radially. Such an AH can be defined in two equivalent ways:

- As the locus where the surface areas of the light fronts of the bundles achieve maxima.
- As the locus where the expansion scalars $\theta = k^{\mu}{}_{;\mu}$ of such bundles become zero.

Both these definitions determine the same hypersurface $R = 2M$. The AH so defined seems to be common to all light emitters. However, in Friedmann models, each observer is central and each one has a differently located AH, see Fig. 18.8. In the left panel the (t, r) coordinates are comoving. The future light cone LC of the present instant of observer O hits the Big Crunch at the r values marked by the vertical strokes. In the right panel the coordinates are (t, R), the BC is a single point and the AH profile is the pair of straight lines. The curves converging at the BC are worldlines of particles of the cosmic dust – in the left panel they would be vertical straight lines. Such structures exist around every comoving observer worldline in any collapsing Friedmann model.

So, it was puzzling where the AH would be for a noncentral observer in an L–T model. This question was investigated and answered in the paper by Krasiński (2021). Since the calculations underlying it are all numerical, only the results are reported here.

It turns out that then the picture differs strongly from both the AH of the central observer in an L–T model and the AH in a Friedmann model. Here are the main differences:

- The loci of $\theta = 0$ no longer coincide with $R = 2M$ and with the locus of maximum R.
- The number of zeros of θ along nonradial rays may be 0, 1, 2 or more depending on the emission point and the initial direction of the ray.
- The nonradial rays reach the BC with $\theta \to +\infty$, i.e. the bundles of nonradial rays *diverge* at the BC. The only rays that reach the BC with $\theta \to -\infty$ are the radial ones.

These results were obtained using the L–T model of Sec. 18.9. The figures show numerically calculated bundles of light rays emitted at different points of worldlines of comoving

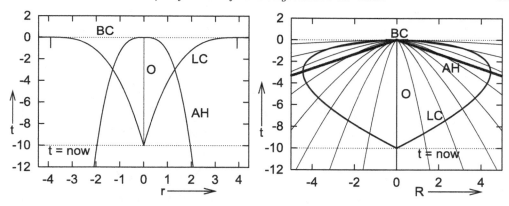

Fig. 18.8 **Left panel:** Profiles of the future apparent horizon AH of a comoving observer O and of the future light cone LC of O's present instant in the collapsing flat Friedmann model (17.33) represented in comoving coordinates. **Right panel:** The same situation in the (t, R) variables. At the AH R becomes maximum along the rays. See the text for more explanation.

observers situated at fixed radial coordinates. In each bundle there are 256 rays with regularly spaced initial directions; ray #0 is radial outward, ray #256 is radial inward.

Figure 18.9 shows the loci of $\theta = 0$ for rays emitted at 6 selected instants on the worldline $(M, \varphi) = (0.005, 0)$, projected on a surface of constant t. Numbers label the emission points, the letters 'f' and 's' refer to first and second zeros of θ. It can be seen that θ has no zeros on some rays, and two or more zeros on other ones. The knot at left is the locus of 3rd, 4th, 5th and 6th θ zeros on the earliest bundle of rays. Compare Fig. 18.7 – the dotted curve in it coincides with the pair (1f, 1s) in Fig. 18.9.

Figure 18.10 shows θ as a function of t along a subset of the bundle emitted at $(t, M) = (5.075, 0.012)$. It can be seen that θ begins at $+\infty$ and ends at $-\infty$ only on the outward radial ray #0. On nonradial rays $\theta(t)$ is not monotonic, and on many of them it has two zeros. The dotted curve shows that $\theta(t)$ has a similar shape even for a ray emitted at $t = 5.1002$, which is later than $R = 2M$.

Figure 18.11 shows the time-ordering of various events along the rays of the bundle emitted at $(t, M) = (5.05, 0.005)$. In the right part of the figure it is seen that this whole branch of the $\theta = 0$ set lies earlier than $R = 2M$, so radial light rays can be sent towards increasing M from the region where $\theta < 0$. This shows that the AH of the central observer, located at $R = 2M$, is a one-way membrane also for noncentral observers, while the locus of $\theta = 0$ plays no such role.

18.14 The influence of inhomogeneities in matter distribution on the cosmic microwave background radiation

Equation (16.20) implies $\nu_o(1 + z) = \nu_e = $ constant along every light ray. The CMB radiation has at present the black-body spectrum (17.89). Assuming that this spectrum is preserved in time, we conclude from (17.89) and (16.20) that the temperature of the

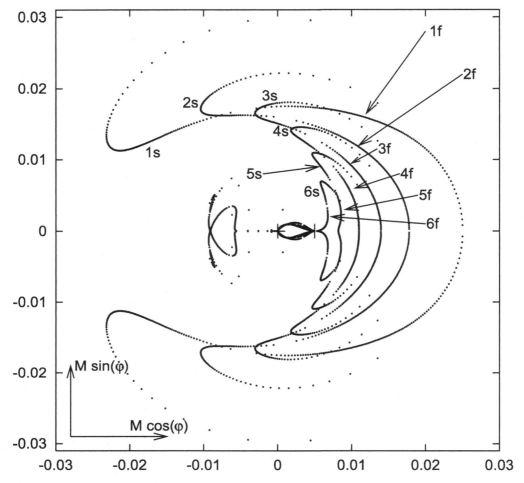

Fig. 18.9 Loci of $\theta = 0$ for rays emitted at 6 instants on the line $(M, \varphi) = (0.005, 0)$ (marked by the vertical stroke) in the L–T model of Sec. 18.9, projected on a surface of constant t.

radiation evolves between the emission instant t_e and the observation instant t_o by the law

$$T(t_o)(1 + z) = T(t_e) = \text{constant}. \tag{18.145}$$

This can be generalised to the L–T model by observing that in small neighbourhoods it behaves like a Friedmann model, so (18.145) can be applied to nearby points on the same null geodesic, with radial coordinates r and $(r + \mathrm{d}r)$. In this way, differences in observed temperature between rays going along different paths can be calculated. The calculated quantity is the **temperature contrast** $\Delta T/T$, where T is the temperature along a ray that has propagated through a Friedmann region all the way and ΔT is the difference between T and the temperature along a neighbouring ray.

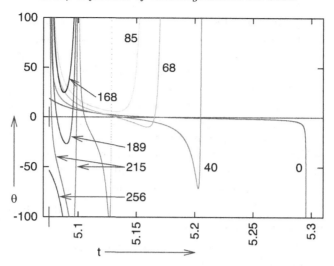

Fig. 18.10 The graphs of $\theta(t)$ along selected rays emitted at $(t, M) = (5.075, 0.012)$ (the continuous curves) and $(5.1002, 0.012)$ (the dotted curve). Ray 0 is radial outward, Ray 256 is radial inward.

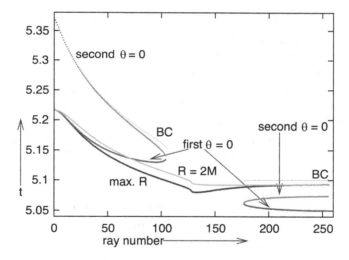

Fig. 18.11 The t coordinates of various events on rays emitted at $(t, M) = (5.05, 0.005)$.

This approach was first applied by Raine and Thomas (1981). They considered a large-scale but small-amplitude condensation in the path of rays of the CMB reaching an observer. The rays were assumed to be emitted and received in a Friedmann region. The equations of null geodesics were integrated numerically in order to calculate the temperature at the reception point and compare it with the temperature at endpoints of rays propagating all the way through a Friedmann medium. The temperature variation was calculated as a function of the direction of observation.

The most comprehensive study by this method was done by Arnau, Fullana, Monreal and Saez (Arnau et al., 1993, 1994; Saez et al., 1993). They computed numerically the dependence of the temperature contrast on the direction of observation of the microwave background radiation in a model consisting of a Friedmann background with the density parameter $\Omega = \rho/\rho_{cr}$ (where ρ_{cr} is the critical density defined by (17.30)) and a localised L–T perturbation superimposed on it. The numerical code allowed them to fit a model with a few free parameters (describing the density profile of the condensation, the background density parameter and the Hubble parameter) to the observed characteristics of the Great Attractor and of the Virgo cluster. The models of the Great Attractor had different density and velocity profiles and the Friedmann background had different density parameters. Specifically, the authors calculated the effects of the following:

1. Various velocity profiles in the attractor model with a fixed background density parameter and fixed distance from the observer.

2. Various background density parameters with a fixed velocity profile and the distance adjusted so as to produce the largest effect.

3. Various distances to the condensation with a fixed velocity profile and a fixed background density parameter.

The result was that the maximum anisotropy to be expected is up to 3×10^{-5} (when $\Omega = 0.15$), at the angular scale of $10°$. This is to be compared with the measurements that gave $\Delta T/T \approx 5 \times 10^{-6}$ (Smoot et al., 1992).

This result invites a repetition of the comment made in Sec. 17.14. The high degree of isotropy of the CMB radiation has frequently been used as an argument in favour of the Friedmann models, and the 'reasoning' was that inhomogeneities in matter distribution would leave an imprint on the radiation. However, for a long time these claims had not been backed up by any calculations. The results of Saez et al. proved that no trace of such an imprint could show up until the precision of measurements of temperature anisotropies reached 10^{-6}, which happened only in 1992 (Smoot et al., 1992; Mather et al., 1993).

18.15 Matching the L–T models to the Schwarzschild and Friedmann solutions

As already observed in Section 18.3, the Schwarzschild solution is the vacuum limit of the L–T model that results when $M_{,r} = 0$. For various purposes, a spacetime model is sometimes considered consisting of an L–T solution in the region $r \leq r_b$ matched to the Schwarzschild solution in the region $r \geq r_b$. The Schwarzschild solution should then be represented in the Lemaître–Novikov coordinates, as in (14.120) – (14.121). As shown in Section 12.17, the matching conditions to (18.16) require that

$$R_{LT}(t, r_b) = R_S(t, r_b), \qquad R_{,r}/\mathcal{N}|_{LT} = R_{,r}/\mathcal{N}|_S, \qquad (18.146)$$

where S stands for 'Schwarzschild' and LT for 'Lemaître–Tolman', and

$$\mathcal{N} = \frac{R_{,r}}{\sqrt{1 + 2E}} \qquad (18.147)$$

in both metrics. As a result, $R_{,r}$ cancels out in the equations, and the matching conditions reduce just to the continuity of E, R and $R_{,t}$, i.e.

$$E(r_b)|_{LT} = E(r_b)|_S, \qquad t_B(r_b)|_{LT} = t_B(r_b)|_S, \qquad M(r_b) = m, \qquad (18.148)$$

where m is the Schwarzschild mass parameter. The result of the matching is in fact a single L–T model with $M(r)$ going over into the constant function $M = m$ at $r = r_b$.

Sometimes, the matching between the L–T and Friedmann spacetimes is considered; for example, when discussing the formation of galaxy clusters or voids (as in Sec. 18.6). It is then assumed that at a certain distance from the centre of the cluster or void the L–T metric goes over into an unperturbed Friedmann model. Equations (18.146) – (18.148) must apply also then, with the subscript 'S' now referring to Friedmann. In particular, the last equation of (18.148) now reads $M_{LT}(r_b) = M_S(r_b)$, where $M_S(r_b)$ is the Friedmann mass function calculated at $r = r_b$. This means that the Friedmann mass removed from within $r = r_b$ must be the same as the L–T mass filling that sphere. This is a necessary and sufficient condition for matching because the other two equations in (18.148) can always be fulfilled: the first determines the value of the Friedmann curvature index k, the second can be fulfilled by time-translation in the Friedmann spacetime.

As observed by Ribeiro (1992a), this equality of masses implies that in such a configuration any region of density higher or lower than in the Friedmann background must be compensated by a region of a lower or higher density, respectively.

It is possible to consider a composite L–T/Friedmann model in which the two masses are not equal. Then, if the mass removed from the Friedmann region is larger than the inserted L–T mass, the L–T/Friedmann boundary will expand outward, pushing a higher-density wall of mass in front of it. If the removed Friedmann mass is smaller than the L–T mass taking its place, the boundary will collapse towards the centre of the L–T region. A comprehensive study of these processes was carried out by Maeda, Sasaki and Sato (1983), Maeda and Sato (1983), Sato and Maeda (1983) and by Sato (1984).

18.16 * The shell focussing singularity

At all points apart from the centre of symmetry, the BB and BC singularities are spacelike. This is verified as follows: since the BB and BC are parts of the $R = 0$ set, we calculate the normal vector field n_μ to any hypersurface $R =$ constant. It is $(R_{,t}, R_{,r}, 0, 0)$, so $g^{\mu\nu} n_\mu n_\nu = g^{00} R_{,t}^2 + g^{11} R_{,r}^2$, which, using (18.14) and (18.16), becomes

$$g^{\mu\nu} n_\mu n_\nu = 2M/R - 1. \qquad (18.149)$$

This is negative for $R > 2M$, zero at the AH and positive for $R < 2M$. Consequently, n_μ is, respectively, spacelike, null or timelike, i.e. the hypersurface $R =$ constant is timelike, null or spacelike. But as we approach the nonsingular centre of symmetry $R \to 0$, the regularity condition (18.31) implies that $2M/R \to 0$, so the nonsingular part of the set $R = 0$ is timelike. As we approach the singularity $R \to 0$ along a line with $M > 0$, the expression (18.149) tends to $+\infty$, so the part of $R = 0$ in which $M > 0$ is spacelike.

The orientation of the $R = 0$ set at the point where the line of centre of symmetry hits the BB or BC is not simple to determine. This is because the comoving coordinates have a singularity there, too. It had taken quite some time to notice that this set is not always a single point, but may be a finite segment of a timelike or null curve. The first to note the possibility of this peculiar behaviour (during a numerical investigation of the $E = 0$ L–T model) were Eardley and Smarr (1979), and they called it **shell focussing**. They stated that the segment on which shell focussing occurs can only be null. However, their criteria for the occurrence of a shell focussing, which they gave without derivation, do not look entirely credible because they refer to the limit of t_B/M as $r \to r_c$, while the value of t_B is coordinate-dependent. C. Hellaby, in his unpublished PhD Thesis,[13] found that the segment in question can be timelike as well.

If the BC singularity is all spacelike, then its intersection with the worldline of the centre of symmetry can be imagined as a single point – no future-directed light ray can leave the singular set. However, if the singularity contains a timelike or null segment, then, as will be shown below, that segment contains vertices of an infinite family of distinct light cones. The conclusion is that the nonspacelike segment of the singularity is an extended arc of a curve that is mapped into a single point in the comoving coordinates.

The discussion below was inspired by the paper of Christodoulou (1984) and the unpublished work of Hellaby and Lake (1988), but is a great simplification of both. Take the $E = 0$ L–T model with the function $M(r)$ chosen as follows:

$$M(r) = M_0 r^3,$$ (18.150)

where M_0 is a constant; this defines the coordinate r. To discuss collapse, we set in (18.20):

$$R(t,r) = \left\{ \frac{9}{2} M(r) \left[t_C(r) - t \right]^2 \right\}^{1/3}$$ (18.151)

so that t increases from an initial value to $t = t_C(r)$ at the BC. Now choose:

$$t_C(r) = ar^2,$$ (18.152)

where $a > 0$ is a constant. This BC starts at $r = 0$ with $t = 0$ and proceeds to greater r as t increases (see Fig. 18.12). Then the equation of an outgoing radial null geodesic is

$$\left(\frac{dt}{dr} \right)_n = \left(\frac{9M_0}{2} \right)^{1/3} \frac{\frac{7}{3}ar^2 - t}{(ar^2 - t)^{1/3}}.$$ (18.153)

The set where $t = 7ar^2/3$ would be a shell crossing, but it lies to the future of the BC, and hence does not belong to the physical spacetime. Along any line that terminates at the BC with nonzero r, the dt/dr becomes infinite, i.e. the geodesics hit the BC vertically in a (t,r) diagram. As we approach any nonsingular point of the centre of symmetry line, $\{r = 0, t < t_C\}$, dt/dr has a finite nonzero limit. In order to find what happens as we approach the intersection of the centre of symmetry with the BC along an arbitrary curve $t = f(r)$, we write (18.153) in the form

$$\left(\frac{dt}{dr} \right)_n = \left(\frac{9M_0}{2} \right)^{1/3} r^{4/3} \frac{\frac{7}{3}a - t/r^2}{(a - t/r^2)^{1/3}},$$ (18.154)

---

[13] We are grateful to Charles Hellaby for giving us access to his Thesis. An abbreviated account of the results of this Thesis is given in the unpublished preprint by Hellaby and Lake (1988).

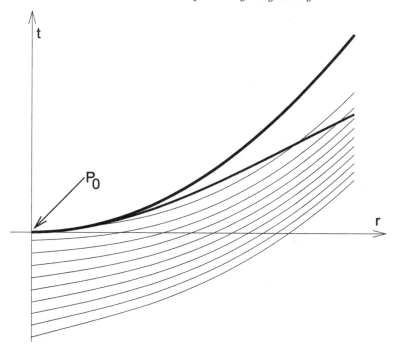

Fig. 18.12 Outgoing radial light rays in the neighbourhood of the Big Crunch in the spacetime defined by (18.151) – (18.152). The vertical arrow is the axis of symmetry, the thicker parabola is the Big Crunch singularity and the lighter curve is the apparent horizon. At all points except at the centre, the rays hit the singularity vertically (the figure does not show this). At all points except at the singularity, the rays hit the axis of symmetry with nonvertical tangents. When the common point of the axis and of the singularity is approached along any curve except the singularity itself, the direction of the outgoing light ray tends to the horizontal. Light rays emanating from the central point of the singularity are tangent to the singular set and to the apparent horizon, but d^2t/dr^2 calculated along those rays at $(t, r) = (0, 0)$ is still zero, while $d^2t_B/dr^2(0) = d^2t_{AH}/dr^2 \neq 0$, so the rays can recede from the singularity out to a finite distance, still remaining before the apparent horizon. The uppermost ray in the figure has this property.

and consider the various possible limits of t/r^2 at $t \to 0, r \to 0$ along different curves. The limit may be zero, finite nonzero or infinite. In the first two cases, (18.154) shows that

$$\lim_{r \to 0, t \to 0} (dt/dr)_n = 0. \tag{18.155}$$

In the third case it is useful to rewrite (18.154) in yet another form

$$\left(\frac{dt}{dr}\right)_n = \left(\frac{9M_0}{2}\right)^{1/3} t^{2/3} \frac{7ar^2/(3t) - 1}{(ar^2/t - 1)^{1/3}}. \tag{18.156}$$

The infinite limit of t/r^2 means zero limit of r^2/t, so (18.155) again holds. Thus, the limiting slope of a light ray emitted from the point P_0 of coordinates $(t, r) = (0, 0)$ consistently comes out to be zero for all nonsingular paths of approach to P_0 (see Fig. 18.12), but when this point is approached along the BC, the tangent to the light ray is vertical. This suggests that there is some structure hidden in P_0.

The equation of the apparent horizon in the spacetime considered is

$$t_{AH}(r) = ar^2 - \frac{4}{3}M_0 r^3. \tag{18.157}$$

Thus, the apparent horizon is tangent to the BC at $r = 0$, and initially proceeds towards larger t; $t_{AH}(r)$ becomes a decreasing function of r only at $r = a/(2M_0) > 0$ (that region is not visible in Fig. 18.12, which shows only $r < 0.2a/M_0$). We thus see that the BC, the apparent horizon and the light rays emitted from P_0 are all tangent at P_0. However, while the second derivatives of both the BC time and the apparent horizon time are equal to $2a$ at P_0, the second derivative of $t = t_n(r)$ along a light ray is still zero, as can be found by differentiating (18.153) and taking the limit $t \to 0, r \to 0$ along the light ray:

$$\left(\frac{d^2 t}{dr^2}\right)_n (P_0) = \lim_{r \to 0} \left[\left(\frac{9M_0}{2}\right)^{1/3} \frac{4ar^{1/3}}{(a - t/r^2)^{4/3}} \left(\frac{7}{9}a - \frac{t}{r^2}\right) \right.$$
$$\left. + \frac{2}{3}\left(\frac{9M_0}{2}\right)^{2/3} r^{2/3} \frac{7a/3 - t/r^2}{(a - t/r^2)^{5/3}} \left(-\frac{a}{3} + \frac{t}{r^2}\right) \right] = 0. \tag{18.158}$$

This is zero no matter what limit t/r^2 has at $(t, r) = (0, 0)$. Thus, light rays that start off from the point P_0 initially recede from the BC and remain earlier than the apparent horizon. Consequently, each such ray will proceed out to a finite r without entering the apparent horizon. In Section 18.18 we will come back to the question whether the ray can escape to infinity. (In the example we consider now, the apparent horizon begins to proceed to earlier times when $r > a/(2M_0)$, so the ray must eventually cross the apparent horizon and then hit the BC.)

Which ray leaving the central point of the singularity is the earliest? Lacking an exact solution of (18.153), it can be identified only by numerical calculation. The best that can be achieved analytically is to prove that such a ray exists. The proofs of existence in another model, with $E < 0$, were given by Christodoulou (1984) and Newman (1986).

Suppose that L is any such ray emitted from $P_0 = (0, 0)$. Outgoing radial rays cannot intersect at $(t, r) \neq (0, 0)$, and cannot be emitted into the spacetime from any other point of the BC. It follows that any radial ray lying to the future of L was also emitted from P_0. Thus there is an infinite family of rays emanating from P_0 and initially receding from the BC and the apparent horizon. Consequently, P_0 only looks to be a single point in the comoving coordinates, in reality it must be a timelike or null set.

We have considered just one example of an L–T model, constructed deliberately so that it shows this unusual behaviour. There are L–T models in which the central point of the BC cannot emit any light ray because that singularity is all spacelike.

A time-reversed behaviour can occur at the central point of the Big Bang.

18.17 * Extending an L–T spacetime through a shell crossing singularity

The shell crossing is considered a less serious breakdown of Einstein's theory than the Big Bang – because it is believed to be an artefact of the zero pressure gradient in the L–T model. In models with nonzero pressure gradients (so far unknown), it is expected that the shell crossing will be replaced by a set of high but finite density.

In this section, we shall present one more reason why a shell crossing is relatively harmless: as observed by Newman (1986), an L–T spacetime can be extended through it. The extension is provided by the Gautreau (1984) coordinates. For a general L–T model these coordinates are $(\tau, R, \vartheta, \varphi)$, where $\tau = t$ and R obeys (18.14). Thus, in (18.16) we have

$$R_{,r} \, dr = dR - R_{,t} \, dt = dR - \ell \sqrt{2E + \frac{2M}{R} - \frac{1}{3} \Lambda R^2} \, d\tau, \qquad \ell = \pm 1, \tag{18.159}$$

$$ds^2 = \frac{1}{1 + 2E} \left[\left(1 - \frac{2M}{R} + \frac{1}{3} \Lambda R^2 \right) d\tau^2 \right.$$
$$\left. + 2\ell \sqrt{2E + \frac{2M}{R} - \frac{1}{3} \Lambda R^2} \, d\tau dR - dR^2 \right] - R^2 \left(d\vartheta^2 + \sin^2 \vartheta d\varphi^2 \right), \tag{18.160}$$

where the properties $M_{,t} = E_{,t} = 0$ translate into the partial differential equations:

$$\frac{\partial M}{\partial \tau} + \ell \sqrt{2E + \frac{2M}{R} - \frac{1}{3} \Lambda R^2} \frac{\partial M}{\partial R} = 0, \tag{18.161}$$

and a similar equation for E. In (18.160) there is no trace of a shell crossing, which means that the metric can be extended through it. However, the derivatives of g_{00}, g_{01} and g_{11} by R will now be singular at the shell crossing because

$$\frac{\partial M}{\partial R} = \frac{\partial M}{\partial r} \frac{\partial r}{\partial R} = \frac{\partial M}{\partial r} \bigg/ R_{,r} \tag{18.162}$$

(all quantities to be taken at $r = r(\tau, R)$). Thus, the extension is of class C^0 (continuous but nondifferentiable), and the curvature tensor still has a singularity at $R_{,r}(\tau, R) = 0$.

The components of the velocity field in the (τ, R) coordinates are

$$u^0 = 1, \qquad u^1 = R_{,t} = \ell \sqrt{2E + \frac{2M}{R} - \frac{1}{3} \Lambda R^2}, \qquad u^2 = u^3 = 0. \tag{18.163}$$

The u^1 is nondifferentiable at the shell crossing, but continuous. Thus the derivative of u^1 (and consequently the Christoffel symbols) will have a finite discontinuity there. The singularity of the flow makes itself visible in the behaviour of the geodesic deviation of the velocity field. As seen from (6.56), to offset the infinite curvature at the shell crossing, the geodesic deviation δx^α must be zero there, which means that different flow lines intersect. However, they continue behind the intersection points. Figures 18.13 and 18.14 show an example of a shell crossing in an $E = 0$ L–T model in the comoving coordinates (Fig. 18.13) and in the Gautreau coordinates (Fig. 18.14). As seen from Fig. 18.14, the region behind the shell crossing contains three superposed flows of dust: the streams that entered through the left wall, through the right wall and through the lower vertex.[14]

---

[14] This kind of extension was considered by Clarke and O'Donnell (1992), but it is not clear from their figure what quantities are represented on the coordinate axes.

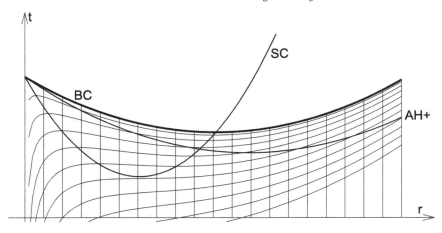

Fig. 18.13 A shell crossing singularity (SC) in an $E = 0$ L–T model in the comoving coordinates. (The model used to draw the figure is the same as will appear in Section 18.18, Eqs. (18.164) – (18.169), but with different parameter values.) The vertical straight lines are flow lines of the dust, terminating in the Big Crunch (BC) singularity. AH$^+$ is the apparent horizon. The other curves in the figure are contours of constant R. In the comoving coordinates, the flow lines cannot intersect in a spacetime diagram, even at shell crossings. A shell crossing is a location where $R_{,r} = 0$, i.e. where the contour of constant R has a horizontal tangent. The figure should help one understand why a necessary condition for avoiding a shell crossing is that the BC is an increasing function of r: note that the centre of symmetry and the BC together form the $R = 0$ contour. Thus, since the contours of small constant R must approximately follow the shape of the $R = 0$ contour, if $t_{BC}(r)$ is not an increasing function, then the other $R = $ constant contours must have a horizontal tangent somewhere. At the point where t_{BC} becomes increasing, the shell crossing set meets the BC and at larger r is later than the BC, i.e. not in the physical region of the spacetime. Compare this with Fig. 18.3, where there are no shell crossings.

18.18 * Singularities and cosmic censorship

The cosmic censorship hypothesis (CCH) does not in fact belong to the field of cosmology. However, the L–T model has been used many times as a testing ground for its various formulations. After reading this section, readers may have a false impression that the subject of CCH is buried under counterexamples. In reality, the debate is far from settled. The emphasis on counterexamples in this text does not reflect the true state of affairs, it resulted from the fact that we are more interested here in applications of the L–T model than in settling the CCH issue.

In its original formulation (Penrose, 1969), the CCH said that relativity does not allow **naked singularities**. All singularities were supposed to be hidden inside horizons. A singularity can be **locally naked**, when light rays proceed from it out to a finite distance and turn back or hit the BC, or **globally naked**, when the light rays it emitted can escape to infinity. The CCH would preferably prohibit both kinds of naked singularity.

It would certainly be desirable if the CCH were correct because a naked singularity could emit matter or radiation in unpredictable quantities at unpredictable times. However, the history of the CCH turned out to be turbulent. A counterexample to this earliest formulation was provided by Yodzis, Seifert and Müller zum Hagen (1973), then the hypothesis was modified to exclude 'nongeneric' cases, then another counterexample to the modified

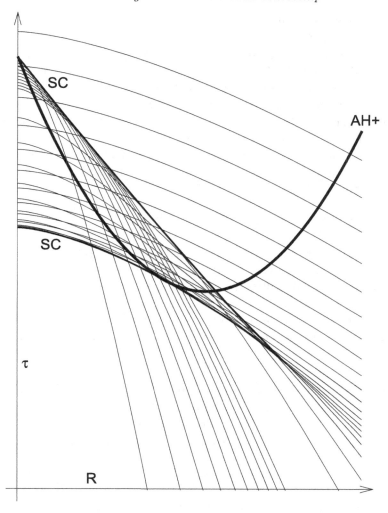

Fig. 18.14 The situation from Fig. 18.13 in the (τ, R) coordinates of Gautreau. The Big Crunch is at the vertical axis, where $R = 0$. The shell crossing (SC) is the wedge-shaped pair of curves; the tip of the wedge is the minimum of SC from Fig. 18.13 and the lower branch of the wedge is the part to the right of the minimum. More flow lines are drawn here than in Fig. 18.13 – the extra lines are the leftmost nine (which would lie, evenly spaced, to the left of the leftmost line in Fig. 18.13) and the uppermost one (which would lie to the right of the rightmost line in Fig. 18.13). Inside the wedge, three streams of matter are seen: the particles that entered through the lower face, the particles that entered through the upper face and the particles that entered through the tip. Note that this diagram does not faithfully represent all the features of Fig. 18.13. For example, the flow lines that intersect the right face of the SC here do not intersect the SC in Fig. 18.13. The function $t(r)$ given by $R(t, r) = $ constant is not one-to-one, i.e. for every point (t_1, r_1) in Fig. 18.13 in the SC to the left of the minimum, with the value R_1 of R, there is a point (t_1, r_2) to the right of the SC that has the same value of R.

version appeared, followed by a new modification, and so the story went. We shall not follow it here; more information can be found in the book by Joshi (1993). Instead, we will describe some of the applications of the L–T model to the discussion of properties of singularities, and to the CCH.

Our example of a naked shell crossing singularity follows the method of Yodzis, Seifert and Müller zum Hagen (1973). However, the functions used here are different because the original example was, apparently needlessly, quite complicated. The idea is to construct a ball of L–T dust of finite radius, matched to the Schwarzschild solution, and then design a shell crossing so that it reaches the surface of the ball before the surface crosses the future apparent horizon at $R = 2M$. At the surface, the apparent horizon coincides with the Schwarzschild event horizon, so the singularity thus constructed will be naked.

We take the collapsing $E = 0 = \Lambda$ L–T model (in which $t < t_C$ and t increases towards $t = t_C$), and we find the locus of $R_{,r} = 0$ in it:

$$t = t_S(r) \overset{\text{def}}{=} t_C(r) + 2\frac{M}{M_{,r}}t_{C,r}. \tag{18.164}$$

Following Yodzis et al. we choose the r-coordinate so that $M(r) = \mu r$, where μ is a constant. The centre of symmetry is at $r = 0$, and the regularity condition (18.31) is fulfilled. The surface of the L–T ball will be at $r = b$, and for $r > b$ we have the Schwarzschild solution with the mass parameter $m = \mu b$ (it must be represented in the Lemaître–Novikov coordinates with $E = 0$ and with the appropriate value of $t_C(b)$).

Next, we choose the function $t_S(r)$ in (18.164) so that it has a minimum at some r:

$$t_S(r) = a(r - c)^2 + d, \tag{18.165}$$

where a, c and d are arbitrary positive parameters, whose values will be chosen later. The minimum of $t_S(r)$ is at $r = c$ and equals d, i.e. for $t < d$ the shell crossing does not exist yet. We equate (18.164) and (18.165) and solve the resulting differential equation for $t_C(r)$. In order to fulfil (18.35), the integration constant must be zero. The result is

$$t_C(r) = \frac{a}{5}r^2 - \frac{2}{3}acr + ac^2 + d. \tag{18.166}$$

The time by which the shell crossing hits the surface of the ball is

$$t_S(b) = a(b - c)^2 + d. \tag{18.167}$$

The time by which the surface of the ball crosses the apparent horizon is found by solving the equation $R = 2M$; it is

$$t_H(b) = t_C(b) - \frac{4}{3}\mu b = \frac{a}{5}b^2 - \frac{2}{3}b(2\mu + ac) + ac^2 + d, \tag{18.168}$$

and this must be greater than $t_S(b)$, which implies

$$c > \frac{3}{5}b + \frac{\mu}{a}. \tag{18.169}$$

This leaves the mass parameter μ, the location of the surface of the ball, $r = b$ and the instant when the shell crossing first appears, $t = d$, still arbitrary – Eq. (18.169) is the only condition to be fulfilled. If it is fulfilled, then the singularity reaches the surface of the ball at $t = t_S(b)$ given by (18.167), while the surface is still outside the Schwarzschild event horizon. Consequently, the singularity can send a light ray out to the future null infinity,

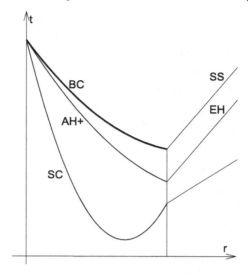

Fig. 18.15 A schematic illustration of a naked shell crossing singularity of the type of Yodzis et al. (1973). The vertical line is the worldline of the surface of the dust ball, the area to the right of it is the Schwarzschild spacetime. BC is the Big Crunch singularity, where $t = t_C(r)$; outside the ball it goes over into the Schwarzschild singularity denoted SS. AH^+ is the future apparent horizon; outside the ball it goes over into the Schwarzschild event horizon EH. SC is the shell crossing singularity that appears at $t = d$ and grows inwards and outwards. The seemingly straight line below EH is the light ray emitted when the shell crossing reaches the surface of the body – it proceeds unimpeded to the future null infinity. For the shell crossing singularity to exist it is necessary that the Big Crunch function inside the body is decreasing with r (for an increasing $t_C(r)$, the shell crossing is later than the Big Crunch).

see Fig. 18.15. For better physical interpretation, it is good to assume $c < b$, i.e. that the minimum of $t_S(r)$ is inside the ball, so that the shell crossing begins within the ball and not at the surface. This can be fulfilled together with (18.169) only if $a > 5\mu/(2b)$. This leaves the parameters μ, b and d still arbitrary.

The Schwarzschild solution represented in the Lemaître–Novikov coordinates must have its $R(t,r)$ equal at $r = b$ to the $R(t,r)$ of (18.164) – (18.166). Hence, $E_S(b) = 0$, $t_{CS}(b) = t_C(b)$ and $m = \mu b$, where the subscript 'S' stands for 'Schwarzschild'. Thus, it is simplest to choose $E_S(r) = 0$ (which, in the Schwarzschild limit, is just a choice of coordinates, see Exercise 10 in Chapter 14), and then $t_{CS}(r)$ can be given any form by a transformation of r alone. A simple form is $t_{CS}(r) = \alpha r$, with $\alpha = t_C(b)/b$. Then, the event horizon, $R = 2m$, is given by $t = \alpha r - 4m/3$, i.e. the singularity $t = t_{CS}(r)$ and the event horizon are parallel straight lines in the (t, r) diagram, as in Fig. 18.15. (With arbitrary r, the event horizon would still be a parallely displaced copy of the singularity, but the graphs of both sets are in general curves.) The null geodesic equation in the Schwarzschild spacetime must then be integrated numerically.

Now we will show that a shell focussing singularity can be globally naked. The proof is similar to the one for shell crossings: the surface of the L–T dust ball has to be chosen at such a location that the ray emitted from the central point of the Big Crunch hits it before the surface crosses the apparent horizon. Then the ray can continue out to infinity in the Schwarzschild region. Figure 18.12 shows that this is possible, since that ray remains before the apparent horizon for some time, but we will prove it formally – using the same example as in Section 18.16, i.e. the model defined by (18.150) – (18.152).

Consider the function $f(r) \overset{\text{def}}{=} t_{AH}(r) - t_n(r)$. We have $f(0) = 0$, $\mathrm{d}f/\mathrm{d}r = 0$, while $\mathrm{d}^2 f/\mathrm{d}r^2(0) = 2a > 0$. From (18.158) we see that $\mathrm{d}^2 t_n/\mathrm{d}r^2$ is continuous at $r = 0$, and so is $\left(\mathrm{d}^2 t_{AH}/\mathrm{d}r^2\right) = 2a - 8M_0 r$. Consequently, $\mathrm{d}^2 f/\mathrm{d}r^2 > 0$ in some range $0 \le r \le r_1$ with $r_1 > 0$. This means that $\mathrm{d}f/\mathrm{d}r$ is a strictly increasing function for $0 < r \le r_1$, and so must be positive for $0 < r \le r_2 > r_1$. This, in turn, means that $f(r)$ is increasing for $0 < r \le r_2$, and so must be positive for $0 < r \le r_3 > r_2$. Thus, $f(r_3) > 0$, i.e. $t_n(r_3) < t_{AH}(r_3)$. Choose the surface of the dust ball at $r = b = r_3$, and match the L–T model to the Schwarzschild spacetime there. The ray emitted from the central point of the Big Crunch will reach the surface of the ball at $r = b$ with $t_n(b) < t_{AH}(b)$, and then will continue unimpeded to infinity in the Schwarzschild region. This was the essential point of Christodoulou's (1984) counterexample to the CCH, except that he used an $E < 0$ L–T model. Figure 18.12 illustrates this: the surface of the ball, $r = b$, can be chosen anywhere between $r = 0$ and the r at which the ray emitted from the singularity intersects the apparent horizon. The exterior region would then look the same as in Fig. 18.15.

The later-found counterexamples proved ever stronger violations of the CCH. We will describe them without going into details, since most of the arguments are highly technical.

Newman (1986) discussed the strength of naked singularities in L–T models. Imagine a nonspacelike geodesic and a set of spacelike vectors orthogonal to it and propagated parallely along it. The set is 3-dimensional at each point of a timelike geodesic and 2-dimensional along a null geodesic. The limiting focussing condition (LFC) means, roughly, that the volume (for timelike geodesics) or area (for null geodesics) of the parallelepiped spanned on the vectors decreases as the singularity is approached. The strong LFC means that the volume/area tends to zero at the singularity. The conclusions of the paper are: (1) radial null geodesics hitting or leaving a shell crossing do not obey the LFC; (2) radial null geodesics hitting or leaving a shell focussing singularity obey the LFC, but not the strong LFC. The first result implies that a shell crossing singularity is weak and thus is not a conclusive counterexample to cosmic censorship.

Waugh and Lake (1988) found that Newman (1986) excluded, through his assumptions about energy density, self-similar configurations.[15] They discussed the self-similar L–T model with $E = 0 = \Lambda$. They found that a globally naked and strong shell focussing singularity may form in it. It is strong in the sense that $\lim_{\lambda \to 0}(\lambda^2 R_{\alpha\beta} k^\alpha k^\beta) \ne 0$, where k^α is the tangent vector to a null geodesic hitting the singularity, λ is the affine parameter on the geodesic with $\lambda = 0$ at the singularity and $R_{\alpha\beta}$ is the Ricci tensor.

Gorini, Grillo and Pelizza (1989) identified another L–T model that was not captured by Newman and showed that it develops a globally naked shell focussing singularity that obeys the strong LFC, so it is to be taken as a serious counterexample to (one particular formulation of) the cosmic censorship postulate.

---

[15] A self-similar spacetime is one that admits a conformal Killing vector field with constant λ in (8.56).

Grillo (1991) showed by an example that a $\Lambda = 0$ L–T model with $E < 0$ and with a certain definite choice of $E(r), M(r)$ and $t_B(r)$ will form a locally naked shell focussing singularity that satisfies the strong LFC.

Dwivedi and Joshi (1992) identified a class of the functions $M(r)$ which generate a strong naked singularity at the centre. In this class, the first and second derivatives of the initial density by r are zero at $r = 0$. Most of the earlier papers considered special cases of this.

In another paper (Joshi and Dwivedi, 1993), the authors showed that in a class of the L–T models there is an at least locally naked strong curvature singularity at the centre. The class includes the configurations of Eardley and Smarr (1979) and of Christodoulou (1984). A continuous family of nonspacelike curves can be emitted from the singularity.

The status of the CCH in 1993 was reviewed by Clarke (1993), and, more extensively, by Joshi (1993).

Deshingkar, Joshi and Dwivedi (1999) defined a shell focussing singularity (SFS) to be strong if there exists a timelike or null geodesic γ terminating in it such that all the geodesic deviation fields V^α along γ obey $\mathcal{V} \stackrel{\text{def}}{=} (s_0 - s)^2 R_{\alpha\beta} V^\alpha V^\alpha = \text{constant} > 0$, where $R_{\alpha\beta}$ is the Ricci tensor, s is the affine parameter along γ and $s = s_0$ is the location of the singularity. By this definition, SFSs are *always strong*. The reason why earlier papers identified some of these singularities as weak was the selection of a special family of curves to calculate $\mathcal{V}$. The property of being strong is stable against perturbations of symmetry.

18.19 Solving the 'horizon problem' without inflation

The 'horizon problem' was explained in Section 17.12 and Fig. 17.7. The inflationary models solve it by postulating that at some time before last scattering the matter in the Universe was a scalar field acting like a repulsive cosmological constant of large value. Consequently, the Universe must have expanded exponentially, by the law $R \propto e^{\Lambda t}$. As observed by Célérier and Szekeres (2002), this not so much *solves*, but rather *postpones* the horizon problem: sufficiently distant volumes are still causally disconnected, but the present observer does not see them. At a time later than now, the observer would again be able to see regions that had not been in causal contact prior to $t = t_{ls}$.

Another solution of this problem was invented by Célérier and Schneider (1998). They assumed that the L–T Big Bang function $t_B(r)$ has a local minimum at $r = 0$ and is increasing with r so that there is a shell crossing at some $t > t_B(r)$ for every $r > 0$. As seen from (18.16), at a shell crossing, where $R_{,r} = 0$, a radial null geodesic has a horizontal tangent in the comoving coordinates. Consequently, all the curves that intersect a shell crossing are timelike or null at the intersection point. The time of shell crossing in the $E = 0$ model is $t \stackrel{\text{def}}{=} t_S(r) = t_B + 2Mt_{B,r}/M_{,r}$, where $t_{B,r} > 0$. Choosing the r coordinate so that $M = M_0 r^3$, and assuming that $t_{S,r} > 0$, i.e. $5t_{B,r} + 2rt_{B,rr} > 0$, we obtain that the curve $t = t_S(r)$ is timelike everywhere.

Now consider a radial null geodesic k_1 (obeying $t = t_n(r)$) sent backwards in time from the observer's position at $t = t_p, r = 0$, see Fig. 18.16. It obeys $(dt/dr)_n = -R_{,r} < 0$ in the whole spacetime region that is later than $t_S(r)$. Since the curve $t = t_S(r)$ started from $r = 0$ at $t < t_p$, and has a strictly positive derivative everywhere, the two curves must intersect somewhere, and $t_n(r)$ will have a zero derivative there. Let the point of

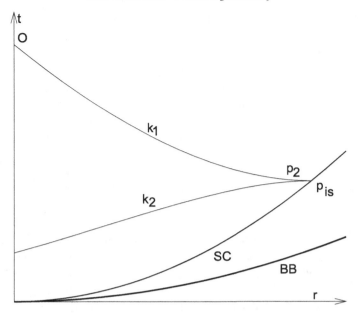

Fig. 18.16 A solution of the 'horizon problem' in an $E = 0$ L–T model with an increasing Big Bang (BB) function $t_B(r)$ (in the picture $t_B(r) = br^2$, where b is constant). The null radial geodesic k_1 sent backwards in time from the observer's position O hits the shell crossing SC at the point p_{is} of coordinates $(t, r) = (t_{is}, r_{is})$. From a point p_2 on k_1, of coordinates $t_2 > t_{is}$, $r_2 < r_{is}$, another radial null geodesic k_2 is sent backwards in time towards the centre. As explained in the text, k_2 must hit the centre at a time later than the BB at $r = 0$ (the shell crossing is in this example tangent to the BB at $r = 0$). Consequently, all areas of the sky that the observer O sees at present had a chance to be in causal contact with a common source in the past. This solves the 'horizon problem' for as long as the observer has in the field of view the shell crossing set $t = t_S(r)$ with $t'_S(r) > 0$; the solution is permanent if $t'_S(r) > 0$ for all $r > 0$.

intersection p_{is} have the coordinates (t_{is}, r_{is}). Consider another radial null geodesic k_2 sent backwards in time from a point p_2 of coordinates (t_2, r_2) with $t_2 > t_{is}$, $r_2 < r_{is}$ towards $r = 0$. The function $t = t_{n2}(r)$ describing that geodesic has a strictly positive derivative at every point at which $t_{n2}(r) > t_S(r)$. Assume that the curves $t_S(r)$ and $t_{n2}(r)$ are both differentiable for $r \in [0, r_2]$ and that the mass density is nowhere zero. It is then a simple exercise to prove that k_2 cannot intersect $t_S(r)$ at any r obeying $0 \leq r \leq r_2$. In brief, the argument is that at an intersection point q of coordinates (t_q, r_q) the derivative $t'_{n2}(r)$ would have to be zero and would have to go down to zero from positive values as r is decreasing. This means, when $t_{n2}(r)$ would be approaching $t_{n2}(r_q)$, the difference $[t_{n2}(r) - t_S(r)]$ would have to be increasing, and the intersection could not occur (see Exercise 12 for more details). Consequently, k_2 will reach the centre at $r = 0$ at a later time than the central point of the Big Bang, $t_B(0)$ (which coincides with the central point of the shell crossing, $t_S(0)$).

Consequently, there was a period in the past of the observer O (with time coordinates $t < T$) such that a signal sent out from the point $(t_1, 0)$ with $t_1 < T$ had enough time to proceed to the neighbourhood of the shell crossing, and then bounce back towards the observer, to reach her at $t = t_p$. Such a value of T exists for every $t_p > t_B(0)$. Hence, all the regions that the observer sees at t_p could have been in causal contact with a common source located in the observer's past. This solution of the 'horizon problem' is not just a postponement; when $t'_S(r) > 0$ for all $r > 0$, the problem is avoided permanently.

The two rays in the construction remain in the region $t > t_S(r)$, i.e. they never hit the shell crossing set. Hence, they remain all the time in the physical region of the spacetime.

The solution of the 'horizon problem' presented in this section assumes that the problem really existed in the form in which it was stated in the original paper (Guth, 1981). However, see the comment in Sec. 17.12 – in that original version this 'problem' was ill-posed and was in fact nonexistent.

18.20 * The evolution of $R(t, M)$ versus the evolution of $\rho(t, M)$

In the Friedmann models, the relation between the areal radius R and the mass density ϵ/c^2 is simple: the larger R, the smaller ϵ. In an L–T model, no such relation exists because of the possible existence of local condensations, as we will prove here.

We rewrite (18.23):

$$R^3(M) - R_0^3 = \frac{6}{\kappa} \int_{M_0}^{M} \frac{du}{\epsilon(u)}, \qquad (18.170)$$

where R_0 is a (possibly time-dependent) constant of integration and M_0 is another constant. Equation (18.170) means that the value of R at any M depends on the values of ϵ in the whole range $[M_0, M]$. Also the inverse relation to (18.170) is nonlocal – to find $\epsilon(t, M)$ we need to know R in an open neighbourhood of the value of M.

Assume now that $\epsilon_2 < \epsilon_1$ over the whole of $[M_0, M]$. Then

$$R_2^3(M) - R_1^3(M) - \left(R_0^3(t_2) - R_0^3(t_1) \right) = \frac{6}{\kappa} \int_{M_0}^{M} \left(\frac{1}{\epsilon_2(u)} - \frac{1}{\epsilon_1(u)} \right) du > 0. \qquad (18.171)$$

Hence, $R_2^3(M) > R_1^3(M)$ if

$$R_0^3(t_2) - R_0^3(t_1) + \frac{6}{\kappa} \int_{M_0}^{M} \left(\frac{1}{\epsilon_2(u)} - \frac{1}{\epsilon_1(u)} \right) du > 0. \qquad (18.172)$$

This will hold if, for example, $R_0(t_i) = 0$, $i = 1, 2$, in addition to $\epsilon_2 < \epsilon_1$.

The converse is not true. This is easy to understand on physical grounds: $R(t_2, M) > R(t_1, M)$ for all $M \in [M_0, M_1]$ means that every shell of constant $M = \widetilde{M}$ has a larger radius at t_2 than it had at t_1. However, the neighbouring shells may have moved closer to $\widetilde{M}$ at t_2 than they were at t_1. If they did, then a local condensation around $\widetilde{M}$ was created, which may result in $\epsilon(t_2, \widetilde{M})$ being larger than $\epsilon(t_1, \widetilde{M})$. This does not happen in the Friedmann limit, where local condensations are excluded by the symmetry assumptions.

A sufficient condition for $\epsilon_2(M) < \epsilon_1(M)$, $M \in [M_0, M_1]$, is

$$\frac{1}{(R_1^3)_{,M}} > \frac{1}{(R_2^3)_{,M}} \qquad \text{for all } M \in [M_0, M_1]. \qquad (18.173)$$

If $R_{,M} > 0$ for all $M \in [M_0, M_1]$ at both t_1 and t_2 (i.e. there are no shell crossings in $[M_0, M_1]$), then this is equivalent to $(R_2{}^3)_{,M} > (R_1{}^3)_{,M}$ for all $M \in [M_0, M_1]$. Incidentally, this implies $R_2 > R_1$ for all $M \in [M_0, M_1]$ if $R_2(M_0) > R_1(M_0)$.

18.21 * Increasing and decreasing density perturbations

Formation of structures in the Universe used to be described by approximate perturbations of the Robertson–Walker models. In that approach, two classes of perturbations had been identified: those that increase with time and those that decrease with time. The same classification can be done in the L–T and Szekeres models (for the latter see Section 20.4). For the L–T model with $\Lambda = 0$, this approach was first applied by Silk (1977). His results are re-derived here, by a somewhat different method.

The quantities used in studying perturbations of the Friedmann models are the density contrast $\epsilon_{,r}/\epsilon$ and the curvature contrast $R^{(3)}_{,r}/R^{(3)}$, where $R^{(3)}$ is the scalar curvature of the hypersurfaces $t = $ constant. Both these functions vanish identically in the Friedmann limit. Since a perturbation with $E = 0$ would be rather exceptional, we assume $E \neq 0$. From (18.15) and (18.25) we find

$$\frac{\epsilon_{,r}}{\epsilon} = \frac{M_{,rr}}{M_{,r}} - 2\frac{R_{,r}}{R} - \frac{R_{,rr}}{R_{,r}}, \qquad R^{(3)} = -\frac{4(RE)_{,r}}{R^2 R_{,r}}, \tag{18.174}$$

$$\Delta K \overset{\text{def}}{=} \frac{R^{(3)}_{,r}}{R^{(3)}} = \frac{E_{,rr}/E - 2R_{,r}{}^2/R^2 - (R_{,rr}/R_{,r})(E_{,r}/E)}{R_{,r}/R + E_{,r}/E}. \tag{18.175}$$

For $E > 0$ we define

$$\alpha(r) \overset{\text{def}}{=} \frac{M_{,r}}{M} - \frac{E_{,r}}{E}, \qquad \beta(r) \overset{\text{def}}{=} \frac{3}{2}\frac{E_{,r}}{E} - \frac{M_{,r}}{M}, \qquad \gamma(r) \overset{\text{def}}{=} \frac{|2E|^{3/2}}{M} t_{B,r}, \tag{18.176}$$

and then find using (18.113) – (18.114) and (18.21):

$$\frac{R_{,rr}}{R_{,r}} = \alpha_{,r}\frac{R}{R_{,r}} + \alpha + \left(\frac{R\beta_{,r}}{R_{,r}} + \beta\right)\frac{\sinh\eta(\sinh\eta - \eta)}{(\cosh\eta - 1)^2}$$

$$+ \frac{R\beta}{R_{,r}}\frac{2\eta + \eta\cosh\eta - 3\sinh\eta}{(\cosh\eta - 1)^3}[\beta(\sinh\eta - \eta) - \gamma] \tag{18.177}$$

$$- \left(\frac{R\gamma_{,r}}{R_{,r}} + \gamma\right)\frac{\sinh\eta}{(\cosh\eta - 1)^2} + \frac{R\gamma}{R_{,r}}\frac{\cosh\eta + 2}{(\cosh\eta - 1)^3}[\beta(\sinh\eta - \eta) - \gamma].$$

We need to know the behaviour of $\epsilon_{,r}/\epsilon$ and ΔK as $\eta \to 0$ and $\eta \to \infty$ in two cases: when $t_{B,r} = 0$ and when $[(2E)^{3/2}/M]_{,r} = 0$. For this purpose, we must know the behaviour in these limits of a few functions present in (18.174) – (18.175). Two of them are $\Phi_3(\eta)$ and $\Phi_4(\eta)$, defined in (18.114); their relevant properties are given in (18.115). The other functions are (a coefficient at ∞ is meant to determine the sign):

$$\lim_{\eta \to 0}\frac{R_{,r}}{R} = \frac{M_{,r}}{3M}, \qquad \lim_{\eta \to \infty}\frac{R_{,r}}{R} = \frac{E_{,r}}{2E} \qquad \text{when } t_{B,r} = 0, \tag{18.178}$$

$$\lim_{\eta \to 0}\frac{R_{,r}}{R} = -t_{B,r} \times \infty, \qquad \lim_{\eta \to \infty}\frac{R_{,r}}{R} = \frac{M_{,r}}{3M} \qquad \text{when } \left[\frac{(2E)^{3/2}}{M}\right]_{,r} = 0, \tag{18.179}$$

$$\psi_3(\eta) \overset{\text{def}}{=} \frac{(2\eta + \eta \cosh \eta - 3 \sinh \eta)(\sinh \eta - \eta)}{(\cosh \eta - 1)^3},$$

$$\lim_{\eta \to 0} \psi_3(\eta) = 0, \qquad \lim_{\eta \to \infty} \psi_3(\eta) = 0, \tag{18.180}$$

$$\psi_4(\eta) \overset{\text{def}}{=} \frac{\cosh \eta + 2}{(\cosh \eta - 1)^3}, \qquad \lim_{\eta \to 0} \psi_4(\eta) = \infty, \qquad \lim_{\eta \to \infty} \psi_4(\eta) = 0, \tag{18.181}$$

$$\lim_{\eta \to 0} \frac{R_{,r}}{R} \frac{(\cosh \eta - 1)^2}{\sinh \eta} = -\frac{(2E)^{3/2}}{M} t_{B,r}, \qquad \lim_{\eta \to 0} \frac{R_{,r}}{R} (\cosh \eta - 1)^2 = 0. \tag{18.182}$$

With $t_{B,r} = 0$, we now find the following limits:

$$\lim_{\eta \to 0} \frac{R_{,rr}}{R_{,r}} = \frac{M_{,rr}}{M_{,r}} - \frac{2}{3} \frac{M_{,r}}{M}, \qquad \lim_{\eta \to \infty} \frac{R_{,rr}}{R_{,r}} = \frac{E_{,rr}}{E_{,r}} - \frac{E_{,r}}{2E}. \tag{18.183}$$

With these, one finds that $\epsilon_{,r} / \epsilon \xrightarrow[\eta \to 0]{} 0$, while

$$\lim_{\eta \to 0} \Delta K = \frac{\dfrac{E_{,rr}}{E} - \dfrac{2M_{,r}{}^2}{9M^2} - \dfrac{E_{,r}}{E} \left(\dfrac{M_{,rr}}{M_{,r}} - \dfrac{2M_{,r}}{3M} \right)}{M_{,r} / (3M) + E_{,r} / E}. \tag{18.184}$$

As $\eta \to \infty$, still with $t_{B,r} = 0$, we obtain

$$\lim_{\eta \to \infty} \frac{\epsilon_{,r}}{\epsilon} = \frac{M_{,rr}}{M_{,r}} - \frac{E_{,rr}}{E_{,r}} - \frac{E_{,r}}{2E}, \qquad \lim_{\eta \to \infty} \Delta K = 0. \tag{18.185}$$

This shows that with $t_{B,r} = 0$ the perturbation of the Friedmann density vanishes at the initial singularity and tends to a finite limit as $t \to \infty$, i.e. it is increasing. The perturbation of curvature displays the exactly opposite behaviour.

With $[(2E)^{3/2}/M]_{,r} = 0$, the limits are

$$\lim_{\eta \to 0} \frac{\epsilon_{,r}}{\epsilon} = +t_{B,r} \times \infty = \lim_{\eta \to 0} \Delta K \tag{18.186}$$

(see Exercise 13 for hints on $\lim_{\eta \to 0} \Delta K$),

$$\lim_{\eta \to \infty} \frac{\epsilon_{,r}}{\epsilon} = 0 = \lim_{\eta \to \infty} \Delta K, \tag{18.187}$$

so the perturbations generated by $t_{B,r}$ are decreasing.

Now we repeat the calculations in (18.174) – (18.187) for the $E < 0$ model. Since infinities will appear both in $R_{,rr}/R_{,r}$ and in $R_{,r}/R$, we must substitute for these quantities before taking the limits. Using (18.15), (18.19), (18.106) – (18.108) and (18.176) we find

$$\begin{aligned} \frac{\epsilon_{,r}}{\epsilon} &= \frac{M_{,rr}}{M_{,r}} - \frac{\alpha_{,r} R}{R_{,r}} - 3\alpha - \left(\frac{R\beta_{,r}}{R_{,r}} + 3\beta \right) \frac{\sin \eta (\eta - \sin \eta)}{(1 - \cos \eta)^2} \\ &\quad - \frac{R\beta}{R_{,r}} \times \frac{-2\eta - \eta \cos \eta + 3 \sin \eta}{(1 - \cos \eta)^3} [\beta(\eta - \sin \eta) - \gamma] \\ &\quad + \left(\frac{R\gamma_{,r}}{R_{,r}} + 3\gamma \right) \frac{\sin \eta}{(1 - \cos \eta)^2} - \frac{\gamma R}{R_{,r}} \frac{\cos \eta + 2}{(1 - \cos \eta)^3} [\beta(\eta - \sin \eta) - \gamma], \end{aligned} \tag{18.188}$$

$$\Delta K = -2 \frac{R_{,r}}{R} + \frac{E_{,rr}/E + (E_{,r}/E)(\epsilon_{,r}/\epsilon - M_{,rr}/M_{,r})}{R_{,r}/R + E_{,r}/E}. \tag{18.189}$$

Here are some useful limits. For $t_{B,r} = 0 = \gamma$:

$$\lim_{\eta \to 0} \frac{R_{,r}}{R} = \frac{M_{,r}}{3M}, \qquad \lim_{\eta \to 2\pi} \frac{R_{,r}}{R} = -\beta \times \infty \equiv -\left(\frac{3E_{,r}}{2E} - \frac{M_{,r}}{M}\right) \times \infty, \qquad (18.190)$$

and for $[(2E)^{3/2}/M]_{,r} = 0 = \beta$:

$$\lim_{\eta \to 0} \frac{R_{,r}}{R} = -t_{B,r} \times \infty, \qquad \lim_{\eta \to 2\pi} \frac{R_{,r}}{R} = t_{B,r} \times \infty. \qquad (18.191)$$

Other functions and limits are:

$$\psi_1(\eta) \stackrel{\text{def}}{=} \frac{(-2\eta - \eta \cos \eta + 3 \sin \eta)(\eta - \sin \eta)}{(1 - \cos \eta)^3},$$

$$\lim_{\eta \to 0} \psi_1(\eta) = 0, \qquad \lim_{\eta \to 2\pi} \psi_1(\eta) = -\infty, \qquad (18.192)$$

$$\psi_2(\eta) \stackrel{\text{def}}{=} \frac{\cos \eta + 2}{(1 - \cos \eta)^3}, \qquad \lim_{\eta \to 0} \psi_2(\eta) = \lim_{\eta \to 2\pi} \psi_2(\eta) = \infty. \qquad (18.193)$$

In calculating the limits with $\gamma = 0$, we observe that $\Phi_1(\eta)/\psi_1(\eta) \xrightarrow[\eta \to 2\pi]{} 0$, so the coefficient of ψ_1 determines the behaviour of the whole $\epsilon_{,r}/\epsilon$ at $\eta \to 2\pi$. Other useful limits are:

$$\lim_{\eta \to 0} \frac{R_{,r}}{R} \frac{(1 - \cos \eta)^2}{\sin \eta} = -\frac{(-2E)^{3/2}}{M} t_{B,r} = -\gamma, \qquad (18.194)$$

$$\lim_{\eta \to 0} \frac{R_{,r}}{R}(1 - \cos \eta)^2 = 0 = \lim_{\eta \to 2\pi} \frac{R_{,r}}{R}(1 - \cos \eta)^2, \qquad (18.195)$$

$$\lim_{\eta \to 2\pi} \frac{R_{,r}}{R} \frac{(1 - \cos \eta)^2}{\sin \eta} = 2\pi\beta - \gamma. \qquad (18.196)$$

Now, when $t_{B,r} = 0$ ($\Longrightarrow \gamma = 0$), we find $\lim_{\eta \to 0} \epsilon_{,r}/\epsilon = 0$ and $\lim_{\eta \to 2\pi} \epsilon_{,r}/\epsilon = +\infty$. Thus, the gradient of $[(-2E)^{3/2}/M]$ generates a perturbation of the Friedmann mass density that vanishes at the Big Bang and increases without limits at the Big Crunch.

With $[(-2E)^{3/2}/M]_{,r} = 0$, the density contrast tends to infinity at both singularities. In this case, the time-difference between the bang and the crunch singularities is constant: $t_C - t_B = 2\pi/B$, where $B \stackrel{\text{def}}{=} (-2E)^{3/2}/M = \text{constant}$. Thus, it is impossible to fulfil both of (18.109) – (18.110) and shell crossings are unavoidable.

For ΔK we obtain the following results. With $t_{B,r} = 0$ ($\gamma = 0$):

$$\lim_{\eta \to 0} \Delta K = -\frac{2M_{,r}}{3M} + \frac{E_{,rr}/E - E_{,r} M_{,rr}/(EM_{,r})}{M_{,r}/(3M) + E_{,r}/E},$$

$$\lim_{\eta \to 2\pi} \Delta K = \beta \times \infty \qquad (18.197)$$

(only the first term in (18.189) is infinite).

With $[(-2E)^{3/2}/M]_{,r} = 0$ ($\beta = 0$), the limits are

$$\lim_{\eta \to 0} \Delta K = -\lim_{\eta \to 2\pi} \Delta K = t_{B,r} \times \infty. \qquad (18.198)$$

Thus, the gradient of $[(-2E)^{3/2}/M]$ generates increasing curvature contrast, while the gradient of t_B generates perturbations that increase without limits at both singularities. As can be seen from (18.174) – (18.177) and (18.113), the terms containing β and γ

do not simply add in $\epsilon_{,r}/\epsilon$ and ΔK: they interact nonlinearly. Hence, unlike with the linearised perturbations of the Friedmann models, conclusions drawn from investigating the cases $\beta = 0$ and $\gamma = 0$ separately do not give a complete picture of the general case.

We recall what had already been said in Sec. 18.12: cosmological models obeying the Einstein equations with dust source do not apply before the epoch of emission of the CMB radiation. Hence, there is no reason to dismiss the L–T models with $dt_B/dr \neq 0$ on the ground that they contain 'huge' inhomogeneities close to the Big Bang.

18.22 Mimicking accelerating expansion of the Universe by inhomogeneities in matter distribution

The accelerating expansion of the Universe was deduced from observations of the type Ia supernovae, see Sec. 17.13. It was *assumed* that for all of them the absolute luminosity at peak is the same. The observed luminosities were inconsistent with the predictions of the $\Lambda = 0$ Friedmann model. For other Friedmann models, the best consistency with observations was achieved when $k = 0$, 32% of the energy density comes from matter (visible or dark) and 68% of the energy density comes from 'dark energy', which plays the role of Λ (Planck, 2014). Thus, the conclusion about accelerating expansion of the Universe followed from the *interpretation of observations via a pre-assumed Friedmann model*, so it is not an objectively registered fact. The example given below (Célérier, 2000; Iguchi et al., 2002; Yoo et al., 2011; Krasiński 2014a) shows how the accelerating expansion can be imitated using an L–T model, without introducing 'dark energy'.

By a reasoning closely analogous to that which led to (17.62) we find that in an L–T geometry, the observer area distance r_O and the luminosity distance D_L between the central observer at $R = 0$ and a light source at (t_e, r_e) are $R(t_e, r_e)|_{\text{ray}}$ and $D_L = (1 + z)^2 R(t_e, r_e)|_{\text{ray}}$, respectively. To calculate D_L as a function of z we need to know the functions $t(r)$ and $z(r)$ along the ray. The first can be found (numerically) as a solution of the null geodesic equation in the metric (18.16) with (18.14). Once this is done, the function $z(r)$ can be found from the Bondi formula (18.126). We then equate the $D_L(z)$ for the L–T model found in this way to the $D_L(z)$ for the ΛCDM model, Eq. (17.69):

$$(1 + z)^2 R(t(z), r(z)) = \frac{1 + z}{\mathcal{H}_0} \int_0^z \frac{\mathrm{d}z'}{\sqrt{\Omega_m(1 + z')^3 + \Omega_\Lambda}}, \tag{18.199}$$

where $\Omega_m = 0.32$, $\Omega_\Lambda = 0.68$ and $c\mathcal{H}_0 = H_0 = 67.1$ km/(s × Mpc) are taken from the current observations (Planck, 2014). Let us assume $E/r^2 = C = \text{constant} > 0$, as in Friedmann models (the solution of (18.199) exists for every C). Then (18.199) implicitly defines $t_B(r)$ via (18.19) – (18.21), so $t_B(r)$ can be numerically calculated.

The $t_B(r)$ thus calculated, together with the assumed $E(r)$, determine the L–T model that has the same function $D_L(z)$ for the central observer as the ΛCDM model, Eq. (17.69). Figure 18.17 shows the comparison of the past light cones of the two models for the same present central observer; it also shows the corresponding Big Bang graphs: the constant one for ΛCDM, and the $t_B(r)$ calculated numerically from (18.199). The $t_B(r)$ graph asymptotically approaches the constant t_{BF} of ΛCDM.

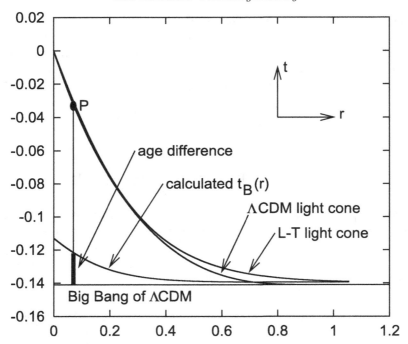

Fig. 18.17 The past light cones of the present central observer in the L–T model that reproduces the $D_L(z)$ from (17.69) *using only* $t_B(r)$ and of the ΛCDM model with the same $D_L(z)$. The vertical straight line is the worldline of a particle of the cosmic fluid. At P it intersects the past light cone of the present central observer, which nearly coincides with the ΛCDM past light cone there. The thick vertical segment shows the difference in age between the ΛCDM and L–T models for the particle observed at P.

This is how the L–T model from Fig. 18.17 mimics the apparent accelerating expansion. In the L–T model, the BB occurs progressively later when the observer position is approached. At the point P the particle in the L–T model is 'younger' than in the corresponding Friedmann model. The age difference increases towards the observer. Therefore, the particle in the L–T model is less decelerated than it would be in a $\Lambda = 0$ Friedmann model that has the BB at $t = t_{BF}$. So, the L–T expansion velocity at P is larger than in that Friedmann model, and the difference in velocities increases towards the observer. Instead of increasing with time, the expansion velocity increases relative to the Friedmann model on approaching the position of the observer. The conclusion is: if an L–T model were used to interpret the observations, no 'accelerating expansion' would appear and there would be no need to introduce the 'dark energy'.

The model discussed above did away with dark energy by using just $t_B(r)$; the function $E(r)$ was assumed the same as in a Friedmann model. The dark energy can be explained away also by assuming t_B constant and adapting $E(r)$ to fulfil (18.199) (Iguchi et al., 2002; Krasiński, 2014b). With $E(r)$ and $t_B(r)$ being both arbitrary one can explain away the acceleration and fit the L–T model to one more set of data (Célérier et al., 2010, see also the more astronomically minded paper by Balazs and Bene, 2018).

The models presented above are too simple to be fitted to the whole set of observations of the type Ia supernovae. They are meant to be a warning that the idea of dark energy may have no basis in reality, similarly to the (now long-forgotten) ideas of aether and continuous creation of matter in steady-state models of the Universe.

The earliest paper in which an inhomogeneous model was used to mimic the accelerating expansion of the Universe was that of Tomita (2000). The author considered a configuration in which a Friedmann region of mass density lower than the large-scale average surrounds the symmetry centre and goes over into another Friedmann region with density larger than average. Because of that historical precedent, many authors now reflexively refer to the L–T models mimicking the accelerating expansion by the term 'void models of acceleration'. But this term is misleading – the entity responsible for the spurious acceleration is an inhomogeneity of mass distribution. The presence of the central void is needless as shown using an explicit example by Célérier et al. (2010).

The following must be stressed: what requires explanation *is not* the accelerating expansion of the Universe, but the measurements of distances to the type Ia supernovae. The latter is a legitimate astronomical task, the former is not an objectively established fact but a model-dependent interpretation of observations that may turn out to be an illusion.

18.23 Drift of light rays

This topic will be presented here briefly and only qualitatively; see Krasiński and Bolejko (2011) for details.[16] It was discussed at a more general level by Korzyński and Kopiński (2018) and by Grasso et al. (2019), see Sec. 16.13.

Null geodesics in the metric (18.16) cannot have constant r over any open segment because (18.135) shows that then also ϑ and φ would be constant, so such a geodesic would be timelike. But $dr/d\lambda = 0$ is possible at isolated points. Thus, r can be used as a (nonaffine) parameter along every segment of a null geodesic on which $dr/d\lambda \neq 0$. Imagine two *nonradial* light rays in an L–T spacetime, the second one emitted by the same source later by τ, both arriving at the same observer. Let the trajectory of the first ray be

$$(t, \vartheta, \varphi) = (T(r), \Theta(r), \Phi(r)). \tag{18.200}$$

Unlike for radial geodesics, we cannot assume that the second ray will proceed through the same values of Θ and Φ as the first one. So, the equation of the second ray will be

$$(t, \vartheta, \varphi) = (T(r) + \tau(r), \Theta(r) + \zeta(r), \Phi(r) + \psi(r)). \tag{18.201}$$

Already at this point, without any calculations, we may conclude the following: the second ray intersects each given hypersurface $r = r_0$ not only later but, in general, at a different comoving location. Consequently, the two rays intersect different sequences of matter worldlines between the source and the observer. The second ray will thus be received by the observer from a different direction in the sky. The result reported below was obtained by linearising the null geodesic equations of the second ray in $\tau(r)$, $\zeta(r)$ and $\psi(r)$.

[16] The paper by Krasiński and Bolejko discussed the drift in Szekeres spacetimes, but the numerical example quoted here was calculated in the L–T limit.

It follows that an observer in an L–T spacetime should see typical light sources drift across the sky. The drift vanishes only for radial directions. The only spacetimes in the L–T family in which there is no drift for all observers and all rays are the Friedmann models (Krasiński and Bolejko, 2011). Observational detection of the drift would thus be evidence of inhomogeneity of the Universe on large scales.

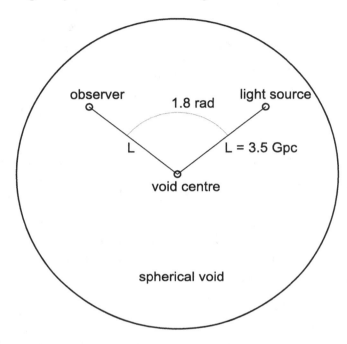

Fig. 18.18 The L–T configuration used in the example. The present distance from the centre of the void to the observer and to the light source is $L = 3.5$ Gpc. The radius of the void is drawn not to scale.

Figures 18.18 and 18.19 show an example of the drift. The values of the parameters in them are only illustrative, they are not related to any real object. The rays propagate through a void of radius ≈ 7 Gpc, the mass density at the centre of the void is $\approx 1/5$ of the average background density, the observer and the light source are at ≈ 3.5 Gpc from the centre of the void, the density profile is $\rho(t_0, x) = \rho_0 \left[1 + \delta - \delta \exp\left(-x^2/\sigma^2\right)\right]$, where t_0 is the current instant, x is defined as $R(t_0, r)$, ρ_0 is the density at the origin, $\delta = 4.05$ and $\sigma = 2.96$ Gpc. We considered rays emitted by still the same source and reaching the same fixed observer. The projections of three rays emitted at (very) different times on the space $t = $ now along the dust flow lines are shown in Fig. 18.19.

The drift rate has not yet been measured for any observed object. For the configuration of Figs. 18.18 and 18.19 it is $\approx 10^{-6}$ arc seconds per year. With the accuracy of the GAIA (Global Astrometric Interferometer for Astrophysics) mission[17] (Gaia, 2020), a few years of monitoring a given source would be needed to detect this effect.

---

[17] At the moment of writing this the predicted parallax standard errors range from 7×10^{-6} arc sec for stars of magnitude 13 to 588×10^{-6} arc sec for stars of magnitude 20.7 (Gaia, 2020).

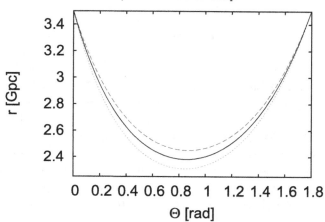

Fig. 18.19 Rays proceeding from the source to the observer in Fig. 18.18 projected on the space $t =$ now along the dust flow lines. **Middle line:** the ray received now. **Upper line:** the ray received 5×10^9 years ago. **Lower line:** the ray received 5×10^9 years in the future.

18.24 * L&T curio shop

In this section, a selection of material will be presented that did not fit under any other heading in this chapter. An overview of papers discussing the L–T models, complete until 1994, can be found in the book by Krasiński (1997).

18.24.1 Lagging cores of the Big Bang

Novikov (1964a) used a similar setup to that of Bonnor (1956) (see Section 18.6) to describe the *lagging cores of Big Bang* (BB). He considered the flat Friedmann model with several Schwarzschild holes matched into it. Inside each hole there was a small expanding Friedmann region that had its local BB later than the BB of the background Friedmann universe. Particles and radiation emitted from such lagging cores of BB should in principle be observable. The hope was that this could explain the source of energy in quasars, but this attempt has had no following except the paper mentioned below.

Neeman and Tauber (1967) considered more possibilities for matter inside the core: it could be either the Friedmann dust or the so-called stiff fluid (equation of state $\epsilon = p$), both with or without Λ. However, in every case there were difficulties to reconcile the parameters of the model with the results of observations in a self-consistent way.

Miller (1976) described an L–T universe with $E = 0 = \Lambda, t_{B,r} < 0$ gradually emerging from the BB singularity. With the BB first occurring far from the centre and proceeding inwards, one can assign mass to the singularity and interpret it as the mass to be emitted. This mass is the function $M(r)$ from (18.14) calculated at $r = r_B(t)$, where $r_B(t)$ is the inverse function to $t_B(r)$; it is well behaved on the singularity. When the BB reaches $r = 0$, the singularity may either disappear or continue to exist with a negative mass and still emitting matter. In the second case it becomes a permanent timelike singularity.

The idea of lagging cores of the BB found its continuation in a series of papers by this author (Krasiński, 2016, 2018 and 2020). In them, the 'lagging cores' were upward 'bumps' on the Big Bang set in the L–T and Szekeres models. It was shown, by numerical calculation of null geodesics in an exact metric, that selected rays of the CMB radiation, when emitted above the slope of the 'bump', would reach the present observers with a strong blueshift $(1+z \leq 10^{-5})$. The parameters of the bumps were such that the blueshifted rays shared several properties with the observed gamma ray bursts. The motivating idea was that blueshifted rays should not be cursed and banned but rather watched for because they may hide in the results of some astronomical observations already made. The gamma ray bursts were chosen for scrutiny because stronger blueshifts are more difficult to produce without contradicting observational facts (like the angular diameter of the observed gamma-ray bursts). Larger (i.e. less blueshifted) $1 + z$ requires smaller amplitude and diameter of the bump on the BB, and then the risk of running into contradiction with observations is smaller. Such less-blueshifted rays may hide among sources of lower-frequency electromagnetic radiation (X or ultraviolet rays). The blueshifted rays are ordinary CMB rays that happened to go off in the preferred directions. In an L–T model all radial directions in the last-scattering epoch (LSE) are preferred in this sense, and in the quasi-spherical Szekeres models the preferred directions are only two, opposite to each other, at every point of the LSE. This last property is especially promising because the observed gamma ray bursts are believed to be strongly collimated. This idea met a violent opposition from astrophysically minded referees and will likely not be further pursued.

18.24.2 Strange or nonintuitive properties of the L–T model

Novikov (1962a) showed that, with some choices of the arbitrary functions in the L–T model, adding some amount of rest mass to a source may decrease the active gravitational mass. The model he used (a Friedmann region going over into the Schwarzschild spacetime through an interpolating L–T region) was specified numerically, and the effect was demonstrated only for a definite amount of mass added. Novikov's interpretation: 'The (negative) potential energy of gravitational interaction between the medium added and the one present previously is greater than the energy corresponding to the mass of the medium added. This leads to a decrease of the total gravitational mass.' The effect is an example of what happens beyond the neck (see Section 18.10): adding more rest mass results in reducing the total active mass. By another numerically specified example Novikov showed that an infinite amount of added rest mass may leave the active mass still finite.

Hellaby and Lake (1985) presented a few more examples that run counter to the intuitions based on the properties of the R–W models:

1. A model with $E > 0$ everywhere that has a globally closed space.
2. A model with $E < 0$ everywhere that has an open space.
3. A region of negative curvature placed between two regions of positive curvature which does not develop shell crossings.

The first situation occurs when, for example:

$$M = M_0\gamma^m, \qquad E = E_0\gamma^{2m/3}, \qquad t_B = -a_0\gamma^p, \qquad (18.202)$$

where M_0, E_0, a_0, m and p are positive constants, and

$$\gamma(r) \overset{\text{def}}{=} 3\sin(\pi r/l) + 2\sin(3\pi r/l), \tag{18.203}$$

$l > 0$ being one more constant. Note that $\gamma(0) = 0 = \gamma(l)$, and $\gamma(r) > 0$ for all $0 < r < l$. For this model, $(2E)^{3/2}/M = (2E_0)^{3/2}/M_0 \overset{\text{def}}{=} C = $ constant and, from (18.19):

$$R(t,r) = \frac{M_0}{2E_0}\gamma^{m/3}(\cosh\eta - 1), \qquad \sinh\eta - \eta = C(t + a_0\gamma^p). \tag{18.204}$$

Thus, for all $t > t_B$, $R = 0$ at $r = 0$ and at $r = l$, and $0 < R < \infty$ for $0 < r < l$. Since $R = 0$ at two different values of r for each t, each space $t = $ constant has two origins and a finite radius. Still, at every r the space expands to infinite size as $t \to \infty$. Hence, the model has closed spaces, but is ever-expanding – a situation that cannot happen in the Friedmann models with $\Lambda = 0$.

The second situation occurs in the spacetime discussed in Section 18.9. Hellaby and Lake's example has similar qualitative properties.

The third situation, an ever-expanding region sandwiched between two recollapsing regions (see Fig. 18.20), occurs with the following functions:

$$\begin{aligned} E &= E_0 r^2(l-r)^2(br-1)[b(l-r)-1], \\ M &= M_0 r^3(l-r)^3, \qquad t_B = -a_0 r(l-r), \end{aligned} \tag{18.205}$$

where the constants E_0, M_0, a_0 and l are all positive and $b > 2/l$. All three functions have zeros at $r = 0$ and $r = l$, while $E(r)$ has two additional zeros at $r = r_1 = 1/b$ and at $r = r_2 = l - 1/b$. For $0 < r < r_1$ and for $r_2 < r < l$, $E(r) < 0$; for $r_1 < r < r_2$ $E(r) > 0$, while $M(r) > 0$ for all $0 < r < l$. It follows that somewhere in (r_1, r_2) E must have at least one local maximum, while in each of the intervals $(0, r_1)$ and (r_2, l) it must have at least one local minimum. The functions M and E have local maxima at $r = l/2$, and E has two other local extrema at

$$\begin{aligned} r = r_{3,4} &= \frac{l}{2} \mp \frac{1}{6b}\sqrt{3\left(3b^2l^2 - 8bl + 8\right)}, \\ 0 &< r_3 < 1/b, \qquad l - 1/b < r_4 < l. \end{aligned} \tag{18.206}$$

They must be local minima. At both of them E has the same value

$$E_{\min} = -\frac{4E_0}{27b^4}(bl-1)^3. \tag{18.207}$$

We have to ensure that $E(r) \geq -1/2$ for all $r \in [0, l]$ and that there are no shell crossings. It suffices to verify $E(r) \geq -1/2$ at the minima of E, the result is

$$E_0 < \frac{27}{8}\frac{b^4}{(bl-1)^3}. \tag{18.208}$$

With the functions (18.205), the no-shell-crossing conditions in the $E > 0$ region are fulfilled (see Sec. 18.10): we have $t_{B,r} < 0$, $M_{,r} > 0$ and $E_{,r} > 0$ for $r < l/2$; $t_{B,r} = M_{,r} = E_{,r} = 0$ for $r = l/2$; and $t_{B,r} > 0$, $M_{,r} < 0$, $E_{,r} < 0$ for $r > l/2$.

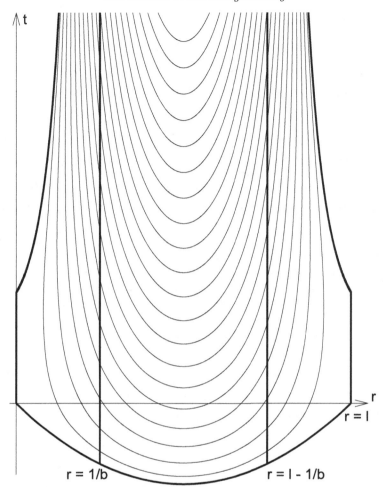

Fig. 18.20 The evolution of the (t, r) surface in the model given by (18.205). The thin lines are contours of constant R. All differences in the value of R between neighbouring contours are the same. The curve at the bottom is the Big Bang, the two curves bordering the surface on the right and on the left together form the Big Crunch. The two straight lines, at $r = r_1 = 1/b$ and at $r = r_2 = l - 1/b$, separate the region with $E(r) > 0$ (between them) from the regions with $E(r) < 0$. For $r < r_1$ and $r > r_2$, the lifetime of the model is finite, but it tends to infinity as r approaches r_1 or r_2. The short straight segments at $r = 0$ and at $r = l$ are the two centres of spherical symmetry.

In the $E < 0$ region, (18.110) has to be fulfilled for $r < l/2$, and its opposite for $r > l/2$. The left-hand side of (18.110) is equivalent to

$$\left[3\pi b^2 - \frac{(2E_0)^{3/2}}{M_0} a_0 \left(b^2 r^2 - b^2 lr + bl - 1 \right)^{5/2} \right] (2r - l). \qquad (18.209)$$

The trinomial in r is positive for $r < r_1$ and for $r > r_2$, i.e. in the whole $E < 0$ interval. In $[0, r_1]$ it has an upper bound at $r = 0$, in $[r_2, l]$ the upper bound is at $r = l$; both bounds are equal to $(bl - 1)$. So, in order to make the expression in square brackets in (18.209) positive it suffices to ensure that it is positive at these bounds. This means

$$M_0 > \frac{(2E_0)^{2/3}}{3\pi b^2} (bl - 1)^{5/2}. \qquad (18.210)$$

Then the factor $(2r - l)$ in (18.209) ensures the correct overall sign of the whole expression in the ranges $r < l/2$ and $r > l/2$.

Just like the spacetime (18.202) – (18.204), the one specified in (18.205) has two centres, at $r = 0$ and at $r = l$. $E(r)$ is negative in the vicinity of each centre and positive around $r = l/2$. Each of the two $E < 0$ regions recollapses onto a different centre, and the ever-expanding region between them can exist without shell crossings being created.

Figure 18.20 shows the evolution of the (t, r) surfaces in this model. At $r = 0$, $r = r_1$, $r = r_2$ and $r = l$ the $r = $ constant shells evolve by the $E = 0$ law. For every $0 < r < r_1$ and for every $r_2 < r < l$, the evolution proceeds by the $E < 0$ law, and that part of the space recollapses after a finite time T, with $T \to \infty$ as $r \to r_1$ or $r \to r_2$. For $r_1 < r < r_2$, the evolution proceeds by the $E > 0$ law and the model has infinite future.

18.24.3 Chances to fit an L–T model to observations

Mészáros (1986) considered the implications of the assumption that the observed part of the Universe is described by a finite-volume subset of the L–T model with a simultaneous BB. There are two ways of verifying this assumption: (1) by observing the outer surface of the L–T sphere; (2) by detecting the *cosine anisotropy* in the matter density and in the Hubble parameter. There is no observational evidence for point (1). As for point (2), Mészáros notes the following: (I) observations of matter density are inconclusive; (II) the Hubble parameter and the temperature of the microwave background radiation do reveal cosine anisotropies. They are usually explained away as resulting from orital motion of the Solar System in the Galaxy. However, the directions and magnitudes of the velocities calculated from the two effects are inconsistent. This inconsistency is a problem in a Robertson–Walker model, but it can be accounted for on the basis of a L–T model. Unfortunately, the precision and interpretation of the observations are questionable. The quadrupole anisotropy of the background radiation is much simpler to explain in an L–T model than in a Friedmann model. Mészáros concluded that there are more arguments for the L–T models than against them, but present observations are inconclusive.

18.24.4 An 'in one ear and out the other' Universe

Hellaby (1987) presented a few illuminating examples of nontrivial L–T geometries and topologies. One of them, called 'In one ear and out the other', has the geometry shown in Fig. 18.21. The coordinate r is allowed to assume negative values. For $r \leq -E_1$, where $E_1 > 0$ is a constant, the model has $E \geq 0$ ($E = 0$ at $r = -E_1$) and collapses from past infinity to the Big Crunch that occurs at a finite t at all $0 > r > -\infty$. For $-E_1 < r < E_2$ (where $E_2 > 0$ is another constant), E is negative, so the model has both a Big Bang and a Big Crunch at finite times. For $r \geq E_2$, the model has $E \geq 0$ again (with $E = 0$

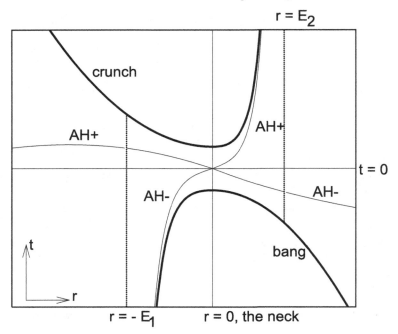

Fig. 18.21 The spacetime diagram for the L–T model of Eqs. (18.211)–(18.214). It has past timelike infinity for $r \leq -E_1 < 0$, a finite lifetime for $-E_1 < r < E_2$ and future timelike infinity for $r \geq E_2 > 0$. There is a neck at $r = 0$, where the past and future apparent horizons touch each other. In the figure, the time along AH^+ tends to $-\infty$ as $r \to -\infty$, and so does the time along AH^- as $r \to +\infty$. Whether these times go to $+\infty$ or $-\infty$ depends on the numerical balance between the various constants; see Exercise 14.

at $r = E_2$), and expands from the Big Bang in the finite past to infinity.[18] In the range $r \leq 0$, the L–T functions are

$$M(r) = \frac{1}{2}\left(M_0 - M_1 r^3\right), \qquad E(r) = -\frac{1}{2}\left(1 - r^2/E_1{}^2\right),$$

$$t_C(r) = \frac{1}{2}\pi M_0 \left(1 + \frac{3}{2}r^2/E_1{}^2\right). \tag{18.211}$$

The function $t_C(r)$ is the crunch time, and the bang time is

$$t_B(r) = t_C(r) - \frac{\pi\left(M_0 - M_1 r^3\right)}{\left(1 - r^2/E_1{}^2\right)^{3/2}}. \tag{18.212}$$

Thus, as seen in Fig. 18.21, the bang time goes down to $-\infty$ as $r \to -E_1$.

In the range $r \geq 0$, the functions are

$$M(r) = \frac{1}{2}\left(M_0 + M_2 r^3\right), \qquad E(r) = -\frac{1}{2}\left(1 - r^2/E_2{}^2\right),$$

$$t_B(r) = -\frac{1}{2}\pi M_0 \left(1 + \frac{3}{2}r^2/E_2{}^2\right). \tag{18.213}$$

[18] The configuration in Fig. 18.21 is a slight generalisation of $(E_1, M_1) = (E_2, M_2)$ used by Hellaby.

The crunch time corresponding to the Big Bang at $t = t_B(r)$ is

$$t_C(r) = t_B(r) + \frac{\pi \left(M_0 + M_2 r^3 \right)}{\left(1 - r^2/E_2{}^2 \right)^{3/2}}, \tag{18.214}$$

and it goes up to $+\infty$ as $r \to E_2$. There is a neck at $r = 0$, where the past and future apparent horizons touch each other.

As we observed in Sec. 18.8, Eq. (18.84), an outgoing future apparent horizon can never be timelike, and can be null only in vacuum. Since we are considering here a model that is not vacuum anywhere, the future apparent horizon in Fig. 18.21 is everywhere spacelike. This observation is sufficient to see that no light ray can be sent from the left part of the spacetime between AH^- and AH^+ to the right part between AH^- and AH^+: a light ray that would touch any point of the future apparent horizon would immediately pass inside it. This is in contrast to the Schwarzschild spacetime, in which, as seen from the Kruskal diagram (Fig. 14.8), it is possible to send light rays along the apparent/event horizon. Thus, as stated by Hellaby (1987), in the L–T spacetime the communication between the opposite sides of the neck is even worse than in the Schwarzschild spacetime. However, it is possible to receive a light ray that originated behind the neck before the AH^-.

18.24.5 A 'string of beads' Universe

Another of Hellaby's (1987) illuminating examples is the 'string of beads' model. This is a chain of recollapsing L–T regions, each of which begins in a separate Big Bang (BB), then they become connected by necks, and then split to end their lives in separate Big Crunches (BCs), see Fig. 18.22.[19] In our example the BC is a mirror-image of the BB. In order to have necks without shell crossings, the extrema of the BB must occur at the same r-s as the extrema of M, E and t_B. The L–T functions in Fig. 18.22 are:

$$t_B(r) = -\frac{\pi M}{(-2E)^{3/2}}, \qquad M(r) = M_0 + M_1 e^{-ar^2} \sin^2(br),$$

$$E(r) = -\frac{1}{2} \left(1 - E_1 e^{-ar^2} \sin^2(br) \right), \tag{18.215}$$

where $0 < M_1 < M_0$, $0 < E_1 < 1$, $a > 0$ and $b > 0$ are arbitrary constants. Note that $-1/2 \le E < 0$ everywhere is guaranteed by (18.215). The corresponding crunch function is $t_C(r) = \pi M/(-2E)^{3/2}$. The space extends infinitely far in all directions. This Universe has no centre of symmetry. It begins its existence in a series of (nonsimultaneous!) explosions, each of which creates a bubble of spacetime isolated from the others. Inside each bubble, the BB is going on for some time, and it ends when the bubble becomes connected to a neighbouring bubble by a neck. At a certain time the BB expires. In our example, the BB ends at the same t in each bubble – this is probably not necessary, but such an example was easier to concoct. After the BB is over, the Universe becomes a chain of regions connected by necks. Later, the BC sets in, also at the same t for each bubble. The

---

[19] Again, our example is a slight generalisation of that presented by Hellaby, who considered the 'beads' being isometric to each other.

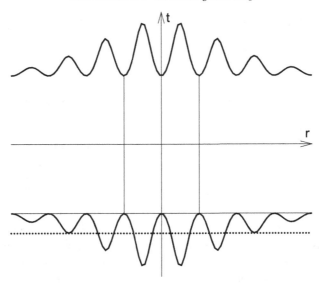

Fig. 18.22 A 'string of beads' Universe. The lower curve is the Big Bang (BB), the upper curve is the Big Crunch (BC). The Universe begins its existence as a chain of unconnected bubbles. The dotted line shows the time by which four bubbles already exist, and for the fifth and sixth the BB just begins. The lower solid horizontal line shows the time by which all the bubbles have completed their BBs. The vertical lines show the positions of three necks. Necks exist only at those local extrema of the BB where $E = -1/2$.

Universe splits again into unconnected bubbles, each of which gradually collapses into its own final singularity.

18.24.6 Uncertainties in inferring the spatial distribution of matter

Partovi and Mashhoon (1984) calculated various cosmologically relevant parameters for two types of spherically symmetric inhomogeneous spacetime: one with a perfect fluid source and a barotropic equation of state, the other with a dust source. The quantities are calculated in terms of successive powers of the redshift around the values in a R–W background. It was found that the leading terms in z are identical to the R–W terms, while the next following terms are formally modified, but in such a way that the inhomogeneous corrections cannot be properly read out from the formulae, and may lead to assignment of incorrect values to the cosmological parameters. The conclusion is that up to those following terms these models are observationally indistinguishable from the R–W models.

Kurki-Suonio and Liang (1992) used the L–T model to demonstrate the ambiguity in reading out the spatial distribution of matter from redshift measurements. What astronomers can directly measure is the number of light sources in a given solid angle and a given redshift interval, i.e. the density of matter in the **redshift space**. Lacking any information about its geometry, the redshift space, in which z is the radial coordinate, is assumed flat. Let the observer be at the centre of symmetry $z = 0$, and let $N(z)$ be the mass contained in the sphere of radius z around the observer. This mass is the sum of rest masses defined in (18.17), and is related to the mass function $M(r)$ from (18.18) by

$N_{,r} = M_{,r} / \sqrt{1 + 2E}$. The mass density in the redshift space, $\widehat{\rho}(z)$, is defined by

$$\frac{c^2 H_0{}^3}{G} N(z) = \int_0^z 4\pi z'^2 \widehat{\rho}(z') dz'. \tag{18.216}$$

The coefficients are there in order to provide convenient units; H_0 is the current value of the Hubble parameter at the observer's position. Thus,

$$\frac{c^2 H_0{}^3}{G} N_{,z} \equiv \frac{c^2 H_0{}^3}{G} \frac{N_{,r}}{z_{,r}} = 4\pi z^2 \widehat{\rho}(z). \tag{18.217}$$

Substituting for $N_{,r}$ from (18.17) – (18.18) and for $z_{,r}$ from (18.125) we obtain

$$\widehat{\rho}(z) = \frac{2 H_0{}^3 M_{,r}}{c^2 \kappa (1 + z) z^2 R_{,tr}} = \frac{H_0{}^3 R^2 R_{,r}}{(1 + z) z^2 R_{,tr}} \rho(r), \tag{18.218}$$

where $\rho(r)$ is the physical mass density. The values of t and r refer to the time and position of emission. Equation (18.218) connects one observed quantity, $\widehat{\rho}(z)(1 + z)z^2$, to two unknown functions, $M(r)$ and $R(t, r)$, and obviously cannot determine both. The authors investigated numerically the dependence of these functions on $\widehat{\rho}(z)$ with various shapes of $t_B(r), E(r)$ and initial density distributions $\rho(t, r)$. They assumed no shell crossings $(R_{,r} > 0)$ and redshifts monotonically growing with distance $(R_{,tr} > 0)$. The results, greatly abbreviated, are as follows:

1. If $t_B(r) = $ constant, then the amplitudes of overdensity in the redshift space are larger than the amplitudes of proper density by 40% or more. The regions of overdensity in proper density have larger sizes than their images in the redshift space.

2. When $M(r)/r^3 = $ constant and $E(r)/r^2 = $ constant, that is, when the bang-time function is the only factor generating inhomogeneity, overdensities in the redshift space translate into *underdensities* in proper density, of larger size and much larger amplitude.

3. Since the two effects influence observations in opposite directions, they can momentarily cancel each other, producing a homogeneous density in the redshift space while the Universe is highly inhomogeneous. Conversely, the Universe may be momentarily very smooth in density, but can appear highly inhomogeneous in the redshift space because of the effects of inhomogeneous velocities. To illustrate this last possibility, the authors fitted the velocity function in the L–T model so that the proper density is constant on the past light cone, while the density in redshift space is the same as in one of the deep-redshift observational surveys. The proper density at the present time then has smaller amplitudes in the peaks and a less regular distribution of the peaks. The authors concluded that using the R–W relation between the redshift and the comoving distance to describe inhomogeneities is 'fundamentally self-inconsistent'. They said further: 'What we see on the past light cone are only momentary "images".'

Kurki-Suonio and Liang's paper contains several graphs and spacetime diagrams that illustrate the conclusions more clearly. In particular, they show how the bang-time non-simultaneity generates decaying inhomogeneities while variations in initial energy generate growing inhomogeneities, even though these processes are not discussed in the paper.

18.24.7 *Is the distribution of matter in our Universe fractal?*

Ribeiro (1992a) defined fractal distribution of matter by requiring that the luminosity distance d_l to the surface of a sphere containing N_c objects obeys $N_c = \sigma d_l{}^D$, where σ is a constant and D is the fractal dimension of the distribution.

In a sequel paper (Ribeiro, 1992b), the author calculated the density versus distance function along the past light cone in the $k = 0$ Friedmann model. This function is not constant and does not seem to agree with the observations. Ribeiro stressed that cosmological models should be tested in precisely this way. For the $k = 0$ Friedmann model this function does not possess any features of a fractal distribution of mass.

In the third paper, Ribeiro (1993) showed that the density versus distance relation measured along a past light cone in the $k \neq 0$ Friedmann model is not homogeneous either, nor is it fractal. However, the functions $E(r), M(r)$ and $t_B(r)$ in the L–T model *can* be chosen so that $N_c = \sigma \sigma d_l{}^D$. Ribeiro found several examples of such functions by numerical experiments. In most examples, either the Hubble law is contradicted or the Hubble constant is smaller than astronomers currently believe it should be. The only example that yields acceptable parameters is $2E(r) = \sinh^2 r, 2M(r) = \alpha r^p, t_0(r) = \ln(e^{\beta_0} + \eta_1 r)$. If $\alpha = 10^{-4}, p = 1.4, \beta_0 = 3.6$ and $\eta_1 = 1000$; then the implied fractal dimension is $D = 1.3$, the Hubble constant is $H = 61$ km s^{-1} Mpc^{-1} and $\sigma = 5.4 \times 10^5$. Even better agreement with observations follows when $\eta_1 = 0$; then $D = 1.4$ and $H = 80$ km s^{-1} Mpc^{-1}. Ribeiro emphasised that no positive observational evidence of spatial homogeneity of the Universe exists. Should it be available, it would actually rule out the Friedmann models because the light cones in them intersect hypersurfaces of different densities.

18.24.8 *General results related to the L–T models*

Novikov (1962b) proved the following theorems for a general spherically symmetric spacetime: 1. If $\lim_{r \to r_0} R(t_0, r) = \infty$ for a certain t_0 (the r_0 may be infinite), and for $r > r_1$ the matter density $\rho \geq A/(8\pi R^2)$, where $A = \text{constant} > 1$, then there exists an r_2 such that the points with $r > r_2$ and $t = t_0$ are in the T-region.[20] 2. For every $r = r_2$ there exist such t_1 and t_2 (dependent on r_2) that the points with (t, r_2) where $t_1 < t < t_2$ are in the T-region.[21] 3. There exists a solution for which the volume of a finite mass goes through a nonsingular minimum and the minimum is simultaneous for the whole space. 4. The same is possible for an infinite mass. 5. There exists a solution in which a volume minimum occurs close to the centre, and, at the same instant, a volume maximum occurs at large distances from the centre. 6. If $M(r) = \int \rho dV$ grows with r faster than the affine radius $l = \int_0^r (-g_{11})^{1/2} dr'$, and does so for a sufficiently large interval of l, then the configuration cannot be static. 7. If $M(r)$ grows faster than $l(r)$ over a sufficiently large interval of monotonic growth of R, then the volumes of different shells cannot go through an extremum simultaneously. 8. If the surface of a spherical ball is in a T-region and is shrinking, then, in a finite proper time, it will shrink to a point; if the surface is in a

---

[20] For the definition of the T- and R-regions, see Section 14.1. This means that with sufficiently high density the spheres $t = \text{constant}, r = \text{constant}$ will not be static for any observer.

[21] This means that, for a time, each matter particle remains in a T-region, i.e. in the vicinity of BB.

T-region and is expanding, then it started its expansion from a point. Finally, Novikov calculated the position of the boundary between the R- and T-regions in a Friedmann model and concluded that the observed radio sources must lie in the T-region. Consequently, they seem to have emerged from a point. These theorems reveal what should be expected from exact solutions generalizing the L–T model for pressure with nonzero spatial gradient.

A later paper by Novikov (1964b) is a continuation of the one described above. One of its new results is that in the Schwarzschild solution the boundary between the R- and T-regions is the $r = 2m$ event horizon, and such boundaries also exist for the Kottler (i.e. Schwarzschild with $\Lambda \neq 0$) and Reissner–Nordström solutions. Another result was discussed in Section 14.12 (Eqs. (14.120), (14.121) and Exercise 10 in Chapter 14).

In a series of papers, Dautcourt (1980, 1983a, b, 1985) developed a new approach to integrating the Einstein equations that is better suited to incorporating observational results into cosmological models. In this approach, the initial data are given on the past light cone of a single event in the spacetime, and then the Einstein equations are integrated off that cone to the future. The first paper (Dautcourt, 1980) is a short exposition of the ideas of the other papers. In the second paper (Dautcourt, 1983b), the author gave a prescription for integrating the Einstein equations in vacuum and for dust off the past light cone of an observer. The Einstein Universe and the Friedmann models were used as illustrations. In the third paper (Dautcourt, 1983a), the author described the ideal observations needed to determine the initial data. In practice, several of those observations are beyond the limits of current technology. In the last paper (Dautcourt, 1985), the author noted that the observable quantities introduced in the preceding papers depend on the 4-velocity of the observer, and proposed a preferred set of coordinates with a simple relation to the observable quantities and a simple transformation law under a Lorentz rotation of the observer. Dautcourt's approach was later developed by Ellis and coworkers; the paper by Stoeger, Ellis and Nel (1992) mentioned in Section 18.3 is one entry in that series.

18.25 Exercises

1. Prove that if the spacetime is spherically symmetric and the metric obeys the Einstein equations with a perfect fluid source, then the rotation of the fluid's velocity is zero.

Hint. See Sec. 14.2 – a perfect fluid energy-momentum tensor inherits all the invariances of the metric, and so do the velocity u^α and acceleration $\dot{u}^\alpha$ fields. Consequently, in our case $\underset{k_{(i)}}{\pounds} u^\alpha = k^\rho_{(i)} u^\alpha_{,\rho} - u^\rho k^\alpha_{(i),\rho} = 0$ for all three generators (8.42), and the same is true for $\dot{u}^\alpha$. Then use the result of Exercise 9 to Chapter 8 – u^α and $\dot{u}^\alpha$ can have only the $\alpha = 0, 1$ components in the metric (8.52) and depend only on t and r. Recall that $\beta = 0$ can be achieved by transforming t and r, see the paragraph under (14.8). Note that only ω_{01} can be nonzero. Write out ω_{01}, then use $g_{\mu\nu} u^\mu u^\nu = 1$ and $(g_{\mu\nu} u^\mu u^\nu)_{,\alpha} = 0$.

2. Prove that $\dot{u}^\alpha = 0$ in (18.1) – (18.2) implies $C_{,r} = 0$.

3. Prove that (18.24) are the necessary and sufficient conditions for the L–T model to reduce to the Friedmann limit.

Hint. Use M as the radial coordinate. With $E = 0$ the thesis follows from (18.20) by direct calculation. With $E \neq 0$ use the second of (18.19) and the second of (18.21) to calculate $\partial \eta / \partial M$, then use the result in calculating $(R^3)_{,M}$ and $(R^3)_{,MM}$. Eliminate $(t - t_B)$ from $(R^3)_{,MM}$ using (18.19) and (18.21). The result will be a polynomial in η, $\sin \eta$ and $\cos \eta$ with coefficients depending on M. Observe that the Jacobian $\partial(M, \eta)/\partial(M, t) = \eta_{,t} \neq 0$. Hence, M and η can now be treated as independent variables. Equate to zero the coefficients of independent functions of η, beginning with the coefficients of η^2. In order to show that (18.24) do indeed define the Friedmann model, change the radial variable in (18.16) to $r = (M/M_0)^{1/3}$ and compare the result with (17.3).

4. Transform the L–T metric to the curvature coordinates.

Hint. Note that, from (18.14),

$$R_{,r}\, dr = dR - R_{,t}\, dt = dR - \ell \sqrt{2E + \frac{2M}{R} - \frac{1}{3}\Lambda R^2}\, dt, \qquad (18.219)$$

where $\ell = +1$ for expanding and $\ell = -1$ for collapsing models. Then transform t by $t = f(T, R)$ and demand that $g_{TR} = 0$. This implies:

$$\frac{\partial f}{\partial R} = -\frac{\ell}{1 - 2M/R + \frac{1}{3}\Lambda R^2} \sqrt{2E + \frac{2M}{R} - \frac{1}{3}\Lambda R^2}. \qquad (18.220)$$

Now M and E are functions of $r\,(T, R)$ and the metric becomes

$$ds^2 = \frac{f_{,T}{}^2}{1 + 2E}\left(1 - \frac{2M}{R} + \frac{1}{3}\Lambda R^2\right) dT^2 - \frac{1}{1 - 2M/R + \frac{1}{3}\Lambda R^2}\, dR^2$$
$$- R^2 \left(d\vartheta^2 + \sin^2 \vartheta d\varphi^2\right). \qquad (18.221)$$

The functions E and M obey the equations

$$0 = M_{,T}\, T_{,t} + M_{,R}\, R_{,t}, \qquad 0 = E_{,T}\, T_{,t} + E_{,R}\, R_{,t}, \qquad (18.222)$$

where $T_{,t} = \partial T / \partial t$ is an element of the matrix inverse to $\begin{bmatrix} f_{,T} & f_{,R} \\ r_{,T} & r_{,R} \end{bmatrix}$. The derivatives $r_{,T}$ and $r_{,R}$ can be found from (18.219) with $dt = f_{,T}\, dT + f_{,R}\, dR$ substituted in it.

5. Prove that (18.46) is a necessary and sufficient condition for the existence of a solution of $\psi_H(x) = 0$ at $x \neq 0$, where ψ_H is given by (18.45). Prove also that only one such value of x exists.

Hint. Calculate $d[\psi_H(x)]/dx$ and consider how it changes. When in trouble, consult Krasiński and Hellaby (2002).

6. Prove that (18.49) are the necessary and sufficient condition for the existence of a solution of $\psi_X(x) = 0$ at $x \neq 0$, where ψ_X is given by (18.47). Prove also that only one such value of x exists.

Hint. Consult Krasiński and Hellaby (2004a). That paper will be helpful also in the next exercise.

7. Verify that (18.63) and (18.66) are the sets of necessary and sufficient conditions for the existence of the appropriate evolutions. Interpret the results (i.e. why the time difference has to be sufficiently *large* in (18.63) and sufficiently *small* in (18.66)).

8. Prove that for a Friedmann model with $\Lambda = 0$ (in which also $p = 0$) the parameter B given by (18.84) is always equal to -2, and so the apparent horizons in it are timelike.

9. Derive (18.103) for the $E > 0$ and $E < 0$ L–T models.

10. Prove that (18.105), (18.109) and (18.110) are sufficient conditions for $R_{,r} > 0$.

Hint. Write (18.106) as follows:

$$f(\eta) = \frac{M_{,r}}{3M} + \left(\frac{3}{2}\frac{E_{,r}}{E} - \frac{M_{,r}}{M}\right)\left[-\frac{2}{3} + \Phi_1(\eta)\right] - \frac{(-2E)^{3/2}}{M}t_{B,r}\Phi_2(\eta). \tag{18.223}$$

The sign of $f(\eta) - M_{,r}/(3M)$ must be considered separately for the ranges $[0, \pi)$ and $[\pi, 2\pi]$. In $(\pi, 2\pi)$, $\Phi_2 < 0$ so the last term above is negative by (18.109), and from (18.110):

$$-\frac{(-2E)^{3/2}}{M}t_{B,r}\Phi_2 > -2\pi\Phi_2\left(\frac{3}{2}\frac{E_{,r}}{E} - \frac{M_{,r}}{M}\right). \tag{18.224}$$

Using this in (18.223) one obtains for $\pi < \eta < 2\pi$:

$$f(\eta) > \frac{M_{,r}}{3M} + \left(\frac{3}{2}\frac{E_{,r}}{E} - \frac{M_{,r}}{M}\right)\left(-\frac{2}{3} + \Phi_1 - 2\pi\Phi_2\right). \tag{18.225}$$

Now let us investigate the factor $\mathcal{F}(\eta) \stackrel{\text{def}}{=} (-2/3 + \Phi_1)$. For $\eta \in [\pi, 2\pi]$, $\Phi_1 \leq 0$, so $\mathcal{F} < 0$. At $\eta = 0$, $\Phi_1 = 2/3$, at $\eta = \pi$, $\Phi_1 = 0$, and for $\eta \in (0, \pi)$

$$\frac{d\Phi_1}{d\eta} = \frac{3\sin\eta - \eta\cos\eta - 2\eta}{(1 - \cos\eta)^2} < 0, \tag{18.226}$$

so for $\eta \in (0, \pi)$ Φ_1 decreases monotonically from $2/3$ to zero. It follows that $\mathcal{F}(\eta) < 0$ for all $0 < \eta \leq 2\pi$. Now observe that $\Phi_1(\eta) - 2\pi\Phi_2(\eta) = \Phi_1(\eta_1)$ where $\eta_1 = 2\pi - \eta$. Thus as η grows from 0 to 2π, $\Phi_1 - 2\pi\Phi_2$ goes through the same set of values as Φ_1, but in reverse order. Consequently, also $(-\frac{2}{3} + \Phi_1 - 2\pi\Phi_2) < 0$ for all $\eta \in (0, 2\pi)$ ($= 0$ at $\eta = 2\pi$). Together with (18.111) this implies

$$\frac{R_{,r}}{R} = f(\eta) > \frac{M_{,r}}{3M} > 0 \qquad \text{for } \eta \in [\pi, 2\pi], \tag{18.227}$$

which means that (18.109) and (18.110) are sufficient conditions for avoiding shell crossings for $\pi \leq \eta \leq 2\pi$.

For $0 \leq \eta \leq \pi$ we have $-2/3 + \Phi_1 \leq 0$ as shown above and $\Phi_2 \geq 0$, which, together with (18.109) and (18.111), leads to (18.227) again.

11. Prove that (18.126) is equivalent to (16.20).

Hint. With the coordinates of (18.16), Eq. (16.20) can be written as $1 + z = (k^0)_e/(k^0)_o$. By rescaling the affine parameter of k^α we can achieve $(k^0)_o = 1$. Keeping the observer at a fixed position, take the logarithmic derivative of $(1 + z)$ by r and compare the result with (18.125). The conclusion is

$$\frac{(k^0)_{,r}}{k^0} = \frac{R_{,tr}(T(r), r)}{\sqrt{1 + 2E(r)}}. \tag{18.228}$$

Integrating this with the initial condition $(k^0)_o = 1$ and using (18.119) we obtain (18.133). Verify that this field is affinely parametrised (note that the integral is calculated along the geodesic tangent to k^α, so the integrand is a function of r only).

12. Prove that the light ray k_2 sent backwards in time towards the centre of symmetry from the point p_2 in Fig. 18.16 can never intersect the shell crossing set if the following assumptions are fulfilled:

(a) The functions $t = t_{n2}(r)$ (describing the light ray k_2) and the shell crossing function $t = t_S(r)$ are differentiable for $r \in [0, r_2]$.
(b) $dt_S/dr > 0$ for $r \in [0, r_2]$.
(c) $R_{,r} > 0$ for all $t > t_S(r)$.

Hint. Here is one suggestion of a proof. Let $t'(r) = dt/dr$. Since $t'_S(r) > 0$ for $r \in [0, r_2]$, there exists a minimum value $T' > 0$ of $t'_S(r)$ in this range, so $t'_S(r) \geq T'$ for $r \in [0, r_2]$.

If k_2 should intersect $t_S(r)$ at some $r_1 \in [0, r_2]$, then $t'_{n2}(r_2) = 0$ and $t'_{n2} \xrightarrow[r \to r_1]{} 0$, while $t'_{n2}(r) > 0$ for $r_1 < r < r_2$ (by virtue of assumption c). Hence, for every $0 < \varepsilon < T'$ there exists such a $\delta > 0$ that $0 \leq t'_{n2}(r) < \varepsilon < T'$ if $r_1 \leq r < r_1 + \delta$.

Since $t_{n2}(r)$ starts from (t_2, r_2) with $t_{n2}(r_2) > t_S(r_2)$, assume that $t_{n2}(r_1 + \delta) > t_S(r_1 + \delta)$ and compare $t_{n2}(r_1)$ with $t_S(r_1)$. By virtue of the statements made we have

$$t_S(r_1) < t_S(r_1 + \delta) - T'\delta, \qquad t_{n2}(r_1) > t_{n2}(r_1 + \delta) - \varepsilon\delta, \qquad (18.229)$$
$$t_{n2}(r_1) - t_S(r_1) > t_{n2}(r_1 + \delta) - t_S(r_1 + \delta) + (T' - \varepsilon)\delta$$
$$> t_{n2}(r_1 + \delta) - t_S(r_1 + \delta). \qquad (18.230)$$

Thus, whenever k_2 would be approaching the horizontal direction, it would be receding from the shell crossing rather than intersecting it.

In order that k_2 can intersect the shell crossing set, it would have to turn back and proceed towards increasing r and decreasing t. To achieve this, it would have to either (1) change abruptly from past-directed ingoing to past-directed outgoing, which is prohibited by assumption (a), or (2) become vertical and then past-directed outgoing. However, at the point where the tangent to k_2 would be vertical, k_2 would become timelike, which cannot happen for a light ray.

Hence, k_2 and $t_S(r)$ cannot intersect between $r = 0$ and $r = r_2$.

13. Verify (18.186), i.e. prove that $\lim_{\eta \to 0} \Delta K = +t_{B,r} \times \infty$ when $[(2E)^{3/2}/M]_{,r} = 0$.
Hint. $[(2E)^{3/2}/M]_{,r} = 0$ is equivalent to $\frac{3}{2} E_{,r}/E - M_{,r}/M = 0$. Since $\Phi_4 \xrightarrow[\eta \to 0]{} \infty$, we have in (18.113):

$$\lim_{\eta \to 0} \frac{R_{,r}}{R} = \lim_{\eta \to 0} \left[-\frac{(2E)^{3/2}}{M} t_{B,r} \Phi_4 \right] = -t_{B,r} \times \infty, \qquad (18.231)$$

so $\lim_{\eta \to 0} R/R_{,r} = 0$. Hence, in (18.175) the terms $E_{,rr}/E$ in the numerator and $E_{,r}/E$ in the denominator become negligible as $\eta \to 0$. What remains is

$$\lim_{\eta \to 0} \Delta K = \lim_{\eta \to 0} \left[-2\frac{R_{,r}}{R} - \frac{E_{,r}}{E} \frac{R_{,rr}}{R_{,r}} \frac{R}{R_{,r}} \right]. \qquad (18.232)$$

The terms multiplied by $\beta_{,r}$ and β in (18.177) are zero by assumption. The terms containing $\alpha_{,r}$ and α will go to zero after being multiplied by $R/R_{,r}$. So the only terms that need to be investigated are those containing $\gamma_{,r}$ and γ. We have, using (18.231),

$$\lim_{\eta \to 0} \left| \frac{R}{R_{,r}} \frac{\sinh \eta}{(\cosh \eta - 1)^2} \right| = \lim_{\eta \to 0} \left| \frac{M}{-(2E)^{3/2} t_{B,r}} \right| < \infty, \tag{18.233}$$

so the contribution from the term containing $(R\gamma_{,r}/R_{,r} + \gamma)$ in (18.177) is negligible compared to $R_{,r}/R$. For the last term in (18.177) we find

$$\lim_{\eta \to 0} \left[\left(\frac{R}{R_{,r}} \right)^2 \frac{\cosh \eta + 2}{(\cosh \eta - 1)^3} \right] = \frac{3}{2} \frac{M^2}{(2E)^3 t_{B,r}{}^2} < \infty, \tag{18.234}$$

which is also negligible compared to $R_{,r}/R$. So the thesis follows from (18.231). $\square$

14. Calculate the equations $t = t_{AH}(r)$ of the apparent horizons for the model (18.211) – (18.214). Show that in the $E < 0$ region $t_{AH-} \to -\infty$ when $r \to -E_1$ and $t_{AH+} \to +\infty$ when $r \to E_2$. Show that in the $E > 0$ region for the past AH $t_{AH-} \xrightarrow[r \to +\infty]{} \varepsilon_r \infty$, where $\varepsilon_r = \pm 1$ and the sign of ε_r is the sign of $(4M_2 E_2{}^3 - 3\pi M_0)$. Show that with $E > 0$ the future AH $t_{AH+} \xrightarrow[r \to -\infty]{} \varepsilon_l \infty$, where $\varepsilon_l = \pm 1$ and its sign is the sign of $(3\pi M_0 - 4M_1 E_1{}^3)$.

15. Verify that the functions defined in (18.215) obey the no-shell-crossing conditions (18.106), (18.110) and (18.111), and their reverses where $M_{,r} < 0$.

16. Let the mass density in the L–T model (18.15) be a differentiable function of r. Prove that the spatial extrema of density in the $E \neq 0$ models are comoving only when the following equations are fulfilled at the extremal values:

$$\left(\frac{|2E|^{3/2}}{M} \right)_{,r} = \left(\frac{|2E|^{3/2}}{M} \right)_{,rr} = t_{B,r} = t_{B,rr} = 0, \qquad \left[\left(\frac{M^3}{E^3} \right)_{,r} \frac{1}{M_{,r}} \right]_{,r} = 0. \tag{18.235}$$

Find the corresponding condition for the $E = 0$ model.

Note: The fact that the extrema are comoving only under the conditions (18.235) means that in general they are not comoving. Thus, they move across the flow lines of dust. This phenomenon is known in cosmology as density waves. Their existence in relativistic models was predicted by Ellis, Hellaby and Matravers (1990) by methods of the linearized Einstein theory. Note also that the first four conditions mean that at the extremum the L–T model has a second-order contact with the Friedmann model that approximates it.

Hint. Choose M as the spatial coordinate. The first four conditions do not change then, but the last one takes the more readable form $(M^3/E^3)_{,MM} = 0$. The extrema are at those values of M where $\epsilon_{,M} = 0$, i.e. $(R^3)_{,MM} = 0$. Then proceed as in Exercise 3. If the extrema are comoving, then the values of M obeying (18.235) do not depend on t.

19

Relativistic cosmology IV: simple generalisations of L–T and related geometries

19.1 The plane- and hyperbolically symmetric spacetimes

The L–T models considered in the previous chapter have 2-spheres as orbits of their symmetry group. They have counterparts: spacetimes whose symmetry orbits are 2-surfaces of zero or negative constant curvature.

A surface of zero curvature has the local geometry of a plane, whose symmetries in the Cartesian coordinates (x, y) are

$$(x', y') = (x + a_1, y), \qquad (x', y') = (x, y + a_2),$$
$$(x', y') = (x \cos a_3 + y \sin a_3, -x \sin a_3 + y \cos a_3), \tag{19.1}$$

where a_1, a_2 and a_3 are the group parameters; a_1 and a_2 are arbitrary, while $0 \leq a_3 \leq 2\pi$. Proceeding just as in Sec. 8.10 we find that a 4-dimensional metric for which (19.1) is the symmetry group is

$$ds^2 = \alpha(t, r)dt^2 + 2\beta(t, r)dtdr + \gamma(t, r)dr^2 + \delta(t, r)\left(dx^2 + dy^2\right). \tag{19.2}$$

Spacetimes with this symmetry group are called **plane symmetric**. As in Sec. 18.1, we find that in such a spacetime the rotation of any velocity field must be zero, so comoving-synchronous coordinates can be introduced, in which the metric becomes

$$ds^2 = e^{C(t,r)}dt^2 - e^{A(t,r)}dr^2 - R^2(t, r)\left(dx^2 + dy^2\right). \tag{19.3}$$

The curvature tensor of a sphere, $R^2\left(d\vartheta^2 + \sin^2 \vartheta d\varphi^2\right)$, $R = \text{constant}$, is

$$R^{\alpha\beta}{}_{\gamma\delta} = R^{-2}\delta^{\alpha\beta}_{\gamma\delta}. \tag{19.4}$$

We can make the curvature negative by the complex transformation $\{\vartheta = i\vartheta', R = iR'\}$. Thereafter, the 2-metric of a symmetry orbit becomes (with primes dropped)

$$ds^2 = R^2\left(d\vartheta^2 + \sinh^2 \vartheta d\varphi^2\right), \tag{19.5}$$

and its curvature tensor is $R^{\alpha\beta}{}_{\gamma\delta} = -R^{-2}\delta^{\alpha\beta}_{\gamma\delta}$. We choose $(x^2, x^3) = (\vartheta, \varphi)$ as coordinates in the spacetime. The Killing fields are then

$$\underset{(1)}{k^\alpha} = \sin\varphi\delta^\alpha{}_2 + \cos\varphi \coth\vartheta\delta^\alpha{}_3,$$

$$\underset{(2)}{k^\alpha} = \cos\varphi\delta^\alpha{}_2 - \sin\varphi \coth\vartheta\delta^\alpha{}_3,$$

$$\underset{(3)}{k^\alpha} = \delta^\alpha{}_3, \tag{19.6}$$

and we require these to be the Killing fields for the whole spacetime. Solving the Killing equations with (19.6) we find the metric

$$ds^2 = \alpha(t,r)dt^2 + 2\beta(t,r)dtdr + \gamma(t,r)dr^2 + \delta(t,r)\left(d\vartheta^2 + \sinh^2\vartheta d\varphi^2\right), \qquad (19.7)$$

where α, β, γ and δ are arbitrary functions. These spacetimes are called **hyperbolically symmetric** because the orbits of their symmetry group have several properties in common with two-sheeted hyperboloids. Once again, we prove that in such a spacetime the rotation of any velocity field must be zero, so we can introduce the coordinates in which

$$ds^2 = e^{C(t,r)}dt^2 - e^{A(t,r)}dr^2 - R^2(t,r)\left(d\vartheta^2 + \sinh^2\vartheta d\varphi^2\right). \qquad (19.8)$$

The surface with the metric (19.5) cannot be embedded in a Euclidean 3-space – the flat 3-metric must have the signature $(+,+,-)$. To verify this, consider (19.5) embedded in a flat 3-space, draw an orthogonal geodesic through each point of the surface, take the affine parameter on the geodesics as the $x^1 = r$-coordinate in space, let $r = R$ on the initial surface and let the surfaces $R \neq r = $ constant have the geometry of (19.5). In such coordinates, the 3-space metric will be

$$ds_3{}^2 = \varepsilon dr^2 + f^2(r)\left(d\vartheta^2 + \sinh^2\vartheta d\varphi^2\right), \qquad (19.9)$$

where $\varepsilon = \pm 1$, to allow for different signatures. The Riemann tensor for the metric (19.9) can vanish only when $\varepsilon = -1$, and $f_{,r}(r) = \pm 1$.

The orbits of spherical, plane- and hyperbolic symmetries are 2-dimensional. However, the symmetry groups themselves are 3-dimensional, so their algebras can be classified into the Bianchi types of Chapter 10. The algebras of spherical, plane and hyperbolic symmetry are of Bianchi types IX, VII$_0$ and VIII, respectively.

It is convenient to represent these spacetimes by the single formula:

$$ds^2 = e^{C(t,r)}dt^2 - e^{A(t,r)}dr^2 - R^2(t,r)\left[d\vartheta^2 + (1/\varepsilon)\sin^2(\sqrt{\varepsilon}\vartheta)d\varphi^2\right], \qquad (19.10)$$

where $\varepsilon = \pm 1, 0$. With $\varepsilon = +1$, the metric is spherically symmetric, with $\varepsilon = -1$ we recover the hyperbolically symmetric metric via the identity $\sin(i\vartheta) = i\sinh\vartheta$, and the limit $\varepsilon \to 0$ of (19.10) is plane symmetric. The three cases will be collectively called G_3/S_2-**symmetric** spacetimes, G_3 standing for 3-dimensional groups and S_2 for 2-dimensional orbits.

We will often use a variant of the notation of (19.10):

$$ds^2 = e^{C(t,r)}dt^2 - e^{A(t,r)}dr^2 - R^2(t,r)\left[d\vartheta^2 + f^2(\vartheta)d\varphi^2\right], \qquad (19.11)$$

where

$$\begin{aligned}
f(\vartheta) &= \sin\vartheta && \text{for spherical symmetry,} \\
f(\vartheta) &= \vartheta && \text{for plane symmetry,} \\
f(\vartheta) &= \sinh\vartheta && \text{for hyperbolic symmetry.}
\end{aligned} \qquad (19.12)$$

19.2 G_3/S_2-symmetric dust solutions with $R_{,r} \neq 0$

Just as for the spherically symmetric models, the case $R_{,r} = 0$ has to be considered separately; we will present it in Section 19.5.

The most general set of solutions for the metric (19.11) – (19.12) with $R_{,r} \neq 0$, a dust source and $\Lambda \neq 0$ was found by Ellis (1967, his case IIaii). They are given by

$$\mathrm{d}s^2 = \mathrm{d}t^2 - \frac{R_{,r}^2(t,r)}{\varepsilon + 2E(r)}\,\mathrm{d}r^2 - R^2(t,r)\left[\mathrm{d}\vartheta^2 + f^2(\vartheta)\mathrm{d}\varphi^2\right], \tag{19.13}$$

where $E(r)$ is an arbitrary function, while $R(t,r)$ is determined by

$$R_{,t}^2 = 2E(r) + \varepsilon - 1 + \frac{2M(r)}{R} - \frac{1}{3}\Lambda R^2. \tag{19.14}$$

The matter density is determined by (18.15). The required signature $(+ - --)$ puts different limits on $E(r)$ for each ε. With $\varepsilon = 0$, $E(r)$ must be nonnegative everywhere, and can vanish only at isolated values of r, provided that at the same values $R_{,r} = 0$ and $R_{,r}^2/(2E)$ has a finite limit. With $\varepsilon = -1$, $E(r)$ must be $\geq 1/2$ everywhere.

In the Friedmann limit $R(t,r) = g(r)S(t)$; then the r-coordinate can be chosen so that $R(t,r) = rS(t)$ and it follows further that $-2E/r^2 = k = $ constant, k being the Friedmann curvature index, $M(r)/r^3 = M_0 = $ constant, M_0 being the Friedmann mass integral. Thus, only $\varepsilon = +1$ allows all three Friedmann limits. The requirement of the Lorentzian signature forces the Friedmann limit to have $k < 0$ when $\varepsilon = 0$ or $\varepsilon = -1$.

19.3 Plane symmetric dust solutions with $R_{,r} \neq 0$

In this section we will consider the metrics (19.13) – (19.14) with $\varepsilon = 0$. It will be more convenient to change to the coordinates $x = \vartheta \cos\varphi, y = \vartheta \sin\varphi$ in which

$$\mathrm{d}s^2 = \mathrm{d}t^2 - \frac{R_{,r}^2(t,r)}{2E(r)}\,\mathrm{d}r^2 - R^2(t,r)\left(\mathrm{d}x^2 + \mathrm{d}y^2\right). \tag{19.15}$$

The interpretation of the plane- and hyperbolically symmetric Ellis solutions is still unclear. A natural idea is that a plane symmetric spacetime should be a family of Euclidean planes, but this leads to difficulties. One of them is this: in a spherically symmetric spacetime, the surfaces P_2 of constant t and r are spheres, and $M(r)$ in (18.14) – (18.15) is the mass inside the sphere of coordinate radius r. But if P_2 are infinite planes, then they do not enclose any volume, so where does the mass $M(r)$ reside?

We now compare the plane symmetric model with its possible Newtonian counterpart. For this purpose, let us note the pattern of expansion in the metric (19.15) obeying (19.14) with $\Lambda = 0$. When $R_{,t} \neq 0$, Eq. (19.14) implies

$$R_{,tt} = -M/R^2. \tag{19.16}$$

Imagine a pair of dust particles, located at (t, r_1, x_0, y_0) and at (t, r_2, x_0, y_0), and consider the affine distance between them in the space of fixed (t, x_0, y_0):

$$\ell_{12}(t) = \int_{r_1}^{r_2} \frac{R_{,r}\,\mathrm{d}r}{\sqrt{2E}} \implies \frac{\mathrm{d}^2\ell_{12}}{\mathrm{d}t^2} = \int_{r_1}^{r_2} \frac{R_{,ttr}\,\mathrm{d}r}{\sqrt{2E}}. \tag{19.17}$$

So, the two particles are receding from each other (or approaching each other if collapse is considered) with acceleration that in general is nonzero.

Take another pair of dust particles, located at (t, r_0, x_1, y_0) and at (t, r_0, x_2, y_0). The distance between them, measured within the symmetry orbit, is

$$\ell_{34}(t) = \int_{x_1}^{x_2} R \, dx \equiv R\,(x_2 - x_1) \implies \frac{d^2 \ell_{34}}{dt^2} = R_{,tt}\,(x_2 - x_1), \qquad (19.18)$$

i.e. the acceleration of the expansion does not in general vanish in this direction. The same result will follow for any direction in the (x, y) surface. Thus, the expansion or collapse in this model proceeds with acceleration in every spatial direction.[1]

At first sight, it seems that a Newtonian model analogous to (19.14) – (19.15) should be dust whose density is constant on parallel (x, y)-planes, and depends only on z. The same should be true for the gravitational potential. But with such a potential, the pattern of expansion is different from that in (19.14) – (19.15). Expansion with acceleration proceeds only in the z-direction, while in the directions orthogonal to z there is no acceleration, or, in a special case, not even any expansion. A solution of the Poisson equation that qualitatively mimics the pattern of expansion of (19.15) is presented in Krasiński (2008). Its (locally plane symmetric) equipotential surfaces are co-axial parallel cones.

The difficulties with localising the active gravitational mass and with the pattern of expansion disappear when the (x, y)-surfaces are flat tori (see Krasiński, 2008, for an in-depth justification). While we identify the set $x = x_0$ with $x_1 = x_0 + 2\pi$, we are still free to carry out the symmetry transformations within $x = x_1$. Thus, the identification can be done with a twist, that will turn a square into a Möbius strip, or with a two-way twist, that will turn it into a projective plane (Borsuk, 1964). This is necessary when one considers the quasiplane Szekeres models which are nonsymmetric generalisations of the plane symmetric dust models, see again Krasiński (2008) and our Sec. 20.3.

The toroidal topology of the (x, y) surfaces neatly explains the pattern of expansion. The circumference of the torus along the x- or y-direction is $2\pi R$ in the coordinates of (19.15). Thus, as R increases with time, the circumference of the torus increases in proportion to R, which causes transversal expansion.

The toroidal topology also solves the problem of where the mass generating the gravitational field resides. As observed by Hellaby and Krasiński (2008), the regularity conditions at an origin $r = r_c$ require that $E/M^{2/3}$ tends to a nonzero constant as $r \to r_c$. The amount of rest mass in an arbitrary volume $\mathcal{V}$, from (19.15) and (18.15), is $N = \int_{\mathcal{V}} \rho \sqrt{g_3} \, d_3 x$, where g_3 is the determinant of the 3-metric of a $t = $ constant subspace of (19.15). Thus,

$$N = \frac{c^2}{4\pi G} \int_{\mathcal{V}} \frac{M_{,r}}{\sqrt{2E}} \, dx \, dy \, dr \equiv \frac{c^2}{4\pi G} \int_{\mathcal{V}} \frac{dM}{\sqrt{2E}} \, dx \, dy. \qquad (19.19)$$

With $0 < E \propto M^{2/3}$ in the vicinity of $M = 0$, the integral with respect to M is finite. With a toroidal topology, the ranges of x and y are finite, so the integrals over x and y give finite values. Thus, the total amount of mass in each space $t = $ constant is finite.

The relation $N_{,r} = M_{,r}/\sqrt{2E}$ that follows from (19.19) is similar to $N_{,r} = M_{,r}/\sqrt{1 + 2E}$ which held in the L–T models, Eq. (18.17). By analogy, we conclude that in a plane symmetric spacetime the factor $1/\sqrt{2E}$ measures the relativistic mass defect/excess, i.e. the discrepancy between the active gravitational mass M and the sum of rest masses N.

---

[1] Since $M \geq 0$, this is in fact deceleration.

19.4 G_3/S_2-symmetric dust in electromagnetic field, the case $R_{,r} \neq 0$

19.4.1 Integrals of the field equations

The most general solution of the Einstein–Maxwell equations for the metric (19.13) with $R_{,r} \neq 0 \neq \Lambda$ was found by Bronnikov and Pavlov (1979). They considered charged dust under the assumption that the metric- and electromagnetic field tensors both have a G_3/S_2 group of symmetries. We shall allow also for magnetic charges.

With the assumed symmetries, in the coordinates of (19.11) – (19.12), the electromagnetic field tensor can have at most two nonzero components, F_{tr} (the electric field) and $F_{\vartheta\varphi}$ (the magnetic field). The $F_{\vartheta\varphi} \neq 0$ is due to a distribution of magnetic monopoles – admittedly a nonrealistic source, but it is useful to know what the theory implies for it.

The Einstein–Maxwell equations are in this case (compare with (13.6), (13.10), (13.12), (13.15) and (13.16)):

$$R_{\alpha\beta} - \frac{1}{2}g_{\alpha\beta}R + \Lambda g_{\alpha\beta} = \frac{8\pi G}{c^4}\epsilon u_\alpha u_\beta + \frac{2G}{c^4}\left(F_\alpha{}^\mu F_{\mu\beta} + \frac{1}{4}g_{\alpha\beta}F_{\mu\nu}F^{\mu\nu}\right), \tag{19.20}$$

$$F^{\mu\nu}{}_{;\nu} = (4\pi/c)\rho_e u^\mu, \tag{19.21}$$

$$*F^{\mu\nu}{}_{;\nu} = (4\pi/c)\rho_m u^\mu. \tag{19.22}$$

Equation (19.20) is (13.16) rewritten for dust. Equation (19.21) is (13.6) with the current composed of electric charges attached to the dust particles. Equation (19.22) is the generalisation of (13.15) for the current of magnetic charges attached to dust particles. The ρ_e and ρ_m are the densities of the electric and magnetic charge, respectively.

Equation (19.21) is equivalent to $(\sqrt{-g}F^{\mu\nu})_{,\nu} = (4\pi/c)\sqrt{-g}\rho_e u^\mu$. Taking $\mu = 1$ and then $\mu = 0$ we find

$$F^{01} = Q_e(r)e^{-(A+C)/2}/R^2, \tag{19.23}$$

$$Q_{e,r} = (4\pi/c)\rho_e e^{A/2}R^2, \tag{19.24}$$

where $Q_e(r)$ is an arbitrary function; it is the electric charge within the r-surface as (19.24) shows. Doing the same with (19.22) we find

$$*F^{01} = Q_m(r)e^{-(A+C)/2}/R^2, \tag{19.25}$$

$$Q_{m,r} = (4\pi/c)\rho_m e^{A/2}R^2, \tag{19.26}$$

where $Q_m(r)$ is the magnetic charge within the r-surface. Equation (19.25) translates into

$$F_{23} = -f(\vartheta)Q_m(r). \tag{19.27}$$

Now the coordinate components of the Einstein equations with the metric (19.11) and the electromagnetic tensor (19.23) – (19.27) become

$$G_{00} = \frac{8\pi G}{c^4}\epsilon e^C + \frac{G}{c^4}\frac{Q_e{}^2 + Q_m{}^2}{R^4}e^C - \Lambda e^C, \tag{19.28}$$

$$G_{01} = 0, \tag{19.29}$$

$$G_{11} = -\frac{G}{c^4}\frac{Q_e{}^2 + Q_m{}^2}{R^4}e^A + \Lambda e^A, \tag{19.30}$$

$$G_{22} \equiv G_{33}/f^2(\vartheta) = \frac{G}{c^4}\frac{Q_e{}^2 + Q_m{}^2}{R^2} + \Lambda R^2. \tag{19.31}$$

For $G_{\alpha\beta}$ we find from (19.11)

$$G_{00} = \frac{\varepsilon e^C}{R^2} + \frac{R_{,t}{}^2}{R^2} + \frac{A_{,t}\,R_{,t}}{R} + e^{C-A}\left(-\frac{R_{,r}{}^2}{R^2} - 2\frac{R_{,rr}}{R} + \frac{A_{,r}\,R_{,r}}{R}\right), \tag{19.32}$$

$$G_{01} = -2\frac{R_{,tr}}{R} + \frac{A_{,t}\,R_{,r}}{R} + \frac{R_{,t}\,C_{,r}}{R}, \tag{19.33}$$

$$G_{11} = -\frac{\varepsilon e^A}{R^2} + \frac{R_{,r}{}^2}{R^2} + \frac{C_{,r}\,R_{,r}}{R} + e^{A-C}\left(-\frac{R_{,t}{}^2}{R^2} - 2\frac{R_{,tt}}{R} + \frac{C_{,t}\,R_{,t}}{R}\right), \tag{19.34}$$

$$G_{22} = \frac{1}{4}e^{-C}\left(-4RR_{,tt} + 2RC_{,t}\,R_{,t} - 2RA_{,t}\,R_{,t} - R^2 A_{,t}{}^2 - 2R^2 A_{,tt} + R^2 C_{,t}\,A_{,t}\right)$$
$$+ \frac{1}{4}e^{-A}\left(4RR_{,rr} + 2RC_{,r}\,R_{,r} - 2RA_{,r}\,R_{,r} + R^2 C_{,r}{}^2 + 2R^2 C_{,rr} - R^2 C_{,r}\,A_{,r}\right). \tag{19.35}$$

In order to derive the two equations below, apply $\left(R^{\alpha\beta} - \frac{1}{2}g^{\alpha\beta}R\right)_{;\beta} = 0$ to the right-hand side of (19.20) and note that (19.22) is equivalent to

$$F_{\mu\nu;\beta} + F_{\nu\beta;\mu} + F_{\beta\mu;\nu} = -\frac{4\pi}{c}\sqrt{-g}\rho_m u^\lambda \epsilon_{\lambda\mu\nu\beta}. \tag{19.36}$$

Equations (19.20) – (19.22) imply the following:

$$\left(\epsilon u^\beta\right)_{;\beta} = 0, \tag{19.37}$$
$$\epsilon u^\alpha{}_{;\beta}\,u^\beta = -(1/c)\left(\rho_e F^\alpha{}_\beta + \rho_m * F^\alpha{}_\beta\right)u^\beta. \tag{19.38}$$

Equation (19.37) follows when $T^{\alpha\beta}{}_{;\beta} = 0$ is contracted with u_α, Eq. (19.38) follows when (19.37) is used in $T^{\alpha\beta}{}_{;\beta} = 0$. The remaining terms cancel out in consequence of (19.36) and the antisymmetry of $F_{\mu\nu}$. Equations (19.37) (the conservation of mass) and (19.38) (the **Lorentz force** acting on charges in motion and pushing them off geodesic trajectories) are general and independent of symmetries of spacetime. Applying (19.38) to our metric (19.11) – (19.12), then using (19.24) and (19.26) we obtain

$$\epsilon C_{,r} = (\rho_e Q_e + \rho_m Q_m)\frac{2e^{A/2}}{cR^2} \equiv \frac{QQ_{,r}}{2\pi R^4}, \tag{19.39}$$

where

$$Q^2 \stackrel{\text{def}}{=} Q_e{}^2 + Q_m{}^2. \tag{19.40}$$

Applying (19.37) to our metric (19.11) – (19.12) we then obtain

$$\frac{\kappa}{2}\epsilon e^{A/2}R^2 = \frac{G}{c^4}N_{,r}, \tag{19.41}$$

where $N_{,r}\,(r)$ is an arbitrary function. In the spherically symmetric case the N so defined is, up to the choice of units, the sum of rest masses within a surface of constant t and r. The coefficient defining the units was chosen so that the notation agrees with that in the pioneering paper (Bronnikov, 1983). From (19.24), (19.26) and (19.41) we now see that the ratios $Q_{e,r}/N_{,r} = \rho_e/(c\epsilon)$ and $Q_{m,r}/N_{,r} = \rho_m/(c\epsilon)$ are both time-independent.

Using now (19.39) we obtain from the equation $G_{01} = 0$

$$2e^{-A/2}R_{,tr} - e^{-A/2}A_{,t}\,R_{,r} = \frac{2R_{,t}}{cR^2}\left(\frac{\rho_e}{\epsilon}Q_e + \frac{\rho_m}{\epsilon}Q_m\right). \tag{19.42}$$

(It is here that the case $R_{,r} = 0$ has to be set aside for separate investigation. With $R_{,r} = 0$, the equation $G_{01} = 0$ reduces to $R_{,t}\,C_{,r} = 0$ and has other consequences, see Sec. 19.5.) Since the expression in parentheses is independent of t, (19.42) implies

$$e^{-A/2}R_{,r} = \Gamma(r) - \frac{\rho_e Q_e + \rho_m Q_m}{c\epsilon R}, \tag{19.43}$$

where $\Gamma(r)$ is an arbitrary function. Using (19.39) and (19.41) we now find

$$\frac{\rho_e Q_e + \rho_m Q_m}{c\epsilon} \equiv \frac{QQ_{,r}}{N_{,r}} = QQ_{,N}. \tag{19.44}$$

With this, (19.43) becomes

$$e^{-A/2}R_{,r} = \Gamma(r) - \frac{QQ_{,N}}{R}, \tag{19.45}$$

and now the first part of (19.39) becomes

$$C_{,r} = 2\frac{e^{A/2}}{R^2}QQ_{,N}. \tag{19.46}$$

Using (19.45) and (19.46) to eliminate $R_{,r}$ and $C_{,r}$, we obtain from (19.34) and (19.30)

$$e^{-C}\left(2\frac{R_{,tt}}{R} + \frac{R_{,t}^2}{R^2} - \frac{C_{,t}\,R_{,t}}{R}\right) - \frac{\Gamma^2 - \varepsilon}{R^2} + \frac{Q^2\left(Q_{,N}^2 - G/c^4\right)}{R^4} + \Lambda = 0. \tag{19.47}$$

Multiplying by $R^2R_{,t}$ we easily find that the integral of this is

$$e^{-C}R_{,t}^2 = \Gamma^2 - \varepsilon + \frac{2M(r)}{R} + \frac{Q^2\left(Q_{,N}^2 - G/c^4\right)}{R^2} - \frac{1}{3}\Lambda R^2, \tag{19.48}$$

where $M(r)$ is an arbitrary function. Comparing this with (18.14) we see that $(\Gamma^2 - \varepsilon)/2$ plays here the role of the L–T energy function $E(r)$.

The function $M(r)$ is the **effective mass** that drives the evolution, but, as we will see, it is a combination of mass and charge that *need not be positive*. In order to see this, we will now compare (19.48) with the Newtonian limit, assuming $\Lambda = 0$. Let $[x]$ denote the physical dimension (unit) of x. The dimensions of the quantities appearing in (19.48) are: $[R] = [\text{length}]$, $[e^{C/2}dt] \equiv [ds] = [c] \times [\text{time}]$, $[M] = [G] \times [\text{mass}]/[c^2]$, $[N] = [\text{mass}] \times [c^2]$ and $[Q] = [\text{charge}] \equiv [\sqrt{\text{mass}}] \times [\text{length}^{3/2}]/[\text{time}]$. The function $(\Gamma^2 - \varepsilon)$ is dimensionless, but, for consistency, must be assumed to have the form $2\mathcal{E}/c^2$, where $[\mathcal{E}] = [\text{velocity}^2]$. In order to find the Newtonian limit of (19.48), we multiply it by c^2 and let $c \to \infty$. Denoting the Newtonian time by τ, we obtain

$$R_{,\tau}^2 = 2\mathcal{E} + 2Gm(r)/R. \tag{19.49}$$

But the Newtonian equation of motion of spherically symmetric charged dust is

$$r_{,\tau}^2 = 2\mathcal{E}(r) + 2\left[G\mathcal{M}(r) - (\rho_e(r)/\rho_\mu(r))\,Q(r)\right]/r, \tag{19.50}$$

where $\mathcal{M}(r)$ is the total mass within the sphere of radius r, $Q(r)$ is the total charge within the same sphere, ρ_e is the charge density and ρ_μ is the mass density. Thus, the $M(r)$ in (19.48) corresponds to the Newtonian $[G\mathcal{M}(r) - (\rho_e(r)/\rho_\mu(r))\, Q(r)]$, and, as announced, does not have to be positive. This relation will appear in other formulae later on.

It remains to verify (19.31) with G_{22} given by (19.35). The verification, tricky and lengthy, is presented in Sec. 24.3. The end result is

$$QQ_{,N}\left[-\frac{G}{c^4}\Gamma N_{,r} + (M + QQ_{,N}\,\Gamma)_{,r}\right] = 0. \tag{19.51}$$

One solution of this is $Q_{,N} = 0$, i.e. a constant total charge. We will deal with this simpler case later. When $Q_{,N} \neq 0$,

$$\frac{G}{c^4}\Gamma N_{,r} = (M + QQ_{,N}\,\Gamma)_{,r}. \tag{19.52}$$

Via (19.52), Γ determines by how much $\mathcal{M} \overset{\text{def}}{=} M + QQ_{,N}\,\Gamma$ increases when a unit of rest mass is added to the source, i.e. Γ is a measure of the gravitational mass defect/excess. Solutions with $\Gamma = 0$ are known; this is the Datt–Ruban class of Section 19.5. Negative Γ can also occur; see the comments on the Novikov (1962a) paper in subsection 18.24.2, $\Gamma < 0$ when the distribution of matter contains necks (see Section 18.10) or fills more than half the volume of a closed 3-space, like in the Friedmann $k > 0$ model matched to the Schwarzschild metric. However, $\Gamma > 0$ corresponds to most ordinary configurations. Note the similarity of $M = \mathcal{M} - QQ_{,N}\,\Gamma$ to the Newtonian relation $Gm(r) = G\mathcal{M}(r) - (\rho_e/\rho_\mu)\,Q$.

Equation (19.28) with (19.32) defines the mass density of the dust. Hints for the calculation are given in Sec. 24.3. The resulting formula is equivalent to (19.41) via (19.45):

$$\kappa\epsilon = \frac{2GN_{,r}}{c^4 R^2 R_{,r}}\left(\Gamma - \frac{QQ_{,N}}{R}\right). \tag{19.53}$$

So finally, the Einstein–Maxwell equations for charged dust reduced to a set that defines $C(t,r)$, $A(t,r)$ and $R(t,r)$ implicitly. The prescription for constructing a solution is this:

1. Choose $M(r)$, $Q_e(r)$, $Q_m(r)$ and $\Gamma(r)$, use (19.40) and (19.44) for $Q_{,N} = Q_{,r}/N_{,r}$ and solve (19.52) to find $N(r)$.
2. Given these, express $e^{A/2}$ through R via (19.45).
3. The set $\{(19.46),\ (19.48)\}$ then defines e^C and R.

In the following, we will assume $Q_m = 0$, i.e. no magnetic charges. This does not influence the solutions in a significant way: with $Q_m = 0$ we have $Q = Q_e$. If, further, $Q_{,N} = 0$, then $Q_e = $ constant, i.e. $\rho_e = 0$ from (19.24). In the spherically symmetric case this means that the whole charge is contained in the point $r = 0$ – the electrically neutral dust is moving in the exterior electric field of a point charge.

Note, from (19.46), that when $Q_{,N} = 0$ the dust moves on geodesics, whereas when $Q_{,N} \neq 0$ it does not. But note from (19.48) and (18.14) that a constant charge Q modifies the geometry and the neutral dust in gravitational & electric field moves on geodesics in a geometry different from that of pure gravitational field. This means that electromagnetic field exerts influence even on electrically neutral matter via the Einstein equations. This is a purely relativistic effect. However, a constant charge does not modify the mass, and with $Q_{,N} = 0$, Eq. (19.53) reduces back to (18.15).

Following Bronnikov and Pavlov (1979) let us note from (19.48) that, with $\Lambda = 0$, the sign of ε puts certain limits on the possible types of evolution of $R(t, r)$. When R increases, the third term and the absolute value of the fourth term on the right of (19.48) decrease. Nothing can prevent R from growing to infinity when $\varepsilon \leq 0$. Thus these two cases cannot contain the recollapsing L–T or Friedmann models as subcases. When $\varepsilon = 0$, $R_{,t}$ will decrease to zero at $R \to \infty$ only if $\Gamma = 0$. However, with $\Gamma = 0$, Eq. (19.52) implies $M_{,r} = 0$, i.e. vacuum in the $Q \to 0$ limit. Thus, the plane symmetric charged models do not contain the $E = 0$ L–T (or $k = 0$ Friedmann) models in a limit, although the evolution of the charged model can have qualitative features of the $k = 0$ model. Only in the spherically symmetric case, $\varepsilon = +1$, are all three types of evolution possible, and all three L–T models are contained as subcases.

Special cases of the Bronnikov–Pavlov class presented here were discovered and discussed by other authors in earlier papers (Krasiński, 1997). Of the more important subcases, the electrically neutral case for all three symmetries was solved by Ellis (1967), the spherically symmetric case with no magnetic charge was solved and discussed by Vickers (1973), the $\Lambda = 0$ subcase of the Vickers case was solved and discussed by Markov and Frolov (1970), the case with zero density of magnetic monopoles and zero cosmological constant for all three symmetries was investigated by Shikin (1974), and the further subcase with zero magnetic charge and zero electric charge density was discussed in an earlier paper by Shikin (1972). In this chapter, we will mention only the most important physical contributions.

With vanishing charges, $Q_e = Q_m = 0$, in the spherically symmetric case $\varepsilon = +1$, (19.38) reduces to the equation of a geodesic, while Eqs. (19.45), (19.46) and (19.48) reduce to those defining the L–T model, where $2E = \Gamma^2 - 1$.

19.4.2 Matching the charged dust metric to the Reissner–Nordström metric

From now on, we will consider spherically symmetric metrics ($\varepsilon = +1$) with $Q_m = 0$.

Matching a charged matter solution to an electrovacuum[2] metric can help in interpreting the arbitrary functions and constants. For spherical symmetry and zero magnetic charge, the appropriate electrovacuum metric is the Reissner–Nordström (R–N) solution[3] presented in Section 14.4. We will allow for a nonzero cosmological constant.

As shown in Exercise 10 to Chapter 14, the R–N metric (14.40) – (14.41) can be transformed to the Lemaître–Novikov-like coordinates in which it becomes

$$ds^2 = dt^2 - \frac{{\mathcal{R}_{,r}}^2 dr^2}{1 + 2E(r)} - \mathcal{R}^2(t, r)\left(d\vartheta^2 + \sin^2 \vartheta d\varphi^2\right), \tag{19.54}$$

where $\mathcal{R}(t, r)$ obeys the equation

$$\mathcal{R}_{,t}^2 = 2E + \frac{2m}{\mathcal{R}} - \frac{e^2}{\mathcal{R}^2} - \frac{1}{3}\Lambda \mathcal{R}^2. \tag{19.55}$$

---

[2] An **electrovacuum metric** is a solution of the Einstein–Maxwell equations such that the mass density ϵ and charge density ρ_e are both zero, but there is an electromagnetic field generated by charges placed outside the region considered.

[3] A generalisation of the Reissner–Nordström solution for magnetic charge q is known (Stephani et al., 2003), but it is geometrically insignificant. The only change is replacing e^2 in the metric by $e^2 + q^2$.

These coordinates are adapted to the hypersurfaces $r = r_b = $ constant, so we may use (7.96) for the second fundamental form (now $X^{n+1} = r$). Continuity of the 3-metric of $r = r_b$ requires that

$$e^{C(t,r_b)} = 1, \qquad R(t, r_b) = \mathcal{R}(t, r_b). \tag{19.56}$$

The transformations that keep the metric (19.11) diagonal are still allowed. Transforming t by $t' = \int e^{C(t,r_b)} dt$ we fulfil the first of (19.56). We will deal with the consequences of the second of (19.56) further on.

The unit vector normal to the hypersurface $r = r_b$ has components

$$X^\mu = \left(0, e^{-A/2}, 0, 0 \right) \equiv (0, (\Gamma - QQ_{,N}/R)/R_{,r}, 0, 0) \tag{19.57}$$

for the interior metric, and, from (19.54),

$$X_{RN}^\mu = \left(0, \sqrt{1 + 2E}/\mathcal{R}_{,r}, 0, 0 \right) \tag{19.58}$$

for the R–N metric. From the continuity of $\Omega_{\alpha\beta}$ we have

$$R_{,r} X^r \big|_{r=r_b} = \mathcal{R}_{,r} X_{RN}^r \big|_{r=r_b} \qquad \text{and} \qquad \left(e^C \right)_{,r} X^r \big|_{r=r_b} = 0. \tag{19.59}$$

The second condition means $C_{,r} = 0$ at $r = r_b$, so (19.46) implies

$$Q_{,N}(r_b) = 0 \implies Q_{,r}(r_b) = 0. \tag{19.60}$$

This, via (19.24), says that the charge density $\rho_e(r_b) = 0$. The first of (19.59) then says

$$\Gamma = \sqrt{1 + 2E}. \tag{19.61}$$

At $r = r_b$, $R(t, r_b)$ and $\mathcal{R}(t, r_b)$ must be the same functions of t by (19.56). Consequently, $R_{,t}(t, r_b)$ and $\mathcal{R}_{,t}(t, r_b)$ must be the same, so, at $r = r_b$, the right-hand side of (19.55) must be the same as the right-hand side of (19.48). This means, since $Q_{,N}(r_b) = 0$, that

$$m = M(r_b), \qquad e^2 = GQ^2(r_b)/c^4. \tag{19.62}$$

This matching was discussed by Vickers (1973), but he overlooked Eq (19.60).

Compare Exercise 6: when you transform the R–N metric (14.40) to new orthogonal coordinates without requiring that the new $g_{t't'} = 1$ and impose (19.48) on $g_{\alpha'\beta'}$, then $g_{t't',r'} = 0$ follows anyway.

19.4.3 Prevention of the Big Crunch singularity by electric charge

The presence of electric charges has a strong influence on the evolution of dust. In discussing this, we shall follow Vickers (1973), with a few corrections. In the present section we will deal only with the Big Bang/Big Crunch (BB/BC) singularities. Shell crossings will be discussed in Section 19.4.6. We assume $\Lambda = 0$ and denote $E(r) = \left(\Gamma^2 - 1 \right)/2$.

The presence or absence of a singularity is detected by investigating the roots of the right-hand side of (19.48), which, for this purpose, is more conveniently written as

$$e^{-C} R^2 R_{,t}^2 = 2E(r)R^2 + 2M(r)R + Q^2 \left(Q_{,N}^2 - G/c^4 \right) \stackrel{\text{def}}{=} W(R). \tag{19.63}$$

The trinomial $W(R)$ has roots only if

$$M^2 \geq 2EQ^2 \left(Q_{,N}{}^2 - G/c^4\right). \tag{19.64}$$

At each root of $W(R)$, the sign of $R_{,t}$ changes, and evolution is possible only in those ranges where $W(R) \geq 0$. The following cases occur:

(a) When $E < 0$
With no roots, $W(R)$ would be negative at all R, so (19.64) is the condition for the existence of a solution of (19.63). With (19.64) fulfilled, the roots are

$$R_\pm = -\frac{M}{2E} \pm \frac{1}{2E}\sqrt{M^2 - 2EQ^2\left(Q_{,N}{}^2 - G/c^4\right)}, \tag{19.65}$$

and $W(R) > 0$ for $R_+ < R < R_-$. Nonsingular solutions of (19.63) exist when both $R_\pm > 0$ (with both $R_\pm < 0$ no solution exists at all, with $R_- R_+ < 0$ the value $R = 0$ is in the allowed range.) This implies

$$Q_{,N}{}^2 < G/c^4 \text{ and } M > 0. \tag{19.66}$$

We will interpret these conditions later in this section.

If there is equality in (19.64), then $R_- = R_+$, $W(R_\pm) = 0$ and $W(R) < 0$ for all $R \neq R_\pm$. Then $R = R_\pm$ is the only solution of (19.63) and the model is static. If, in addition, $\Gamma(r_b) = 0$, then $E(r_b) = -1/2$, and in this case the exterior R–N metric is the extreme one, with $e^2 = m^2$, as seen from (19.60), (19.64) and (19.62).

With (19.66) fulfilled, R oscillates between R_+ and $R_- > R_+$, never going down to zero.

(b) When $E = 0$
The singularity is avoided if and only if $M > 0$ and $Q_{,N}{}^2 < G/c^4$. Collapse is then halted and reversed once and for all, with the turning point at $R_{\min} = Q^2(G/c^4 - Q_{,N}{}^2)/(2M)$. At the outer surface of a charged sphere, $r = r_b$, this is $R_{\min} = GQ^2/(2Mc^4)\big|_{r=r_b}$, which by (19.62) means $R_{\min} = e^2/(2m) < m - \sqrt{m^2 - e^2}$. Thus, the bounce of a whole charged sphere occurs inside the inner R–N event horizon.

(c) When $E > 0$
In this case $W(R) > 0$ either everywhere (if it has no roots) or outside the range $[R_-, R_+]$. There are no roots when $M^2 < 2EQ^2\left(Q_{,N}{}^2 - G/c^4\right)$, in which case $W(R) > 0$ for all R including $R = 0$, and the model can run into the singularity. Thus, (19.64) is here one of the necessary conditions for the existence of nonsingular solutions. With (19.64) fulfilled, $W(R)$ has two roots, and at least one of them has to be positive if singularity is to be avoided. With $M > 0$, we have $R_- < 0$ always and $R_+ > 0$ if and only if $Q_{,N}{}^2 < G/c^4$. With $M < 0$, $R_+ > 0$ and $R_- < R_+$ always, so nonsingular solutions exist with no further conditions, provided $R > R_+$ initially. Collapse is then halted and reversed as in case (b).

Now we will interpret the conditions (19.66). The inequality $Q_{,N}{}^2 < G/c^4$, with $Q_m = 0$ and $Q_e = Q$, translates into $|\rho_e| < \sqrt{G}\epsilon/c$.

When $M < 0$, the BB/BC singularity is avoided only if $E > 0$ and (19.64) is fulfilled, with no further conditions on ρ_e. This is a nonrelativistic bounce, since it occurs also in Newton's theory, under the same conditions ($M = G\mathcal{M} - (\rho_e/\rho_\mu)Q < 0$ in (19.50)).

With $M > 0$, a nonsingular bounce will occur only if $|\rho_e| < \sqrt{G}\epsilon/c$, i.e. if the absolute value of the charge density is sufficiently *small* (but nonzero) compared with the mass density. This is a purely relativistic effect that does not occur in the Newtonian limit: with $M > 0$, $R = 0$ is always in the allowed range of Newtonian solutions. The physical interpretation of the relativistic bounce is this: as seen from (19.48), the charges provide a correction to the effective mass M, so that $\overline{M} = M + (1/2)Q^2\left(Q_{,N}{}^2 - G/c^4\right)/R$ drives the evolution. This correction is negative at small charge density $(Q_{,N}{}^2 < G/c^4)$, so it weakens gravitation, thus helping the dust to bounce. However, at large charge density $(Q_{,N}{}^2 > G/c^4)$, charges enhance the effective mass and thus oppose bounce. Nevertheless, in the latter case, the Newtonian electrostatic repulsion can prevail, provided that $M < 0$ at the same time. Recall a similar phenomenon that we encountered while discussing the motion of particles in the Reissner–Nordström spacetime (Section 14.16), where electric charge in the source of the gravitational field creates effective antigravitation, provided the charge is small enough compared to the mass.

With $Q_{,N} = 0$ the singularity is avoided in every case when a solution exists. Thus, for neutral dust moving in an exterior electric field, the BB/BC singularity never occurs. This was first found by Shikin (1972). This is a purely relativistic effect.

We have proven that in some cases a collapsing solution of (19.48), for which $R \neq 0$ initially, does not go down to 0. However, if the charged dust occupies a volume around the centre of symmetry $R = 0$, then, at any time, there exist dust particles with all values of R, including $R = 0$. We will find in the next section the conditions for the centre to be nonsingular. Thus, the inner turning points given by (19.65) will exist arbitrarily close to the centre. This has consequences for the shell crossings discussed in Section 19.4.6.

If $EM < 0$ and $Q_{,N}{}^2 = G/c^4$, then (19.63) has a time-independent solution $R = -M/E$. In this case, the electrostatic repulsion just balances the gravitational attraction and the whole configuration is static – but unstable. Any small perturbation of this value of R will send the dust either into collapse that will terminate at $R = 0$ or into expansion $R \to \infty$. Which case occurs depends on the sign of E: with $E > 0$ only the perturbation towards larger R is possible, with $E < 0$ the opposite is true.

At the surface of a charged sphere, where $Q_{,N}(r_b) = 0$ by (19.60), Eq. (19.63) will coincide with (14.180) (up to a transformation of time) if $J_0 = 0$ (radial motion), $m = M - qe\Gamma/\mu$ and $e^2\left(1 - q^2/\mu^2\right) = GQ^2/c^4$ (this last equation can hold only if $(q/\mu)^2 < 1$). Consequently, the surface of a charged dust sphere obeys the equation of radial motion of a charged particle in the Reissner–Nordström (R–N) spacetime. As shown in Section 14.16, for such a particle the reversal of fall to escape can occur only inside the inner R–N horizon, at $R < r_- = m - \sqrt{m^2 - e^2}$. Thus, the surface of a collapsing charged dust sphere must continue to collapse until it crosses the inner horizon at $R = r_-$, and can bounce at $R < r_-$. Then, as seen from Fig. 14.12, it cannot re-expand into the same spacetime region I from which it collapsed, since this would require motion backwards in time. The surface would thus pass through the tunnel between the singularities and re-expand into another copy of the asymptotically flat region.

The bounce at small charge density $(Q_{,N}{}^2 < G/c^4)$ would be interesting physically, since the real Universe has no detectable net charge, so only small charges could exist in it. We saw that *an arbitrarily small uncompensated charge can prevent the BB/BC singularity.* Unfortunately, Ori (1990, 1991) proved that precisely in this case a shell crossing is unavoidable, and it will block the passage through the tunnel. We will derive this result in Section 19.4.6. Thus, a nonsingular bounce through the R–N tunnel is impossible when $Q_{,N}{}^2 < G/c^4$ everywhere.

19.4.4 * Charged dust in curvature and mass-curvature coordinates

It is instructive to transform the metric given by (19.11) with $f(\vartheta) = \sin\vartheta$, (19.45), (19.46) and (19.48) to coordinates in which the function $R(t,r)$ is the radial coordinate.

We note that $R_{,r}\,dr = dR - R_{,t}\,dt$, and then we take t to be a function of the new coordinates: $t = f(\tau, R)$. Thus

$$R_{,r}\,dr = dR - R_{,t}\,(f_{,\tau}\,d\tau + f_{,R}\,dR)\,, \qquad dt = f_{,\tau}\,d\tau + f_{,R}\,dR, \tag{19.67}$$

and the new metric components, using (19.48), are found to be

$$g_{\tau\tau} = e^C f_{,\tau}{}^2\Delta/u^2, \tag{19.68}$$

$$g_{\tau R} = \left[e^C f_{,\tau}\,f_{,R}\,\Delta + f_{,\tau}\,R_{,t} \right]/u^2, \tag{19.69}$$

$$g_{RR} = \left[e^C f_{,R}{}^2\Delta - 1 + 2f_{,R}\,R_{,t} \right]/u^2, \tag{19.70}$$

where we have defined

$$\Delta \overset{\text{def}}{=} 1 - \frac{2\mathcal{M}}{R} + \frac{GQ^2}{c^4 R^2} + \frac{1}{3}\Lambda R^2, \tag{19.71}$$

$$u \overset{\text{def}}{=} \Gamma - QQ_{,N}/R, \tag{19.72}$$

$$\mathcal{M} \overset{\text{def}}{=} M + QQ_{,N}\,\Gamma. \tag{19.73}$$

Note, from (19.67), that the transformation for collapsing dust ($R_{,t} < 0$) is different from that for expanding dust ($R_{,t} > 0$). The transformation inverse to (19.67) is analogous to introducing, in the Reissner–Nordström region $r > r_+ = m + \sqrt{m^2 - e^2}$, coordinates comoving with the congruence of charged particles that are radially collapsing or expanding, respectively. Figure 14.12 shows that in this way we obtain a coordinate map that covers regions I and II for the collapsing congruence or regions I and IV for the expanding congruence. The extension is to the future or to the past, respectively.

This can now be specialised in two ways. One possibility is to choose the proper curvature coordinates, in which $g_{\tau R} = 0$, so

$$f_{,R} = -e^{-C}R_{,t}/\Delta. \tag{19.74}$$

Then, substituting it in (19.70) we obtain

$$g_{RR} = -1/\Delta. \tag{19.75}$$

The other possible specialisation of (19.68) – (19.70) is to choose the $\mathcal{M}$ defined in (19.73) as the new τ coordinate. These **mass-curvature coordinates** were first introduced by Ori (1990). The surfaces $\mathcal{M} = $ constant are timelike, so none of the coordinates is time and the metric cannot be diagonal. Since $\mathcal{M}$, Q, Γ and N depend only on r in the original coordinates, we have $Q = Q(\mathcal{M})$, $\Gamma = \Gamma(\mathcal{M})$ and $N = N(\mathcal{M})$. The Jacobi matrices of the transformations $(t, r) \leftrightarrow (\mathcal{M}, R)$ are

$$\frac{\partial(t, r)}{\partial(\mathcal{M}, R)} = \begin{bmatrix} f_{,\mathcal{M}}, & f_{,R} \\ r_{,\mathcal{M}}, & r_{,R} \end{bmatrix}, \tag{19.76}$$

$$\frac{\partial(\mathcal{M}, R)}{\partial(t, r)} = \begin{bmatrix} 0 & \mathcal{M}_{,r} \\ R_{,t}, & R_{,r} \end{bmatrix}. \tag{19.77}$$

These matrices must be inverse to each other. Hence:

$$f_{,\mathcal{M}} = -\frac{R_{,r}}{R_{,t}\,\mathcal{M}_{,r}}, \qquad f_{,R} = 1/R_{,t}, \qquad r_{,\mathcal{M}} = 1/\mathcal{M}_{,r}, \qquad r_{,R} = 0. \tag{19.78}$$

In the coordinates $(x^0, x^1) \overset{\text{def}}{=} (\mathcal{M}, R)$ the velocity field u^α still has only one component:

$$\begin{aligned} u^R &= e^{-C/2} R_{,t} = \pm\sqrt{\Gamma^2 - 1 + \frac{2M}{R} + \frac{Q^2\left(Q_{,N}{}^2 - G/c^4\right)}{R^2} - \frac{1}{3}\Lambda R^2} \\ &\equiv \pm\sqrt{u^2 - \Delta}. \end{aligned} \tag{19.79}$$

(the $+$ sign for expansion, $-$ for collapse).

We define the auxiliary quantities

$$\alpha \overset{\text{def}}{=} e^{C/2}/u, \qquad \alpha f_{,\mathcal{M}} = -\frac{e^{C/2}}{u}\frac{R_{,r}}{R_{,t}\,\mathcal{M}_{,r}} \overset{\text{def}}{=} F(\mathcal{M}, R), \tag{19.80}$$

and, using (19.71), (19.79) and (19.48), we obtain from (19.68) – (19.70):

$$g_{\mathcal{M}\mathcal{M}} = F^2\Delta \overset{\text{def}}{=} A, \tag{19.81}$$

$$g_{\mathcal{M}R} = F\alpha u^2/R_{,t} = Fu/u^R \overset{\text{def}}{=} B, \tag{19.82}$$

$$g_{RR} = 1/(u^R)^2 \overset{\text{def}}{=} C. \tag{19.83}$$

The function $F(\mathcal{M})$ is to be found from the field equations. Equations (19.81) – (19.83) differ from those of Ori (1990) only by notation. They imply

$$AC - B^2 = -F^2 \implies \det(g) = -\left(FR^2\sin\vartheta\right)^2. \tag{19.84}$$

Using (19.78), (19.23) and (19.80) we find that the only nonvanishing components of the electromagnetic tensor in the $(\mathcal{M}, R)$ coordinates are

$$F^{\mathcal{M}R} = -F^{R\mathcal{M}} = \frac{Q}{FR^2}, \qquad F_{\mathcal{M}R} = -F_{R\mathcal{M}} = -\frac{FQ}{R^2}. \tag{19.85}$$

Further, using (19.24), (19.72), (19.45), (19.79), (19.77), (19.80) and (19.78) for the first equation below, and (19.41), (19.45), (19.72), (19.79), (19.52), (19.77), (19.80) and (19.78) for the second one, we find for the charge density and energy density

$$\frac{4\pi\rho_e}{c} = \frac{uQ_{,r}}{R^2 R_{,r}} \equiv -\frac{uQ_{,\mathcal{M}}}{R^2 f_{,\mathcal{M}} R_{,t}} \equiv -\frac{Q_{,\mathcal{M}}}{R^2 F u^R}, \tag{19.86}$$

$$\kappa\epsilon = \frac{2u\mathcal{M}_{,r}}{\Gamma R^2 R_{,r}} \equiv -\frac{2\alpha u}{\Gamma R^2 F R_{,t}} \equiv -\frac{2}{\Gamma R^2 F u^R}. \tag{19.87}$$

Equations (19.81) – (19.83) and (19.85) – (19.87) show that $F(\mathcal{M}, R)$ is the only unknown function (because $\mathcal{M}$ and R are now coordinates and (M, Γ, Q) are arbitrary functions of $\mathcal{M}$). In these coordinates we obtain

$$u_{\mathcal{M}} \ = \ uF, \qquad u_R = 1/u^R, \tag{19.88}$$

$$g^{\mathcal{M}\mathcal{M}} \ = \ -\frac{1}{(Fu^R)^2}, \qquad g^{\mathcal{M}R} = \frac{u}{Fu^R}, \qquad g^{RR} = -\Delta. \tag{19.89}$$

Dividing (19.52) by $\mathcal{M}_{,r}$ and using the definition of $\mathcal{M}$, Eq. (19.73), we find that in the $(\mathcal{M}, R)$ coordinates Eq. (19.52) reads

$$\frac{G}{c^4}\Gamma N_{,\mathcal{M}} = 1. \tag{19.90}$$

The function F can be found from (19.38). With $\rho_m = Q_m = 0$, using (19.44), they read

$$u^{\alpha}{}_{;\beta}\, u^{\beta} = -\frac{\rho_e}{c\epsilon}\, F^{\alpha\beta}u_{\beta} \equiv -Q_{,N}\, F^{\alpha\beta}u_{\beta}, \tag{19.91}$$

and, using (19.81) – (19.83) and (19.88) – (19.89), imply just one equation:

$$F_{,R} = -\frac{u^R{}_{,\mathcal{M}}}{u\,(u^R)^2}. \tag{19.92}$$

Using (19.72), (19.79) and (19.90), we transform (19.92) to

$$F_{,R} = -\frac{1}{(u^R(\mathcal{M}, R))^3}\left\{\Gamma_{,\mathcal{M}} + \frac{1}{R\Gamma}\left[1 - \frac{c^4}{G}\left(Q_{,N}{}^2 + QQ_{,NN}\right)\right]\right\}. \tag{19.93}$$

This coincides, except for notation, with Ori's (1990) result. Note that we have *not* assumed $\Lambda = 0$. As Ori (1990) stressed, Eqs. (19.81) – (19.83), (19.88) and (19.93) determine the metric explicitly, in contrast to the representation by Vickers used in Section 19.4.1, where the Einstein–Maxwell equations were reduced to a set of two differential equations.

With $\Lambda = 0$, the integral of (19.93) is of the form $\int \frac{(ax+b)x^2}{(cx^2+dx+e)^{3/2}}\mathrm{d}x$ and is elementary, though complicated. The full list of results is given in Ori's paper.

It can now be verified that the Einstein–Maxwell equations are all fulfilled. The identity $Q_{,N} \equiv Q_{,\mathcal{M}}/N_{,\mathcal{M}} = G\Gamma Q_{,\mathcal{M}}/c^4$, resulting from (19.90), is helpful in the calculations.

Note, from (19.79) – (19.80), that Fu^R is insensitive to the sign of u^R, so, by (19.81) – (19.83) and (19.87), the metric and the mass density are also insensitive to it. As explained in the remark after (19.73), $u^R > 0$ and $u^R < 0$ correspond to different maps with different domains. Thus, integrating (19.93) from R_1 to $R > R_1$ with $u^R > 0$ we integrate forward in time, while calculating the same integral with $u^R < 0$, we integrate backwards in time.

19.4.5 Regularity conditions at the centre

Just as in the L–T model, the set $R = 0$ in charged dust consists of the BB/BC singularity (which we showed to be avoidable) and the centre of symmetry, which may or may not be singular. We will now derive the conditions for the absence of the central singularity.

Let $r = r_c$ correspond to the centre of symmetry. From (19.11) – (19.12) and (19.41) we see that $N(r) = \int_{\mathcal{V}} \epsilon\sqrt{g_3}\mathrm{d}_3x$, where $\mathcal{V}$ is a sphere centred at $r = r_c$ in a $t =$ constant space. Thus, if ϵ has no singularity of the type of the Dirac delta at the centre, $N(r)$ must obey $N(r_c) = 0$. Similarly, (19.11) – (19.12) and (19.24) show that if there is no delta-type singularity of ρ_e at the centre, then the electric charge must obey $Q(r_c) = 0$. In addition, we assume that $\epsilon(r_c) \neq 0$ (motivated by physical intuition). Then, with both ϵ and ρ_e being nonsingular at r_c, the ratio $\rho_e(r_c)/\epsilon(r_c)$ is nonsingular, and (19.24) with (19.41) show that $\lim_{r \to r_c} Q/N = \lim_{r \to r_c} Q_{,N} = \rho_e(r_c)/[c\epsilon(r_c)]$ is finite (possibly zero). Then, (19.90) implies that $\Gamma(r_c) \neq 0$ and $\lim_{r \to r_c} N/M = \text{constant} \neq 0$. Thus, we can use M instead of N in calculating the behaviour of functions in the vicinity of $r = r_c$.

Since $R(t, r_c) = 0$ and $N(r_c) = 0$, we find from (19.53)

$$\lim_{r \to r_c} \frac{R^3}{N} = \lim_{r \to r_c} \left(\frac{3R^2 R_{,r}}{N_{,r}} \right) = \lim_{r \to r_c} \left[\frac{3}{4\pi\epsilon} \left(\Gamma - \frac{QQ_{,N}}{R} \right) \right]. \tag{19.94}$$

This limit will be finite if $\lim_{r \to r_c}(QQ_{,N}/R) < \infty$. We already know that $Q_{,N}(r_c) \overset{\text{def}}{=} \tilde{q}_0 = \text{constant}$, so $Q = \tilde{q}_0 N + O_1(N)$, using the notation of Section 18.4. Thus, $Q = q_0 M + O_1(M)$ and $\lim_{r \to r_c}(QQ_{,N}/R) < \infty$ if $R = \text{constant} \times M^\gamma + O_\gamma(M)$, where $\gamma < 1$. Then, (19.94) imposes the further condition that, in the neighbourhood of r_c,

$$R = \beta(t)M^{1/3} + O_{1/3}(M). \tag{19.95}$$

Since $R(t, r_c) = 0$ at all times, we have $R_{,t}(t, r_c) = 0$. All other terms in (19.48) except $(\Gamma^2 - \varepsilon)$ vanish at $r = r_c$, so, in the spherically symmetric case $\varepsilon = +1$, we must have

$$\lim_{r \to r_c} \Gamma^2(r) = 1 \implies \lim_{r \to r_c} E(r) = 0. \tag{19.96}$$

Note that this does not exclude $\Gamma < 0$.

19.4.6 * Shell crossings in charged dust

As can be seen from (19.80), $F = 0$ is equivalent to $R_{,r} = 0$, so $F = 0$ is a locus of shell crossings. Then, from (19.87) we see that $\Gamma F u^R$ must be negative for the density of dust to be positive. Since $u^R = dR/ds < 0$ during collapse, ΓF must be positive.

Now let us write the solution of (19.93) as follows:

$$F = - \int_{R_1}^{R} \frac{1}{(u^R(\mathcal{M}, x))^3} \left\{ \Gamma_{,\mathcal{M}} + \frac{1}{x\Gamma} \left[1 - \frac{c^4}{G} (Q_{,N}^2 + QQ_{,NN}) \right] \right\} dx + g(\mathcal{M}), \tag{19.97}$$

where $g(\mathcal{M})$ is an arbitrary function and R_1 is the initial value of R. When approaching the bounce in the central region, we have $R_1 > R$ and $u^R > 0$, so the integral term is positive when the expression in curly braces is negative. From now on we assume $\Lambda = 0$.

We saw in Section 19.4.3 that there are dust particles with all values of R, including arbitrarily small values. At the turning point in collapse $u^R \to 0$. The integrand is of the form $\left[(ax^2 + bx + c)^{-3/2} (a_2 x^3 + a_3 x^2) \right]$, and the trinomial has real zeros, so the integral is unbounded (which shows that $u^R = 0$ is a coordinate singularity). From (19.97) it is seen that as we approach $\{R = 0, \mathcal{M} = 0\}$, the term with $1/(x\Gamma)$ will dominate. We know from the regularity conditions that $\lim_{\mathcal{M} \to 0} \Gamma^2 = 1$, $\lim_{\mathcal{M} \to 0} Q = 0$ and $\lim_{\mathcal{M} \to 0} Q_{,N} = \text{constant} < \infty$. Thus, as long as $Q_{,N}^2 < G/c^4$, the sign of $x\Gamma$ in the vicinity of $R = 0$ will determine the sign of the infinity in F. Now it turns out that the sign of F will necessarily change to the opposite during collapse. If $\Gamma > 0$, then $F > 0$ can be achieved at $R = R_1$ by a choice of $\Gamma_{,\mathcal{M}}$ and $g(\mathcal{M})$, but $F \to -\infty$ as $R \to 0$. If $\Gamma < 0$, then $F < 0$ can be achieved at $R = R_1$, but $F \to +\infty$ as $R \to 0$. This means that there is necessarily a shell crossing somewhere at $R > 0$ if $Q_{,N}{}^2 < G/c^4$ holds all the way down to $\mathcal{M} = 0$. This is the theorem proved by Ori (1990, 1991).[4]

The infinity in F can be avoided altogether if the term in curly braces in (19.97) is zero at the same x at which $u^R = 0$. The zero of u^R is given by (19.65); it is $R = R_+$. In that case F is finite at $R = R_+$ and, by (19.87), ϵ becomes infinite, i.e. $R = R_+$ becomes a true curvature singularity. Thus, also in this case the charged dust cannot tunnel through the Reissner–Nordström throat.

The only situations in which both the BB/BC and shell crossing singularities could possibly be avoided are the following:

1. When $Q_{,N}{}^2 > G/c^4$, $E > 0$ and $M < 0$. Then, as shown in Section 19.4.3, a nonsingular bounce is possible, and shell crossings will not occur.

2. When $\lim_{\mathcal{M} \to 0} Q_{,N}{}^2 = G/c^4$. Then, because of $\lim_{\mathcal{M} \to 0} Q/R = 0$, the term $\Gamma_{,\mathcal{M}}$ in (19.97) has a chance to outbalance the other one and secure the right sign of F everywhere.

At the time of the first edition of this book, no examples of such solutions were known. Here is an update.

The first case, $Q_{,N}{}^2 > G/c^4$, $E > 0$ and $M < 0$, had reportedly been investigated by Ori and coworkers, but never published.[5]

The second case was followed in-depth by Krasiński and Bolejko (2006 and 2007), but the results were rather disappointing and partly mysterious. Here is a collection of quotations that summarise the results in the 2007 paper:

'In our previous paper ... We found that shell crossings are still unavoidable when the energy function $E \geq 0$, but the conditions did not lead to a contradiction when $E < 0$. ... However, ... the energy density ... was becoming negative in a vicinity of the center of symmetry, for a brief period around the bounce instant. Then, exact consideration showed that ... a weakly charged spherically symmetric dust ball will always have such a negative-energy-density region if the charge and mass densities become equal in absolute value at the center. ... In the general case treated by Ori this problem can be avoided by an appropriate choice of the arbitrary functions.

   The explicit example we gave in paper I had a problem ... there was a permanent central singularity in it (the limit of the mass density at the center was actually $-\infty$). The permanent infinity is easily cured .... but a direction-dependent point singularity at the center necessarily appears at the instant of maximal compression.

   In the present paper we ... show that ... The conditions (2.14)' [these are (19.66) in this book] 'guarantee that a particle that had $R > 0$ initially will not hit the set $R = 0$ in the future or in the past. But ... the configurations obeying (2.14) contain a cleverly hidden singularity of a type hitherto unknown

---

[4] We thank Amos Ori for an extended correspondence on this point. The discussion helped to clarify several other points in this section.

[5] A. Ori, private communication from around 2007.

in dust solutions. On the worldline of the center of symmetry, where $R(t, r) = 0$ permanently, there is a point in which $\epsilon \to +\infty$ for a single instant. This instant is the limit at $R \to 0$ of the hypersurface $S_{\min}$ consisting of the instants in which the mass shells with $R > 0$ attain their minimal sizes. However, if we approach the same spacetime location along the hypersurface $S_{\min}$, then $\epsilon \to -\infty$. ...

The conclusion of this paper is the following: the weakly charged spherically symmetric dust distribution considered here ($Q,_N^2 < G/c^4$ at $N > 0$ and $Q,_N^2 = G/c^4$ at $N = 0$) must contain at least one of the following features:

(1) A big bang/big crunch singularity;

(2) A permanent central singularity;

(3) A shell crossing singularity in a vicinity of the center;

(4) A finite time interval around the bounce instant in which the energy density becomes negative, and a transient momentary singularity of infinite energy density at a single point on the worldline of the center of symmetry. ...

A permanently nonsingular pulsating configuration of spherically symmetric charged dust does not exist. At most, a single full cycle of nonsingular bounce can occur, and shell crossings will necessarily appear during the second collapse phase. The ... bounce occurs at $R > 0$, but the momentary isolated singularity at the center of symmetry is still there.

The possibility of ϵ going negative in the presence of electric charges does not seem to have been noticed and may need further work on its interpretation. If the negative-energy-density region existed permanently in some part of the space, with comoving boundary, then we might suspect that this is a consequence of a bad choice of parameters that implies unphysical properties in that region. However, here we have the case in which the energy density is positive for some time, and then these same matter particles acquire negative energy density in a time interval around the bounce instant. This suggests that there is some physical process involved in this, which should be further investigated.

... The uncharged limit of the family of configurations ... is the LT model. ... For the latter, shell crossings can be avoided. ... Why, then, are they unavoidable with $Q \neq 0$?

The answer is this: in the LT model, the BB or BC are unavoidable .... In the cases that are colloquially called "free of shell crossings", in reality the shell crossings are not removed, but shifted to the epoch before the BB or after the BC, or both. Thus, the shell crossings are no longer in the domain of physical applicability of the model. When the BB/BC singularities are replaced with a smooth bounce in charged dust, the shell crossings that were hidden on the other side of BB/BC become physically accessible, and they terminate the evolution of the configuration.'

The remarks quoted above are related to the closing comment of Sec. 18.10.

19.5 The Datt–Ruban solution

Now we will deal with the case $R,_r = 0$ in (19.11) – (19.12). Equations (19.28) – (19.38) are still valid. Equations (19.29) and (19.33) imply $R,_t C,_r = 0$. With $R,_t = 0$, taking the combination $e^{-C} G_{00} + e^{-A} G_{11}$, we obtain $\kappa \epsilon = 0$, i.e. an (electro-) vacuum solution, which we will not consider. Thus $C,_r = 0$, which means that the dust is moving on geodesics. A transformation of time can then be used to achieve $C = 0$. Using $Q^2 \overset{\text{def}}{=} Q_e^2 + Q_m^2$, we find from (19.30) and (19.34)

$$\frac{\epsilon}{R^2} + \frac{R,_t^2}{R^2} + 2\frac{R,_{tt}}{R} - \frac{GQ^2}{c^4 R^4} + \Lambda = 0. \tag{19.98}$$

Since $R,_r = 0$, Q must be constant. According to (19.24) and (19.26), this means that the densities of electric and magnetic charge must be zero. Thus, the only kind of electromagnetic field that is compatible with the geometry we are now considering is the free field; the dust is uncharged and moves in the exterior electric field.

Multiplying (19.98) by $R^2 R_{,t}$ and integrating we obtain

$$R_{,t}^{\;2} = -\varepsilon + \frac{2M}{R} - \frac{GQ^2}{c^4 R^2} - \frac{1}{3}\Lambda R^2, \tag{19.99}$$

where M is a constant. Then, (19.31) and (19.35) imply

$$RR_{,tt} + \frac{1}{2}RA_{,t}\, R_{,t} + \frac{1}{4}R^2 A_{,t}^{\;2} + \frac{1}{2}R^2 A_{,tt} + \frac{GQ^2}{c^4 R^2} + \Lambda R^2 = 0. \tag{19.100}$$

This is much simplified by the substitution $A = 2\ln u(t,r)$ and then $u = R_{,t}\, w(t,r)$. For the function w the following equation results:

$$2RR_{,tt}\, w_{,t} + RR_{,t}\, w_{,tt} + R_{,t}^{\;2} w_{,t} = 0. \tag{19.101}$$

Its general solution is $w = X(r) \int \left[1/\left(RR_{,t}^{\;2}\right)\right] dt + Y(r)$, where $X(r)$ and $Y(r)$ are arbitrary functions. Thus,

$$e^{A/2} = R_{,t} \left[X(r) \int \frac{1}{RR_{,t}^{\;2}} dt + Y(r)\right]. \tag{19.102}$$

The expression for the metric is

$$ds^2 = dt^2 - e^A dr^2 - R^2(t)\left[d\vartheta^2 + (1/\varepsilon)\sin^2\left(\sqrt{\varepsilon}\vartheta\right) d\varphi^2\right], \tag{19.103}$$

and the following is found from (19.28) and (19.32) with use of (19.99) and (19.102):

$$\kappa\epsilon = \frac{2X}{R^2 e^{A/2}}. \tag{19.104}$$

This solution of the Einstein–Maxwell equations was first semi-published by Ruban (1972), and then mentioned in a later paper (Ruban, 1983).

The Ruban spacetime is globally a T-region:[6] the gradient of R is timelike everywhere. Consequently, as mentioned in Section 14.1, the curvature coordinates do not exist in this case. Since R depends only on t, the spaces $t = $ constant do not contain their centres of symmetry: $R \neq 0$ at all points of such a space, except when $R = 0$ at a given value of t, but this is a singularity. In fact, (19.99) shows that with $\Lambda = 0$ and $Q \neq 0$ no Big-Bang-like singularity can occur: the term with the electric charge is negative and tends to $-\infty$ when $R \to 0$, while the right-hand side of this equation must be positive for R to exist. (Note that with $\Lambda = 0$ and $\varepsilon = +1$, Eq. (19.99) has no solutions at all if $M^2 < GQ^2/c^4$, and has only the static solution $R = M$ when $M^2 = GQ^2/c^4$. However, we have shown at the beginning of this section that the solution with $R_{,t} = 0$ is electrovacuum.)

---

[6] For the definition of the T- and R-regions, see Section 14.1.

The radius R is constant in each 3-space $t =$ constant. The geometry of this 3-space with $\varepsilon = +1$ is that of a 3-dimensional cylinder whose sections of constant r are spheres, all of the same radius, and the coordinate r measures the position along the generator. The space is inhomogeneous along the r-direction, and the electric field has its only component also in the r-direction. The radius of the cylinder evolves according to (19.99).

The subcase $Q = 0$, $\varepsilon = +1$ of this solution, i.e. the case of zero electric field and spherical symmetry, was found by Ruban in an earlier paper (Ruban, 1968) and discussed in an illuminating way in yet another paper (Ruban, 1969); we will report on that discussion below. The further subcase $\Lambda = 0$ appeared in a paper by Datt published as early as 1938 (Datt, 1938), but the author arbitrarily dismissed it as being of 'little physical significance'. In this subcase the explicit formulae for R and $e^{A/2}$ may be given. The solution for $R(t)$ is the same as for the $k = +1$ Friedmann model, while

$$e^{A/2} = 2X(r)(1 - Z \cot Z) + Y(r) \cot Z, \qquad Z \overset{\text{def}}{=} \arcsin \sqrt{R/(2M)}. \qquad (19.105)$$

From (19.102) and (19.104) it can be seen that the Ruban solution becomes spatially homogeneous ($\epsilon_{,r} = 0$) when $X/Y = \overline{C} =$ constant. Then, by the transformation $r' = \int Y(r)dr$ the factor $e^{A/2}$ is made independent of r, which shows that in this case an additional Killing field exists. This is the metric with the Kantowski–Sachs symmetry introduced in Section 10.7. When, further, $\overline{C} = 0$, the solution becomes vacuum and corresponds to that part of the Schwarzschild manifold that is not covered by the curvature coordinates, i.e. inside the event horizon, as mentioned in Section 14.4.

Various generalisations of these subcases were found by several authors, see Krasiński (1997) for a complete list. Among them are solutions with the Kantowski–Sachs geometry and various sources more general than just perfect fluid. Here we mention only one generalisation: Korkina and Martinenko (1975) worked out the case when the source in the Einstein equations for the metric (19.103) (with $\varepsilon = +1$) is a general perfect fluid, with nonzero, time-dependent, pressure. With no specific equation of state, the Einstein equations cannot be solved to the end, and are then just reduced to a single ordinary differential equation that contains an arbitrary function of time (the pressure).

Now we come to the interpretation of the spherically symmetric Ruban solution, $\varepsilon = +1$. Note that the matter density in this solution, (19.104), depends on r and is everywhere positive if $X > 0$. Thus, the amount of rest mass contained inside a sphere $r = r_0 =$ constant does depend on the value of r_0, and is an increasing function of r. Nevertheless, as seen from (19.99), the active gravitational mass M that drives the evolution is constant. It looks as if all matter added to the source loses its ability to gravitate, and the active gravitational mass is just a parameter of the space on which infalling matter has no influence. Ruban (1969) interpreted this property as follows: the gravitational mass defect of any matter added exactly cancels its contribution to the active mass.

Equation (19.99) with $Q = 0$ is the same that governs the Friedmann models, Eq. (17.28). Note that the constant $\varepsilon = \pm 1, 0$ in (19.99) and (19.102) that determines the type of symmetry fixes the type of evolution. Thus with $\varepsilon = +1$ (spherical symmetry), the model is necessarily the recollapsing one ($k = +1$). Now compare (19.99) with the law of evolution of the L–T model, (18.14) – the Datt–Ruban (D–R) model has $E = -1/2$ and $R_{,r} = 0$ permanently. Thus, it behaves like a neck in the L–T model (see Section 18.10).

Finally, take (19.99) with $Q = 0$ and compare it with the Schwarzschild solution in the Lemaître–Novikov coordinates, (14.120) – (14.121). It is clear that the two solutions can be matched across $r = r_b$ if $E(r_b) = -1/2$ and $m = M$, see Exercise 7 and Fig. 19.1.

In the special case $Q = \Lambda = 0$, the D–R model explodes out of a singularity at $R = 0$ at $t = t_B$, expands until $R = 2M$ is reached at $t = t_B + \pi M$ and then collapses back to $R = 0$ at $t = t_B + 2\pi M$ (Exercise 8). Thus, the hypersurface across which the D–R model is matched to the Schwarzschild solution lies inside the Schwarzschild event horizon, except at the instant of maximal expansion, when the D–R sphere touches the Schwarzschild horizon from inside (Ruban, 1983), see Fig. 19.1 again.

The D–R model and its generalisations have no analogues in the Newtonian theory and do not show up in linear approximations to Einstein's theory.

Krasiński and Giono (2012) made an attempt to use the charged Ruban solution as a source for the maximally extended R–N solution, and were again unsuccessful. The matching of the two metrics works only for a limited time and shell crossings do not allow the Ruban sphere to pass through the tunnel between asymptotically flat regions of R–N. Here is an extended abstract of that paper, partly composed of literal quotations.

Let us take the Reissner–Nordström metric in the standard form (14.40) – (14.41) with $e^2 < m^2$ and let us for a while rename its coordinates: $(t, r) \to (\rho, \mathcal{R})$. Let $F \stackrel{\text{def}}{=} e^{2\nu}$. Let us then transform the $\mathcal{R}$ coordinate in the region $F < 0$ to $\tau(\mathcal{R})$ defined by $(\mathrm{d}\mathcal{R}/\mathrm{d}\tau)^2 = -F$. After this, the R–N metric in this region becomes

$$\mathrm{d}s^2 = \mathrm{d}\tau^2 - [-F(\mathcal{R}(\tau))]\,\mathrm{d}\rho^2 - \mathcal{R}^2(\tau)\left(\mathrm{d}\vartheta^2 + \sin^2\vartheta\mathrm{d}\varphi^2\right). \tag{19.106}$$

When $\Lambda = 0$, the equation $(\mathrm{d}\mathcal{R}/\mathrm{d}\tau)^2 = -F$ can be integrated as follows:

$$\tau - \tau_0 = \mu\sqrt{-\mathcal{R}^2 + 2m\mathcal{R} - e^2} + \mu m \arcsin\left(\frac{\mathcal{R} - m}{\sqrt{m^2 - e^2}}\right), \tag{19.107}$$

where $\mu = \pm 1$ and τ_0 is an arbitrary constant. Equation (19.107) may be represented in the parametric form with the parameter η defined by $(\mathcal{R} - m)/\sqrt{m^2 - e^2} = -\cos\eta$:

$$\mathcal{R} = m - \sqrt{m^2 - e^2}\cos\eta, \qquad \tau - \widetilde{\tau}_0 = \mu\left(m\eta - \sqrt{m^2 - e^2}\sin\eta\right), \tag{19.108}$$

where $\widetilde{\tau}_0 \stackrel{\text{def}}{=} \tau_0 - \mu m\pi/2$. As τ increases, $\mathcal{R}$ oscillates between the roots of F.

We will show that the R–N metric in the coordinates of (19.106) and the Ruban metric defined by (19.99) and (19.102) – (19.103) can be matched across a hypersurface H of constant $\rho = r$. On each such H the 3-metrics are identical when $\tau = t$ and

$$m = M, \qquad e^2 = \overline{Q}^2 \stackrel{\text{def}}{=} GQ^2/c^4, \tag{19.109}$$

since then $\mathcal{R}(\tau)$ and the $R(t)$ of (19.103) obey the same equation $(\mathrm{d}\mathcal{R}/\mathrm{d}\tau)^2 = -F$ and (19.99) with $\varepsilon = +1$, respectively. The coincidence of the second fundamental forms requires that at H the 3-dimensional metrics h_{ij} of H obey

$$\left.\frac{1}{\mathcal{N}}\frac{\mathrm{d}h_{ij}}{\mathrm{d}r}\right|_{\text{Ruban}} = \left.\frac{1}{\mathcal{N}}\frac{\mathrm{d}h_{ij}}{\mathrm{d}\rho}\right|_{\text{RN}}, \tag{19.110}$$

where $i, j = 0, 2, 3$ and $\mathcal{N}$ is the r-component of the unit normal vector to H, thus $|g_{rr}|\mathcal{N}^2 = 1$ on each side of H. But the relevant components of the Ruban metric do not depend on r, and the relevant components of the R–N metric do not depend on ρ, so (19.110) is fulfilled in a trivial way: it reduces to $0 = 0$. Note, however, that the matching is possible only in that region of the R–N manifold where $F < 0$. In the subcase $\Lambda = 0$, this is the region between the two event horizons.

Equation (19.108) applies also in the Ruban region, with $m = M$ and $e = \overline{Q}$. It is independent of r, so each constant-r shell evolves by the same law. This means that all shells, including the outer surface of the Ruban region, oscillate between $R = R_-$ and $R = R_+$, the roots of F. This also means that at each $R = R_-$ the surface of the Ruban region touches the inner R–N event horizon, and at each $R = R_+$ it touches the outer R–N event horizon, see Fig. 19.1.

For the rest of this section we assume $\Lambda = 0$. In this case the Ruban metric is (19.103) with $\varepsilon = 1$ and the solution of (19.99) can be written as

$$R = M - \sqrt{M^2 - \overline{Q}^2} \cos\eta, \qquad t - t_B = \mu\left(M\eta - \sqrt{M^2 - \overline{Q}^2} \sin\eta\right), \qquad (19.111)$$

where $\mu = +1$ in the expansion phase and $\mu = -1$ in the collapse phase. This solution exists only when $\overline{Q}^2 < M^2$.[7] We use (19.111) in (19.102), noting that

$$\int \frac{\mathrm{d}t}{R R_{,t}^2} \equiv \int \frac{\mathrm{d}R}{R R_{,t}^3}. \qquad (19.112)$$

Then we obtain in (19.102)

$$e^{A/2} = X(r)\left[2 + \overline{Q}^2\frac{1 - M/R}{M^2 - \overline{Q}^2} - \sqrt{-1 + \frac{2M}{R} - \frac{\overline{Q}^2}{R^2}}\,\frac{\overline{Q}^2}{R^2}\arcsin\left(\frac{R - M}{\sqrt{M^2 - \overline{Q}^2}}\right)\right]$$

$$+ \mu Y(r)\sqrt{-1 + \frac{2M}{R} - \frac{\overline{Q}^2}{R^2}}. \qquad (19.113)$$

A shell crossing is where $e^{A/2} = 0$. From (19.104) one sees that this is a curvature singularity. The question we seek to answer is now: can the functions $X(r)$ and $Y(r)$ in (19.113) be chosen in such a way that $e^{A/2} \neq 0$ everywhere? Unfortunately, the answer given by the formulae is a definitive 'no', i.e. shell crossings are inevitable. To see this, let us first recall that R changes between the values

$$R_- = M - \sqrt{M^2 - \overline{Q}^2} \qquad \text{and} \qquad R_+ = M + \sqrt{M^2 - \overline{Q}^2} \qquad (19.114)$$

(this is a copy of (14.143) rewritten in the notation of the Ruban model). At $R = R_\pm$ the expression under the square root in (19.113) is zero, and in the whole range $0 < R_- < R < R_+$ the function $e^{A/2}$ given by (19.113) is continuous. Now, at $R = R_-$ we have

$$e^{A/2}\bigg|_{R=R_-} = -X(r)\frac{R_-}{\sqrt{M^2 - \overline{Q}^2}}, \qquad (19.115)$$

---

[7] With $\overline{Q}^2 = M^2$ and $\Lambda = 0$, Eq. (19.99) has the static solution $R = M = |\overline{Q}|$, which is a coordinate transform of the Robinson (1959) metric, see Krasiński and Giono (2012).

while at $R = R_+$ we have

$$e^{A/2}\Big|_{R=R_+} = +X(r)\frac{R_+}{\sqrt{M^2 - \overline{Q}^2}}. \tag{19.116}$$

Thus, whatever the sign of $X(r)$, the signs of $e^{A/2}$ at the ends of the range $[R_-, R_+]$ are opposite, i.e. $e^{A/2} = 0$ for some value of R within this range. This is a shell crossing. Note that it appears every time when R traverses this range, which means that the shell crossings will not allow R to go through even half a cycle of oscillation.

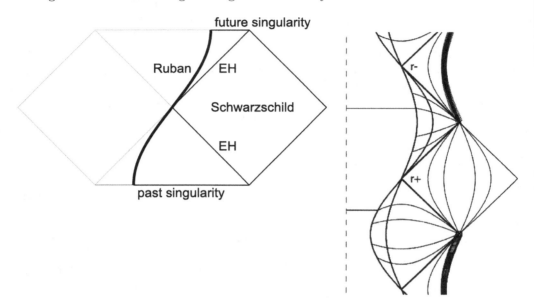

Fig. 19.1 **Left panel:** A schematic Penrose diagram of the (uncharged) Ruban model (to the left of the thick curve) matched to the Schwarzschild spacetime. EH is the Schwarzschild event horizon. **Right panel:** The maximally extended Reissner–Nordström spacetime of Fig. 14.12 with a charged Ruban region matched to it. The two long wavy curves show possible matching surfaces at two different values of the radial coordinate, which is the r of (19.103) and the ρ of (19.106). Note: on these curves r and ρ are constant. It is R that pulsates between the R_- and R_+ of (19.114), and it changes *along* these curves. The lines of constant R are the horizontal hyperbola-like arches. Shell crossings appear within each half-cycle of oscillation, the possible positions of two of them are marked with thin horizontal straight lines.

The geometrical nature of this shell crossing is different than in the L–T and Szekeres models of Chapter 20. In the latter, the spheres that collide are one within the other. At a shell crossing, the smaller sphere catches up with the larger one during expansion (or vice versa during collapse). In the Ruban model, the spheres are surfaces of constant r in the 3-dimensional cylinder $t = $ constant, and they move up and down along the generators as the cylinder expands or collapses. It turns out they will collide before the cylinder manages to proceed all the way between the minimum and maximum value of R or vice versa.

Kantowski and Sachs (1966) noted the possible occurrence of this singularity in the subcase $Q = Y(r) = 0$, where they said 'The singularities for the closed models are of two kinds. ... In one kind of a singularity the cylinder squashes to a disk, in the other it contracts to a line.' What we called shell crossing here is the 'squashing to a disk.'

The shell crossings could possibly be prevented by pressure with nonzero gradient in the r-direction, but such solutions are not yet known. Then, assuming that the pressure and its gradient would be zero at the surface of the charged fluid ball, its surface would follow a timelike geodesic in the R–N spacetime.

When $\overline{Q} = 0$, the charged Ruban solution goes over into the Datt–Ruban neutral dust solution (19.105). Then from (19.111) we obtain

$$R = M(1 - \cos \eta), \qquad t - t_B = \mu M(\eta - \sin \eta). \tag{19.117}$$

This shows that now R starts from 0 at $t = t_B$, then increases to $2M$ at $t = t_B + \mu M\pi$ and decreases to 0 again at $t = t_B + 2\mu M\pi$. At $R = 0$ the model has a Big Bang/Crunch type singularity. Equation (19.113) then simplifies to the form equivalent to (19.102):

$$
\begin{aligned}
e^{A/2} &= \sqrt{-1 + \frac{2M}{R}} \left\{ X(r) \left[\frac{2}{\sqrt{-1 + \frac{2M}{R}}} - \arcsin \left(\frac{R}{M} - 1 \right) \right] + \mu Y(r) \right\} \\
&\stackrel{\text{def}}{=} \sqrt{-1 + \frac{2M}{R}} \mathcal{F}(t, r).
\end{aligned} \tag{19.118}
$$

This model exists only for the finite time $T = 2\pi M$ between the Big Bang and Big Crunch, but we will show that, unlike in the charged case, the functions $X(r)$ and $Y(r)$ can be chosen so that within this interval $\mathcal{F} > 0$ and there are no shell crossings. Note that at both singularities $e^{A/2} \to \infty$, so the space achieves infinite size along the r direction.

With $R = 0$ at the start of the expansion phase, where $\mu = +1$, we have

$$\mathcal{F}|_{R=0} = X\frac{\pi}{2} + Y. \tag{19.119}$$

Thus, $X > 0$ and $Y > 0$ will guarantee that $\mathcal{F}|_{R=0, \mu=+1} > 0$. We also have

$$\frac{\mathrm{d}\mathcal{F}}{\mathrm{d}R} = \frac{X}{R\left(\frac{2M}{R} - 1\right)^{3/2}}, \tag{19.120}$$

which will be positive for all $0 < R < 2M$ if $X > 0$, becoming $+\infty$ as $R \to 2M$. Since also $\mathcal{F} \to +\infty$ as $R \to 2M$, the switch from expansion to collapse, i.e. from $\mu = +1$ to $\mu = -1$, will keep $\mathcal{F} > 0$ at the beginning of collapse. So we only have to guarantee that $\mathcal{F} > 0$ at the end of the collapse phase, when $R \to 0$ again. This will be the case when

$$\mathcal{F}|_{R=0, \mu=-1} = X\frac{\pi}{2} - Y > 0. \tag{19.121}$$

Consequently, both (19.119) and (19.121) will be positive during the whole cycle when $0 < Y < \pi X/2$ at all values of r. The fact that the shell crossings may disappear when $\overline{Q} = 0$ shows that the limiting transition $\overline{Q} \to 0$ is discontinuous; just as it was in the R–N $\to$ Schwarzschild transition. The uncharged Ruban spacetime has a finite time of existence, just as the Kruskal–Szekeres manifold between its two singularities.

19.6 Exercises

1. Find the Killing vector fields for the transformation (19.1). Then solve the Killing equations for the components of the metric tensor in spacetime and verify that the solution is indeed given by (19.3).

2. Verify that $-\delta^{\alpha\beta}_{\gamma\delta}/R^2$ is the curvature tensor for the metric (19.5).

3. Find the Killing fields for the metric (19.5) and verify that they are given by (19.6). Then solve the Killing equations in four dimensions with the Killing fields (19.6) and verify that the resulting metric is (19.7).

4. Calculate the Riemann tensor for (19.9) and verify that it can vanish only with $\varepsilon = -1$. Then, (19.9) is flat when $\mathrm{d}f(r)/\mathrm{d}r = 1$.

5. Find the Bianchi types for the algebras of Killing fields given by (8.42) and (19.6) and for the algebra defined by the transformations (19.1).

6. Let the coordinates of the Reissner–Nordström metric (14.40 – 14.41) be $(\tau, \mathcal{R})$. and let $e^{2\nu} = F$ as in (14.181). Transform (14.40) to new coordinates (t, r) and demand that the transformed metric is diagonal ($g_{tr} = 0$), but do not demand that $g_{tt} = 1$. The $g_{tr} = 0$ requirement means that

$$F\tau_{,t}\,\tau_{,r} - \frac{1}{F}\,\mathcal{R}_{,t}\,\mathcal{R}_{,r} = 0. \tag{19.122}$$

Then the R–N metric acquires the form (19.11) with

$$e^{\overline{C}} = F\tau_{,t}^2 - \frac{1}{F}\,\mathcal{R}_{,t}^2, \tag{19.123}$$

$$e^{\overline{A}} = F\tau_{,r}^2 - \frac{1}{F}\,\mathcal{R}_{,r}^2, \tag{19.124}$$

$$F = 1 - \frac{2m}{\mathcal{R}} + \frac{e^2}{\mathcal{R}^2} + \frac{1}{3}\Lambda\mathcal{R}^2. \tag{19.125}$$

So, it is now a special case of the charged dust metric with $Q_{,N} = 0$ and $N_{,r} = M_{,r} = 0$ in (19.52) – (19.53). Consequently, it must obey (19.48). Solve (19.122) – (19.123) for $\tau_{,t}$ and $\tau_{,r}$, then impose the integrability condition $\tau_{,tr} = \tau_{,rt}$. Assuming $e^2 = GQ^2/c^4$ and $m = M$ show, using (19.48), that this leads to $\overline{C}_{,r} = 0$. Then $e^{\overline{C}} = 1$ can be achieved by a simple transformation of t.

Hint. In the primary coordinates $F_{,\tau} = 0$, so $F_{,t} = F_{,\mathcal{R}}\,\mathcal{R}_{,t}$, $F_{,r} = F_{,\mathcal{R}}\,\mathcal{R}_{,r}$.

7. Show that the Datt–Ruban solution (19.99) with $Q = 0$ and (19.103) can be matched to the Schwarzschild solution in the Lemaître–Novikov coordinates, (14.120) – (14.121).

8. Solve the equation of evolution of the Datt–Ruban model, (19.99), in the case $Q = \Lambda = 0$. Show that the model expands from a singularity $R = 0$ at $t = t_B$ to the maximal size $R = 2M$ at $t = t_B + \pi M$, and then recollapses in a time-symmetric manner.

9. Show that the subcase $X = 0$ of the Datt–Ruban model (19.102) – (19.103) with $\varepsilon = +1$ is isometric to the Schwarzschild solution, (14.40) with (14.42), taken in the black hole region $r < 2m$.

20

Relativistic cosmology V: the Szekeres geometries

20.1 The Szekeres–Szafron family of metrics

Consider the metric

$$ds^2 = dt^2 - e^{2\alpha}dz^2 - e^{2\beta}\left(dx^2 + dy^2\right), \tag{20.1}$$

where α and β are functions of (t, x, y, z) to be determined from the Einstein equations. The source is taken to be a perfect fluid, and the coordinates of (20.1) are assumed to be comoving so that the velocity field $u^\mu = \delta^\mu{}_0$. This implies that the acceleration $\dot{u}^\mu = 0$ and that pressure depends only on time.

Several parametrisations are in use for the solutions of the Einstein equations with the metric (20.1), and some notations are conflicting. We shall follow Szafron's (1977) exposition. The orthonormal tetrad components of the Einstein tensor for (20.1) are

$$
\begin{aligned}
G_{00} = {}& e^{-2\beta}\left(-\alpha_{,x}{}^2 - \alpha_{,y}{}^2 - \alpha_{,xx} - \beta_{,xx} - \alpha_{,yy} - \beta_{,yy}\right) \\
& + e^{-2\alpha}\left(-3\beta_{,z}{}^2 - 2\beta_{,zz} + 2\alpha_{,z}\,\beta_{,z}\right) + 2\alpha_{,t}\,\beta_{,t} + \beta_{,t}{}^2,
\end{aligned} \tag{20.2}
$$

$$G_{01} = e^{-\alpha}\left(-2\beta_{,tz} + 2\alpha_{,t}\,\beta_{,z} - 2\beta_{,t}\,\beta_{,z}\right), \tag{20.3}$$

$$G_{02} = e^{-\beta}\left(-\alpha_{,tx} - \beta_{,tx} - \alpha_{,t}\,\alpha_{,x} + \beta_{,t}\,\alpha_{,x}\right), \tag{20.4}$$

$$G_{03} = e^{-\beta}\left(-\alpha_{,ty} - \beta_{,ty} - \alpha_{,t}\,\alpha_{,y} + \beta_{,t}\,\alpha_{,y}\right), \tag{20.5}$$

$$G_{11} = e^{-2\alpha}\beta_{,z}{}^2 + e^{-2\beta}\left(\beta_{,xx} + \beta_{,yy}\right) - 3\beta_{,t}{}^2 - 2\beta_{,tt}, \tag{20.6}$$

$$G_{12} = e^{-\alpha-\beta}\left(-\beta_{,zx} + \beta_{,z}\,\alpha_{,x}\right), \tag{20.7}$$

$$G_{13} = e^{-\alpha-\beta}\left(-\beta_{,zy} + \beta_{,z}\,\alpha_{,y}\right), \tag{20.8}$$

$$
\begin{aligned}
G_{22} = {}& e^{-2\beta}\left(\alpha_{,y}{}^2 + \alpha_{,x}\,\beta_{,x} + \alpha_{,yy} - \alpha_{,y}\,\beta_{,y}\right) \\
& + e^{-2\alpha}\left(\beta_{,z}{}^2 + \beta_{,zz} - \alpha_{,z}\,\beta_{,z}\right) \\
& - \alpha_{,t}\,\beta_{,t} - \alpha_{,t}{}^2 - \beta_{,t}{}^2 - \alpha_{,tt} - \beta_{,tt},
\end{aligned} \tag{20.9}
$$

$$G_{23} = e^{-2\beta}\left(-\alpha_{,xy} - \alpha_{,x}\,\alpha_{,y} + \alpha_{,x}\,\beta_{,y} + \beta_{,x}\,\alpha_{,y}\right), \tag{20.10}$$

$$
\begin{aligned}
G_{33} = {}& e^{-2\beta}\left(\alpha_{,x}{}^2 + \alpha_{,xx} - \alpha_{,x}\,\beta_{,x} + \alpha_{,y}\,\beta_{,y}\right) + e^{-2\alpha}\left(\beta_{,z}{}^2 + \beta_{,zz} - \alpha_{,z}\,\beta_{,z}\right) \\
& - \alpha_{,t}\,\beta_{,t} - \alpha_{,t}{}^2 - \beta_{,t}{}^2 - \alpha_{,tt} - \beta_{,tt}.
\end{aligned} \tag{20.11}
$$

With perfect fluid source, the Einstein equations are $G_{00} = \kappa\epsilon$ and $G_{11} = G_{22} = G_{33} = \kappa p$ (the cosmological constant can be taken into account by $\kappa p \to \kappa p + \Lambda$, $\kappa\epsilon \to \kappa\tilde{\epsilon} - \Lambda$).

The equations $G_{01} = G_{12} = G_{13} = 0$ say, respectively,

$$\left(e^{\beta-\alpha} \beta_{,z} \right)_{,t} = 0, \qquad \left(e^{-\alpha} \beta_{,z} \right)_{,x} = \left(e^{-\alpha} \beta_{,z} \right)_{,y} = 0. \tag{20.12}$$

The last two equations imply that

$$\beta_{,z} = u(t, z) e^{\alpha}. \tag{20.13}$$

The cases $u = 0$ and $u \neq 0$ must be considered separately because the integration proceeds differently in each case, and the limit $\beta_{,z} \to 0$ of the solution with $\beta_{,z} \neq 0$ can be taken only after a nontrivial reparametrisaton (see Section 20.2.3). The relation between these two cases is analogous to that between the Datt–Ruban and L–T models. We will see later that the L–T and Datt–Ruban models are the spherically symmetric limits of the $\beta_{,z} \neq 0$ and $\beta_{,z} = 0$ subfamily, respectively.

20.1.1 *The* $\beta_{,z} = 0$ *subfamily*

The equations $G_{01} = G_{12} = G_{13} = 0$ are fulfilled identically. In solving the other equations, we can assume that $\beta_{,tx} = 0 = \beta_{,ty}$ because otherwise the equations have no perfect fluid solutions; the proof is given in Section 24.7. Then

$$\beta = \ln \Phi(t) + \nu(x, y), \tag{20.14}$$

where Φ and ν are unknown functions, while $G_{02} = G_{03} = 0$ imply

$$e^{\alpha-\beta} \alpha_{,x} = \tilde{\sigma}_1(z, x, y), \qquad e^{\alpha-\beta} \alpha_{,y} = \tilde{\sigma}_2(z, x, y), \tag{20.15}$$

where $\tilde{\sigma}_1$ and $\tilde{\sigma}_2$ are other unknown functions. Using (20.14) and denoting $\sigma_i(z, x, y) \overset{\text{def}}{=} \tilde{\sigma}_i e^{\nu}$, $i = 1, 2$, we obtain

$$e^{\alpha} \alpha_{,x} = \sigma_1(z, x, y) \Phi(t), \qquad e^{\alpha} \alpha_{,y} = \sigma_2(z, x, y) \Phi(t). \tag{20.16}$$

The integrability condition $(e^{\alpha} \alpha_{,x})_{,y} = (e^{\alpha} \alpha_{,y})_{,x}$ implies $\sigma_{1,y} = \sigma_{2,x}$, so a function $\sigma(z, x, y)$ exists such that $\sigma_1 = \sigma_{,x}$, $\sigma_2 = \sigma_{,y}$. Then (20.16) can be integrated to give

$$e^{\alpha} = \Phi(t) \sigma(z, x, y) + \lambda(t, z), \tag{20.17}$$

where $\lambda(t, z)$ is an unknown function. Substituting for β and α from (20.14) and (20.17) in (20.6) we obtain in the equation $G_{11} = \kappa p$

$$e^{-2\nu} \left(\nu_{,xx} + \nu_{,yy} \right) = 2\Phi\Phi_{,tt} + \Phi_{,t}^2 + \kappa p \Phi^2 \overset{\text{def}}{=} -k = \text{constant}, \tag{20.18}$$

because ν depends only on x and y while Φ and p depend only on t.

Now it is convenient to introduce the complex variables

$$\xi \overset{\text{def}}{=} x + iy, \qquad \bar{\xi} = x - iy, \tag{20.19}$$

in which the first part of (20.18) becomes

$$-4e^{-2\nu} \nu_{,\xi\bar{\xi}} = k. \tag{20.20}$$

The expression $-8e^{-2\nu}\nu_{,\xi\bar{\xi}}$ is the curvature scalar of the metric

$$ds_2{}^2 = e^{2\nu}d\xi d\bar{\xi}, \tag{20.21}$$

and (20.20) says that this curvature is constant. Thus, depending on the sign of k, (20.21) is equivalent to one of the 2-metrics in (19.11) – (19.12). We will now transform it.

The derivative of (20.20) by ξ is equivalent to $(\nu_{,\xi\xi} - \nu_{,\xi}{}^2)_{,\bar{\xi}} = 0$, whose solution is

$$\nu_{,\xi\xi} - \nu_{,\xi}{}^2 = \tau(\xi), \tag{20.22}$$

$\tau(\xi)$ being an arbitrary function. A transformation of the form $\xi = f(\xi')$ $(\to \bar{\xi} = \bar{f}(\bar{\xi}'))$ is a conformal symmetry of (20.21); the new metric has a different ν:

$$\tilde{\nu}(\xi', \bar{\xi}') = \nu(\xi, \bar{\xi}) + \frac{1}{2}\ln(f_{,\xi'}) + \frac{1}{2}\ln\left(\bar{f}_{,\bar{\xi}'}\right). \tag{20.23}$$

After such a transformation

$$\tilde{\nu}_{,\xi'\xi'} - \tilde{\nu}_{,\xi'}{}^2 = \tau(f(\xi'))f_{,\xi'}{}^2 + \frac{1}{2}(\ln f_{,\xi'})_{,\xi'\xi'} - \frac{1}{4}(\ln f_{,\xi'})_{,\xi'}{}^2 \overset{\text{def}}{=} \tilde{\tau}(\xi'). \tag{20.24}$$

We can choose f so that $\tilde{\tau} = 0$, after which the new ν will obey

$$\nu_{,\xi\xi} - \nu_{,\xi}{}^2 = 0. \tag{20.25}$$

This implies $(e^{-\nu})_{,\xi\xi} = 0 = (e^{-\nu})_{,\bar{\xi}\bar{\xi}}$ since ν is real. Hence,

$$e^{-\nu} = a\xi\bar{\xi} + B\xi + \bar{B}\bar{\xi} + c, \tag{20.26}$$

where a, B, $\bar{B}$ and c are arbitrary constants (a and c being real), and (20.20) implies

$$ac - B\bar{B} = k/4. \tag{20.27}$$

Lemma 20.1 *If $k \neq 0$, then (20.21) may be transformed to*

$$ds_2{}^2 = \frac{dx^2 + dy^2}{\left[1 + \frac{1}{4}k\left(x^2 + y^2\right)\right]^2}. \tag{20.28}$$

Proof: Let $k > 0$. Then $ac > 0$ from (20.27). Carry out the following chain of transformations: (1) $\xi = \xi' - \bar{B}/a$; (2) $\xi' = 1/(a\xi'')$; (3) $\xi'' = x + iy$. The resulting metric is (20.28).

If $k < 0$ and $a \neq 0$, then the same chain of transformations will do the job. If $a = 0$ and $c \neq 0$, then $\xi = 1/\xi'$ restores $a \neq 0$, with B and $\bar{B}$ interchanged. If $a = c = 0$, then $a \neq 0$ is restored by the Haantjes transformation (8.81) which, in the variables $(\xi, \bar{\xi})$, is

$$\xi = \frac{\xi' - C\xi'\bar{\xi}'}{1 - \bar{C}\xi' - C\bar{\xi}' + C\bar{C}\xi'\bar{\xi}'}, \tag{20.29}$$

where $C = C_1 + iC_2$. The transformed metric is

$$ds_2{}^2 = \frac{d\xi d\bar{\xi}}{\left[B\xi' + \bar{B}\bar{\xi}' - (BC + \bar{B}\bar{C})\xi'\bar{\xi}'\right]^2}, \tag{20.30}$$

and (20.28) is achieved in the already-known way. $\square$

Lemma 20.2 *Equation (20.28) applies also when $k = 0$.*

Proof:

If $a \neq 0$, then $c = B\overline{B}/a$ from (20.27), and the job is done by the chain of transformations (1) $\xi = \xi' - \overline{B}/a$; (2) $\xi' = 1/(a\xi'')$.

If $a = 0$, then, with $k = 0$, Eq. (20.27) implies $B = \overline{B} = 0$, and then the metric is reduced to (20.28) by $\xi = c\xi'$. $\square$

For the proof that (20.28) is equivalent to the 2-dimensional metrics of constant curvature contained in (19.11) − (19.12) see Exercise 1.

Conclusion

The coordinates (x, y) may be chosen so that $B = 0$, $c = 1$, $a = k/4$, so, from (20.14)

$$e^{\beta} = \Phi e^{\nu} = \frac{\Phi(t)}{1 + \frac{1}{4}k(x^2 + y^2)} = \frac{\Phi(t)}{1 + \frac{1}{4}k\xi\overline{\xi}}. \tag{20.31}$$

In the coordinates (20.19) the equations $G_{22} - G_{33} = 0$ and $G_{23} = 0$, using (20.9) − (20.11), imply $P + \overline{P} = 0$ and $P - \overline{P} = 0$, where $P \overset{\text{def}}{=} \alpha_{,\xi\xi} + \alpha_{,\xi}{}^2 - 2\alpha_{,\xi}\beta_{,\xi}$. So,

$$\alpha_{,\xi\xi} + \alpha_{,\xi}{}^2 - 2\alpha_{,\xi}\beta_{,\xi} = 0, \tag{20.32}$$

and the same holds for the complex conjugate of (20.32). After substituting from (20.17) and (20.31), and using (20.25), Eq. (20.32) implies

$$\left(e^{-\nu}\sigma\right)_{,\xi\xi} = \left(e^{-\nu}\sigma\right)_{,\overline{\xi}\overline{\xi}} = 0. \tag{20.33}$$

Since σ depends on z, the integration constants will be arbitrary functions of z. In the (x, y) coordinates, the solution of (20.33) is[1]

$$\sigma = e^{\nu}\left[\frac{1}{2}U(z)(x^2 + y^2) + V_1(z)x + V_2(z)y + 2W(z)\right]. \tag{20.34}$$

With this, the only equations not yet satisfied are $G_{22} = \kappa p$ and $G_{00} = \kappa \epsilon$. Substituting (20.34), (20.17), (20.31), (20.18) and $\beta_{,z} = 0$ in (20.9) and taking $G_{22} = \kappa p$ we obtain

$$\lambda_{,tt}\,\Phi + \lambda_{,t}\,\Phi_{,t} + \lambda\Phi_{,tt} + \lambda\Phi\kappa p = U(z) + kW(z). \tag{20.35}$$

Substituting all the results in (20.2) we obtain for the matter density

$$\kappa\epsilon = 2e^{-\alpha}(\lambda\Phi_{,tt}/\Phi - \lambda_{,tt}) + 3\Phi_{,t}{}^2/\Phi^2 + 3k/\Phi^2. \tag{20.36}$$

Now all the Einstein equations are solved. Collecting all the information together:

$$\begin{aligned}
ds^2 &= dt^2 - e^{2\alpha}dz^2 - e^{2\beta}\left(dx^2 + dy^2\right), \\
e^{\beta} &= \Phi(t)e^{\nu}, \qquad e^{\alpha} = \Phi(t)\sigma(z, x, y) + \lambda(t, z), \\
\sigma &= e^{\nu}\left[\frac{1}{2}U(z)(x^2 + y^2) + V_1(z)x + V_2(z)y + 2W(z)\right], \\
e^{-\nu} &= 1 + \frac{1}{4}k(x^2 + y^2),
\end{aligned} \tag{20.37}$$

---

[1] The coefficients at U and W were added in order to make the subsequent formulae consistent with those of Szafron (1977). Szafron's Eqs. (2.23) and (2.24) are inconsistent between them.

where k is an arbitrary constant, $\Phi(t)$ is determined by the equation

$$2\Phi\Phi_{,tt} + \Phi_{,t}^2 + k + \kappa p \Phi^2 = 0, \tag{20.38}$$

while $\lambda(t,z)$ and the matter density are determined by (20.35) and (20.36).

This is as far as one can get without assuming any equation of state, which is necessary to determine p and then solve (20.38). However, since the p in (20.38) may depend only on t, the barotropic equation of state $\epsilon = \epsilon(p)$ makes matter density spatially homogeneous.

The uncharged $(Q = 0)$ Datt–Ruban class results from (20.37) – (20.38) as the limit $V_1 = V_2 = 0$, $U = kW$, $\Phi = R$, $\kappa p = \Lambda$, $\alpha = A/2$, $k = \epsilon$. Then (20.38) coincides with the subcase $Q = 0$ of (19.98), and (20.37) reproduces the subcase $Q = 0$ of (19.103). See Exercise 1 for a hint on how to deal with the 2-metric $e^{2\nu}\left(dx^2 + dy^2\right)$.

All the Robertson–Walker models are contained here as the limit $\lambda = 0$, $U = -kW$, in the form (17.99) – (17.100). They result in a simpler form if $U = W = V_2 = 0$, $V_1 = 1$ in addition (which just means a more specific choice of coordinates in the R–W limit):

$$ds^2 = dt^2 - \frac{\Phi^2(t)}{\left[1 + \frac{1}{4}k\left(x^2 + y^2\right)\right]^2}\left(x^2 dz^2 + dx^2 + dy^2\right), \tag{20.39}$$

and then the following transformation will reduce (20.39) to (17.1):

$$\Phi = R, \qquad x = r\sin\vartheta, \qquad y = r\cos\vartheta, \qquad z = \varphi. \tag{20.40}$$

Equation (20.38) is identical to (17.25) that governs the evolution of the R–W models (the cosmological constant can be taken into account here by redefining p and ϵ).

The reparametrisation $U = \tilde{u} + kW$, $\lambda = \tilde{\lambda} - 2\Phi W$ has the same result as if $W = 0$; $\tilde{u}$ then replaces U and $\tilde{\lambda}$ replaces λ in the equations. If $k \neq 0$, then the reparametrisation $\lambda = \tilde{\lambda} - \Phi(U/k + W)$, $W = 2\widetilde{W} + U/k$ leads to the same result as if $U = -kW$, i.e. as if the right-hand side of (20.35) were zero. Both these parametrisations have been used in various papers; see Krasiński (1997) and Section 20.4.

In general, this family of spacetimes has no symmetry (Bonnor, Sulaiman and Tomimura, 1977 and Exercise 2). When $U = kW$ and $V_1 = V_2 = 0$, it acquires a 3-dimensional symmetry group acting on 2-dimensional orbits; the symmetry is spherical, plane or hyperbolic when $k > 0$, $k = 0$ or $k < 0$, respectively. In the general case, a certain quasi-symmetry is present: the surfaces $\{t = \text{constant}, z = \text{constant}\}$ have constant curvature proportional to k. The lack of symmetry in the spaces $t = \text{constant}$ is due to the spheres being placed nonconcentrically (when $k > 0$) and to the 'planes' being nonparallel (when $k = 0$).

The orbits of the $O(3)$ group of (20.39) have nothing to do with the 2-surfaces of constant curvature present in the general Szafron spacetime: the former are the spheres on which $x^2 + y^2 = \text{constant}$, the latter are surfaces of constant z. In contrast to this, the Datt–Ruban limit results in a natural way: the 2-surfaces of constant curvature become orbits of the symmetry group; for example, the spheres become concentric when $k > 0$ and the 'planes' become parallel when $k = 0$.

An explicit solution of (20.38) and (20.35) corresponding to $\kappa p = \Lambda$ was given by Barrow and Stein-Schabes (1984); it involves elliptic functions.

When $p = 0$, Eq. (20.38) has the same solutions as those in Sec. 17.4. Equation (20.35) can then be solved explicitly, too. We shall come back to these solutions in Section 20.2.

20.1.2 The $\beta_{,z} \neq 0$ subfamily

We go back to Eq. (20.13) and consider the case when $u(t,z) \neq 0$. Then

$$e^\alpha = \frac{\beta_{,z}}{u(t,z)}, \tag{20.41}$$

and the first of (20.12) becomes $(e^\beta u)_{,t} = 0$, whose solution is

$$e^\beta = \Phi(t,z)e^{\nu(z,x,y)}, \tag{20.42}$$

where $\Phi \overset{\text{def}}{=} 1/u$. The remaining equations in (20.12) are already fulfilled by (20.13). An arbitrary factor dependent on z can be introduced in e^α by the transformation $z = f(z')$:

$$e^\alpha = h(z)\Phi(t,z)\beta_{,z}, \tag{20.43}$$

where $h(z)$ is an arbitrary function. This will simplify the transition to the R–W limit.

The equation $G_{11} = \kappa p$, in consequence of (20.41), (20.42) and (20.6), becomes

$$e^{-2\nu}(\nu_{,xx} + \nu_{,yy}) + 1/h^2 = 2\Phi\Phi_{,tt} + \Phi_{,t}{}^2 + \kappa p\Phi^2 \overset{\text{def}}{=} - k(z); \tag{20.44}$$

the last part of the equation follows because the left-hand side does not depend on t, while the middle part does not depend on x and y. The function $k(z)$ is arbitrary.

We can now repeat the reasoning in (20.19) – (20.27), with the modification that the functions ν, τ, $\tilde{\nu}$, $\tilde{\tau}$, f and $\overline{f}$ now depend on z. The metric (20.21) and its curvature were only an auxiliary construction that helped us guess the substitution (20.23) that reduced (20.22) to (20.25). However, the reasoning that led to (20.28) is no longer valid because it involved coordinate transformations in the (x,y) subspace. Thus, $e^{-\nu}$ will have a form similar to (20.26), but the coefficients will be arbitrary functions of z:

$$e^{-\nu} = A(z)\left(x^2 + y^2\right) + 2B_1(z)x + 2B_2(z)y + C(z), \tag{20.45}$$

while $\Phi(t,z)$ is defined by the equation

$$2\frac{\Phi_{,tt}}{\Phi} + \frac{\Phi_{,t}{}^2}{\Phi^2} + \frac{k(z)}{\Phi^2} + \kappa p(t) = 0. \tag{20.46}$$

This shows that $\Phi(t,z)$ can be redefined by $\Phi \to \Phi f(z)$, which will result only in rescaling $k(z)$. In consequence of (20.44), $A(z)$, $B_1(z)$, $B_2(z)$, $C(z)$, $h(z)$ and $k(z)$ must obey

$$4\left(AC - B_1{}^2 - B_2{}^2\right) = 1/h^2(z) + k(z) \overset{\text{def}}{=} \mathcal{G}(z). \tag{20.47}$$

With α and β given by (20.43), (20.42), (20.45), (20.46) and (20.47) the Einstein equations corresponding to (20.3) – (20.11) are now satisfied. The $G_{01} = G_{12} = G_{13} = 0$ were solved in obtaining (20.41) and (20.42), the $G_{11} = \kappa p$ led to (20.44), the remaining ones just happen to be already fulfilled. The verification is more or less elementary with the exception of $G_{22} = G_{33} = \kappa p$. Hints for verifying the latter are given in Sec. 24.4. The equation $G_{00} = \kappa \epsilon$ defines the matter density. In order to make the transition to the L–T limit easier, we will represent the solution of (20.46) by the formal integral

$$\Phi_{,t}{}^2 = \frac{2M(z)}{\Phi} - k(z) - \frac{\kappa}{3\Phi}\int p\left(\frac{\partial\Phi^3}{\partial t}\right)dt, \tag{20.48}$$

and then the energy density is given by

$$
\kappa\epsilon = (\Phi_{,z} + \Phi\nu_{,z})^{-1} \left\{ \frac{2M_{,z}}{\Phi^2} \right.
$$

$$
\left. + \frac{6M\nu_{,z}}{\Phi^2} - \frac{\kappa}{3\Phi^2} \int p \left(\frac{\partial^2 \Phi^3}{\partial t \partial z} \right) dt - \frac{\kappa\nu_{,z}}{\Phi^2} \int p \left(\frac{\partial \Phi^3}{\partial t} \right) dt \right\}, \tag{20.49}
$$

where the first line is all that remains in the L–T limit; see further on. Equation (20.49) is easier to verify than $G_{22} = G_{33} = \kappa p$, but hints given in the same Sec. 24.4 may be useful.

Collecting all the information together, the metric we obtained is given by

$$
\begin{aligned}
ds^2 &= dt^2 - e^{2\alpha}dz^2 - e^{2\beta}\left(dx^2 + dy^2\right), \\
e^{\beta} &= \Phi(t,z)e^{\nu(z,x,y)}, \\
e^{\alpha} &= h(z)\Phi(t,z)\beta_{,z} \equiv h(z)\left(\Phi_{,z} + \Phi\nu_{,z}\right), \\
e^{-\nu} &= A(z)\left(x^2 + y^2\right) + 2B_1(z)x + 2B_2(z)y + C(z), \tag{20.50}
\end{aligned}
$$

with $\Phi(t,z)$ obeying (20.46) and the arbitrary functions of (20.45) obeying (20.47). Equation (20.46) can be integrated once $p(t)$ has been specified.

This subfamily, like the preceding one, has in general no symmetry (Exercise 2), and acquires a G_3 with 2-dimensional orbits when A, B_1, B_2 and C are all constant (then $\nu_{,z} = 0$). If $\kappa p = \Lambda$ in addition, then the L–T model results, with $z = r$ and $\Phi = R$; the metric of the spheres $S_{t,z}$ of constant (t,z) can then be transformed to the standard form as explained in Lemmas 20.1 and 20.2 and Exercise 1. For the interpretation of the coordinates in (20.50) see the next section. The sign of $\mathcal{G}(z)$ in (20.47) determines the geometry of the $S_{t,z}$ surfaces in the same way as the sign of k did for (20.36) – (20.38). However, with A, B_1, B_2 and C being functions of z, different surfaces of constant z may have different geometries within a single space $t = $ constant (i.e. they can be spheres in one part of the space and surfaces of constant negative curvature elsewhere, the curvature being zero at the boundary). Moreover, $\mathcal{G}(z)$ and $k(z)$ are different functions, so the relation between the geometries of these surfaces and the type of evolution of the spacetime is not as rigid as in the $\beta_{,z} = 0$ subfamily. However, a connection exists: as seen from (20.47), $\mathcal{G} \le 0$ is possible only with $k < 0$, while all signs of k may co-exist with $\mathcal{G} > 0$.

The Robertson–Walker limit in the form (17.96) – (17.98) follows when $\Phi(t,z) = f(z)R(t)$, $k = k_0 f^2$ and $k_0 = $ constant. When $B_1 = B_2 = 0$, $C = 4A = 1$ and $f = z$ in addition, the R–W limit is 'natural'; its $O(3)$ orbits are the spheres from the Szafron $\beta_{,z} \ne 0$ spacetime, made concentric. (The additional specialisations of the arbitrary functions amount to a more specific choice of coordinates in the R–W limit.)

20.1.3 Interpretation of the Szekeres–Szafron coordinates

The transition from the metric of the 2-surface $S_{t,r}$ of constant t and r in (19.11) – (19.12) to (20.28) is called **stereographic projection**. With $f(\vartheta) = \sin\vartheta$, the surface $S_{t,r}$ in (19.11) is a sphere of radius R. With $f(\vartheta) = \vartheta$, $S_{t,r}$ has the metric of a Euclidean plane. With $f(\vartheta) = \sinh\vartheta$, the metric can be visualised as that of a two-sheeted hyperboloid, but in a 3-space with indefinite metric (Exercise 3). In the $\beta_{,z} = 0$ subfamily, where the metric of the 2-surface is (20.28), the transformation is as in Exercise 1:

$$(x, y) = \begin{cases} (2/\sqrt{-k}) \coth(\vartheta/2)(\cos\varphi, \sin\varphi) & \text{for } k < 0, \\ (2/\sqrt{k}) \cot(\vartheta/2)(\cos\varphi, \sin\varphi) & \text{for } k > 0. \end{cases} \tag{20.51}$$

In the $\beta_{,z} \neq 0$ subfamily, the metric in the 2-surface of constant (t, z) is not as simple, since the functions A, B_1, B_2 and C depend on z. By the same method as used in Lemma 20.1, $A \neq 0$ can be restored in (20.50) by coordinate transformations in the (x, y) surface when $A = 0$ initially. Suppose that $\mathcal{G} \neq 0$ in (20.47). Then, writing $A = \sqrt{|\mathcal{G}|}/(2S)$, $B_1 = -\sqrt{|\mathcal{G}|}P/(2S)$, $B_2 = -\sqrt{|\mathcal{G}|}Q/(2S)$, $\varepsilon \overset{\text{def}}{=} \mathcal{G}/|\mathcal{G}|$ and using (20.47) we find

$$C = \frac{1}{2} \sqrt{|\mathcal{G}|} S \left(\varepsilon + \frac{P^2 + Q^2}{S^2} \right). \tag{20.52}$$

Furthermore, defining $\widetilde{\Phi} = \Phi/\sqrt{|\mathcal{G}|}$ and $\widetilde{k} = k/|\mathcal{G}|$, we can represent the metric (20.50) as[2]

$$e^{-\nu} = \sqrt{|\mathcal{G}|}\mathcal{E}, \qquad \mathcal{E} \overset{\text{def}}{=} \mathcal{E} = \frac{S}{2} \left[\left(\frac{x - P}{S} \right)^2 + \left(\frac{y - Q}{S} \right)^2 + \varepsilon \right],$$

$$ds^2 = dt^2 - \frac{(\Phi_{,z} - \Phi\mathcal{E}_{,z}/\mathcal{E})^2}{\varepsilon - k(z)} dz^2 - \frac{\Phi^2}{\mathcal{E}^2} (dx^2 + dy^2). \tag{20.53}$$

When $\mathcal{G} = 0$, the transition from (20.50) to (20.53) is $(A, B_1, B_2) = (1, -P, -Q)/(2S)$ and Φ unchanged. Then $C = (P^2 + Q^2)/(2S)$ and (20.53) applies with $\varepsilon = 0$.

In the parametrisation of (20.53) (invented by Hellaby, 1996a), Eq. (20.47) was used up to calculate C, so P, Q and S are arbitrary and independent. However, it is no longer visible that all three signs of $\mathcal{G}(z)$ may occur within the same space of constant t.

In all cases, within each single surface of constant (t, z), the transformation from the (ϑ, φ) coordinates to the (x, y) coordinates is

$$\frac{1}{S}(x - P, y - Q) = \begin{cases} \cot(\vartheta/2)(\cos\varphi, \sin\varphi) & \text{when } \varepsilon = +1, \\ (2/\vartheta)(\cos\varphi, \sin\varphi) & \text{when } \varepsilon = 0, \\ \coth(\vartheta/2)(\cos\varphi, \sin\varphi) & \text{when } \varepsilon = -1. \end{cases} \tag{20.54}$$

The quantity ε determines whether the (x, y) 2-surfaces are spherical ($\varepsilon = +1$), pseudo-spherical ($\varepsilon = -1$) or planar ($\varepsilon = 0$). The geometric interpretation of the stereographic projection (20.54) in the cases $\varepsilon = \pm 1$ is shown in Fig. 20.1.

With $\varepsilon = 0$, the interpretation of the (ϑ, φ) coordinates in (20.50) is most easily seen when they are transformed to the Cartesian coordinates $X = \vartheta \cos\varphi$, $Y = \vartheta \sin\varphi$. Then, using (20.54), we find that the transformation $(x, y) \to (X, Y)$ is

$$(X, Y) = \frac{2S}{(x - P)^2 + (y - Q)^2}(x - P, y - Q), \tag{20.55}$$

which is an inversion in the sphere of radius $\sqrt{2S}$ centred at $(x, y) = (P, Q)$.

---

[2] The tildes were dropped in (20.53) for better readability. The Φ in (20.53) is $\widetilde{\Phi}$ and the $k(z)$ is $\widetilde{k}(z)$.

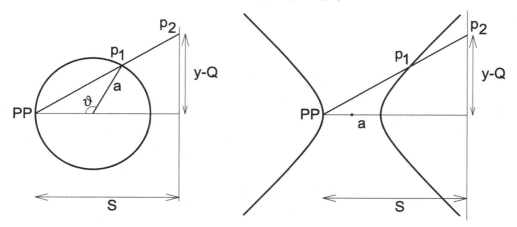

Fig. 20.1 The stereographic projection from the (ϑ, φ) coordinates to the Szekeres–Szafron coordinates (x, y) for a sphere (left) and for a two-sheeted hyperboloid (right). The radius of the sphere is $a = 1$, a is also half the distance between the vertices of the hyperboloid. A point p_1 in the surface with the (ϑ, φ) coordinates is mapped on the point p_2 in the plane with the (x, y) coordinates related to (ϑ, φ) by (20.54). The point PP is the pole of projection; p_2 lies on the straight line PP–p_1. S is the distance from the pole to the plane. On the hyperboloid, the coordinate ϑ has no simple metric interpretation, but, similar to the spherical ϑ-coordinate, it determines the distance of the point p_2 from the projection of PP on the plane. The figure is the cross-section of the sphere–plane setup with the $x = P$ plane. The coordinate φ in both surfaces is the azimuthal angle around the axis of symmetry that passes through PP.

The (x, y) coordinates in the cases $\varepsilon = +1$ and $\varepsilon = 0$ have the range $(-\infty, +\infty)$. With $\varepsilon = -1$, x and y vary in two ranges in which $\mathcal{E}$ has a constant sign; $\mathcal{E}$ is zero when

$$(x - P)^2 + (y - Q)^2 = S^2, \tag{20.56}$$

so $\mathcal{E} > 0$ for x and y outside this circle, and $\mathcal{E} < 0$ for x and y inside it. Figure 20.1 suggests that with $\varepsilon = -1$ we should rather take $(-\mathcal{E})$ as the metric function so that x and y have finite ranges. However, both the $\mathcal{E} > 0$ and the $\mathcal{E} < 0$ regions are Szekeres spacetimes because they are mapped onto one another by $(x, y) = (x', y')/\left(x'^2 + y'^2\right)$, which interchanges the roles of A and C in the metric (20.50). The metric (20.53) is independent of the sign of $\mathcal{E}$.

The surface area of a $\{t = \text{constant}, r = \text{constant}\}$ surface is finite only in the $\varepsilon = +1$ case, for which it equals $4\pi R^2$. In the other two cases, it is infinite.

20.1.4 Common properties of the two subfamilies

Two common properties have already been mentioned: the lack of any symmetry in general and the existence of the surfaces of constant curvature $\{t = \text{constant}, z = \text{constant}\}$. Other properties in common are the following:

The Weyl tensor of the Szafron spacetimes has its magnetic part with respect to the velocity field of the source equal to zero (Szafron and Collins, 1979; Barnes and Rowlingson, 1989) and is in general of Petrov type D (Szafron, 1977). It vanishes in the R–W limit only. Rotation and acceleration of the fluid source are zero, the expansion is nonzero, the shear tensor is

$$\sigma^\alpha{}_\beta = \frac{1}{3}\Sigma \, \text{diag} \, (0, 2, -1, -1), \qquad \text{where}$$

$$\Sigma = \frac{\Phi_{,tz} - \Phi_{,t} \, \Phi_{,z}/\Phi}{\Phi_{,z} - \Phi\mathcal{E}_{,z}/\mathcal{E}} \qquad \text{for } \beta_{,z} \neq 0,$$

$$\Sigma = \frac{\lambda_{,t} - \lambda\Phi_{,t}/\Phi}{\lambda + \Phi\sigma} \qquad \text{for } \beta_{,z} = 0. \tag{20.57}$$

Two eigenvalues of the 3-dimensional Ricci tensor, $^{(3)}R_{AB}$, of the slices $t = $ constant are also equal, and its eigenframe coincides with that of shear (Collins and Szafron, 1979). The eigenspaces corresponding to the degenerate eigenvalues of shear and of $^{(3)}R_{AB}$ are the surfaces of constant curvature mentioned above.

The slices $t = $ constant of these spacetimes are conformally flat (Berger, Eardley and Olson, 1977), i.e. their Cotton–York tensor (7.50) is zero.

The curvature of the (x, y)-surfaces is a global constant only in the $\beta_{,z} = 0$ subfamily, and there it is equal to the curvature index of the $t = $ constant slices in the resulting R–W limit. In the $\beta_{,z} \neq 0$ subfamily, the curvature of the (x, y)-surfaces is determined by $\mathcal{G}(z) = 4(AC - B_1{}^2 - B_2{}^2)$ and is independent of the curvature index of the R–W limit, determined by $k(z)$. Both $k(z)$ and $\mathcal{G}(z)$ are constant only within each (x, y)-surface and can vary within a single $t = $ constant slice; the sign of k can vary, too. The global constancy of k is a peculiarity of the R–W class.

In the Szafron spacetimes with arbitrary $p(t)$, the evolution equations are not yet solved. The missing element is the equation of state. The favourite equation of state for astronomers used to be the barotropic one, $f(\epsilon, p) = 0$. When $\epsilon = \epsilon(p)$ and $p = p(t)$, the Szafron spacetimes trivialise: those with $\beta_{,z} \neq 0$ become R–W, and those with $\beta_{,z} = 0$ acquire either a R–W or a K–S geometry, or the plane and hyperbolic counterparts of the latter (Spero and Szafron, 1978). The only equation of state that produces nontrivial results is $p = $ constant, in particular $p = 0$, and this case will be discussed in Section 20.2.

The discussion of the thermodynamical interpretation and of the existence of a thermodynamical scheme (see Section 15.5) applies to the Szafron metrics. As shown by Krasiński, Quevedo and Sussman (1997), in the $\beta_{,z} \neq 0$ subfamily Eqs. (15.55) and (15.58) necessarily either make the thermodynamics trivial (because p is constant) or else imply a symmetry group with at least 2-dimensional orbits. Hence, the general $\beta_{,z} \neq 0$ Szafron spacetimes require an interpretation in terms of a more complicated source than a single-component perfect fluid – it could be a mixture in which chemical reactions occur, or a mixture of two fluids, such as was first introduced by Letelier (1980).

In the $\beta_{,z} = 0$ subfamily, there is a subset of solutions that obey (15.55) and (15.58) while still having no symmetry.[3]

---

[3] We omit the derivation of these results because the formulae involved are rather complicated, while the subcase that admits a thermodynamical scheme without a symmetry has no clear physical interpretation. We refer the reader to the original paper (Krasiński, Quevedo and Sussman, 1997).

20.1.5 * The invariant definitions of the Szekeres–Szafron metrics

Several results in this subsection are quoted without proofs. While they are important and worth knowing, the proofs are rather technical and laborious. Some of them are the subjects of Exercises 4–7. Interested readers are advised to consult the original papers.

In Section 20.1 we began with the historically earliest definition, through the metric (20.1). Later, a few different coordinate-independent definitions were discovered.

The first invariant definition of the Szekeres limit $p = 0$ of the whole family was given by Wainwright (1977), and then extended by Szafron (1977) to the general case. The general Szafron spacetimes follow when the following properties are assumed:

1. The velocity field of the fluid is geodesic and irrotational.

2. The Weyl tensor is of type D, and the velocity vector of the fluid at every point of the spacetime lies in the 2-plane spanned by the two principal null directions.

3. A vector orthogonal to both principal null directions is an eigenvector of shear.

4. The 2-surfaces tangent to the principal null directions admit orthogonal 2-surfaces.

It follows that the principal null directions are orthogonal to the surfaces of constant (t, z).

The principal null directions are not in general geodesic. They become geodesic if and only if the spacetimes acquire local rotational symmetry, that is, a G_3/S_2 symmetry group.

Szafron and Collins (1979) worked out another invariant definition. The following conditions must be obeyed simultaneously:

1. The metric is a solution of the Einstein equations with a perfect fluid source.

2. The flow lines of the fluid source are geodesic and nonrotating (the last property implies that the flow lines are orthogonal to a family S_t of spacelike hypersurfaces).

3. Each hypersurface of the family S_t is conformally flat.

4. Two of the eigenvalues of the Ricci tensor $^{(3)}R_{AB}$ of the hypersurfaces S_t are equal.

5. Two of the eigenvalues of the shear tensor $\sigma_{\alpha\beta}$ are equal.

For spacetimes that have a G_3/S_2 symmetry group properties 4, 5 and $w = 0$ are automatically fulfilled, and property 3 results from $\dot{u}^\alpha = 0$ via the field equations. Hence, these subcases of the Szekeres–Szafron spacetimes are defined by property 1 and $\dot{u}^\alpha = 0$.

If conditions 1–5 are fulfilled, then there exist coordinates in which the metric has the form (20.1). These coordinates are defined up to the following transformations:

$$t = t' + C, \qquad z = f(z'), \qquad \xi = g(\xi'), \qquad \overline{\xi} = \overline{g}(\overline{\xi}'),$$
$$\xi \stackrel{\text{def}}{=} x + iy, \qquad \xi' \stackrel{\text{def}}{=} x' + iy', \tag{20.58}$$

where C is an arbitrary constant, f is an arbitrary real function and g is an arbitrary complex function.

The proper Szekeres solutions result from the two definitions when $p = 0$.

One more invariant definition was given by Barnes and Rowlingson (1989): the Szafron spacetimes result from the Einstein equations if the following assumptions are made:

1. The source is a geodesic and rotation-free perfect fluid.

2. The Weyl tensor is purely electric and of Petrov type D.

3. The shear tensor has two equal eigenvalues and its degenerate eigensurface coincides with that of the Weyl tensor.

20.2 The Szekeres solutions and their properties

20.2.1 The $\beta_{,z} = 0$ subfamily

When $\kappa p = \Lambda = $ constant, (20.38) is integrated to

$$\Phi_{,t}{}^2 = -k + \frac{2M}{\Phi} - \frac{1}{3}\Lambda\Phi^2 \overset{\text{def}}{=} \mathcal{U}(\Phi) \tag{20.59}$$

and, using (20.38), Eq. (20.35) can be integrated to (see Exercise 8)

$$\lambda_{,t}\,\Phi\Phi_{,t} + \frac{\lambda M}{\Phi} + \frac{1}{3}\Lambda\lambda\Phi^2 = (U + kW)\Phi + X(z), \tag{20.60}$$

where M is an arbitrary constant and $X(z)$ is an arbitrary function (the integral (20.60) is valid only when $\Phi_{,t} \neq 0$, but this will be always assumed, otherwise the solution is static). The solution of (20.60) can be formally represented as

$$\lambda = \sqrt{\mathcal{U}}\left[\int \frac{(U + kW)\Phi + X(z)}{\Phi \mathcal{U}^{3/2}}d\Phi + Y(z)\right], \tag{20.61}$$

where $Y(z)$ is another arbitrary function.

If we want to interpret Λ as the cosmological constant, then the energy density has to be redefined by $\kappa\epsilon = \kappa\tilde{\epsilon} - \Lambda$. Then, from (20.36), (20.59) and (20.60), we obtain

$$\kappa\tilde{\epsilon} = \frac{2X + 6M\sigma}{\Phi^2 e^\alpha}. \tag{20.62}$$

With $\Lambda \neq 0$, the solutions of (20.59) involve elliptic functions. Each solution will contain an additional arbitrary constant which defines the initial instant of evolution. The evolution may or may not begin with a Big Bang singularity, see Fig. 17.2 – a sufficiently large $-\Lambda = \lambda > 0$ together with a sufficiently large initial R (here Φ) can prevent the BB. But if the BB occurs, then it is necessarily simultaneous in the comoving time t.

An additional singularity occurs where (and if) $e^\alpha = 0$. This is a shell crossing, discussed for the $\beta_{,z} \neq 0$ subfamily in Subsection 20.3.4 and for the L–T model in Section 18.10.

Equation (20.59) is identical to the Friedmann equation (17.28) and so are its solutions. Note that M becomes the mass integral in the Friedmann limit, so it should be assumed positive for 'physical reasons'. However, a solution of (20.59) exists when $M < 0$ and $\{k < 0 \text{ or } \Lambda < 0\}$, and it calls for a deeper consideration – which is still missing.

One nonconstant function in the set $\{U, V_1, V_2, W, X, Y\}$ can be specified by a choice of the coordinate z. Hence, the general solution depends on two arbitrary constants (k and M) and five arbitrary functions of z.

In the parametrisation in which $U = -kW$ (see the text after (20.40)), Eq. (20.61) reproduces (19.102), with $(\Phi, z, Q, k) = (R, r, 0, \varepsilon)$.

The solution of (20.59) – (20.60) with $\Lambda \neq 0 = k$ involves elementary functions only and was given by Barrow and Stein-Schabes (1984). When $M \neq 0$, one can redefine $\lambda = \tilde{\lambda} + X\Phi/(3M)$, $W = \widetilde{W} - X/(3M)$, and the result is as if $X = 0$. Thus, $X = 0$ can be assumed without loss of generality whenever $M \neq 0$, and the Barrow–Stein-Schabes result is given in such a parametrisation. (We shall keep X for the sake of comparison with other solutions later.) That solution tends asymptotically to the de Sitter metric.

Equations (20.59) and (20.60) have elementary solutions when $\Lambda = 0$; they define the Szekeres (1975a) models with $\beta_{,z} = 0$. The solutions of (20.59) with $\Lambda = 0$ are identical to (17.32) – (17.34); then the integral in (20.61) can be explicitly calculated, see Szekeres (1975a) and Bonnor and Tomimura (1976). We shall avoid displaying the explicit solutions because then the various possible signs of k and M have to be considered separately (but see Section 20.4). Bonnor and Tomimura discussed the evolution of each model. Depending on the case, the metric or the matter density may or may not approach homogeneity near the initial singularity, at the timelike infinity or at a final singularity. The growing and decaying inhomogeneities are both present, just as in the L–T model, see Section 18.21.

In the $k = +1$ model, the spheres of constant t and z expand and recollapse just as in the Friedmann $k > 0$ model – because the function R is here the same. Along the z-direction the space has infinite extent at the instant of the Big Bang, then collapses to a minimum size, and expands to infinite extent at the final singularity (Barrow and Silk, 1981). The maximum of expansion of the spheres and the minimum of collapse in the z-direction do not in general occur simultaneously. This discussion does not take into account shell crossings, which can be avoided in the Datt–Ruban limit – see Sec. 19.5.

20.2.2 The $\beta_{,z} \neq 0$ subfamily

When $\kappa p = \Lambda = $ constant, (20.46) has the integral

$$\Phi_{,t}^{2} = -k(z) + \frac{2M(z)}{\Phi} - \frac{1}{3}\Lambda\Phi^{2}; \tag{20.63}$$

the only difference from (20.59) being the dependence of k, M and Φ on z. As before, the density is modified to $\kappa\widetilde{\epsilon} = \kappa\epsilon - \Lambda$, and

$$\kappa\widetilde{\epsilon} = \frac{\left(2Me^{3\nu}\right)_{,z}}{e^{2\beta}\left(e^{\beta}\right)_{,z}}. \tag{20.64}$$

Again, with $\Lambda \neq 0$ the solutions of (20.63) involve elliptic functions. A general formal integral of (20.63) was presented by Barrow and Stein-Schabes (1984). Any solution of (20.63) will contain one more arbitrary function $t_B(z)$ in the combination $(t - t_B(z))$. When $\Lambda = 0$, the instant $t = t_B(z)$ is a singularity at which $\Phi = 0$, and it goes over into the Big Bang in the Friedmann limit. When $t_{B,z} \neq 0$ (that is, in general) the instant of singularity is position-dependent, just as it was in the L–T model.

As before, another singularity may occur where $\left(e^{\beta}\right)_{,z} = 0$ (if this equation has solutions). This is a shell crossing, but now it is qualitatively different from that in the L–T models. In L–T, whole spherical shells stick together at it. Here, as can be seen from (20.50), when a shell crossing exists, its intersection with a $t = $ constant space will be a circle, or, in exceptional cases, a single point, not a sphere (see Section 20.3.4).

Equation (20.63) is formally identical with the Friedmann equation, but, just as in the L–T limit, each surface $z = $ constant evolves independently of the others.

The models defined by (20.50) and (20.63) contain eight functions of z, but only six of them are arbitrary; one can be specified by a choice of z (still arbitrary up to now), the other is determined by (20.47).

The elementary solution of (20.63) that results when $k(z) = 0$ was presented by Barrow and Stein-Schabes (1984). The solutions of (20.63) with $\Lambda = 0$ are formally identical with Friedmann's, and define the Szekeres (1975a) models with $\beta_{,z} \neq 0$. We will present them in the Goode–Wainwright representation in Section 20.4.

The subcase $AC - B_1{}^2 - B_2{}^2 > 0$, in which the surfaces of constant t and z are spheres, is the most important one physically. It is called the **quasi-spherical Szekeres model**. Its properties are discussed in Section 20.3, but two little curiosities are listed below.

The mass within the sphere $z = r$ at the time $t = t_0$

$$m(t_0, r) = \int_{t=t_0} \epsilon\sqrt{-g_3}\mathrm{d}_3x = 4\pi \int_0^r 2M_{,z}\, h(z)\mathrm{d}z = m(r), \qquad (20.65)$$

does not depend on t and is the same as in a L–T model (Szekeres, 1975a).

A general (nonsymmetric) quasi-spherical Szekeres metric can be matched to the Schwarzschild metric across a $z = $ constant hypersurface (Bonnor, 1976); see Sec. 20.3.9. With $\beta_{,z} = 0$, the matching is possible only inside the Schwarzschild event horizon.

20.2.3 * The $\beta_{,z} = 0$ family as a limit of the $\beta_{,z} \neq 0$ family

In integrating the Einstein equations for the metric (20.1) the cases $\beta_{,z} = 0$ and $\beta_{,z} \neq 0$ had to be considered separately. Having derived the solutions, we can show now that the $\beta_{,z} = 0$ family is a limit of the $\beta_{,z} \neq 0$ family. The consideration below follows that of Hellaby (1996b), with notation adapted to ours and with some corrections.[4]

In the $\beta_{,z} = 0$ family the parameter $\mathcal{G}$ determining the type of quasi-symmetry, defined by (20.47), coincides with the parameter k determining the type of evolution, and they are constant. Thus, the limit will be $k \to \varepsilon = $ constant and $\mathcal{G} \to \varepsilon$. We assume that k tends to $\mathcal{G}$ sufficiently fast so that $1/h^2 = \mathcal{G} - k$ has the property $\lim_{k \to \varepsilon}(1/h^2)_{,z}/\sqrt{\varepsilon - k} = 0$.[5] We choose the arbitrary functions in (20.50) and (20.63) as follows

$$A = \int \frac{1}{2}U(z)\sqrt{\varepsilon - k}\,\mathrm{d}z + \frac{1}{4}\varepsilon, \qquad C = \int 2W(z)\sqrt{\varepsilon - k}\,\mathrm{d}z + 1,$$

$$B_1 = \int \frac{1}{2}V_1(z)\sqrt{\varepsilon - k}\,\mathrm{d}z, \qquad B_2 = \int \frac{1}{2}V_2(z)\sqrt{\varepsilon - k}\,\mathrm{d}z,$$

$$M = -\int X(z)\sqrt{\varepsilon - k}\,\mathrm{d}z + M_0, \qquad t_B = t_1 \int \sqrt{\varepsilon - k}\,\mathrm{d}z + t_0, \qquad (20.66)$$

where M_0, t_1 and t_0 are constants, and in addition we define

$$\mathcal{E}_1 \stackrel{\mathrm{def}}{=} 1 + \frac{1}{4}\varepsilon\left(x^2 + y^2\right),$$

$$\mathcal{E}_0 \stackrel{\mathrm{def}}{=} \frac{1}{2}U(z)\left(x^2 + y^2\right) + V_1(z)x + V_2(z)y + 2W(z). \qquad (20.67)$$

Assuming that each integral in (20.66) becomes zero in the limit $k \to \varepsilon$ (it is up to us to choose the integration constants appropriately) we find in (20.50) and (20.66)

---

[4] The formulae in Hellaby (1996b) reproduce only the subcase in which U, V_1, V_2 and W are constant.
[5] This can be achieved by taking $\mathcal{G} = k + zv^2$, $\varepsilon = k + \alpha v + v^2$ ($\alpha = $ constant) and taking the limit $v \to 0$.

$$M \rightarrow M_0, \qquad t_B \rightarrow t_0, \qquad k \rightarrow \varepsilon, \qquad \frac{M_{,z}}{\sqrt{\varepsilon - k}} \rightarrow -X(z),$$

$$\frac{t_{B,z}}{\sqrt{\varepsilon - k}} \rightarrow t_1, \qquad \frac{(1/h^2)_{,z}}{\sqrt{\varepsilon - k}} \rightarrow 0, \qquad \frac{k_{,z}}{\sqrt{\varepsilon - k}} \rightarrow 2(U + \varepsilon W), \qquad (20.68)$$

$$\lim_{k \to \varepsilon} e^{-\nu} = \mathcal{E}_1, \qquad \lim_{k \to \varepsilon} \left(-\frac{\nu_{,z}}{\sqrt{\varepsilon - k}} \right) = \frac{\mathcal{E}_0}{\mathcal{E}_1}. \qquad (20.69)$$

The last limit in (20.68) follows, via (20.66), from (20.47) differentiated by z.

Up to this point, the conditions at $k \to \varepsilon$ were imposed on functions of z that were arbitrary, so the conditions just meant giving the functions a definite form. Now let us take a look at the Φ that obeys (20.63). Once the limits of M and t_B at $k \to \varepsilon$ were defined by (20.66), the form of Φ in this limit was also fixed. Since the new M, k and t_B are independent of z, $\lim_{k \to \varepsilon} \Phi(t, z)$ also became a function of t only. We take this limit to be the $\Phi(t)$ of the $\beta_{,z} = 0$ subfamily and denote

$$\lim_{k \to \varepsilon} \Phi(t, z) \stackrel{\text{def}}{=} \Phi_1(t). \qquad (20.70)$$

Then Φ_1 obeys (20.59) with $k = \varepsilon$ and $M = M_0$, so $\lim_{k \to \varepsilon} \Phi_{1,t}$ is now also determined.

Since (20.63) is the same equation as (18.14), we can use here (18.103), with $r = z$, $R = \Phi$ and $E = -k/2$, to write in the $\Lambda = 0$ case:

$$\Phi_{,z} = \left(\frac{M_{,z}}{M} - \frac{k_{,z}}{k} \right) \Phi + \left[\left(\frac{3k_{,z}}{2k} - \frac{M_{,z}}{M} \right) (t - t_B) - t_{B,z} \right] \Phi_{,t}. \qquad (20.71)$$

The limits of M, $(M_{,z}/\sqrt{\varepsilon - k})$, k, $(k_{,z}/\sqrt{\varepsilon - k})$, t_B, $(t_{B,z}/\sqrt{\varepsilon - k})$, Φ and $\Phi_{,t}$ at $k \to \varepsilon$ are now well-defined, so $\lim_{k \to \varepsilon} (\Phi_z/\sqrt{\varepsilon - k})$ is also well-defined by (20.71). So we define

$$\lim_{k \to \varepsilon} \frac{\Phi_{,z}}{\sqrt{\varepsilon - k}} \stackrel{\text{def}}{=} -\lambda(t, z). \qquad (20.72)$$

Now we differentiate (20.63) by z, divide the result by $\sqrt{\varepsilon - k}$ and take the limit $k \to \varepsilon$. Using (20.68) and (20.72) we then find that $\lambda(t, z)$ obeys (20.60).

Using (20.66) – (20.72) in (20.50) we finally obtain (20.37) in the now-used notation

$$ds^2 = dt^2 - (\lambda + \Phi_1 \mathcal{E}_0/\mathcal{E}_1)^2 dz^2 - (\Phi_1/\mathcal{E}_1)^2 (dx^2 + dy^2). \qquad (20.73)$$

Thus, the $\beta_{,z} = 0$ Szekeres solutions are a limit of those with $\beta_{,z} \neq 0$, also when $\Lambda \neq 0$.

Using (20.71) and (20.72) we can verify that $e^\alpha = -\lambda - \Phi_1 \mathcal{E}_0/\mathcal{E}_1$ implied via (20.37) coincides in the spherically symmetric case ($U = kW$, $V_1 = V_2 = 0$, $k \to \varepsilon = 1$) with (19.105), with (X, Y) of (19.105) replaced by $(X + 6W M_0, t_1)$; see Exercise 13.

20.3 Properties of the quasi-spherical Szekeres solutions with $\beta_{,z} \neq 0 = \Lambda$

This subclass of the Szekeres models is best investigated and physically the most important. Except for Subsection 20.3.6, this section follows Hellaby and Krasiński (2002).

In the quasi-spherical Szekeres metrics $\varepsilon = +1$, but we keep ε in the equations in order to be able to make comments on the $\varepsilon = 0$ and $\varepsilon = -1$ cases wherever appropriate.

We will represent the quasi-spherical Szekeres solution with $\beta_{,z} \neq 0$ in the parametrisation introduced in (20.53). The formula for energy density in these variables is[6]

$$\kappa\epsilon = \frac{2\left(M_{,z} - 3M\mathcal{E}_{,z}/\mathcal{E}\right)}{\Phi^2\left(\Phi_{,z} - \Phi\mathcal{E}_{,z}/\mathcal{E}\right)}. \tag{20.74}$$

20.3.1 Basic physical restrictions

We choose $\Phi \geq 0$ ($\Phi = 0$ is the origin, bang or crunch; in no case is continuation to $\Phi < 0$ possible) and $M(z) \geq 0$, so that any vacuum exterior has positive Schwarzschild mass.

We require the metric to be nondegenerate and nonsingular, except at the bang or crunch. Since $(\mathrm{d}x^2 + \mathrm{d}y^2)/\mathcal{E}^2$ is the metric of the unit sphere, plane or pseudosphere, $|S(r)| \neq 0$ is necessary for a sensible mapping, so $S > 0$ is a reasonable choice. In the cases $\varepsilon = 0$ and $\varepsilon = -1$, $\mathcal{E}$ necessarily goes to zero at certain (x, y) values; these sets do not belong to the spacetime. For a well-behaved z-coordinate, we need to require

$$0 < \frac{(\Phi_{,z} - \Phi\mathcal{E}_{,z}/\mathcal{E})^2}{\varepsilon - k} < \infty, \tag{20.75}$$

i.e. $\varepsilon - k > 0$ except where $(\Phi_{,z} - \Phi\mathcal{E}_{,z}/\mathcal{E})^2 = 0$. Quasi-pseudo-spherical regions, $\varepsilon = -1$, then require $k \leq -1$, hence they must evolve hyperbolically. Similarly, quasi-planar regions, $\varepsilon = 0$, may evolve only parabolically or hyperbolically, $k \leq 0$, whereas quasi-spherical regions exist for all $k \leq 1$. In the Lemaître–Tolman limit ($\mathcal{E}_{,z} = 0$, $\varepsilon = 1$), the equality $1 - k = 0 = (\Phi_{,z})^2$ can occur in closed models where $\Phi = R$ on a spatial section is at a maximum, or in wormhole models where it is minimum, $\Phi_{,z}(t, r_m) = 0 \; \forall \; t$. These can only occur at constant z and must hold for all (x, y) values. We will consider maxima and minima in Section 20.3.5.

The density must be positive and finite, which adds

$$\text{either} \quad M_{,z} - 3M\mathcal{E}_{,z}/\mathcal{E} \geq 0 \quad \text{and} \quad \Phi_{,z} - \Phi\mathcal{E}_{,z}/\mathcal{E} \geq 0, \tag{20.76}$$

$$\text{or} \quad M_{,z} - 3M\mathcal{E}_{,z}/\mathcal{E} \leq 0 \quad \text{and} \quad \Phi_{,z} - \Phi\mathcal{E}_{,z}/\mathcal{E} \leq 0. \tag{20.77}$$

If $\Phi_{,z} - \Phi\mathcal{E}_{,z}/\mathcal{E} = 0$ anywhere other than at a regular extremum, we have a shell crossing.

20.3.2 The significance of $\mathcal{E}$

The Szekeres metric is covariant with the transformations $z = g(\widetilde{z})$, where g is an arbitrary function. Hence, if $\Phi_{,z} < 0$ in some neighbourhood, we can take $g = 1/\widetilde{z}$ and obtain $\mathrm{d}\Phi/\mathrm{d}\widetilde{z} > 0$. Therefore, $\Phi_{,z} > 0$ can be assumed to hold in a neighbourhood of any $z = z_0$. However, if $\Phi_{,z}$ changes sign somewhere, then this is a coordinate-independent property.

As seen from (20.53), $\mathcal{E}$ is always nonzero when $\varepsilon = +1$. Since the sign of $\mathcal{E}$ is not defined by the metric, we can assume that $\mathcal{E} > 0$. Can $\mathcal{E}_{,z}$ change sign?

$$\mathcal{E}_{,z} = \frac{1}{2} S_{,z} \left[-\frac{(x - P)^2 + (y - Q)^2}{S^2} + \varepsilon\right] - \frac{(x - P)P_{,z} + (y - Q)Q_{,z}}{S}. \tag{20.78}$$

---

[6] The M and Φ in (20.74) are in fact $\widetilde{M} = M/|\mathcal{G}|^{3/2}$ and $\widetilde{\Phi} = \Phi/\sqrt{|\mathcal{G}|}$; the tildes were dropped for clarity.

The discriminant of this with respect to $(x - P)$ is

$$\Delta_x = \frac{1}{S^2} \left[-\frac{S_{,z}^2}{S^2}(y - Q)^2 - 2\frac{S_{,z}}{S}(y - Q)Q_{,z} + P_{,z}^2 + \varepsilon S_{,z}^2 \right]. \tag{20.79}$$

The discriminant of Δ_x with respect to $(y - Q)$ is

$$\Delta_y = \frac{4S_{,z}^2}{S^6} \left(P_{,z}^2 + Q_{,z}^2 + \varepsilon S_{,z}^2 \right). \tag{20.80}$$

Thus, $\Delta_y \geq 0$ when $\varepsilon = +1$, so $\mathcal{E}_{,z} = 0$ will have in general two solutions and one in exceptional situations. The exceptional situations are when $\Delta_y = 0$. They are as follows:

(i) $S_{,z} = 0$ at $z = z_0$. Then $\mathcal{E}_{,z}(z_0) = 0$ defines a straight line in the (x, y) plane. This will be dealt with below (see after (20.90)).

(ii) $S_{,z} = P_{,z} = Q_{,z} = 0$ at $z = z_0$. Then $\mathcal{E}_{,z} \equiv 0$ at $z = z_0$, and (20.74) shows that ϵ is spherically symmetric there.

When $\Delta_y > 0$, Δ_x will change sign at

$$y_{1,2} = Q + \frac{S}{S_{,z}} \left(-Q_{,z} \pm \sqrt{P_{,z}^2 + Q_{,z}^2 + \varepsilon S_{,z}^2} \right). \tag{20.81}$$

For every y such that $y_1 < y < y_2$ there will be two values of x (one when $y = y_1$ or $y = y_2$) such that $\mathcal{E}_{,z} = 0$. They are

$$x_{1,2} = P - P_{,z}\frac{S}{S_{,z}} \pm S\sqrt{-\left(\frac{y - Q}{S} + \frac{Q_{,z}}{S_{,z}}\right)^2 + \frac{P_{,z}^2 + Q_{,z}^2}{S_{,z}^2} + \varepsilon}. \tag{20.82}$$

The regions where $\mathcal{E}_{,z}$ is positive and negative depend on the sign of $S_{,z}$. If $S_{,z} > 0$, then $\mathcal{E}_{,z} > 0$ for $x < x_1$ and for $x > x_2$, if $S_{,z} < 0$, then $\mathcal{E}_{,z} > 0$ for $x_1 < x < x_2$. For $x = x_1$ and $x = x_2$ we have $\mathcal{E}_{,z} = 0$, but note that x_1 and x_2 are members of a continuous family labelled by y. All the values of x and y from (20.81) – (20.82) lie on the circle

$$\left[x - \left(P - \frac{P_{,z}S}{S_{,z}}\right)\right]^2 + \left[y - \left(Q - \frac{Q_{,z}S}{S_{,z}}\right)\right]^2 = S^2 \left(\frac{P_{,z}^2 + Q_{,z}^2}{S_{,z}^2} + \varepsilon\right). \tag{20.83}$$

The centre of this circle is at the point

$$(x, y) = \left(P - \frac{P_{,z}S}{S_{,z}}, Q - \frac{Q_{,z}S}{S_{,z}}\right), \tag{20.84}$$

and its radius is

$$L_{(\mathcal{E}_{,z}=0)} = S\sqrt{\left(P_{,z}^2 + Q_{,z}^2\right)/S_{,z}^2 + \varepsilon}. \tag{20.85}$$

Thus, with $S_{,z} > 0$, $\mathcal{E}_{,z} < 0$ inside the circle, $= 0$ on the circle and > 0 outside it. Note that this radius can never be zero when $\varepsilon = +1$. It can be infinite when $S_{,z} = 0$, in which case the circle becomes a straight line. Even then, $\mathcal{E}_{,z}$ will have different signs on the opposite sides of the line – see the paragraph below (20.90). Consequently, in the quasi-spherical Szekeres metrics $\mathcal{E}_{,z}$ cannot have the same sign throughout the (x, y) sphere.

We will consider the variation of $\mathcal{E}(z, x, y)$ around the spheres of constant t and z. Setting $\varepsilon = +1$ and applying the transformation (20.54) to $\mathcal{E}$ and to its derivatives gives

$$\mathcal{E} = S/(1 - \cos \vartheta), \tag{20.86}$$

$$\mathcal{E}_{,z} = -\frac{S_{,z} \cos \vartheta + \sin \vartheta \, (P_{,z} \cos \varphi + Q_{,z} \sin \varphi)}{1 - \cos \vartheta}, \tag{20.87}$$

$$\mathcal{E}_{,zz} = -\frac{S_{,zz} \cos \vartheta + \sin \vartheta \, (P_{,zz} \cos \varphi + Q_{,zz} \sin \varphi)}{1 - \cos \vartheta}$$
$$+ 2\frac{S_{,z}}{S} \frac{S_{,z} \cos \vartheta + \sin \vartheta \, (P_{,z} \cos \varphi + Q_{,z} \sin \varphi)}{1 - \cos \vartheta} + \frac{S_{,z}{}^2 + P_{,z}{}^2 + Q_{,z}{}^2}{S}. \tag{20.88}$$

The locus of $\mathcal{E}_{,z} = 0$ is

$$S_{,z} \cos \vartheta + P_{,z} \sin \vartheta \cos \varphi + Q_{,z} \sin \vartheta \sin \varphi = 0. \tag{20.89}$$

Writing $z = \cos \vartheta$, $y = \sin \vartheta \cos \varphi$ and $x = \sin \vartheta \sin \varphi$, we see that (x, y, z) is on a unit sphere centred at $(0, 0, 0)$, and (20.89) becomes $S_{,z} z + P_{,z} x + Q_{,z} y = 0$, which is the equation of an arbitrary plane through $(0, 0, 0)$. Such planes all intersect the unit sphere along great circles, therefore $\mathcal{E}_{,z} = 0$ is a great circle, with locus

$$\tan \vartheta = -S_{,z} / \, (P_{,z} \cos \varphi + Q_{,z} \sin \varphi) \, . \tag{20.90}$$

The plane has unit normal $(P_{,z}, Q_{,z}, S_{,z}) / \sqrt{P_{,z}{}^2 + Q_{,z}{}^2 + S_{,z}{}^2}$.

Now it is easy to understand the meaning of the special case $S_{,z} = 0$ mentioned after (20.80). As seen from (20.90), with $S_{,z} = 0$ we have $\vartheta = 0$, which means that the great circle defined by $\mathcal{E}_{,z} = 0$ passes through the pole of stereographic projection. In this case, the image of the circle $\mathcal{E}_{,z} = 0$ on the (x, y) plane is a straight line passing through $(x, y) = (P, Q)$, as indeed follows from (20.78). (See also Figs. 20.2 and 20.3 in Sec. 20.3.7). Then $\mathcal{E}_{,z}$ has a different sign on each side of the straight line. The other special case, $\mathcal{E}_{,z} \equiv 0$, corresponds to the spherically symmetric subcase. Then, the positions of the great circle from (20.89) and of the poles from (20.96) below are undetermined.

From (20.87) and (20.86) we find

$$\mathcal{E}_{,z} / \mathcal{E} = -(1/S) \, [S_{,z} \cos \vartheta + \sin \vartheta \, (P_{,z} \cos \varphi + Q_{,z} \sin \varphi)] \, , \tag{20.91}$$

thus

$$\mathcal{E}_{,z} / \mathcal{E} = \text{constant} \Rightarrow S_{,z} z + P_{,z} x + Q_{,z} y = S \times \text{constant}, \tag{20.92}$$

which is a plane parallel to the $\mathcal{E}_{,z} = 0$ plane, implying that all loci of $\mathcal{E}_{,z} / \mathcal{E} = \text{constant}$ are circles parallel to the $\mathcal{E}_{,z} = 0$ great circle.

The locations of the extrema of $\mathcal{E}_{,z} / \mathcal{E}$ are found as follows

$$\frac{\partial (\mathcal{E}_{,z} / \mathcal{E})}{\partial \varphi} = \frac{\sin \vartheta \, (P_{,z} \sin \varphi - Q_{,z} \cos \varphi)}{S} = 0 \Rightarrow \tan \varphi_e = \frac{Q_{,z}}{P_{,z}}$$

$$\Rightarrow \cos \varphi_e = \varepsilon_1 \frac{P_{,z}}{\sqrt{P_{,z}{}^2 + Q_{,z}{}^2}}, \qquad \varepsilon_1 = \pm 1, \tag{20.93}$$

$$\frac{\partial (\mathcal{E}_{,z} / \mathcal{E})}{\partial \vartheta} = \frac{S_{,z} \sin \vartheta - \cos \vartheta \, (P_{,z} \cos \varphi + Q_{,z} \sin \varphi)}{S} = 0$$

$$\Rightarrow \tan \vartheta_e = \frac{P_{,z} \cos \varphi_e + Q_{,z} \sin \varphi_e}{S_{,z}} = \varepsilon_1 \frac{\sqrt{P_{,z}{}^2 + Q_{,z}{}^2}}{S_{,z}}, \tag{20.94}$$

$$\Rightarrow \cos \vartheta_e = \varepsilon_2 \frac{S_{,z}}{\sqrt{S_{,z}{}^2 + P_{,z}{}^2 + Q_{,z}{}^2}}, \qquad \varepsilon_2 = \pm 1. \tag{20.95}$$

The extreme value is then

$$(\mathcal{E}_{,z}/\mathcal{E})_{\text{extreme}} = -(\varepsilon_2/S)\sqrt{S_{,z}{}^2 + P_{,z}{}^2 + Q_{,z}{}^2}. \tag{20.96}$$

Since

$$(\cos \vartheta_e, \sin \vartheta_e \cos \varphi_e, \sin \vartheta_e \sin \varphi_e) = \frac{\varepsilon_2(S_{,z}, P_{,z}, Q_{,z})}{\sqrt{S_{,z}{}^2 + P_{,z}{}^2 + Q_{,z}{}^2}}, \tag{20.97}$$

the remark under (20.90) shows that an extremum of $\mathcal{E}_{,z}/\mathcal{E}$ lies on the unit normal to the plane of $\mathcal{E}_{,z} = 0$, i.e. at a pole to the $\mathcal{E}_{,z} = 0$ great circle.

Clearly, $\mathcal{E}_{,z}/\mathcal{E}$ has a dipole variation around each constant-z sphere, changing sign when we go over to the antipodal point: $(\vartheta, \varphi) \to (\pi - \vartheta, \varphi + \pi)$. Writing, from (20.91),

$$\Phi_{,z} - \Phi \mathcal{E}_{,z}/\mathcal{E} = \Phi_{,z} + (\Phi/S)\left[S_{,z} \cos \vartheta + \sin \vartheta \left(P_{,z} \cos \varphi + Q_{,z} \sin \varphi \right) \right] \tag{20.98}$$

we see that $\Phi \mathcal{E}_{,z}/\mathcal{E}$ is the correction to the radial separation $\Phi_{,z}$ of constant-z shells, due to their being nonconcentric. In particular, $\Phi S_{,z}/S$ is the forward ($\vartheta = 0$) displacement, and $\Phi P_{,z}/S$ and $\Phi Q_{,z}/S$ are the two sideways displacements ($\vartheta = \pi/2$, $\varphi = 0$) and ($\vartheta = \pi/2$, $\varphi = \pi/2$). The shortest 'radial' distance is where $\mathcal{E}_{,z}/\mathcal{E}$ is maximum.

It will be shown in Subsection 20.3.4 that when $\Phi_{,z} > 0$ and $M_{,z} > 0$, conditions for avoiding shell crossings require, among other things, that $\Phi_{,z}/\Phi > M_{,z}/(3M)$, see (20.139). This implies that the energy density given by (20.74), as a function of $x \stackrel{\text{def}}{=} \mathcal{E}_{,z}/\mathcal{E}$

$$\kappa \epsilon = \frac{2 \left(M_{,z} - 3Mx \right)}{\Phi^2 \left(\Phi_{,z} - \Phi x \right)} \tag{20.99}$$

has a negative derivative by x:

$$\kappa \epsilon_{,x} = -\frac{6M \left[\Phi_{,z}/\Phi - M_{,z}/(3M) \right]}{\Phi \left(\Phi_{,z} - \Phi x \right)^2} < 0, \tag{20.100}$$

so the density is minimum where $\mathcal{E}_{,z}/\mathcal{E}$ is maximum.

20.3.3 Conditions of regularity at the origin

In this subsection we investigate those Szekeres spacetimes in which at least one origin exists. An origin is an analogue of the centre of symmetry in the L–T models, but it need not exist. Examples of spacetimes that do not contain an origin are the Datt–Ruban and Kantowski–Sachs models of Sec. 19.5. Simple 2-dimensional analogues are a one-sheeted hyperboloid and a cylinder.

For reference, we write out the evolution equations of the Szekeres models with $\Lambda = 0$, even though they are identical to those of the L–T and Friedmann models. The function $\Phi = \Phi(t, z)$ satisfies (20.63):

$$\Phi_{,t}{}^2 = 2M(z)/\Phi - k(z), \tag{20.101}$$

where $M(z)$ and $k(z)$ are arbitrary functions. It follows that

$$\Phi_{,tt} = -M/\Phi^2 < 0. \tag{20.102}$$

The function $M(z)$ is the active gravitational mass within the sphere of coordinate radius z. We assume $M \geq 0$ and $\Phi \geq 0$. In (20.101), $(-k(z))$ represents twice the energy per unit mass of the particles in the shell of matter at constant z, but in the metric (20.53) it also determines the geometry of the spatial sections $t = $ constant. The solution of (20.101) depends on the sign of k; it can be

(1) hyperbolic (when $k < 0$):

$$\Phi = -\frac{M}{k}(\cosh \eta - 1), \qquad \sinh \eta - \eta = \frac{1}{M}(-k)^{3/2}\sigma(t - t_B), \tag{20.103}$$

(2) parabolic (when $k = 0$):

$$\Phi = \left[9M(t - t_B)^2/2\right]^{1/3}, \tag{20.104}$$

(3) elliptic (when $k > 0$):

$$\Phi = \frac{M}{k}(1 - \cos \eta), \qquad \eta - \sin \eta = \frac{k^{3/2}}{M}\sigma(t - t_B), \tag{20.105}$$

where $t_B(z)$ is the arbitrary Big Bang/Crunch function and $\sigma = \pm 1$ permits time reversal. More correctly, the three types of evolution correspond to $k/M^{2/3} < 0, = 0$ and > 0, since $k = 0$ at a spherical type origin for all three evolution types (see further on in this subsection). The behaviour of $\Phi(t, z)$ is identical to that of $R(t, r)$ in the L–T model, and is unaffected by the dependence of the Szekeres solutions on the (x, y) coordinates.

When $\varepsilon = +1$, $\Phi = 0 \forall\, t$ where the 2-spheres have zero radius (which we assume to be at $z = 0$; this can be satisfied by a transformation of z). Similarly, $\Phi_{,t}(t, 0) = 0 = \Phi_{,tt}(t, 0)$ $\forall\, t$. We also require $M(0) = 0$ (no point mass at the origin). Thus, by (20.105), (20.103) and (20.101) – (20.102), for constant η

$$\lim_{z \to 0}(M/k) = 0, \qquad \lim_{z \to 0} k = 0. \tag{20.106}$$

The energy density (20.74) and the scalar polynomial invariants of the Riemann tensor must be finite at the origin. The nonzero tetrad components of the Riemann tensor in the orthonormal tetrad connected with the metric (20.53) are

$$\begin{aligned}
R_{0101} &= 2\mathcal{B}_1 - \mathcal{B}_2, & R_{1212} = R_{1313} = \mathcal{B}_1 - \mathcal{B}_2, \\
R_{0202} &= R_{0303} = \frac{1}{2} R_{2323} = -\mathcal{B}_1,
\end{aligned} \tag{20.107}$$

where

$$\mathcal{B}_1 \overset{\text{def}}{=} \frac{M}{\Phi^3}, \qquad \mathcal{B}_2 \overset{\text{def}}{=} \frac{M_{,z} - 3M\mathcal{E}_{,z}/\mathcal{E}}{\Phi^2(\Phi_{,z} - \Phi\mathcal{E}_{,z}/\mathcal{E})}. \tag{20.108}$$

To ensure that all scalar polynomial invariants of the Riemann tensor are finite at the origin it suffices to ensure that (20.107) – (20.108) are finite there. Note that $\mathcal{B}_2 = \kappa\epsilon/2$, so we only need to ensure the proper behaviour of $\kappa\epsilon$ and of $\mathcal{B}_1$. Thus we require

$$0 < \kappa\epsilon(0) = \lim_{z \to 0}\left[\frac{2M_{,z}}{\Phi^2\Phi_{,z}} \times \frac{1 - 3M\mathcal{E}_{,z}/(M_{,z}\,\mathcal{E})}{1 - \Phi\mathcal{E}_{,z}/(\Phi_{,z}\,\mathcal{E})}\right] = \lim_{z \to 0}\frac{2M_{,z}}{\Phi^2\Phi_{,z}} < \infty, \tag{20.109}$$

$$0 < \mathcal{B}_1(0) = \lim_{z\to 0} \frac{M}{\Phi^3} = \lim_{z\to 0} \frac{M_{,z}}{3\Phi^2\Phi_{,z}} < \infty \Rightarrow \lim_{z\to 0} \frac{3\Phi_{,z}M}{\Phi M_{,z}} = 1. \qquad (20.110)$$

We see from (20.103) and (20.105) that

$$d\eta/dt = \sqrt{|k|}/\Phi, \qquad (20.111)$$

so, using (20.109)

$$\lim_{z\to 0} \frac{d\eta}{dt} = \lim_{z\to 0} \frac{\sqrt{|k|}}{\Phi} = \left(\lim_{z\to 0} \frac{\sqrt{|k|}}{M^{1/3}}\right)\left(\lim_{z\to 0} \frac{M^{1/3}}{\Phi}\right) = \left(\lim_{z\to 0} \frac{\sqrt{|k|}}{M^{1/3}}\right)\left[\lim_{z\to 0}(\kappa\epsilon/2)\right]^{1/3}. \qquad (20.112)$$

By (20.109), $\epsilon(0)$ is not zero and not permanently infinite. We require $0 < \lim_{z\to 0}(d\eta/dt) < \infty$, otherwise η would fail to measure the pace of evolution at the centre. Consequently, $0 < \lim_{z\to 0}\left(\sqrt{|k|}/M^{1/3}\right) < \infty$, which we prefer to write as

$$0 < \lim_{z\to 0}\left(|k|^{3/2}/M\right) < \infty. \qquad (20.113)$$

Using de l'Hôpital's rule, this gives

$$0 < \lim_{z\to 0}\left(\frac{|k|^{3/2}}{M}\right) = \lim_{z\to 0}\left(\frac{3\sqrt{|k|}|k|_{,z}}{M_{,z}}\right) < \infty \Longrightarrow \lim_{z\to 0}\left(\frac{3M|k|_{,z}}{2M_{,z}|k|}\right) = 1. \qquad (20.114)$$

Equation (20.110) implies $0 < \Phi^2\Phi_{,z}/M_{,z} < \infty$. This must be considered separately for $k > 0$ and for $k < 0$. For $k > 0$ and $\sigma = +1$ we have from (20.71) and (20.105):

$$\Phi^2\frac{\Phi_{,z}}{M_{,z}} = \frac{M^2}{k^3}(1-\cos\eta)\left[\left(1 - \frac{Mk_{,z}}{M_{,z}k}\right)(1-\cos\eta)^2\right.$$
$$\left. + \left(\frac{3Mk_{,z}}{2M_{,z}k} - 1\right)\sin\eta(\eta - \sin\eta) - \frac{k^{3/2}t_{B,z}}{M_{,z}}\sin\eta\right]. \qquad (20.115)$$

In consequence of (20.114) the following happens at $z = 0$:

The factor M^2/k^3 is positive and finite because of (20.113).

The first term in the second line is zero.

The first term in square brackets increases from zero at $\eta = 0$ to $2/3$ at $\eta = \pi$ and decreases back to zero at $\eta = 2\pi$. The last term in square brackets changes from zero at $\eta = 0$ to $(-k^{3/2}t_{B,z}/M_{,z})$ at $\eta = \pi/2$ and back to zero at $\eta = \pi$, but changes sign to the opposite there, then proceeds to $(+k^{3/2}t_{B,z}/M_{,z})$ at $\eta = 3\pi/2$ and to zero again at $\eta = 2\pi$. As $\eta \to 0$ from above, both the first and last term tend to zero, but the last one dominates because $\left[\sin\eta/(1-\cos\eta)^2\right] \to +\infty$. Thus, we must require

$$\lim_{z\to 0} Mt_{B,z}/M_{,z} < \infty. \qquad (20.116)$$

This will ensure $\Phi^2\Phi_{,z}/M_{,z} > 0$ for $0 < \eta < \pi$ and for some part of, but not for the whole range $\pi \le \eta \le 2\pi$ because $\sin\eta < 0$ for $\eta \in (\pi, 2\pi)$. As $\eta \to 2\pi$ from below with (20.116) fulfilled, the last term in (20.115) becomes negative and dominant. This means that the energy density will become negative on approaching the Big Crunch – recall that we reported the same phenomenon for collapsing charged dust in Sec. 19.4.6.

For $k < 0$ and $\sigma = +1$, Eq. (20.115) holds with k replaced by $(-k)$, $(1-\cos\eta)$ replaced by $(\cosh\eta - 1)$, $(\eta - \sin\eta)$ replaced by $(\sinh\eta - \eta)$ and $\sin\eta$ is replaced by $\sinh\eta$. Here, the first term in the second line of (20.115) is again zero at $z = 0$, but the first and the last one do not change sign for any $\eta > 0$, so (20.116) ensures $\Phi^2\Phi_{,z}/M_{,z} > 0$ for all $\eta > 0$.

Lastly, the metric must be well behaved, so $0 < \mathcal{E} < \infty$ must hold. We also need to ensure that the rate of change of the proper radius $\mathcal{R} \overset{\text{def}}{=} \int\sqrt{-g_{zz}}dz$ with respect to the areal radius Φ is finite. That rate of change is $d\mathcal{R}/d\Phi$, so we require $0 < (d\mathcal{R}/d\Phi)^2 = -g_{zz}/(\Phi_{,z})^2 < \infty$, i.e., using (20.109) and the second of (20.106),

$$0 < \lim_{z\to 0} \frac{(\Phi_{,z} - \Phi\mathcal{E}_{,z}/\mathcal{E})^2}{(1-k)\Phi_{,z}{}^2} < \infty \Rightarrow 0 < \lim_{z\to 0}\left(1 - \frac{3M\mathcal{E}_{,z}}{M_{,z}\mathcal{E}}\right)^2 < \infty$$

$$\Rightarrow \lim_{z\to 0}\left|\frac{M\mathcal{E}_{,z}}{M_{,z}\mathcal{E}}\right| < \infty \text{ and } \lim_{z\to 0}\left|\frac{M\mathcal{E}_{,z}}{M_{,z}\mathcal{E}}\right| \neq \frac{1}{3}. \tag{20.117}$$

This should hold for all (x,y), i.e. all (ϑ, φ). Thus, (20.91) gives

$$\lim_{z\to 0}\left|\frac{MS_{,z}}{M_{,z}S}\right| < \infty, \qquad \lim_{z\to 0}\left|\frac{MP_{,z}}{M_{,z}S}\right| < \infty, \qquad \lim_{z\to 0}\left|\frac{MQ_{,z}}{M_{,z}S}\right| < \infty, \tag{20.118}$$

all three limits being different from $1/3$.

All of the above shows that, near an origin,

$$M \sim \Phi^3, \qquad |k| \sim \Phi^2, \qquad S \sim \Phi^n, \qquad P \sim \Phi^n, \qquad Q \sim \Phi^n, \qquad n \geq 0. \tag{20.119}$$

The condition $\mathcal{E}_{,z}/\mathcal{E} \leq M_{,z}/(3M)$ that will be obtained in the next subsection implies, via (20.110) and (20.119), that $n \leq 1$ near an origin.

20.3.4 Shell crossings

A shell crossing, if it exists, is the locus of zeros of the function $\chi \overset{\text{def}}{=} \mathcal{E}\Phi_{,z}/\Phi - \mathcal{E}_{,z}$, see (20.53). Observe that $\Phi_{,z} > 0$ and $\chi < 0$ cannot hold for all x and y. This would lead to $\mathcal{E}_{,z} > \mathcal{E}\Phi_{,z}/\Phi > 0$, and we know that $\mathcal{E}_{,z}$ cannot be positive at all x and y (see the text below (20.85)). Hence, with $\Phi_{,z} > 0$, there must be a region in which $\chi > 0$. By a similar argument, $\Phi_{,z} < 0$ and $\chi > 0$ cannot hold for all x and y, so with $\Phi_{,z} < 0$ there must be a region in which $\chi < 0$.

Assuming $\Phi_{,z} > 0$, can χ be positive for all x and y? Writing

$$\chi = \frac{1}{2S}\left(\frac{S_{,z}}{S} + \frac{\Phi_{,z}}{\Phi}\right)[(x-P)^2 + (y-Q)^2]$$

$$- \frac{1}{2}\varepsilon S\left(\frac{S_{,z}}{S} - \frac{\Phi_{,z}}{\Phi}\right) + \frac{1}{S}[(x-P)P_{,z} + (y-Q)Q_{,z}], \tag{20.120}$$

the discriminant of this with respect to $(x - P)$ is

$$\Delta_x = \frac{P_{,z}{}^2}{S^2} - \frac{1}{S^2}\left(\frac{S_{,z}}{S} + \frac{\Phi_{,z}}{\Phi}\right)\left[\left(\frac{S_{,z}}{S} + \frac{\Phi_{,z}}{\Phi}\right)(y-Q)^2\right.$$

$$\left. + 2(y-Q)Q_{,z} - \varepsilon S^2\left(\frac{S_{,z}}{S} - \frac{\Phi_{,z}}{\Phi}\right)\right], \tag{20.121}$$

and the discriminant of the above with respect to $(y - Q)$ is

$$\Delta_y = 4 \frac{1}{S^2} \left(\frac{S_{,z}}{S} + \frac{\Phi_{,z}}{\Phi} \right)^2 \left[\frac{P_{,z}^2 + Q_{,z}^2 + \varepsilon S_{,z}^2}{S^2} - \varepsilon \frac{\Phi_{,z}^2}{\Phi^2} \right]. \tag{20.122}$$

Note that in the quasi-plane model $\varepsilon = 0$ it is impossible to have $\Delta_y < 0$, i.e. χ will have zeros ($\Rightarrow$ there will be shell crossings), unless the range of variability of x and y is suitably limited so that the zeros are left outside. This question was considered by Krasiński (2008) and the proposed solution was this: one has to identify certain pairs of lines so that the (x, y) surface becomes finite in size and the shell crossings are left outside it. However, the metric in the nonsymmetric case permits the identifications only with a twist (the left side of the elementary rectangle is identified with the right side mirror-reflected, the same for the pair upper side–lower side). This introduces a new difficulty because such an object is a projective plane (Borsuk, 1964) – a one-sided surface. To avoid this problem, one has to assemble four elementary rectangles, suitably turned, into a surface sheet that has two sides and becomes the new elementary rectangle whose opposite edges are identified. We do not quote the details because the discussion is long and technical.

The discussion below applies only to the case $\varepsilon = +1$. Then χ will have the same sign for all x and y (there will be no shell crossings) when $\Delta_y < 0$, that is, if and only if

$$\frac{\Phi_{,z}^2}{\Phi^2} > \varepsilon \frac{P_{,z}^2 + Q_{,z}^2 + \varepsilon S_{,z}^2}{S^2} \stackrel{\text{def}}{=} \Psi^2(z). \tag{20.123}$$

If $\Phi_{,z}^2 / \Phi^2 = \Psi^2$, then $\Delta_y = 0$, so $\Delta_x = 0$ at just one value of $y = y_{SS}$. With this value of y, $\chi = 0$ at one value of $x = x_{SS}$. In this case, the shell crossing is a single point in the constant-(t, z) surface, i.e. a curve in a space of constant t and a 2-surface in spacetime.

If $\Phi_{,z}^2 / \Phi^2 < \Psi^2$, then the locus of $\chi = 0$ is in general a circle (a straight line when $S_{,z}/S = -\Phi_{,z}/\Phi$) in the (x, y) plane. The straight line is a projection on the (x, y) plane of a circle on the sphere of constant t and z, so is in fact not a special case.

When $\Delta_y > 0$ ($\Phi_{,z}^2 / \Phi^2 < \Psi^2$), the two values of y at which Δ_x changes sign are

$$y_{1,2} = Q + \frac{-Q_{,z} \pm \sqrt{\delta}}{S_{,z}/S + \Phi_{,z}/\Phi},$$

$$\delta \stackrel{\text{def}}{=} P_{,z}^2 + Q_{,z}^2 + \varepsilon \left(S_{,z}^2 - S^2 \Phi_{,z}^2 / \Phi^2 \right), \tag{20.124}$$

and then for every y such that $y_1 < y < y_2$ there are two values of x (only one if $y = y_1$ or $y = y_2$) such that $\chi = 0$. These are

$$x_{1,2} = P + \frac{-P_{,z} \pm \sqrt{- \left[(S_{,z}/S + \Phi_{,z}/\Phi)(y - Q) + Q_{,z} \right]^2 + \delta}}{S_{,z}/S + \Phi_{,z}/\Phi}. \tag{20.125}$$

The x and y obeying (20.125) lie on a circle with the centre at

$$(x_{SC}, y_{SC}) = \left(P - \frac{P_{,z}}{S_{,z}/S + \Phi_{,z}/\Phi}, Q - \frac{Q_{,z}}{S_{,z}/S + \Phi_{,z}/\Phi} \right), \tag{20.126}$$

and with the radius $L_{SC} = \sqrt{\delta}/(S_{,z}/S + \Phi_{,z}/\Phi)$. This is in general a different circle from the one defined by $\mathcal{E}_{,z} = 0$. By the definition of χ, the shell crossing set intersects with the surface of constant t and z along the line $\mathcal{E}_{,z}/\mathcal{E} = \Phi_{,z}/\Phi = $ constant. As noted after (20.92), this circle lies in a plane parallel to the $\mathcal{E}_{,z} = 0$ great circle. It follows that the $\mathcal{E}_{,z} = 0$ and SC circles cannot intersect unless they coincide.

Now we will consider the conditions for avoiding shell crossings. These were worked out by Szekeres (1975b) and improved upon by Hellaby and Krasiński (2002); the account below is based on the latter.

For positive density, (20.74) shows that $\chi_1 \overset{\text{def}}{=} (M_{,z} - 3M\mathcal{E}_{,z}/\mathcal{E})$ and χ must have the same sign. Consider the case when both are positive (when $\chi_1 \leq 0$ and $\chi < 0$, the inequalities in all the following should be reversed).

The equality $\chi_1 = \chi = 0$ can happen for a particular (x,y) value if $M_{,z}/3M = \Phi_{,z}/\Phi$, but the latter can hold for all time only in the Friedmann limit (because then shear $= 0$ by (20.57)). So, $\chi_1 = \chi = 0$ can hold for all (x,y) only if $M_{,z} = 0$, $\mathcal{E}_{,z} = 0$ and $\Phi_{,z} = 0$ at some z value, which requires $M_{,z} = k_{,z} = t_{B,z} = S_{,z} = P_{,z} = Q_{,z} = 0$ at this z.

Let $\chi_1 > 0$. This must hold for all $\mathcal{E}_{,z}/\mathcal{E}$, including the maximum defined by (20.96), so

$$\frac{M_{,z}}{3M} > \left.\frac{\mathcal{E}_{,z}}{\mathcal{E}}\right|_{\max} = \frac{\sqrt{(S_{,z})^2 + (P_{,z})^2 + (Q_{,z})^2}}{S} \quad \forall\, z. \tag{20.127}$$

This is also sufficient and implies $M_{,z} \geq 0$ for all z.

By the same method as used above, $\chi > 0$ for all $\mathcal{E}_{,z}/\mathcal{E}$ is equivalent to

$$\frac{\Phi_{,z}}{\Phi} \geq \left.\frac{\mathcal{E}_{,z}}{\mathcal{E}}\right|_{\max} > 0 \quad \forall\, z. \tag{20.128}$$

Now the three types of evolution have to be considered separately.

Hyperbolic evolution, $k < 0$. For hyperbolic models we rewrite (20.71) in analogy to (18.113) $-$ (18.114) using (20.103):

$$\frac{\Phi_{,z}}{\Phi} = \frac{M_{,z}}{M}(1 - \Phi_3) + \frac{k_{,z}}{k}\left(\frac{3}{2}\Phi_3 - 1\right) - \frac{(-k)^{3/2}t_{B,z}}{M}\Phi_4, \tag{20.129}$$

where

$$\Phi_3 \overset{\text{def}}{=} \frac{\sinh\eta(\sinh\eta - \eta)}{(\cosh\eta - 1)^2}, \qquad \Phi_4 \overset{\text{def}}{=} \frac{\sinh\eta}{(\cosh\eta - 1)^2}. \tag{20.130}$$

When $\eta \to 0$, we have $\Phi_3 \to 2/3$ and $\Phi_4 \to +\infty$. Thus the term with Φ_4 dominates and goes to $\pm\infty$, its sign being determined by $[-(-k)^{3/2}t_{B,z}/M]$. Consequently, $\chi > 0$ implies

$$t_{B,z} < 0 \quad \forall\, z. \tag{20.131}$$

Similarly, where $\eta \to \infty$ we have $\Phi_3 \to 1$ and $\Phi_4 \to 0$, so $\Phi_{,z}/\Phi \to k_{,z}/(2k)$ and

$$\Phi_{,z}/\Phi - \mathcal{E}_{,z}/\mathcal{E} > 0 \quad \Rightarrow \quad k_{,z}/(2k) - \mathcal{E}_{,z}/\mathcal{E} > 0. \tag{20.132}$$

Following the above analysis of $M_{,z} - 3M\mathcal{E}_{,z}/\mathcal{E} \geq 0$ we obtain

$$k_{,z}/(2k) > \sqrt{(S_{,z})^2 + (P_{,z})^2 + (Q_{,z})^2}/S \quad \forall\, z, \tag{20.133}$$

which implies $k_{,z} < 0\ \forall\, z$ and is sufficient for (20.132).

Parabolic evolution, $k = 0$. When $k = 0$ we obtain from (20.104):

$$\frac{\Phi_{,z}}{\Phi} = \frac{M_{,z}}{3M} - \frac{2}{3}\frac{t_{B,z}}{t - t_B}.$$ (20.134)

Requiring that $\Phi_{,z}/\Phi - \mathcal{E}_{,z}/\mathcal{E} > 0$ at $t \to t_B$ and at $t \to \infty$ we obtain (20.131) and (20.127), respectively.

Elliptic evolution, $k > 0$. Here we rewrite (20.71) in analogy to (18.106) – (18.108) using (20.105):

$$\frac{\Phi_{,z}}{\Phi} = \frac{M_{,z}}{M}(1 - \Phi_1) + \frac{k_{,z}}{k}\left(\frac{3}{2}\Phi_1 - 1\right) - \frac{k^{3/2}t_{B,z}}{M}\Phi_2,$$ (20.135)

where

$$\Phi_1 \overset{\text{def}}{=} \frac{\sin\eta(\eta - \sin\eta)}{(1 - \cos\eta)^2}, \qquad \Phi_2 \overset{\text{def}}{=} \frac{\sin\eta}{(1 - \cos\eta)^2}.$$ (20.136)

When $\eta \to 0$, $\Phi_1 \to 2/3$ and $\Phi_2 \to +\infty$. Thus the term containing Φ_2 dominates and (20.131) is needed to keep $\Phi_{,z}/\Phi > 0$. At $\eta \to 2\pi$, $\Phi_1 \to -\infty$, $\Phi_2 \to -\infty$ and $\Phi_2/\Phi_1 \to 1/(2\pi)$. Consequently, $\Phi_{,z}/\Phi > 0$ now gives

$$\frac{2\pi M}{k^{3/2}}\left(\frac{M_{,z}}{M} - \frac{3k_{,z}}{2k}\right) + t_{B,z} > 0 \quad \forall\, z.$$ (20.137)

From (20.105), the crunch time is

$$t_C = t_B + 2\pi M/k^{3/2},$$ (20.138)

so (20.137) says that the crunch time must increase with z.

The proof that (20.131) and (20.137) are sufficient to keep

$$\Phi_{,z}/\Phi > M_{,z}/(3M) > 0$$ (20.139)

is given in Exercise 10 to Chapter 18. Then (20.127) implies $\Phi_{,z}/\Phi > \mathcal{E}_{,z}/\mathcal{E}$, which ensures no shell crossings.

20.3.5 Regular maxima and minima

Certain topologies imply extrema in Φ, i.e. $\Phi_{,z}(t, z_m) = 0$, $\forall\, t$. For example, closed spatial sections have a maximum areal radius, and wormholes have a minimum areal radius.

Suppose that $\varepsilon - k = 0$ in (20.53) at some $z = z_m$. To keep $|g_{rr}| < \infty$, we need at z_m

$$\chi \overset{\text{def}}{=} \Phi_{,z} - \Phi\mathcal{E}_{,z}/\mathcal{E} = 0,$$ (20.140)

so, from (20.74), $(M_{,z} - 3M\mathcal{E}_{,z}/\mathcal{E})_{z=z_m} = 0 \forall\, (t, x, y)$ to keep $\epsilon < \infty$. More specifically, along any given spatial slice away from the bang or crunch, we want

$$\chi/\sqrt{\varepsilon - k} \to L, \qquad\qquad 0 < L < \infty,$$ (20.141)
$$(M_{,z} - 3M\mathcal{E}_{,z}/\mathcal{E})/\chi \to N, \qquad\qquad 0 \le N < \infty.$$ (20.142)

In consequence of (20.91), (20.129) and (20.135), $\chi = 0$ implies

$$M_{,z} = k_{,z} = t_{B,z} = S_{,z} = P_{,z} = Q_{,z} = 0. \tag{20.143}$$

The limits (20.141) and (20.142) must hold good for all t and for all (x, y), so

$$\frac{M_{,z}}{\sqrt{\varepsilon - k}}, \qquad \frac{k_{,z}}{\sqrt{\varepsilon - k}}, \qquad \frac{t_{B,z}}{\sqrt{\varepsilon - k}}, \qquad \frac{\Phi_{,z}}{\sqrt{\varepsilon - k}}, \tag{20.144}$$

$$\frac{S_{,z}}{\sqrt{\varepsilon - k}}, \qquad \frac{P_{,z}}{\sqrt{\varepsilon - k}}, \qquad \frac{Q_{,z}}{\sqrt{\varepsilon - k}}, \qquad \frac{\mathcal{E}_{,z}}{\sqrt{\varepsilon - k}} \tag{20.145}$$

must all have finite limits at $z \to z_m$. Using de l'Hôpital's rule, each of the above limits can be expressed in the form

$$L_{(M,z)} \overset{\text{def}}{=} \lim_{k \to \varepsilon} \frac{M_{,z}}{\sqrt{\varepsilon - k}} = -\frac{2M_{,zz}\sqrt{\varepsilon - k}}{k_{,z}} \overset{\text{def}}{=} -\frac{2M_{,zz}}{L_{(k,z)}}. \tag{20.146}$$

20.3.6 The mass dipole

In the $\beta_{,z} \neq 0$ Szekeres solution, the distribution of mass over each single sphere of constant t and z has the form of a mass-dipole superposed with a monopole. This was first noted by Szekeres (1975b), then generalised and explained in detail by de Souza (1985). The presentation here is based on the latter reference, but somewhat modified.

The basic idea is to separate the expression for matter density, (20.74), into a spherically symmetric part ϵ_s, depending only on t and z, and a nonsymmetric $\Delta\epsilon$. Without additional requirements, this can be done in an infinite number of ways. De Souza invented a meaningful unique separation. To begin with, add and subtract $H(t, z)/\Phi^2$ on the right-hand side of (20.74), where $H(t, z)$ is an arbitrary function, obtaining

$$\epsilon = \epsilon_s(t, z) + \Delta\epsilon(t, z, x, y), \tag{20.147}$$

where

$$\kappa\epsilon_s = \frac{H}{\Phi^2}, \qquad \kappa\Delta\epsilon = \frac{\mathcal{E}\left(2M_{,z} - H\Phi_{,z}\right) - \mathcal{E}_{,z}\left(6M - H\Phi\right)}{\Phi^2\left(\mathcal{E}\Phi_{,z} - \Phi\mathcal{E}_{,z}\right)}. \tag{20.148}$$

Additional requirements on H will now make the splitting unique. Transform (x, y) to spherical polar coordinates on a sphere of radius 1 that is tangent to the (x, y) plane at $(x, y) = (0, 0)$. The transformation is (20.54) with $S = 1$ and $P = Q = 0$, thus

$$x = \cot(\vartheta/2)\cos\varphi, \qquad y = \cot(\vartheta/2)\sin\varphi. \tag{20.149}$$

For ease of calculation we will now rewrite $\mathcal{E}$ defined in (20.53) into a notation similar to that of (20.50).[7] Using (20.149) for x and y we obtain

$$\mathcal{E} = \mathcal{A}\cot^2(\vartheta/2) + 2\mathcal{B}_1\cot(\vartheta/2)\cos\varphi + 2\mathcal{B}_2\cot(\vartheta/2)\sin\varphi + \mathcal{C}. \tag{20.150}$$

---

[7] Conceptually, the notation used below is different from that of (20.50) – the functions $\mathcal{A}, \mathcal{B}_1, \mathcal{B}_2, \mathcal{C}$ are *defined* by (20.150).

We substitute this in (20.148) and consider the equation $\Delta\epsilon = 0$. We embed the sphere in a Euclidean 3-space, and express (ϑ, φ) through the Cartesian coordinates in the space

$$X = \sin\vartheta \cos\varphi, \qquad Y = \sin\vartheta \sin\varphi, \qquad Z = \cos\vartheta. \qquad (20.151)$$

Using (20.150) and (20.151) in $\Delta\epsilon = 0$ we obtain

$$(6M - H\Phi)\left[\mathcal{A}_{,z}(1 + Z) + 2\mathcal{B}_{1,z}X + 2\mathcal{B}_{2,z}Y + \mathcal{C}_{,z}(1 - Z)\right] =$$
$$(2M_{,z} - H\Phi_{,z})\left[\mathcal{A}(1 + Z) + 2\mathcal{B}_1 X + 2\mathcal{B}_2 Y + \mathcal{C}(1 - Z)\right]. \qquad (20.152)$$

Now we require that the surface of $\Delta\epsilon = 0$ passes through the centre of the sphere, i.e. that (20.152) is fulfilled at $X = Y = Z = 0$, thus

$$(6M - H\Phi)(\mathcal{A}_{,z} + \mathcal{C}_{,z}) = (2M_{,z} - H\Phi_{,z})(\mathcal{A} + \mathcal{C}). \qquad (20.153)$$

Since $(\mathcal{A}, \mathcal{C}, M, \Phi)$ depend only on t and z, this can be solved for H:

$$H = \frac{2M_{,z}(\mathcal{A} + \mathcal{C}) - 6M(\mathcal{A} + \mathcal{C})_{,z}}{\Phi_{,z}(\mathcal{A} + \mathcal{C}) - \Phi(\mathcal{A} + \mathcal{C})_{,z}}. \qquad (20.154)$$

This solution makes sense except when $[(\mathcal{A} + \mathcal{C})/\Phi]_{,z} \equiv 0$. But then shear is zero (see (20.57)) and the Szekeres model degenerates into the Friedmann model. Hence, (20.154) applies as long as the Szekeres metric is inhomogeneous.

With H given by (20.154), Eqs. (20.148) become

$$\kappa\epsilon_s = \frac{2M_{,z}(\mathcal{A} + \mathcal{C}) - 6M(\mathcal{A} + \mathcal{C})_{,z}}{\Phi^2\left[\Phi_{,z}(\mathcal{A} + \mathcal{C}) - \Phi(\mathcal{A} + \mathcal{C})_{,z}\right]}, \qquad (20.155)$$

$$\kappa\Delta\epsilon = \frac{(\mathcal{A} + \mathcal{C})_{,z} - (\mathcal{A} + \mathcal{C})\mathcal{E}_{,z}/\mathcal{E}}{\Phi_{,z} - \Phi\mathcal{E}_{,z}/\mathcal{E}} \times \frac{6M\Phi_{,z} - 2M_{,z}\Phi}{\Phi^2\left[\Phi_{,z}(\mathcal{A} + \mathcal{C}) - \Phi(\mathcal{A} + \mathcal{C})_{,z}\right]}. \qquad (20.156)$$

Now, $\Delta\epsilon = 0$ has two solutions:

$$(\mathcal{A} + \mathcal{C})_{,z} - (\mathcal{A} + \mathcal{C})\mathcal{E}_{,z}/\mathcal{E} = 0 \qquad (20.157)$$

and $(2M/\Phi^3)_{,z} = 0$. The second one defines a hypersurface that depends on t, that is, it is not comoving except when $(2M/\Phi^3)_{,z} \equiv 0$, but then the Friedmann model results again. The first hypersurface, call it H_1, has its equation independent of t – it is a world-sheet of a comoving surface, and $\Delta\epsilon$ changes sign when H_1 is crossed. Hence, $\Delta\epsilon$ is a dipole-like contribution to matter density. The separation (20.155) – (20.156) is global, but the orientation of the dipole axis is different on every sphere of constant t and z. Illustrations showing the positions of the dipole axis in explicit examples of the Szekeres models can be viewed in the paper by Krasiński (2018a), .

We will now verify that H_1 intersects every sphere of constant t and z along a circle, unless $P_{,z} = Q_{,z} = S_{,z} = 0$ $(= \mathcal{A}_{,z} = \mathcal{C}_{,z})$, in which case the dipole component of density is zero. The intersection of H_1 with a sphere of constant t and z is a circle parallel to the great circle $\mathcal{E}_{,z} = 0$, as follows by comparing (20.157) with (20.92). It will coincide with the $\mathcal{E}_{,z} = 0$ circle at those points where $\mathcal{A}_{,z} + \mathcal{C}_{,z} = 0$ (if they exist).

The solution of (20.157) will exist when

$$\left(\frac{\mathcal{E}_{,z}}{\mathcal{E}}\right)_{\min} \leq \frac{\mathcal{A}_{,z} + \mathcal{C}_{,z}}{\mathcal{A} + \mathcal{C}} \leq \left(\frac{\mathcal{E}_{,z}}{\mathcal{E}}\right)_{\max}. \qquad (20.158)$$

Since $(\mathcal{E}_{,z}/\mathcal{E})_{\min} = -(\mathcal{E}_{,z}/\mathcal{E})_{\max}$ (see (20.96)), Eq. (20.158) is equivalent to

$$\frac{(\mathcal{A}_{,z}+\mathcal{C}_{,z})^2}{(\mathcal{A}+\mathcal{C})^2} \leq \left(\frac{\mathcal{E}_{,z}}{\mathcal{E}}\right)^2_{\text{extreme}} = \frac{P_{,z}^{\,2}+Q_{,z}^{\,2}+S_{,z}^{\,2}}{S^2}. \tag{20.159}$$

We have

$$\mathcal{A}+\mathcal{C} = \left(S^2+T^2\right)/(2S), \qquad \text{where} \quad T^2 \stackrel{\text{def}}{=} P^2+Q^2+1,$$

$$\mathcal{A}_{,z}+\mathcal{C}_{,z} = \frac{S_{,z}\left(S^2-T^2\right)}{2S^2} + \frac{PP_{,z}+QQ_{,z}}{S}. \tag{20.160}$$

Substituted in (20.159), this leads to

$$4S^2T^2S_{,z}^{\,2} - 4SS_{,z}\left(PP_{,z}+QQ_{,z}\right)\left(S^2-T^2\right)$$
$$+ \left(P_{,z}^{\,2}+Q_{,z}^{\,2}\right)\left(S^2+T^2\right)^2 - 4S^2\left(PP_{,z}+QQ_{,z}\right)^2 \geq 0. \tag{20.161}$$

The discriminant of this with respect to $S_{,z}$ is

$$\Delta = -16S^2\left(S^2+T^2\right)^2 \left[\left(PQ_{,z}-QP_{,z}\right)^2 + P_{,z}^{\,2} + Q_{,z}^{\,2}\right], \tag{20.162}$$

and is negative unless $P_{,z} = Q_{,z} = 0$. Thus, with $(P_{,z},Q_{,z}) \neq (0,0)$, the left-hand side of (20.161) is strictly positive. Even when $P_{,z} = Q_{,z} = 0$, it is still strictly positive unless $S_{,z} = 0$. However, $P_{,z} = Q_{,z} = S_{,z} = 0$ implies $\mathcal{A}_{,z} = \mathcal{C}_{,z} = 0$ and $\mathcal{E}_{,z} = 0$ on the whole sphere, and then $\Delta\epsilon = 0$; i.e. on such a sphere the density is spherically symmetric. Hence, apart from the spherically symmetric subcase, (20.158) is fulfilled, with sharp inequalities in both places. This means that the $\Delta\epsilon = 0$ hypersurface H_1 intersects every sphere of constant t and z along a circle parallel to the $\mathcal{E}_{,z} = 0$ circle (see the remark after (20.92)). The fact that both inequalities in (20.158) are sharp means that H_1 cannot be tangent to the sphere at a single point, the intersection must be a circle.

20.3.7 * The absolute apparent horizon

This subsection and the next two are a composite of Subsection 19.7.6 from the first edition of this book and of the paper by Krasiński and Bolejko (KB12, 2012a). The old version had to be modified because, following Hellaby and Krasiński (HK02, 2002), it applied the term 'apparent horizon' to a construction different from the commonly accepted one. Following KB12, we now try to put the terminology in order.

The notion of apparent horizon (AH) in a collapsing object can be carried over from the L–T to the quasispherical Szekeres models in two ways: 1. As the boundary of the region, in which every bundle of null geodesics has negative expansion scalar (following Szekeres, 1975b). 2. As the locus, at which null lines that are as nearly radial as possible are turned towards decreasing areal radius Φ (this was the approach of HK02). These lines are in general nongeodesic. The name 'absolute apparent horizon' (AAH) was proposed for this locus by KB12. The AH and AAH coincide in the L–T models. In the quasispherical Szekeres models, the AH is different from (but not disjoint with) the AAH. Properties of the AAH and the relations between the AAH and the AH will be discussed here and illustrated using an explicit Szekeres metric, which is a simple generalisation of the L–T model of Sec. 18.9. It turns out that some observers who are already within the AH are, for some time, not yet within the AAH. Nevertheless, no light signal can be sent through the AH from the inside.

With $\varepsilon = +1$ in (20.53), a null vector field $k^\alpha = dx^\alpha/dt$ obeys

$$\frac{\chi^2}{1-k} \left(\frac{dz}{dt}\right)^2 = 1 - \frac{\Phi^2}{\mathcal{E}^2} \left[\left(\frac{dx}{dt}\right)^2 + \left(\frac{dy}{dt}\right)^2\right], \tag{20.163}$$

where χ was defined in (20.140). The dz/dt is maximum on rays with $k^x = 0 = k^y$. We will call them 'nearly radial rays' (NRR), in general they are not geodesic.[8] The equation $dt/dz|_n = j\chi/\sqrt{1-k}$, $j = \pm 1$, when solved, would give $t = t_n(z)$ along the NRR. The areal radius along a NRR is $\Phi_n = \Phi(t_n(z), z)$. These rays stop proceeding towards larger Φ when the total derivative $D\Phi_n/dz$ defined by

$$\frac{D\Phi_n}{dz} = \Phi_{,t}\ (t_n)_{,z} + \Phi_{,z} = \ell j \chi \frac{\sqrt{2M/\Phi - k}}{\sqrt{1-k}} + \Phi_{,z}, \qquad \ell = \pm 1. \tag{20.164}$$

becomes zero; in the above $\Phi_{,t} = \ell\sqrt{2M/\Phi - k}$ was used. Geodesic light rays going off with $dx/dt = dy/dt = 0$ will not remain in the surface of constant x and y. Since the NRRs seem to be the fastest possible escape routes, we will call the locus of $D\Phi_n/dz = 0$ **absolute apparent horizon** (AAH) – 'absolute' because intuition suggests that along a NRR the light signal should escape farther from the origin than along any other path. An example given further on (see Fig. 20.4 and the text under (20.189)) will show that this is only partly true. Along some directions, the NRRs can indeed proceed towards larger values of Φ than geodesic rays. But along other directions the reverse happens: the NRRs are redirected to decreasing values of Φ where geodesic rays can still proceed outwards.

Assuming that Φ is increasing with z on constant-t slices ($\Phi_{,z} > 0$), for a solution of $D\Phi_n/dz = 0$ to exist we must require $\ell j = -1$, i.e. either $\{j = +1, \ell = -1\}$ (outgoing rays in a collapsing phase $\Longleftrightarrow$ future AAH), or $\{j = -1, \ell = +1\}$ (incoming rays in an expanding phase $\Longleftrightarrow$ past AAH). By 'outgoing' we mean moving away from $\Phi = 0$. A ray passing through $\Phi = 0$ would change from incoming to outgoing, and, since $\Phi_{,z}$ flips sign there, j would also have to flip there.

Now define

$$\mathcal{D} \overset{\text{def}}{=} \sqrt{\frac{1-k}{2M/\Phi - k}} - 1. \tag{20.165}$$

Then $(\mathcal{D} > 0) \Longleftrightarrow (\Phi > 2M)$. By (20.164) the equation $D\Phi_n/dz = 0$ becomes

$$\Phi\mathcal{E}_{,z} + \mathcal{E}\mathcal{D}\Phi_{,z} = 0, \tag{20.166}$$

which, explicitly, is

$$(S_{,z}/S - \mathcal{D}\Phi_{,z}/\Phi)\left[(x-P)^2 + (y-Q)^2\right]$$
$$+2\left[(x-P)P_{,z} + (y-Q)Q_{,z}\right] - S^2 \left(\frac{S_{,z}}{S} + \mathcal{D}\frac{\Phi_{,z}}{\Phi}\right) = 0. \tag{20.167}$$

---

[8] Radial light rays cannot be defined in general Szekeres spacetimes (Nolan and Debnath, 2007). A nongeodesic light ray is guided by mirrors or optical fibres.

The discriminant of this with respect to $(x - P)$ is

$$\Delta_x = 4P_{,z}{}^2 - 4\left(\frac{S_{,z}}{S} - \mathcal{D}\frac{\Phi_{,z}}{\Phi}\right)\left[\left(\frac{S_{,z}}{S} - \mathcal{D}\frac{\Phi_{,z}}{\Phi}\right)(y - Q)^2\right.$$
$$\left. + 2(y - Q)Q_{,z} - S^2\left(\frac{S_{,z}}{S} + \mathcal{D}\frac{\Phi_{,z}}{\Phi}\right)\right]. \tag{20.168}$$

The discriminant of the above with respect to $(y - Q)$ is

$$\Delta_y = 64\left(\frac{S_{,z}}{S} - \mathcal{D}\frac{\Phi_{,z}}{\Phi}\right)^2\left(P_{,z}{}^2 + Q_{,z}{}^2 + S_{,z}{}^2 - S^2\mathcal{D}^2\frac{\Phi_{,z}{}^2}{\Phi^2}\right). \tag{20.169}$$

If $\Delta_y < 0$ for all t, z, then $\Delta_x < 0$ for all y and there is no x obeying (20.166), i.e. the AAH does not intersect this particular surface of constant (t, z).

If $\Delta_y = 0$, then $\Delta_x < 0$ for all y except one value $y = y_0$, at which $\Delta_x = 0$. At the corresponding $x = x_0$, (20.166) has a solution, so the intersection of the AAH with this one constant-(t, z) surface is a single point. Note that the situation when the AAH touches the whole 3-dimensional $t = \text{constant}$ hypersurface at a certain value of t is exceptional; this requires, from (20.166), that $P_{,z} = Q_{,z} = S_{,z} = \Phi_{,z} = 0$ at this t. The first three functions being zero mean just spherical symmetry, but the fourth one defines a special location. These equations hold in the Datt–Ruban solution, see Section 19.5.

If $\Delta_y > 0$, then $\Delta_x > 0$ for every y such that $y_1 < y < y_2$, where

$$y_{1,2} = Q + \frac{-Q_{,z} \pm \sqrt{\delta}}{S_{,z}/S - \mathcal{D}\Phi_{,z}/\Phi},$$

$$\delta \stackrel{\text{def}}{=} P_{,z}{}^2 + Q_{,z}{}^2 + S_{,z}{}^2 - S^2\mathcal{D}^2\frac{\Phi_{,z}{}^2}{\Phi^2} \tag{20.170}$$

and then a solution of (20.166) exists given by

$$x_{1,2} = P + \frac{-P_{,z} \pm \sqrt{-\left[\left(\frac{S_{,z}}{S} - \mathcal{D}\frac{\Phi_{,z}}{\Phi}\right)(y - Q) + Q_{,z}\right]^2 + \delta}}{S_{,z}/S - \mathcal{D}\Phi_{,z}/\Phi}. \tag{20.171}$$

Except for the special case when $S_{,z}/S = \mathcal{D}\Phi_{,z}/\Phi$, these values lie on the circle in the (x, y) coordinate plane, with the centre at

$$(x_{AAH}, y_{AAH}) = \left(P - \frac{P_{,z}}{S_{,z}/S - \mathcal{D}\Phi_{,z}/\Phi}, Q - \frac{Q_{,z}}{S_{,z}/S - \mathcal{D}\Phi_{,z}/\Phi}\right), \tag{20.172}$$

and with the radius $L_{AAH} = \sqrt{\delta}/(S_{,z}/S - \mathcal{D}\Phi_{,z}/\Phi)$. The special case $S_{,z}/S = \mathcal{D}\Phi_{,z}/\Phi$ (when the locus of AAH in the (x, y) plane is a straight line) is an artefact of the stereographic projection because this straight line is an image of a circle on the sphere.

In summary, the intersection of AAH with the (x, y) coordinate plane is
(1) nonexistent when $\Phi_{,z}{}^2/\Phi^2 > \Psi^2/\mathcal{D}^2$ (this is the Ψ defined in (20.123));
(2) a single point when $\Phi_{,z}{}^2/\Phi^2 = \Psi^2/\mathcal{D}^2$;
(3) a circle or a straight line when $\Phi_{,z}{}^2/\Phi^2 < \Psi^2/\mathcal{D}^2$.

The condition $\Phi_{,z}{}^2/\Phi^2 < \Psi^2/\mathcal{D}^2$ is consistent with (20.123) when $|\mathcal{D}| < 1$. Then, in case (2) a shell crossing is automatically excluded by (20.123).

From (20.166) and from $\Phi > 0$, $\mathcal{E} > 0$ and $\Phi_{,z} > 0$ we have

$$\text{sign } \mathcal{D} = -\text{sign } \mathcal{E}_{,z}, \qquad (\mathcal{D} = 0) \Longleftrightarrow (\mathcal{E}_{,z} = 0). \qquad (20.173)$$

But $\mathcal{D} > 0$ and $\mathcal{D} < 0$ define regions independent of x and y. Hence, $\mathcal{E}_{,z}$ has the opposite sign to $\mathcal{D}$ on the whole of AAH.

By the argument that was used following (20.92) the two circles, $\mathcal{E}_{,z} = 0$ and the one where the AAH intersects the sphere of constant (t, z), cannot intersect unless they coincide. Indeed, these circles lie in parallel planes: the line on the $(t, z) = \text{constant}$ sphere defined by (20.166) has the property $\mathcal{E}_{,z} / \mathcal{E} = -\mathcal{D}\Phi_{,z} / \Phi = \text{constant}$, so it must be a circle in a plane parallel to the $\mathcal{E}_{,z} = 0$ great circle.

On the (x, y) coordinate plane, when the $\mathcal{E}_{,z} = 0$ and AAH circles are disjoint, they may be either one inside the other or each one outside the other. However, when projected on the sphere, these two situations are topologically equivalent: depending on the position of the pole of projection the same two circles may project on the plane either as one circle inside the other or as two separate circles; see Figs. 20.2 and 20.3.

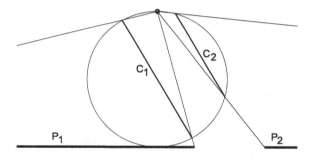

Fig. 20.2 The projections of circles C_1 and C_2 on a sphere (seen here edge on) to the (x, y) coordinate plane (seen as the horizontal line) will be the circles P_1 and P_2 that are outside each other (only parts of P_1 and P_2 are visible). Circle C_1 is the $\mathcal{E}_{,z} = 0$ set, C_2 is the AAH circle.

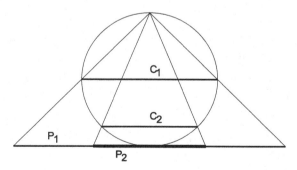

Fig. 20.3 The same circles as in Fig. 20.2 projected on a plane from a different pole will be one inside the other. The transition from the situation of Fig. 20.2 to the one above is continuous and occurs when the sphere is rotated, but the pole and the plane are not moved. Then, when a clockwise rotation is applied to Fig. 20.2, C_1 will pass through the pole at one value $\varphi = \varphi_0$ of the rotation angle. Its image on the plane acquires a larger and larger radius when approaching φ_0, until it becomes a straight line at $\varphi = \varphi_0$. When φ increases further, the straight line bends in the opposite direction and surrounds P_2.

20.3.8 * The apparent horizon and its relation to the AAH

At $\Phi = 2M$, Eq. (20.164) becomes

$$\frac{D\Phi_n}{dz} = (1 + \ell j)\Phi_{,z} - \ell j\Phi\mathcal{E}_{,z} /\mathcal{E}, \tag{20.174}$$

so $\Phi = 2M$ does not coincide with the AAH ($D\Phi_n/dz = 0$) except where $\mathcal{E}_{,z} = 0$.

The construction of the AAH was done with the tacit assumption that the 'nearly radial rays' originated from a common source. The whole reasoning makes sense only when the common source is at the origin. Consequently, we will now compare the AAH with the ordinary AH around the origin for which Szekeres (1975b) found that it is located at $\Phi = 2M$. The method he used was nearly the same as the one we used in Sec. 18.8 (indeed, *we* copied *his* method), so we do not quote the derivation.

For a nearly radial ray, defined under (20.163), we have

$$\left.\frac{dt}{dz}\right|_n = \frac{j}{\sqrt{1-k}}\left(\Phi_{,z} - \frac{\Phi\mathcal{E}_{,z}}{\mathcal{E}}\right), \qquad j = \pm 1, \tag{20.175}$$

where $j = +1$ for outgoing rays, and $j = -1$ for ingoing rays. The AAH is where (20.164) holds. We consider it in the collapse phase ($\ell = -1$), so it is defined by the outgoing rays ($j = +1$). Then, (20.164) becomes

$$\Phi_{,z}\left(\frac{\sqrt{1-k}}{\sqrt{2M/\Phi - k}} - 1\right) + \Phi\frac{\mathcal{E}_{,z}}{\mathcal{E}} = 0. \tag{20.176}$$

For $\Phi_{,z}$ we will use (20.135) – (20.136) rewritten as follows:

$$\frac{\Phi_{,z}}{\Phi} = \left(\frac{M_{,z}}{M} - \frac{k_{,z}}{k}\right) + \left(\frac{3}{2}\frac{k_{,z}}{k} - \frac{M_{,z}}{M}\right)\frac{\sin\eta(\eta - \sin\eta)}{(1 - \cos\eta)^2} - \frac{k^{3/2}}{M}t_{B,z}\frac{\sin\eta}{(1 - \cos\eta)^2}. \tag{20.177}$$

From (20.101) and (20.105) with $\sigma = +1$ and $\pi \le \eta \le 2\pi$, where $\Phi_{,t} < 0$, we have

$$\sqrt{2M/\Phi - k} = -\sqrt{k}\frac{\sin\eta}{1 - \cos\eta}. \tag{20.178}$$

We substitute (20.177) and (20.178) in (20.176), then rearrange the resulting equation for later convenience, and obtain:

$$\begin{aligned}
\Psi(\eta) \stackrel{\text{def}}{=} &\left[\left(\frac{M_{,z}}{M} - \frac{k_{,z}}{k}\right)(1 - \cos\eta)^{3/2} + \left(\frac{3}{2}\frac{k_{,z}}{k} - \frac{M_{,z}}{M}\right)\frac{\sin\eta(\eta - \sin\eta)}{\sqrt{1 - \cos\eta}}\right. \\
&\left. - \frac{k^{3/2}}{M}t_{B,z}\frac{\sin\eta}{\sqrt{1 - \cos\eta}}\right] \times \left[\sqrt{1-k}\sqrt{1 - \cos\eta} + \frac{\sqrt{k}\sin\eta}{\sqrt{1 - \cos\eta}}\right] \\
&- \sqrt{k}\sin\eta(1 - \cos\eta)\frac{\mathcal{E}_{,z}}{\mathcal{E}} = 0.
\end{aligned} \tag{20.179}$$

We will now show that, given $M(z)$, $k(z)$ and $t_B(z)$, Eq. (20.179) determines η at the AAH as a function of (z, x, y). With η at the AAH found, we calculate from (20.105):

$$t(M, x, y)_{\text{AAH}} = \frac{M}{k^{3/2}}(\eta - \sin\eta)_{\text{AAH}} + t_B. \tag{20.180}$$

We assume that shell crossings are absent, so (20.131), (20.137) and their implication

$$M_{,z}/M - k_{,z}/k > 0 \tag{20.181}$$

are fulfilled. We observe that

$$\lim_{\eta \to 2\pi} \frac{\sin \eta}{\sqrt{1 - \cos \eta}} = -\sqrt{2} = -\lim_{\eta \to 0} \frac{\sin \eta}{\sqrt{1 - \cos \eta}} \tag{20.182}$$

and verify using (20.179) that

$$\lim_{\eta \to \pi} \Psi(\eta) = 4\sqrt{1 - k} \left(\frac{M_{,z}}{M} - \frac{k_{,z}}{k} \right) > 0, \tag{20.183}$$

being positive in consequence of (20.181), and

$$\lim_{\eta \to 2\pi} \Psi(\eta) = 2\sqrt{-2E} \left[2\pi \left(\frac{3}{2} \frac{E_{,z}}{E} - \frac{M_{,z}}{M} \right) - \frac{(-2E)^{3/2}}{M} t_{B,z} \right] < 0, \tag{20.184}$$

being negative in consequence of (20.137). Thus, at every $M > 0$ there exists an $\eta_0 \in (\pi, 2\pi)$ at which $\Psi(\eta_0) = 0$, and it is unique (see Sec. 24.5). In passing, we have proved that each particle at $M > 0$ in a recollapsing quasispherical Szekeres model must cross the AAH before it hits the Big Crunch at $\eta = 2\pi$.

As an illustration, we take the model analogous to the L–T model of Sec. 18.9,[9] in which

$$t_B(M) = -bM^2 + t_{B0}, \tag{20.185}$$
$$t_C(M) = aM^3 + T_0 + t_{B0}, \tag{20.186}$$

where M is used as the z coordinate, $t_B(M)$ is the bang time, $t_C(M)$ is the crunch time, a, b, t_{B0} and T_0 are arbitrary constants; t_{B0} is the time-coordinate of the central point of the Big Bang and T_0 is the time between the Big Bang and Big Crunch measured along the central line $M = 0$. Then, from (20.105), since $\eta = 2\pi$ at $t = t_C$:

$$k(M) = \frac{(2\pi M)^{2/3}}{(aM^3 + bM^2 + T_0)^{2/3}}. \tag{20.187}$$

As shown in Sec. 20.3.2, the extreme values of $\mathcal{E}_{,z}/\mathcal{E}$ are

$$D_e \stackrel{\text{def}}{=} \left. \frac{\mathcal{E}_{,z}}{\mathcal{E}} \right|_{\text{extreme}} = \pm \frac{\sqrt{S_{,z}^2 + P_{,z}^2 + Q_{,z}^2}}{S}. \tag{20.188}$$

In choosing (P, Q, S) precaution must be taken not to make D_e too large. If it is too large, then either the numerator or the denominator of (20.74) becomes negative in some region of space, thus rendering the energy density negative there (and infinite where the denominator is zero). Physically, this means that in those regions the dipole component (20.156) dominates over the monopole. We thus first choose such a value of D_e that will make the difference between the graphs of AH and of AAH visible at the scale of a figure, and then we choose such P, Q and S that will imply the chosen value of D_e. To maximise

---

[9] The values of a, b, T_0 and t_{B0} used here are different from those in Sec. 18.9. This model is meant to be an illustration to various geometrical possibilities; it is not supposed to describe any real object.

D_e, at least one of the derivatives $P_{,z}$, $Q_{,z}$, $S_{,z}$ has to be large. Experiments showed that the following choice will yield the desired result:

$$a = 0.1, \qquad b = 5000, \qquad T_0 = 12.5, \qquad t_{B0} = 0,$$
$$S = M^{0.29}, \qquad P = 0.5M^{0.29}, \qquad Q = 0. \tag{20.189}$$

A cross-section of the resulting AAH is shown in Fig. 20.4. (This figure is an extension of Fig. 18.1 – it shows the graphs extended to the opposite side of the $M = 0$ axis.) It shows $t(M)$ on the AAH for two points in the (x, y) surface: the one where the D_e given in (20.188) is maximum (positive) and where it is minimum (negative). These two curves are compared with the ordinary AH and with the crunch time function $t_C(M)$. See Sec. 24.6 for the proof that all four curves indeed have a common origin at $M = 0$.

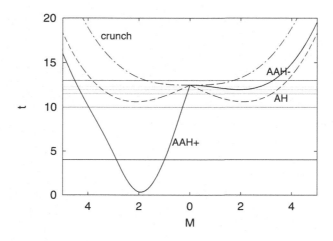

Fig. 20.4 A comparison of the future absolute apparent horizon (AAH) with the ordinary future apparent horizon (AH) in the model defined by (20.185) and (20.186). Curve AAH− is the AAH along the direction where the contribution from $\mathcal{E}_{,z}/\mathcal{E}$ is maximum and curve AAH+ is the AAH along the direction where this contribution is minimum. The dashed-dotted line represents the Big Crunch singularity. Horizontal solid lines show the instants for which the graphs in Figs. 20.5, 20.6 and 20.7 are drawn, these are: $t = 4.0, 10.0, 11.5, 11.9772, 12.3$ and 13.0.

Figure 20.4 shows that the contribution from $\mathcal{E}_{,z}/\mathcal{E}$ can either enlarge or diminish the region where the accelerated rays are forced towards the Big Crunch, depending on the direction. In the direction where this contribution is maximum ($\mathcal{E}_{,z}/\mathcal{E} > 0$ – curve AAH−), the AAH appears later than the ordinary AH, and the term $\mathcal{E}_{,z}/\mathcal{E}$ causes that the accelerating ray can still proceed towards increasing Φ in a region where a geodesic bundle already converges. In the direction where D_e is minimum ($\mathcal{E}_{,z}/\mathcal{E} < 0$ – curve AAH+), the AAH appears earlier than the AH, and the term $\mathcal{E}_{,z}/\mathcal{E}$ causes that the accelerating ray is turned inward where a geodesic bundle is still diverging.

Note what this means physically. When the AAH has a smaller radius than the AH, the observer who has already fallen into the AH still has a chance to send a message, using nongeodesic rays, to observers occupying loci with larger Φ. However, the nongeodesic ray has no chance to escape from inside the AH and will be turned towards the Big Crunch as well – see Subsection 20.3.9. Even this is possible only in some of the directions; in other directions the AAH is outside the AH and no ray within the AH, geodesic or not, can proceed towards larger Φ (Fig. 20.6 illustrates this point, too.) Intuitively, one may explain this as follows: in a direction in which the distance (as measured by z) between the constant-Φ shells is small, the nearly radial nongeodesic route of escape is advantageous, but in a direction where the distance is large, the nearly radial route is not advantageous – the ray will get to larger Φ sooner by veering off the nearly radial direction.

The coordinates (x, y) are not very intuitive. To better visualize the AAH let us first use the stereographic projection (20.54) to transform (x, y) to (ϑ, φ), and then map the AAH into an abstract Euclidean space with the coordinates (ξ, ψ, ζ) defined as follows

$$\xi = M_{AAH}(\vartheta, \varphi) \sin \vartheta \cos \varphi, \qquad \psi = M_{AAH}(\vartheta, \varphi) \sin \vartheta \sin \varphi,$$
$$\zeta = M_{AAH}(\vartheta, \varphi) \cos \vartheta. \tag{20.190}$$

We now use these coordinates to present the shape of the AAH. As seen from (20.91), when $Q_{,z} = 0$ the extreme values of $\mathcal{E}_{,z}/\mathcal{E}$ with respect to φ are at $\varphi = 0$ and $\varphi = \pi$, which, as follows from (20.190), implies $\psi = 0$. Figure 20.5 presents the intersection of the AAH with the plane $(\xi, \psi = 0, \zeta)$ at six different time instants.

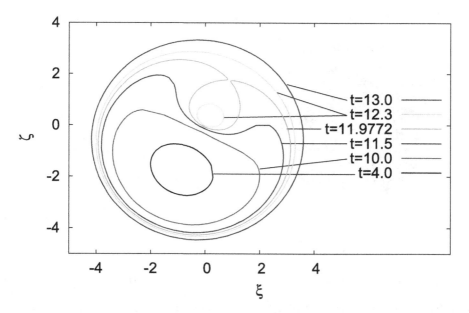

Fig. 20.5 The AAH at six different time instants (shown in Fig. 20.4 as horizontal lines). The curves shown are located at different times in the plane $(\xi, \psi = 0, \zeta)$. This figure can be imagined as a view of Fig. 20.4 by an observer sitting high on the t-axis and looking down; with one spatial dimension added. The value of M_{AAH} at (ξ_0, ζ_0) is the distance between (ξ_0, ζ_0) and the origin $\Phi = 0$ (which is at $(\xi, \zeta) = (0, 0)$ inside the smallest contour in the upper right area).

The origin, where $\Phi = 0$ ($\xi = 0$, $\zeta = 0$), is inside the smallest contour in Fig. 20.5. The AAH+ first appears off the origin (inside the closed curve in the lower left part of the figure). At this instant, most rays will miss it. Then it increases in diameter and encroaches on the origin. At the instant corresponding to the lowest point of AAH− in Fig. 20.4, the cross-section is still connected, but consists of two tangent rings, one inside the other (see the third curve from outside in Fig. 20.5, which presents the AAH just moments before this instant, $t = 11.9772$). The point of tangency lies at the minimum of AAH−. From that moment on, the cross-section splits into two disjoint contours, the smaller of which becomes progressively smaller with increasing t, and shrinks to a point at the instant of the minimum of the Big Crunch (inside the smallest ring in Fig. 20.5).

For better readability, Fig. 20.5 does not show the cross-sections of the ordinary AH. It first appears shortly after $t = 10.0$ and, at the moment of first appearance, would show in Fig. 20.5 as a single circle with the centre at $(\xi, \zeta) = (0, 0)$ and radius slightly larger than $M = 2$. From this instant on, the cross-section of the AH splits into two concentric circles. The smaller circle has its radius decreasing as t increases, and shrinks to a point at $t = T_0 + t_{B0} = 12.5$ (the smaller contour of the AAH shrinks to a point at the same instant). The larger circle of the AH keeps increasing, and intersects the larger contour of the AAH (which is not a circle) at two points at every instant.

Three-dimensional surface-plots of the AAH and AH at $t = 11.5$ and $t = 13$ are shown in Figs. 20.6 and 20.7. At $t = 11.5$ the Big Crunch singularity has not yet appeared, the

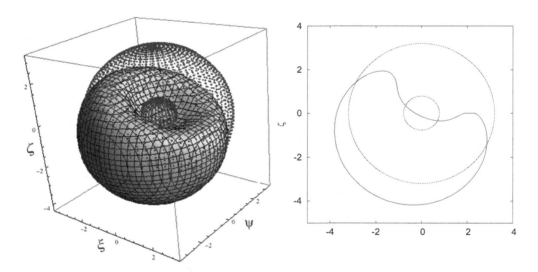

Fig. 20.6 **Left**: The M-coordinate on the AAH (solid surface) and AH (dotted surface) represented as a function of ϑ and φ in the space $t = 11.5$. The value of $M(\vartheta, \varphi)$ is the distance of a point on the surface shown from the point $(0, 0, 0)$, in the direction specified by (ϑ, φ). The axes in this picture are in an abstract Euclidean space with coordinates (ξ, ψ, ζ), see (20.190). The (M, ϑ, φ) are spherical polar coordinates in this space. The AH consists of two disjoint spheres – one with radius $M_{AH} = 0.783$ and the other with $M_{AH} = 3.199$. The origin $(\xi = 0, \psi = 0, \zeta = 0)$ is inside the smaller AH surface. **Right**: Intersections of the AAH (solid line), and the inner and outter AH (dotted lines) with the plane $(\xi, \psi = 0, \zeta)$ (analogous to Fig. 20.5).

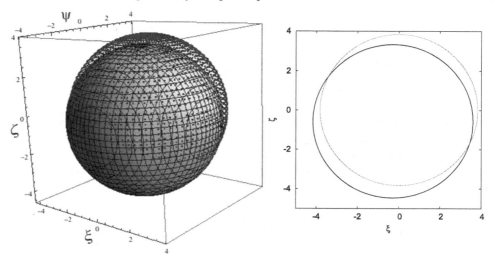

Fig. 20.7 The analogue of Fig. 20.6 at the later instant $t = 13.0$, when the singularity already exists at $\Phi = 0$. The small-radius components of the AH and AAH no longer exist.

AAH is still a connected surface and the AH consists of two disjoint parts – one has the radius $M_{AH} = 0.783$ and the other has $M_{AH} = 3.199$. They are the dotted surfaces in Fig. 20.6, and, as seen, each one is partly inside and partly outside the AAH (shown as the solid surface). Each part of the AH is a sphere in the (ξ, ψ, ζ) coordinates, while the AAH has the shape of a ping-pong ball depressed on one side. At $t = 13.0$ the singularity already exists at $\Phi = 0$. Each of AH and AAH is a single surface, which surrounds the singularity. As before, the AAH is partly outside and partly inside the AH.

Between $t = 11.5$ and $t = 13.0$ there is a period when each of AH and AAH is split into two disjoint parts. We do not provide an illustration for this configuration because it would be unreadable. One can imagine it as the object from Fig. 20.7 that contains a small copy of itself inside. The object inside does not intersect with the large one.

Figures 20.5–20.7 demonstrate that the AH and AAH do not in fact reveal the whole truth about the future fate of the light rays. There exists a region between the $M = 0$ axis and the AAH− in Fig. 20.4, in which future-directed rays are only formally not yet in the black hole. They are still able to proceed outwards, but only for a short while. They have no way to avoid intersecting the AAH and the AH in near future, and so they are doomed to hit the Big Crunch.

20.3.9 * Which is the true horizon – the AH or the AAH?

To get insight into this question we recall (see the end of Sec. 20.2.2) that a quasispherical Szekeres metric can be matched to the Schwarzschild metric across a $z =$ constant hypersurface (Bonnor, 1976). For the matching, the Schwarzschild metric must be transformed to the Lemaître–Novikov coordinates of Sec. 14.12. Then it has the form

$$ds^2 = dt^2 - \frac{R_{,r}^{\ 2}}{1 + 2E(r)} dr^2 - R^2(t, r) \left(d\vartheta^2 + \sin^2 \vartheta d\varphi^2 \right), \tag{20.191}$$

where $R(t, r)$ is determined by the equation

$$R_{,t}^{\ 2} = 2E(r) + \frac{2m}{R}, \tag{20.192}$$

m being the Schwarzschild mass and $E(r)$ being an arbitrary function. In this form, the Schwarzschild metric emerges as the limit $M_{,r} = 0$ of an L–T model, and the limit of constant M, P, Q, S of a quasispherical Szekeres solution.

Furthermore, the coordinates used on a sphere of constant (t, r) in (20.191) must be transformed to those used in (20.53). Suppose the matching is to be done at $r = z = b =$ constant. Then the transformation is

$$\vartheta = 2 \arctan \left\{ \frac{\sqrt{[x - P(b)]^2 + [y - Q(b)]^2}}{S(b)} \right\}, \quad \varphi = \arctan \left[\frac{y - Q(b)}{x - P(b)} \right]. \tag{20.193}$$

The transformed metric (20.191) is

$$ds^2 = dt^2 - \frac{R_{,r}^{\ 2}}{1 + 2E(r)} dr^2 - \frac{R^2(t, r)}{\mathcal{E}_1^{\ 2}} \left(dx^2 + dy^2 \right), \tag{20.194}$$

where

$$\mathcal{E}_1 \overset{\text{def}}{=} \frac{S(b)}{2} \left\{ \left[\frac{x - P(b)}{S(b)} \right]^2 + \left[\frac{y - Q(b)}{S(b)} \right]^2 + 1 \right\}. \tag{20.195}$$

Now the matching conditions between (20.194) – (20.195) and (20.53) are fulfilled at any $r = z = b =$ constant, provided that $2E(r)$ of (20.194) and the $(-k(z))$ of (20.53) have the same value at $r = z = b$, and that $R(t, b)$ of (20.191) is the same function of t as $\Phi(t, b)$ of (20.53). The latter condition implies

$$M(b) = m, \tag{20.196}$$

where M is the function from (20.63).

To answer the question asked in the title of this section we need to verify whether the AH is spacelike or otherwise. For the L–T models, this analysis was done in Sec. 18.8, with the result that the ingoing[10] part of the AH (close to $M = 0$ in Fig. 20.4) can be any, while the outgoing part can be spacelike and pointwise null, but never timelike. The corresponding analysis for the exemplary Szekeres model considered here was done by KB12, and the result was nearly the same: the outgoing part of the AH can only be spacelike, while the ingoing part can be any.

As Fig. 20.4 shows, even if a segment of the ingoing branch of the AH is timelike, a ray going out of the AH will be trapped in the funnel around $\Phi = 0$. Whether it later crosses the AAH or not, it will be forced to hit the Big Crunch within a finite segment of its affine parameter. Where the AH is spacelike, a null line that crossed it once cannot cross it again without being redirected towards the past. This shows that in its outgoing

---

[10] 'Ingoing' ('outgoing') mean 'Φ decreases (increases) as we proceed along the AH towards greater t'.

part the AH acts as a black hole surface, even in that region where the AAH is inside it. The conclusion is that *the true horizon is the AH rather than the AAH*.

This conclusion is strengthened by the following fact. If a portion of the Szekeres spacetime, of finite spatial diameter, is matched to the Schwarzschild spacetime, then, from (20.196), the AH matches to the Schwarzschild event horizon. Thus, no signal can escape to infinity if it was within the Szekeres AH while crossing the $z = b$ surface. The intersection of the AAH with this surface leaves no trace in the Schwarzschild metric. In particular, this happens in that part of the Szekeres region, where the AAH is earlier than the AH (in the left half of Fig. 20.4).

20.4 * The Goode–Wainwright representation of the Szekeres solutions

Goode and Wainwright (1982) introduced a description of the Szekeres solutions in which many properties of the two subfamilies can be considered at one go.[11] The metric is

$$ds^2 = dt^2 - S^2 \left[e^{2\nu} \left(dx^2 + dy^2 \right) + H^2 W^2 dz^2 \right],$$ (20.197)

where $S(t, z)$ is defined by

$$S_{,t}^2 = -k + 2\mathcal{M}/S,$$ (20.198)

$k = 0, \pm 1$, and $\mathcal{M}(z)$ is an arbitrary function,

$$H = A(x, y, z) - \beta_+ f_+ - \beta_- f_-,$$ (20.199)

A, e^ν and W will be defined below (they differ for each subfamily), $\beta_+(z)$ and $\beta_-(z)$ are functions of z defined below, and $f_+(t, z)$ and $f_-(t, z)$ are the two linearly independent solutions of the equation

$$F_{,tt} + 2(S_{,t}/S)F_{,t} - (3\mathcal{M}/S^3)F = 0.$$ (20.200)

Those solutions of (20.198) for which $S_{,t} > 0$ are

$$S = \mathcal{M}g_{,\eta}(\eta), \qquad t - T(z) = \mathcal{M}g(\eta),$$ (20.201)

where $T(z)$ is an arbitrary function (the bang time) and $g(\eta)$ is

$$g(\eta) = \begin{cases} \eta - \sin \eta & \text{when } k = +1, \\ \sinh \eta - \eta & \text{when } k = -1, \\ \eta^3/6 & \text{when } k = 0. \end{cases}$$ (20.202)

With $k = 0$, S can be rescaled so that $\mathcal{M} = 1$ (see later), and this choice of S will be assumed. The corresponding solutions of (20.200) are

$$f_+ = \begin{cases} (6\mathcal{M}/S) \left[1 - (\eta/2) \cot(\eta/2) \right] - 1 & \text{when } k = +1, \\ (6\mathcal{M}/S) \left[1 - (\eta/2) \coth(\eta/2) \right] + 1 & \text{when } k = -1, \\ \eta^2/10 & \text{when } k = 0, \end{cases}$$ (20.203)

---

[11] The G–W notation became a standard that is partly in conflict with the notation widely used for the L–T models. It is used only in this section.

$$f_- = \begin{cases} (6\mathcal{M}/S)\cot(\eta/2) & \text{when } k = +1, \\ (6\mathcal{M}/S)\coth(\eta/2) & \text{when } k = -1, \\ 24/\eta^3 & \text{when } k = 0. \end{cases} \tag{20.204}$$

The coefficients in $f_\pm$ were chosen for later convenience. The solutions with $S_{,t} < 0$ are time reverses of (20.203) – (20.204), and $\mathcal{M} > 0$ is assumed for correspondence of the results with the Friedmann models. With these assumptions, f_+ are the solutions that increase with time and f_- are the solutions that decrease with time.

The two subfamilies are now defined as follows:

$$\underline{\beta_{,z} \neq 0}$$

$T_{,z}{}^2 + \mathcal{M}_{,z}{}^2 \neq 0$, so $S_{,z} \neq 0$, and then:

$$e^{-\nu} = \frac{1}{f(z)} \left[a(z)\left(x^2 + y^2\right) + 2b(z)x + 2c(z)y + d(z) \right], \tag{20.205}$$

where $f(z)$ is arbitrary and a, b, c, d are functions of z obeying

$$ad - b^2 - c^2 = \varepsilon/4, \qquad \varepsilon = 0, \pm 1, \tag{20.206}$$

$$W^2 = (\varepsilon - kf^2)^{-1}, \tag{20.207}$$

$$\beta_+ = -kf\mathcal{M}_{,z}/(3\mathcal{M}), \qquad \beta_- = fT_{,z}/(6\mathcal{M}), \tag{20.208}$$

$$A = f\nu_{,z} - k\beta_+, \tag{20.209}$$

and it is understood that $\mathcal{M} = 1$ when $k = 0$ (see later).

$$\underline{\beta_{,z} = 0}$$

$\mathcal{M}_{,z} = T_{,z} = 0$, so $S_{,z} = 0$ and:

$$e^{-\nu} = 1 + \frac{1}{4}k\left(x^2 + y^2\right), \qquad W = 1, \tag{20.210}$$

$$A = \begin{cases} e^\nu \left\{ a(z)\left[1 - \frac{1}{4}k\left(x^2 + y^2\right)\right] + b(z)x + c(z)y \right\} - k\beta_+ & \text{for } k = \pm 1, \\ a(z) + b(z)x + c(z)y - \frac{1}{2}\beta_+\left(x^2 + y^2\right) & \text{for } k = 0, \end{cases} \tag{20.211}$$

a, b, c, β_+ and β_- being arbitrary functions of z.

The remarkable thing about the Goode–Wainwright (G–W) parametrization is that (20.197) – (20.204) all hold for both subfamilies, the $\beta_{,z} = 0$ subfamily differing from the other one only by $\mathcal{M}$ and T being constant. In both subfamilies the subcase $\beta_+ = \beta_- = 0$ gives the Friedmann models, but represented in untypical coordinates. That this subcase is indeed a coordinate transform of the Friedmann models can be seen by applying the criteria of Section 17.15.

With $\beta_{,z} \neq 0$, the G–W representation arises from (20.50) as follows. Let

$$k(z) = K\phi^2(z), \tag{20.212}$$

where $k(z)$ is the function from (20.44), and $K = 0, \pm 1$. When $k \neq 0$, $\phi = |k|^{1/2}$; when $k = 0$, $\phi(z)$ is taken to be a new arbitrary function. We define $S(t, z)$ by

$$\Phi = \phi S, \tag{20.213}$$

so that S obeys

$$2\frac{S_{,tt}}{S} + \frac{S_{,t}{}^2}{S^2} + \frac{K}{S^2} = 0,$$

(20.214)

resulting from (20.46) with (20.212), (20.213) and $p = 0$. Equation (20.198) is the integral of (20.214). Comparing (20.63) with (20.198) and using (20.213) we find that

$$\mathcal{M} = M/\phi^3.$$

(20.215)

Equation (20.47) now becomes

$$AC - B_1{}^2 - B_2{}^2 = \frac{1}{4}\left[\frac{1}{h^2(z)} + K\phi^2\right].$$

(20.216)

Let us define $G(z)$ by

$$1/h^2(z) + K\phi^2 = \varepsilon G^2,$$

(20.217)

where $\varepsilon = 0, \pm 1$ so that $G = \left|1/h^2 + K\phi^2\right|^{1/2}$ when $1/h^2 + K\phi^2 \neq 0$ and $0 \neq G(z)$ is an arbitrary function otherwise. Then let us define a, b, c, d, f and W in terms of the functions from (20.45) and (20.47) by

$$(A, B_1, B_2, C, \phi) = (a, b, c, d, f)G, \qquad h = W/G;$$

(20.218)

the functions a, b, c, d will then obey (20.206). In the new variables, from (20.42), (20.45) and (20.213), we obtain

$$e^\beta = Se^\nu,$$

(20.219)

where ν is given by (20.205) and is related to the one from (20.45) (which we for a moment denote $\bar\nu$) by

$$e^{\bar\nu} = e^\nu/\phi = e^\nu/(fG).$$

(20.220)

From (20.50), (20.213) and (20.218) – (20.219) we obtain

$$e^\alpha = WSf\left(S_{,z}/S + \nu_{,z}\right)$$

(20.221)

and so, from (20.197),

$$H = fS_{,z}/S + f\nu_{,z}.$$

(20.222)

When $k = 0$, the functions $\phi(z)$ and $f(z)$ are not yet defined; then let us define ϕ by

$$\phi = M^{1/3},$$

(20.223)

where M is the function from (20.63). Then, from (20.215), $\mathcal{M} = 1$.

Now it can be verified case by case using (20.199), (20.202) – (20.204), (20.208), (20.209) and (20.222) that

$$F \stackrel{\text{def}}{=} \beta_+ f_+ + \beta_- f_- = A - H = -fS_{,z}/S - k\beta_+$$

(20.224)

obeys (20.200) (note that with $k = 0$ necessarily $\beta_+ = 0$ from (20.208)). In the case now considered, the formula for matter density (20.64), using (20.220), (20.219), (20.215), (20.222), (20.208) and (20.224), becomes[12]

$$\kappa\epsilon = 6\mathcal{M}(1 + F/H)/S^3 \equiv 6\mathcal{M}A/(HS^3). \tag{20.225}$$

In the $\beta_{,z} = 0$ subfamily, when $k \neq 0$ in (20.37), it can be rescaled to $+1$ or -1 by simple rescalings of x, y, ϕ, U, V_1 and V_2. We will therefore assume that $k = 0, \pm 1$. The G–W representation is then introduced in a different way when $k = \pm 1$ and when $k = 0$.

When $k = \pm 1$ we redefine U, W and λ by

$$U = (u - kw)/2, \qquad W = (ku + w)/2, \qquad \lambda = \widetilde{\lambda} - ku\Phi, \tag{20.226}$$

where u and w depend only on z, and then $\widetilde{\lambda}$ obeys (20.35) with $p = 0$ and $U + kW = 0$ by virtue of (20.18). This can be assumed with no loss of generality – see the comments after (20.40). The G–W representation then arises from (20.37) by

$$
\begin{aligned}
\Phi &= S, \qquad e^\beta = Se^\nu, \\
W &= a/2, \qquad U = -ka/2, \qquad V_1 = b, \qquad V_2 = c, \\
e^\alpha &= (A + k\beta_+)S + \lambda \overset{\text{def}}{=} HS, \qquad \beta_+ = -kX(z)/(3M),
\end{aligned}
\tag{20.227}
$$

where e^ν and A are given by (20.210) – (20.211) and $X(z)$ is the function from (20.60). Hence,

$$F = A - H = -\lambda/S - k\beta_+, \tag{20.228}$$

and it can be verified, case by case again, using (20.61), that $F = \beta_+ f_+ + \beta_- f_-$ obeys (20.200), where f_+ and f_- are given by (20.203) – (20.204) and $\beta_-(z)$ is arbitrary.

When $k = 0$, we first observe that when $\Lambda = k = 0$, one can assume $X(z) = 0$ in (20.60) with no loss of generality because $X = 0$ results after the reparametrisation $\lambda = \widetilde{\lambda} + X\Phi/3$ and using (20.198) with $\mathcal{M} = 1$ (then $e^\alpha = \widetilde{\lambda} + \Phi\sigma$, as before). Then, we take $\Phi = S = e^\beta$ and $e^\alpha = \lambda + S\sigma$, where σ is given by (20.34) and S is given by (20.201) – (20.202). Next, we find λ from (20.61) in the case $k = \Lambda = 0$. Finally, we define

$$F \overset{\text{def}}{=} -\lambda/S = \beta_+ f_+ + \beta_- f_- + A, \tag{20.229}$$

where $\beta_+ = -U$, $b = V_1$, $c = V_2$, $a = 2W$ and $\beta_-(z)$ is arbitrary. The functions f_+ and f_- are as in (20.203) – (20.204), while A is as in (20.211). Using the information given in this paragraph, one can verify that such an F obeys (20.200). The matter density is given by (20.225) for the $\beta_{,z} = 0$ subfamily as well.

As Goode and Wainwright observed, (20.200) is linear and has the same form as the equation derived for the density perturbation in the linearised perturbation scheme around the Friedmann background, if the perturbed solution is also dust. This holds both in relativity and in Newtonian theory; see Eqs. (15.9.23) and (15.10.57) in the Weinberg (1972) textbook and Eqs. (12.10) and (12.31) in the textbook by Raychaudhuri (1979). However, this is only a formal similarity. The following differences should be noted:

---

[12] In deriving (20.225) one must note that $k = 1/k$ when $k = \pm 1$, and when $k = 0$ the term with β_+ does not arise.

1. In the corresponding equation of the linearised perturbation scheme, the coefficients $(2S_{,t}/S)$ and $3\mathcal{M}/S^3 = \kappa\rho/2$ are taken from the background Friedmann model, whereas here they come from the perturbed model. Assuming F small, this means, not surprisingly, that the equation of the linearised perturbation is the linear approximation to (20.200).

2. In the linearised perturbation scheme, the solution of (20.200) is $\delta = (\epsilon_p - \epsilon_b)/\epsilon_b$, where ϵ_p is the perturbed density and ϵ_b is the Friedmann background density. In the Szekeres models, linearizing about the background, one obtains from (20.225) $F = A(\epsilon_p - \epsilon_b)/\epsilon_b$, that is, $F \cong \delta$ only if $A \cong 1$. In general, F has no clear physical meaning, although it does define a certain deformation on the Friedmann background.

3. In the linearised perturbation scheme, every solution of (20.200) generates a perturbed model. In the $\beta_{,z} \neq 0$ Szekeres models the coefficients β_+ and β_- are provided by other data (see (20.208)), and when $k = 0$ the growing mode is necessarily absent ($\beta_+ \equiv 0$). In the $\beta_{,z} = 0$ subfamily β_- is truly arbitrary, but β_+ is again provided from elsewhere.

Still, the G–W representation is enlightening in several ways. From (20.208) one can see that the growing mode of perturbation is generated by the spatial inhomogeneity in the mass distribution ($\mathcal{M}_{,z} \neq 0$), and the decaying mode is generated by the nonsimultaneity of the Big Bang ($T_{,z} \neq 0$) – a result that required much labour to obtain in the standard representation of the L–T model, see Section 18.21.

The expansion and shear of the dust source are in both cases:

$$\theta = 3S_{,t}/S - F_{,t}/H, \qquad 2\sigma^1{}_1 = 2\sigma^2{}_2 = -\sigma^3{}_3 = 2F_{,t}/(3H). \tag{20.230}$$

From the above and from (20.203) – (20.204) one can now show that (20.200) is the Raychaudhuri equation (15.46), which, surprisingly, becomes linear for the Szekeres models in the G–W representation.

In general, there can be two scalar polynomial curvature singularities, they occur where $S = 0$ and $H = 0$. The first is the Big Bang. It is pointlike when $\beta_- = 0$ or cigar-type when $\beta_- \neq 0$. In the $\beta_{,z} \neq 0$ subfamily, $\beta_- = 0$ implies a simultaneous Big Bang; in the $\beta_{,z} = 0$ subfamily, the Big Bang is always simultaneous. The second singularity is a shell crossing. When $\beta_-(z_0) > 0$, the shell crossing along the flow lines with $z = z_0$ will occur later than the Big Bang. Whenever it occurs, it is of the pancake type.

For a list of papers in which the G–W representation was used see Krasiński (1997).

20.5 Selected interesting subcases of the Szekeres–Szafron family

20.5.1 The Szafron–Wainwright model

Szafron and Wainwright (1977) considered the subcase of the Szafron $\beta_{,z} = 0$ class in which $k = 0 = W$ and $\kappa p = \alpha/t^2$, $\alpha = $ constant. Then, from (20.38), $Q(t) \stackrel{\text{def}}{=} [\Phi(t)]^{3/2}$ obeys

$$(4/3)Q_{,tt} + (\alpha/t^2)\, Q = 0, \tag{20.231}$$

and (20.35) reduces to

$$(4/3)\chi_{,tt} + (\alpha/t^2)\, \chi = 4U(z)/\left(3Q^{1/3}\right), \tag{20.232}$$

where $\chi \overset{\text{def}}{=} \lambda Q^{1/3}$. Note that (20.231) is the homogeneous part of (20.232). The general solution of (20.231) is

$$Q(t) = C_1 t^{1-q} + C_2 t^q, \tag{20.233}$$

where C_1 and C_2 are arbitrary constants and $\alpha = (4/3)q(1-q)$; q will be real only when $\alpha \leq 1/3$. With $\alpha = 1/3$ ($\Rightarrow q = 1/2$), the two basis solutions in (20.233) become linearly dependent, and there exists an additional solution. With q complex, the constants C_1 and C_2 must be complex, too, and the real solutions for Q have different properties than those with real q. All these cases have to be considered separately. Szafron and Wainwright considered only the case $\alpha < 1/3$ ($q \neq 1/2$ and real).

A. The case $\alpha < 1/3$

With reference to (20.37), the solution is

$$\Phi(t) = [Q(t)]^{2/3}, \qquad Q(t) = C_1 t^{1-q} + C_2 t^q,$$

$$\lambda(t,z) = \chi/[Q(t)]^{1/3}, \qquad \chi = \lambda_1 t^{1-q} + \lambda_2 t^q,$$

$$\lambda_1 = B(z) + \frac{U(z)}{1-2q} \int \frac{t^q}{Q^{1/3}} dt, \qquad \lambda_2 = A(z) - \frac{U(z)}{1-2q} \int \frac{t^{1-q}}{Q^{1/3}} dt, \tag{20.234}$$

where $A(z)$ and $B(z)$ are arbitrary functions.

This model begins with a (simultaneous) Big Bang at $t = 0$ and then expands forever. It reduces to a $k = 0$ Robertson–Walker metric when $U = V_1 = V_2 = 0$, $B = C_1$ and $A = C_2$, but the equation of state in the limit is none of the astrophysicists' favourites. The density in the R–W limit is $\kappa\epsilon = (4/3)(Q_{,t}/Q)^2$.

B. The case $\alpha = 1/3$ ($q = 1/2$)

Only the expressions for Q and χ are different from the previous case:

$$\chi = A(z)\sqrt{t}\ln t + B(z)\sqrt{t} + U(z)Q \int \left[\frac{1}{Q^2} \left(\int Q^{2/3} dt \right) dt \right],$$

$$Q(t) = C_1\sqrt{t}\ln t + C_2\sqrt{t}; \tag{20.235}$$

see Exercise 14. The evolution of this model is similar to the previous one: a simultaneous Big Bang followed by expansion forever. The R–W limit results in the same way.

C. The case $\alpha > 1/3$

In this case, q is complex:

$$q_{\pm} = 1/2 \pm iq_2, \qquad q_2 = \sqrt{3\alpha - 1}/2 \tag{20.236}$$

(note that $q_- = 1 - q_+$), and the solution is

$$Q(t) = C_1\sqrt{t}\cos(q_2 \ln t) + C_2\sqrt{t}\sin(q_2 \ln t),$$

$$\chi = A(z)\sqrt{t}\sin(q_2 \ln t) + B(z)\sqrt{t}\cos(q_2 \ln t) \tag{20.237}$$

$$+ \frac{U(z)\sqrt{t}}{q_2} \left[\sin(q_2 \ln t) \int \frac{\sqrt{t}\cos(q_2 \ln t)}{Q^{1/3}} dt - \cos(q_2 \ln t) \int \frac{\sqrt{t}\sin(q_2 \ln t)}{Q^{1/3}} dt \right].$$

The function Φ begins from $\Phi = 0$ at $t = 0$, increases to a maximum and then decreases back to zero a finite time later. Unlike in typical recollapsing R–W models, the duration of the cycle increases with increasing t. The period between bang and crunch tends to zero as $t \to 0$ and to infinity as $t \to \infty$, and the same is true for the maximum value of Φ in each cycle; see Fig. 20.8. The pressure $\kappa p = \alpha/t^2$ acts like a cosmological constant decreasing with time, and is responsible for this behaviour, which persists also in the R–W limit (achieved in the same way as in case A).

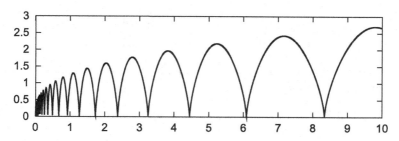

Fig. 20.8 A graph of the function $\Phi(t)$ for a model of the Szafron–Wainwright class with $\alpha > 1/3$. The graph is self-similar: it looks the same for every scale on the horizontal axis.

The value of Φ is the radius of the cylindrical space $t =$ constant. Along the generator of the cylinders, the variation of λ with time is still more complicated. This model has never been investigated from the point of view of cosmology.

The solutions described above have their exact counterparts in the $\beta_{,z} \neq 0$ family. With $p = \alpha/t^2$, the solution for Φ is of the same algebraic form, only its constants become arbitrary functions of z. The one corresponding to $\alpha < 1/3$ was found by Szafron (1977); the other two have never been investigated.

20.5.2 The toroidal universe of Senin

Senin (1982) found a solution whose local geometry (not just topology!) is that of a 3-dimensional torus. We shall re-derive it here.

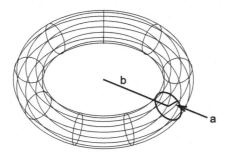

Fig. 20.9 A 2-torus of large radius b and small radius a.

The parametric equations of a 2-dimensional torus embedded in a Euclidean 3-space, with large radius b and small radius a (Fig. 20.9) are

$$x = (a \cos \psi + b) \cos \varphi, \qquad y = (a \cos \psi + b) \sin \varphi, \qquad z = a \sin \psi, \qquad (20.238)$$

where φ is the azimuthal angle around the vertical axis in Fig. 20.9, while ψ is the angle around the small circle, both angles vary in the range $[0, 2\pi]$. We will make the torus 3-dimensional by allowing that the subset $\varphi = $ constant is a sphere rather than a circle. The generalisation is straightforward, the appropriate parametric equations being:

$$
\begin{aligned}
x &= (a \cos \psi + b) \cos \varphi, \qquad y = (a \cos \psi + b) \sin \varphi, \\
z &= a \sin \psi \sin \zeta, \qquad u = a \sin \psi \cos \zeta, \qquad (20.239)
\end{aligned}
$$

the 3-torus being now embedded in a 4-dimensional Euclidean space with the metric $dx^2 + dy^2 + dz^2 + du^2$. The metric of the 3-torus is

$$ds^2 = a^2 \left[d\psi^2 + (\cos \psi + b/a)^2 d\varphi^2 + \sin^2 \psi d\zeta^2 \right]. \qquad (20.240)$$

In this 3-space, the surfaces $\zeta = $ constant are 2-dimensional tori, and the surfaces $\varphi = $ constant have the local geometry of 2-spheres. Note, however, that although on a surface $\varphi = $ constant the coordinate ψ plays the role of the lateral angle (normally denoted by ϑ), it varies from 0 to 2π, not from 0 to π. Moreover, in the 3-space (20.240) the points with $\{\psi = \psi_1 = \pi + \psi_0, \zeta = \zeta_0\}$ *are not* equivalent to points with $\{\psi = \psi_2 = \pi - \psi_0, \zeta = \pi + \zeta_0\}$, as they would be on a sphere, because the coefficient of $d\varphi^2$ does not return to its initial value after ψ is increased by π. We conclude from this that each surface $\varphi = $ constant is actually a pair of spheres that touch each other at one pole, and their opposite poles are identified, as in Fig. 20.10.

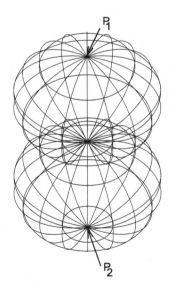

Fig. 20.10 The subspace $\varphi = $ constant of the 3-torus with the metric (20.240). The two spheres are tangent at one pole, the poles P_1 and P_2 are identified. The coordinate ψ has the value 0 at P_1, π at the common point of the spheres and 2π at P_2; it increases along the meridians of the spheres.

The next step is to make the 3-torus (20.240) a subspace $t = $ constant of a 4-dimensional spacetime, and to allow the torus parameters to evolve with time. The simplest idea is to assume that the two radii a and b are momentary values of functions of time, so

$$ds^2 = a^2(t) \left\{ dt^2 - d\psi^2 - \sin^2 \psi d\xi^2 - [\cos \psi + b(t)/a(t)]^2 d\varphi^2 \right\}. \tag{20.241}$$

Somewhat miraculously, this metric does allow perfect fluid solutions. The tetrad components of the Einstein tensor, in the orthonormal tetrad defined by (20.241), are

$$
\begin{aligned}
G_{00} &= \frac{a_{,t}^2}{a^4} + \frac{2 \cos \psi a_{,t}^2}{a^4 S} + \frac{2 a_{,t} b_{,t}}{a^4 S} + \frac{2 \cos \psi}{a^2 S} + \frac{1}{a^2}, \\
G_{11} &= G_{22} = \frac{a_{,t}^2}{a^4} - \frac{a_{,tt}}{a^3} - \frac{\cos \psi a_{,tt}}{a^3 S} - \frac{b_{,tt}}{a^3 S} - \frac{\cos \psi}{a^2 S}, \\
G_{33} &= \frac{a_{,t}^2}{a^4} - \frac{2 a_{,tt}}{a^3} - \frac{1}{a^2},
\end{aligned}
\tag{20.242}
$$

where $S \overset{\text{def}}{=} a \cos \psi + b$. To allow a perfect fluid source, only one equation, $G_{22} = G_{33}$, must be imposed on the set (20.242), and it is equivalent to

$$a_{,tt}/a = b_{,tt}/b - 1. \tag{20.243}$$

Hence, one of the radii of the torus can be chosen arbitrarily, while the other one will follow from (20.243). Since the metric depends on just two variables, t and ψ, the existence of a thermodynamical scheme is guaranteed (see Section 15.5). One sensible choice would then be to define an equation of state, which would provide a second equation connecting a and b, and then solve all equations. For a test, Senin chose $a_{,tt} = -a$, and then

$$a = a_0 \sin t, \qquad b = C_1 t + C_2; \tag{20.244}$$

a_0, C_1 and C_2 being arbitrary constants. This is a perfect fluid solution with

$$
\begin{aligned}
u^\alpha &= \delta_0^\alpha/a, \qquad \kappa p = a_0^2/a^4, \\
\kappa \epsilon &= \frac{2 C_1 \cos t + 3 a_0 \cos \psi + (C_1 t + C_2)/\sin t}{a_0^2 \sin^3 t (a \cos \psi + b)}.
\end{aligned}
\tag{20.245}
$$

Each surface of constant t and ξ has the local geometry of a torus with large radius b and small radius a; the two radii evolve according to different laws. The model has a finite duration. At $t = 0$, it has a Big Bang singularity; the space $t = $ constant begins its expansion from a ring of radius C_2. The large radius of the torus keeps growing from C_2 to $(\pi C_1 + C_2)$; the small radius grows from zero at $t = 0$ to the maximum a_0 at $t = \pi/2$, then collapses again to zero at the final singularity $t = \pi$. The final singularity is again a ring, of radius $(2\pi C_1 + C_2)$. With $C_2 = 0$, the Big Bang is pointlike. The energy density will be positive throughout the evolution when $C_2 = 0$ if C_1/a_0 is sufficiently large.

When $C_1 = C_2 = b = 0$, the solution reduces to the $k > 0$ R–W model in the form (17.93) – (17.94) with the equation of state $\epsilon = 3p$.

In order to relate the Senin solution to other ones, it is convenient to transform the coordinates as follows:

$$t = \arccos\left[(a_0 - T)/a_0\right], \qquad \psi = 2 \arctan(r/2). \tag{20.246}$$

Then the metric becomes

$$ds^2 = dT^2 - \frac{a^2(T)}{(1+r^2/4)^2} \left(dr^2 + r^2 d\zeta^2\right) - \left[a(T)\frac{1-r^2/4}{1+r^2/4} + \frac{1}{2}b(T)\right]^2 d\varphi^2, \quad (20.247)$$

$$a^2(T) = a_0{}^2 - (a_0 - T)^2, \qquad b(T) = C_1 \arccos\left[(a_0 - T)/a_0\right] + C_2, \quad (20.248)$$

and is the following limit of the Szafron $\beta_{,z} = 0$ metric given by (20.37) with (20.35) and (20.38) (see Exercise 15):

$$k = 1, \qquad U = -1/2, \qquad V_1 = V_2 = 0, \qquad W = 1/2. \quad (20.249)$$

20.6 Selected further reading on the Szekeres models

Walters and Hellaby (2012) generalised the construction of models evolving between data given at two different times (see Sec. 18.6) to the quasi-spherical Szekeres models.

Sussman and Delgado Gaspar (2015) provided a numerical example (based on exact formulae) of a set of four thick-pancake-like overdensities surrounding a central void, constructed from a single quasi-spherical Szekeres region. In theory, the number of pancakes could be arbitrarily large. This shows that the Szekeres metrics are sufficiently general to provide 'a fully relativistic nonperturbative coarse-grained modelling of cosmic structure at all scales' (this is a quotation from the authors' abstract). This multiple structure evolves without shell crossings in ever-expanding models in their full range of validity.

The paper also contains several other noteworthy results, for example, the discussion of the positions of 3-space extrema of mass density, expansion $\theta = u^\alpha{}_{;\alpha}$ and the scalar curvature $^{(3)}R$ of the constant-t hypersurfaces (extremum in the profile along the z coordinate not always coincides with the extremum in space). Also, the introductory description of the geometry of the quasi-spherical Szekeres models is instructive and illuminating.

The paper by Buckley and Schlegel (2020) contains an excellent pedagogical presentation, with several instructive illustrations, of the quasi-spherical Szekeres models and some of their geometrical properties. It is especially recommended to those readers who found the formal presentation in this book too terse.

In the paper by Krasiński and Bolejko (2011) it was shown that in a general quasi-spherical Szekeres (QSS) model the objects seen by an observer necessarily drift across the sky. This was done by parametrising the light rays not by the affine parameter, but by the quasi-radial coordinate r (called z in this book). In such a parametrisation, it was clear that rays emitted from the same source at different times and hitting the same observer will in general not intersect always the same intermediate particles comoving with the cosmic medium. By taking the linear approximation to the geodesic equations, the small parameters being the differences between the t, x and y coordinates on two rays at a fixed r, it was shown that the only metrics in the QSS family in which the drift is zero for all sources and all observers are the Friedmann ones. The rate of the drift was numerically calculated for nonradial rays in an exemplary L–T model, see Sec. 18.23.

The rarely discussed and still incompletely understood quasi-plane and quasi-hyperbolic Szekeres models were investigated in the papers by Hellaby and Krasiński (2008), Krasiński (2008) and Krasiński and Bolejko (2012b).

Here is a summary of selected results found by Krasiński (2008):

1. The pattern of decelerated expansion in the plane symmetric model disagrees with the Newtonian analogues and intuitions. An exemplary Newtonian potential that gives a similar pattern of expansion has co-axial parallel cones as its equipotential surfaces.

2. Embeddings of the constant (t, z) surfaces in the Euclidean space suggest that the flat surfaces contained in the plane symmetric model can be rather naturally interpreted as flat tori whose circumferences are proportional to the value of $\Phi(t, z)$, and thus vary with time. Such a topology immediately explains the pattern of expansion and implies that the total mass contained within a $z = $ constant surface is finite.

3. The quasi-plane and quasi-hyperbolic models are permanently trapped, so no apparent horizons exist in them. Consequently, these models cannot be used to describe the formation of black holes (the whole Universe is one black hole all the time).

4. The toroidal interpretation can be extended to the nonsymmetric quasi-plane model. With that interpretation, the nonsymmetric model can be free of shell crossings, unlike the model with infinite spaces.

5. For both interpretations (toroidal and infinite), the quasi-plane model cannot describe the formation of structures that collapse to high densities. Just as in the ever-expanding Lemaître–Tolman and quasi-spherical Szekeres models, the density perturbations tend to finite values in the asymptotic future. Thus, the quasi-plane model may describe only the formation of moderate-density condensations and of voids.

7. With the toroidal interpretation, the mass function $M(z)$ is proportional to the active gravitational mass contained within a solid torus of 'radial' coordinate z.

The quasi-plane model with toroidal spaces may be a testing ground for the idea of a 'small Universe', proposed by Ellis (1984). A small Universe is one with compact spatial sections, in which the present observer has already seen several times around the space. Several papers were devoted to checking this idea against the observational data (a conclusive proof of any nontrivial topology is, unfortunately, still lacking). However, the background geometry has always been a homogeneous isotropic Robertson–Walker metric with identifications in the underlying manifold. The quasi-plane toroidal Szekeres model has a less general topology (identifications in it may exist only in 2-dimensional surfaces, in the z-direction the space is infinite), but is inhomogeneous and has no symmetry.

Here is a summary of selected results of Krasiński and Bolejko (2012b):

1. Although there exists no origin, where Φ would be zero permanently, a set where $M = 0$ can exist. At this location, $\Phi_{,t}$ is constant.

2. The whole spacetime is both future- and past- globally trapped.

3. The geometrical interpretation of the (x, y) coordinates in a constant-(t, z) surface was clarified. Contrary to an earlier claim, this surface consists of just one sheet, doubly covered by the (x, y) map.

4. The surfaces $(z = $ constant, $\varphi = 0)$ are locally isometric to surfaces of revolution in the Euclidean space (in special cases to a plane and a cone) when $E \geq 1$, but cover the latter an infinite number of times. When $1/2 < E < 1$, they cannot be embedded in a Euclidean space even locally. The values $E \leq 1/2$ are prohibited by the spacetime signature.

20.7 Exercises

1. Show that (20.28) is a coordinate transform of the 2-dimensional metrics resulting from (19.11) – (19.12) when t and r are constant.

Hint. With $k > 0$, apply the transformation $x = (2/\sqrt{k}) \cot(\vartheta/2) \cos\varphi$, $y = (2/\sqrt{k})$ $\cot(\vartheta/2) \sin\varphi$ to (20.28). With $k < 0$, the transformation is $x = (2/\sqrt{-k}) \coth(\vartheta/2) \cos\varphi$, $y = (2/\sqrt{-k}) \coth(\vartheta/2) \sin\varphi$.

2. Prove that, in the most general case, the Szafron metrics given by (20.36) – (20.38) and (20.49) – (20.50) have no symmetry.

Hint. Solve the Killing equations and verify that with no further limitations on the metric the solution is $k^\mu \equiv 0$. Note: this is a laborious exercise. You may use the paper by Bonnor et al. (1977) as an aid.

3. Consider a surface given by the equation $r = a = $ constant in the coordinates (r, ϑ, φ) related to the Cartesian coordinates (x, y, z) by

$$x = r \sinh\vartheta \cos\varphi, \qquad y = r \sinh\vartheta \sin\varphi, \qquad z = r \coth\vartheta \qquad (20.250)$$

in the pseudo-Euclidean flat 3-space with the metric $ds^2 = dx^2 + dy^2 - dz^2$. Show that the surface obeys the equation of a two-sheeted hyperboloid $(z/a)^2 - (x^2 + y^2)/a^2 = 1$ and that its internal metric coincides with that of the surface $(t = $ constant, $\Phi = a)$ in the metric (19.11) with $f(\vartheta) = \sinh\vartheta$.

4. Verify that the magnetic part of the Weyl tensor with respect to the velocity field in the Szafron spacetimes (20.50) and (20.37) is zero.

5. Prove that (1) two eigenvalues of the 3-dimensional Ricci tensor, $^{(3)}\Phi_{AB}$, of the slices $t = $ constant are equal (in both classes of Szafron's spacetimes); (2) the eigenframe of this Ricci tensor coincides with that of shear; (3) the eigenspaces corresponding to the degenerate eigenvalues of shear and of $^{(3)}\Phi_{AB}$ are the surfaces of constant curvature defined by the Szekeres geometry.

6. Prove that the slices $t = $ constant of the Szafron spacetimes are conformally flat, i.e. that their Cotton–York tensor (7.50) is zero.

7. Verify that the barotropic equation of state $\varepsilon = \varepsilon(p)$ reduces the $\beta_{,z} \neq 0$ Szafron spacetimes to the Robertson–Walker ones, and the $\beta_{,z} = 0$ spacetimes to either R–W or Kantowski–Sachs-type spacetimes, or the plane and hyperbolically symmetric counterparts of the latter.

Note: this is a very laborious exercise. It is advisable to use the paper by Spero and Szafron (1978) as an aid.

8. Verify that (20.60) is the integral of (20.35) when Φ obeys (20.38) and (20.59).

Hint. (1) Multiply (20.35) by $\Phi_{,t}$. (2) After the term with $\Phi_{,t}{}^2$ add and subtract $\lambda_{,t} \Phi\Phi_{,tt}$. (3) In the term $(-\lambda_{,t} \Phi\Phi_{,tt})$ substitute for $\Phi\Phi_{,tt}$ from (20.38). (4) In the term $\frac{1}{2}\lambda\Phi_{,t}{}^2$ substitute for $\Phi_{,t}{}^2$ from (20.59). (5) In the term $\lambda\Phi_{,t} \Phi_{,tt}$ substitute for $\Phi_{,t} \Phi_{,tt}$ from the derivative of (20.59) by t. (6) After cancelling out whatever possible you should recognise that the integral of the final expression is (20.60).

9. Calculate the integral in (20.61) when $\Lambda = 0$ and verify that along the z-direction the space has infinite extent at the instant of the Big Bang, then collapses to a minimum size, and expands to infinite extent as the final singularity is reached. Show that the maximum of expansion of the spheres and the minimum of collapse in the z-direction do not in general occur simultaneously.

10. Verify (20.65).

11. Verify that the class I $(\beta_{,z} \neq 0)$ Szekeres solution with $AC - B_1{}^2 - B_2{}^2 > 0$ can be matched to the Schwarzschild solution. Verify that for the class II Szekeres solution with $k = +1$ the matching hypersurface must lie inside the Schwarzschild horizon.

12. Verify that the metric (20.53) results from (20.50) after the reparametrisation indicated is carried out.

13. Verify that the explicit solution for λ that results from (20.71) – (20.72) in the limit $k \to \varepsilon = 1$, $U = kW$, $V_1 = V_2 = 0$ coincides with (19.105).

Hint. Note that with $2E = -k = -1$ the first of (18.19) can be written as $R = 2M \sin^2(\eta/2)$, so $\eta = 2 \arcsin \sqrt{R/(2M)} = 2Z$. Note also that with $U = kW$, $V_1 = V_2 = 0$ we get $\sigma = 2W$ from (20.37), so $e^\alpha = 2W\Phi + \lambda$ and $e^{A/2} = e^\alpha$ from (20.37) and (19.103). Then divide (20.71) by $\sqrt{\varepsilon - k}$, take the limit $k \to \varepsilon = 1$ and use (20.68) – (20.69).

14. Show that the first of (20.235) can be written in the equivalent form

$$\chi = A(z)\sqrt{t}\ln t + B(z)\sqrt{t} + U(z)\sqrt{t}\left[(\ln t)\int \frac{\sqrt{t}}{Q^{1/3}}dt - \int \frac{(\ln t)\sqrt{t}}{Q^{1/3}}dt\right]. \qquad (20.251)$$

Show that the term containing $U(z)$ in (20.251) follows from the one in (20.234) as the limit $q \to 1/2$.

Hint. Note that

$$\mathcal{L} \overset{\text{def}}{=} t^{1-q}\int \frac{t^q}{Q^{1/3}}\,dt - t^q\int \frac{t^{1-q}}{Q^{1/3}}\,dt \xrightarrow[q\to 1/2]{} 0, \qquad (20.252)$$

so to calculate $\lim_{q\to 1/2}[\mathcal{L}/(1-2q)]$ you need to apply the de l'Hôpital formula. Note that you will be calculating $\partial/\partial q$, not $\partial/\partial t$.

15. Verify that the metric (20.247) – (20.249) is in the (20.37) – (20.38) family with $a = \Phi$ and $b/2 = \lambda$. Verify that $b/2 = \lambda$ obeys (20.35) and $a = \Phi$ obeys (20.38). Note that T in (20.247) – (20.248) is the same coordinate as t in (20.37).

21

The Kerr metric

21.1 The Kerr–Schild metrics

In the first two sections we shall briefly describe the way in which Kerr (1963) found his solution. In the next sections, we shall introduce the derivation by Carter (1973), and will derive its most important properties. Sections 21.1 and 21.2 are based on the papers by Boyer and Lindquist (1967) and by Kerr and Schild (1965).

The starting point is the consideration of metrics of the form

$$g_{\mu\nu} = \eta_{\mu\nu} - l_\mu l_\nu, \tag{21.1}$$

where $\eta_{\mu\nu}$ is the flat (Minkowski) metric in any coordinates and l_μ is a null vector field. The reasons given for studying this metric vary from paper to paper. Kerr and Schild (1965) justified it by saying that it allows for an easy conversion between covariant and contravariant components (see below). Boyer and Lindquist (1967) gave another reason: the Schwarzschild solution has this form (see Exercise 1), which inspired the search for further metrics with the same property. The vector field l_μ is required to be null with respect to the metric $g_{\mu\nu}$, but then it follows that it is null with respect to $\eta_{\mu\nu}$ as well and that, consequently, it does not matter which metric is used to raise and lower the index of l. It follows also that the inverse metric has the form

$$g^{\mu\nu} = \eta^{\mu\nu} + l^\mu l^\nu. \tag{21.2}$$

The sign of $l_\mu l_\nu$ is invariant. Depending on it, one or another coordinate becomes time. For example, take the coordinates (u, v, y, z) in which the flat background metric is

$$\begin{bmatrix} 0 & 1 & 0 & 0 \\ 1 & 0 & 0 & 0 \\ 0 & 0 & -1 & 0 \\ 0 & 0 & 0 & -1 \end{bmatrix}$$

and take $l^\mu = (1, 0, 0, 0)$. Consider both possible signs of $l_\mu l_\nu$, thus take $\varepsilon l_\mu l_\nu$ with $\varepsilon = \pm 1$. Then

$$ds^2 = 2dudv + \varepsilon dv^2 - dy^2 - dz^2.$$

With $\varepsilon = +1$ we have

$$ds^2 = d(u + v)^2 - du^2 - dy^2 - dz^2,$$

whereas with $\varepsilon = -1$ we have

$$ds^2 = du^2 - d(u - v)^2 - dy^2 - dz^2.$$

442

To be consistent with the existing literature, we must take the sign as in (21.1) (note that we use a different signature from that in the papers cited).

The following formulae will help in the calculations further on:

$$\left\{{\alpha \atop \beta\gamma}\right\}l^\beta l^\gamma = \left\{{\alpha \atop \beta\gamma}\right\}(\eta)l^\beta l^\gamma, \qquad \left\{{\alpha \atop \beta\gamma}\right\}l_\alpha l^\gamma = \left\{{\alpha \atop \beta\gamma}\right\}(\eta)l_\alpha l^\gamma,$$

$$\left\{{\alpha \atop \beta\alpha}\right\} = \left\{{\alpha \atop \beta\alpha}\right\}(\eta), \tag{21.3}$$

$$\left\{{\alpha \atop \beta\gamma}\right\}l^\gamma = \left\{{\alpha \atop \beta\gamma}\right\}(\eta)l^\gamma - \frac{1}{2}l^\rho\left(l^\alpha l_\beta\right)_{;\rho},$$

$$\left\{{\alpha \atop \beta\gamma}\right\}l_\alpha = \left\{{\alpha \atop \beta\gamma}\right\}(\eta)l_\alpha + \frac{1}{2}l^\rho\left(l_\beta l_\gamma\right)_{;\rho}, \tag{21.4}$$

where $\left\{{\alpha \atop \beta\gamma}\right\}(\eta)$ denotes the Christoffel symbols calculated from the flat metric $\eta_{\alpha\beta}$ (recall: $\eta_{\alpha\beta}$ may be expressed in non-Lorentzian coordinates, so the Christoffel symbols do not have to be zero). In consequence of (21.3), the covariant derivatives in (21.4) can be taken with respect to any of the two metrics, so let us assume that they are with respect to $\eta_{\alpha\beta}$.

Now let us consider the vacuum Einstein equations, $R_{\mu\nu} = 0$, and take the component $R_{\mu\nu}l^\mu l^\nu = 0$. Using (21.3) – (21.4) and, along the way, the identity $k_{\alpha,\beta} - k_{\beta,\alpha} = k_{\alpha;\beta} - k_{\beta;\alpha}$, after a lot of algebra we find that $R_{\mu\nu}l^\mu l^\nu = 0$ implies

$$\dot{l}^\beta l_\beta = 0, \tag{21.5}$$

where

$$\dot{l}^\alpha \stackrel{\text{def}}{=} l^\mu l^\alpha{}_{;\mu}. \tag{21.6}$$

In consequence of (21.3) it does not matter which of the two metrics is used to calculate the covariant derivative in (21.6), both definitions of $\dot{l}^\alpha$ give the same result. Since $\eta_{\mu\nu}l^\mu l^\nu = 0$, we have $\eta_{\mu\nu}l^\mu \dot{l}^\nu = 0$, so $\dot{l}^\mu$ is orthogonal to the null vector l^μ and is itself null by (21.5). Two null vectors can be orthogonal only when they are collinear. Consequently,

$$l^\mu l^\alpha{}_{;\mu} = \sigma l^\alpha. \tag{21.7}$$

Thus, the vacuum Einstein equations imply that l^α must be tangent to geodesics, not affinely parametrised when $\sigma \neq 0$. To obtain an affine parametrisation, we define

$$k^\mu \stackrel{\text{def}}{=} l^\mu/\sqrt{2H}, \tag{21.8}$$

where the function H is defined by $H_{,\nu}\,l^\nu/\sqrt{2H} = \sigma$. Then

$$k^\mu{}_{;\nu}\,k^\nu = 0; \qquad g_{\mu\nu} = \eta_{\mu\nu} - 2Hk_\mu k_\nu, \qquad g^{\mu\nu} = \eta^{\mu\nu} + 2Hk^\mu k^\nu. \tag{21.9}$$

In terms of k^α, Eqs. (21.3) do not change, whereas, in consequence of k^α being geodesic and affinely parametrised, Eqs. (21.4) simplify to

$$\left\{{\alpha \atop \beta\gamma}\right\}k^\gamma = \left\{{\alpha \atop \beta\gamma}\right\}(\eta)k^\gamma - k^\rho H_{,\rho}\,k^\alpha k_\beta,$$

$$\left\{{\alpha \atop \beta\gamma}\right\}k_\alpha = \left\{{\alpha \atop \beta\gamma}\right\}(\eta)k_\alpha + k^\rho H_{,\rho}\,k_\beta k_\gamma. \tag{21.10}$$

In the next step, the formula given below will be needed (in deriving it, one must use

$k_{\alpha,\beta} - k_{\beta,\alpha} = k_{\alpha;\beta} - k_{\beta;\alpha}$ and $k^\rho k_{\alpha;\rho} = 0 = k^\rho k_{\rho;\alpha}$, the last two equations stemming from k^α being null and geodesic):

$$\left\{{\alpha \atop \beta\gamma}\right\} = \left\{{\alpha \atop \beta\gamma}\right\}(\eta) - k^\alpha k_\beta H_{,\gamma} - H k^\alpha k_{\beta;\gamma} - H k^\alpha{}_{;\gamma} k_\beta - H_{,\beta} k^\alpha k_\gamma$$
$$- H k^\alpha k_{\gamma;\beta} - H k^\alpha{}_{;\beta} k_\gamma + \eta^{\alpha\rho} H_{,\rho} k_\beta k_\gamma + \eta^{\alpha\rho} H k_{\beta;\rho} k_\gamma$$
$$+ \eta^{\alpha\rho} H k_\beta k_{\gamma;\rho} + 2 H k^\rho H_{,\rho} k^\alpha k_\beta k_\gamma. \tag{21.11}$$

The covariant derivatives are with respect to $\eta_{\alpha\beta}$. This formula shows that the difference of two Christoffel symbols is a tensor, in agreement with (4.23). A direct calculation with use of (21.3) rewritten in terms of k^μ, of (21.10) – (21.11) and of the identities mentioned above, gives for the Riemann tensor

$$R_{\alpha\beta\gamma\delta} k^\beta k^\delta = H_{;\mu\nu} k^\mu k^\nu k_\alpha k_\gamma. \tag{21.12}$$

In vacuum, the Weyl tensor will obey the same equation. But then, it will obey also (11.55), and so we have derived

Lemma 21.1 *The vacuum Kerr–Schild metrics cannot be of the general Petrov type, they are **algebraically special**, i.e. of Petrov type II or simpler. The Kerr–Schild vector field k^α is at the same time the degenerate Debever vector of the Weyl tensor.*

From the Goldberg–Sachs theorem 16.4 it follows then immediately that the field k^α is not only geodesic but also shearfree.

21.2 The derivation of the Kerr metric by the original method

The Kerr solution has no clear-cut invariant definition. In order to understand how it came about, it is most instructive to follow the historical path. The presentation in this section is based on the paper by Kerr and Schild (1965). (The derivation by Debney, Kerr and Schild (1969) used a different tetrad and a different notation.)

In the Minkowski space, let us choose the real null coordinates (u, v) and the complex null coordinates $(\xi, \bar\xi)$, related to the Lorentzian (t, x, y, z) by

$$(u, v, \xi, \bar\xi) = (t + x, t - x, y + iz, y - iz)/\sqrt{2}. \tag{21.13}$$

In these coordinates, the Minkowski metric is

$$[\eta_{\alpha\beta}] = [\eta^{\alpha\beta}] = \begin{bmatrix} 0 & 1 & 0 & 0 \\ 1 & 0 & 0 & 0 \\ 0 & 0 & 0 & -1 \\ 0 & 0 & -1 & 0 \end{bmatrix}. \tag{21.14}$$

A real null direction field in the Minkowski spacetime is in these coordinates given by

$$k_\mu dx^\mu = du + Y\bar{Y}dv + \bar{Y}d\xi + Yd\bar\xi \iff k^\mu = (Y\bar{Y}, 1, -Y, -\bar{Y}), \tag{21.15}$$

where $Y(u, v, \xi, \bar\xi)$ is an arbitrary complex function. If this field is to be geodesic and affinely parametrised, then the function Y must obey

$$k^\rho Y_{,\rho} = k^\rho \bar{Y}_{,\rho} = 0, \tag{21.16}$$

this follows from $k^\rho k^\alpha{}_{;\rho} = 0$ using (21.3). For the metric (21.14), $\left\{{\alpha \atop \beta\gamma}\right\}(\eta) = 0$.

Let us now consider a Kerr–Schild metric (21.9) with (21.14) as the background Minkowski metric and (21.15) as k^α. Let us introduce the double-null tetrad of Sec. 16.6 with $e_0{}^\alpha = k^\alpha$. The simplest choice of the other tetrad vectors is[1]

$$
\begin{aligned}
e_1{}^\alpha &\stackrel{\text{def}}{=} \ell^\alpha = \delta^\alpha{}_0 + Hk^\alpha, & \ell_\alpha dx^\alpha &= dv - Hk_\alpha dx^\alpha \\
e_2{}^\alpha &\stackrel{\text{def}}{=} m^\alpha = (\overline{Y}, 0, -1, 0), & m_\alpha dx^\alpha &= \overline{Y}dv + d\overline{\xi}, \\
e_3{}^\alpha &\stackrel{\text{def}}{=} \overline{m}^\alpha = (Y, 0, 0, -1), & \overline{m}_\alpha dx^\alpha &= Ydv + d\xi.
\end{aligned}
\tag{21.17}
$$

Then, by (16.52), the field (21.15) will be shearfree if Y and $\overline{Y}$ obey in addition

$$
\overline{m}^\rho Y{}_{,\rho} = m^\rho \overline{Y}{}_{,\rho} = 0 \quad \Longrightarrow \quad m^\rho m^\alpha{}_{,\rho} = \overline{m}^\rho \overline{m}^\alpha{}_{,\rho} = 0.
\tag{21.18}
$$

By virtue of (16.55) we also have

$$
Z = \theta + i\omega \equiv \Gamma^1{}_{23} = \overline{m}^\sigma \overline{Y}{}_{,\sigma}.
\tag{21.19}
$$

In what follows we will assume that $Z = \theta + i\omega \neq 0$.

The tetrad (21.15) – (21.17) is not the special one we used in Section 16.8 following (16.96). Nevertheless, some of the Ricci rotation coefficients vanish independently of the property $\Gamma^1{}_{21} = 0$; they are listed in the last line of (21.20). The other ones vanish in consequence of the form of the tetrad (21.15) – (21.17):[2]

$$
\begin{aligned}
\Gamma^a{}_{b0} &= \Gamma^0{}_{02} = \Gamma^1{}_{12} = \Gamma^0{}_{03} = \Gamma^1{}_{13} = \Gamma^a{}_{22} = \Gamma^a{}_{33} = \Gamma^2{}_{23} = \Gamma^3{}_{32} = 0, \\
\Gamma^a{}_{00} &= \Gamma^1{}_{a0} = \Gamma^1{}_{22} = \Gamma^1{}_{33} = 0.
\end{aligned}
\tag{21.20}
$$

With (21.14), the background Christoffel symbols vanish and the covariant derivatives in (21.11) become partial derivatives. Using (21.11) thus simplified, we find

$$
\begin{aligned}
\Gamma^0{}_{01} &= -\Gamma^1{}_{11} = -k^\rho H{}_{,\rho}, & \Gamma^1{}_{23} &= \Gamma^3{}_{03} = Z, \\
\Gamma^0{}_{21} &= \Gamma^3{}_{11} = -m^\rho H{}_{,\rho} - H\overline{Y}{}_{,u}, & \Gamma^1{}_{21} &= \Gamma^3{}_{01} = \overline{Y}{}_{,u}, \\
\Gamma^0{}_{23} &= \Gamma^3{}_{13} = H\overline{Z}, & \Gamma^2{}_{21} &= H(Z - \overline{Z})
\end{aligned}
\tag{21.21}
$$

(plus the Γ-s complex conjugate to those above that result by interchanging $2 \leftrightarrow 3$).

In calculating the $R^i{}_{jkl}$ below one must keep in mind the nonobvious identities like $\Gamma^2{}_{03} = -\Gamma_{303} = \Gamma_{033} = \Gamma^1{}_{33} = 0$ (the last step by (21.20)). If a certain component $\Gamma^i{}_{jk}$ is not found in the list (21.19) – (21.21), then such identities should be employed.

Substituting (21.21) in the equation $0 = R_{23} = R^0{}_{203} + R^1{}_{213} + R^2{}_{223}$, then using (9.21), (21.16) – (21.19) and (16.72) we find

$$
k^\rho H{}_{,\rho}(Z + \overline{Z}) + H\left(Z^2 + \overline{Z}^2\right) = 0.
\tag{21.22}
$$

Since Z obeys $k^\rho Z{}_{,\rho} = -Z^2$ (from (16.72)), the above says that $k^\rho[H/(Z + \overline{Z})]{}_{,\rho} = 0$, so

$$
H = e^{3P}(Z + \overline{Z}),
\tag{21.23}
$$

where P is a real function such that $k^\rho P{}_{,\rho} = 0$.

---

[1] The tetrad (21.17) obeys the orthogonality relations (16.45) with respect to both $g_{\alpha\beta}$ and $\eta_{\alpha\beta}$.
[2] In this section, the indices of the Riemann and Ricci tensors are tetrad indices.

In further calculations, the two equations given below and their complex conjugates will be helpful. They follow from (21.19) with use of (21.16) – (21.18):

$$m^\rho Z_{,\rho} = (\overline{Z} - Z)\overline{Y}_{,u}, \tag{21.24}$$

$$\ell^\rho Z_{,\rho} = Y_{,u}\overline{Y}_{,u} - HZ^2 + \overline{m}^\rho \overline{Y}_{,u\rho}. \tag{21.25}$$

In consequence of (21.10) – (21.11) and (21.19) – (21.23) $R_{10} = 0$ is fulfilled identically (see Exercise 7), while $R_{12} = 0$ and $R_{13} = \overline{R_{12}} = 0$ give (see Exercise 8)

$$m^\rho P_{,\rho} = -\frac{\overline{Z}}{Z}\,\overline{Y}_{,u} \equiv -\frac{\overline{Z}}{Z}\,\ell^\rho\overline{Y}_{,\rho}\,, \qquad \overline{m}^\rho P_{,\rho} = -\frac{Z}{\overline{Z}}\,Y_{,u} \equiv -\frac{Z}{\overline{Z}}\,\ell^\rho Y_{,\rho}. \tag{21.26}$$

These two equations imply the integrability condition (16.106) that says $\overline{m}^\rho\,(m^\sigma P_{,\sigma})_{,\rho} - m^\rho\,(\overline{m}^\sigma P_{,\sigma})_{,\rho} = 2\Gamma^s{}_{[23]}e_s{}^\rho P_{,\rho}$. Explicitly, it reads

$$S \overset{\text{def}}{=} \frac{Z}{\overline{Z}^2}Y_{,u}\,m^\rho\overline{Z}_\rho - \frac{Z}{\overline{Z}}m^\rho Y_{u\rho} - \frac{\overline{Z}}{Z^2}\overline{Y}_{,u}\,\overline{m}^\rho Z_\rho + \frac{\overline{Z}}{Z}\overline{m}^\rho\overline{Y}_{,u\rho}$$

$$+ \left(\frac{Z}{\overline{Z}} - \frac{\overline{Z}}{Z}\right)Y_{,u}\,\overline{Y}_{,u} - \ell^\rho P_{,\rho}\,(\overline{Z} - Z) = 0. \tag{21.27}$$

The equation $R_{11} = 0$ is the most difficult to handle. The tricky point is to note that $R_{11} = R^2{}_{121} + R^3{}_{131} \equiv -2R_{1213} = -2R^0{}_{213}$, this reduces the number of $R^i{}_{jkl}$ to calculate. See Sec. 24.8 for hints on how to verify the equation $R^0{}_{213} = 0$ given below:

$$\ell^\rho P_\rho \overline{Z}(Z + \overline{Z}) \;+\; \frac{1}{Z}\,(Z + \overline{Z})^2 Y_{,u}\,\overline{Y}_{,u} + Zm^\rho Y_{,u\rho} - \frac{\overline{Z}^2}{Z}\overline{m}^\rho\overline{Y}_{,u\rho} + \frac{\overline{Z}^2}{Z^2}Y_{,u}\,\overline{m}^\rho Z_\rho$$

$$- \frac{\overline{Z}}{Z}Y_{,u}\,m^\rho\overline{Z}_\rho = 0. \tag{21.28}$$

Taking the imaginary part of the above we obtain $(Z + \overline{Z})S = 0$, where S is defined in (21.27) and $S = 0$ is already imposed.

At this point, the situation with the $R_{ij} = 0$ equations is this:

- $R_{00} = 0$ is the null Raychaudhuri equation (16.72) which became $k^\gamma Z_{,\gamma} = -Z^2$ (since Lemma 21.1 implied $\sigma = 0$), so it has already been used.
- $R_{10} \equiv 0$ as stated below (21.25).
- $R_{12} = R_{13} = 0$ were used to derive (21.26).
- $R_{02} = R_{03} = R_{22} = R_{33} \equiv 0$ by virtue of (21.20) – (21.21).
- $R_{11} = 0$ was used to derive (21.28), and we found that the imaginary part of this equation is fulfilled by virtue of (21.27).

So, it only remains to take into account the other part of $R_{11} = 0$. Using (21.27) in (21.28) to eliminate the $Y_{,u}\,m^\rho Z_{,\rho}$ term we obtain the last equation of the $R_{ij} = 0$ set:

$$\ell^\rho P_{,\rho} = -\,(1/Z + 1/\overline{Z})\,Y_{,u}\,\overline{Y}_{,u}. \tag{21.29}$$

The functions Y and $\overline{Y}$ are independent – see Exercise 9. Each of $(P, Y, \overline{Y})$ obeys

$$k^\rho\phi_{,\rho} = 0, \qquad v^\rho\phi_{,\rho} = 0, \qquad v^\rho \overset{\text{def}}{=} \ell^\rho - \frac{\ell^\sigma Y_{,\sigma}}{\overline{Z}}m^\rho - \frac{\ell^\sigma\overline{Y}_{,\sigma}}{Z}\overline{m}^\rho, \tag{21.30}$$

where ϕ stands for any of $(P, Y, \overline{Y})$. The vectors k^α and v^α are linearly independent. This means that the gradients of P, Y and $\overline{Y}$ are orthogonal to two linearly independent vectors, i.e. that they lie in the same 2-dimensional plane within the tangent vector space at each point of the manifold. Consequently, the gradients $P_{,\alpha}$, $Y_{,\alpha}$ and $\overline{Y}_{,\alpha}$ are linearly dependent, so there exists a functional relation of the form $\psi(P, Y, \overline{Y}) = 0$, and hence P is a function of Y and $\overline{Y}$. Then, Eqs. (21.26) become by virtue of (21.28):

$$P_{,Y} = -\overline{Y}_{,u}/Z, \qquad P_{,\overline{Y}} = -Y_{,u}/\overline{Z}. \tag{21.31}$$

With this, and with $P = P(Y, \overline{Y})$, (21.29) is satisfied identically. Calculating the directional derivative of the first equation in (21.31) along m^α, using (21.19), (21.24) and $m^\alpha \overline{Y}_{,u\alpha} = -m^\alpha_{,u} \overline{Y}_{,\alpha}$ (by (21.28)) $= -\overline{Y}_{,u}^2$ (by (21.17)) we obtain $\overline{Z} P_{,YY} = \overline{Z} P_{,Y}^2$. Then, calculating $\overline{m}^\alpha \partial/\partial x^\alpha$ of the second equation in (21.31) and carrying out the analogous operations we obtain $Z P_{,\overline{YY}} = Z P_{,\overline{Y}}^2$. These two equations are equivalent to

$$\left(e^{-P}\right)_{,YY} = 0 = \left(e^{-P}\right)_{,\overline{YY}}. \tag{21.32}$$

The solution of this is $e^{-P} = B + A_1 \overline{Y} + A_2 Y + CY\overline{Y}$, where A_1, A_2, B and C are constants. Since e^{-P} must be real, it follows that B and C are arbitrary real, while $A \stackrel{\text{def}}{=} \overline{A}_1 = A_2$ can be complex (they are real when $A_1 = A_2$). Thus,

$$e^{-P} = B + \overline{A}\overline{Y} + AY + CY\overline{Y}. \tag{21.33}$$

Each of the functions Y, $\overline{Y}$, Z and $\overline{Z}$ satisfies the equation (see Exercise 10)

$$\mathcal{K}^\rho \phi_{,\rho} \stackrel{\text{def}}{=} B\phi_{,u} + C\phi_{,v} + A\phi_{,\overline{\xi}} + \overline{A}\phi_{,\xi} = 0, \tag{21.34}$$

where ϕ is any of $(Y, \overline{Y}, Z, \overline{Z})$. It follows that P, H and all components of k_μ obey the same equation, so $\mathcal{K}^\rho g_{\alpha\beta,\rho} = 0$ for the whole metric. The components of $\mathcal{K}^\alpha$ are constant, so by (8.12) $\mathcal{K}^\alpha$ is a Killing field – for both $g_{\alpha\beta}$ and $\eta_{\alpha\beta}$.

The coordinates $(u, v, \xi, \overline{\xi})$ are not yet defined uniquely – we are free to transform them so as to preserve the form of $\eta_{\alpha\beta}$. Let us then assume that the Killing field $\mathcal{K}^\alpha$ is timelike[3] with respect to $\eta^{\mu\nu}$, and let us carry out such a transformation after which $\mathcal{K}^\alpha$ becomes collinear with the time axis of the background Minkowski space, thus

$$A = 0, \qquad B = C > 0 \Longrightarrow \mathcal{K}^\alpha = B(1, 1, 0, 0). \tag{21.35}$$

Then, equations (21.16), (21.18) and (21.34) imply

$$Y_{,\overline{\xi}} = YY_{,u}, \qquad Y_{,\xi} = Y_{,v}/Y, \qquad Y_{,u} + Y_{,v} = 0. \tag{21.36}$$

Each of these is a quasi-linear partial differential equation of first order and can be solved by standard methods (Courant and Hilbert, 1965; Sneddon, 1957). Combining the three solutions we obtain

$$F(Y, \xi, \overline{\xi}, u - v) \stackrel{\text{def}}{=} Y^2\overline{\xi} - \xi + (u - v)Y + \Phi(Y) = 0, \tag{21.37}$$

where Φ is an arbitrary complex function of one variable.

---

[3] Kerr and Schild (1965) and Debney et al. (1969) considered also the cases when $\mathcal{K}^\alpha$ is spacelike or null.

Calculating $0 = m^\rho F_{,\rho} \equiv \overline{Y} F_{,u} - F_{,\xi}$ we find

$$\overline{Z} = -(1 + Y\overline{Y})/F_{,Y} \qquad (21.38)$$

(where $F_{,Y} = 2Y\overline{\xi} + u - v + \Phi_{,Y}$). Then, calculating $0 = F_{,u} \equiv F_{,Y} Y_{,u} + Y$ we find, with use of (21.31) and (21.38),

$$Y_{,u} = -Y/F_{,Y} = -\overline{Z} P_{,\overline{Y}} = (1 + Y\overline{Y}) P_{,\overline{Y}}/F_{,Y} . \qquad (21.39)$$

Hence $P_{,\overline{Y}} = -Y/(1 + Y\overline{Y})$, so $\mathrm{e}^{-P} = \text{constant} \times (1 + Y\overline{Y})$ and

$$\mathrm{e}^{3P} = \sqrt{2} m/(1 + Y\overline{Y})^3, \qquad (21.40)$$

where m is an arbitrary constant.

Equations (21.9), (21.15), (21.23), (21.40), (21.37) and (21.38) implicitly define the most general vacuum Kerr–Schild metric in which the (necessarily existing) Killing field is timelike. The definition is implicit because Y has not been given as a function of the coordinates. In order to find it, we must specify the function $\Phi(Y)$ in (21.37). It is here that we shall make a noncovariant assumption: we assume that $\Phi(Y) = \alpha Y^2 + \beta Y - \overline{\alpha}$.

We adapted the coordinates to $\mathcal{K}^\alpha$, but we are still free to carry out the transformations preserving the direction of $\mathcal{K}^\alpha$. These are the translations and rotations in the 3-space orthogonal to $\mathcal{K}^\alpha$. By $\overline{\xi} = \overline{\xi}' - \alpha$ we then cancel the term $\alpha Y^2 - \overline{\alpha}$ in $\Phi(Y)$, and by one of the translations $u = u' - \mathrm{Re}(\beta)$ or $v = v' + \mathrm{Re}(\beta)$ we cancel the real part of β. Finally, we obtain $\Phi = -\mathrm{i}\sqrt{2} a Y$, where $\mathrm{Im}(\beta) = -\sqrt{2} a$ was assumed for later convenience.

With this Φ, we transform back to the real coordinates (t, x, y, z) by

$$\xi = \frac{1}{\sqrt{2}}(x + \mathrm{i}y), \qquad u = \frac{1}{\sqrt{2}}(t + z), \qquad v = \frac{1}{\sqrt{2}}(t - z), \qquad (21.41)$$

and F becomes

$$F = (x - \mathrm{i}y)Y^2/\sqrt{2} + \sqrt{2}(z - \mathrm{i}a)Y - (x + \mathrm{i}y)/\sqrt{2}, \qquad (21.42)$$

the function $Y(x, y, z)$ being defined by $F = 0$. In order to simplify the final expression, we introduce the real function $r(x, y, z) > 0$ by[4]

$$\frac{x^2 + y^2}{r^2 + a^2} + \frac{z^2}{r^2} = 1. \qquad (21.43)$$

The solution of $F = 0$ is now[5]

$$Y = \frac{(r - z)(r + \mathrm{i}a)}{r(x - \mathrm{i}y)} . \qquad (21.44)$$

Substituting this where necessary, we obtain

$$F_{,Y} = \frac{\sqrt{2}}{r}\left(r^2 - \mathrm{i}az\right), \qquad Y\overline{Y} = \frac{r - z}{r + z}, \qquad \overline{Z} = -\frac{\sqrt{2}r^2\left(r^2 + \mathrm{i}az\right)}{(r + z)\left(r^4 + a^2 z^2\right)},$$

---

[4] Interpreting x, y and z as Cartesian coordinates, the surfaces $r = \text{constant}$ are confocal ellipsoids of revolution, a being the radius of the focal ring and r being the semiminor axis of an ellipsoid.

[5] We discard the other solution of $F = 0$ obtained by $r \to -r$ because we assumed $r > 0$. The metric *can* be extended to $r < 0$, but the extension involves nontrivial problems – see Sec. 21.9.

$$e^{3P} = \frac{\sqrt{2}m(r+z)^3}{8r^3}, \qquad H = -\frac{mr(r+z)^2}{2\left(r^4 + a^2 z^2\right)}, \tag{21.45}$$

and then substituting (21.41) in (21.15) we obtain in (21.9)

$$\begin{aligned}
ds^2 &= dt^2 - dx^2 - dy^2 - dz^2 - \frac{2mr^3}{r^4 + a^2 z^2}\left[dt + \frac{z}{r}dz\right.\\
&\left. + \frac{r}{r^2 + a^2}(x\,dx + y\,dy) + \frac{a}{r^2 + a^2}(x\,dy - y\,dx)\right]^2.
\end{aligned} \tag{21.46}$$

This is the form in which the Kerr metric first appeared in the literature,[6] but Kerr (1963) did not give any hint on how it had been derived. Nothing in that paper presaged the prominence that the Kerr metric acquired later. It is the simplest exact solution of Einstein's equations that describes the exterior field of a stationary rotating body or black hole. Because of this, it became the basis of a great many papers discussing astrophysical aspects of black holes. Also, it is believed to be the universal asymptotic state towards which all nonstationary uncharged black holes should evolve. This statement is based on the still disputed **cosmic censorship hypothesis** (see Section 18.18). Consequently, the Kerr metric frequently appears in papers on geometrical aspects of black holes.

21.3 Basic properties

The form (21.46) becomes the Minkowski metric when $m = 0$. Also, it is approximately flat at large r – the Kerr–Schild term is then negligible compared with the Lorentzian part. Thus, at large r (21.46) obeys the assumptions the weak-field approximation to relativity discussed in Sec. 12.18. The component

$$g_{00} = 1 - \frac{2mr^3}{r^4 + a^2 z^2} \approx 1 - \frac{2m}{r} + \frac{2ma^2 z^2}{r^5} + O(1/r^5), \tag{21.47}$$

when compared with (12.155), shows that m is the mass of the source. The g_{0I} components, however, are not exactly of the form (12.156) because of the terms of order $1/r$:

$$g_{tx} = -\frac{2mr^4 x - 2mar^3 y}{(r^2 + a^2)(r^4 + a^2 z^2)}, \qquad g_{ty} = -\frac{2mr^4 y + 2mar^3 x}{(r^2 + a^2)(r^4 + a^2 z^2)},$$

$$g_{tz} = -\frac{2mr^2 z}{r^4 + a^2 z^2}. \tag{21.48}$$

The unwanted terms are $2mr^4 x$ and $2mr^4 y$ in g_{tx} and g_{ty}, and the whole of g_{tz}. We guess that we need a transformation (12.130) to transform the metric as in (12.152). The simplest choice is $b_I = 0$ and a time-independent b_0. The following b_0 will do the job:

$$b_{0,x} = -\frac{2mr^4 x}{(r^2 + a^2)(r^4 + a^2 z^2)}, \qquad b_{0,y} = -\frac{2mr^4 y}{(r^2 + a^2)(r^4 + a^2 z^2)},$$

$$b_{0,z} = -\frac{2mr^2 z}{r^4 + a^2 z^2}. \tag{21.49}$$

---

[6] Actually, the signature in Kerr's paper was $(-+++)$.

These derivatives of b_0 identically obey the integrability conditions $b_{0,xy} = b_{0,yx}$ and $b_{0,xz} = b_{0,zx}$ (by virtue of (21.43)), so such a b_0 does exist, and the transformed g_{0I} are:

$$\tilde{g}_{tx} = \frac{2mar^3 y}{(r^2 + a^2)(r^4 + a^2 z^2)} \approx \frac{2may}{r^3} + O(1/r^3),$$

$$\tilde{g}_{ty} = -\frac{2mar^3 x}{(r^2 + a^2)(r^4 + a^2 z^2)} \approx -\frac{2max}{r^3} + O(1/r^3), \qquad \tilde{g}_{tz} = 0. \qquad (21.50)$$

These are of the form (12.156) with $P^I = (0, 0, -ma)$ (where $(x^1, x^2, x^3) = (x, y, z)$). Thus, $(-a)$ is the total angular momentum of the source per unit mass. This much was said in the original paper of Kerr (1963), and the physical interpretation became clear.

The coordinates of (21.46) are useful for interpreting the parameters, but are not convenient in other calculations. Often, it is more practical to use r as one of the coordinates. One such coordinate system, introduced by Kerr (1963), is related to (21.46) by

$$x = (r \cos \varphi + a \sin \varphi) \sin \vartheta \equiv \sqrt{r^2 + a^2} \sin \vartheta \cos[\varphi - \arctan(a/r)],$$

$$y = (r \sin \varphi - a \cos \varphi) \sin \vartheta \equiv \sqrt{r^2 + a^2} \sin \vartheta \sin[\varphi - \arctan(a/r)],$$

$$z = r \cos \vartheta, \qquad (21.51)$$

where $r \geq 0$, $\vartheta \in [0, \pi]$ and $\varphi \in [0, 2\pi]$. The transformed metric is

$$ds^2 = dt^2 - dr^2 - 2a \sin^2 \vartheta dr d\varphi - \Sigma d\vartheta^2 - (r^2 + a^2) \sin^2 \vartheta d\varphi^2$$

$$- \frac{2mr}{\Sigma} (dt + dr + a \sin^2 \vartheta d\varphi)^2, \qquad \Sigma \overset{\text{def}}{=} r^2 + a^2 \cos^2 \vartheta. \qquad (21.52)$$

In these coordinates the metric is independent of $x^3 = \varphi$, which shows that one more Killing vector field exists in addition to the timelike $\underset{(1)}{k}{}^\alpha = \delta^\alpha_0$; it is $\underset{(2)}{k}{}^\alpha = \delta^\alpha_3$. The Abelian group generated by $\underset{(1)}{k}{}^\alpha$ and $\underset{(2)}{k}{}^\alpha$ is the complete symmetry group of the Kerr metric, as can be verified by solving the Killing equations. The coordinate φ becomes the azimuthal angle in the limit $m = 0 = a$, so we say that the Kerr metric is axially symmetric.

At infinity, the (x, y, z) coordinates of (21.46) become Cartesian rectangular. Then, as already stated, the surfaces of constant r are confocal ellipsoids of revolution whose foci lie on the ring $\{z = 0, x^2 + y^2 = a^2\}$. On this ring $r = 0$ and $\vartheta = \pi/2$; it is a singularity of the coordinates of (21.52), and, as we shall see below, a singularity of the spacetime as well. It is seen, using (21.51), that the following holds

$$\frac{x^2 + y^2}{a^2 \sin^2 \vartheta} - \frac{z^2}{a^2 \cos^2 \vartheta} = 1, \qquad (21.53)$$

which shows that the surfaces $\vartheta = $ constant are one-sheeted hyperboloids of revolution, with foci on the same ring. It follows from (21.43) that the hyperboloid corresponding to $\vartheta = \pi/2$ is degenerated to the plane $z = 0$ with the interior of the ring $r = 0$ removed. Figure 21.1 shows the axial cross-section through the family of the constant-(r, ϑ) surfaces.

The geometry of the $\varphi = $ constant surfaces is more complicated. To see it, let us change the variables in (21.51) to

$$\xi = x \cos \varphi + y \sin \varphi, \qquad \eta = -x \sin \varphi + y \cos \varphi. \qquad (21.54)$$

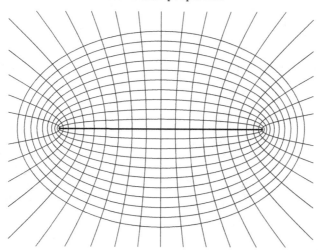

Fig. 21.1 A cross-section through the space t = constant by a plane containing the axis of symmetry. The figure shows the surfaces of constant r (the ellipses) and the surfaces of constant ϑ (the hyperbolae). The ellipsoids and the hyperboloids all have their foci on the ring of radius a, seen here as the thick horizontal line. The (r, ϑ, φ) coordinates are singular on this ring; it has the equation $r = 0$, $\vartheta = \pi/2$. When $\vartheta \to \pi/2$, the hyperboloid degenerates to the $z = 0$ plane with the interior of the disc $r = 0$ removed.

It follows then from (21.51) that

$$(\xi = r \sin \vartheta, \ \eta = -a \sin \vartheta) \iff (a/\eta)^2 - (z/\xi)^2 = 1. \tag{21.55}$$

The coordinates (ξ, η) are thus obtained by rotating the x- and y-axes by the angle φ and at $\varphi = 0$ they coincide with (x, y). Since $r \geq 0$ and $\vartheta \in [0, \pi]$, we have $\xi \geq 0$ and $\eta/a \leq 0$. The surface (21.55) consists of two sheets, $z \geq 0$ and $z \leq 0$ that are mirror images of each other; it is shown in Fig. 21.2. Other φ = constant surfaces are obtained by rotating the one in Fig. 21.2 around the z-axis. The surface contains the straight half-lines $\{z = \xi \cot \vartheta, \vartheta = \text{constant}\}$ along which it intersects the ϑ = constant hyperboloids.

A still more readable form of the Kerr metric results when the **Boyer–Lindquist** (1967) (B–L) **coordinates** are introduced. They are related to those of (21.52) by[7]

$$t \ = \ t' + 2m \int \frac{r\mathrm{d}r}{\Delta_r(r)}, \qquad \varphi = -\varphi' - a \int \frac{\mathrm{d}r}{\Delta_r(r)},$$

$$\Delta_r(r) \ \overset{\text{def}}{=} \ r^2 - 2mr + a^2, \qquad r = r', \quad \vartheta = \vartheta'. \tag{21.56}$$

The resulting metric, with primes dropped, is (see Exercise 12)

$$ds^2 \ = \ \left(1 - \frac{2mr}{\Sigma}\right) dt^2 + \frac{4mra \sin^2 \vartheta}{\Sigma} dt d\varphi - \frac{\Sigma}{\Delta_r} dr^2 - \Sigma d\vartheta^2$$

$$- \left(\frac{2mra^2 \sin^2 \vartheta}{\Sigma} + r^2 + a^2\right) \sin^2 \vartheta d\varphi^2. \tag{21.57}$$

[7] The φ' in (21.56) and the term with $dt d\varphi$ in (21.57) have opposite signs to those implied by Boyer and Lindquist (1967). The changed form is consistent with the papers referred to in subsequent sections.

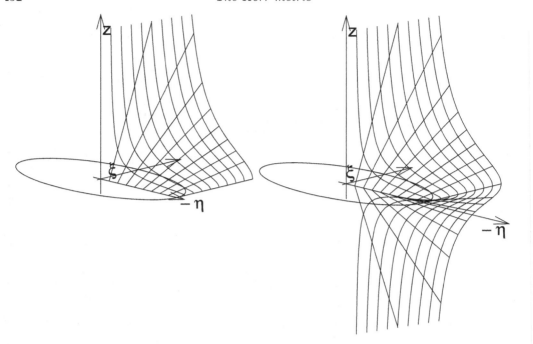

Fig. 21.2 A surface of constant φ in the coordinates of (21.52). The tangent planes to this surface become vertical at the right edge (where $\eta = -a$) and at $z \to \pm\infty$. The distance from the z-axis to the right edge of the surface is $|a|$. The surface contains the straight half-lines $\{z = \xi \cot \vartheta, \xi \geq 0\}$ seen in the figure, they are intersections with the surfaces of constant ϑ. The left panel shows the upper half of this surface. The other half is its mirror reflection in the $z = 0$ plane; the right panel shows both halves. Other surfaces of constant φ are obtained by rotating this one around the z-axis. The η-axis is drawn in reverse, so the coordinate on it is $(-\eta)$. The circle is the singular ring $\{r = |a|, \vartheta = \pi/2\}$.

The limit $a = 0$ of this is seen to be the Schwarzschild metric.

The Schwarzschild metric has a spurious singularity at $r = 2m$, where $g_{rr} \to \infty$ and $g_{00} = 0$. The coordinate t in the Schwarzschild metric (and also in the Kerr metric in the B–L coordinates) coincides with the proper time of an observer at infinity. The ratio of the proper time interval of the observer at rest in the set $g_{00} = 0$ to the corresponding proper time interval of infinity is zero, which means that light emitted from $g_{00} = 0$ arrives to a distant observer with an infinite redshift. In the Kerr metric, the set $g_{rr} = \infty$ does not coincide with $g_{00} = 0$. The first one, given by $r^2 - 2mr + a^2 = 0$, exists only when $a^2 \leq m^2$ and will be seen later to be a spurious singularity, removable by a coordinate transformation. With $a^2 < m^2$, the spurious singularity consists of two disjoint sets

$$r = r_\pm \overset{\text{def}}{=} m \pm \sqrt{m^2 - a^2}. \tag{21.58}$$

The set $r^2 - 2mr + a^2 \cos^2 \vartheta = 0$ is sometimes called an *infinite redshift hypersurface*, by analogy with the Schwarzschild metric. However, as rightly pointed out by Carter (1973), this name is misleading. An observer at rest in the B–L coordinates has the coordinates r, ϑ and φ all constant. On the hypersurface $g_{00} = 0$ and inside it (where $g_{00} < 0$), for such an 'observer' the interval $ds^2 \leq 0$, which means that she/he would have to be moving with the velocity of light or faster just to remain at rest. This means that the state of rest with respect to infinity is impossible where $g_{00} \leq 0$ – stationary observers do not exist in that region. This is why this hypersurface has also a second, more appropriate name: the **stationary limit hypersurface**.

It is seen from (21.57) that with $a^2 < m^2$, the coordinate r becomes time in those regions where $\Delta_r < 0$. Consequently, just as in the Schwarzschild metric, no material body can stay at constant r there. In those regions where $\Delta_r > 0$ and $g_{00} \leq 0$, null vectors must necessarily have a nonzero φ-component. The minimal value of the φ-component is found from $g_{\mu\nu} v^\mu v^\nu = 0$ with $v^r = v^\vartheta = 0$ using (21.57); it is

$$v_{\min}^\varphi = v^0 \frac{2mra - \varepsilon \Sigma \sqrt{\Delta_r}/\sin\vartheta}{2mra^2\sin^2\vartheta + \Sigma(r^2 + a^2)}, \tag{21.59}$$

where v^0 is the t-component and $\varepsilon \overset{\text{def}}{=} a/|a|$. (As expected, $v_{\min}^\varphi \to 0$ when $g_{00} \to 0$, i.e. when $\Sigma \to 2mr$ and $\Delta_r \to a^2\sin^2\vartheta$. The other solution of $ds^2 = 0$ at $dr = d\vartheta = 0$, with $+$ in the numerator, does not have this property.) This is an example of the phenomenon of **frame dragging** in the gravitational field of rotating bodies, which was mentioned in Section 12.18; see after (12.158).

When ϑ is near $\pi/2$, while $r < 0$ and $|r|$ is sufficiently small, the term $(2mra^2\sin^2\vartheta/\Sigma)$ in the component $g_{\varphi\varphi}$ of (21.57) becomes negative and large in absolute value, which means that φ becomes a timelike coordinate. Hence, the curves of constant t, ϑ and r in that region are timelike. If we require that, by continuity, these lines are closed with period 2π also in that region, then *closed timelike curves exist in the $r < 0$ sheet of the extended Kerr manifold*. This applies to all three varieties of the Kerr solution. This observation was reported by Boyer and Lindquist (1967), and attributed to a remark by Carter.

The existence or nonexistence of regions with $\Delta_r \leq 0$, their geometry and their relation to the regions with $g_{00} \leq 0$ will be discussed in Section 21.5. Their physical meaning will emerge gradually later.

The singularities of the Kerr metric are best recognised when one calculates the tetrad components of the Riemann tensor in the following orthonormal tetrad (Carter, 1973):

$$e^0 = \sqrt{\frac{\Delta_r}{\Sigma}}\left(dt - a\sin^2\vartheta d\varphi\right), \qquad e^1 = \sqrt{\frac{\Sigma}{\Delta_r}}dr,$$

$$e^2 = \sqrt{\Sigma}d\vartheta, \qquad e^3 = \sin\vartheta \frac{adt - (r^2 + a^2)d\varphi}{\sqrt{\Sigma}}. \tag{21.60}$$

The tetrad components of the Weyl tensor (equal to the Riemann tensor) are then

$$\begin{aligned}
C_{0101} &= -C_{2323} = 2I_1, \qquad C_{0123} = 2C_{0213} = -2C_{0312} = 2I_2, \\
C_{0202} &= C_{0303} = -C_{1212} = -C_{1313} = -I_1, \\
I_1 &\overset{\text{def}}{=} mr \frac{r^2 - 3a^2\cos^2\vartheta}{\Sigma^3}, \qquad I_2 \overset{\text{def}}{=} ma\cos\vartheta \frac{3r^2 - a^2\cos^2\vartheta}{\Sigma^3}. \tag{21.61}
\end{aligned}$$

The singularity is located on the ring $\{r = 0, \vartheta = \pi/2\}$, where $\Sigma = 0$. Since the interior of the ring is nonsingular, one can extend the metric through it to $r < 0$. From (21.51) we see that the (t, x, y, z) coordinates cover this extension. With $a^2 \leq m^2$, the spurious singularities lie in the $r > 0$ region, so the extension does not result in just another copy of the same set. We will come back to this in Section 21.9.

Equations (21.61) imply that the Kerr metric is of Petrov type D.

21.4 * Derivation of the Kerr metric by Carter's method – from the separability of the Klein–Gordon equation

The Kerr metric emerged in Section 21.2 after imposing coordinate-dependent assumptions of unknown interpretation on the vacuum Einstein equations. It would be good to have an invariant definition that would single out this metric by geometric properties. Carter (1973) came one step towards this goal by considering generalisations of the Schwarzschild metric in which the Klein–Gordon equation, in appropriately chosen coordinates, is still separable. Although this method is not unique and not invariant either, and makes ample use of the expected result, its bonus is that it led to the generalisation the Kerr metric for the cosmological constant and the electric/magnetic charge in a rather simple way. We will present this derivation now.[8]

The Klein–Gordon equation in curved space has the form

$$\frac{1}{\psi} \frac{\partial}{\partial x^\alpha} \left(\sqrt{-g} g^{\alpha\beta} \frac{\partial \psi}{\partial x^\beta} \right) - m_0{}^2 \sqrt{-g} = 0, \tag{21.62}$$

where m_0 is not to be confused with the mass parameter m of the Kerr metric. Equation (21.62) is called separable when the function ψ is a product of functions of one variable, $\psi = \prod_i \psi_i(x^i)$. Such a form of ψ causes the left-hand side of (21.62) to become a sum of four ordinary differential equations, each involving just one coordinate.

In considering the Klein–Gordon equation, it is convenient to use the metric co-form $(\partial/\partial s)^2 = g^{\alpha\beta} (\partial/\partial x^\alpha)(\partial/\partial x^\beta)$ instead of $g_{\alpha\beta}$. For the Schwarzschild metric, introducing the coordinate $\mu \stackrel{\text{def}}{=} \cos \vartheta$, the co-form is

$$\left(\frac{\partial}{\partial s} \right)^2 = \frac{1}{r^2} \left[-(1 - \mu^2) \left(\frac{\partial}{\partial \mu} \right)^2 - \frac{1}{1 - \mu^2} \left(\frac{\partial}{\partial \varphi} \right)^2 - \Delta_r \left(\frac{\partial}{\partial r} \right)^2 + \frac{Z_r{}^2}{\Delta_r} \left(\frac{\partial}{\partial t} \right)^2 \right], \tag{21.63}$$

where $\Delta_r \stackrel{\text{def}}{=} r^2 - 2mr$ and $Z_r \stackrel{\text{def}}{=} r^2$, these quantities being introduced for later convenience. The corresponding Klein–Gordon equation is

$$\frac{r^2}{Z_r} \left\{ -\frac{1}{\psi} \frac{\partial}{\partial \mu} \left[(1 - \mu^2) \frac{\partial \psi}{\partial \mu} \right] - \frac{1}{\psi (1 - \mu^2)} \frac{\partial^2 \psi}{\partial \varphi^2} - m_0{}^2 r^2 \right\}$$
$$- \frac{1}{\psi} \frac{\partial}{\partial r} \left(\frac{r^2 \Delta_r}{Z_r} \frac{\partial \psi}{\partial r} \right) + \frac{1}{\psi} \frac{r^2 Z_r}{\Delta_r} \frac{\partial^2 \psi}{\partial t^2} = 0. \tag{21.64}$$

The substitution $\psi = \prod_i \psi_i(x^i)$ and multiplication by the appropriate functions allows one

---

[8] Carter (1968a) showed that the separability of the Klein–Gordon equation implies the separability of the Hamilton–Jacobi equation for geodesics. This implication will not be exploited here.

now to separate out the term that depends only on φ, then the term that depends only on μ, and then the terms depending only on r and t.

Now we would like to generalise the co-form (21.63) so as to preserve the separability property, but allow the Kerr metric as a subcase. We thus keep the assumption that the metric functions are independent of t and φ. We also keep the property that the co-form is proportional to the sum of a term independent of r and a term independent of μ. We must take it into account that the Kerr metric is not diagonal and contains the $g_{t\varphi}$ terms. Consequently, we take as our first hypothesis

$$\left(\frac{\partial}{\partial s}\right)^2 = \frac{1}{Z}\left[-\Delta_\mu\left(\frac{\partial}{\partial\mu}\right)^2 - \frac{1}{\Delta_\mu}\left(Z_\mu\frac{\partial}{\partial t} + Q_\mu\frac{\partial}{\partial\varphi}\right)^2\right]$$
$$+ \frac{1}{Z}\left[-\Delta_r\left(\frac{\partial}{\partial r}\right)^2 + \frac{1}{\Delta_r}\left(Z_r\frac{\partial}{\partial t} + Q_r\frac{\partial}{\partial\varphi}\right)^2\right], \tag{21.65}$$

where $\Delta_\mu(\mu)$, $Z_\mu(\mu)$, $Q_\mu(\mu)$, $\Delta_r(r)$, $Z_r(r)$ and $Q_r(r)$ are arbitrary functions, and Z will be determined later. The factors Δ_μ and Δ_r could be made equal to 1 by coordinate transformations and renaming the other functions, but it is more convenient to keep them and make use of the freedom later in another way. The determinant of the metric is

$$\sqrt{-g} = Z^2/|Z_rQ_\mu - Z_\mu Q_r|. \tag{21.66}$$

The Klein–Gordon equation in the metric determined by (21.65) is

$$\frac{1}{\psi}\left[-\frac{\partial}{\partial\mu}\left(\frac{\sqrt{-g}}{Z}\Delta_\mu\frac{\partial\psi}{\partial\mu}\right) - \frac{\partial}{\partial r}\left(\frac{\sqrt{-g}}{Z}\Delta_r\frac{\partial\psi}{\partial r}\right) - \frac{\sqrt{-g}}{Z}m_0{}^2 Z\psi\right]$$
$$- \frac{1}{\psi}\left(Z_\mu\frac{\partial}{\partial t} + Q_\mu\frac{\partial}{\partial\varphi}\right)\left[\frac{1}{\Delta_\mu}\frac{\sqrt{-g}}{Z}\left(Z_\mu\frac{\partial\psi}{\partial t} + Q_\mu\frac{\partial\psi}{\partial\varphi}\right)\right]$$
$$+ \frac{1}{\psi}\left(Z_r\frac{\partial}{\partial t} + Q_r\frac{\partial}{\partial\varphi}\right)\left[\frac{1}{\Delta_r}\frac{\sqrt{-g}}{Z}\left(Z_r\frac{\partial\psi}{\partial t} + Q_r\frac{\partial\psi}{\partial\varphi}\right)\right] = 0. \tag{21.67}$$

The factor $\sqrt{-g}/Z$ present in each term may depend only on r and μ in consequence of the assumed symmetry. But to achieve separation, it must be a product of a function of r by a function of μ. Then, after dividing (21.67) by $\sqrt{-g}/Z$, we can transform $r = f_1(r')$, $\mu = f_2(\mu')$, and redefine Δ_μ, Δ_r, Z_μ, Q_μ, Z_r and Q_r so as to achieve the result as if $\sqrt{-g}/Z = 1$, which, by (21.66), means that we choose

$$Z = Z_rQ_\mu - Z_\mu Q_r. \tag{21.68}$$

However, this is only a necessary, not yet a sufficient, condition for separability; the term that may still cause problems is the one containing m_0. It will allow separability if Z has the form $Z = U_\mu(\mu) + U_r(r)$, where U_μ and U_r are functions of one variable, as yet undetermined. The derivative $Z_{,r\mu}$ must then vanish, which means, from (21.68)

$$\frac{\mathrm{d}Z_r}{\mathrm{d}r}\frac{\mathrm{d}Q_\mu}{\mathrm{d}\mu} - \frac{\mathrm{d}Z_\mu}{\mathrm{d}\mu}\frac{\mathrm{d}Q_r}{\mathrm{d}r} = 0. \tag{21.69}$$

This can be satisfied in three ways: (1) at least one of (Z_r, Q_μ, Z_μ, Q_r) is zero; (2) Z_r and Z_μ are constant; (3) Q_r and Q_μ are constant. We shall believe Carter (1973) that case (1) leads to subcases of (2) and (3). Cases (2) and (3) are clearly equivalent, so, for better correspondence with the Schwarzschild limit, we take

$$Q_r = C_r = \text{constant}, \qquad Q_\mu = C_\mu = \text{constant}. \tag{21.70}$$

In this way, we have arrived at the following metric form:

$$ds^2 = \frac{\Delta_r \left(C_\mu dt - Z_\mu d\varphi\right)^2 - \Delta_\mu \left(C_r dt - Z_r d\varphi\right)^2}{C_\mu Z_r - C_r Z_\mu} - \left(C_\mu Z_r - C_r Z_\mu\right)\left(\frac{dr^2}{\Delta_r} + \frac{d\mu^2}{\Delta_\mu}\right). \tag{21.71}$$

Now we take into account the electromagnetic field. The Klein–Gordon equation with the electromagnetic 4-potential included is

$$\frac{1}{\psi}\left(\frac{\partial}{\partial x^\alpha} - ieA_\alpha\right)\left[\sqrt{-g}\,g^{\alpha\beta}\left(\frac{\partial\psi}{\partial x^\beta} - ieA_\beta\psi\right)\right] - m_0{}^2\sqrt{-g} = 0. \tag{21.72}$$

For simplicity, we assume that only the t- and φ-components of the 4-potential are nonzero, just as in the Schwarzschild case, and we assume the same symmetry as for the Kerr metric, so the nonzero components depend only on r and μ. Thus

$$A_\alpha dx^\alpha = A_0(r,\mu)dt + A_3(r,\mu)d\varphi. \tag{21.73}$$

Substituting this in (21.72), and taking into account the simplifications achieved on the way to (21.71), we obtain

$$\frac{1}{\psi}\left[-\frac{\partial}{\partial\mu}\left(\Delta_\mu\frac{\partial\psi}{\partial\mu}\right) - \frac{\partial}{\partial r}\left(\Delta_r\frac{\partial\psi}{\partial r}\right) - m_0{}^2 Z\psi\right]$$

$$-\frac{1}{\psi}\left(Z_\mu\frac{\partial}{\partial t} + C_\mu\frac{\partial}{\partial\varphi} + ieX_\mu\right)\left[\frac{1}{\Delta_\mu}\left(Z_\mu\frac{\partial\psi}{\partial t} + C_\mu\frac{\partial\psi}{\partial\varphi} + ieX_\mu\psi\right)\right]$$

$$+\frac{1}{\psi}\left(Z_r\frac{\partial}{\partial t} + C_r\frac{\partial}{\partial\varphi} - ieX_r\right)\left[\frac{1}{\Delta_r}\left(Z_r\frac{\partial\psi}{\partial t} + C_r\frac{\partial\psi}{\partial\varphi} - ieX_r\psi\right)\right] = 0, \tag{21.74}$$

where the following abbreviations were introduced:

$$A_0 Z_r + A_3 C_r \overset{\text{def}}{=} X_r, \qquad A_0 Z_\mu + A_3 C_\mu \overset{\text{def}}{=} -X_\mu. \tag{21.75}$$

Equation (21.74) will be separable if X_r depends only on r and X_μ depends only on μ. Assuming this, the 4-potential is

$$ZA_\alpha dx^\alpha = X_r\left(C_\mu dt - Z_\mu d\varphi\right) + X_\mu\left(C_r dt - Z_r d\varphi\right). \tag{21.76}$$

The constants C_μ and C_r, if nonzero, can be rescaled by the coordinate transformation $t = \alpha t'$, $r = \gamma r'$, accompanied by redefinitions of Δ_r, Δ_μ, Z_μ, X_r and X_μ. Both C_μ and C_r could be rescaled to 1 in this way. However, in anticipation of the result we aim at, we rescale only C_μ and rename C_r as follows: $C_\mu = 1$, $C_r = a$. We thus leave out the case $C_\mu = 0$. With this, we now proceed to the Einstein–Maxwell equations. We include the cosmological constant and magnetic charge. We will calculate the Einstein tensor using the following orthonormal tetrad of differential forms:

$$e^0 = \sqrt{\frac{\Delta_r}{Z}}\,(dt - Z_\mu d\varphi), \qquad e^1 = \sqrt{\frac{Z}{\Delta_r}}\,dr, \qquad e^2 = \sqrt{\frac{Z}{\Delta_\mu}}\,d\mu,$$

$$e^3 = \sqrt{\Delta_\mu/Z}\,(a\,dt - Z_r d\varphi). \tag{21.77}$$

With A_α given by (21.76), the electromagnetic field tensor $F_{\alpha\beta}$ has only two nonzero tetrad components in the tetrad (21.77):

$$\begin{aligned}
F_{01} &= [X_{r,r}\,(Z_r - aZ_\mu) - Z_{r,r}\,(X_r + aX_\mu)]\,/Z^2, \\
F_{23} &= -[X_{\mu,\mu}\,(Z_r - aZ_\mu) + Z_{\mu,\mu}\,(X_r + aX_\mu)]\,/Z^2
\end{aligned} \tag{21.78}$$

(see Exercise 15). Using (13.10) we now find for the tetrad components of $T_{\alpha\beta}$:

$$T_{00} = -T_{11} = T_{22} = T_{33} = \left(F_{01}{}^2 + F_{23}{}^2\right)/(8\pi). \tag{21.79}$$

For the tetrad components of the Einstein tensor we find

$$\begin{aligned}
G_{00} = & -\frac{1}{2Z}\Delta_{\mu,\mu\mu} - \frac{a}{2Z^2}\Delta_{\mu,\mu}Z_{\mu,\mu} - \frac{a^2}{4Z^3}\Delta_\mu Z_{\mu,\mu}{}^2 + \frac{3}{4Z^3}\Delta_r Z_{\mu,\mu}{}^2 \\
& -\frac{a^2}{4Z^3}\Delta_\mu Z_{r,r}{}^2 - \frac{1}{Z^2}\Delta_r Z_{r,rr} + \frac{3}{4Z^3}\Delta_r Z_{r,r}{}^2 - \frac{1}{2Z^2}\Delta_{r,r}Z_{r,r},
\end{aligned} \tag{21.80}$$

$$G_{03} = \frac{1}{2Z^2}\sqrt{\Delta_r \Delta_\mu}\,(aZ_{r,rr} + Z_{\mu,\mu\mu}), \tag{21.81}$$

$$\begin{aligned}
G_{11} = & \frac{1}{2Z}\Delta_{\mu,\mu\mu} + \frac{a}{2Z^2}\Delta_{\mu,\mu}Z_{\mu,\mu} + \frac{a^2}{4Z^3}\Delta_\mu Z_{\mu,\mu}{}^2 - \frac{1}{4Z^3}\Delta_r Z_{\mu,\mu}{}^2 \\
& +\frac{a^2}{4Z^3}\Delta_\mu Z_{r,r}{}^2 - \frac{1}{4Z^3}\Delta_r Z_{r,r}{}^2 + \frac{1}{2Z^2}\Delta_{r,r}Z_{r,r},
\end{aligned} \tag{21.82}$$

$$\begin{aligned}
G_{22} = & -\frac{a^2}{4Z^3}\Delta_\mu Z_{\mu,\mu}{}^2 + \frac{1}{4Z^3}\Delta_r Z_{\mu,\mu}{}^2 - \frac{a}{2Z^2}\Delta_{\mu,\mu}Z_{\mu,\mu} \\
& -\frac{a^2}{4Z^3}\Delta_\mu Z_{r,r}{}^2 + \frac{1}{2Z}\Delta_{r,rr} - \frac{1}{2Z^2}\Delta_{r,r}Z_{r,r} + \frac{1}{4Z^3}\Delta_r Z_{r,r}{}^2,
\end{aligned} \tag{21.83}$$

$$\begin{aligned}
G_{33} = & -\frac{a}{Z^2}\Delta_\mu Z_{\mu,\mu\mu} - \frac{a}{2Z^2}\Delta_{\mu,\mu}Z_{\mu,\mu} - \frac{3a^2}{4Z^3}\Delta_\mu Z_{\mu,\mu}{}^2 + \frac{1}{4Z^3}\Delta_r Z_{\mu,\mu}{}^2 \\
& -\frac{3a^2}{4Z^3}\Delta_\mu Z_{r,r}{}^2 + \frac{1}{2Z}\Delta_{r,rr} - \frac{1}{2Z^2}\Delta_{r,r}Z_{r,r} + \frac{1}{4Z^3}\Delta_r Z_{r,r}{}^2.
\end{aligned} \tag{21.84}$$

We are now going to solve the equations $G_{ij} + \Lambda g_{ij} = \kappa T_{ij}$.

The $G_{03} = 0$ equation is a sum of two terms, one of which depends only on μ and the other only on r. The solution is easily found to be

$$Z_r = Cr^2 + C_1 r + C_2, \qquad Z_\mu = -aC\mu^2 + C_3\mu + C_4, \tag{21.85}$$

where the C and C_i are arbitrary constants. Substituting these in $G_{22} - G_{33} = 0$ and using $Z = Z_r - aZ_\mu$ (from (21.68)) we obtain

$$C_4 = C_2/a - \left(C_1{}^2 + C_3{}^2\right)/(4aC). \tag{21.86}$$

Substituting this in (21.85) we see that the transformation $\mu = \mu' + C_3/(2aC)$ and the redefinition $C_2 = aC_2' + C_1{}^2/(4C)$ have the same result as if $C_3 = 0$. Thus

$$Z_r = C\left[r + C_1/(2C)\right]^2 + aC_2', \qquad Z_\mu = C_2' - aC\mu^2. \tag{21.87}$$

The transformation $t = t' + C_2'\varphi - aC\varphi$ does not change the combination $(Z_r - aZ_\mu)$ and has the same result as if $C_2' = aC$. Then, a translation of r makes $C_1 = 0$, and the rescaling $\varphi = \varphi'/C$, $\Delta_\mu = C\Delta_\mu'$, $\Delta_r = C\Delta_r'$ has the same result as if $C = 1$. Thus finally, we have

$$Z_r = r^2 + a^2, \qquad Z_\mu = a\left(1 - \mu^2\right). \tag{21.88}$$

The equation $G_{00} + G_{11} = 0$ is now fulfilled identically.

We now solve the Maxwell equations. The equations $F_{[\alpha\beta,\gamma]} = 0 = \left(\sqrt{-g}F^{1\beta}\right)_{,\beta} = \left(\sqrt{-g}F^{2\beta}\right)_{,\beta}$ (coordinate indices) are already fulfilled. The remaining ones are

$$(Z_r F_{01})_{,r} - (Z_\mu F_{23})_{,\mu} = 0, \tag{21.89}$$

$$aF_{01,r} - F_{23,\mu} = 0, \tag{21.90}$$

the F_{ij} above being tetrad components (see Exercise 17). The solution of (21.90) is, using (21.78) and (21.88),

$$X_r = Dr^4 + \mathcal{E}r - aD_1, \qquad X_\mu = -a^3 D\mu^4 + \mathcal{Q}\mu + D_1, \tag{21.91}$$

where $\mathcal{E}$ and $\mathcal{Q}$ are arbitrary constants. Equation (21.89) results in $D = 0$. The constant D_1 does not enter the electromagnetic field tensor, and can thus be assumed zero with no loss of generality. The final solution of the Maxwell equations is then $X_r = \mathcal{E}r$, $X_\mu = \mathcal{Q}\mu$. With these, and with the Z_r and Z_μ of (21.88), the expression in (21.79) becomes

$$F_{01}{}^2 + F_{23}{}^2 = \left(\mathcal{E}^2 + \mathcal{Q}^2\right)/Z^2. \tag{21.92}$$

We now go back to the Einstein equations. The equation $G_{00} - G_{22} = -2\Lambda$ is equivalent to $\Delta_{r,rr} - 4\Lambda r^2 = -\Delta_{\mu,\mu\mu} + 4\Lambda a^2\mu^2$, which means that both sides must be constant. Calling this constant $2E$ and integrating we obtain

$$\Delta_r = \frac{1}{3}\Lambda r^4 + Er^2 - 2mr + E_2, \qquad \Delta_\mu = \frac{1}{3}\Lambda a^2\mu^4 - E\mu^2 + E_3\mu + E_4, \tag{21.93}$$

where m and E_i are arbitrary constants. The last equation to solve is, using (21.92),

$$G_{22} - \Lambda = \left(\mathcal{E}^2 + \mathcal{Q}^2\right)/\left(8\pi Z^2\right) \stackrel{\text{def}}{=} \left(e^2 + q^2\right)/Z^2, \tag{21.94}$$

and it leads to $E_2 - a^2 E_4 = e^2 + q^2$.

With this, the Einstein–Maxwell equations are solved. However, a few conditions have to be imposed on the solution in order that the metric is physically reasonable. One of them is $E_3 = 0$ – otherwise, with the term proportional to $\mu = \cos\vartheta$, the metric would not be mirror-symmetric with respect to the equatorial plane. To avoid a singularity at the axis $\vartheta = 0$ that would appear after the transformation $\mu = \cos\vartheta$ we require $E = E_4 + \Lambda a^2/3$, and to obtain the correct Schwarzschild limit we require $E_4 = 1$. Finally then,

$$\Delta_\mu = \left(1 - \frac{1}{3}\Lambda a^2\mu^2\right)\left(1 - \mu^2\right),$$

$$\Delta_r = \frac{1}{3}\Lambda r^2\left(r^2 + a^2\right) + r^2 - 2mr + a^2 + e^2 + q^2. \tag{21.95}$$

The final result for the metric is

$$ds^2 = \frac{\Delta_r}{\Sigma}\left(dt - a\sin^2\vartheta d\varphi\right)^2 - \frac{\sin^2\vartheta\left(1 - \frac{1}{3}\Lambda a^2\cos^2\vartheta\right)}{\Sigma}\left[adt - \left(r^2 + a^2\right)d\varphi\right]^2$$

$$- \Sigma \left(\frac{\mathrm{d}r^2}{\Delta_r} + \frac{\mathrm{d}\vartheta^2}{1 - \frac{1}{3}\Lambda a^2 \cos^2 \vartheta} \right), \qquad \Sigma = Z = r^2 + a^2 \cos^2 \vartheta. \tag{21.96}$$

This differs from the result of Carter (1973) by a constant factor in the first two terms[9] that can be recovered by rescaling t and φ. A metric equivalent to (21.96) under a coordinate transformation was found by Frolov (1974).[10] The algebraic structure of the Weyl tensor for (21.96) is still the same as in (21.61) but with I_1 and I_2 changed to

$$I_1 = \frac{1}{\Sigma^3} \left[mr \left(r^2 - 3a^2 \cos^2 \vartheta \right) - \left(e^2 + q^2 \right) \left(r^2 - a^2 \cos^2 \vartheta \right) \right],$$

$$I_2 = \frac{1}{\Sigma^3} \left[ma \cos \vartheta \left(3r^2 - a^2 \cos^2 \vartheta \right) - 2 \left(e^2 + q^2 \right) ar \cos \vartheta \right]. \tag{21.97}$$

so (21.96) is still of Petrov type D.

Comparing (21.96) with the spherically symmetric limit ($a = 0$, Section 14.4) and the proper Kerr limit ($\Lambda = 0 = e = q$, Eq. (21.57)) we recognise e and q as the electric and magnetic charges, respectively,[11] and a as the angular momentum per unit mass while m is the mass of the source and Λ is the cosmological constant.

The (so far) farthest-reaching generalisation of the Kerr metric was obtained by Debever, Kamran and McLenaghan (1983, 1984). It contains 13 arbitrary constants, for most of which no physical or geometrical interpretation has been provided. In addition to the parameters contained in (21.96), it includes the acceleration of the source (Stephani et al., 2003). It is still of Petrov type D, and has the same Abelian symmetry group. The paper by Debever, Kamran and McLenaghan (1984) and the book by Stephani et al. (2003) contain lists of specialisations of this metric that recover the earlier-known solutions.

Historically, the generalisation of the Kerr metric for electric charge was first found by E. Newman et al. (1965) by a rather mysterious procedure. They allowed the constants and coordinates in the Reissner–Nordström metric to be complex and applied such a complex coordinate transformation to it that the result was real – and was the subcase $\Lambda = q = 0$ of (21.96).[12] It is called the **Kerr–Newman metric**. The first author to generalise the Kerr metric for the cosmological constant was Carter (1968a); that paper contains a 10-parameter electrovacuum solution and a list of possible specialisations of it.

21.5 The event horizons and the stationary limit hypersurfaces

We come back now to the proper Kerr metric, with $\Lambda = e = q = 0$, and we will consider this case to the end of this chapter.

We have already noted that the hypersurfaces given by (21.58), if they exist, play a special role. We will see later that they are **event horizons**. We also noted that the stationary limit hypersurfaces, given by

$$r^2 - 2mr + a^2 \cos^2 \vartheta = 0 \Longrightarrow r = m \pm \sqrt{m^2 - a^2 \cos^2 \vartheta}, \tag{21.98}$$

[9] Another difference is that Carter used the signature $(+ + +-)$.
[10] We are grateful to Valeri Frolov for demonstrating the equivalence in a letter to A. K., long ago.
[11] As remarked in Section 14.4, the magnetic charge may be removed by a duality rotation.
[12] The trick is known to work also in other cases, leading from given solutions of the Einstein or Einstein–Maxwell equations to other solutions, but nobody knows why it works, and why only in these particular cases – see Stephani et al. (2003) for more information.

play a special role, too. We will now consider the shapes of these two families of hyper-surfaces and their relation to each other.

When $a^2 < m^2$, there are two event horizons, the one at $r = r_-$ is contained inside that at $r = r_+$. There are also two stationary limit hypersurfaces in this case, see Fig. 21.3. The outer one, at $r = m + \sqrt{m^2 - a^2 \cos^2 \vartheta}$, envelops the outer event horizon, and is tangent to it only at the axis where $\vartheta = 0$ or $\vartheta = \pi$. The inner stationary limit surface, at $r = m - \sqrt{m^2 - a^2 \cos^2 \vartheta}$, lies all within the inner event horizon and is tangent to it also only at the axis. It is tangent to the disc $r = 0$ at its singular edge.

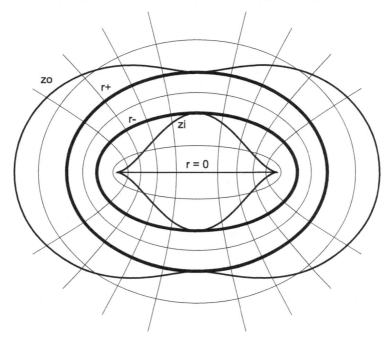

Fig. 21.3 An axial cross-section through the space $t = $ constant in the Kerr metric, in the Boyer–Lindquist coordinates, in the case $a^2 < m^2$. The surfaces $r = r_+$ and $r = r_-$ are disjoint; the outer stationary limit surface ('zo' in the figure) is tangent to $r = r_+$ at the axis of symmetry. Likewise, the inner stationary limit surface (zi) is tangent to $r = r_-$ at the axis. The disc $r = 0$ is seen as the horizontal line; the surface zi has a sharp edge, tangent to the disc. When $|a|/m$ decreases, the surface $r = r_-$ approaches the disc $r = 0$, whereas the surface $r = r_+$ recedes from the disc and becomes more spherical. When $a \to 0$, the disc $r = 0$ and the surfaces zi and $r = r_-$ all collapse to a single point, while the surfaces zo and $r = r_+$ coalesce at $r = 2m$ and become spherical.

As $a \to 0$, the Kerr metric tends to the Schwarzschild metric. Then, the inner stationary limit surface and the inner event horizon both shrink to a point, together with the disc $r = 0$. The outer stationary limit surface and the outer event horizon coalesce and go over into the Schwarzschild horizon at $r = 2m$.

As $a^2 \to m^2$, the two event horizons approach each other to meet at $r = m$ when $a^2 = m^2$. The concave regions of the outer stationary limit surface shrink to points, and the surface becomes conical in their neighbourhoods, the vertices of the cones touching the event horizon. Similarly, the neighbourhoods of those points of the inner stationary limit surface that lie on the axis of symmetry become conical, the vertices being common with the outer surface, see Fig. 21.4. With $a^2 = m^2$, the metric (21.57) has no Schwarzschild limit. When $a \to 0$ it becomes the Minkowski metric in spherical coordinates.

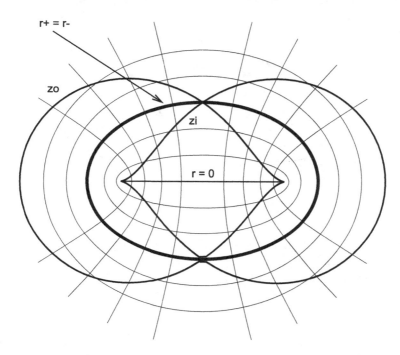

Fig. 21.4 An axial cross-section through the space $t = $ constant in the Kerr metric, in the Boyer–Lindquist coordinates, in the case $a^2 = m^2$. There is now a single surface $r = r_+ = r_-$, the surfaces zo and zi both touch it at the axis (but are not tangent to it – they approach it at nonzero angle). As $a \to m$, the concave areas on the zo surface from Fig. 21.3 shrink to points, and the neighbourhoods of those points become conical. The zi surface becomes conical at the axis as well.

When $a^2 > m^2$, the event horizon disappears. This case, similarly to the previous one, has no Schwarzschild limit. The conical points of the stationary limit surfaces change to open holes, and the two stationary limit surfaces coalesce into one surface that has the topology of a torus. The hole in the surface around the axis of symmetry is the larger, the greater the difference $a^2 - m^2$. The surface still has an edge at $r = 0$, where it is tangent to the disc $r = 0$, but the 'outer' and the 'inner' part of it join smoothly at two rings, given by $(r = m, \vartheta = \arccos(m/a))$ and $(r = m, \vartheta = \pi - \arccos(m/a))$. The geometry of this case is shown in Fig. 21.5.

A comment must be added here. We have already alluded to the possibility of extending the Kerr spacetime to negative values of the Boyer–Lindquist r-coordinate (in the footnote to Eq. (21.44)), and such an extension will be discussed in Sec. 21.9. The following question then arises: where should the regions with $r < 0$ be placed in Figs. 21.3–21.5? In the theory of analytic functions, one deals with the problem

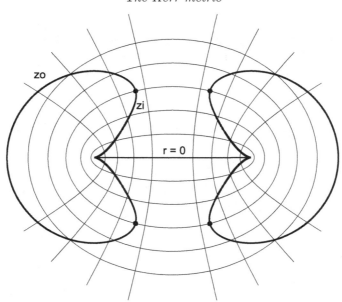

Fig. 21.5 An axial cross-section through the space $t = $ constant in the Kerr metric, in the Boyer–Lindquist coordinates, in the case $a^2 > m^2$. The spurious singularities at $r = r_+$ and $r = r_-$ have disappeared, and the two stationary limit surfaces merged into one that has the topology of a torus. The points mark the places where zo (the surface with the plus sign in (21.98)) and zi meet at $(r = m, \vartheta = \vartheta_0 \stackrel{\text{def}}{=} \arccos(m/a))$ and $(r = m, \vartheta = \pi - \vartheta_0)$.

of multi-valued mappings by introducing *Riemann surfaces*: multi-sheeted manifolds such that to each value of a given function there corresponds a single point of the Riemann surface (see, for example, Knopp (1996)). In our present case, one has to imagine that each of the planes of Figs. 21.3–21.5 branches out into two sheets at the disc $r = 0$. By going through the interior of the ring $\{r = 0, \vartheta = \pi/2\}$ from above, one does not reach the lower half of the figure, but a sheet with $r < 0$. There are two such sheets. To enter the second one from the first, one has to go back up through the interior of the ring, go around the ring on the right or on the left, and then go through the disc $r = 0$ from below. Note that by going around the ring one intersects the inner horizon $r = r_-$ (in Fig. 21.3) or *the* horizon $r = m$ (in Fig. 21.4) twice, each time a different half of it. This agrees with the picture of the maximally extended Kerr manifold shown in Fig. 21.14: there are two disjoint $r < 0$ sheets, and to proceed from one to the other one has to go through the $r = r_-$ horizon at least twice.

Figure 21.6 shows the 3-dimensional subspace of the Kerr spacetime with $a^2 < m^2$, given by $\vartheta = \pi/2$ in (21.57); the vertical axis is time. The top part of the figure shows a perspective view; the bottom part shows the view from above. The direction of rotation is clockwise ($a < 0$), so that the X-axis in the top figure would move towards the viewer. The outermost ring is the stationary limit surface at $r = 2m$, the two middle rings are the event horizons $r_\pm$ and the innermost ring is the inner stationary limit surface that, because of $\vartheta = \pi/2$, coincides with the ring singularity at $r = 0$. Several future light cones are shown. For $r \gg 2m$ they look like tilted Minkowski light cones. At $r = 2m$, the light cone has one generator parallel to the t-axis: no timelike vector at that point can

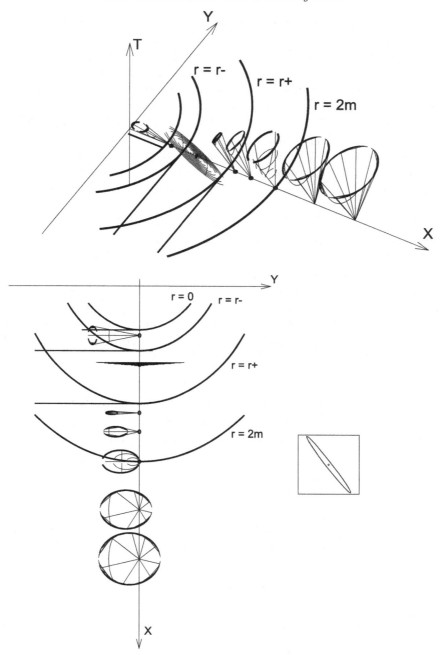

Fig. 21.6 A general perspective view (top graph) and a view along the time axis from above (bottom graph) of the Kerr subspace $\vartheta = \pi/2$ in the coordinates of (21.57). Explanation is given in the text. Large dots mark the positions of the vertices of the cones in the $T = 0$ plane. The inset is an $X =$ constant intersection of the light cone from the sector $r \in (r_-, r_+)$.

have a zero φ-component.[13] As we move from $r = 2m$ towards $r = r_+$, the cones lean forward further and become thinner in all directions. The limit $r \to r_+$ is discontinuous. As $r \to r_+$ from the $r > r_+$ side, the cones tend to a single beam along the Y-direction in the $T = 0$ plane. As $r \to r_+$ from the $r < r_+$ side, the cones tend to the whole plane $X = \sqrt{r_+^2 + a^2} = $ constant (not shown in Fig. 21.6 for the sake of better clarity). With $r > r_+$, the intersections of the cones with a $T = $ constant plane are ellipses that recede to $Y \to -\infty$ as $r \to r_+$. With $r_- < r < r_+$, the intersections of the cones with the $X = $ constant planes are ellipses elongated and rotated as shown in the inset; their axes both become infinite as $r \to r_+$. A similar discontinuity exists at $r = r_-$; the intersections of the cones with a $T = $ constant plane again become ellipses in the region $r < r_-$.

Since at $r = r_+$ the cone degenerates to the plane $X = \sqrt{r_+^2 + a^2} = $ constant, no timelike vector attached there can have a zero r-component; in fact r takes over as the time-coordinate here. Inside $r = r_+$, it is impossible for an ingoing timelike or null curve to turn back towards increasing r without becoming spacelike along an arc or having a non-differentiable reflection. This shows that $r = r_+$ is an event horizon. In Fig. 21.6, ingoing and outgoing null curves cross at $r = r_+$, but the B–L coordinates give a false picture here. In reality, there are two event horizons at $r = r_+$. The Kruskal diagram (Fig. 14.8) gives a good picture of the situation, and we will deal with the corresponding extension for the Kerr manifold in Section 21.9. For $r < r_-$, the cones become similar to those in the sector $(r_+, 2m)$. As $r \to 0$, the cones squeeze to the straight line $\{X = a, T = Y\}$. See the last exercise in this chapter for directions on how to draw this figure.

21.6 The Hamiltonian and the Poisson bracket

When looking for an extremum of the functional $\mathcal{L} \overset{\text{def}}{=} \int L(q^i(t), \dot{q}^i(t), t)dt$ (where $\dot{q}^i \overset{\text{def}}{=} dq^i/dt$) with respect to the functions $q^i(t), i = 1, \ldots, m$, it is often useful to express L as a function of the variables q^i and $p_i \overset{\text{def}}{=} \partial L/\partial \dot{q}^i$, and then form the Hamiltonian

$$H \overset{\text{def}}{=} \sum_{l=1}^{m} p_l \dot{q}^l(q, p, t) - L. \tag{21.99}$$

The functions $q^i(t)$ and $p_i(t)$ are called positions and momenta, respectively, and treated as independent variables in the Hamiltonian. In terms of H, the Euler–Lagrange equations that determine $q^i(t)$,

$$\frac{d}{dt}\left(\frac{\partial L}{\partial \dot{q}^i}\right) - \frac{\partial L}{\partial q^i} = 0, \tag{21.100}$$

become the Hamilton equations for the functions $q^i(t), p_i(t)$:

$$\dot{q}^i = \frac{\partial H}{\partial p_i}, \qquad \dot{p}_i = -\frac{\partial H}{\partial q^i}. \tag{21.101}$$

This formalism was invented for mechanics, but is useful also for solving the geodesic equations in differential geometry.

---

[13] Because of $g_{t,\varphi} \neq 0$, the time axis is not orthogonal to the $t = $ constant hypersurface, but it is drawn as orthogonal for better readability.

The Poisson bracket for the functions $F(q^i, p_i)$ and $G(q^i, p_i), i = 1, \ldots, m$, is defined as

$$\{F, G\} \overset{\text{def}}{=} \sum_{l=1}^{m} \left(\frac{\partial F}{\partial q^l} \frac{\partial G}{\partial p_l} - \frac{\partial F}{\partial p_l} \frac{\partial G}{\partial q^l} \right). \tag{21.102}$$

From the Hamilton equations it follows that for any function $E(q^i, p_i, t)$,

$$\frac{\mathrm{d}E}{\mathrm{d}t} = \sum_{l=1}^{m} \left(\frac{\partial E}{\partial q^l} \frac{\mathrm{d}q^l}{\mathrm{d}t} + \frac{\partial E}{\partial p_l} \frac{\mathrm{d}p_l}{\mathrm{d}t} \right) + \frac{\partial E}{\partial t} = \{E, H\} + \frac{\partial E}{\partial t}. \tag{21.103}$$

So, if E does not depend on t directly (i.e. depends on t only via q^l and p_l) and has zero Poisson bracket with the Hamiltonian, then it is constant along the curve $\{q^i(t), p_i(t)\}$ that obeys the Hamilton equations (21.101).

The Poisson bracket has the following properties:

$$\begin{aligned} \{F, G\} &= -\{G, F\}, & \{F_1 + F_2, G\} &= \{F_1, G\} + \{F_2, G\}, \\ \{F_1 F_2, G\} &= F_1 \{F_2, G\} + F_2 \{F_1, G\}. \end{aligned} \tag{21.104}$$

21.7 General geodesics

For each Killing field k^α there exists a first integral of the geodesic equations. Let p^α be a vector field tangent to an affinely parametrised geodesic. Then, along the geodesic, $\mathrm{d}(k_\alpha p^\alpha)/\mathrm{d}s = D(k_\alpha p^\alpha)/\mathrm{d}s = k_{\alpha;\beta} p^\alpha p^\beta + k_\alpha p^\alpha{}_{;\beta} p^\beta = 0$, the first term being zero because only $k_{(\alpha;\beta)} = 0$ enters the formula, the second one being zero because p^α is geodesic.

For the Killing fields in the Kerr metric represented in the Boyer–Lindquist coordinates, $k^\alpha_{(t)} = \delta^\alpha_0$ and $k^\alpha_{(\varphi)} = \delta^\alpha_3$, there are two first integrals of the geodesic equations:

$$p_\alpha k^\alpha_{(t)} = p_0 \overset{\text{def}}{=} E, \qquad p_\alpha k^\alpha_{(\varphi)} = p_3 \overset{\text{def}}{=} -L_z. \tag{21.105}$$

Moreover, the first integral $g_{\alpha\beta} p^\alpha p^\beta = $ constant exists in every metric. For timelike geodesics we choose the affine parameter so that

$$g_{\alpha\beta} p^\alpha p^\beta = \mu_0{}^2, \tag{21.106}$$

where μ_0 is the mass of the orbiting particle. This parametrisation allows for an easy specialisation to null geodesics, for which $\mu_0 = 0$.

For the Kerr metric, there exists a fourth first integral, discovered by Carter (1973). The geodesic equations are the Euler–Lagrange equations for the following Lagrangian:

$$L = \frac{1}{2} g_{\alpha\beta} \frac{\mathrm{d}x^\alpha}{\mathrm{d}s} \frac{\mathrm{d}x^\beta}{\mathrm{d}s} \overset{\text{def}}{=} \frac{1}{2} g_{\alpha\beta} \dot{x}^\alpha \dot{x}^\beta. \tag{21.107}$$

The momentum associated to $\dot{x}^\alpha$ is $p_\alpha = g_{\alpha\beta} \dot{x}^\beta$. The corresponding Hamiltonian is

$$H \overset{\text{def}}{=} p_\alpha \dot{x}^\alpha - L = \frac{1}{2} g_{\alpha\beta} \dot{x}^\alpha \dot{x}^\beta = L \left(p_\alpha, q^\beta \right). \tag{21.108}$$

Lemma 21.2 *Let the Hamiltonian have the form*

$$H = \frac{1}{2} \frac{H_r + H_\mu}{U_r + U_\mu}, \tag{21.109}$$

where U_r depends only on r, U_μ depends only on μ, H_r depends only on r and p_r and H_μ depends only on μ and H_μ. Then the quantity

$$-K \overset{\text{def}}{=} -\frac{U_r H_\mu - U_\mu H_r}{U_r + U_\mu} \tag{21.110}$$

has a vanishing Poisson bracket with the Hamiltonian, and thus is a constant of the motion.

Proof:

By the assumptions made, H_r commutes with H_μ, and obviously H_r commutes with itself. Consequently, for H in the form (21.109), by the rules (21.104), we obtain

$$\{H_r, H\} = \frac{1}{2}(H_r + H_\mu)\left\{H_r, \frac{1}{U_r + U_\mu}\right\} = \frac{1}{2}\frac{H_r + H_\mu}{(U_r + U_\mu)^2}\frac{dU_r}{dr}\frac{\partial H_r}{\partial p_r}$$

$$= \frac{H}{U_r + U_\mu}\{U_r, H_r\}. \tag{21.111}$$

Since U_r and U_μ do not depend on the momenta, they have zero Poisson brackets among them. So, using the assumptions of the lemma, we have

$$\{U_r, H\} = \frac{1}{2(U_r + U_\mu)}\{U_r, H_r + H_\mu\} = \frac{1}{2(U_r + U_\mu)}\frac{dU_r}{dr}\frac{\partial H_r}{\partial p_r} = \frac{1}{2(U_r + U_\mu)}\{U_r, H_r\}. \tag{21.112}$$

From (21.112) and (21.111) we see that

$$\{H_r, H\} = 2H\{U_r, H\}. \tag{21.113}$$

Now it is easy to verify that

$$-K \overset{\text{def}}{=} 2U_r H - H_r \equiv \frac{U_r H_\mu - U_\mu H_r}{U_r + U_\mu} \tag{21.114}$$

commutes with H, and so is a constant of the motion. $\square$

For applying the Hamilton method to the geodesic equations in the Kerr metric, it is convenient to rewrite it in the form equivalent to (21.57):

$$ds^2 = \frac{\Delta_r}{\Sigma}\left[dt - a(1 - \mu^2)\,d\varphi\right]^2 - \frac{1 - \mu^2}{\Sigma}\left[a\,dt - (r^2 + a^2)\,d\varphi\right]^2 - \Sigma\left(\frac{dr^2}{\Delta_r} + \frac{d\mu^2}{1 - \mu^2}\right), \tag{21.115}$$

where Σ and Δ_r are defined in (21.52) and (21.56) and $\mu = \cos\vartheta$ as in (21.63). The Lagrangian and the Hamiltonian are found from (21.108) using (21.115):

$$H = L = \frac{1}{2}\left\{\frac{\Delta_r}{\Sigma}\left[\dot{t} - a(1 - \mu^2)\dot{\varphi}\right]^2 - \frac{1 - \mu^2}{\Sigma}\left[a\dot{t} - (r^2 + a^2)\dot{\varphi}\right]^2 - \Sigma\left(\frac{\dot{r}^2}{\Delta_r} + \frac{\dot{\mu}^2}{1 - \mu^2}\right)\right\}. \tag{21.116}$$

From here we find the momenta using the definition $p_i \overset{\text{def}}{=} \partial L/\partial\dot{q}^i$:

$$p_t = \left(1 - \frac{2mr}{\Sigma}\right)\dot{t} + \frac{2mra\sin^2\vartheta}{\Sigma}\dot{\varphi}, \tag{21.117}$$

$$p_\varphi = \frac{2mra\sin^2\vartheta}{\Sigma}\dot{t} - \left(\frac{2mra^2\sin^2\vartheta}{\Sigma} + r^2 + a^2\right)\sin^2\vartheta\dot{\varphi}, \qquad (21.118)$$

$$p_r = -\frac{\Sigma}{\Delta_r}\dot{r}, \qquad p_\mu = -\frac{\Sigma}{1-\mu^2}\dot{\mu}. \qquad (21.119)$$

The Hamiltonian (21.116) rewritten as a functional of the momenta (21.117) – (21.119), with terms grouped as in (21.109), is

$$H = \frac{-\Delta_\mu p_\mu{}^2 - \Delta_\mu{}^{-1}(a\Delta_\mu p_t + p_\varphi)^2}{2\Sigma} + \frac{-\Delta_r p_r{}^2 + \Delta_r{}^{-1}\left[(r^2+a^2)p_t + ap_\varphi\right]^2}{2\Sigma}, \qquad (21.120)$$

where $\Delta_\mu \stackrel{\text{def}}{=} 1 - \mu^2 \equiv \sin^2\vartheta$, and $\Sigma \equiv r^2 + a^2\mu^2$.

Comparing (21.120) with (21.109) we identify the parts of the Hamiltonian as follows:

$$H_r = -\Delta_r p_r{}^2 + \Delta_r{}^{-1}\left[(r^2+a^2)p_t + ap_\varphi\right]^2, \qquad (21.121)$$

$$H_\mu = -\Delta_\mu p_\mu{}^2 - \Delta_\mu{}^{-1}(a\Delta_\mu p_t + p_\varphi)^2, \qquad (21.122)$$

$$U_r = r^2 + a^2, \qquad U_\mu = -a^2(1-\mu^2) \equiv -a^2\sin^2\vartheta \qquad (21.123)$$

(the splitting of Σ into U_r and U_μ was done so that it agrees with Carter's (1973) notation). Using the above in (21.110) we obtain for the fourth constant of the motion

$$K = \frac{(r^2+a^2)\left\{\Delta_\mu p_\mu{}^2 + \Delta_\mu^{-1}[a\Delta_\mu p_t + p_\varphi]^2\right\}}{\Sigma}$$

$$+ \frac{a^2\Delta_\mu\left\{\Delta_r p_r{}^2 - \Delta_r^{-1}[(r^2+a^2)p_t + ap_\varphi]^2\right\}}{\Sigma} \qquad (21.124)$$

$$= a^2\sin^2\vartheta\left[\frac{\Sigma}{\Delta_r}\dot{r}^2 - \frac{\Delta_r}{\Sigma}\left(a\sin^2\vartheta\dot{\varphi} - \dot{t}\right)^2\right]$$

$$+ (r^2+a^2)\left\{\Sigma\dot{\vartheta}^2 + \frac{\sin^2\vartheta}{\Sigma}\left[(r^2+a^2)\dot{\varphi} - a\dot{t}\right]^2\right\}. \qquad (21.125)$$

The constants defined in (21.105) and (21.106) are

$$E = \frac{r^2 - 2mr + a^2\cos^2\vartheta}{\Sigma}\dot{t} + \frac{2mra\sin^2\vartheta}{\Sigma}\dot{\varphi}, \qquad (21.126)$$

$$L_z = -\frac{2mra\sin^2\vartheta}{\Sigma}\dot{t} + \frac{\left[-\Delta_r a^2\sin^2\vartheta + (r^2+a^2)^2\right]\sin^2\vartheta}{\Sigma}\dot{\varphi}, \qquad (21.127)$$

$$\mu_0{}^2 = -\Sigma\left(\frac{\dot{r}^2}{\Delta_r} + \dot{\vartheta}^2\right) - \frac{\sin^2\vartheta}{\Sigma}\left[(r^2+a^2)\dot{\varphi} - a\dot{t}\right]^2 + \frac{\Delta_r}{\Sigma}\left(a\sin^2\vartheta\dot{\varphi} - \dot{t}\right)^2. \qquad (21.128)$$

These equations can now be solved for the velocities, which allows for some general discussion. We introduce the following new functions:

$$R(r) \stackrel{\text{def}}{=} -K\Delta_r - \mu_0{}^2(r^2+a^2)\Delta_r + \left[(r^2+a^2)E - aL_z\right]^2, \qquad (21.129)$$

$$\Theta(\vartheta) \stackrel{\text{def}}{=} K + \mu_0{}^2 a^2\sin^2\vartheta - \sin^2\vartheta\left(aE - \frac{L_z}{\sin^2\vartheta}\right)^2. \qquad (21.130)$$

The components of the velocity are then

$$\Sigma^2 \dot{r}^2 \;=\; R(r), \qquad \Sigma^2 \dot{\vartheta}^2 = \Theta(\vartheta), \tag{21.131}$$

$$\Sigma \dot{\varphi} \;=\; \left(\frac{1}{\sin^2 \vartheta} - \frac{a^2}{\Delta_r} \right) L_z + \frac{2mra}{\Delta_r} E, \tag{21.132}$$

$$\Sigma \dot{t} \;=\; -\frac{2mra}{\Delta_r} L_z + \left[\frac{\left(r^2 + a^2\right)^2}{\Delta_r} - a^2 \sin^2 \vartheta \right] E. \tag{21.133}$$

This can be disentangled still further. From (21.131) $\Sigma \dot{\vartheta}/\sqrt{\Theta(\vartheta)} = \Sigma \dot{r}/\sqrt{R(r)} = 1$. We take the quotient of these two equations to obtain (21.134), and any one of them using (21.134) to obtain (21.135) below. The result is

$$\int \frac{d\vartheta}{\sqrt{\Theta}} \;=\; \int \frac{dr}{\sqrt{R}} = Q = \text{constant}, \tag{21.134}$$

$$\lambda - \lambda_0 \;=\; \int \frac{a^2 \cos^2 \vartheta d\vartheta}{\sqrt{\Theta}} + \int \frac{r^2 dr}{\sqrt{R}}, \tag{21.135}$$

where λ is the affine parameter. These equations implicitly determine $r(\lambda)$ and $\vartheta(\lambda)$.

In order to disentangle (21.132) – (21.133) in a similar way, we note from (21.131) that $1/\Sigma = \dot{\vartheta}/\sqrt{\Theta(\vartheta)} = \dot{r}/\sqrt{R(r)}$, and use one or the other of these equations to obtain an integrand that is a function of r only, or one that is a function of ϑ only. The result is

$$\varphi - \varphi_0 \;=\; \int \frac{L_z d\vartheta}{\sin^2 \vartheta \sqrt{\Theta}} + \int \frac{2mraE - a^2 L_z}{\Delta_r \sqrt{R}} \, dr, \tag{21.136}$$

$$t - t_0 \;=\; \int \frac{\left(r^2 + a^2\right)^2 E - 2mraL_z}{\Delta_r \sqrt{R}} \, dr - \int \frac{a^2 \sin^2 \vartheta E}{\sqrt{\Theta}} \, d\vartheta. \tag{21.137}$$

In this form, the geodesic equations can be solved and investigated numerically. The equations of null geodesics follow as the subcase $\mu_0 = 0$.

Without numerical integration, only a few qualitative conclusions may be drawn. For example, if an orbit lies in the equatorial plane ($\vartheta = \pi/2$ and $\dot{\vartheta} = 0$ on the whole orbit) or is tangent to that plane at one point ($\vartheta = \pi/2$ and $\dot{\vartheta} = 0$ only at that point), then

$$C \stackrel{\text{def}}{=} K + \mu_0{}^2 a^2 - (aE - L_z)^2 = 0. \tag{21.138}$$

For the orbits that cross the equatorial plane at nonzero angle, $C > 0$ (because $R \geq 0$ and $\Theta \geq 0$ from (21.131)). For the orbits that never go through $\vartheta = \pi/2$, C may be negative. For the orbits that run along the symmetry axis, $L_z = 0$.

Equation (21.138) being fulfilled all along a geodesic is a necessary and sufficient condition for it to lie in an equatorial plane. The necessity was shown above, for the sufficiency see Exercise 18. A detailed investigation of timelike geodesics in the Kerr solution, also in its generalisation for Λ, was carried out by Kraniotis (2004–2014).

21.8 Geodesics in the equatorial plane

This section is based on the papers by Bardeen (1973) and by Boyer and Lindquist (1967).

In the spherically symmetric case each timelike or null geodesic lies in a coordinate plane, and the coordinates can be chosen so that it is the equatorial surface. In the Kerr metric the equatorial surface is $\vartheta = \pi/2$, and a given geodesic either does or does not lie in it. Equations (21.131)–(21.137) simplify when it does, and we will consider this case now.

Consider (21.131) in the case $\vartheta = \pi/2$, $\dot{\vartheta} = 0$. The ranges of r accessible for motion are determined by $R(r) \geq 0$. It is more convenient to treat $R(r)$ as a function of E. For motions in the equatorial surface, $C = 0$ in (21.138), which allows us to express K through the other constants. Substituting such K in (21.129) we obtain

$$\frac{1}{r} R(r) = \left[r \left(r^2 + a^2 \right) + 2ma^2 \right] E^2 - 4amL_z E - (r - 2m){L_z}^2 - r\Delta_r {\mu_0}^2. \tag{21.139}$$

The discriminant with respect to E of the right-hand side of the above is

$$\delta = 4\Delta_r \left\{ r^2 {L_z}^2 + {\mu_0}^2 r \left[r \left(r^2 + a^2 \right) + 2ma^2 \right] \right\}, \tag{21.140}$$

and for each $r > 0$ the signs of δ and Δ_r are the same. Hence, R has zeros only in those regions where $\Delta_r \geq 0$. Before we find out what this means, note that for the tangent vector p^α to a timelike geodesic necessarily $p^0 \propto dx^0/ds > 0$, the opposite would mean that x^0 runs backward with respect to s. From (21.105) and (21.57) we obtain

$$p^0 = g^{00} E - g^{03} L_z = B \left(E - \omega L_z \right) / \left(\Delta_r \Sigma \right), \tag{21.141}$$

where

$$B \overset{\text{def}}{=} \left(r^2 + a^2 \right) \Sigma + 2mra^2 \sin^2 \vartheta, \qquad \omega \overset{\text{def}}{=} 2amr/B. \tag{21.142}$$

Hence, $p^0 > 0 \iff E > \omega L_z$, which in the surface $\vartheta = \pi/2$ reduces to

$$E > 2amL_z/D, \qquad D \overset{\text{def}}{=} r \left(r^2 + a^2 \right) + 2ma^2. \tag{21.143}$$

The two roots of the equation $R(r) = 0$ are

$$E_\pm = \frac{1}{D} \left[2amL_z \pm \sqrt{\Delta_r \left(r^2 {L_z}^2 + {\mu_0}^2 rD \right)} \right]. \tag{21.144}$$

The root E_- does not obey (21.143), so the motion in $\vartheta = \pi/2$ can occur only at $E \geq E_+$.

Now let us consider the sign of Δ_r. If $a^2 > m^2$, then $\Delta_r > 0$ for all r, and there exists no hypersurface analogous to the Schwarzschild horizon. This means that there are no obstacles to communication with the singular ring at $\{r = 0, \vartheta = \pi/2\}$. This ring is then a **naked singularity**. The prevailing opinion among astrophysicists is that real astronomical objects, when they are about to collapse, must get rid of the 'excess' angular momentum to achieve $a^2 < m^2$ and become a black hole. Nevertheless, our familiar celestial bodies, like the Sun and the Earth, have $a^2 > m^2$ (verification of this is left as an exercise for the reader), so $a^2 > m^2$ is not necessarily untypical.

If $a^2 < m^2$, then $\Delta_r > 0$ for $r < r_-$ and for $r > r_+$, where the $r_\pm$ are given by (21.58). Between r_- and r_+, the discriminant (21.140) is negative, which means that $R(r)$ cannot have any zeros there, which in turn means that $dr/ds \neq 0$. Hence, there exist no circular orbits in that region and no turning points for other orbits. A body that enters the region (r_-, r_+) from the side of $r > r_+$ must keep going towards smaller r until it flies through $r = r_-$. Unlike in the Schwarzschild case, however, it does not have to hit the singularity, since there can be a turning point at $r < r_-$. Conversely, a body that entered the region (r_-, r_+) from the side of $r < r_-$ must keep going towards larger r until it flies through $r = r_+$. Thus, the region $r_- < r < r_+$ is analogous to the region $r < 2m$ in the Schwarzschild spacetime and to the corresponding region in the Reissner–Nordström spacetime. The analogy with the R–N spacetime is far-reaching, see Section 21.9.

The value of $E_{\min}(r) = E_+(r)$ depends on the sign of aL_z. When $aL_z > 0$, the orbital angular momentum of the body on the orbit and the internal angular momentum of the source of the gravitational field have the same sense. Such orbits are called **direct**. When $aL_z < 0$, the two angular momenta have opposite senses. Such orbits are called **retrograde**. The difference between these orbits is a relativistic effect: the Newtonian gravitational field does not 'feel' in which direction the central body is rotating. It is only sensitive to the asymmetries (polar flattening) in the central body caused by the centrifugal force. Retrograde orbits in Newton's theory are the same as direct orbits.

In relativity, this difference is quite pronounced, as can be seen from Fig. 21.7. Note from (21.144) that with $aL_z < 0$ and r sufficiently close to r_+ (i.e. with $\Delta_r \approx 0$) $E_{\min} < 0$, so E can be negative. This is the total energy 'at infinity', with the energy equivalent of the rest mass included.[14] Thus, $E < 0$ means that the whole energy contained in the particle is insufficient to actually send it to infinity. In other words, the energy that the particle must have lost on entering such an orbit was greater than its rest energy. This effect does not appear for direct orbits, or in the Schwarzschild limit $a = 0$.

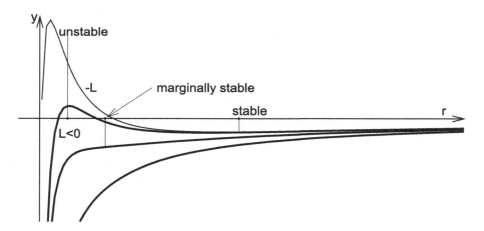

Fig. 21.7 Graphs of $y(r) \overset{\text{def}}{=} E_{\min}(r)/\mu_0 - 1$ for different values of L_z. For every L_z, $y(r) \to 0$ as $r \to \infty$. The lowest graph corresponds to $L_z = 0$, then $y(r)$ is all monotonic. The two upper graphs correspond to large $|L_z|$ – the function $y(r)$ then has a maximum that determines the position of the unstable circular orbit and a minimum that determines the stable circular orbit. Note the large difference between the direct and retrograde orbits: the uppermost curve has $L = aL_z > 0$, the second curve has $L < 0$ with the same $|L|$. The third curve from above approximately corresponds to such L_z at which the maximum and the minimum coalesce into a single point where $\mathrm{d}^2y/\mathrm{d}r^2 = 0$. The value of r at this point determines the radius of the marginally stable circular orbit. On all four curves $r \geq r_+$; the y-axis is drawn at $r \approx r_+$.

---

[14] The gravitational field 'at infinity' vanishes, hence special relativity applies there, in which $p^0 = E$ is the total energy of the orbiting particle.

More information about the orbits follows from the graph of the function $E_{\min}(r) = E_+(r)$; see Fig. 21.7. We consider only the case $a^2 < m^2$ and the region $r > r_+$. Each orbit has a constant $E \geq E_{\min}(r)$, so the area accessible for motion is above the curve $E_{\min}(r)$, and the value of E determines the allowed range of r on the orbit. Note that $E_{\min}/\mu_0 \xrightarrow[r\to\infty]{} 1$ independently of the value of L_z. When $L_z = 0$, $\mathrm{d}(E_{\min}/\mu_0)/\mathrm{d}r > 0$ for all $r \geq r_+$, so $E_{\min}/\mu_0 < 1$ for all $r \geq r_+$. However, when $|L_z|$ is sufficiently large, then there exists an interval (r_1, r_2), with $r_+ \leq r_1 < r_2$, in which $E_{\min}/\mu_0 \geq 1$ while $E_{\min}/\mu_0 < 1$ for $r > r_2$, for each sign of L_z. For retrograde orbits the values of r_1 and r_2 can be explicitly calculated, see Exercise 19. Thus, with $|L_z|$ sufficiently large, $E_{\min}/\mu_0$ has a local maximum at some $r = r_u > r_+$ and a local minimum at $r = r_s > r_u$, as shown in Fig. 21.7. The region where $E_{\min}/\mu_0 < 1$ is the locus of bound orbits, and $r = r_s$ is the radius of the stable circular orbit (i.e. there exists one for each sufficiently large value of $|L_z|$, for each sign of L_z). A circular orbit also exists at $r = r_u$, but is unstable. The smallest r on each curve is $r = r_+$, and $\Delta_r(r_+) = 0$. Equation (21.144) shows that $E_{\min}(r_+) > 0$ on direct orbits and < 0 on retrograde orbits.

Where can the negative-energy orbits occur? Note that the momentum vector of a particle must be timelike, $g^{\alpha\beta}p_\alpha p_\beta > 0$. In an orthonormal tetrad we have for the tetrad components of the momentum

$$p_{\hat{0}}{}^2 - p_{\hat{1}}{}^2 - p_{\hat{2}}{}^2 - p_{\hat{3}}{}^2 = \mu_0{}^2 > 0 \Longrightarrow |p_{\hat{3}}| < p_{\hat{0}}. \tag{21.145}$$

For the metric (21.57), we choose the following orthonormal tetrad:

$$e^0 = e^\nu \mathrm{d}t, \qquad e^1 = e^\lambda \mathrm{d}r, \qquad e^2 = \sqrt{\Sigma}\,\mathrm{d}\vartheta, \qquad e^3 = e^\psi(\mathrm{d}\varphi - \omega \mathrm{d}t), \tag{21.146}$$

where B and ω were defined in $(21.142)^{15}$ and

$$e^{2\nu} = \Delta_r\Sigma/B, \qquad e^{2\lambda} = \Sigma/\Delta_r, \qquad e^{2\psi} = B\sin^2\vartheta/\Sigma. \tag{21.147}$$

The tetrad components of the momentum, defined in (21.105), are then

$$p_{\hat{0}} = e^{-\nu}(E - \omega L_z), \qquad p_{\hat{3}} = -e^{-\psi}L_z. \tag{21.148}$$

From the first equation we have $E = e^\nu p_{\hat{0}} - \omega e^\psi p_{\hat{3}}$. But $p_{\hat{0}} = p^{\hat{0}} > 0$, as explained after (21.140). Hence, for E to be negative, $(-\omega e^\psi p_{\hat{3}})$ must be negative. Since $|p_{\hat{3}}| < p_{\hat{0}}$, it follows that $\omega^2 e^{2\psi} > e^{2\nu}$, which means that $g_{00} < 0$. Thus, orbits with negative energy can exist only between the stationary limit hypersurfaces.

This was a necessary condition for the existence of orbits of negative energy. What is a sufficient condition? Suppose that $p_{\hat{1}} = 0$ at an initial point of the orbit. Then $p_{\hat{0}}$ is bounded because $p_{\hat{0}}{}^2 = \mu_0{}^2 + p_{\hat{2}}{}^2 + p_{\hat{3}}{}^2$, while $p_{\hat{2}}{}^2 = \Sigma^2\dot{\vartheta}^2$ and $p_{\hat{3}}{}^2 = e^{-2\psi}L_z{}^2$ are both bounded in the region $r > r_+$; see (21.130) – (21.131) and (21.142). Consequently, $e^\nu p_{\hat{0}}$ can be made as small as we wish by taking r sufficiently close to r_+. Then, since $E = e^\nu p_{\hat{0}} + \omega L_z$, and ω is seen to be nonzero at $r = r_+$, it suffices to take $aL_z < 0$ and such r that $e^\nu p_{\hat{0}} < |\omega L_z|$ to achieve $E < 0$.

Equation (21.144) simplifies for photon orbits, for which $\mu_0 = 0$:

$$E^\nu_{\min} \stackrel{\text{def}}{=} E_+|_{\mu_0=0} = |L_z|\left(r\sqrt{\Delta_r} \pm 2am\right)/D, \tag{21.149}$$

---

15 The verification that the metric defined by such a tetrad coincides with (21.57) is rather laborious. The first thing to verify is $\Delta_r\Sigma^2 - 4a^2m^2r^2\sin^2\vartheta = B\left(r^2 - 2mr + a^2\cos^2\vartheta\right)$.

where '+' corresponds to $L_z > 0$ and '−' to $L_z < 0$. For any sign, the function $F(r) \overset{\text{def}}{=} E^\nu_{\min}/|L_z|$ has exactly one maximum and no minima for $r \in (r_+, \infty)$ (see Exercise 20). Thus, a photon orbit in the equatorial surface can have only one turning point in this range of r, and there are no stable circular orbits. For each value of L_z, there exists a circular orbit that lies on the maximum of $F(r)$, and so is unstable. The same conclusion follows in the Schwarzschild limit $a = 0$ (see Misner, Thorne and Wheeler (1973) for more details on this subcase). A representative graph of the function $F(r)$ is shown in Fig. 21.8.

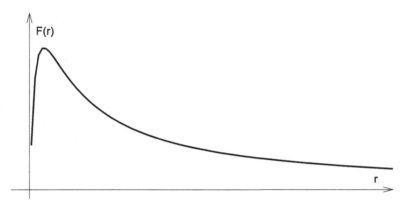

Fig. 21.8 The analogue of the graph from Fig. 21.7 for null geodesics in the equatorial plane. The curve $F(r)$ begins at $r = r_+$. The graph has the same general shape for all values of L_z.

Still more information about equatorial null geodesics can be drawn from the equation of radial motion. As seen from (21.131) and (21.139), for null geodesics (where $\mu_0 = 0$) the affine parameter may be redefined by $s' = Es$, so that only the ratio L_z/E enters the equation. Consequently, it may be assumed without loss of generality that $E = 1$. We redefine some of the variables as follows:

$$\rho \overset{\text{def}}{=} r/(2m), \qquad \lambda \overset{\text{def}}{=} -L_z/(2m), \qquad \alpha \overset{\text{def}}{=} a/(2m). \qquad (21.150)$$

Then, for equatorial null geodesics, (21.131) and (21.139) imply

$$\dot{r}^2 = \left[\rho^3 + (\alpha^2 - \lambda^2)\rho + (\alpha + \lambda)^2\right]/\rho^3 \overset{\text{def}}{=} \psi(\rho)/\rho^3. \qquad (21.151)$$

The turning points of the null geodesics are at the zeros of $\psi(\rho)$, and the motion can take place in those ranges of ρ where $\psi(\rho)/\rho^3 > 0$. Since we will extend the Kerr manifold in the next section to include $r < 0$, we will take it into account here.

Apart from the case $\alpha = -\lambda$, i.e. $a = L_z$, the last term in ψ is positive, so $\psi(0) > 0$, while $\psi < 0$ for sufficiently large $\rho < 0$. Thus, ψ typically has a root at $\rho = \rho_1 < 0$ and is positive for some $0 > \rho > \rho_1$. Any ray coming from $r = -\infty$ will eventually return to $r = -\infty$ (the ray with $\alpha = -\lambda$ hits the $r = 0$ singularity). If $\lambda^2 \leq \alpha^2$ while $\lambda \neq -\alpha$, then there are no positive roots of ψ, and any ray coming from the side of positive ρ will hit $r = 0$. However, if λ^2 is sufficiently large, then ψ has two positive roots, $\rho_2 > \rho_1 > 0$. In that case, there are rays that come in from $r = +\infty$ and turn back at ρ_2, possibly after circling the central body several times in a spiral, and rays that come from the side of

$r = 0$ and turn back towards $r = 0$ after reaching ρ_1. The situation is shown in Fig. 21.9 for the case $m^2 > a^2$. Figures 21.10 and 21.11 show the situations when $a^2 = m^2$ and $a^2 > m^2$, respectively. The contours separating the allowed and prohibited areas are the graphs of $\lambda(\rho)$ found by solving the equation $\psi(\rho) = 0$ for λ; from (21.151) the solution is

$$\lambda = \left(-\alpha \pm |\rho| \sqrt{\rho^2 - \rho + \alpha^2}\right) / (1 - \rho) \overset{\text{def}}{=} \lambda_\pm(\rho). \tag{21.152}$$

In comparing (21.152) with the figures note well how the presence of $|\rho|$ influences the relation between the different branches of $\lambda(\rho)$. Namely, in the $\rho < 0$ area of Fig. 21.9, the upper branch corresponds to $\lambda_-(\rho)$, and the lower branch to $\lambda_+(\rho)$, while the opposite is true in the $\rho > 0$ area. Note also that it is ψ/ρ^3 that must be positive in the allowed region; hence in the $\rho < 0$ area the allowed region is where $\psi < 0$.

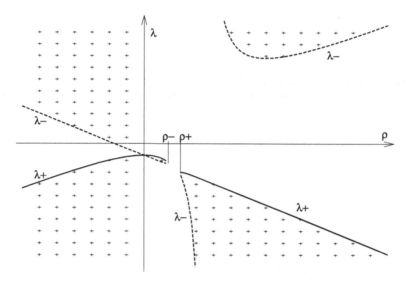

Fig. 21.9 The allowed ranges of ρ for null geodesics in the case $a^2 < m^2$ (the areas covered with crosses are prohibited). On the $\rho < 0$ side, most rays coming from $r = -\infty$ turn back at some $\rho < 0$, but those with $\lambda = -\alpha$ hit $r = 0$. For $\rho > 0$ and large λ^2, the rays can either emanate from $\rho = 0$ and turn back at a finite ρ_1 or come in from $\rho = +\infty$ and turn back at a finite ρ_2. For small λ^2, on the $\rho > 0$ side, the rays mostly hit the singularity at $\rho = 0$, except for a small range of λ, in which there is the little prohibited 'peninsula'. The right tip of the peninsula is at $r = r_-$, the inner event horizon. The left tip of the prohibited wedge in the $\rho > 0$ region is at $r = r_+$, the outer event horizon. With decreasing α, the peninsula shrinks and its left tip moves towards the ρ-axis (and so does the tip of the white wedge on the left), the upper prohibited area on the right moves down and left, while the lower prohibited area moves down and right. In the limit $a = 0$ the peninsula disappears, and the whole graph becomes mirror-symmetric with respect to the ρ-axis. For what happens when a increases, see Figs. 21.10 and 21.11.

For timelike geodesics, we can choose the affine parameter so that $\mu_0 = 1$. We define $\Gamma \overset{\text{def}}{=} E^2 - 1$, and, using (21.150), we obtain from (21.131) and (21.139) (with $\vartheta = \pi/2$):

$$\dot{r}^2 = \frac{1}{\rho^3} \left[\Gamma \rho^3 + \rho^2 + \left(\alpha^2 \Gamma - \lambda^2\right) \rho + (\alpha E + \lambda)^2\right] \overset{\text{def}}{=} \frac{1}{\rho^3} \tilde{\psi}(\rho). \tag{21.153}$$

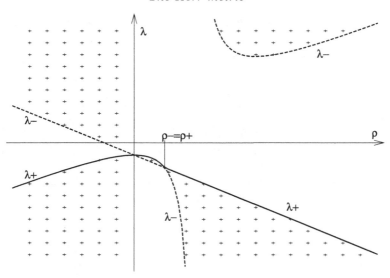

Fig. 21.10 The allowed ranges of ρ for null geodesics in the case $a^2 = m^2$. The little peninsula from Fig. 21.9 has extended to touch the tip of the lower prohibited area. The contact point has $\rho = \rho_- = \rho_+$.

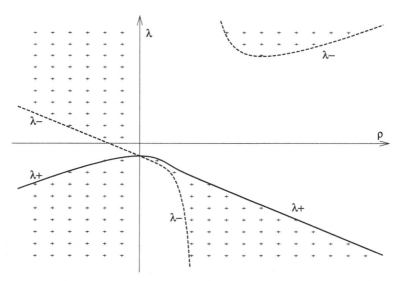

Fig. 21.11 The allowed ranges of ρ for null geodesics in the case $a^2 > m^2$. The two prohibited areas at $\{\rho > 0, \lambda < 0\}$ merged into one. As a^2 increases, the lower prohibited wedge becomes gradually wider and its tip moves down, while the upper prohibited area on the right moves up and to the right.

The allowed range of r is given by $\widetilde{\psi}(\rho)/\rho^3 \geq 0$. The zeros of $\widetilde{\psi}(\rho)$ exist only for those ρ at which the discriminant of $\widetilde{\psi}(\rho)$ treated as a function of λ is nonnegative, thus

$$\delta = \rho\,(\rho - \rho_-)\,(\rho - \rho_+)\,(\Gamma\rho + 1) \geq 0, \tag{21.154}$$

where $\rho_{\pm}$ are the event horizons; $\rho_- < 1/2$, $\rho_- < \rho_+ < 1$. Note that $\Gamma \geq -1$ by definition. The following cases have to be considered:

Case 1: $\Gamma < 0$.

Then $|\Gamma| \leq 1$, and $\delta \geq 0$ for $0 \leq \rho \leq \rho_-$ and for $\rho_+ \leq \rho \leq 1/|\Gamma|$. By considering the signs of δ and of $(1 - \rho)$ (the coefficient of λ^2), the following subcases are discerned:

(*i*) $\rho < 0$ – this area is all prohibited;

(*ii*) $0 \leq \rho \leq \rho_-$ – orbits exist for $\lambda \leq \lambda_-$ and for $\lambda \geq \lambda_+$, where

$$\lambda_{\pm}(\rho) = \left(-\alpha E \pm \sqrt{\delta}\right) / (1 - \rho) \tag{21.155}$$

(note that $\lambda_- < \lambda_+$ when $\rho < 1$ and $\lambda_+ < \lambda_-$ when $\rho > 1$);

(*iii*) $\rho_- < \rho < \rho_+$ – all values of λ are allowed;

(*iv*) $\rho_+ \leq \rho < 1$ – orbits exist for $\lambda \leq \lambda_-$ and for $\lambda \geq \lambda_+$;

(*v*) $\rho = 1$ – orbits exits for $\lambda \geq \lambda_c$ when $\alpha E > 0$ and for $\lambda \leq \lambda_c$ when $\alpha E < 0$;
$\lambda_c \stackrel{\text{def}}{=} -\left[\Gamma\left(2\alpha^2 + 1\right) + \alpha^2 + 1\right] / (2\alpha E)$;

(*vi*) $1 < \rho \leq 1/|\Gamma|$ – orbits exist for $\lambda_+ \leq \lambda \leq \lambda_-$;

(*vii*) $\rho > 1/|\Gamma|$ – this area is all prohibited.

Figure 21.12 shows Case 1 with $a > 0$. The curves $\lambda(\rho)$ determined by $\widetilde{\psi}(\rho) = 0$ have vertical tangents at $\rho = 0$, $\rho = \rho_{\pm}$, $\rho \to 1$ and at $\rho = -1/\Gamma$. At $\rho > 1$, the upper part of each 'belly' in the main figure is the $\lambda_-(\rho)$ branch, the lower part is $\lambda_+(\rho)$. At $\rho < 1$, the roles of λ_- and λ_+ reverse; λ_- goes to $+\infty$ or $-\infty$ as $\rho \to 1^{\pm}$, respectively. Curve a corresponds to $\Gamma = -1$, curves b–d correspond to increasing values of Γ. The information about the allowed ranges of ρ can be briefly summarised by saying that the allowed area in the main panel is the surface sheet between the curve being considered and the $\rho = 0$ vertical line. The inset shows a magnified view of the region $0 \leq \rho \leq \rho_-$, where small prohibited 'peninsulas', similar to the one seen in Fig. 21.9, exist.

Case 2: $\Gamma > 0$

Then there are the following ranges of ρ to consider: (a) $\rho \leq -1/\Gamma$, (b) $-1/\Gamma < \rho < 0$, (c) $0 \leq \rho \leq \rho_-$, (d) $\rho_- < \rho < \rho_+$, (e) $\rho_+ \leq \rho \leq 1$ and (f) $\rho > 1$.

In cases (a), (c) and (e) the allowed range is $\lambda \leq \lambda_-$ and $\lambda \geq \lambda_+$, in case (f) the allowed range is $\lambda_- \leq \lambda \leq \lambda_+$, in the remaining two cases all values of λ are allowed.

Case 3: $\Gamma = 0$.

Details of Case 2 and the whole Case 3 are left as an exercise for the readers.

Bound orbits exist in those cases where a $\lambda = $ constant line has a segment that runs in the allowed region and has both ends on the same $\lambda(\rho)$ curve. In Fig. 21.12 such orbits exist for curves c and d; two of them are shown.

21.9 * The maximal analytic extension of the Kerr metric

This section is based on the paper by Boyer and Lindquist (1967), who were the first to construct a complete extension of the case $a^2 < m^2$. Important contributions to this

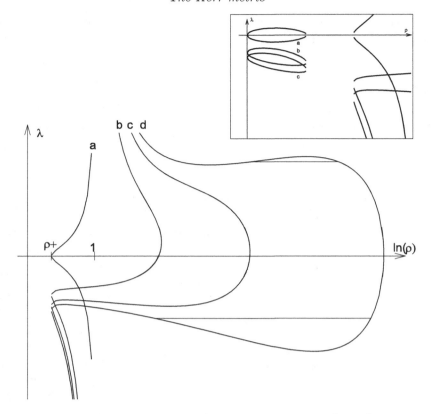

Fig. 21.12 The allowed ranges of ρ for timelike geodesics in the case $a^2 < m^2$, $\Gamma < 0$ for different values of Γ. All the curves correspond to $a > 0$; Γ increases from curve (a) (on which $\Gamma = -1 \Longleftrightarrow E = 0$) to curve (d). The allowed areas are between each curve and the $\ln(\rho) = 0$ line. With $\Gamma < 0$, the whole region $\rho < 0$ is prohibited. Little prohibited peninsulas similar to the one from Fig. 21.9 are present also here, they are shown in the inset (where curve (d) is omitted). The scale on the ρ-axis is logarithmic in the main figure and linear in the inset. Horizontal lines show the allowed ranges of ρ on two bound orbits. Since $a > 0$ for all curves, $\lambda > 0$ corresponds to retrograde orbits and $\lambda < 0$ to direct orbits; the figure clearly shows their inequivalence. There exist allowed regions also with $\rho < 0$ when $\Gamma > 0$, see the text.

subject were published also by Carter, who first constructed such an extension along the symmetry axis (Carter, 1966b), and then extended the Boyer–Lindquist (B–L) construction to the charged case (Carter, 1968b) and to the critical case $a^2 = m^2$ (Carter, 1973).

The maximal extension of the Kerr metric is qualitatively similar to that of the R–N metric in Sec. 14.15. In the Schwarzschild metric, the singularity at $r = 0$ lies in the region where the hypersurfaces $r = $ constant are spacelike, i.e. transversal to all causal curves. Hence, any causal curve running in that region must hit the singularity at a finite value of the affine parameter. In the R–N and Kerr metrics, the singularity at $r = 0$ lies in the region where the hypersurfaces of constant r are timelike. Consequently, a causal curve in that region may steer clear of the singularity and continue infinitely far to the future.

However, in consequence of the geometric arrangement of the event horizons, such a curve cannot return to the same asymptotically flat region from which it entered – this would require travelling backward in time. This is how the infinite chain structure arises. We will see a similar thing happen in the Kerr spacetime.

The B–L coordinates are useful for many calculations, but not for considering the extensions across spurious singularities, since they had in fact *introduced* these singularities. Therefore, we will have to begin with other coordinates.

Take the Kerr metric in the form (21.52) and find its inverse. It is

$$
\left(\frac{\partial}{\partial s}\right)^2 = \left(\frac{\partial}{\partial t}\right)^2 - \frac{1}{\Sigma}\left[(r^2 + a^2)\left(\frac{\partial}{\partial r}\right)^2 - 2a\,\frac{\partial}{\partial r}\frac{\partial}{\partial \varphi}\right.
$$

$$
+\ \frac{1}{\sin^2 \vartheta}\left(\frac{\partial}{\partial \varphi}\right)^2 + \left.\left(\frac{\partial}{\partial \vartheta}\right)^2\right] + \frac{2mr}{\Sigma}\left(\frac{\partial}{\partial t} - \frac{\partial}{\partial r}\right)^2 . \tag{21.156}
$$

This has the Kerr–Schild form (21.9), with the flat metric (the limit $m = 0$ of (21.156)) expressed in spheroidal coordinates. The Kerr–Schild vector field k^α is proportional to

$$
k^\alpha = \left(k^t, k^r, k^\vartheta, k^\varphi\right) = (1, -1, 0, 0). \tag{21.157}
$$

This is *the* affinely parametrised Kerr–Schild field – it obeys $k^\rho k^\alpha{}_{;\rho} = 0$. Since $k^\alpha = \mathrm{d}x^\alpha/\mathrm{d}s$, the coordinate t or r can be chosen as the affine parameter. Since $k^t > 0$, the vector field k^α is future-pointing, and, since $k^r < 0$, it is ingoing. It is tangent to the surfaces ($\vartheta = \text{constant}, \varphi = \text{constant}$). With $\vartheta \neq \pi/2$, the null geodesics tangent to k^α proceed towards smaller r until they cross the $r = 0$ disc and continue into the $r < 0$ region. The null geodesics lying in $\vartheta = \pi/2$ hit the ring singularity at $\{r = 0, \vartheta = \pi/2\}$.

The Kerr metric is of Petrov type D, so it defines a second degenerate Debever congruence (call it ℓ^α), which, by virtue of the Goldberg–Sachs theorem 16.4, must also be geodesic and shearfree. There is no reason why one of these congruences should be preferred, so it should be possible to express the Kerr metric in the Kerr–Schild form (21.9) also with respect to ℓ^α. The B–L coordinates of (21.57) are helpful in identifying ℓ^α. The metric expressed in them is invariant under the substitution $\mathrm{d}r \to -\mathrm{d}r$, which changes the ingoing field (21.157) into an outgoing one, and so puts an outgoing field into the position of k^α. That outgoing field must thus be a Kerr–Schild field, too. To find that second field we transform k^α to the B–L coordinates, obtaining

$$
k^\alpha = \left[(r^2 + a^2)/\Delta_r, -1, 0, a/\Delta_r\right], \tag{21.158}
$$

which implies that

$$
\ell^\alpha = \left[(r^2 + a^2)/\Delta_r, +1, 0, a/\Delta_r\right], \tag{21.159}
$$

and then we transform ℓ^α back to the coordinates of (21.52):

$$
\ell^\alpha = \left[(r^2 + a^2 + 2mr)/\Delta_r, +1, 0, -2a/\Delta_r\right]. \tag{21.160}
$$

This field is geodesic and affinely parametrised, too. The coordinate r is still an affine parameter on the geodesics tangent to ℓ^α, but t is not. The expressions (21.159) and

(21.160) become singular where $\Delta_r = 0$. To find ℓ^α on the horizons, we note that it is proportional to the vector field (in the B–L coordinates)

$$\widetilde{\ell}^\alpha = \left(r^2 + a^2, \Delta_r, 0, a \right). \tag{21.161}$$

This one is geodesic, but no longer affinely parametrised. At $r = r_\pm$, still being null, it becomes $\widetilde{\ell}^\alpha_\pm = (2mr_\pm, 0, \ 0, a)$, i.e. it becomes tangent to the $r = r_\pm$ hypersurfaces. Thus, the hypersurfaces $r = r_\pm$ are tangent to the light cones, i.e. matter and light can cross them in one direction only (they are crossed by the null curves tangent to k^α, which is an ingoing field, so the allowed direction is from outside in). This confirms that they are event horizons. Moreover, note that $\widetilde{\ell}^\alpha_\pm$ is affinely parametrised. We will consider the null geodesics tangent to $\widetilde{\ell}^\alpha_\pm$ later on (see the paragraph around (21.174)).

In the coordinates of (21.52), the fields k^α and ℓ^α appear nonsymmetrically, while they play similar roles in the metric. This is because the coordinates of (21.52) are adapted to k^α. Since in the B–L coordinates the roles of k^α and ℓ^α are interchanged by the substitution $dr \to -dr$, we find the coordinates adapted to ℓ^α in the following way: (1) carry out the transformation inverse to (21.56) on (21.157) with dr replaced by $-dr$ (the hint to Exercise 12 will help here) – the result will be (21.52) with dr replaced by $-dr$; (2) carry out the same transformation on the components of the fields k^α and ℓ^α. The result is

$$k^\alpha = \left[\left(r^2 + a^2 + 2mr \right) / \Delta_r, -1, 0, -2a/\Delta_r \right], \qquad \ell^\alpha = (1, +1, 0, 0), \tag{21.162}$$

i.e. the roles of k^α and ℓ^α are now interchanged, except that k^α is still ingoing, while ℓ^α is still outgoing. In these coordinates it is seen that k^α at the horizons is tangent to them.

This looks mysterious, but the coordinates of (21.157) (call them, after Boyer and Lindquist (1967), E frame) and of (21.162) (call them E' frame) do not have the same domain, and the horizon tangent to k^α is not the same one that is tangent to ℓ^α. To see this, note that both vector fields are future-pointing, k^α is ingoing and ℓ^α is outgoing. Begin at a point in the $r > r_+$ region in the E frame and proceed along the null geodesics tangent to k^α. Since it is ingoing, by moving to the future you proceed towards smaller r, and you will eventually cross the surface $r = r_+$ (at which k^α has no singularity). Not so with the ℓ^α field – if you follow its integral lines to the future from any point in $r > r_+$, then you proceed towards larger r and you will never reach any of the horizons. In order to meet the $r = r_+$ hypersurface, you would have to move along ℓ^α to the past. Thus, the extension of the region $r > r_+$ along the k field in the E frame is not the same as the extension along the ℓ field in the E' frame. This is an analogy to the Kruskal extension of the Schwarzschild manifold.

The transformation from E to E' is regular for all $r > m$ only when $a^2 > m^2$, but then no horizons exist. When $a^2 < m^2$, the domains of the two frames overlap only partially. The transformation is

$$t' = t - \frac{m}{\sqrt{m^2 - a^2}} \left[(r_+) \times \ln \left| \frac{r - r_+}{2m} \right| - (r_-) \times \ln \left| \frac{r - r_-}{2m} \right| \right],$$

$$\varphi' = \varphi - \frac{a}{2\sqrt{m^2 - a^2}} \ln \left| \frac{r - r_+}{r - r_-} \right|, \tag{21.163}$$

and it is singular at both horizons.

These inequivalent extensions now have to be put together into a single extended manifold. Let us begin with the region $r > r_+$ and the E' frame, and let us move back in time along the ℓ field. We can carry out the transformation from the E' frame to the B–L coordinates separately in each of the regions $r > r_+$, $r_- < r < r_+$ and $-\infty < r < r_-$. Hence, this E' frame can be understood to be the common extension of these three regions; we will call them regions (0), (−1) and (−2), respectively. Now let us go back to the $r > r_+$ region, let us transform the coordinates to the E frame and let us move forward in time along the k field. Again, we can carry out the transformation from the E frame to the B–L coordinates in each of the regions $r > r_+$, $r_- < r < r_+$ and $-\infty < r < r_-$. The first one is the same (0) as before, but the other two are different from (−1) and (−2) since they lie to the future of (0). We shall call them (+1) and (+2), respectively. The result of these extensions is schematically shown in Fig. 21.13 on the left.

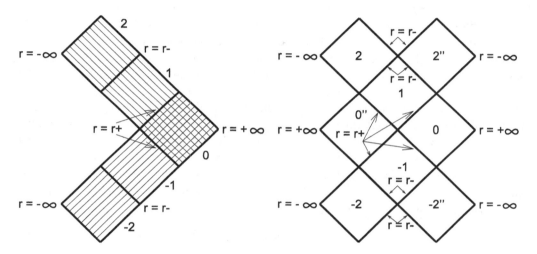

Fig. 21.13 **Left:** The coordinate patches −1 and −2 obtained by extending the region $r > r_+$ along the ℓ field (in the E' frame), and the patches 1 and 2 obtained by extending the same region along the k field (in the E frame). **Right:** The patches 0'' and −2'' were added by extending the −1 patch along the k field, and then the patches +1'' and +2'' were added by extending the 0'' patch along the ℓ field. At this point there is no justification yet for identifying the patches 1'' and 1, and not even for placing 0'' side by side with 1, but a proof that this is correct is provided further on in the text. The straight segments of the left and right edges of the figure are null infinities. Spatial infinities are the extreme points on the left and on the right. The extreme points at mid-height are $+\infty$; the others are $-\infty$.

But then, in the (−1) region the field k exists together with ℓ. Since $\Delta_r < 0$ there, we see from (21.162) that a future-pointing field is actually $(-k^\alpha)$, not k^α itself, and $(-k^\alpha)$ is *outgoing* there, just like ℓ^α. We will thus hit the $r = r_+$ horizon irrespective of whether we move along $(-k^\alpha)$ or along ℓ^α. However, the horizon that we meet when moving along $(-k^\alpha)$ is not the same as that which we meet when moving along ℓ^α. At the first one, ℓ^α becomes tangent to it, while $(-k^\alpha)$ crosses it smoothly (as can be seen by transforming to the E frame); at the second one $(-k^\alpha)$ becomes tangent and ℓ^α crosses it. Thus, just like we did in the Kruskal extension of the Schwarzschild spacetime, we must assume that there are two ways back to the future from region (-1) and that there exists a second copy of region (0) that we will call $(0'')$. Similarly, when we move from region (-1) along $(-k)$ back in time across $r = r_-$, we obtain a second copy of region (-2) that we will call $(-2'')$.

Let us then move from region (-1) to $(0'')$ along the $(-k)$ field.[16] We are then back in a region where $\Delta_r > 0$, so k^α itself is future-pointing and ingoing, while ℓ^α is future-pointing and outgoing. By moving along k^α to the future across $r = r_+$ and then across $r = r_-$, we create second copies of regions $(+1)$ and $(+2)$ that we call $(+1'')$ and $(+2'')$, respectively. By analogy with the Kruskal extension of the Schwarzschild spacetime, we expect that we can identify regions $(+1)$ and $(+1'')$ – after all, there are timelike curves inside the light cone tangent to the k and ℓ fields with vertex in the region (-1), and their future should be continuously predictable. This (at this point hypothetical) identification is shown in Fig. 21.13 on the right. However, in order to prove that such an identification is correct, we must find a coordinate map that will cover regions $(+1)$, (0), (-1), $(0'')$ and $(+1'')$ simultaneously and will show explicitly that $(+1)$ and $(+1'')$ coincide.

There is a difficulty here. In the Schwarzschild case, the $a \to 0$ limits of k^α and ℓ^α are tangent to the surfaces $\{\vartheta = \text{constant}, \varphi = \text{constant}\}$, and the Kruskal diagram shows one such surface. Here, the fields k^α and ℓ^α are not surface-forming, so will not be tangent to a common surface. We will choose such coordinates u and v that the integral curves of k^α and ℓ^α will lie in the hypersurfaces $u + v = \text{constant}$ and $u - v = \text{constant}$, respectively.

We take the B–L coordinates, since in them both null fields are treated on a nearly equal footing – see (21.158) and (21.159). As we approach $r = r_+$ along these fields, they approach the direction $(2mr_+, 0, 0, a)$, which is tangent to the helix in spacetime with the equation $\mathrm{d}\varphi/\mathrm{d}t = a/(2mr_+)$. In the first step, we transform φ to $\varphi' \overset{\text{def}}{=} w$ so that k^α and ℓ^α approach the horizon along the hypersurface $w = \text{constant}$:

$$w = \varphi - at/(2mr_+).$$ (21.164)

This makes $k^w(r_+) = 0$, as desired.

The null curves tangent to the k and ℓ fields obey

$$\mathrm{d}t/\mathrm{d}r = \varepsilon \left(r^2 + a^2\right)/\Delta_r, \qquad \varepsilon = \mp 1$$ (21.165)

(plus the equation that determines $w(r)$). The solution of (21.165) is

$$\mathrm{e}^{\varepsilon r - t} \left(\frac{r - r_+}{2m}\right)^{\varepsilon \sigma_+} \left(\frac{r - r_-}{2m}\right)^{-\varepsilon \sigma_-} = C_\pm = \text{constant},$$

$$\sigma_\pm \overset{\text{def}}{=} mr_\pm / \sqrt{m^2 - a^2}$$ (21.166)

and we want each of the constants $C_\pm$ to be a value of a function $\widetilde{F}$ of the new variables u and v such that $C_\pm = \widetilde{F}(u \pm v)$. Taking each side of (21.166) to the power ε/σ_+ we obtain

$$\exp\left(\frac{r - \varepsilon t}{\sigma_+}\right) \frac{r - r_+}{2m} \left(\frac{r - r_-}{2m}\right)^{-r_-/r_+} = F(u \pm v).$$ (21.167)

We now want to cancel the spurious singularity at $r = r_+$ (we will deal with the one at

---

[16] Actually, after we have crossed the $r = r_+$ hypersurface, we have to move on along k again.

$r = r_-$ later). Using the extension of the R–N metric in Sec. 14.15 as a guide, inspired by a loose similarity of (21.166) to (14.149) (see Exercise 24) and following Boyer and Lindquist (1967), we guess that the right choice might be

$$F(u \pm v) = (u \pm v)^2. \tag{21.168}$$

Then we obtain

$$u \pm v = \sqrt{\frac{r - r_+}{2m}} \left(\frac{r - r_-}{2m}\right)^{-r_-/(2r_+)} \exp\left(\frac{r \pm t}{2\sigma_+}\right), \tag{21.169}$$

$$\Psi(r) \overset{\text{def}}{=} \frac{r - r_+}{2m} \left(\frac{r - r_-}{2m}\right)^{-r_-/r_+} e^{r/\sigma_+} = u^2 - v^2, \tag{21.170}$$

$$t = \sigma_+ \ln\left|\frac{u+v}{u-v}\right| \equiv \sigma_+ \operatorname{artanh}\left(\frac{2uv}{u^2+v^2}\right). \tag{21.171}$$

We have $\Psi(r) \underset{r\to r_-}{\longrightarrow} -\infty$, $\Psi(r_+) = 0$ and $\Psi(r) \underset{r\to+\infty}{\longrightarrow} +\infty$. The function $\Psi(r)$ is well defined and analytic on the whole plane, so $r(u,v)$ exists for all values of u and v. The function $t(u,v)$ is analytic everywhere except on the straight lines $u^2 = v^2$, where t becomes infinite. In spite of that, the Kerr metric transformed to the (u, v, ϑ, w) coordinates[17]

$$
\begin{aligned}
ds^2 = \frac{1}{\Sigma \Delta_r} &\left\{ \left[\frac{2\sigma_+ f\Sigma_+}{r_+{}^2 + a^2} (vdu - udv) + a\sin^2\vartheta \Delta_r dw\right]^2 \right. \\
&\left. - \left[\frac{2\sigma_+ f\Sigma}{r^2 + a^2} (vdu - udv)\right]^2\right\} + \frac{4\sigma_+{}^2 f\Sigma}{(r^2 + a^2)^2} (dv^2 - du^2) - \Sigma d\vartheta^2 \\
&- \frac{\sin^2\vartheta}{\Sigma} \left[\frac{af(r + r_+)}{\sqrt{m^2 - a^2}(r - r_-)} (vdu - udv) - (r^2 + a^2) dw\right]^2,
\end{aligned}
$$

$$\Sigma_\pm \overset{\text{def}}{=} \Sigma(r_\pm, \vartheta),$$

$$f(r) \overset{\text{def}}{=} \frac{\Delta_r}{\Psi(r)} = 4m^2 \left(\frac{r - r_-}{2m}\right)^{2m/r_+} e^{-r/\sigma_+} \tag{21.172}$$

is analytic for all values of u and v, including the two straight lines, and also for all values of $w \in [0, 2\pi]$ and of $\vartheta \in [0, \pi]$. The only term that might raise suspicion is the one with Δ_r in the denominator because $\Delta_r = 0$ at $r = r_+$. However, the coefficient of $(vdu - udv)^2$ between the {} is proportional to $\Sigma_+^2/(r_+{}^2 + a^2)^2 - \Sigma^2/(r^2 + a^2)^2$, which contains the factor $(r - r_+)$ that cancels the one in Δ_r. Since $r = r_+$ is no longer a singularity, (21.172) applies throughout regions $(+1)$, (0), (-1) and $(0'')$ in Fig. 21.13, so we have shown that the identifications made in the figure are indeed allowed. For $r > r_-$ the metric (21.172) is a coordinate transform of (21.57), see Exercise 25.

But recall that $u^2 - v^2 = -\infty$ corresponds to $r = r_-$, so the extension we have just found has dealt with the spurious singularity at $r = r_+$, and not with the one at $r = r_-$. In order to remove the spurious singularities at $r = r_-$, we would have to choose

---

[17] Boyer and Lindquist (1967) worked in the signature $(+++-)$. A simple rewrite of their Eq. (3.8) into our signature $(+---)$ gives an incorrect result, though. The signs in front of both terms containing dw had to be reversed. The metric (21.172) was verified to be a coordinate transform of (21.57).

$F(u \pm v) = (u \pm v)^{-2r_-/r_+}$ in (21.168), with a result analogous to (21.171). Then the range $u \in (-\infty, +\infty)$, $v \in (-\infty, +\infty)$ would cover the region $r \in (-\infty, r_+)$, but the true singularity at $\{r(u,v) = 0, \vartheta = \pi/2\}$ would still be there. The coordinate system thus obtained would apply throughout the regions $\{-2, -1, -2''\}$ and $\{+2, +1, +2''\}$, plus in the regions $(+3)$ and (-3) that we are going to construct now.

We can continue to extend the spacetime in the same way both upwards and downwards, see Fig. 21.14. Thus, we can generate regions $(+3)$ and $(+4)$ by adapting the coordinates to the ℓ field in region $(+2)$ and proceeding to the future, and we can generate regions $(+3)$ and $(+4'')$ by adapting coordinates to the k field in region $(+2'')$ and proceeding to the future. Likewise, we can extend region (-2) along the k field to the past and thus generate regions (-3) and (-4), and extend region $(-2'')$ along the ℓ field to the past, thus generating regions (-3) and $(-4'')$. Exactly as in the case of the R–N spacetime (see Section 14.15) we must then decide whether we wish to continue the process *ad infinitum*, or to identify two regions that are isometric, for example (-1) and $(+3)$.

The arcs of hyperbolae in Fig. 21.14 represent the $r = $ constant hypersurfaces; they are timelike in even-numbered areas where $r > r_+$ and $r < r_-$, and spacelike in odd-numbered areas where $r_- < r < r_+$. The thick arcs represent the sets $r = 0$; they are nonsingular except at $\vartheta = \pi/2$. This is the only difference from the corresponding diagram for the R–N spacetime (Fig. 14.12): here, null and timelike curves can be continued through the open disc $\{r = 0, \vartheta \neq \pi/2\}$ to $r \to -\infty$. The left and right edges of Fig. 21.14 are null infinities; those with numbers divisible by 4 are $+\infty$, the remaining ones are $-\infty$. The extreme points of the edges are spatial infinities. Thick straight segments are the event horizons, alternately r_- and r_+. Identifications can be made such that an odd-numbered area n is identified with the area $(n + 4k)$, where k is an integer number.

There remains now the question of *geodesic completeness* of the mosaic manifold represented by Fig. 21.14. We call a manifold **geodesically complete** if every geodesic in it can be continued to an arbitrarily large absolute value of the affine parameter. We already know that the extended Kerr manifold is not geodesically complete in the strict sense: there are null geodesics that hit the ring singularity $\{r = 0, \vartheta = \pi/2\}$ at a finite value of the affine parameter and cannot be continued beyond it. But we can consider the geodesics that do not run into this singularity and then ask whether they can be continued indefinitely. It turns out that in the extended Kerr geometry they can.

Consider the equations (21.131) – (21.133). We applied them to timelike and null geodesics, but spacelike geodesics will be included if we replace $\mu_0{}^2$ with a negative quantity. The infinite values that appear as $r \to \pm\infty$ are no problem, since we know that the spacetime becomes approximately flat there, and so complete. The components of tangent vectors to geodesics become infinite at $\Sigma = 0$, $\sin\vartheta = 0$ and $\Delta_r = 0$. The set $\Sigma = 0$ is the singularity that we know exists. The set $\sin\vartheta = 0$ is the axis of symmetry. As seen from (21.131), a geodesic can run only in the regions where $\Theta \geq 0$, but $\Theta \to -\infty$ as $\vartheta \to 0$ unless $L_z = 0$. Thus the axis cannot be intersected by any geodesic that has $L_z \neq 0$, and for those with $L_z = 0$ the set $\sin\vartheta = 0$ is nonsingular. Thus, we have to take care only about the infinities in $\dot{t}$ and $\dot{\varphi}$ that appear on the horizons $\Delta_r = 0$.

Before we continue, note that a timelike or null geodesic is ingoing when along it,

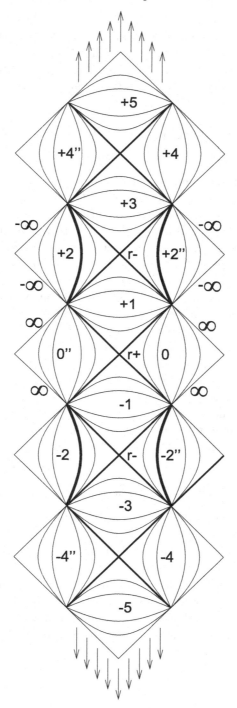

Fig. 21.14 The compactified (u, v) surface of the maximally extended Kerr spacetime. The arcs of hyperbolae are the $r = $ constant lines, the thick arcs are the $r = 0$ lines. See the text for more explanation.

simultaneously, either $(dt/ds > 0$ and $dr/ds < 0)$ or $(dt/ds < 0$ and $dr/ds > 0)$. This is taken care of by the single inequality $dr/dt < 0$. By looking at (21.129), (21.131) and (21.133) we see that a geodesic is ingoing at $r_\pm$ if and only if $\dot{r}(r_\pm)$ and $(2mr_\pm E - aL_z)$ have opposite signs. Now transform (21.132) and (21.133) by the transformation inverse to (21.56) (in which, recall, t' and φ' are the B–L coordinates). The result is (the nonprimed coordinates are now those of (21.52)):

$$\Sigma\dot{t} = -Ea^2\sin^2\vartheta + \left\{\left(r^2 + a^2\right)^2 E - 2mraL_z + 2mr\dot{r}\right\}/\Delta_r,$$
$$\Sigma\dot{\varphi} = L_z/\sin^2\vartheta + a\left\{2mrE - aL_z + \dot{r}\right\}/\Delta_r. \tag{21.173}$$

If $\dot{r}$ and $(2mr_\pm E - aL_z)$ have opposite signs, then the terms in braces vanish at both horizons, and their quotients by Δ_r have finite limits, which means that we have removed the singularities by a coordinate transformation. But why does this work only for ingoing geodesics? This is because the coordinates of (21.52) are adapted to the ingoing k field. In order to deal with outgoing geodesics, we must employ coordinates adapted to the outgoing ℓ field, i.e. related to those of (21.173) and (21.52) by $dr \to -dr$ (see the paragraph containing (21.162)). Indeed, after this transformation the terms containing $\dot{r}$ in (21.173) change signs, so these coordinates remove the singularities at $\Delta_r = 0$ for outgoing geodesics. Thus, by switching between coordinate systems we can carry all geodesics through both horizons. If a geodesic does not strike the true singularity at $\Sigma = 0$, it can be continued to arbitrarily large absolute values of its affine parameter.

The foregoing discussion concerned geodesics intersecting $r = r_\pm$, but the conclusion applies also to geodesics that have turning points at $r = r_\pm$ or approach $r = r_\pm$ asymptotically (by winding around the horizon). The key point is that $\dot{t}$ and $\dot{\varphi}$ remain finite as $r \to r_\pm$, asymptotically or not.

What remains are the null geodesics that run along the horizons – we left them after (21.161) for separate consideration. On such geodesics $r = r_\pm =$ constant ($\dot{r} = 0$), $\Delta_r = 0$ and, as seen from (21.161), $\dot{\vartheta} = 0$. Solving the equation $dx^\alpha/ds = \tilde{\ell}^\alpha_\pm$ we obtain[18]

$$t = s + s_1, \qquad \varphi = a\left(s + s_2\right)/\left(2mr_\pm\right), \tag{21.174}$$

where s is an affine parameter and s_1 and s_2 are arbitrary constants. Such a geodesic can obviously be continued to $s \to \pm\infty$.

Finally, then, all geodesics that do not strike the ring singularity $\Sigma = 0$ can be continued to infinite absolute values of their affine parameters.

We would like now to draw a figure analogous to 14.13 for a surface of constant t and $\vartheta = \pi/2$ in (21.57), but the choice of the t coordinate in the Kerr metric is not unique: the t-coordinate lines are not orthogonal to the spaces of constant t. We wish to choose such a t-coordinate that will minimise the consequences of rotation. A promising choice is $t' = t - a\varphi$, then the surfaces of constant t' and $\vartheta = \pi/2$ have the metric

$$ds_2{}^2 = \frac{r^2}{\Delta_r}\,dr^2 + r^2 d\varphi^2. \tag{21.175}$$

---

[18] The result obtained by Boyer and Lindquist (1967) in a strangely roundabout way is equivalent to (21.174), but their parameter is nonaffine.

For the sake of comparison let us take $a = e$, then (21.175) is the same as (14.162), so Fig. 14.13 applies unchanged.

In the extreme case $a^2 = m^2$ we use as a guide the $e^2 = m^2$ case of R–N (Eqs. (14.167) – (14.170)) and the procedure applied earlier in this section. The considerations about the null Kerr–Schild fields k and ℓ do not depend on whether $a^2 < m^2$ or $a^2 = m^2$. So, in the present case, one of the fields passes through the hypersurface $r = m$ smoothly, while the other becomes tangent to it. The transformation (21.164) that 'untwists' the helical field is here $w = \varphi - t/(2m)$, and the solution of (21.165) is

$$\pm t = \xi(r) \overset{\text{def}}{=} r - m - \frac{2m^2}{r - m} + 2m \ln \left(\frac{r}{m} - 1 \right). \tag{21.176}$$

We now introduce the null coordinates p and q such that the fields k^α and ℓ^α obeying (21.165) are tangent to the hypersurfaces $p \pm q = $ constant; they are $p = t + \xi$, $q = t - \xi$. The transformed metric (21.57) with $a = m$ is

$$
\begin{aligned}
ds^2 \;=\; & \frac{1}{4} \left(1 - \frac{2mr}{\Sigma} \right) (dp + dq)^2 - \frac{\Sigma(r - m)^2}{4\,(r^2 + m^2)} \, (dp - dq)^2 - \Sigma d\vartheta^2 \\
& + \frac{2m^2 r \sin^2 \vartheta}{\Sigma} \, (dp + dq) \left(dw + \frac{dp}{4m} + \frac{dq}{4m} \right) \\
& - \left(\frac{2m^3 r \sin^2 \vartheta}{\Sigma} + r^2 + m^2 \right) \left(dw + \frac{dp}{4m} + \frac{dq}{4m} \right)^2,
\end{aligned}
\tag{21.177}
$$

and it is nonsingular everywhere except at the ring $\Sigma = 0$. The coordinates (p, q) have infinite ranges, so in order to compactify the space and be able to draw a diagram like Fig. 21.14, we transform (p, q) just like we did in (14.170), $p = \tan P$, $q = \tan Q$. The resulting picture is again easy to construct, by analogy with Fig. 14.16. We can extend the region $r > m$ across $r = m$ either along the ingoing k field to the future or along the outgoing ℓ field to the past. Once we have crossed the $r = m$ hypersurface, nothing can prevent us from going to $r \to -\infty$, unless we hit the $\{r = 0, \vartheta = \pi/2\}$ singularity. But on the other side of the horizon, when we used the k field to cross it, we can use the ℓ field to go to the past, and when we crossed the horizon along the ℓ field, we can use the k field to go to the future. What results looks like Fig. 21.14, but with the odd-numbered regions all squeezed to lines, together with their lower-left and upper-right neighbours; see Fig. 21.15. All $r = $ constant surfaces are now either timelike or null. This extension was first constructed by Carter (1973), see there for a brief presentation and references to earlier work.

Just like in the extremal R–N case, the invariant spatial distance from any point in the $r > m$ region to the horizon at $r = m$ along a curve of constant (t, ϑ, φ) is infinite. The metric of the surface $\{t - m\varphi = \text{constant}, \vartheta = \pi/2\}$ is identical to the metric of constant t and $\varphi = \pi/2$ in (14.166), so the embedding of its $r > m$ part in a Euclidean space looks the same as the left part of Fig. 14.17.

21.10 * The Penrose process

Penrose (1969) contemplated a process by which, *in principle*, the rotational energy of a Kerr black hole can be extracted. The idea is based on the observation we made after

Fig. 21.15 The (P,Q) surface of the maximally extended Kerr spacetime with $a^2 = m^2$. The whole left edge represents $r = -\infty$, the whole right edge represents $r = +\infty$. The thick straight segments form the horizon $r = m$, the thick hyperbolae are the $r = 0$ sets. More explanation is given in the text.

(21.148), that a body on a retrograde orbit inside the stationary limit hypersurface can have a negative total energy if its turning point $\dot{r} = 0$ is close enough to the event horizon $r = r_+$. In brief, Penrose's idea was this: put two masses at the ends of a sufficiently strong spring, squeeze the spring and bind the masses together. Then send the composite on an orbit that enters the region between the stationary limit hypersurface and the event horizon $r = r_+$. That region is called the **ergosphere**, and the meaning of this term will become clear at the end of this section. Design the orbit so that it has its turning point very close to $r = r_+$ (how close will become clear below). When the composite object is at the turning point, release the spring in such a direction that one of the masses is sent, with $\dot{r} = 0$ at the initial point, on a retrograde orbit with a negative energy and with $|L_z|$ sufficiently small that it falls through the event horizon. It follows from the reasoning between (21.148) and (21.149) that this is possible: we direct the ejected mass so that $aL_z < 0$, we make $|L_z|$ small enough that the mass is sure to go through $r = r_+$ (see Fig. 21.7), and the orbit has to be pre-designed so that at the turning point $e^\nu p_{\hat{0}} < |\omega L_z|$.

Since the mass dropped into the black hole carried away some negative energy, the other mass acquires additional energy and additional momentum by recoil so that it returns to the outside of the stationary limit hypersurface having a greater energy than it had at the beginning of the journey. This trick can be applied for as long as the stationary limit hypersurface and the horizon exist. The logical conclusion is that the extra energy of the returning mass was gained at the expense of the rotational energy of the black hole: the slower the black hole rotates, the smaller $|a|$ becomes, and the smaller is the volume of the ergosphere. However, this is a speculation that goes beyond the applicability of the Kerr metric. In order to discuss this energy-extraction process in a correct way, we would have to use a nonstationary solution in which the angular momentum of the source of the gravitational field depends on time.

The ergosphere was named after the Greek word $\varepsilon\rho\gamma o$, meaning 'work' – because, as shown, thanks to its existence, a rotating black hole is in principle able to do some work.

21.11 Stationary–axisymmetric spacetimes and locally nonrotating observers

We call a spacetime **stationary** when its metric tensor allows a timelike Killing field (if the Killing field is hypersurface-orthogonal, then the spacetime is called **static**). We call it **axisymmetric** when there exists a Killing field whose integral lines are closed. For the stationary–axisymmetric spacetimes it is assumed that the two Killing fields commute, so they are surface-forming (the Kerr metric (21.57) is an example). Then, coordinates can be chosen so that the metric is independent of $x^0 = t$ (where $\mathrm{d}x^\alpha/\mathrm{d}t = \underset{(0)}{k}{}^\alpha$ – the timelike Killing field) and of $x^3 = \varphi$ (where $\mathrm{d}x^\alpha/\mathrm{d}\varphi = \underset{(3)}{k}{}^\alpha$ – the Killing field connected with axial symmetry). Let the other two coordinates be x^1 and x^2.

At this point, it is assumed in addition that the surfaces generated by the Killing fields admit orthogonal surfaces, i.e. that in the coordinates adapted to the Killing fields $g_{01} = g_{02} = g_{13} = g_{23} = 0$ (this property is called **orthogonal transitivity**). This is equivalent to the requirement that the metric, and the motion of matter if any is present, is invariant under the discrete transformation $(t, \varphi) \to (-t, -\varphi)$.[19] Several theorems were proven in which the property $\left[\underset{(0)}{k}, \underset{(3)}{k}\right] = 0$ and the orthogonal transitivity follow from other assumptions. The intention of those theorems (see Stephani et al. (2003) for a brief listing) was to show that spacetimes that do not possess these properties are rare, unimportant or weird in some sense. The fact is, though, that not much is known about the cases left out.

In a stationary–axisymmetric spacetime that is orthogonally transitive, coordinates in the (x^1, x^2) surfaces can be chosen so that $g_{12} = 0$ – since we know that every 2-dimensional metric is conformally flat, and the property $g_{12} = 0$ is even weaker. Thus

$$\mathrm{d}s^2 = g_{00}\mathrm{d}t^2 + 2g_{03}\mathrm{d}t\mathrm{d}\varphi + g_{33}\mathrm{d}\varphi^2 + g_{11}\mathrm{d}(x^1)^2 + g_{22}\mathrm{d}(x^2)^2. \tag{21.178}$$

When two Killing fields exist and nothing is assumed about them, the basis of the space of Killing fields can be chosen arbitrarily. The transformation of the basis of the form

$$\underset{(0)}{k}{}' = C_0 \underset{(0)}{k} + D_0 \underset{(3)}{k}, \qquad \underset{(3)}{k}{}' = C_3 \underset{(0)}{k} + D_3 \underset{(3)}{k} \tag{21.179}$$

induces a transformation of the coordinates adapted to the Killing fields; the coordinates

[19] An example of a configuration that does not obey this is a rotating gaseous body, inside which the gas circulates in the meridional planes.

(t', φ') adapted to $\underset{(0)}{k}'$ and $\underset{(3)}{k}'$ are related to (t, φ) by

$$t' = C_0 t + C_3 \varphi, \qquad \varphi' = D_0 t + D_3 \varphi. \tag{21.180}$$

Such a transformation preserves the independence of the metric tensor of t and φ and the orthogonal transitivity; it only reshuffles the components g_{00}, g_{03} and g_{33} among themselves. Note, however, that the Killing fields are in fact *unique* if they correspond to stationarity and axial symmetry, and it is assumed in addition that the spacetime is asymptotically flat. Then the integral lines of $\underset{(3)}{k}$ are closed, and the coordinate φ defined by this field is periodic with the period 2π. This excludes those transformations (21.180) in which $C_3 \neq 0$, or else the strange behaviour illustrated in Fig. 21.16 would occur: after increasing φ by $2\pi C_3 D_0 / (C_0 D_3 - C_3 D_0)$ we would land at the same t' line from which we started, but with the t'-coordinate increased by $\Delta t' \overset{\text{def}}{=} -2\pi C_3$, see Exercise 26. The orbits of the $\underset{(3)}{k}$ field would thereby be disrupted and changed into infinite helices. The time coordinate would thus fail to be a continuous function of the spacetime point.[20]

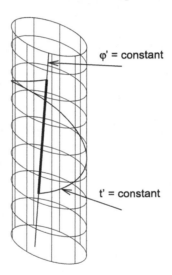

φ' = constant

t' = constant

Fig. 21.16 An orbit of axial symmetry. A coordinate transformation of the form (21.180) would result in disrupting the orbits of the $\underset{(3)}{k}$ field and in a discontinuous time-coordinate. The thick helix segment shows the jump in t' along the $\varphi' = 0$ line (given in the text).

The constant D_0 has to be zero for a different reason. Suppose that the coordinates of the stationary–axisymmetric spacetime that is asymptotically flat go over, asymptotically, into the cylindrical coordinates:

$$ds^2 = dt^2 - r^2 d\varphi^2 - dr^2 - dz^2. \tag{21.181}$$

---

[20] However, exactly this kind of time coordinate is used on the surface of the Earth – the discontinuity occurs across the line of change of date that runs through the middle of the Pacific. This discontinuity played a dramatic role in one novel by Jules Verne (1874).

This means that asymptotically g_{03} must go to zero, while g_{33}/r^2 goes to -1. After the transformation (21.180) with $D_0 \neq 0$, in the coordinates adapted to the transformed Killing fields, the g_{03} component becomes

$$g'_{03} = (g_{03} - D_0 g_{33}/D_3) / (D_3 C_0),$$

(21.182)

and would fail to go to zero asymptotically at infinity. Consequently, $D_0 = 0$ in consequence of the assumed asymptotic flatness.

Finally, $D_3 = 1$ in order that the period of φ is 2π. Thus, the only freedom remaining is the choice of the unit of time connected with C_0.

This uniqueness allows us to introduce the notion of *locally nonrotating observers* (Bardeen, 1970). Write the metric (21.178) in the form

$$ds^2 = e^{2\nu} dt^2 - e^{2\psi} (d\varphi - \omega dt)^2 - e^{2\lambda} dr^2 - e^{2\mu} d\vartheta^2.$$

(21.183)

Now imagine an observer moving with the 4-velocity U^α such that $U^r = U^\vartheta = 0$, i.e. circulating within the (t, φ) surface. Define $\Omega \stackrel{\text{def}}{=} U^\varphi/U^t \equiv d\varphi/dt$ – this is the angular velocity of the observer. Imagine that she/he has set up mirrors all along her/his orbit so that light rays emitted by her/his carry-on source can travel around the same circle in both directions and come back to her/him. Let the time of the round-trip of the light ray be T_ε, with $\varepsilon = +1$ for the ray rotating forward, and $\varepsilon = -1$ for the ray rotating backward with respect to the observer. The observer will during the time T_ε move from φ_1 to $\varphi_2 = \varphi_1 + \Omega T_\varepsilon$. The light ray, during its round trip, will move from φ_1 to $\varphi_3 = \varphi_1 + \Omega T_\varepsilon + 2\pi\varepsilon$. Along the ray $ds^2 = 0$, so $\varepsilon e^\nu dt = e^\psi (d\varphi - \omega dt)$, and consequently

$$dt = \frac{e^\psi}{\omega e^\psi + \varepsilon e^\nu} d\varphi.$$

(21.184)

Integrating this over the whole round-trip we obtain

$$T_\varepsilon = \frac{\Omega T_\varepsilon + 2\pi\varepsilon}{\omega + \varepsilon e^{\nu-\psi}}.$$

(21.185)

Solving this for T_ε we obtain

$$T_\varepsilon = \frac{2\pi}{e^{\nu-\psi} - \varepsilon(\Omega - \omega)}.$$

(21.186)

Along the observer's path $d\varphi = \Omega dt$, so her/his proper time s is related to t by

$$ds^2 = \left[e^{2\nu} - e^{2\psi} (\Omega - \omega)^2 \right] dt^2.$$

(21.187)

Hence, the proper time of the round-trip of the ray measured by the observer is

$$S_\varepsilon = \sqrt{e^{2\nu} - e^{2\psi}(\Omega - \omega)^2} T_\varepsilon \equiv 2\pi e^\psi \sqrt{\frac{1 + \varepsilon e^{\psi-\nu}(\Omega - \omega)}{1 - \varepsilon e^{\psi-\nu}(\Omega - \omega)}}.$$

(21.188)

This time is different for the forward ray and for the backward ray ($S_1 \neq S_{-1}$), except when $\Omega = \omega$. Since we showed that in an asymptotically flat spacetime the (t, φ) coordinates of (21.183) are unique, ω is uniquely determined by the geometry, so *there is a uniquely determined set of observers for whom the effect of rotation, ($S_1 - S_{-1}$), disappears.* They are called **locally nonrotating observers** (Bardeen, 1970). Their worldlines are orthogonal to the $t = $ constant hypersurfaces, so their rotation tensor defined by (15.33) is zero.

The quantities present in (21.183) were defined for the Kerr metric in (21.147) and (21.142). The Kerr metric is stationary, axisymmetric and asymptotically flat, with the period of φ equal to 2π, so it is an example of a spacetime that allows the existence of locally nonrotating observers.

21.12 * Ellipsoidal spacetimes

Together with locally nonrotating observers, one can define another set of observers in the Kerr spacetime – those for whom the locally orthogonal subspaces are composed of confocal ellipsoids of revolution. In order to define these observers, we first introduce the notion of an *ellipsoidal spacetime*. It is defined in analogy to the spherically symmetric spacetimes that are composed of concentric spheres. The ellipsoidal spacetimes are composed of concentric ellipsoids in a similar manner.[21]

Consider a Euclidean space E_3 with rectangular Cartesian coordinates (x, y, z) and the metric $ds^2 = dx^2 + dy^2 + dz^2$. Introduce such (r, ϑ, Φ) coordinates that $r = $ constant defines an ellipsoid of revolution:

$$x = g(r)\sin\vartheta\cos\Phi, \qquad y = g(r)\sin\vartheta\sin\Phi, \qquad z = r\cos\vartheta. \tag{21.189}$$

They will be orthogonal when $g^2(r) - r^2 = $ constant, i.e. when

$$g(r) = \sqrt{r^2 + a^2}, \qquad a = \text{constant}. \tag{21.190}$$

This implies that the ellipsoids corresponding to different values of r are confocal, and a is the radius of their common focal ring $\{r = 0, \vartheta = \pi/2\}$. The surfaces $\vartheta = $ constant are one-sheeted hyperboloids of revolution with their foci on the same ring. The metric in these coordinates becomes

$$ds^2 = \frac{r^2 + a^2\cos^2\vartheta}{r^2 + a^2}\, dr^2 + \left(r^2 + a^2\cos^2\vartheta\right)d\vartheta^2 + \left(r^2 + a^2\right)\sin^2\vartheta d\Phi^2. \tag{21.191}$$

The space with the metric (21.191) is still flat (and note that it coincides with the space of constant t in the limit $m = 0$ of the Kerr metric (21.57)). We make it curved by assuming that $g_{rr} = f^2(r, \vartheta)$ is an arbitrary function – in analogy to a curved spherically symmetric space that results when the Cartesian $g_{rr} = 1$ is replaced by $g_{rr} = f^2(r)$.

We now define an **ellipsoidal spacetime** as follows: there should exist a congruence of timelike lines in it, with the tangent vector field

$$u^\alpha = U\delta^\alpha{}_0 + V\delta^\alpha{}_3 \tag{21.192}$$

(where $x^0 = t$ and $x^3 = \varphi$) such that the metric

$$ds_3{}^2 = f^2(r, \vartheta)dr^2 + \left(r^2 + a^2\cos^2\vartheta\right)d\vartheta^2 + \left(r^2 + a^2\right)\sin^2\vartheta d\Phi^2 \tag{21.193}$$

---

[21] In this section, we will consider confocal ellipsoids of revolution. In principle, it should be possible to generalise this construction to other types and arrangements of the ellipsoids, but so far those other possibilities have not been explored.

coincides with the metric of the space locally orthogonal to u^{α}, i.e.

$$h_{\alpha\beta} = g_{\alpha\beta} - u_{\alpha}u_{\beta}. \tag{21.194}$$

From the normalisation $g_{\alpha\beta}u^{\alpha}u^{\beta} = 1$ we have

$$g_{00}U^2 + 2g_{03}UV + g_{33}V^2 = 1. \tag{21.195}$$

Since we intend to apply this construction to the Kerr metric, we assume that the ellipsoidal spacetime is stationary and axially symmetric, and thus has the metric (21.178). Equating (21.193) with (21.194) we obtain

$$
\begin{aligned}
- \ h_{\alpha\beta}dx^{\alpha}dx^{\beta} &= -g_{11}dr^2 - g_{22}d\vartheta^2 - \left(g_{00}g_{33} - g_{03}{}^2\right)U^2\left(d\varphi - \frac{V}{U}dt\right)^2 \\
&= f^2(r,\vartheta)dr^2 + \left(r^2 + a^2\cos^2\vartheta\right)d\vartheta^2 + \left(r^2 + a^2\right)\sin^2\vartheta d\Phi^2.
\end{aligned} \tag{21.196}
$$

Thus,

$$
\begin{aligned}
g_{00}g_{33} - g_{03}{}^2 &= -\left(r^2 + a^2\right)\sin^2\vartheta/U^2, \qquad g_{11} = -f^2(r,\vartheta), \\
g_{22} &= -\left(r^2 + a^2\cos^2\vartheta\right), \qquad d\Phi = d\varphi - (V/U)dt.
\end{aligned} \tag{21.197}
$$

The set (21.195) – (21.197) cannot be uniquely solved for $g_{\alpha\beta}$ because we defined only the 3-dimensional projection of the metric, $h_{\alpha\beta}$. Hence, $g_{\alpha\beta}$ will contain one arbitrary function, which we choose to be

$$k^2 \stackrel{\text{def}}{=} g_{33} + \left(r^2 + a^2\right)\sin^2\vartheta. \tag{21.198}$$

The resulting 4-dimensional metric is then

$$
\begin{aligned}
ds^2 &= \left(\frac{1 \mp kV}{U}dt \pm kd\varphi\right)^2 - f^2(r,\vartheta)dr^2 - \left(r^2 + a^2\cos^2\vartheta\right)d\vartheta^2 \\
&\quad - \left(r^2 + a^2\right)\sin^2\vartheta\left[d\varphi - (V/U)dt\right]^2.
\end{aligned} \tag{21.199}
$$

The conditions (21.195) and (21.197) define metric components, so they are coordinate-dependent. Thus, the definition of ellipsoidality in fact says that *coordinates should exist in which the metric obeys (21.195) and (21.197)*. In order to show that the Kerr metric is ellipsoidal, the Boyer–Lindquist coordinates of (21.57) have to be transformed as follows:

$$t = t' + a\varphi, \qquad (r, \vartheta, \varphi) = (r', \vartheta', \varphi'). \tag{21.200}$$

This transformation is of the 'forbidden' type (21.180) that leads to a discontinuous time-coordinate (we have already used it in Sec. 21.9 to derive (21.175) and in the very last paragraph). The resulting metric is

$$
\begin{aligned}
ds^2 &= \left(1 - \frac{2mr}{\Sigma}\right)dt^2 + 2a\left(1 - \frac{2mr\cos^2\vartheta}{\Sigma}\right)dtd\varphi - \frac{\Sigma}{\Delta_r}dr^2 - \Sigma d\vartheta^2 \\
&\quad + \left[a^2 - \frac{2mra^2\cos^4\vartheta}{\Sigma} - \left(r^2 + a^2\right)\sin^2\vartheta\right]d\varphi^2.
\end{aligned} \tag{21.201}
$$

This is seen to coincide with (21.199) when

$$k^2 = a^2\left(1 - 2mr\cos^4\vartheta/\Sigma\right), \qquad V = \left(\pm k - g_{03}U\right)/g_{33},$$

$$U^2 = \left(r^2 + a^2\right)/\Delta_r, \tag{21.202}$$

where the metric components refer to (21.201).

The ellipsoidal spacetimes were introduced by Krasiński (1978) with the hope that the construction would help in the search for a perfect fluid source of the Kerr metric. It did not (see Section 21.14), and the ellipsoidal form turned out to be rather difficult to use in calculations. So far, it has found no practical application and very little following (Racz, 1992; Racz and Süveges, 1998; Zsigrai, 2003; Papakostas, 2009).

21.13 A Newtonian analogue of the Kerr solution

We found in Sec. 12.11 that Newton's gravitation theory results as the $c \to \infty$ limit of relativity if the metric developed in a series of powers of $1/c^2$ has the form (12.46). The corresponding expression for the Kerr metric is (21.47). Since the $r = $ constant surfaces in the Kerr metric are confocal ellipsoids of revolution, its Newtonian limit might have such ellipsoids as equipotential surfaces. However, the potential implied in (21.47) is not necessarily the exact desired one – it may be an approximation to the correct formula. Thus, in order to find the correct Newtonian potential we have to consider the complete set of Newtonian field equations and equations of motion.

One source of such a potential was found by Chasles (1840).[22] It is a homoeoid, i.e. an ellipsoid of revolution with a uniform 2-dimensional distribution of matter over it. The equipotential surfaces are ellipsoids confocal to the source. In the spheroidal coordinates defined in (21.189) – (21.190), the exterior potential is (Krasiński, 1980)

$$V_e(r) = -(GM/a)\arctan(a/r), \tag{21.203}$$

where M is the total mass of the homoeoid. Knowing that the potential depends only on r, it is easy to guess a continuous density distribution that will generate the same exterior potential; it is (Krasiński, 1980)

$$\rho(r, \vartheta) = f(r)/\left(r^2 + a^2\cos^2\vartheta\right), \tag{21.204}$$

where $f(r)$ is an arbitrary function. The total mass of the source is then

$$M = 4\pi \int_0^{r_0} f(r')\mathrm{d}r', \tag{21.205}$$

and the surface of the source is at $r = r_0$. The field (21.203) is the unique solution of the Laplace equation that depends only on r (recall that this is the spheroidal r!) and obeys the boundary conditions $\lim_{r\to\infty} V = 0$ and $\lim_{r\to\infty} r^2 \mathrm{d}V/\mathrm{d}r = GM$.

Some properties of this source were investigated by Krasiński (1980) and by Bażański, Kaczyńska and Krasiński (1985). Among other things, the interior potential was found, and the equations of motion were solved to find the distribution of velocity of matter and of pressure. The equipotential surfaces are ellipsoids of the same (confocal) family also inside the source. If $\rho_{,r} < 0$, then the surfaces of constant matter density are more oblate than

---

[22] Information from Webster (1949). The exact references were tracked by Marie-Noëlle Célérier (to whom A. K. expresses his gratitude) in the library of the Meudon Observatory in Paris.

the equipotential surfaces; they do not coincide in any case. The pressure is not constant on the equipotential surfaces either. The focal ring of the ellipsoids, $\{r = 0, \vartheta = \pi/2\}$, is in general a singularity; in order to make it nonsingular the density would have to *increase* with increasing r, at least in some neighbourhood of the ring. We refer the reader to the two papers cited above for a more extended list of properties of this configuration. Let us only note a few curiosities here:

1. The exterior field is determined only by the mass and angular momentum of the body; it does not depend on the body's size.

2. At a point P inside the body, the gravitational potential is determined only by the mass contained inside the equipotential ellipsoid passing through P; the matter exterior to that ellipsoid gives no contribution.

3. Although the configuration is rather special, in the limit $a \to 0$ it reproduces *all possible* spherically symmetric configurations (because of the arbitrary function $f(r)$).

21.14 A source of the Kerr field?

The Kerr solution has been known for more than 60 years now, and from the very beginning its existence provoked the simple question: what material body could generate such a vacuum field around it? Several authors tried hard to find a model of the source, but so far without success. The most promising positive result is that of Roos (1976), who investigated the Einstein equations with a perfect fluid or an anisotropic nondissipating fluid distribution, with the boundary condition that the Kerr metric is matched to it. Roos' result was that the equations form an integrable set. All attempts so far to find an explicit solution have failed.[23] The sources known are rather artificial (see Krasiński (1978) for references): a 2-dimensional disc spanned on the singular ring of the Kerr metric; a body with anisotropic stresses, sometimes enveloped in a crust of another kind of anisotropic matter. The continuing lack of success prompted some authors to spread the suspicion that a perfect fluid source might not exist, rumours about this suspicion were then taken as a serious suggestion. The opinion of one of the present authors (A. K.) is that a bright new idea is needed, as opposed to routine standard tricks tested so far. Just a look at the supposedly simple but difficult to handle Newtonian models with homogeneous density distribution (Chandrasekhar, 1969) allows one to realise that the corresponding problem in relativity must be at least as difficult.

21.15 Exercises

1. Prove that the Schwarzschild metric has the property (21.1).

Hint. The Schwarzschild metric must first be transformed from the curvature coordinates (14.40) − (14.42) by

$$t = t' - 2m \int \frac{dr'}{r' - 2m}, \qquad (r, \vartheta, \varphi) = (r', \vartheta', \varphi') \qquad (21.206)$$

---

[23] See Krasiński (1978) for an overview of the attempts prior to 1976; all the later attempts can be classified according to the same scheme and none has moved the matter forward.

after which it becomes

$$ds^2 = dt'^2 - dr'^2 - r'^2 \left(d\vartheta'^2 + \sin^2 \vartheta' d\varphi'^2 \right) - \frac{2m}{r'} \left(dt' + dr' \right)^2 . \tag{21.207}$$

This is the Kerr–Schild form with $l_0 = l_1 = \sqrt{2m/r'}$. Now find $g^{\alpha\beta}$ in these coordinates, verify that $g^{\alpha\rho} l_\rho = \eta^{\alpha\rho} l_\rho$ and that $g_{\alpha\beta} l^\alpha l^\beta = \eta_{\alpha\beta} l^\alpha l^\beta = 0$.

2. Prove that the vector field l_μ in (21.1) is null with respect to $g_{\mu\nu}$ if and only if it is null with respect to $\eta_{\mu\nu}$. Prove then that $l^\mu \overset{\text{def}}{=} g^{\mu\nu} l_\nu \equiv \eta^{\mu\nu} l_\nu$ and verify that (21.2) holds.

3. Verify (21.3) and (21.4).

Hint. The following indentities may be useful:

$$l_{\mu,\nu} - l_{\nu,\mu} = l_{\mu;\nu} - l_{\nu;\mu}, \qquad l^\rho l_{\rho;\mu} = 0 \qquad \text{and} \qquad \eta_{\alpha\beta;\gamma} = 0. \tag{21.208}$$

4. Verify (21.12) (it is rather laborious and requires several changes of names of indices in order to detect simplifications).

5. Verify that (21.16) and (21.18) are consistent and admit nonzero Y-s.

Hint. Solve (21.16) and (21.18) for $Y_{,\xi}$ and $Y_{,\overline{\xi}}$, and then check the integrability condition $Y_{,\xi\overline{\xi}} = Y_{,\overline{\xi}\xi}$.

6. Verify (21.24) and (21.25).

7. Compare the formula for $R_{10} = R^1{}_{110} + R^2{}_{120} + R^3{}_{130}$ (mentioned under (21.25)) with $R_{01} = R^0{}_{001} + R^2{}_{021} + R^3{}_{031}$ and note how much simpler the first one comes out. They are equivalent, both lead to $R_{01} = 0 = R_{10}$ after using (21.10) − (21.11) and (21.19) − (21.23), but R_{01} needs a larger number of simplifications.

8. Verify (21.26).

Hint. Be sure to calculate R_{12}, not R_{21} (the reason is the same as in Exercise 7). Equation (21.24) will be helpful in that.

9. Prove that the assumption $Z \neq 0$ in (21.19) implies that Y and $\overline{Y}$ in (21.15) obey $dY \wedge d\overline{Y} \neq 0$ ($\Longleftrightarrow Y_{,[\alpha} \overline{Y}_{,\beta]} \neq 0$).

Hint. Assume that $Y_{,[\alpha} \overline{Y}_{,\beta]} = 0$, which implies $\overline{Y}_{,\alpha} = \mathcal{A} Y_{,\alpha}$, where $\mathcal{A}$ is a function, substitute this in (21.19) and use (21.18). It will follow that $Z = 0$.

10. Verify that the functions Y, $\overline{Y}$, Z and $\overline{Z}$ obey (21.34).

Hint. To prove that Y obeys (21.34), solve $k^\rho Y_{,\rho} = 0 = \overline{m}^\rho Y_{,\rho}$ and $\overline{Z} = Y Y_{,u} - Y_{,\xi}$ for $Y_{,v}$, $Y_{,\xi}$ and $Y_{,\overline{\xi}}$, then substitute for $Y_{,u}$ from (21.31). After substituting all of these in (21.34) recognise $\left(e^{-P} \right)_{,\overline{Y}}$ in the result. For $\overline{Y}$ the thesis follows by complex conjugation. To prove (21.34) for Z note that the directional derivatives along $\mathcal{K}^\alpha$ and along $\overline{m}^\alpha$ commute, and use $\mathcal{K}^\rho \overline{Y}_{,\rho} = 0$.

11. Verify that if x, y and z are interpreted as Cartesian coordinates, then the surfaces $r = $ constant defined by (21.43) are confocal ellipsoids of revolution. Since they are axially symmetric, the foci of cross-sections through the axis all lie on a circle. Verify that the radius of that circle is a and that $2r$ is the smallest diameter of the ellipsoid $r = $ constant.

12. Verify that the transformation (21.56) turns (21.52) into (21.57).

Hint. Substitute the transformation formulae (21.56) in (21.52). The most difficult part of this exercise is simplifying the coefficient of the new dr^2 (we omit primes here). Begin by substituting $8m^3 r^3 = 4m^2 r^2 \left(\Sigma + a^2 \sin^2 \vartheta - \Delta_r \right)$, then $4m^2 r^2 = 2mr \left(\Sigma + a^2 \sin^2 \vartheta - \Delta_r \right)$; the rest should be easy. Similar tricks help in verifying that there is no $dr d\varphi$ term in (21.57).

13. Verify (21.61)

Remark: This looks like a difficult computational task – when done by hand. My (A. K.'s) computer algebra program Ortocartan (Krasiński, 2001b) did this calculation in 10 *milli*seconds, and as a bonus verified that the Ricci tensor of the tetrad (21.60) equals zero. Preparing and verifying the data, and designing the chain of simplifications took about 2 hours. (Of course, it helped much that I knew the expected result.)

14. Verify, by any method, that the Kerr metric is of Petrov type D.

15. Verify that with the electromagnetic potential A_α given by (21.76), the electromagnetic field tensor has only two tetrad components in the tetrad (21.77), given by (21.78).

Hint. Calculate first the tensor components $F_{\alpha\beta} = A_{\alpha,\beta} - A_{\beta,\alpha}$. Then you will need the following formulae for the inverse tetrad $e_i{}^\alpha$:

$$e_{\widehat{0}} = \frac{1}{\sqrt{Z\Delta_r}} \left(Z_r \partial/\partial t + a\partial/\partial\varphi\right), \qquad e_{\widehat{1}} = \sqrt{\Delta_r/Z} \, \partial/\partial r,$$

$$e_{\widehat{2}} = \sqrt{\Delta_\mu/Z} \, \partial/\partial\mu, \qquad e_{\widehat{3}} = -\frac{1}{\sqrt{Z\Delta_\mu}} \left(Z_\mu \partial/\partial t + \partial/\partial\varphi\right). \tag{21.209}$$

16. Verify that the only Maxwell equations that are not identically fulfilled up to (21.88) are $\left(\sqrt{-g}F^{0\beta}\right)_{,\beta} = 0 = \left(\sqrt{-g}F^{3\beta}\right)_{,\beta}$.

17. Verify that the equations shown in Exercise 16 are equivalent to (21.89) – (21.90).

Hint. Recall from the paragraph above (21.68) that $\sqrt{-g} = Z$. Then use the relation $F^{\alpha\beta} = e_i{}^\alpha e_j{}^\beta F^{ij}$ between the tensor components $F^{\alpha\beta}$ and the tetrad components F^{ij}. The tetrad vectors are given by (21.209). Use the fact that the only nonzero F^{ij} are $F^{\widehat{0}\widehat{1}} = -F_{\widehat{0}\widehat{1}}$ and $F^{\widehat{2}\widehat{3}} = F_{\widehat{2}\widehat{3}}$. In this way you will find that F^{01} (tensor component) $= -(Z_r/Z)F_{\widehat{0}\widehat{1}}$ (tetrad component), and similarly $F^{02} = (Z_\mu/Z)F_{\widehat{2}\widehat{3}}$, $F^{31} = -(a/Z)F_{\widehat{0}\widehat{1}}$, $F^{32} = (1/Z)F_{\widehat{2}\widehat{3}}$.

18. Show that if (21.138) is fulfilled all along a geodesic, then the geodesic lies in the equatorial plane.

Hint: A geodesic lies in the equatorial plane when $\vartheta = \pi/2$ all along it, so $\dot{\vartheta} \equiv 0$. This means that $\Theta(\vartheta) \equiv 0$ in (21.131). So, one must find the solutions of $\Theta(\vartheta) = 0$ with K obeying (21.138). There are two solutions: one is $\cos\vartheta = 0$, the other is

$$\sin^2\vartheta = \frac{L_z{}^2}{a^2\left(E^2 - \mu_0{}^2\right)}. \tag{21.210}$$

Calculate $\ddot{\vartheta}$ for both of these solutions by differentiating $\Sigma\dot{\vartheta} = \pm\sqrt{\Theta(\vartheta)}$, which follows from (21.131). Note that $\dot{\vartheta}/\sqrt{\Theta(\vartheta)}$, which will appear in the calculation, has a finite limit at $\dot{\vartheta} \to 0$, equal to $\pm 1/\Sigma$, in both solutions of $\Theta(\vartheta) = 0$. For $\cos\vartheta = 0$, $\ddot{\vartheta} = 0$, which means that $\cos\vartheta = 0$ holds all along the orbit. For (21.210), $\ddot{\vartheta} \neq 0$, so it holds only at one point of the orbit, and ϑ has an extremum there.

19. Prove that with $a^2 < m^2$ and a sufficiently large value of $|L_z|$, there exists a range (r_1, r_2), with $r_+ \leq r_1 < r_2$, such that $E_{\min}(r)/\mu_0 > 1$ for $r_1 < r < r_2$ and $E_{\min}(r)/\mu_0 < 1$ for $r > r_2$ in (21.144).

Hint. Substitute (21.144) with the + sign in $E_{\min}(r)/\mu_0 > 1$ to obtain

$$\sqrt{\Delta_r\left(r^2 L_z{}^2 + \mu_0{}^2 rD\right)} > \mu_0 D - 2amL_z. \tag{21.211}$$

If $aL_z > 0$ and, at a given r, $aL_z > \mu_0 D/(2m)$, then the right-hand side above is negative and the inequality is trivially satisfied. This proves that indeed $E_{\min}(r)/\mu_0 > 1$ at every r for a sufficiently large $aL_z > 0$.

When $aL_z < 0$, and also at such r where $\mu_0 D - 2amL_z > 0$ in spite of $aL_z > 0$, more follows. Note that $r^2 \Delta_r - 4a^2m^2 = (r - 2m)D$. Equation (21.211) implies then $\Delta_r \left(r^2 L_z{}^2 + \mu_0{}^2 rD \right) > (\mu_0 D - 2amL_z)^2$, which, in consequence of $D > 0$, simplifies to

$$- 4mr^2 + (L_z/\mu_0)^2 r - 2m \left(\frac{L_z}{\mu_0} - a \right)^2 > 0. \tag{21.212}$$

If $|L_z|$ is large enough, then the discriminant of this with respect to r is positive, hence $E_{\min}(r)/\mu_0 > 1$ for $r_1 < r < r_2$, with

$$r_{1,2} = \frac{1}{4m} \left[\left(\frac{L_z}{\mu_0} \right)^2 \mp \sqrt{ \left(\frac{L_z}{\mu_0} \right)^4 - 16m^2 \left(\frac{L_z}{\mu_0} - a \right)^2 } \right], \tag{21.213}$$

and $E_{\min}(r)/\mu_0 < 1$ for $r > r_2$. Then verify that $r_1 \geq r_+$ for any L_z.

20. Prove that the function $F(r) \overset{\text{def}}{=} E_{\min}^\nu / |L_z|$ defined in (21.149) has only one maximum and no minima in the range $r \in (r_+, \infty)$.

Hint. One method to do it is as follows: 1. Note that $F = (r - 2m)/ \left(r\sqrt{\Delta_r} \mp 2am \right)$. Note also that $F \neq 0$ at $r = 2m$ with one or the other sign (depending on the sign of a) because the denominator vanishes there, too, and the limit of the whole expression is finite. The sign is $-$ when $a > 0$, and we follow this case. 2. Calculate $F' \overset{\text{def}}{=} dF/dr$ and observe that it is $+\infty$ at $r = r_+$ and goes to zero from below as $r \to \infty$. Hence it has at least one zero in the range $r \in (r_+, \infty)$. 3. The equation $F' = 0$ is equivalent to $2am\sqrt{\Delta_r} = $ (polynomial of degree 3 in r). 4. Square this equation and simplify, r will factor out. The resulting fifth-degree equation looks bad, but it factorises and becomes $(r - 2m)^2 \left(r^3 - 6mr^2 + 9m^2r - 4ma^2 \right) \overset{\text{def}}{=} (r - 2m)^2 H(r) = 0$. The $r = 2m$ is not a zero of F': with $a > 0$ the limit of F' at $r \to 2m$ is finite; with the plus sign, F' has a nonzero value there. Consequently, the zeros of F' are a subset of the zeros of H. 5. Observe that $H(r)$ has two zeros in the range $r \in (r_+, \infty)$; however, only one of them is also a zero of F'. This is done as follows: $H(r)$ is positive at $r = r_+$ and at $r = 4m$, but negative at its minimum at $r = 3m$. Thus one of its zeros is in $(r_+, 3m)$ and the other in $(3m, 4m)$. Now, with $a > 0$, $F'(r)$ changes sign an odd number of times in $(r_+, 3m)$ (because $F' > 0$ at r_+ and $F' < 0$ at $3m$), but an even number of times in $(3m, 4m)$, hence the second zero of H is not a zero of F'. (With $a < 0$, the opposite is true: $F'(r)$ changes sign an even number of times in $(r_+, 3m)$, but an odd number of times in $(3m, 4m)$, so the first zero of H is not a zero of F'.) Consequently, F' has only one zero in (r_+, ∞). That this corresponds to a maximum of F follows from point 2 above: F is increasing in the neighbourhood of $r = r_+$ and decreasing as $r \to \infty$.

21. Draw graphs analogous to Fig. 21.12 for the cases $\Gamma > 0$ and $\Gamma = 0$ (by using a plotting computer program).

22. Verify that the vector field k^α in (21.157) is geodesic and affinely parametrised.

Hint. The components of $g_{\alpha\beta}$ and of $g^{\alpha\beta}$ needed to calculate the Christoffel symbols can be read out from (21.52) and (21.156).

23. Verify that the vector fields k^α and ℓ^α given by (21.157) and (21.160) (in the coordinates of (21.52)) and (21.158) – (21.159) (in the B–L coordinates of (21.57)) are *not* surface-forming.

24. The analogue of the function ϕ from (14.144) in the Kerr metric is Δ_r. The r^* defined under (14.147) with $\phi = \Delta_r$ is

$$e^{r^*} = \left(\frac{r - r_+}{2m}\right)^{\sigma_+/(2m)} \left(\frac{r - r_-}{2m}\right)^{-\sigma_-/(2m)}, \tag{21.214}$$

where $\sigma_\pm$ are defined in (21.166). This shows the similarity between (21.166) and (14.149).

25. Verify that the metric (21.172) is equivalent to (21.57) under the coordinate transformation defined in (21.169) – (21.171).

Hint. The integral in (21.165) is

$$\tilde{\chi} \stackrel{\text{def}}{=} \int \frac{\varepsilon \left(r^2 + a^2\right) dr}{\Delta_r} = \varepsilon r + \ln\left[\left(\frac{r - r_+}{2m}\right)^{\varepsilon\sigma_+} \left(\frac{r - r_-}{2m}\right)^{-\varepsilon\sigma_-}\right]. \tag{21.215}$$

Denote $\chi = e^{\tilde{\chi}}$. The function ψ of (21.170) was defined as $\psi = \chi^{\varepsilon/\sigma_+}$. Remembering that $\chi_{,r}/\chi = \tilde{\chi}_{,r} = \varepsilon \left(r^2 + a^2\right)/\Delta_r$ we find

$$\psi_{,r} = \frac{\psi \left(r^2 + a^2\right)}{\sigma_+ \Delta_r}. \tag{21.216}$$

From (21.169) we have

$$u = \sqrt{\psi} \cosh\left[t/\left(2\sigma_+\right)\right], \qquad v = \sqrt{\psi} \sinh\left[t/\left(2\sigma_+\right)\right]. \tag{21.217}$$

Using the above we calculate

$$vdu - udv = -\psi dt/\left(2\sigma_+\right), \tag{21.218}$$

$$dv^2 - du^2 = -\frac{\psi \left(r^2 + a^2\right)^2}{4\sigma_+^2 \Delta_r^2} dr^2 + \frac{\psi}{4\sigma_+^2} dt^2. \tag{21.219}$$

Using this in (21.172) we obtain (21.57). The calculation is horrible and it is advisable to use a computer algebra program for it. In calculating the g_{tt} component it is simpler to verify that $\Sigma\mathcal{H} - \Sigma + 2mr = 0$, where $\mathcal{H}$ is the coefficient of dt^2 that comes out of (21.172).

26. Verify the reasoning under (21.180): after increasing φ by $2\pi C_3 D_0/\left(C_0 D_3 - C_3 D_0\right)$ along a line of constant t' we land on the line of constant φ' with t' increased by $(-2\pi C_3)$.

Hint. The equation of the $t' = 0$ line (call it Line I) is $t = -C_3\varphi/C_0$, the equation of the $\varphi' = 0$ line (call it Line II) is $t = -D_3\varphi/D_0$. The two lines intersect at $t_1 = \varphi_1 = 0 = t_1' = \varphi_1'$. They will intersect for a second time at the point where φ_2^I on Line I is equal to $(2\pi + \varphi_2^{II})$. The corresponding t coordinates of the intersection point must be equal, so

$$\frac{C_3}{C_0} \left(2\pi + \varphi_2^{II}\right) = \frac{D_3}{D_0} \varphi_2^{II} \implies \varphi_2^{II} = \frac{2\pi C_3 D_0}{C_0 D_3 - C_3 D_0}. \tag{21.220}$$

The t corresponding to φ_2^{II} on Line I is

$$t_2 = -\frac{C_3}{C_0} \left(2\pi + \varphi_2^{II}\right) = -2\pi \frac{C_3 D_3}{C_0 D_3 - C_3 D_0}. \tag{21.221}$$

The (t', φ') corresponding to φ_2^{II} and t_2 are $(-2\pi C_3, 0)$.

27. Prove that for the locally nonrotating observers defined in Sec. 21.11 the rotation tensor is zero and that their worldlines are orthogonal to the $t = $ constant hypersurfaces.

28. Find the equations of the light cones in Fig. 21.6.

Hint. The metric of the subspace $\vartheta = \pi/2$ in (21.57) is

$$ds^2 = \left(1 - \frac{2m}{r}\right) dt^2 + \frac{4ma}{r} dtd\varphi - \left(\frac{2ma^2}{r} + r^2 + a^2\right) d\varphi^2 - \frac{r^2}{\Delta_r} dr^2. \tag{21.222}$$

The equation of the light cones is $ds^2 = 0$. The relation between the B–L coordinates (r, φ) and the (X, Y) spatial coordinates in the figure is $X = \sqrt{r^2 + a^2} \cos\varphi$, $Y = \sqrt{r^2 + a^2} \sin\varphi$. The cones are drawn at the points of the plane $\varphi = 0$, where $dr = \left(\sqrt{r^2 + a^2}/r\right) dX$ and $d\varphi = dY/\sqrt{r^2 + a^2}$. Thus the equation of a cone with vertex at $(T, X, Y) = (0, 0, X_0)$ is

$$\left(1 - \frac{2m}{r}\right) T^2 + \frac{4ma}{r\sqrt{r^2 + a^2}} TY - \left[\frac{2ma^2}{r(r^2 + a^2)} + 1\right] Y^2$$
$$- \frac{r^2 + a^2}{\Delta_r} (X - X_0)^2 = 0. \tag{21.223}$$

The vertical axis in the figure is T, the (X, Y) axes are as shown. The equations for drawing the graphs are obtained by solving the above for T or for X, but a few special cases must be treated separately. The limit $r = 2m$ (the outer stationary limit hypersurface) in the above equation is nonsingular. Before taking the limits $r \to r_\pm$, the equation must be multiplied by Δ_r. These limits are discontinuous, as described at the end of Section 21.5. On approaching $r = r_\pm$ from inside the segment (r_-, r_+), the equation becomes $X = X_0(r_\pm)$, which is a vertical plane. On approaching these values from outside that segment, the equation becomes singular, but by calculating the curves along which the cones intersect with a $T = $ constant plane one finds that they are ellipses whose both axes shrink to a point, while the ellipses themselves recede to $Y \to -\infty$ as $r \to r_\pm$. Before taking the limit $r \to 0$, the equation must be multiplied by r; in the limit it becomes $-2m[T - (a/|a|)Y]^2 = 0$, which is one of the straight lines $T = Y$ (when $a > 0$) or $T = -Y$ (when $a < 0$, the case shown in the figure).

22

Relativity enters technology: the Global Positioning System

This chapter is based on the articles by Neil Ashby (1998–2015).[1]

The US-based Global Positioning System (GPS) is by today one of a few Global Navigation Satellite Systems (GNSS) that are working or are being developed and prepared for operation. The other ones are the Russian GLONASS (Global Navigation Satellite System), the European Galileo and the Chinese Beidou. The GPS was the first (launched in 1977), and by now is the longest-working and most widely used, therefore this overview will be focussed on it. It was originally intended for military use only, but is accessible to civilian users since 1989. The relativistic effects are present in all GNSS.

22.1 Purpose and setup

The GPS provides space and time coordinates of portable receivers on the surface of the Earth and above it. It consists of a constellation of satellites that carry atomic clocks, several ground-based monitoring stations and a Master Control Station in Colorado Springs. Information between the satellites and the ground stations is transmitted by electromagnetic waves at two frequencies (Ashby, 1998).

The orbital planes of the GPS satellites are inclined at 55° to the equatorial plane. At this angle the perturbations of the satellite paths caused by the Earth's quadrupole moment have zero time-average. The basic system includes 6 such orbits that intersect the equatorial plane in 12 points at angular separations of 30°, and 4 satellites on each orbit. In addition, there are several spare satellites and clocks. This distribution of satellites in space ensures that at least 4 of them are at nearly all times in the field of view for every point on Earth's surface. Each satellite carries a cesium clock of accuracy 5×10^{-14}, i.e. 4 nanoseconds per 24 hours.

The accuracy in time of 1 ns implies the accuracy in space of 30 cm. The relativistic effects are much larger than this, as will be seen from the following. Accuracies of about 1 mm are technically feasible by today, but are accessible only for US military use.

Each satellite sends a stream of circularly polarised electromagnetic waves that carries information about its time-coordinate and position in space. The signals are received by a network of monitoring stations on the ground and forwarded to the Master Control

---

[1] Consultation by Neil Ashby and the permission to use his figures are gratefully acknowledged; I (A.K.) am also grateful to Naresh Dadhich for the permission on behalf of the publisher of Ashby (1998).

Station. There, the state of the whole constellation is analysed and predicted for the next few hours. Based on the predictions, the clocks in the satellites are periodically adjusted.

The users carry locators that determine the user's time, position and velocity using the information transmitted in the signal – see below. Until 2000, civilians were shielded from the highest precision: the two carrier frequencies were intentionally fluctuated (high-precision positioning required access to undisturbed signals). From 2000 on, this limitation does not exist, and new signals are being added to increase the precision even further.[2]

Without the relativistic corrections, the GPS would be useless – see the end of this chapter. This system could be used for experimental tests of relativity, but its precision is not better than that of dedicated experiments.

22.2 The principle of position determination

Let $\vec{r}_i$ be the positions of clocks, $i = 1, 2, 3, 4$, given at times t_i. Let $\vec{r}$ and t be the unknown position and time of the receiver; all the times and positions are referred to a local inertial frame. Consider the signals sent from $\vec{r}_i$ at t_i that are simultaneously received at $\left(\vec{r}, t\right)$. Since the velocity of light c is constant, we have

$$c^2 \left(t - t_i\right)^2 = \left|\vec{r} - \vec{r}_i\right|^2. \tag{22.1}$$

These are 4 equations, to be solved for the 4 unknowns $\left(\vec{r}, t\right)$. But, in consequence of the high precision of the system, the reference frame with respect to which the times and positions are calculated must be carefully defined – see the next sections.

The timing signals are places in the wave trains where the phase of the electromagnetic wave changes sign. At those events, momentarily $F_{\mu\nu} = 0$, and this is independent of the reference system.

The following reference systems will be used in the calculations:
ECEF = Earth-centred, Earth-fixed. This system rotates with the Earth.
ECI = Earth-centred inertial. This system, like the previous one, has its origin at the centre of the Earth, but does not rotate.
The metre and the second of the SI are defined in the ECEF frame.

22.3 The reference frames and the Sagnac effect

Equation (22.1) applies in a local inertial frame, while navigators on the Earth use the noninertial ECEF frame. Transformations between them are needed in using the GPS.

Ignore for a moment the gravitational field of the Earth. In an inertial frame, in cylindrical coordinates, the Minkowski metric is

$$- \mathrm{d}s^2 = -(c\mathrm{d}t)^2 + \mathrm{d}r^2 + r^2 \mathrm{d}\varphi^2 + \mathrm{d}z^2. \tag{22.2}$$

---

[2] See https://www.gps.gov/systems/gps/modernization/civilsignals/ for the currently available menu of signals.

The transformation equations to the ECEF with the coordinates (t', r', φ', z') are

$$(t, r, z) = (t', r', z'), \qquad \varphi = \varphi' + \omega_E t', \tag{22.3}$$

where $\omega_E = 7.292115 \times 10^{-5}$ rad/s is the angular velocity of rotation of the Earth. The Minkowski metric in the ECEF frame is thus

$$- ds^2 = - \left(1 - \omega_E^2 r'^2/c^2\right) (cdt')^2 + 2\omega_E r'^2 d\varphi' dt' + dr^2 + r^2 d\varphi'^2 + dz'^2. \tag{22.4}$$

So, in a rotating frame, the time coordinate t' is not the proper time of the comoving observer (which is $d\tau = \sqrt{1 - (\omega_E r'/c)^2 dt'})$, but the proper time of the ECI observer.

Suppose that a set of observers on the rotating Earth use the Einstein synchronisation method (by sending light rays between them and registering the emission/detection times). Since $\omega_E r'/c \approx 1.5 \times 10^{-6}$, $(\omega_E r'/c)^2 \approx 2.25 \times 10^{-12} \ll \omega_E r'/c$ and can be neglected. The equation of the light signal is then

$$0 = -ds^2 = -(cdt')^2 + 2 \left(\omega_E/c\right) r'^2 d\varphi' (cdt') + d\sigma'^2, \tag{22.5}$$

where

$$d\sigma'^2 \overset{\text{def}}{=} dr^2 + r^2 d\varphi'^2 + dz'^2. \tag{22.6}$$

The solution of (22.5), up to $\omega_E r'/c$, is

$$cdt' = d\sigma' + \frac{1}{c} \omega_E r'^2 d\varphi' = d\sigma' + 2\omega_e dA_z'/c, \tag{22.7}$$

see Fig. 22.1 for the definition of dA_z'. Thus, the light pulse traverses a path $\mathcal{P}$ in the time

$$T = \int_{\mathcal{P}} dt' = \int_{\mathcal{P}} \frac{1}{c} d\sigma' + 2\frac{\omega_E}{c^2} \int_{\mathcal{P}} dA_z'. \tag{22.8}$$

The first term applies in an inertial frame, the second one is the correction due to rotation.

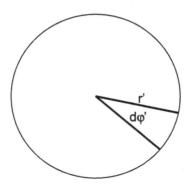

Fig. 22.1 The surface area of the wedge is $dA_z' = \frac{1}{2} r'^2 d\varphi$. This is the area swept by the radius-vector of the head of the light ray, projected on the equatorial plane.

Now suppose that we try to synchronise clocks on the equator by sending a light ray eastward, all around the Earth. Then the path $\mathcal{P}$ is the whole equator, and

$2\omega_E/c^2 = 1.6227 \times 10^{-21}$ s/m^2,

$r' = a_1 = 6\ 378\ 137$ m is the equatorial radius of the Earth,

$2\frac{\omega_E}{c^2}\int_{\mathcal{P}}\mathrm{d}A'_z = 2\frac{\omega_E}{c^2} \times \frac{1}{2}\pi a_1{}^2 = 207.4$ ns is the additional time needed for the light ray to

catch up with its point of origin.

The last quantity is the error in synchronisation caused by Earth's rotation; it is called the **Sagnac effect.**[3] For a ray sent around the equator in the westward direction, the error would have the same magnitude, but opposite sign – the ray would meet its point of origin by 207.4 ns earlier than if the Earth were not rotating.

One might think that the error would be avoided by moving a portable reference clock along the equator, with negligible velocity. This does not help. The proper time of the clock would be, from (22.4),

$$\mathrm{d}\tau^2 = (\mathrm{d}s/c)^2 = \mathrm{d}t'^2\left[1 - \left(\frac{\omega_E r'}{c}\right)^2 - 2\frac{\omega_E r'^2}{c^2}\frac{\mathrm{d}\varphi'}{\mathrm{d}t'} - \frac{1}{c^2}\left(\frac{\mathrm{d}\sigma'}{\mathrm{d}t'}\right)^2\right]. \tag{22.9}$$

Assuming slow motion, $c^{-1}\mathrm{d}\sigma'/\mathrm{d}t' \ll 1$, and $(\omega_E r'/c)^2 \ll \omega_E r'/c$ again, we obtain $\mathrm{d}\tau \approx \mathrm{d}t' - \omega_E r'^2\mathrm{d}\varphi'/c^2$, so for the travel time all around the equator eastward we again obtain the same formula (22.8), and the same synchronisation error. The conclusion is:

In synchronising clocks all over the Earth, constant reference must be made to the synchronisation in the underlying inertial frame, ECI. This is the only self-consistent frame.

22.4 Earth's gravitation and the SI time units

We assume the following:

(i) The mass distribution in the Earth is static.

(ii) The origin of the local inertial frame is at the Earth's centre of mass.

Then the linearised solution of the Einstein equations that describes the Earth's gravitational field is, in agreement with (12.155) – (12.157),

$$-\mathrm{d}s^2 = -\left(1 + \frac{2V}{c^2}\right)(c\mathrm{d}t)^2 + \left(1 - \frac{2V}{c^2}\right)(\mathrm{d}r^2 + r^2\mathrm{d}\vartheta^2 + r^2\sin^2\vartheta\mathrm{d}\varphi^2), \tag{22.10}$$

where

$$V = -\frac{GM}{r}\left[1 - J_2\left(\frac{a_1}{r}\right)^2 P_2(\cos\vartheta)\right] \tag{22.11}$$

is the Newtonian gravitational potential of the Earth (calculated up to terms of order r^{-3}), M is the mass of the Earth, a_1 is, as before, the equatorial radius of the Earth and

- $GM = 3.98600418 \times 10^{14}$ (m^3/s^2),

---

[3] Georges Sagnac (1913) was an opponent of special relativity, and he interpreted his effect as a proof of existence of the aether. The correct explanation of this effect was first given by Paul Langevin (1921). See Frauendiener (2018) for more on this story.

- $J_2 = 1.0826300 \times 10^{-3}$ is the quadrupole moment of the Earth,
- $P_2(\cos \vartheta) = \frac{1}{2} \left(3 \cos^2 \vartheta - 1 \right)$ is the second Legendre polynomial.

We transform the metric (22.10) to the rotating ECEF frame by (22.3), and neglect terms $(\omega r'/c)^n$ where $n > 2$. The result is

$$-ds^2 = - \left(1 + \frac{2\Phi}{c^2} \right) (cdt')^2 + 2\frac{\omega_E}{c} \, r'^2 \sin^2 \vartheta' cdt' d\varphi'$$
$$+ \left(1 - \frac{2V}{c^2} \right) \left(dr'^2 + r'^2 d\vartheta'^2 + r'^2 \sin^2 \vartheta' d\varphi'^2 \right), \qquad (22.12)$$

where

$$\Phi \overset{\text{def}}{=} V - \frac{1}{2} \left(\omega_E r' \sin \vartheta' \right)^2 \qquad (22.13)$$

is the **effective potential** – the sum of the gravitational and centrifugal potentials. In (22.12), t' is the proper time of observers at rest at infinity.

The SI time units are defined by atomic clocks at rest in the ECEF system at the mean sea level. These clocks rotate with the Earth and, because of rotational flattening, are at different distances from the Earth's centre. So, for precise synchronisation, we need the equation of the idealised Earth's surface (called **geoid**) and the values of Φ all over it. The geoid is defined as the surface of constant effective potential, thus from (22.11) and (22.13)

$$-\frac{2GM}{r'} \left[1 - J_2 \left(\frac{a_1}{r'} \right)^2 \frac{3\cos^2 \vartheta' - 1}{2} \right] - \left(\omega_E r' \sin \vartheta' \right)^2 \overset{\text{def}}{=} 2\frac{\Phi_0}{c^2} = \text{constant.} \qquad (22.14)$$

Models of the geoid that include higher multipoles are known. However, the sum of the first 20 multipoles gives a correction to Φ_0/c^2 of the order of 10^{-16}, so it is negligible.

To find the value of Φ_0/c^2, we take it on the equator, where $\vartheta' = \pi/2$ and $r' = a_1$:

$$\frac{\Phi_0}{c^2} = -6.9693 \times 10^{-10}. \qquad (22.15)$$

It is seen from (22.12) that τ – the proper time of a clock at rest on the geoid – is related to the inertial time at infinity t' by

$$d\tau = ds/c = dt' \left(1 + \Phi_0/c^2 \right). \qquad (22.16)$$

Thus, the atomic clocks on the Earth *run slow* compared to the (fictional, idealised) clocks at rest at infinity (because $\Phi_0 < 0$). The rate of the two frequencies is $\approx -7 \times 10^{-10}$, which is $\approx 10,000$ times the fractional frequency stability of the best cesium clocks.

However, since Φ_0 is constant, all atomic clocks at the sea level beat at the same rate. Consequently, the time coordinate on the Earth may be rescaled by a constant factor so that it coincides with the proper time of the standard clocks,

$$t'' = \left(1 + \Phi_0/c^2 \right) t'. \qquad (22.17)$$

Up to terms of order $1/c^2$, the transformation (22.17) changes the metric (22.12) as follows:

$$-ds^2 = - \left(1 + 2\frac{\Phi - \Phi_0}{c^2} \right) (cdt'')^2 + 2\frac{\omega_E}{c} \, r'^2 \sin^2 \vartheta' d\varphi' cdt''$$

$$+ \left(1 - \frac{2V}{c^2}\right) \left(\mathrm{d}r'^2 + r'^2 \mathrm{d}\vartheta'^2 + r'^2 \sin^2 \vartheta' \mathrm{d}\varphi'^2\right). \tag{22.18}$$

On the geoid, where $\Phi = \Phi_0$, the time coordinate t'' is the proper time. We will drop the primes on t'' from now on. The ECI metric results from (22.18) when $\omega_E = 0$ and $\Phi = V$.

Now apply the transformation (22.17) to the metric (22.10). The result is

$$- \mathrm{d}s^2 = - \left(1 + 2\frac{V - \Phi_0}{c^2}\right) (c\mathrm{d}t)^2 + \left(1 - \frac{2V}{c^2}\right) \left(\mathrm{d}r^2 + r^2 \mathrm{d}\vartheta^2 + r^2 \sin^2 \vartheta \mathrm{d}\varphi^2\right). \tag{22.19}$$

This metric is in the ECI reference system. The time-coordinate in (22.19) is the well-defined time on the geoid, so it can be used as a universal basis for synchronisation in the whole GPS. To take into account the time-dilation and gravitational frequency shifts, let us factor out $(c\mathrm{d}t)^2$ from (22.19):

$$\mathrm{d}s^2 = \left[1 + 2\frac{V - \Phi_0}{c^2} - \left(1 - \frac{2V}{c^2}\right)\frac{v^2}{c^2}\right] (c\mathrm{d}t)^2, \tag{22.20}$$

where

$$v^2 \stackrel{\text{def}}{=} \left(\frac{\mathrm{d}r}{\mathrm{d}t}\right)^2 + r^2 \left(\frac{\mathrm{d}\vartheta}{\mathrm{d}t}\right)^2 + r^2 \sin^2 \vartheta \left(\frac{\mathrm{d}\varphi}{\mathrm{d}t}\right)^2 \tag{22.21}$$

is the velocity of the clock in the ECI system. We can neglect the second V/c^2 in (22.20) since it is multiplied by $(v/c)^2$. Neglecting also other terms of the order $(v/c)^4$, the proper time of a moving clock is related to its time-coordinate t by

$$\mathrm{d}s/c \stackrel{\text{def}}{=} \mathrm{d}\tau = \left(1 + \frac{V - \Phi_0}{c^2} - \frac{v^2}{2c^2}\right) \mathrm{d}t. \tag{22.22}$$

Solving this for t, again only up to terms of order $(v/c)^2$, we obtain

$$\int_{\text{path}} \mathrm{d}t = \int_{\text{path}} \left(1 - \frac{V - \Phi_0}{c^2} + \frac{v^2}{2c^2}\right) \mathrm{d}\tau. \tag{22.23}$$

This is the set of corrections to be applied in synchronising the GPS clocks. The last term in (22.23) is often called the **transverse** or second order **Doppler effect**.

22.5 Selected corrections of the orbits of the GPS satellites

22.5.1 Corrections for gravity and velocity

The term Φ_0 includes the Earth's quadrupole moment, but the correction for quadrupole in Φ_0/c^2 is $GMJ_2/(2c^2 a_1) \approx 3.76 \times 10^{-13}$. At the satellites' positions this would be $\approx 10^{-14}$, and we will neglect it for a while.

We also assume that the GPS satellites follow Keplerian orbits. This is not the case for low orbits, but the GPS orbits have radii of $\approx 25,000$ km, and for them the assumption is fulfilled with sufficient accuracy. With the quadrupole term neglected, the satellites move in the potential $V = -GM/r$. Let a be the semimajor axis of the orbit, and e its eccentricity. The orbits are then ellipses, and their equation is

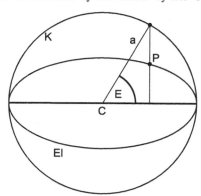

Fig. 22.2 Definition of the eccentric anomaly E for the point P on the ellipse El with semimajor axis a. The circle K has radius a and is tangent to El at both ends of its longer axis. (Imagine the ellipse being stretched in the vertical direction by the ratio a/b, where b is its semiminor axis. After this, El will coincide with K. E is the azimuthal coordinate of the point on K where P will land after the stretching.)

$$r = a \frac{1 - e^2}{1 + e \cos \varphi}, \qquad e = \sqrt{1 - (b/a)^2}, \tag{22.24}$$

where φ is the azimuthal coordinate (also called **true anomaly**) of the satellite in the polar coordinates whose origin coincides with the centre of the Earth and with the focus of the ellipse; $\varphi = 0$ at the perigee. Another useful description of the orbit is via the **eccentric anomaly** E defined in Fig. 22.2. Using E, the equation of the orbit is

$$r = a(1 - e \cos E). \tag{22.25}$$

The relation between the two coordinates is

$$\cos \varphi = \frac{\cos E - e}{1 - e \cos E} \tag{22.26}$$

(the proof of this is left as an exercise for the reader). The dependence of E on the Newtonian time is (see Exercise 1)

$$E - e \sin E = \sqrt{\frac{GM}{a^3}} \, (t - t_p), \tag{22.27}$$

where t_p is the time of passage of the satellite through the perigee, $E = 0$ in Fig. 22.2.

Using these formulae, the total energy per unit mass of a satellite is (see Exercise 2)

$$\frac{v^2}{2} - \frac{GM}{r} = -\frac{GM}{2a}, \tag{22.28}$$

so it is constant on the orbit, as expected. Using this in (22.23) to eliminate v^2 we obtain

$$\int_{\text{path}} dt = \int_{\text{path}} \left[1 + \frac{3GM}{2ac^2} + \frac{\Phi_0}{c^2} - \frac{2GM}{c^2} \left(\frac{1}{a} - \frac{1}{r} \right) \right] d\tau. \tag{22.29}$$

The term in parentheses is zero on circular orbits, while the two constant terms, on a GPS satellite orbit, sum up to

$$\frac{\Delta t_{\text{GPS}}}{\Delta \tau} \overset{\text{def}}{=} \frac{3GM}{2ac^2} + \frac{\Phi_0}{c^2} = -4.465 \times 10^{-10}. \tag{22.30}$$

Since this is < 0, a clock on a circular orbit with a the same as for GPS satellites goes faster than a clock on the Earth. So, in order that the receivers on the Earth see the right frequency of the clock signal, 10.23 MHz, the clocks on satellites are adjusted before launch to go at the corrected frequency:

$$\nu_{\text{corr}} = \left(1 - 4.465 \times 10^{-10}\right) \times 10.23 \text{ MHz} = 10.22999999543 \text{ MHz}. \tag{22.31}$$

Figure 22.3[4] shows the frequency shift $\Delta \nu = \nu - \nu_{\text{corr}}$ resulting from (22.30) as a function of a. On low orbits (for example the Space Shuttle orbit) $\Delta \nu < 0$ because a in (22.30) is smaller, so $3GM/(2ac^2) > 0$ is large, and makes $\Delta t/\Delta \tau > 0$, so $\Delta \nu/\nu < 0$. On high orbits $\Delta \nu/\nu > 0$. The two effects cancel out at $a \approx 9545$ km.

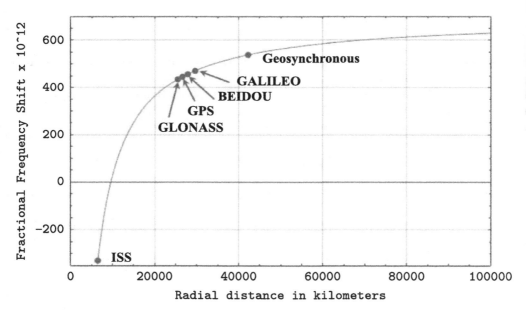

Fig. 22.3 Fractional frequency shifts at various distances from the Earth's centre. The distance is the a in (22.30). ISS = International Space Station.

At the launch of the first GPS satellite on 23 June 1977 some engineers did not believe that relativistic effects were meaningful. But, just in case, a frequency synthesiser was built into the satellite clock system so that corrections could be switched on if relativity turned out to be correct. The uncorrected system ran for 20 days, during which the frequency

[4] Figure 22.3 was supplied by N. Ashby by private communication to A. K.

measured for the satellite clock was by 442.5×10^{-12} larger than for the clocks on the ground. The prediction of relativity was 446.5×10^{-12}. This was in fact an experiment confirming the combination of two relativistic effects (transverse Doppler and gravitational redshift). After 20 days, the frequency correction was switched on. In later-generation satellites, the correction is built in into the hardware.

22.5.2 The eccentricity correction

The last term in (22.29) can be integrated exactly if we observe, from (22.27), that

$$\frac{dE}{dt} = \frac{1}{1 - e \cos E} \sqrt{\frac{GM}{a^3}}, \tag{22.32}$$

$$\frac{1}{r} - \frac{1}{a} = \frac{e \cos E}{a(1 - e \cos E)} \equiv \frac{e}{a} \sqrt{\frac{a^3}{GM}} \cos E \frac{dE}{dt}. \tag{22.33}$$

This last term in (22.29) contains the factor $1/c^2$, and the difference between dt and $d\tau$ is also of order $1/c^2$. So, if we replace $d\tau$ by dt in the integral, the error will be of order $1/c^4$ – negligible at our assumed accuracy. Consequently, using (22.32) we obtain

$$\Delta t = \frac{2GM}{c^2} \int \left(\frac{1}{a} - \frac{1}{r} \right) dt = \frac{2\sqrt{GMa}}{c^2} e(\sin E - \sin E_0). \tag{22.34}$$

The additive constant $\propto \sin E_0$ only resets the origin of time. The variable term is

$$\Delta t = + \left(4.4428 \times 10^{-10} \frac{\text{s}}{\sqrt{\text{m}}} \right) e\sqrt{a} \sin E. \tag{22.35}$$

This correction could have been applied in the satellite clocks. However, insufficient computing power available in the 1970s dictated the decision to apply (22.35) in the receivers. Now it is too late to reverse this decision because of the large investment already made by manufacturers into the devices existing on the market. This means that each hand-held GPS locator applies this relativistic correction.

The eccentricity effect (22.35) allows for one more test of relativity. It was carried out because an official document distributed in 1997 by the Aerospace Corporation claimed that the correction (22.35) could be neglected for a receiver on the orbit. This error, if actually implemented in the system, would have catastrophic consequences: the next generation of the GPS satellites is able to work independently of ground control for periods of up to 180 days. They determine their positions by picking up signals from other GPS satellites. Figures 22.4 and 22.5[5] show the results of this experiment.

22.6 The 9 largest relativistic effects in the GPS

The relativistic effects influence the rate of time flow in the components of the system. Because of the large value of c, small discrepancies in timing translate into large differences in calculated positions of the receivers. If the relativistic corrections were neglected, the

---

[5] These figures were copied from Ashby (1998) and used here with the permissions of N. Ashby and N. Dadhich.

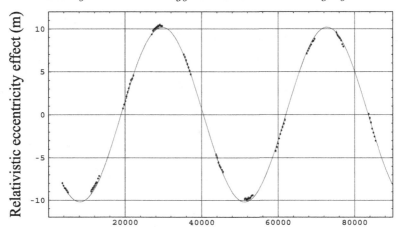

Time from beginning of day Oct 22 1995 (s)

Fig. 22.4 The relativistic eccentricity effect: Comparison of experimental results with the prediction of the theory for the GPS satellite with the largest $e = 0.01486$. The mean difference in positions between theory and experiment is 22 cm, which is 2.2% of the amplitude, 10.2 m.

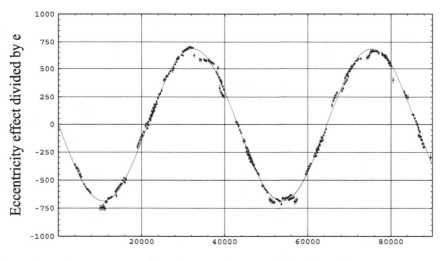

Time from perigee Oct 22 1995 (s)

Fig. 22.5 The relativistic eccentricity effect: $\Delta t/e$ measured experimentally vs. the prediction of theory for 5 GPS satellites with the largest values of e. Here the agreement between the theory and experiment was at the level of 2.5%.

errors would accumulate over time. An illustrative measure of the importance of those corrections is the error in determining the position of the receiver if the corrections were neglected for 24 hours. The table (based on Ashby, 1996) shows the 9 largest effects.

The influence of relativistic effects on the precision of location in the GPS (Adapted from Ashby, 1996)

Effect	Error after 24 hours
Influence on the clocks in the receivers	
Earth's gravitational field	**18 km**
Oblateness of the Earth	9.7 m
Altitude of the clock (e.g. 10 km)	28 m
Rotation of the Earth (on the equator)	31 m
Velocity of the clock (e.g. 600 km/h)	10 m
Synchronisation on the rotating Earth (Sagnac effect)	up to 62 m
Influence on the clocks in the satellites	
Earth's gravitational field	**4.3 km**
Orbital velocity of the satellite	**2.2 km**
Influence on the propagation of the electromagnetic waves	
Rotation of the Earth	up to 41 m

So, every time when we determine our position using a GPS receiver and obtain a correct result, we carry out an experiment that confirms the predictions of general relativity. Note the new quality of this test: the earlier tests in laboratory or on an orbit measured tiny effects, invisible in everyday situations. The effect of relativity in the GPS is very visible, and neglecting it would knock out the whole system.

22.7 Exercises

1. Verify (22.27).

Hint. Standard courses of mechanics tell us that the angular momentum of the satellite is related to the parameters of the orbit by

$$J^2 = GMm^2a\left(1 - e^2\right), \tag{22.36}$$

and the angular momentum first integral is

$$mr^2 d\varphi/dt = J, \tag{22.37}$$

where M and m are the masses of the Earth and of the satellite. Then from (22.26)

$$\sin\varphi \frac{d\varphi}{dt} = \frac{\left(1 - e^2\right)\sin E}{\left(1 - e\cos E\right)^2}\frac{dE}{dt} \quad \text{and} \quad \sin\varphi = \frac{\sqrt{1 - e^2}\sin E}{1 - e\cos E}, \tag{22.38}$$

so

$$\frac{d\varphi}{dt} = \frac{\sqrt{1 - e^2}}{1 - e\cos E}\frac{dE}{dt}. \tag{22.39}$$

Substituting (22.25) and (22.39) in (22.37) we obtain $(1 - e \cos E) \mathrm{d}E/\mathrm{d}t = \sqrt{GM/a^3}$, whose solution is (22.27).

2. Verify (22.28).

Hint. Adapt the (r, ϑ, φ) coordinates to the orbital plane of the satellite so that $\vartheta = \pi/2$ all along the orbit. Then, from (22.21)

$$v^2 = \left(\frac{\mathrm{d}r}{\mathrm{d}t}\right)^2 + r^2 \left(\frac{\mathrm{d}\varphi}{\mathrm{d}t}\right)^2 . \tag{22.40}$$

Substitute in this for r from (22.25), for $\mathrm{d}\varphi/\mathrm{d}t$ from (22.39), for $\mathrm{d}E/\mathrm{d}t$ from (22.27), then calculate v^2 from (22.40) and $(v^2/2 - GM/r)$ using again (22.25) for r.

23
Subjects omitted from this book

As stated in the introduction, it would not be possible to include the whole of relativity in a book of manageable size. We chose to go into several selected topics in-depth, but omitted some other topics completely or nearly so. This short chapter is a list of the topics we omitted, with some suggestions to the reader for further reading.

The following topics were covered inadequately or not at all:

1. **Gravitational waves.** Speaking most generally, a gravitational wave (GW) is any gravitational field that propagates through space independently of matter. In theory, it may or may not be periodic, but the actually detected GWs are short impulses, generated by pairs of black holes or neutron stars that spiral into each other and coalesce. There exists a large collection of exact solutions of Einstein's equations describing GWs, for these see Stephani et al. (2003). There exists also an elaborate theory of nearly linear waves, a relatively good source for it is the book by Ohanian and Ruffini (1994). The theory of generation and detection of GWs is worked out rather well (see Jaranowski and Krolak, 2009). A network of detectors is working and keeps discovering new sources of GW impulses, so frontline knowledge can be gained only from papers. The pioneer of the search for gravitational waves was Joseph Weber (1961); his small book can be recommended to readers interested in the history of the subject.

2. **The Cauchy problem.** In each coordinate system, the set of Einstein's equations can be separated into those equations that contain at most the first-order time derivatives of the metric components and those that are of second order in time. The former are limitations imposed on the initial data, the latter are the dynamical evolution equations. This approach makes it possible to discuss such problems as the global existence or nonexistence of solutions, horizons in general spacetimes and general principles of propagation of gravitational waves. The pioneering paper in this field was that by Arnowitt, Deser and Misner (1962), which gave the name ADM to the whole approach. Misner, Thorne and Wheeler (1973) and Joshi (1993) are also good sources for this topic, and Hawking and Ellis (1973) make elaborate use of it.

3. **Generating new stationary–axisymmetric solutions out of known solutions.** This is a very large field of activity. We saw in Section 21.2 that the Kerr metric resulted from the Einstein equations in consequence of a set of assumptions, not all of which had a clear interpretation. By changing these assumptions one can obtain other solutions. The

first important step in this direction was made by Ernst (1968a), who reduced the Einstein equations for stationary–axisymmetric vacuum spacetimes to one equation for one complex function, now called the **Ernst potential** (in a subsequent paper, this approach was extended to the Einstein–Maxwell equations (Ernst, 1968b)). Then it was discovered that these equations admit transformations of variables that lead from one solution to another, with a geometry that is not a coordinate transform of the initial one. On the basis of this finding, several **generating techniques** for obtaining new solutions were introduced. Meanwhile, the most famous among the next-generation stationary–axisymmetric metrics, the Tomimatsu–Sato (1972) solution, was found. An overview of the generating techniques can be found in Stephani et al. (2003), a more extended overview is the book by Belinskii and Verdaguer (2001).

4. **The Penrose transform.** We mentioned this approach when discussing the maximal analytic extensions of the Reissner–Nordström spacetime in Section 14.15, and of the Kerr metric in Section 21.9. More on it can be found in the books by Misner, Thorne and Wheeler (1973) and by Hawking and Ellis (1973). The main advantage of this approach is that it maps points at infinity into finite points of another manifold, which in turn allows one to discuss values of functions instead of limits. This is a powerful tool, even though few spacetimes are known for which the Penrose transform has been constructed explicitly.

5. **Cosmic censorship.** We mentioned this subject in Section 18.18, but our brief account does not fairly represent the activity behind it. This is a lively paradigm, on which perhaps the best source is the book by Joshi (1993).

6. **Experimental tests.** Apart from the very few classical and most basic tests, we did not really do justice to this subject. This is now a science in itself, with large groups of physicists involved in projects lasting many years. As a historical introduction to the subject, the old volume of proceedings of the Fermi school from 1972 can be recommended (Bertotti, 1974), a discussion of results and their meaning for the theory can be found in the book by Will (2018).

7. **Spinor methods.** Our Chapter 11 is a very concise introduction to this subject, but spinors can do much more than we described there. Fortunately, a large monograph (Penrose and Rindler, 1984) is available.

8. **Relativistic astrophysics.** We did not give a fair representation of the classical applications of relativity to astrophysics because we concentrated on the conceptual basis of relativity. The most extended course on this subject are still the two volumes by Zel'dovich and Novikov (1971, 1974); briefer accounts can be found in Misner, Thorne and Wheeler (1973) and Weinberg (1972).

9. **The history of relativity.** This subject is treated rather superficially in most textbooks, and ours is no exception. That history is important and can be exciting is best attested by the classic book by Pais (1982) – a detailed account of Einstein's life and scientific activities. Other important sources are the book by Mehra (1974) that explains how relativity had been taking shape step by step, and the collection of original papers in which special and general relativity were created (Einstein et al., 1923). Some important bits of history can be found in the monograph by Dicke (1964).

10. **Special relativity.** We omitted this subject altogether because we assumed that

special relativity is now part of all courses on electrodynamics and should be familiar to anyone setting out to study general relativity. Should any reader need to learn, we recommend the following sources: Synge's (1965) book is an expert-level complete textbook. The book by Kopczyński and Trautman (1992) is only in a small part devoted to special relativity, but it presents an enlightening geometric approach that simplifies many problems. Equally enlightening is the textbook by Rindler (1980). Finally, special relativity in the context of electrodynamics is presented in a readable way by Jackson (1975).

24

Comments to selected exercises and calculations

24.1 Exercise 1 to Chapter 14

We introduce the following symbols for parts of (14.1):

$$\vartheta = \arctan(\sqrt{N}/A), \tag{24.1}$$

$$\varphi = \arctan(B). \tag{24.2}$$

The following relations exist between these functions:

$$N + A^2 = 1, \tag{24.3}$$

$$N = \sin^2 \vartheta' \cos^2 \varphi' \left(1 + B^2\right), \tag{24.4}$$

$$\sin^2 \vartheta = N. \tag{24.5}$$

With these symbols and relations, the following is true:

$$d\vartheta = \frac{1}{2A \sin \vartheta' \cos \varphi' \sqrt{1 + B^2}} \left(N_{,\vartheta'} \, d\vartheta' + N_{,\varphi'} \, d\varphi'\right), \tag{24.6}$$

$$d\varphi = \frac{1}{1 + B^2} \left(B_{,\vartheta'} \, d\vartheta' + B_{,\varphi'} \, d\varphi'\right). \tag{24.7}$$

The 2-dimensional metric of a sphere should not change under this transformation, so the following should hold:

$$d\vartheta^2 + \sin^2 \vartheta d\varphi^2 = d\vartheta'^2 + \sin^2 \vartheta' d\varphi'^2. \tag{24.8}$$

This implies

$$\frac{1}{A^2 N} \left(\frac{1}{4} N_{,\vartheta'}{}^2 + A^2 \sin^4 \vartheta' \cos^4 \varphi' B_{,\vartheta'}{}^2\right) = 1, \tag{24.9}$$

$$\frac{1}{4} N_{,\vartheta'} N_{,\varphi'} + A^2 \sin^4 \vartheta' \cos^4 \varphi' B_{,\vartheta'} B_{,\varphi'} = 0, \tag{24.10}$$

$$\frac{1}{A^2 N} \left(\frac{1}{4} N_{,\varphi'}{}^2 + A^2 \sin^4 \vartheta' \cos^4 \varphi' B_{,\varphi'}{}^2\right) = \sin^2 \vartheta'. \tag{24.11}$$

For verifying on a computer, the first and third equations are more convenient in the following equivalent form:

$$\frac{1}{4} N_{,\vartheta'}{}^2 + A^2 \sin^4 \vartheta' \cos^4 \varphi' B_{,\vartheta'}{}^2 - A^2 N = 0, \tag{24.12}$$

$$\frac{1}{4} \, N_{,\varphi'}{}^2 + A^2 \sin^4 \vartheta' \cos^4 \varphi' B_{,\varphi'}{}^2 - A^2 N \sin^2 \vartheta' = 0. \tag{24.13}$$

24.2 Exercise 14 to Chapter 14

Throughout the calculation we assume $q^2/\mu^2 < 1$.

From (14.180) we have at $dr/ds = 0$:

$$\left(\Gamma^2 - 1\right) r^2 + 2\left(m + qe\Gamma/\mu\right) r - e^2 \left(1 - q^2/\mu^2\right) = 0. \tag{24.14}$$

The discriminant Δ of this is

$$\Delta/4 = \left(m + qe\Gamma/\mu\right)^2 + e^2 \left(\Gamma^2 - 1\right) \left(1 - q^2/\mu^2\right) \stackrel{\text{def}}{=} \delta. \tag{24.15}$$

For the beginning, we assume $\Gamma^2 > 1$. Then $\delta > 0$ and (24.14) has two roots, r_1 and $r_2 > r_1$. For the greater root we assume $r_2 > r_+$, i.e.

$$\frac{1}{\Gamma^2 - 1} \left[-\left(m + qe\Gamma/\mu\right) + \sqrt{\delta}\right] > m + \sqrt{m^2 - e^2} = r_+. \tag{24.16}$$

Now we multiply (24.16) by $\left(\Gamma^2 - 1\right)$. After some further manipulation the result becomes

$$\begin{aligned}
\sqrt{\delta} \; &> \; r_+ \left(\Gamma^2 - 1\right) + m + qe\Gamma/\mu \\
&= \; \left(m + \sqrt{m^2 - e^2}\right) \Gamma^2 + (qe/\mu)\Gamma - \sqrt{m^2 - e^2} \stackrel{\text{def}}{=} W(\Gamma).
\end{aligned} \tag{24.17}$$

The discriminant Δ_W of W with respect to Γ is evidently positive, so $W(\Gamma)$ has two roots:

$$\Gamma_\pm = \frac{1}{r_+} \left(-qe/\mu \pm \sqrt{\Delta}\right), \qquad \Delta = q^2 e^2/\mu^2 + 4 r_+ \sqrt{m^2 - e^2}, \tag{24.18}$$

and $W > 0$ for $\Gamma < \Gamma_-$ and for $\Gamma > \Gamma_+$. Now it can be verified that $-1 < \Gamma_- < 0$ and $0 < \Gamma_+ < 1$; for the verification one must observe that $m + qe/\mu > 0$ because $|e| < m$ and $|q/\mu| < 1$. Therefore, with $\Gamma^2 > 1$ assumed, Γ is in the range that guarantees $W(\Gamma) > 0$ in (24.17), and taking the square of both sides in (24.17) we obtain an inequality that is a necessary condition for (24.17). After some reordering it results in

$$\left(\Gamma^2 - 1\right) \left(\Gamma r_+ + qe\Gamma/\mu\right)^2 < 0, \tag{24.19}$$

which is a clear-cut contradiction. The assumption $r_2 > r_+$ led to this, so the conclusion is that with $\Gamma^2 > 1$ the greater root of (24.14) cannot be greater than r_+. Then it follows from the text under (14.180) that $r < r_-$.

Now we turn to the case $\Gamma^2 < 1$ in (24.14). Then δ in (24.15) is not explicitly guaranteed to be positive. However, again from the text under (14.180), we know that with $q^2/\mu^2 < 1$, (24.14) must have at least one positive root – otherwise, the radial trajectory discussed there would have no turning points and could hit $r = 0$, which we know is impossible. So we just assume $\delta > 0$, which, using (24.15), translates to

$$0 < 1 - \Gamma^2 < \frac{\left(m + qe\Gamma/\mu\right)^2}{e^2 \left(1 - q^2/\mu^2\right)}. \tag{24.20}$$

With this assumption, we proceed to consider (24.16) again. This time, because of $\Gamma^2 - 1 < 0$, we obtain the opposite of (24.17):

$$\sqrt{\delta} < W(\Gamma) = r_+ \left(\Gamma^2 - 1\right) + m + qe\Gamma/\mu. \tag{24.21}$$

Since $\sqrt{\delta} > 0$, we can safely take the square of both sides of the above to obtain the necessary condition: the left-hand side of (24.19) must be *positive*. But this time $\Gamma^2 - 1 < 0$, so this is again a contradiction. Consequently, $r_2 > r_+$ is impossible also with $\Gamma^2 < 1$.

In the remaining case, $\Gamma^2 = 1 \Longrightarrow \Gamma = \varepsilon_1 = \pm 1$. Then the turning point at $r = r_{\rm tp} > 0$ will exist only when $m + qe\varepsilon_1/\mu > 0$, and we again consider the Ansatz $r_{\rm tp} > r_+$, i.e.

$$\left(1 - q^2/\mu^2\right) e^2 > 2r_+ \left(m + qe\varepsilon_1/\mu\right). \tag{24.22}$$

Let us take $e = \alpha m$, where $|\alpha| < 1$. Then (24.22) becomes

$$W_2 \stackrel{\rm def}{=} q^2/\mu^2 + 2\varepsilon_1 \left(1 + \sqrt{1 - \alpha^2}\right) q/\mu + 1 + 2\sqrt{1 - \alpha^2} < 0. \tag{24.23}$$

(When $q = 0$ we have a contradiction already here.) The roots of W_2 are

$$(q/\mu)_\pm = -\varepsilon_1 + (-\varepsilon_1 \pm 1) \sqrt{1 - \alpha^2}, \tag{24.24}$$

and (24.23) will be fulfilled if

$$(q/\mu)_- < q/\mu < (q/\mu)_+. \tag{24.25}$$

It turns out that this cannot be fulfilled for $|q/\mu| < 1$. This is because

- With $\varepsilon_1 = +1$ the roots are $(q/\mu)_+ = -1$ and $(q/\mu)_- = -1 - 2\sqrt{1 - \alpha^2}$, and any q/μ between them will obey $q/\mu < -1$, contrary to the assumption.
- With $\varepsilon_1 = -1$ the roots are $(q/\mu)_+ = 1 + 2\sqrt{1 - \alpha^2}$ and $(q/\mu)_- = 1$, so any q/μ between them will obey $q/\mu > +1$, again contrary to the assumption.

So, finally, the requirement that the greater root of (24.14) obeys $r_2 > r_+$ when $q^2/\mu^2 < 1$ leads to a contradiction in all three cases: $\Gamma^2 > 1$, $\Gamma^2 < 1$ and $\Gamma^2 = 1$. Consequently, $r_2 < r_-$. $\square$

24.3 Verifying Eqs. (19.35) with (19.31) and (19.32) with (19.28)

The transformations of G_{22} described below are time-consuming and tedious unless you use a computer-algebra program. If you do, begin by letting the computer calculate G_{22} out of the metric (19.11).

Denote the right-hand side of (19.48) by $\mathcal{K}$, so

$$e^{-C} R_{,t}{}^2 = \mathcal{K}. \tag{24.26}$$

Step 1

Substitute in (19.35) for $C_{,rr}$ and $C_{,r}$ using (19.46). The result is

$$G_{22} = \frac{1}{4} e^{-C} \left(-4RR_{,tt} + 2RC_{,t} R_{,t} - 2RA_{,t} R_{,t} - R^2 A_{,t}{}^2 - 2R^2 A_{,tt} + R^2 C_{,t} A_{,t}\right)$$
$$+ e^{-A/2} \left(QN_{,r} Q_{,NN} - QR_{,r} Q_{,N} /R + N_{,r} Q_{,N}{}^2\right)$$

$$+ e^{-A} \left(RR_{,rr} - \frac{1}{2} RA_{,r} R_{,r} \right) + Q^2 Q_{,N}{}^2 / R^2. \tag{24.27}$$

Step 2

Rewrite the equation $G_{01} = 0$ using (19.33) in the following form:

$$A_{,t} = \frac{e^C}{R_{,t} R_{,r}} \left(2e^{-C} R_{,t} R_{,tr} - e^{-C} R_{,t}{}^2 C_{,r} \right). \tag{24.28}$$

The expression in parentheses is the r-derivative of the $\mathcal{K}$ from (24.26). So, substitute in (24.27) for $A_{,t}$ and $A_{,tt}$ using

$$A_{,t} = \frac{e^C}{R_{,t} R_{,r}} \mathcal{K}_{,r}, \tag{24.29}$$

without substituting for $\mathcal{K}$ yet. The result is

$$G_{22} = R^{-2} Q^2 Q_{,N}{}^2 - \frac{1}{2} R^2 R_{,t}{}^{-1} R_{,r}{}^{-1} \mathcal{K}_{,tr} - \frac{1}{4} e^C R^2 R_{,t}{}^{-2} R_{,r}{}^{-2} \mathcal{K}_{,r}{}^2$$
$$+ \left(-\frac{1}{2} R_{,t} + \frac{1}{2} RR_{,tt} R_{,t}{}^{-1} + \frac{1}{2} RR_{,tr} R_{,r}{}^{-1} - \frac{1}{4} RC_{,t} \right) \frac{R}{R_{,t} R_{,r}} \mathcal{K}_{,r}$$
$$+ e^{-A/2} \left(QN_{,r} Q_{,NN} - R^{-1} QR_{,r} Q_{,N} + N_{,r} Q_{,N}{}^2 \right)$$
$$+ e^{-A} \left(-\frac{1}{2} RA_{,r} R_{,r} + RR_{,rr} \right) + \frac{1}{2} e^{-C} RC_{,t} R_{,t} - e^{-C} RR_{,tt}. \tag{24.30}$$

Step 3

Denote

$$u = \Gamma - QQ_{,N} / R, \tag{24.31}$$

and then substitute for $A_{,r}$ using (19.45):

$$A_{,r} = -2e^{A/2} (u/R_{,r})_{,r}. \tag{24.32}$$

Step 4

Eliminate all occurrences of $e^{-A/2}$ and e^{-A} using (19.45) in the form

$$e^{-A/2} = u/R_{,r}. \tag{24.33}$$

The result is

$$G_{22} = R^{-2} Q^2 Q_{,N}{}^2 - R^{-1} uQQ_{,N} + RuR_{,r}{}^{-1} u_{,r} + uR_{,r}{}^{-1} N_{,r} QQ_{,NN}$$
$$+ uR_{,r}{}^{-1} N_{,r} Q_{,N}{}^2 - \frac{1}{2} R^2 R_{,t}{}^{-1} R_{,r}{}^{-1} \mathcal{K}_{,tr} - \frac{1}{4} e^C R^2 R_{,t}{}^{-2} R_{,r}{}^{-2} \mathcal{K}_{,r}{}^2$$
$$+ \left(-\frac{1}{2} R_{,t} + \frac{1}{2} RR_{,tt} R_{,t}{}^{-1} + \frac{1}{2} RR_{,tr} R_{,r}{}^{-1} - \frac{1}{4} RC_{,t} \right) \frac{R}{R_{,t} R_{,r}} \mathcal{K}_{,r}$$
$$- e^{-C} RR_{,tt} + \frac{1}{2} e^{-C} RC_{,t} R_{,t}. \tag{24.34}$$

Step 5

Express the derivatives of (24.26) by t and by r through the derivatives of $\mathcal{K}$. Solve the resulting equations for $R_{,tr}$ and $R_{,tt}$, obtaining

$$R_{,tr} = \frac{1}{2R_{,t}} e^{C} \left(K_{,r} + e^{-C} C_{,r} R_{,t}{}^{2} \right), \tag{24.35}$$

$$R_{,tt} = \frac{1}{2R_{,t}} e^{C} \left(K_{,t} + e^{-C} C_{,t} R_{,t}{}^{2} \right). \tag{24.36}$$

Then substitute for $R_{,tr}$ and $R_{,tt}$ using (24.35) – (24.36). The result is

$$G_{22} = RuR_{,r}{}^{-1}u_{,r} + R_{,r}{}^{-1}N_{,r} \, QQ_{,NN} \, u + R_{,r}{}^{-1}N_{,r} \, Q_{,N}{}^{2}u - R^{-1}QQ_{,N} \, u$$

$$+ R^{-2}Q^{2}Q_{,N}{}^{2} - \frac{1}{2} \, R^{2}R_{,t}{}^{-1}R_{,r}{}^{-1}K_{,tr} + \frac{1}{4} \, e^{C}R^{2}R_{,t}{}^{-3}R_{,r}{}^{-1}K_{,t} \, K_{,r}$$

$$- \frac{1}{2} \, RR_{,t}{}^{-1}K_{,t} - \frac{1}{2} \, RR_{,r}{}^{-1}K_{,r} + \frac{1}{4} \, R^{2}C_{,r} \, R_{,r}{}^{-2}K_{,r} \,. \tag{24.37}$$

Step 6

In Step 5, $C_{,r}$ reappeared. Eliminate it using (19.46) and (19.45):

$$C_{,r} = 2QQ_{,N} \, R_{,r} \, / \, \left(R^{2}u \right), \tag{24.38}$$

then eliminate $u_{,r}$ using (24.31) and $QQ_{,N} \, /u$ using

$$QQ_{,N} \, /u = R(\Gamma/u - 1). \tag{24.39}$$

The result is

$$G_{22} = R^{-2}Q^{2}Q_{,N}{}^{2} - \frac{1}{2} \, R^{2}R_{,t}{}^{-1}R_{,r}{}^{-1}K_{,tr} + \frac{1}{4} \, e^{C}R^{2}R_{,t}{}^{-3}R_{,r}{}^{-1}K_{,t} \, K_{,r}$$

$$- \frac{1}{2} \, RR_{,t}{}^{-1}K_{,t} + \frac{1}{2} \, R\Gamma u^{-1}R_{,r}{}^{-1}K_{,r} - RR_{,r}{}^{-1}K_{,r} + RuR_{,r}{}^{-1}\Gamma_{,r} \,. \tag{24.40}$$

Step 7

Now it is time to substitute for K. This must be done step by step to avoid explosion of size of the formula for G_{22}. The order of actions is this:

(7a) Substitute for $K_{,tr}$, **(7b)** substitute for $K_{,t}$, **(7c)** substitute, using (24.35), for $R_{,tr}$ that reappeared in Step (7b), **(7d)** substitute for $K_{,r}$, **(7e)** substitute again for $C_{,r}$ using (24.38), **(7f)** substitute for $Q_{,N}{}^{3}/u$ and for $Q_{,N} \, /u$, but not for $Q_{,N}{}^{2}/u$, using (24.39). The result is

$$G_{22} = \frac{G}{c^{4}} \left(\Gamma R_{,r}{}^{-1}N_{,r} + R^{-2}Q^{2} - \Gamma^{2}u^{-1}R_{,r}{}^{-1}N_{,r} \right) + \Lambda R^{2} - 2R\Gamma R_{,r}{}^{-1}\Gamma_{,r}$$

$$+ RuR_{,r}{}^{-1}\Gamma_{,r} + R\Gamma^{2}u^{-1}R_{,r}{}^{-1}\Gamma_{,r} - \Gamma R_{,r}{}^{-1}N_{,r} \, QQ_{,NN}$$

$$- \Gamma R_{,r}{}^{-1}N_{,r} \, Q_{,N}{}^{2} + \Gamma^{2}u^{-1}R_{,r}{}^{-1}N_{,r} \, QQ_{,NN} + \Gamma^{2}u^{-1}R_{,r}{}^{-1}N_{,r} \, Q_{,N}{}^{2}$$

$$- R_{,r}{}^{-1}M_{,r} + \Gamma u^{-1}R_{,r}{}^{-1}M_{,r} \,. \tag{24.41}$$

Step 8

Add $\left[-\Lambda R^{2} - GQ^{2}/(c^{4}R^{2}) \right]$ to the result of Step 7. This should be equal to zero by (19.31) and (19.35), but is not yet fully simplified. Factor out $1/(RuR_{,r})$ to obtain

$$G_{22} - \Lambda R^{2} - GQ^{2}/(c^{4}R^{2}) = \frac{1}{RuR_{,r}} \left[\frac{G}{c^{4}} \, R\Gamma N_{,r} \, (u - \Gamma) + RM_{,r} \, (\Gamma - u) \right.$$

$$- R\Gamma u N_{,r} \, QQ_{,NN} - R\Gamma u N_{,r} \, Q_{,N}{}^{2} + R\Gamma^{2}N_{,r} \, QQ_{,NN} + R\Gamma^{2}N_{,r} \, Q_{,N}{}^{2}$$

$$\left. - 2R^{2}\Gamma u\Gamma_{,r} + R^{2}\Gamma^{2}\Gamma_{,r} + R^{2}u^{2}\Gamma_{,r} \right], \tag{24.42}$$

then use the following consequences of (24.31):

$$\Gamma^2 = \Gamma \left(u + QQ_{,N}/R\right), \qquad u^2 = u\left(\Gamma - QQ_{,N}/R\right), \tag{24.43}$$

and use (24.31) itself. The final result is equivalent to (19.51).

To derive (19.53)

(1) Use (24.29) for $A_{,t}$; (2) use (24.32) for $A_{,r}$; (3) use (24.33) for e^{-A}; (4) use (24.26) for $R_{,t}{}^2$; (5) factor out e^C; (6) substitute for $\mathcal{K}$ the right-hand side of (19.48); (7) use (24.31) for u; (8) use (19.52) for $M_{,r}$.

24.4 Verifying the Einstein equations (20.2), (20.9) and (20.11)

The verification of $G_{22} = G_{33} = \kappa p$ using (20.9) and (20.11) is complicated and is best done using an algebraic computer program. The consecutive substitutions listed below are the same for G_{22} and G_{33}. The intermediate results are given only for $G_{22} = \kappa p$.

Step 1

Substitute in G_{22} from (20.42) – (20.43) for the derivatives of $\alpha = \ln h + \ln \Phi + \beta_{,z}$ and of $\beta = \ln \Phi + \nu$, but do not substitute for $\beta_{,z}$ (it will later factor out). Then substitute for $e^{-2\beta}$ and $e^{-2\alpha}$ using (20.42) and (20.43). The result is

$$\begin{aligned}
G_{22} - \kappa p &= k\Phi^{-2} + h^{-2}\Phi^{-2} - h^{-3}\Phi^{-2}h_{,z}\,\beta_{,z}{}^{-1} - h^{-2}\Phi^{-3}\Phi_{,z}\,\beta_{,z}{}^{-1} \\
&\quad + \Phi^{-3}\Phi_{,t}{}^2\Phi_{,z}\,\beta_{,z}{}^{-1} + \Phi^{-2}\Phi_{,tt}\,\Phi_{,z}\,\beta_{,z}{}^{-1} - \Phi^{-2}\Phi_{,tz}\,\Phi_{,t}\,\beta_{,z}{}^{-1} \\
&\quad - \Phi^{-1}\Phi_{,ttz}\,\beta_{,z}{}^{-1} + e^{-2\nu}\Phi^{-2}\nu_{,zx}\,\nu_{,x}\,\beta_{,z}{}^{-1} + e^{-2\nu}\Phi^{-2}\nu_{,zyy}\,\beta_{,z}{}^{-1} \\
&\quad - e^{-2\nu}\Phi^{-2}\nu_{,zy}\,\nu_{,y}\,\beta_{,z}{}^{-1}.
\end{aligned} \tag{24.44}$$

Step 2

Substitute for $\Phi_{,tt}$ using (20.46), then calculate $\Phi_{,ttz}$ and substitute for it, too. Factor out $\beta_{,z}{}^{-1}$, then in the coefficient of $\beta_{,z}{}^{-1}$ substitute $\beta_{,z} = \Phi_{,z}/\Phi + \nu_{,z}$. The result is

$$\begin{aligned}
\beta_{,z}\left(G_{22} - \kappa p\right) &= k\Phi^{-2}\nu_{,z} + h^{-2}\Phi^{-2}\nu_{,z} + \frac{1}{2}\Phi^{-2}k_{,z} - h^{-3}\Phi^{-2}h_{,z} \\
&\quad + e^{-2\nu}\Phi^{-2}\nu_{,zx}\,\nu_{,x} + e^{-2\nu}\Phi^{-2}\nu_{,zyy} - e^{-2\nu}\Phi^{-2}\nu_{,zy}\,\nu_{,y}.
\end{aligned} \tag{24.45}$$

Step 3

Denote $e^{-\nu} = Q$ (this is the polynomial in (20.45)), substitute $e^{-2\nu} = Q^2$ and by differentiating $e^{-\nu}$ find the following:

$$\begin{aligned}
\nu_{,zxx} &= -\nu_{,z}\,\nu_{,x}{}^2 + \nu_{,z}\,\nu_{,xx} + 2\nu_{,x}\,\nu_{,zx} - Q_{,zxx}/Q, \\
\nu_{,zx} &= \nu_{,z}\,\nu_{,x} - Q_{,zx}/Q, \qquad \nu_{,x} = -Q_{,x}/Q
\end{aligned} \tag{24.46}$$

plus the analogous expressions for $\nu_{,zyy}$, $\nu_{,zy}$ and $\nu_{,y}$, then substitute $\nu_{,z}\,Q = -Q_{,z}$. Do not substitute for $\nu_{,z}$. The result is

$$\begin{aligned}
\beta_{,z}\left(G_{22} - \kappa p\right) &= \frac{1}{2}\Phi^{-2}k_{,z} - h^{-3}\Phi^{-2}h_{,z} + k\Phi^{-2}\nu_{,z} + h^{-2}\Phi^{-2}\nu_{,z} \\
&\quad - \Phi^{-2}QQ_{,zyy} + \Phi^{-2}Q_{,zx}\,Q_{,x} + \Phi^{-2}Q_{,zy}\,Q_{,y} + \Phi^{-2}Q_{,x}{}^2\nu_{,z} \\
&\quad + \Phi^{-2}Q_{,y}{}^2\nu_{,z} + \Phi^{-2}Q_{,z}\,Q_{,yy}.
\end{aligned} \tag{24.47}$$

Step 4

Now substitute for Q the polynomial from (20.45). The expression for G_{22} will explode a bit, and will include terms containing x^2 and y^2; they were born out of the terms $Q_{,x}{}^2$, $Q_{,zx}\,Q_{,x}$ and the like that were present in Step 3. Eliminate them using the following consequence of (20.45):

$$x^2 = \frac{1}{A}\left(Q - 2B_1 x - 2B_2 y - C\right) - y^2. \tag{24.48}$$

The factors $\nu_{,z}\,Q$ and Q reappear after (24.48). Substitute for them again. The result is

$$
\begin{aligned}
\beta_{,z}\left(G_{22} - \kappa p\right) = {}& \frac{1}{2}\,\phi^{-2}k_{,z} - h^{-3}\phi^{-2}h_{,z} + k\phi^{-2}\nu_{,z} + h^{-2}\phi^{-2}\nu_{,z} \\
& - 4\phi^{-2}AC\nu_{,z} - 2\phi^{-2}AC_{,z} - 2\phi^{-2}CA_{,z} \\
& + 4\phi^{-2}B_1 B_{1,z} + 4\phi^{-2}B_2 B_{2,z} + 4\phi^{-2}B_1{}^2\nu_{,z} + 4\phi^{-2}B_2{}^2\nu_{,z}. \tag{24.49}
\end{aligned}
$$

Step 5

Calculate k from (20.47), then substitute for it in (24.49). The computer will now tell you that $\beta_{,z}\left(G_{22} - \kappa p\right) = \beta_{,z}\left(G_{33} - \kappa p\right) = 0$. $\square$

Now come hints for verifying (20.49), also best used in an algebraic computer program.

Step 1

Substitute for all the derivatives of α in (20.2) using (20.43). After this, terms containing $\beta_{,zxx}$ and $\beta_{,zyy}$ will appear. Substitute for the derivatives of β except for $\beta_{,z}$. Substitute for $e^{-2\alpha}$ and $e^{-2\beta}$ using (20.42) and (20.43). The result is

$$
\begin{aligned}
G_{00} = {}& 3\Phi^{-2}\Phi_{,t}{}^2 - 3h^{-2}\Phi^{-2} + 2h^{-3}\Phi^{-2}h_{,z}\,\beta_{,z}{}^{-1} + 2h^{-2}\Phi^{-3}\Phi_{,z}\,\beta_{,z}{}^{-1} \\
& - 2\Phi^{-3}\Phi_{,t}{}^2\Phi_{,z}\,\beta_{,z}{}^{-1} + 2\Phi^{-2}\Phi_{,tz}\,\Phi_{,t}\,\beta_{,z}{}^{-1} - e^{-2\nu}\Phi^{-2}\nu_{,zxx}\,\beta_{,z}{}^{-1} \\
& - e^{-2\nu}\Phi^{-2}\nu_{,zyy}\,\beta_{,z}{}^{-1} - e^{-2\nu}\Phi^{-2}\nu_{,xx} - e^{-2\nu}\Phi^{-2}\nu_{,yy}. \tag{24.50}
\end{aligned}
$$

Step 2

Factor out $(\Phi\beta_{,z})^{-1}$ – this is the coefficient of the long expression in (20.49). Eliminate $\nu_{,xx}$ and $\nu_{,zxx}$ using (20.44) in the form

$$\nu_{,xx} = -\left(k + 1/h^2\right)e^{2\nu} - \nu_{,yy} \tag{24.51}$$

and the derivative by z of the above. Do not eliminate $\nu_{,z}$ and $\beta_{,z}$. The result is

$$
\begin{aligned}
\Phi\beta_{,z}\,G_{00} = {}& \Phi^{-1}k_{,z} + k\Phi^{-1}\beta_{,z} + 2k\Phi^{-1}\nu_{,z} + 2h^{-2}\Phi^{-2}\Phi_{,z} + 3\Phi^{-1}\Phi_{,t}{}^2\beta_{,z} \\
& - 2h^{-2}\Phi^{-1}\beta_{,z} + 2h^{-2}\Phi^{-1}\nu_{,z} - 2\Phi^{-2}\Phi_{,t}{}^2\Phi_{,z} + 2\Phi^{-1}\Phi_{,tz}\,\Phi_{,t}. \tag{24.52}
\end{aligned}
$$

Step 3

Eliminate $\Phi_{,t}{}^2$ using (20.48) and $\Phi_{,t}\,\Phi_{,tz}$ using the derivative of (20.48) by z. Then substitute $\beta_{,z} = \Phi_{,z}/\Phi + \nu_{,z}$. The result will be (20.49).

24.5 Equation (20.179) defines η at the AAH uniquely

We prove here that (20.179) is fulfilled by only one value of $\eta \in (\pi, 2\pi)$ for each set of values of (z, x, y). We begin by recalling the following:

1. Equation (20.179) that determines $\eta(z, x, y)$ on the AAH is derived from (20.164), and so all quantities in it are calculated along the 'nearly radial ray' (NRR) that obeys (20.163) with x and y being constant, i.e.

$$\frac{(\Phi_{,z} - \Phi\mathcal{E}_{,z}/\mathcal{E})^2}{1-k}\left(\frac{dz}{dt}\right)^2 = 1. \tag{24.53}$$

2. Note, from (20.105) with $\sigma = +1$, that

$$\frac{\partial t}{\partial \eta} = \frac{M}{k^{3/2}}(1 - \cos\eta) > 0 \qquad \text{for } \eta \in (0, 2\pi), \tag{24.54}$$

at every fixed (z, x, y), so, in this range, $t(\eta)$ is a monotonically increasing function i.e. $(\eta_i < \eta_j) \Longrightarrow (t(\eta_i) < t(\eta_j))$ for every fixed (z, x, y).

Now suppose that (20.179) has more than one solution for η at a given (z, x, y), and call the solutions $(\eta_1, \ldots, \eta_k)$, with $\eta_1 < \cdots < \eta_k$. Then there would be k instants $t_1 < \cdots < t_k$, at which the given NRR would intersect the AAH, all the $t_i, i = 1, \ldots, k$ corresponding to the same (z, x, y) in (20.179). Then the function $z(t, x, y)$ would have the property $z(t_1, x, y) = z(t_2, x, y)$. Since this function is continuous (even differentiable, see (24.53)), this means that for some $\bar{t} \in (t_1, t_2)$ we would have $dz/dt|_{t=\bar{t}} = 0$ while (24.53) holds. This is possible only when $k \xrightarrow[t \to \bar{t}]{} 1$ at the right speed or $(\Phi_{,z} - \Phi\mathcal{E}_{,z}/\mathcal{E}) \xrightarrow[t \to \bar{t}]{} \infty$ at each t_i. The first locus is a neck – an analogue of the Kruskal–Szekeres wormhole in the Schwarzschild solution, see Sec. 18.10. At the second locus the energy density is zero by (20.74). Both loci are special locations in spacetime that may or may not exist.

Thus, for every NRR obeying (24.53) that does not traverse a neck or a point with $\epsilon = 0$, $dz/dt \neq 0$ everywhere along it. This means that $z(t_1, x, y) = z(t_2, x, y)$ cannot happen, i.e. that (20.179) has only one solution for η at each given (z, x, y). $\square$

24.6 The four curves in Fig. 20.4 meet at one point

For the numerical calculation we need to know the value of the function $t(M)_{\text{AAH}}$ in (20.180) at $M = 0$. This has to be calculated exactly because numerical programs are unreliable in calculating limits. From (20.185) and (20.187) we have

$$\lim_{M \to 0} t_B(M) = t_{B0}, \qquad \lim_{M \to 0} \frac{M}{k^{3/2}} = \frac{T_0}{2\pi}, \tag{24.55}$$

so

$$\lim_{M \to 0} t(M)_{\text{AAH}} = t_{B0} + \frac{T_0}{2\pi}\lim_{M \to 0}(\eta - \sin\eta)\Big|_{\text{AAH}}. \tag{24.56}$$

To calculate $\lim_{M \to 0} \eta$ along the AAH we use (20.179). Equation (20.182) shows that $\sin\eta/\sqrt{1 - \cos\eta}$ is finite in the full range $\eta \in [0, 2\pi]$. However, with M being the z coordinate, the function Ψ as given by (20.179) contains terms proportional to $1/M$, also the extreme values of $\mathcal{E}_{,M}/\mathcal{E}$ from (20.188) are proportional to $1/M$ with (P, Q, S) given by (20.189). Consequently, to find $\lim_{M\to 0} \eta$ along the AAH we have to multiply (20.179) by M and only then take the limit $M \to 0$. The result is

$$\lim_{M\to 0} (M\Psi) = (1/3)(1 - \cos\eta)^2, \qquad (24.57)$$

and this is zero at $\eta = 0$ (i.e., at the BB) and $\eta = 2\pi$ (at the BC). Note that this limit is independent of x and y because the term containing $\mathcal{E},_z /\mathcal{E}$ in (20.179) disappears in the limit $M \to 0$ after being multiplied by M. With $\eta = 2\pi$, Eq. (24.56) becomes

$$\lim_{M\to 0} t(M)_{\mathrm{AAH}} = t_{B0} + T_0 = \lim_{M\to 0} t_C(M), \qquad (24.58)$$

i.e. the AAH at $M = 0$ coincides with the Big Crunch in the model now considered. $\square$

24.7 The discarded case in Eqs. (20.2)–(20.11)

In Section 20.1 we omitted the case when $\beta,_z = 0$ and $(\beta,_{tx} \neq 0$ or $\beta,_{ty} \neq 0)$ in (20.2) – (20.11). We show here that it contains no perfect fluid solutions.

Lemma 24.1 *If $\beta,_z = 0$ and $\beta,_{tx} \neq 0$, and the metric (20.1) is to obey the Einstein equations with a perfect fluid source, then coordinates may be chosen so that $\alpha,_z = 0$.*

Proof:

For a perfect fluid source the G_{ij} components in (20.4) – (20.5) and (20.10) must be zero. We rewrite them as follows:

$$\alpha,_{tx} = -\beta,_{tx} -\alpha,_t \alpha,_x +\beta,_t \alpha,_x , \qquad (24.59)$$

$$\alpha,_{ty} = -\beta,_{ty} -\alpha,_t \alpha,_y +\beta,_t \alpha,_y \qquad (24.60)$$

$$\alpha,_{xy} = -\alpha,_x \alpha,_y +\alpha,_x \beta,_y +\beta,_x \alpha,_y . \qquad (24.61)$$

We apply to these the integrability conditions $\alpha,_{tx,y} -\alpha,_{ty,x} = 0$, $\alpha,_{ty,x} -\alpha,_{xy,t} = 0$ and $\alpha,_{tx,y} -\alpha,_{xy,t} = 0$ and use (24.59) – (24.61) in the result. The resulting equations are, respectively,

$$\alpha,_x \beta,_{ty} -\alpha,_y \beta,_{tx} = 0, \qquad (24.62)$$

$$-\beta,_{txy} = 2\alpha,_x \beta,_{ty} -\beta,_y \beta,_{tx} -\beta,_x \beta,_{ty} , \qquad (24.63)$$

$$-\beta,_{txy} = 2\alpha,_y \beta,_{tx} -\beta,_y \beta,_{tx} -\beta,_x \beta,_{ty} . \qquad (24.64)$$

Now we differentiate (24.64) by z. Since $\beta,_z = 0$, the result is $\alpha,_{yz} \beta,_{tx} = 0$. Then, since $\beta,_{tx} \neq 0$ by assumption, the implication is

$$\alpha,_{yz} = 0. \qquad (24.65)$$

But from (24.62) with $\beta,_{tx} \neq 0$ we have

$$\alpha,_y = \beta,_{ty} \alpha,_x /\beta,_{tx} . \qquad (24.66)$$

Differentiating this by z and using $\beta,_z = 0$ and (24.65) we obtain $\alpha,_{xz} \beta,_{ty} = 0$. We thus have to consider two cases separately.

Case 1: $\beta_{,ty} = 0$.

Then from (24.62) we have immediately $\alpha_{,y} = 0$, and from (24.63) $\beta_{,y} = 0$. We substitute this in (20.9) and (20.11) and take the equation $G_{22} - G_{33} = 0$, the result is

$$\alpha_{,xx} + \alpha_{,x}^2 - 2\alpha_{,x}\,\beta_{,x} = 0. \tag{24.67}$$

We differentiate this by t and use (24.59) to eliminate $\alpha_{,tx}$ and $\alpha_{,txx}$. The result is

$$-\beta_{,txx} - 2\alpha_{,x}\,\beta_{,tx} + 2\beta_{,x}\,\beta_{,tx} = 0. \tag{24.68}$$

We differentiate this by z. Since $\beta_{,tx} \neq 0 = \beta_{,z}$, the conclusion $\alpha_{,xz} = 0$ follows anyway. Thus we go back to consider

Case 2: $\alpha_{,xz} = 0$.

Equations (24.65) and the above imply that

$$\alpha = \alpha_1(t, x, y) + \alpha_2(t, z). \tag{24.69}$$

We substitute this in (24.59) and differentiate the result by z. The resulting equation is $\alpha_{1,x}\alpha_{2,tz} = 0$. But with $\alpha_{1,x} = 0$ we would have $\alpha_{,x} = 0$, and then $\beta_{,tx} = 0$ from (24.59) – contrary to the assumption. Hence, $\alpha_{2,tz} = 0$, which means that $\alpha_2 = \alpha_3(t) + \alpha_4(z)$. Then, the coordinate transformation $z' = \int e^{\alpha_4}dz$ makes the metric independent of z.

Lemma 24.2 *If $\beta_{,z} = 0$ and $\beta_{,tx} \neq 0$, and the metric (20.1) is to obey the Einstein equations with a perfect fluid source, then coordinates may be chosen so that $\alpha_{,y} = \alpha_{,z} = \beta_{,y} = 0$, i.e. so that the metric depends only on t and x.*

Proof:

By Lemma 24.1 no generality is lost on assuming that $\alpha_{,z} = 0$. We substitute (24.66) in (24.60) and use (24.59) to eliminate $\alpha_{,tx}$. Using (24.62), the result is $\alpha_{,x}\,(\beta_{,ty}/\beta_{,tx})_{,t} = 0$. Since $\alpha_{,x} = 0$ would lead to $\beta_{,tx} = 0$ (from (24.59)), this implies that

$$\beta_{,ty}/\beta_{,tx} = f(x, y) \implies \alpha_{,y} = f(x, y)\alpha_{,x}, \tag{24.70}$$

where $f(x, y)$ is an arbitrary function independent of t, and (24.62) was used again.

With $\alpha_{,z} = \beta_{,z} = 0$, we carry out the variable transformation in (20.9) − (20.11):

$$\xi = x + iy, \qquad \bar{\xi} = x - iy, \tag{24.71}$$

and consider the equations $G_{22} - G_{33} = 0$ and $G_{23} = 0$. As shown under (20.31) they imply (20.32):

$$\alpha_{,\xi\xi} + \alpha_{,\xi}^2 - 2\alpha_{,\xi}\,\beta_{,\xi} = 0 \tag{24.72}$$

and its complex conjugate. The solution of (24.72) is

$$e^{\alpha - 2\beta}\alpha_{,\xi} = h(t, \bar{\xi}), \tag{24.73}$$

where $h(t, \bar{\xi})$ is an arbitrary function, independent of ξ. Since α and β are real functions, by taking the complex conjugate of (24.73) we obtain

$$e^{\alpha - 2\beta}\alpha_{,\bar{\xi}} = \bar{h}(t, \xi), \tag{24.74}$$

which is the general solution of the complex conjugate equation to (24.72). Note that $h \neq 0$, or else α becomes a function of t only, which leads to $\beta_{,tx} = 0$ via (24.59) – contrary to the assumption.

Now we do the transformation (24.71) in (24.70) and calculate

$$f(x,y) = \frac{\alpha_{,y}}{\alpha_{,x}} \equiv i \frac{\alpha_{,\xi} - \alpha_{,\bar{\xi}}}{\alpha_{,\xi} + \alpha_{,\bar{\xi}}} \equiv i \frac{\alpha_{,\xi}/\alpha_{,\bar{\xi}} - 1}{\alpha_{,\xi}/\alpha_{,\bar{\xi}} + 1}. \tag{24.75}$$

This shows that $\alpha_{,\xi}/\alpha_{,\bar{\xi}} = h/\bar{h}$ does not depend on t. Hence

$$\frac{\alpha_{,\xi}}{\alpha_{,\bar{\xi}}} = \frac{h(t,\bar{\xi})}{\bar{h}(t,\xi)} = \frac{h(t_0,\bar{\xi})}{\bar{h}(t_0,\xi)}, \tag{24.76}$$

where t_0 is some fixed instant. We define the function $g(\xi)$ by

$$1/\bar{h}(t_0,\xi) = dg/d\xi. \tag{24.77}$$

Then (24.76) may be written as

$$\frac{\alpha_{,\xi}}{\alpha_{,\bar{\xi}}} = \frac{dg/d\xi}{d\bar{g}/d\bar{\xi}}. \tag{24.78}$$

Let us take $g(\xi)$ and $\bar{g}(\bar{\xi})$ as the new variables. Then (24.78) is equivalent to $\partial\alpha/\partial g = \partial\alpha/\partial\bar{g}$, whose solution is

$$\alpha = \alpha(t, g + \bar{g}). \tag{24.79}$$

From (24.73) and (24.77) we have, using (24.79),

$$e^{2\beta} = e^{\alpha}\alpha_{,\xi} \frac{1}{h(\bar{\xi})} = e^{\alpha}\alpha' g_{,\xi}\, \bar{g}_{,\bar{\xi}}, \tag{24.80}$$

where α' is the derivative of α by its argument $(g + \bar{g})$. Using the transformation (24.71) we now see that (24.80) implies

$$e^{2\beta}\left(dx^2 + dy^2\right) = e^{2\beta}d\xi\, d\bar{\xi} = e^{\alpha}\alpha' g_{,\xi}\, \bar{g}_{,\bar{\xi}}\, d\xi d\bar{\xi} = e^{\alpha}\alpha'\, dg d\bar{g}. \tag{24.81}$$

Equation (24.79) shows that if g and $\bar{g}$ are chosen as the new complex coordinates $\xi' = g = x' + iy'$, $\bar{\xi}' = \bar{g} = x' - iy'$, then α will depend only on t and $x' = \frac{1}{2}(g + \bar{g})$ and (24.81) shows that the transformed $e^{2\bar{\beta}} = e^{\alpha}\alpha'$ will also depend only on t and x', so both α and $\tilde{\beta}$ will be independent of y'. $\square$

Conclusion. *If $\beta_{,ty} \neq 0$ and $\beta_{,z} = 0$, and the metric (20.1) is to obey the Einstein equations with a perfect fluid source, then coordinates may be chosen so that the metric depends only on t and y.*

Proof:
The transformation $(x, y) = (y', x')$ is an isometry of (20.1). The rest of the proof follows by interchanging x and y in the proof above. $\square$

Since the cases $\beta_{,tx} \neq 0$ and $\beta_{,ty} \neq 0$ are thus equivalent, we will follow only the case $\beta_{,tx} \neq 0 \Longrightarrow \alpha = \alpha(t,x)$, $\beta = \beta(t,x)$.

Lemma 24.3 *If the functions α and β in the metric (20.1) depend only on t and x, while $\beta_{,tx} \neq 0$, then the metric does not obey the Einstein equations with a perfect fluid source.*

Proof:

We substitute $\alpha_{,y} = \alpha_{,z} = \beta_{,y} = \beta_{,z} = 0$ in (20.2) – (20.11) and take the equation $G_{22} - G_{33} = 0$. Equation (24.67) follows again, and its solution is analogous to (24.73):

$$e^\alpha \alpha_{,x} = h(t)e^{2\beta} \implies 2\beta = \alpha + \ln \alpha_{,x} - \ln h, \tag{24.82}$$

where $h(t)$ is an arbitrary function. We assume $h \neq 0$, or else (24.59) would imply $\beta_{,tx} = 0$, then we use (24.82) to eliminate β from $G_{02} = 0$; the result is

$$\alpha_{,tx} + \frac{\alpha_{,txx}}{2\alpha_{,x}} - \frac{\alpha_{,xx}\,\alpha_{,tx}}{2\alpha_{,x}{}^2} + \frac{1}{2}\alpha_{,t}\,\alpha_{,x} + \frac{h_{,t}}{2h}\alpha_{,x} = 0. \tag{24.83}$$

This is integrated with respect to x, yielding

$$e^\alpha\,(\alpha_{,t} + h_{,t}/h) = k_1(t)\alpha + k_2(t), \tag{24.84}$$

where $k_1(t)$ and $k_2(t)$ are arbitrary functions. Using (24.82) and (24.84) in the integrability condition $(e^\alpha \alpha_{,t})_{,x} = (e^\alpha \alpha_{,x})_{,t}$ we obtain

$$\beta_{,t} = k_1 e^{-\alpha}/2 - h_{,t}/h. \tag{24.85}$$

Now we take the equation $G_{11} - G_{22} = 0$, use (24.82), (24.84) and (24.85) to eliminate β, $\alpha_{,t}$, $\alpha_{,tt}$, $\beta_{,t}$ and $\beta_{,tt}$, and multiply the result by $\alpha_{,x}$, obtaining

$$he^{-\alpha}\,(\beta_{,xx} - \alpha_{,x}\,\beta_{,x}) + \left[-\frac{1}{2}\,k_{1,t}e^{-\alpha} + (k_{1,t}\alpha + k_{2,t})\,e^{-\alpha} \right.$$
$$\left. - 2\frac{h_{,t}}{h}\,(k_1\alpha + k_2)\,e^{-\alpha} - \frac{1}{2}k_1{}^2 e^{-2\alpha} + 2k_1\,(k_1\alpha + k_2)\,e^{-2\alpha} \right]\alpha_{,x} = 0. \tag{24.86}$$

The integral of this is

$$he^{-\alpha}\beta_{,x} - \frac{1}{2}k_{1,t}e^{-\alpha} - (k_{1,t}\alpha + k_{2,t})\,e^{-\alpha} + 2\frac{h_{,t}}{h}\,(k_1\alpha + k_2)\,e^{-\alpha}$$
$$+ 2k_1\frac{h_{,t}}{h}e^{-\alpha} - \frac{1}{4}k_1{}^2 e^{-2\alpha} - k_1\,(k_1\alpha + k_2)\,e^{-2\alpha} + \ell_1(t) = 0, \tag{24.87}$$

where $\ell_1(t)$ is an arbitrary function. We substitute here for β from (24.82) and multiply the resulting equation by $e^{2\alpha}\alpha_{,x}$. This can be integrated by x again with the result

$$\frac{1}{2}he^\alpha \alpha_{,x} + \frac{1}{2}k_{1,t}e^\alpha - (k_{1,t}\alpha + k_{2,t})\,e^\alpha + 2\frac{h_{,t}}{h}\,(k_1\alpha + k_2)\,e^\alpha$$
$$- \frac{1}{4}k_1{}^2\alpha - k_1\left(\frac{1}{2}k_1\alpha^2 + k_2\alpha \right) + \frac{1}{2}\ell_1(t)e^{2\alpha} + \ell_2(t) = 0. \tag{24.88}$$

On this equation, the following sequence of operations should now be carried out: (1) replace $e^\alpha \alpha_{,x}$ from (24.82); (2) differentiate the result by t; (3) use (24.85) to eliminate $\beta_{,t}$; (4) use (24.84) to eliminate $\alpha_{,t}$; (5) multiply (24.88) by $k_1 e^{-\alpha}$ and subtract the product from the equation obtained in the previous step. The term containing β will cancel out, and a polynomial in α, $e^{-\alpha}$ and e^α will remain. Since $\alpha_{,x} \neq 0$ and the other functions do not depend on x, the coefficients of independent functions of α must vanish separately. The coefficient of $\alpha^2 e^{-\alpha}$ will vanish only when $k_1 = 0$. However, then we have $\beta_{,t} = -h_{,t}/h$ in

(24.85), which makes $\beta_{,tx} = 0$, contrary to our assumption. Consequently, no perfect fluid solutions exist for (20.2) – (20.11) when $\beta_{,z} = 0$ and ($\beta_{,tx} \neq 0$ or $\beta_{,ty} \neq 0$). $\square$

In this way, we have confirmed the concluding sentence of the Szekeres (1975a) paper: 'A tedious and uninstructive computation reveals that in fact there are no solutions to these equations' (by 'these equations' he meant the Einstein equations of the $\beta_{,z} = 0$ subfamily with $\beta_{,tx} \neq 0$ or $\beta_{,ty} \neq 0$).

24.8 Hints for verifying Eq. (21.28)

The formula for $R^0{}_{213}$ obtained by substituting the various $\Gamma^i{}_{jk}$ in (9.21) and taking into account (21.20) – (21.21) is

$$R^0{}_{213} = \left(m^\rho H_{,\rho} + H\overline{Y}_{,u}\right)_{,\alpha} \overline{m}^\alpha + (H\overline{Z})_{,\alpha} \ell^\alpha + H^2 \overline{Z}^2 + \left(m^\rho H_{,\rho} + H\overline{Y}_{,u}\right) Y_{,u}$$
$$- k^\rho H_{,\rho} H\overline{Z}. \tag{24.89}$$

In simplifying the above one must take into account, several times, each of (21.17) – (21.19) and (21.23) – (21.26) (remembering that $k^\rho P_{,\rho} = 0$) and their following consequences:

$$m^\rho H_{,\rho} + H\overline{Y}_{,u} = -3e^{3P}\left(\overline{Z}^2/Z\right)\overline{Y}_{,u} - e^{3P}\overline{ZY}_{,u} + e^{3P} m^\rho \overline{Z}_{,\rho}, \tag{24.90}$$

$$\overline{m}^\alpha\left(m^\rho \overline{Z}_{,\rho}\right)_{,\alpha} = 2\overline{m}^\alpha m^\rho_{,\alpha} m^\sigma Y_{,\sigma\rho} - 2m^\rho m^\sigma \overline{m}^\alpha_{,\rho} Y_{,\alpha\sigma} - m^\rho m^\sigma \overline{m}^\alpha_{,\sigma\rho} Y_{,\alpha}. \tag{24.91}$$

To derive the last equation above one must skilfully handle the term $\overline{m}^\alpha m^\rho m^\sigma Y_{,\sigma\rho\alpha} \equiv \overline{m}^\alpha m^\rho m^\sigma Y_{,\alpha\sigma\rho}$. The chain of equations is this:

$$\overline{m}^\alpha m^\rho m^\sigma Y_{,\alpha\sigma\rho} = m^\rho m^\sigma\left[(\overline{m}^\alpha Y_{,\alpha\sigma})_{,\rho} - \overline{m}^\alpha_{,\rho} Y_{,\alpha\sigma}\right], \tag{24.92}$$

$$\overline{m}^\alpha Y_{,\alpha\sigma} = (\overline{m}^\alpha Y_{,\alpha})_{,\sigma} - \overline{m}^\alpha_{,\sigma} Y_{,\alpha} = -\overline{m}^\alpha_{,\sigma} Y_{,\alpha}, \tag{24.93}$$

the last equality by (21.18). Other useful equations are these:

$$m^\rho \overline{Z}_{,\rho} = m^\rho m^\sigma Y_{,\sigma\rho}, \tag{24.94}$$

$$\overline{m}^\alpha m^\rho_{,\alpha} m^\sigma Y_{,\sigma\rho} = \overline{m}^\alpha \overline{Y}_{,\alpha} m^\sigma Y_{,\sigma u} = Z m^\sigma Y_{,\sigma u}, \tag{24.95}$$

$$m^\rho m^\sigma \overline{m}^\alpha_{,\rho} Y_{,\alpha\sigma} = m^\rho Y_{,\rho} m^\sigma Y_{,\sigma u} = \overline{Z} m^\sigma Y_{,\sigma u}, \tag{24.96}$$

the first equation above from (21.18)–(21.19), the other two from (21.17) and (21.19).

References

Allen, C. W. (1973). *Astrophysical Quantities*. The Athlone Press, University of London. [**174**]

Alpher, R. A. and Herman, R. C. (1948). Evolution of the Universe, *Nature* **162**, 774. [**283**]

Aluri, P. K., Cea, P., Chingangbam, P., Chu, M.-C., Clowes R. G., Hutsemékers, D., Kochappan, J. P., Lopez, A. M., Liu, L., Martens, N. C. M., Martins, C. J. A. P., Migkas, K., Colgáin, E. Ó., Pranav, P., Shamir, L., Singal, A. K., Sheikh-Jabbari, M. M., Wagner, J., Wang, S.-J., Wiltshire, D. L., Yeung, S., Yin, L. and Zhao, W. (2023). Is the observable Universe consistent with the cosmological principle? *Class. Quant. Grav.* **40**, 094001. [**224**]

Appelquist, T. and Chodos, A. (1983). Quantum effects in Kaluza–Klein theories, *Phys. Rev. Lett.* **50**, 141. [**159**]

Arnau, J. V., Fullana, M., Monreal, L. and Saez, D. (1993). On the microwave background anisotropies produced by nonlinear voids, *Astrophys. J.* **402**, 359. [**330**]

Arnau, J. V., Fullana, M. and Saez, D. (1994). Great attractor-like structures and large-scale anisotropy, *Mon. Not. Roy. Astr. Soc.* **268**, L17. [**330**]

Arnowitt, R., Deser, S. and Misner, C. W. (1962). The dynamics of general relativity, in *Gravitation: An Introduction to Current Research*. Edited by L. Witten. Wiley Interscience, New York, pp. 227 – 264. Reprinted in *Gen. Relativ. Gravit.* **40**, 1997 (2008), with an editorial note by J. Pullin, *Gen. Relativ. Gravit.* **40**, 1989 (2008). See also Krasiński, A., Ellis, G. F. R., and MacCallum, M. A. H. (2013). *Golden Oldies in General Relativity, Hidden Gems*, pp. 41 - 79. [**511**]

Ashby, N. (1996). Relativity in the palm of your hand, *Mercury* **25**, 23. [**508, 509**]

Ashby, N. (1998). Relativistic effects in the Global Positioning System. In: *Gravitation and relativity at the turn of the millenium. Proceedings of the 15th International Conference on General Relativity and Gravitation in Pune, India, 1997*. Edited by N. Dadhich and J. V. Narlikar. Inter-University Centre for Astronomy and Astrophysics, Pune, p. 231. [**499, 507**]

Ashby, N. (2014). Relativity in GNSS. Chapter 24 in *Springer handbook of spacetime*, Springer, Heidelberg, pp. 509 – 525. [**499**]

Ashby, N. (2015). GNSS and other applications of general relativity. In: *General Relativity: the most beautiful of theories*. Edited by C. Rovelli, de Gruyter, Berlin. Studies in Mathematical Physics Vol. 28. [**499**]

Balazs, B. and Bene, G. (2018). Simplified model of voids able to mimic accelerating expansion at high z without dark energy, *Gravitation and Cosmology* **24**, 331. [**348**]

Bardeen, J. M. (1970). A variational principle for rotating stars in general relativity, *Astrophys. J.* **162**, 71. [**489, 490**]

Bardeen, J. M. (1973). Timelike and null geodesics in the Kerr metric, in: *Black Holes – les astres occlus*. Edited by C. de Witt and B. S. de Witt. Gordon and Breach, New York, London, Paris, pp. 219 – 239. [**468**]

Barnes, A. (1970). On gravitational collapse against a cosmological background, *J. Phys.* **A3**, 653. [**306, 317**]

Barnes, A. (1973). On shearfree normal flows of a perfect fluid, *Gen. Relativ. Gravit.* **4**, 105. [**218**]

Barnes, A. (1984). Shear-free flows of a perfect fluid, in *Classical General Relativity*. Edited by W. B. Bonnor, J. N. Islam and M. A. H. MacCallum. Cambridge University Press, pp. 15 – 23. [**65**]

Barnes, A. and Rowlingson, R. R. (1989). Irrotational perfect fluids with a purely electric Weyl tensor, *Class. Quant. Grav.* **6**, 949. [**399, 401**]

Barrow, J. D. and Silk, J. (1981). The growth of anisotropic structures in a Friedmann Universe, *Astrophys. J.* **250**, 432. [**403**]

Barrow, J. D. and Stein-Schabes, J. A. (1984). Inhomogeneous cosmologies with cosmological constant, *Phys. Lett.* **A103**, 315. [**395, 402, 403, 404**]

Bażański, S., Kaczyńska, R. and Krasiński, A. (1985). Physical properties of the extended Chasles equilibrium figure, *Phys. Lett.* **A115**, 33. [**492**]

Belinskii, V. and Verdaguer, E. (2001). *Gravitational Solitons.* Cambridge University Press, Cambridge. [**141, 512**]

Berger, B. K., Eardley, D. M. and Olson, D. W. (1977). Note on the spacetimes of Szekeres, *Phys. Rev.* **D16**, 3086. [**400**]

Bergmann, P. G. (1968). Comments on the scalar–tensor theory, *Int. J. Theor. Phys.* **1**, 25. [**145**]

Bertotti, B. (1959). Uniform electromagnetic field in the theory of general relativity, *Phys. Rev.* **116**, 1331. [**165**]

Bertotti, B. (editor) (1974). *Experimental Gravitation. Proceedings of the International School of Physics 'Enrico Fermi', Course 56.* Academic Press, New York and London. [**512**]

Bianchi, L. (1898). Sugli spazi a tre dimensioni che ammettono un gruppo continuo di movimenti [On the three-dimensional spaces which admit a continuous group of motions], *Memorie di Matematica e di Fisica della Società Italiana delle Scienze* **11**, 267. English translation in *Gen. Relativ. Gravit.* **33**, 2171 (2002), with an editorial note by R. Jantzen, *Gen. Relativ. Gravit.* **33**, 2157 (2002). [**94, 96**]

Bini, D., and Jantzen R. T. (2002). Circular holonomy, clock effects and gravitoelectromagnetism: Still going around in circles after all these years..., *Nuovo Cim.* **117B**, 983. [**246, 248**]

Bisogni, S., Risaliti, G. and Lusso, E. (2018). A Hubble Diagram for Quasars, *Front. Astron. Space Sci.* **4**, article 68. [**181**]

Boesgaard, A. M. and Steigman, G. (1985). Big bang nucleosynthesis – theories and observations, *Ann. Rev. Astron. Astrophys.* **23**, 319. [**283**]

Bolejko, K., Krasiński, A., Hellaby, C. and Célérier, M.-N. (2010). *Structures in the Universe by Exact Methods – Formation, Evolution, Interactions.* Cambridge University Press, Cambridge. [**301, 302**]

Bona, C. and Coll, B. (1985). On the Stephani Universe, *C. R. Acad. Sci. Paris* **301**, 613. [**221**]

Bona, C. and Coll, B. (1988). On the Stephani Universes, *Gen. Relativ. Gravit.* **20**, 297. [**221**]

Bondi, H. (1947). Spherically symmetrical models in general relativity, *Mon. Not. Roy. Astr. Soc.* **107**, 410; reprinted in *Gen. Relativ. Gravit.* **31**, 1783 (1999), with an editorial note by A. Krasiński, *Gen. Relativ. Gravit.* **31**, 1777 (1999). [**249, 292, 293, 306, 309, 321**]

Bonnor, W. B. (1956). The formation of the nebulae, *Z. Astrophysik* **39**, 143; reprinted in *Gen. Relativ. Gravit.* **30**, 1113 (1998), with an editorial note by A. Krasiński, *Gen. Relativ. Gravit.* **30**, 1111 (1998). [**297, 351**]

Bonnor, W. B. (1976a). Non-radiative solutions of Einstein's equations for dust, *Commun. Math. Phys.* **51**, 191. [**404, 427**]

Bonnor, W. B. (1976b). Do freely falling bodies radiate? *Nature* **263**, 301. [**404, 427**]

Bonnor, W. B. (1985). An open recollapsing cosmological model with $\Lambda = 0$, *Mon. Not. Roy. Astr. Soc.* **217**, 597. [**310**]

Bonnor, W. B., Sulaiman, A. H. and Tomimura, N. (1977). Szekeres's space-times have no Killing vectors, *Gen. Relativ. Gravit.* **8**, 549. [**395, 440**]

Bonnor, W. B. and Tomimura, N. (1976). Evolution of Szekeres's cosmological models, *Mon. Not. Roy. Astr. Soc.* **175**, 85. [**403**]

Borsuk, K. (1964). *Multidimensional analytic geometry* (in Polish). Second edition. Polish Scientific Publishers, Warsaw. [**369, 413**]

Bower, G. C. (Project Scientist) for the Event Horizon Telescope Collaboration (2022). Focus on first Sgr A* results from the Event Horizon Telescope, *Astrophys. J. Lett.* **930**, L12 – L21. [**192**]

Boyer, R. H. and Lindquist, R. W. (1967). Maximal analytic extension of the Kerr metric, *J. Math. Phys.* **8**, 265. [**442, 451, 453, 468, 475, 478, 481, 484**]

Brans, C. and Dicke, R. H. (1961). Mach's principle and a relativistic theory of gravitation, *Phys. Rev.* **124**, 925. [**144**]

Bronnikov, K. A. (1983). Some exact models of nonspherical collapse. II, *Gen. Relativ. Gravit.* **15**, 823. [**371**]

Bronnikov, K. A. and Pavlov, N. V. (1979). Relyativistskiye raspredeleniya zaryazhenoy pyli s ploskoy, sfericheskoy i psevdosfericheskoy simmetriyami [Relativistic distributions of charged dust with flat, spherical and pseudospherical symmetries], in: *Diskusyonnye voprosy teorii otnositelnosti i gravitatsii [Controversial Questions of the Theory of Relativity and Gravitation]*. Nauka, Moscow, p. 59. [**370, 374**]

Buckley, R. G. and Schlegel, E. M. (2020). Physical geometry of the quasispherical Szekeres models, *Phys. Rev.* **D101**, 023511. [**438**]

Cahen, M. and Defrise, L. (1968). Lorentzian 4 dimensional manifolds with "local isotropy", *Commun. Math. Phys.* **11**, 56. [**168**]

Capozziello, S. and De Laurentis, M. (2011). Extended theories of gravity, *Phys. Rep.* **509**, 167. [**130**]

Carter, B. (1966a). The complete analytic extension of the Reissner–Nordström metric in the special case $e^2 = m^2$, *Phys. Lett.* **21**, 423. [**202, 205**]

Carter, B. (1966b). Complete analytic extension of the symmetry axis of Kerr's solution of Einstein's equations, *Phys. Rev.* **141**, 1242. [**476**]

Carter, B. (1968a). Hamilton–Jacobi and Schrödinger separable solutions of Einstein's equations, *Commun. Math. Phys.* **10**, 280. [**454, 459**]

Carter, B. (1968b). Global structure of the Kerr family of gravitational fields, *Phys. Rev.* **174**, 1559. [**476**]

Carter, B. (1973). Black hole equilibrium states. Part I: Analytic and geometric properties of the Kerr solutions, in: *Black Holes – les astres occlus*. Edited by C. de Witt and B. S. de Witt. Gordon and Breach, New York, London, Paris, p. 61. Reprinted in *Gen. Relativ. Gravit.* **41**, 2874 (2009), with an editorial note by N. Kamran and A. Krasiński, *Gen. Relativ. Gravit.* **41**, 2867 (2009). [**201, 202, 205, 442, 452, 453, 454, 455, 459, 465, 467, 476, 485**]

Célérier, M.-N. (2000). Do we really see a cosmological constant in the supernovae data? *Astron. Astrophys.* **353**, 63. [**282, 347**]

Célérier, M.-N., Bolejko, K. and Krasiński, A. (2010). A (giant) void is not mandatory to explain away dark energy with a Lemaître – Tolman model, *Astronomy and Astrophysics* **518**, A21. [**348, 349**]

Célérier, M.-N. and Schneider, J. (1998). A solution to the horizon problem: a delayed big bang singularity, *Phys. Lett.* **A249**, 37. [**341**]

Célérier, M.-N. and Szekeres, P. (2002). Timelike and null focusing singularities in spherical symmetry: A solution to the cosmological horizon problem and a challenge to the cosmic censorship hypothesis, *Phys. Rev.* **D 65**, 123516. [**341**]

Chandrasekhar, S. (1969). *Ellipsoidal Figures of Equilibrium*. Yale University Press, New Haven and London, p. 46. [**493**]

Chasles, M. (1840). Nouvelle solution du problème de l'attraction d'un ellipsoïde hétérogène sur un point extérieur [A new solution of the problem of attraction of a heterogeneous ellipsoid at an exterior point], *Journal de Mathématiques Pures et Appliquées* **5**, pp. 465 – 488. A briefer account of the same topic, under the same title: *Comptes Rendus Hebdomadaires des Séances de l'Académie des Sciences* **6** *Séance du Lundi 25 Juin 1838.*, pp. 902 – 915. [**492**]

Christodoulou, D. (1984). Violation of cosmic censorship in the gravitational collapse of a dust cloud, *Commun. Math. Phys.* **93**, 171. [**332, 334, 340, 341**]

Clarke, C. J. S. (1993). A review of cosmic censorship, *Class. Quant. Grav.* **10**, 1375. [**341**]

Clarke, C. J. S. and O'Donnell, N. (1992). Dynamical extension through a space-time singularity, *Rendiconti del Seminario Matematico della Università e Politecnico di Torino* **50**(1), 39. [**335**]

Coles P. and Ellis, G. F. R. (1997). *Is the Universe Open or Closed? The Density of Matter in the Universe*. Cambridge University Press, Cambridge. [**261, 281**]

Coll, B. and Ferrando, J. J. (1989). Thermodynamic perfect fluid. Its Rainich theory, *J. Math. Phys.* **30**, 2918. [**221**]

Collins, C. B. (1979). Intrinsic symmetries in general relativity, *Gen. Relativ. Gravit.* **10**, 925. [**99**]

Collins, C. B. and Ellis, G. F. R. (1979). Singularities in Bianchi cosmologies, *Phys. Rep.* **56** no 2, 65. [**100**]

Collins, C. B. and Szafron, D. A. (1979). A new approach to inhomogeneous cosmologies: intrinsic symmetries. I, *J. Math. Phys.* **20**, 2347. [**400**]

Collins, C. B. and Wainwright, J. (1983). On the role of shear in general relativistic cosmological and stellar models, *Phys. Rev.* **D27**, 1209. [**100**]

Courant, R. and Hilbert, D. (1965). *Methods of Mathematical Physics, vol. II: Partial Differential Equations.* Wiley Interscience, New York. [**447**]

Damour, T., Jaranowski, P. and Schäfer, G. (2016). Conservative dynamics of two-body systems at the fourth post-Newtonian approximation of general relativity, *Phys. Rev.* **D93**, 084014. [**175**]

Datt, B. (1938). Über eine Klasse von Lösungen der Gravitationsgleichungen der Relativität [On a class of solutions of the gravitation equations of relativity], *Z. Physik* **108**, 314. English translation *Gen. Relativ. Gravit.* **31**, 1619 (1999), with an editorial note by A. Krasiński, *Gen. Relativ. Gravit.* **31**, 1615 (1999). [**104, 385**]

Dautcourt, G. (1980). The cosmological problem as initial value problem on the observer's past light cone, in: *9th International Conference on General Relativity and Gravitation*, University of Jena, p. 315. [**361**]

Dautcourt, G. (1983a). The cosmological problem as initial value problem on the observer's past light cone: observations, *Astron. Nachr.* **304**, 153. [**361**]

Dautcourt, G. (1983b). The cosmological problem as initial value problem on the observer's past light cone: geometry, *J. Phys.* **A16**, 3507. [**361**]

Dautcourt, G. (1985). Observer dependence of past light cone initial data in cosmology, *Astron. Nachr.* **306**, 1. [**361**]

de la Macorra, A., Garrido, J. and Almaraz, E. (2022). Towards a solution to the H_0 tension, *Phys. Rev.* **D105**, 023526. [**261**]

de Sitter, W. (1916). On Einstein's theory of gravitation and its astronomical consequences. Second paper, *Mon. Not. Roy. Astr. Soc.* **77**, 155 (1916–17). [**168**]

de Sitter, W. (1917). On Einstein's theory of gravitation and its astronomical consequences. Third paper, *Mon. Not. Roy. Astr. Soc.* **78**, 3. [**87**]

de Souza, M. M. (1985). Hidden symmetries of Szekeres quasi-spherical solutions, *Revista Brasileira de Física* **15**, 379. [**416**]

Debever, R. (1959). Sur le tenseur de super-énergie [On the super-energy tensor], *C. R. Acad. Sci. Paris* **249**, 1324; Tenseur de super-énergie, tenseur de Riemann: cas singuliers [The super-energy tensor, the Riemann tensor: the singular cases], *C. R. Acad. Sci Paris* **249**, 1744. [**65, 116**]

Debever, R. (1964). Le rayonnement gravitationnel: le tenseur de Riemann en relativité génerale [Gravitational radiation: the Riemann tensor in general relativity], *Cahiers Phys.* **18**, 303. [**116**]

Debever, R., Kamran N. and McLenaghan, R. G. (1983). A single expression for the general solution of Einstein's vacuum and electrovac field equations with cosmological constant for Petrov type D admitting a nonsingular aligned Maxwell field, *Phys. Lett.* **93A**, 399; also reported in *10th International Conference on General Relativity and Gravitation, Padova 1983 – contributed papers.* Edited by B. Bertotti, F. de Felice and A. Pascolini. Consiglio Nazionale delle Ricerche, Roma 1983, pp. 216 – 218. [**459**]

Debever, R., Kamran, N. and McLenaghan, R. G. (1984). Exhaustive integration and a single expression for the general solution of the type D vacuum and electrovac field equations with cosmological constant for a nonsingular aligned Maxwell field, *J. Math. Phys.* **25**, 1955. [**459**]

Debney, G. C., Kerr, R. P. and Schild, A. (1969). Solutions of the Einstein and Einstein–Maxwell equations, *J. Math. Phys.* **10**, 1842. [**444, 447**]

Deshingkar, S. S., Joshi, P. S. and Dwivedi, I. H. (1999). Physical nature of the central singularity in spherical collapse, *Phys. Rev.* **D59**, 044018. [**341**]

Dicke, R. H. (1964). *The Theoretical Significance of Experimental Relativity.* Gordon and Breach, New York. [**3, 144, 174, 512**]

Dicke, R. H., Peebles, P. J. E., Roll, P. G. and Wilkinson, D. T. (1965). Cosmic black-body radiation, *Astrophys. J.* **142**, 414. [**283**]

Dixon, W. G. (2015). The new mechanics of Myron Mathisson and its subsequent development, in *Equations of Motion in Relativistic Gravity*. Edited by D. Puetzfeld, C. Lämmerzahl and B. Schutz, Springer, pp. 1 – 66. [**169**]

Droste, J. (1917). The field of a single centre in Einstein's theory of gravitation, and the motion of a particle in that field, *Koninklijke Nederlandsche Akademie van Wetenschappen Proceedings* **19**, 197; reprinted, in *Gen. Relativ. Gravit.* **34**, 1545 (2002), with an editorial note by T. Rothman, *Gen. Relativ. Gravit.* **34**, 1541 (2002). [**168**]

Dwivedi, I. H. and Joshi, P. S. (1992). Cosmic censorship violation in non-self-similar Tolman–Bondi models, *Class. Quant. Grav.* **9**, L69. [**341**]

Dyson, F. W., Eddington, A. S. and Davidson, C. (1920). A determination of the deflection of light by the Sun's gravitational field, from observations made at the total eclipse of May 29, 1919, *Phil. Trans. Roy. Soc. London* A220, 291; cited after Will (2018). [**178**]

Eardley, D. M. (1974). Self-similar spacetimes: geometry and dynamics, *Commun. Math. Phys.* **37**, 287. [**100**]

Eardley, D. M. and Smarr, L. (1979). Time functions in numerical relativity: Marginally bound dust collapse, *Phys. Rev.* **D19**, 2239. [**332, 341**]

Ehlers, J. (1961). Beiträge zur relativistischen Mechanik kontinuierlicher Medien [Contributions to the relativistic mechanics of continuous media], *Abhandlungen der Mathematisch-Naturwissenschaftlichen Klasse der Akademie der Wissenschaften und Literatur Mainz*, No 11; English translation: *Gen. Relativ. Gravit.* **25**, 1225 (1993). See also Krasiński, A., Ellis, G. F. R., and MacCallum, M. A. H. (2013), pp. 275 – 323. [**216, 217**]

Ehlers, J. (1981). Über den Newtonschen Grenzwert der Einsteinschen Gravitationstheorie [On the Newtonian limit of Einstein's gravitation theory]. In: Nitsch, J., Pfarr, J., Stachow, E.-W. (eds.) Grundlagenprobleme der modernen Physik: Festschrift für Peter Mittelstaedt zum 50. Geburtstag, pp. 65 – 84. Bibliographisches Institut, Mannheim, Wien, Zürich (1981). English translation: *Gen. Relativ. Gravit.* **51**, 163 (2019), with an editorial note by T. Buchert and T. Mädler, *Gen. Relativ. Gravit.* **51**, 162 (2019). [**133**]

Ehlers, J. and Kundt, W. (1962). Exact solutions of the gravitational field equations, in *Gravitation, an introduction to current research*. Edited by L. Witten. Wiley, New York, pp. 49 – 101. [**65, 114**]

EHTC Collaboration (The Event Horizon Telescope Collaboration) (2019). First M87 Event Horizon Telescope Results I – V, *Astrophys. J. Lett.* **875**, L1 – L5. [**192**]

Einstein, A. (1916). Die Grundlage der allgemeinen Relativitätstheorie [The foundation of the general theory of relativity], *Ann. Physik* **49**, 769; English translation: Einstein et al. (1923, pp. 109 – 164). [**123**]

Einstein, A. (1925). Einheitliche Feldtheorie von Gravitation und Elektrizität [Unified field theory of gravitation and electricity], *Sitzungsberichte Preuss. Akad. Wiss.* p. 414. [**133**]

Einstein, A., Lorentz, H. A., Weyl, H. and Minkowski, H. (1923). *The Principle of Relativity. A Collection of Original Papers on the Special and General Theory of Relativity*. Dover Publications. [**4, 129, 512**]

Einstein, A. and Straus, E. G. (1945). The influence of the expansion of space on the gravitation fields surrounding the individual stars, *Rev. Mod. Phys.* **17**, 120. [**303**]

Einstein, A. and Straus, E. G. (1946). Corrections and additional remarks to our paper: The influence of the expansion of space on the gravitation fields surrounding the individual stars, *Rev. Mod. Phys.* **18**, 148. [**303**]

Eisenhart, L. P. (1940). *An Introduction to Differential Geometry with Use of Tensor Calculus*. Princeton University Press, Princeton. [**45, 51**]

Eisenhart, L. P. (1964). *Riemannian Geometry*. Princeton University Press, Princeton. [**45, 58**]

Ellis, G. F. R. (1967). Dynamics of pressure-free matter in general relativity, *J. Math. Phys.* **8**, 1171. [**368, 374**]

Ellis, G. F. R. (1971). Relativistic cosmology, in: *Proceedings of the International School of Physics 'Enrico Fermi', Course 47: General Relativity and Cosmology*. Edited by R. K. Sachs. Academic Press, New York and London, pp. 104 – 182. Reprinted in *Gen. Relativ. Gravit.* **41**, 581 (2009), with an editorial note by W. Stoeger, *Gen. Relativ. Gravit.* **41**, 575 (2009). [**216, 217, 224, 240, 241, 246**]

Ellis, G. F. R. (1973). Relativistic cosmology, in: *Cargèse Lectures in Physics*, vol. 6. Edited by E. Schatzman. Gordon and Breach, New York, p. 1. [**224**]

Ellis, G. F. R. (1984). Relativistic cosmology: its nature, aims and problems, in: *General Relativity and Gravitation*. Edited by B. Bertotii, F. de Felice, A. Pascolini. D. Reidel, Dordrecht, p. 215. [**439**]

Ellis, G. F. R., Hellaby, C. and Matravers, D. R. (1990). Density waves in cosmology, *Astrophys. J.* **364**, 400. [**365**]

Ellis, G. F. R. and MacCallum, M. A. H. (1969). A class of homogeneous cosmological models, *Commun. Math. Phys.* **12**, 108. [**94, 106**]

Ernst, F. J. (1968a). New formulation of the axially symmetric gravitational field problem, *Phys. Rev.* **167**, 1175. [**512**]

Ernst, F. J. (1968b). New formulation of the axially symmetric gravitational field problem II, *Phys. Rev.* **168**, 1415. [**512**]

Estabrook, F. B., Wahlquist, H. D. and Behr, C. G. (1968). Dyadic analysis of spatially homogeneous world models, *J. Math. Phys.* **9**, 497. [**94**]

Etherington, I. M. H. (1933). On the definition of distance in general relativity, *Phil. Mag., ser.* 7 **15**, 761. Reprinted in *Gen. Relativ. Gravit.* **39**, 1055 (2007), with an editorial note by G. F. R. Ellis, *Gen. Relativ. Gravit.* **39**, 1047 (2007). [**240, 241**]

Everitt, C. W. F. (1974). The gyroscope experiment I: General description and analysis of gyroscope performance. In *Experimental Gravitation. Proceedings of the International School of Physics "Enrico Fermi", course 56*. Edited by B Bertotti, Academic Press, New York 1974, p. 331. [**154**]

Everitt, C. W. F. et al. (2008). Gravity Probe B data analysis status and potential for improved accuracy of scientific results, *Class. Quant. Grav.* **25**, 114002. [**154**]

Everitt, C. W. F. et al. (2011). Gravity Probe B: final results of a space experiment to test general relativity, *Phys. Rev. Lett.* **106**, 221101. [**154**]

Everitt, C. W. F. et al. (2015). The Gravity Probe B test of general relativity, *Class. Quant. Grav.* **32**, 224001. [**154**]

Fermi, E. (1922), Sopra i fenomeni che avvengono in vicinanza di una linea oraria [On phenomena occurring close to a world line], *Atti della Reale Accademia Nazionale dei Lincei, serie 5, Rendiconti Classe di scienze fisiche, matematiche e naturali*, **31**, 1° **semestre** 21 – 23, 51 – 52, 101 – 103. English translation: https://www34.homepage .villanova.edu/robert.jantzen/mg/fermi/bookfiles/fermiandastrophys/pdf/ [**126, 246**]

Ferraris, M., Francaviglia, M. and Reina, C. (1982). Variational formulation of general relativity from 1915 to 1925. 'Palatini's method' discovered by Einstein in 1925, *Gen. Relativ. Gravit.* **14**, 243. [**133**]

Finkelstein, D. (1958). Past–future asymmetry of the gravitational field of a point particle, *Phys. Rev.* **110**, 965. [**183**]

Flanders, H. (1963). *Differential Forms with Applications to Physical Sciences*. Academic Press, New York and London. Republished in 1989 by Dover Books. [**89, 289**]

Fomalont, E. B. and Sramek, R. (1975). A confirmation of Einstein's general theory of relativity by measuring the bending of microwave radiation in the gravitational field of the Sun, *Astrophys. J.* **199**, 749. [**179**]

Fomalont, E. B. and Sramek, R. (1976). Measurements of the solar gravitational deflection of radio waves in agreement with general relativity, *Phys. Rev. Lett.* **36**, 1475. [**179**]

Fomalont, E. B. and Sramek, R. (1977). The deflection of radio waves by the Sun, *Comments Astrophys.* **7**, 19. [**179**]

Fowler, W. A. (1967). *Nuclear Astrophysics*. American Philosophical Society, Philadelphia. [**283**]

Foyster, J. M. and McIntosh, C. B. G. (1972). A class of solutions of Einstein's equations which admit a 3-parameter group of isometries, *Commun. Math. Phys.* **27**, 241. [**164**]

Frauendiener, J. (2018). Notes on the Sagnac effect in General Relativity, *Gen. Relativ. Gravit.* **50**, 147. [**502**]

Friedmann, A. A. (1922). Über die Krümmung des Raumes [On the curvature of space], *Z. Physik* **10**, 377 (1922). See Friedmann (1924) for information on translations of both Friedmann papers in later years. [**105, 224, 256, 260, 268, 282**]

Friedmann, A. A. (1924). Über die Möglichkeit einer Welt mit konstanter negativer Krümmung des Raumes [On the possibility of a world with constant negative curvature of space], *Z. Physik* **21**, 326 (1924). English translation of both Friedmann papers *Gen. Relativ. Gravit.* **31**, 1991 and 2001 (1999) with an editorial note by A. Krasiński and G. F. R. Ellis *Gen. Relativ. Gravit.* **31**, 1985 (1999), see also addendum: *Gen. Relativ. Gravit.* **32**, 1937 (2000). [**105, 256, 268**]

Frolov, V. P. (1974). Resheniya tipa Kerra – Nyumana – Unti – Tamburino uravneniy Eynshteina s kosmologicheskim chlenom [Solutions of Einstein's equations of the Kerr – Newman – Unti – Tamburino type with the cosmological term], *Teor. Mat. Fiz.* **21**, 213; English translation: *Theor. Math. Phys.* **21**, 1088 (1974). [**459**]

Frolov, V. P. and Novikov, I. D. (1998). *Black Hole Physics: Basic Concepts & New Developments.* Kluwer, Amsterdam, 770 pp. [**192**]

Fronsdal, C. (1959). Completion and embedding of the Schwarzschild solution, *Phys. Rev.* **116**, 778. [**188**]

Gaia Mission Science Performance (2020). https://www.cosmos.esa.int/web/gaia/science-performance#astrometric%20performance [**350**]

Gamow, G. (1948). The evolution of the Universe, *Nature* **162**, 680. [**283**]

Gautreau, R. (1984). Imbedding a Schwarzschild mass into cosmology, *Phys. Rev.* **D29**, 198. [**294, 304, 335**]

Gödel, K. (1949). An example of a new type of cosmological solutions of Einstein's field equations of gravitation, *Rev. Mod. Phys.* **21**, 447; reprinted in *Gen. Relativ. Gravit.* **32**, 1409 (2000), with an editorial note by G. F. R. Ellis, *Gen. Relativ. Gravit.* **32**, 1399 (2000). [**94**]

Goldberg, J. and Sachs, R. K. (1962). A theorem on Petrov types, *Acta Phys. Polon* **22 (supplement)**, 13. Reprinted in *Gen. Relativ. Gravit.* **41**, 433 (2009), with an editorial note by A. Krasiński and M. Przanowski, *Gen. Relativ. Gravit.* **41**, 421 (2009). See also Krasiński, A., Ellis, G. F. R., and MacCallum, M. A. H. (2013), pp. 247 – 274. [**239**]

Goode, S. W. and Wainwright, J. (1982). Singularities and evolution of the Szekeres cosmological models, *Phys. Rev.* **D26**, 3315. [**429**]

Gorini, V., Grillo, G. and Pelizza, M. (1989). Cosmic censorship and Tolman–Bondi spacetimes, *Phys. Lett.* **A135**, 154. [**340**]

Grasso, M., Korzyński, M. and Serbenta, J. (2019). Geometric optics in general relativity using bilocal operators, *Phys. Rev.* **D99**, 064038. [**253, 349**]

Graves, J. C. and Brill, D. R. (1960). Oscillatory character of Reissner–Nordström metric for an ideal charged wormhole, *Phys. Rev.* **120**, 1507. [**187, 198, 202, 205**]

Gregory, S. A. and Thompson L. A. (1978). The Coma/A1367 supercluster and its environs, *Astrophys. J.* **222**, 784. [**296**]

Grillo, G. (1991). On a class of naked strong-curvature singularities, *Class. Quant. Grav.* **8**, 739. [**341**]

Grishchuk, L. P. (1967). Cosmological models and spatial homogeneity criteria, *Astron. Zh.* **44**, 1097; English translation: *Sov. Astron. A. J.* **11**, 881 (1968). [**106**]

Guth, A. H. (1981). Inflationary Universe: a possible solution to the horizon and flatness problems, *Phys. Rev.* **D23**, 347. [**277, 279, 280, 343**]

Haantjes, J. (1937). Conformal representations of an *n*-dimensional euclidean space with a non-definite fundamental form on itself, *Koninklijke Nederlandsche Akademie van Wetenschappen Proceedings* **40**, 700. [**86**]

Haantjes, J. (1940). Die Gleichberechtigung gleichförmig beschleunigter Beobachter für die elektromagnetischen Erscheinungen [Equivalence of uniformly accelerated observers with respect to electromagnetic phenomena], *Koninklijke Nederlandsche Akademie van Wetenschappen Proceedings* **43**, 1288. [**86**]

Harness, R. S. (1982). Space-times homogeneous on a time-like hypersurface, *J. Phys.* **A15**, 135. [**99**]

Hasse, W. and Perlick, V. (1988). Geometrical and kinematical characterization of parallax-free world models, *J. Math. Phys.* **29**, 2064. [**249**]

Hawking, S. W. and Ellis, G. F. R. (1973). *The Large-scale Structure of Spaceetime*. Cambridge University Press, Cambridge. [**219, 511, 512**]

Hellaby, C. (1987). A Kruskal-like model with finite density, *Class. Quant. Grav.* **4**, 635. [**318, 355, 357**]

Hellaby, C. (1996a). The nonsimultaneous nature of the Schwarzschild $R = 0$ singularity, *J. Math. Phys.* **37**, 2892. [**312, 398**]

Hellaby, C. (1996b). The null and KS limits of the Szekeres model, *Class. Quant. Grav.* **13**, 2537. [**404**]

Hellaby, C. and Krasiński, A. (2002). You cannot get through Szekeres wormholes or regularity, topology and causality in quasi-spherical Szekeres models, *Phys. Rev.* **D66**, 084011. [**318, 405, 414, 418**]

Hellaby, C. and Krasiński, A. (2008). Physical and geometrical interpretation of the $\epsilon \leq 0$ Szekeres models, *Phys. Rev.* **D77**, 023529. [**369, 438**]

Hellaby, C. and Lake, K. (1984). The redshift structure of the Big Bang in inhomogeneous cosmological models. I. Spherical dust solutions, *Astrophys. J.* **282**, 1; + erratum *Astrophys. J.* **294**, 702 (1985). [**291, 323, 324, 325**]

Hellaby, C. and Lake, K. (1985). Shell crossings and the Tolman model, *Astrophys. J.* **290**, 381 + erratum *Astrophys. J.* **300**, 461 (1985). [**310, 317, 352**]

Hellaby, C. and Lake, K. (1988). *The singularity of Eardley, Smarr and Christodoulou*. Preprint 88/7, Institute of Theoretical Physics and Astrophysics, University of Cape Town. [**332**]

Hubble, E. P. (1929). A relation between distance and radial velocity among extragalactic nebulae, *Proc. Nat. Acad. Sci. USA* **15**, 169. [**141, 256, 282**]

Hubble, E. P. (1936). *The Realm of the Nebulae* (New Haven: Yale University Press, 1982), 207 (Republication; First published 1936). [**256, 282**]

Hubble, E. P. (1953). The law of the redshifts, *Mon. Not. Roy. Astr. Soc.* **113**, 658. [**141, 256**]

Iguchi, H., Nakamura, T. and Nakao, K. (2002). Is dark energy the only solution to the apparent acceleration of the present Universe? *Progr. Theor. Phys.* **108**, 809. [**347, 348**]

Jackson, J. D. (1975). *Classical Electrodynamics*. Second edition. J. Wiley & Sons, Inc. [**150, 157, 513**]

Jantzen, R. (2001). Editor's note: On the three-dimensional spaces which admit a continuous group of motions, *Gen. Relativ. Gravit.* **33**, 2157. [**94**]

Jaranowski, P. and Krolak, A. (2009). *Analysis of Gravitational-Wave Data*. Cambridge University Press, Cambridge. [**511**]

Joshi, P. S. (1993). *Global Aspects in Gravitation and Cosmology*. Clarendon Press, Oxford. [**316, 337, 341, 511, 512**]

Joshi, P. S. and Dwivedi, I. H. (1993). Naked singularities in spherically symmetric inhomogeneous Tolman–Bondi dust cloud collapse, *Phys. Rev.* **D47**, 5357. [**341**]

Kaluza, T. (1921). Zum Unitätsproblem der Physik [On the problem of unity of physics], *Sitzungsber. Preuss. Akad. Wiss. Berlin, Math. Phys. Kl.* p. 966. [**159**]

Kantowski, R. (1965). Some relativistic cosmological models (PhD Thesis). Reprinted in *Gen. Relativ. Gravit.* **30**, 1665 (1998), with an editorial note by A. Krasiński, *Gen. Relativ. Gravit.* **30**, 1663 (1998). [**104**]

Kantowski, R. and Sachs, R. K. (1966). Some spatially homogeneous anisotropic relativistic cosmological models, *J. Math. Phys.* **7**, 443. [**80, 104, 184, 388**]

Kasner, E. (1921). Geometrical theorems on Einstein's cosmological equations, *Amer. J. Math.* **43**, 217. Reprinted in *Gen. Relativ. Gravit.* **40**, 872 (2008), with an editorial note by J. Wainwright, *Gen. Relativ. Gravit.* **40**, 865 (2008). [**144**]

Kerr, R. P. (1963). Gravitational field of a spinning mass as an example of algebraically special metrics, *Phys. Rev. Lett.* **11**, 237. [**442, 449, 450**]

Kerr, R. P. and Schild, A. (1965). A new class of vacuum solutions of the Einstein field equations, in *Atti del Convegno sulla Relatività Generale: Problemi dell'Energia e Onde Gravitazionali*. G. Barbèra, Florence. Reprinted in *Gen. Relativ. Gravit.* **41**, 2485 (2009), with an editorial note by A. Krasiński, E. Verdaguer and R. P. Kerr, *Gen. Relativ. Gravit.* **41**, 2469 (2009). See also Krasiński, A., Ellis, G. F. R., and MacCallum, M. A. H. (2013), pp. 439 – 472. [**442, 444, 447**]

Kinnersley, W. (1969). Type D vacuum metrics, *J. Math. Phys.* **10**, 1195. [**168**]

Klein, O. (1926). Quantentheorie und fünfdimensionale Relativitätstheorie [Quantum theory and five-dimensional relativity], *Z. Physik* **37**, 895 (1926); The atomicity of electricity as a quantum theory law, *Nature* **118**, 516 (1926). [**159**]

Knopp, K. (1996). *Theory of Functions, Part II.* New York: Dover. [**462**]

Kompaneets, A. S. and Chernov, A. S. (1964). Solutions of the gravitation equations for a homogeneous anisotropic model, *Zh. Eksper. Teor. Fiz.* **47**, 1939; English translation: *Sov. Phys. JETP* **20**, 1303 (1965). [**104**]

Kopczyński, W. (1972). A non-singular universe with torsion, *Phys. Lett. A* **39**, 219. [**146**]

Kopczyński, W. and Trautman, A. (1992). *Spacetime and Gravitation.* Państwowe Wydawnictwo Naukowe, Warszawa, and J. Wiley, Chichester – New York – Brisbane – Toronto – Singapore. [**513**]

Korkina, M. P. and Martinenko, V. G. (1975). Dovilni 'T-kuli' v zagalniy teorii vidnosti [Arbitrary 'T-spheres' in the general theory of relativity], *Ukr. Fiz. Zh.* **20**, 626. [**385**]

Korzyński, M. and Kopiński, J. (2018). Optical drift effects in general relativity, *J. Cosm. Astropart. Phys.* **03**, 012. [**xvii, 249, 251, 253, 349**]

Kottler, F. (1918). Über die physikalischen Grundlagen der Einsteinschen Gravitationstheorie [On the physical foundations of Einstein's gravitation theory], *Ann. Physik* **56**, 410. [**168**]

Kraniotis, G. V. (2004). Precise relativistic orbits in Kerr and Kerr-(anti) de Sitter spacetimes, *Class. Quant. Grav.* **21**, 4743. [**468**]

Kraniotis, G. V. (2005). Frame dragging and bending of light in Kerr and Kerr-(anti) de Sitter spacetimes, *Class. Quant. Grav.* **22**, 4391. [**468**]

Kraniotis, G. V. (2007). Periapsis and gravitomagnetic precessions of stellar orbits in Kerr and Kerr-de Sitter black hole spacetimes, *Class. Quant. Grav.* **24**, 1775. [**468**]

Kraniotis, G. V. (2011). Precise analytic treatment of Kerr and Kerr-(anti) de Sitter black holes as gravitational lenses, *Class. Quant. Grav.* **28**, 085021. [**468**]

Kraniotis, G. V. (2014). Gravitational lensing and frame dragging of light in the Kerr-Newman and the Kerr-Newman-(anti) de Sitter black hole spacetimes, *Gen. Relativ. Gravit.* **46**, 1818. [**468**]

Krasiński, A. (1974). Solutions of the Einstein field equations for a rotating perfect fluid. Part I – Presentation of the flow-stationary and vortex-homogeneous solutions, *Acta. Phys. Polon.* **B5**, 411 (1974); Part II – Properties of the flow-stationary and vortex-homogeneous solutions, *Acta. Phys. Polon.* **B6**, 223 (1975). [**99**]

Krasiński, A. (1978). Ellipsoidal spacetimes, sources for the Kerr metric, *Ann. Phys.* **112**, 22. [**492, 493**]

Krasiński, A. (1980). A Newtonian model of the source of the Kerr metric, *Phys. Lett.* **A80**, 238. [**492**]

Krasiński, A. (1981). Spacetimes with spherically symmetric hypersurfaces, *Gen. Relativ. Gravit.* **13**, 1021. [**99**]

Krasiński, A (1983). Symmetries of the Riemann tensor, in *10th International Conference on General Relativity and Gravitation. Abstracts of Contributed Papers.* Edited by F. de Felice and A. Pascolini. University of Padua, p. 290. [**73, 74**]

Krasiński, A. (1997). *Inhomogeneous Cosmological Models.* Cambridge University Press, Cambridge. [**xiv, 141, 193, 286, 292, 297, 298, 303, 351, 374, 385, 395, 433**]

Krasiński, A. (1998). Rotating dust solutions of Einstein's equations with 3-dimensional symmetry groups; Part 1: Two Killing fields spanned on u^α and w^α, *J. Math. Phys.* **39**, 380; Part 2: One Killing field spanned on u^α and w^α, *J. Math. Phys.* **39**, 401; Part 3: All Killing fields linearly independent of u^α and w^α, *J. Math. Phys.* **39**, 2148. [**99**]

Krasiński, A. (1999). [Editorial note to the Nariai 1950 papers] *Gen. Relativ. Gravit.* **31**, 945. See also Krasiński, A., Ellis, G. F. R., and MacCallum, M. A. H. (2013), pp. 350 – 353. [**165, 256**]

Krasiński, A. (2001a). Rotating Bianchi type V dust models generalizing the $k = -1$ Friedmann models, *J. Math. Phys.* **42**, 355; Friedmann limits of hypersurface-homogeneous rotating dust models, *J. Math. Phys.* **42**, 3628. [**99**]

Krasiński, A. (2001b). The newest release of the Ortocartan set of programs for algebraic calculations in general relativity, *Gen. Relativ. Gravit.* **33**, 145. [**119, 495**]

Krasiński, A. (2008). Geometry and topology of the quasi-plane Szekeres model, *Phys. Rev.* **78**, 064038. [**369, 413, 438, 439**]

Krasiński, A. (2014a). Accelerating expansion or inhomogeneity? A comparison of the ΛCDM and Lemaître – Tolman models, *Phys. Rev.* **D89**, 023520; erratum: *Phys. Rev.* **D89**, 089901(E) (2014). [**347**]

Krasiński, A. (2014b). Accelerating expansion or inhomogeneity? Part 2: Mimicking acceleration with the energy function in the Lemaître – Tolman model, *Phys. Rev.* **D90**, 023524. [**348**]

Krasiński, A. (2014c). Blueshifts in the Lemaître – Tolman models, *Phys. Rev.* **D90**, 103525. [**324**]

Krasiński, A. (2016a). Cosmological blueshifting may explain the gamma ray bursts, *Phys. Rev.* **D93**, 043525. [**279, 324, 352**]

Krasiński, A. (2016b). Existence of blueshifts in quasispherical Szekeres spacetimes, *Phys. Rev.* **D94**, 023515. [**324, 352**]

Krasiński, A. (2018a). Properties of blueshifted light rays in quasi-spherical Szekeres metrics, *Phys. Rev.* **D97**, 064047. [**324, 352, 417**]

Krasiński, A. (2018b). Short-lived flashes of gamma radiation in a quasi-spherical Szekeres metric, arXiv 1803.10101 [not to be published in consequence of extremely adverse reaction of the referees]. [**324, 352**]

Krasiński, A. (2020). Gamma radiation from areal radius minima in a quasi-spherical Szekeres metric, *Acta Phys. Polon.* **B51**, 483. [**324, 352**]

Krasiński, A. (2021). Expansion of bundles of light rays in the Lemaître – Tolman models, *Rep. Math. Phys.* **88**, 203. [**324, 326**]

Krasiński, A. (2023). Spacetimes with no position drift, *Acta Phys. Polon.* **B54**, 2-A.1 (2023). [**249**]

Krasiński, A., Behr, C. G., Schücking, E., Estabrook, F. B., Wahlquist, H. D., Ellis, G. F. R., Jantzen, R. and Kundt, W. (2003). The Bianchi classification in the Schücking–Behr approach, *Gen. Relativ. Gravit.* **35**, 475. See also Krasiński, A., Ellis, G. F. R., and MacCallum, M. A. H. (2013), p. 109. [**94**]

Krasiński, A. and Bolejko, K. (2006). Avoidance of singularities in spherically symmetric charged dust, *Phys. Rev.* **D73**, 124033 + erratum *Phys. Rev.* **D75**, 069904 (2007). Fully corrected text: arXiv:gr-qc/0602090. [**382**]

Krasiński, A. and Bolejko, K. (2007). Can a charged dust ball be sent through the Reissner – Nordström wormhole? *Phys. Rev.* **D76**, 124013. [**382**]

Krasiński, A. and Bolejko, K. (2011). Redshift propagation equations in the $\beta' \neq 0$ Szekeres models, *Phys. Rev.* **D83**, 083503. [**249, 349, 350, 438**]

Krasiński, A. and Bolejko, K. (2012a). Apparent horizons in the quasi-spherical Szekeres models, *Phys. Rev.* **D85**, 124016. [**418**]

Krasiński, A. and Bolejko, K. (2012b). Geometry of the quasi-hyperbolic Szekeres models, *Phys. Rev.* **D86** 104036. [**438, 439**]

Krasiński, A. and Ellis, G. F. R. (1999). Editor's note: On the curvature of space; On the possibility of a world with constant negative curvature of space, *Gen. Relativ. Gravit.* **31**, 1985. [**141, 256**]

Krasiński, A., Ellis, G. F. R. and MacCallum, M. A. H. (2013). *Golden Oldies in General Relativity, Hidden Gems*. Springer Verlag, Berlin Heidelberg.

Krasiński, A. and Giono, G. (2012). The charged dust solution of Ruban – matching to Reissner–Nordström and shell crossings, *Gen. Relativ. Gravit.* **44**, 239. [**386, 387**]

Krasiński, A. and Hellaby, C. (2002) Structure formation in the Lemaître–Tolman model, *Phys. Rev.* **D65**, 023501. [**297, 298, 299, 300, 301, 362**]

Krasiński, A. and Hellaby, C. (2004a). More examples of structure formation in the Lemaître–Tolman model, *Phys. Rev.* **D69**, 023502. [**297, 298, 300, 301, 362**]

Krasiński, A. and Hellaby, C. (2004b). Formation of a galaxy with a central black hole in the Lemaître–Tolman model, *Phys. Rev.* **D69**, 043502. [**301, 318**]

Krasiński, A. and Hellaby, C. (2006). Structure formation in the Lemaître–Tolman cosmological model (a non-perturbative approach). In: *Topics in mathematical physics, general relativity and cosmology, in honor of Jerzy Plebański. Proceedings of 2002 international conference.* Edited by H. Garcia-Compean, B. Mielnik, M. Montesinos and M. Przanowski. World Scientific, New Jersey, London, Singapore, Beijing, Shanghai, Hong Kong, Taipei, Chennai, p. 279. [**298, 301**]

Krasiński, A., Hellaby, C., Bolejko, K. and Célérier, M.-N. (2010). Imitating accelerated expansion of the Universe by matter inhomogeneities – corrections of some misunderstandings, *Gen. Relativ. Gravit.* **42**, 2453. [**307**]

Krasiński, A. and Plebański, J. (1980). *N*-dimensional complex Riemann–Einstein spaces with $O(n-1, \mathbb{C})$ as the symmetry group, *Rep. Math. Phys.* **17**, 217. [**165**]

Krasiński, A., Quevedo, H. and Sussman, R. (1997). On thermodynamical interpretation of perfect fluid solutions of the Einstein equations with no symmetry, *J. Math. Phys.* **38**, 2602; more detailed version available as a preprint. [**221, 400**]

Kristian, J. and Sachs, R. K. (1966). Observations in cosmology, *Astrophys. J.* **143**, 379; reprinted in *Gen. Relativ. Gravit.* **43**, 337 (2011), with an editorial note by G. F. R. Ellis, *Gen. Relativ. Gravit.* **43**, 331 (2011). [**224**]

Kruskal, M. (1960). Maximal extension of Schwarzschild metric, *Phys. Rev.* **119**, 1743. [**183**]

Kundt, W. (2003). The spatially homogeneous cosmological models, *Gen. Relativ. Gravit.* **35**, 491. [**94**]

Kurki-Suonio, H. and Liang, E. (1992). Relation of redshift surveys to matter distribution in spherically symmetric dust Universes, *Astrophys. J.* **390**, 5. [**358**]

Lake, K. and Pim, R. (1985). Development of voids in the thin-wall approximation. I General characteristics of spherical vacuum voids, *Astrophys. J.* **298**, 439. [**304**]

Lang, K. (1974). *Astrophysical Formulae.* Springer, Berlin – Heidelberg – New York. [**174, 261**]

Lang, K. (1999). *Astrophysical Formulae. Volume I: Radiation, Gas Processes and High Energy Astrophysics; Volume II: Space, Time, Matter and Cosmology.* Springer, Berlin – Heidelberg – New York. [**284**]

Langevin, P. (1921). Sur la théorie de relativité et l'expérience de M. Sagnac [On the theory of relativity and the experiment of Mr. Sagnac], *C. R. Acad. Sci. Paris* **173**, 831. [**502**]

Laplace, P. S. (1795). *Exposition du Système du Monde [Presentation of the World System]*; cited after Schneider, Ehlers and Falco (1992). [**192**]

Lazkoz, R., Leanizbarrutia, I. and Salzano, V. (2018). Forecast and analysis of the cosmological redshift drift, *European Phys. J. C*, **78**, 11. [**272**]

Lemaître, G. (1927). Un univers homogène de masse constante et de rayon croissant, rendant compte de la vitesse radiale de nébuleuses extra-galactiques [A homogeneous Universe of constant mass and increasing radius accounting for the radial velocity of the extra-galactic nebulae], *Ann. Soc. Sci. Bruxelles* **A47**, 49. See Lemaître (1931) for a somewhat edited English translation and Lemaître (2013) for a faithful translation with historical comments. [**224, 256, 282**]

Lemaître, G. (1931). A homogeneous Universe of constant mass and increasing radius accounting for the radial velocity of the extra-galactic nebulae, *Mon. Not. R. Astr. Soc.* **91**, 483. [**256**]

Lemaître, G. (1933a). L'Univers en expansion [The expanding Universe], *Ann. Soc. Sci. Bruxelles* **A53**, 51; English translation: *Gen. Relativ. Gravit.* **29**, 641 (1997), with an editorial note by A. Krasiński, *Gen. Relativ. Gravit.* **29**, 637 (1997). [**183, 194, 291, 292, 293**]

Lemaître, G. (1933b). La formation des nébuleuses dans l'univers en expansion [Formation of nebulae in the expanding Universe], *C. R. Acad. Sci. Paris* **196**, 1085. [**297**]

Lemaître, G. (2013). A homogeneous Universe of constant mass and increasing radius accounting for the radial velocity of the extra-galactic nebulae, *Gen. Relativ. Gravit.* **45**, 1635; with an editorial note by J. P. Luminet, *Gen. Relativ. Gravit.* **45**, 1619 (2013). [**256**]

Lense, J. and Thirring, H. (1918). Über den Einfluß der Eigenrotation der Zentralkörper auf die Bewegung der Planeten und Monde nach der Einsteinschen Gravitationstheorie [On the influence of the proper rotation of central bodies on the motions of planets and moons according to Einstein's theory of gravitation], *Phys. Zeitschr.* **19**, 156. English translation: *Gen. Relativ. Gravit.* **16**, 711 (1984), combined with comments by B. Mashhoon, F. W. Hehl and D. S. Theiß. [**154**]

Letelier, P. S. (1980). Anisotropic fluids with two perfect-fluid components, *Phys. Rev.* **D22**, 807. [**400**]

Li, J. Z. and Liang, C. B. (1985). A new plane-symmetric solution of Einstein–Maxwell equations, *Chin. Phys. Lett.* **2**, 23. [**165**]

Livio, M. (2011). Mystery of the missing text solved, *Nature* **479**, 171 (2011). [**256**]

Loeb, A. (1998). Direct measurement of cosmological parameters from the cosmic deceleration of extragalactic objects, *Astrophys. J.* **499**, L111. [**272**]

Luminet, J. P. (2013). Editorial note to the Lemaître (1927) paper, *Gen. Relativ. Gravit.* **45**, 1619 (2013). [**256**]

MacCallum, M. A. H. (1979). Anisotropic and inhomogeneous relativistic cosmologies, in *General Relativity, an Einstein Centenary Survey*, Edited by S. W. Hawking and W. Israel, Cambridge University Press, pp. 552–553. [**132**]

MacCallum, M. A. H. (2018). Computer algebra in gravity research, *Liv. Rev. in Relativ.* **21**, 6. [**93**]

MacCallum, M. A. H. and Ellis, G. F. R. (1970). A class of homogeneous cosmological models, *Commun. Math. Phys.* **19**, 31. [**94**]

Maeda, K., Sasaki, M. and Sato, H. (1983). Voids in the closed Universe, *Progr. Theor. Phys.* **69**, 89. [**297, 331**]

Maeda, K. and Sato, H. (1983a). Expansion of a thin shell around a void in expanding Universe, *Progr. Theor. Phys.* **70**, 772. [**297, 331**]

Maeda, K. and Sato, H. (1983b). Expansion of a thin shell around a void in expanding Universe. II, *Progr. Theor. Phys.* **70**, 1276. [**331**]

Manzano, M. and Mars, M. (2021). Null shells: general matching across null boundaries and connection with cut-and-paste formalism, *Class. Quant. Grav.* **38**, 155008. [**146**]

Maoz, D. (2016). *Astrophysics in a Nutshell, 2nd ed.* Princeton University Press, Princeton, NJ and Oxford, UK. [**278, 279, 284**]

Markov, M. A. and Frolov, V. P. (1970). Metrika zakrytogo mira Fridmana, vozmushchennaya elektricheskim zaryadom (k teorii elektromagnitnykh 'Fridmonov') [The metric of a closed Friedman world perturbed by an electric charge (towards a theory of electromagnetic 'Friedmons')], *Teor. Mat. Fiz.* **3**, 3; English translation: *Theor. Math. Phys.* **3**, 301 (1970). [**374**]

Mather, J. C., Bennett, C. L., Boggess, N. W., Hauser, M. G., Smoot, G. F. and Wright, E. L. (1993). Recent results from COBE, in: *General Relativity and Gravitation 1992. Proceedings of the 13th International Conference on General Relativity and Gravitation at Cordoba 1992.* Edited by R. J. Gleiser, C. N. Kozameh and O. M. Moreschi. Institute of Physics Publishing, Boston and Philadelphia, pp. 151 – 158. [**330**]

Mattig, W. (1958). Über den Zusammenhang zwischen Rotverschiebung und scheinbarer Helligkeit [On the connection between redshift and apparent luminosity], *Astron. Nachr.* **284**, 109. [**264**]

Maurin, K. (1973). *Analysis, Part I: Elements.* D. Reidel Publishing Company, Warsaw 1973. *Analysis, Part II: Integration, Distributions, Holomorphic Functions, Tensor and Harmonic Analysis.* Springer, Dordrecht 1980. [**89**]

Mehra, J. (1974). *Einstein, Hilbert and the Theory of Gravitation.* D. Reidel, Dordrecht. [**4, 129, 130, 512**]

Mészáros, A. (1986). New model of the metagalaxy, *Acta Phys. Hung.* **60**, 75. [**355**]

Miller, B. D. (1976). Negative-mass lagging cores of the Big Bang, *Astrophys. J.* **208**, 275. [**351**]

Milne, E. A. (1934). A Newtonian expanding Universe, *Quart. J. Math. Oxford* **5**, 64 (1934); McCrea, W. H. and Milne, E. A. Newtonian Universes and the curvature of space, *Quart. J. Math. Oxford* **5**, 73 (1934); both papers reprinted, in *Gen. Relativ. Gravit.* **32**, 1939 and 1949 (2000), with an editorial note by A. Krasiński, *Gen. Relativ. Gravit.* **32**, 1933 (2000). [**141, 266**]

Milne, E. A. (1948). *Kinematic Relativity.* Clarendon Press, Oxford. [**266**]

Misner, C. W. and D. H. Sharp (1964). Relativistic equations for adiabatic, spherically symmetric gravitational collapse, *Phys. Rev.* **B136**, 571. [**291**]

Misner, C. W., Thorne, K. S. and Wheeler, J. A. (1973). *Gravitation.* Freeman, San Francisco. [**136, 141, 284, 472, 511, 512**]

Musgrave, P. https://github.com/grtensor/grtensor [**66**]

Mustapha, N. and Hellaby, C. (2001). Clumps into voids, *Gen. Relativ. Gravit.* **33**, 455. [**296, 298**]

Nariai, H. (1950). On some static solutions of Einstein's gravitational field equations in a spherically symmetric case, *Scientific Reports of the Tôhoku University* **34**, 160 (1950); On a new cosmological solution of Einstein's field equations of gravitation, *Scientific Reports of the Tôhoku University* **35**, 46 (1951); both papers reprinted in *Gen. Relativ. Gravit.* **31**, 951 and 963 (1999), with an editorial note by A. Krasiński, *Gen. Relativ. Gravit.* **31**, 945 (1999). See also Krasiński, A., Ellis, G. F. R., and MacCallum, M. A. H. (2013), pp. 349 – 374. [**165, 290**]

Neeman, Y. and Tauber, G. (1967). The lagging-core model for quasi-stellar sources, *Astrophys. J.* **150**, 755. [**192, 351**]

Newman, E. T., Couch, E., Chinnapared, K., Exton, A., Prakash, A. and Torrence, R. (1965). Metric of a rotating charged mass, *J. Math. Phys.* **6**, 918. [**459**]

Newman, R. P. A. C. (1986). Strengths of naked singularities in Tolman–Bondi spacetimes, *Class. Quant. Grav.* **3**, 527. [**334, 335, 340**]

Nolan, B. C. and Debnath, U. (2007). Is the shell-focusing singularity of Szekeres space-time visible? *Phys. Rev.* **D76**, 104046. [**419**]

Nordström, G. (1918). On the energy of the gravitational field in Einstein's theory, *Koninklijke Nederlandsche Akademie van Wetenschappen Proceedings* **20**, 1238. [**168**]

Novikov, I. D. (1962a). O nekotorykh svoystvakh resheniy uravneniy Eynshteina dlya sfericheski simmetrichnykh polyey tyagotyeniya (I) [On some properties of solutions of Einstein's equations for spherically symmetric gravitational fields (I)], *Vestn. Mosk. Univ.* no. 5, 90. [**352, 373**]

Novikov, I. D. (1962b). O povedenii sfericheski-symmetrichnykh raspredeleniy mass v obshchey teorii otnositelnosti [On the behaviour of spherically symmetric mass distributions in the general relativity theory], *Vestn. Mosk. Univ.* no. 6, 66. [**164, 360**]

Novikov, I. D. (1964a). Zaderzhka vzryva chasti Fridmanovskogo mira i sverkhzvezdy [Delay of explosion of a part of the Fridman Universe and quasars], *Astron. Zh.* **41**, 1075; English translation: *Sov. Astr. A. J.* **8**, 857 (1965). [**192, 351**]

Novikov, I. D. (1964b). R- i T-oblasti v prostranstve-vremeni so sfericheski-simetrichnym prostranstvom [R- and T-regions in a spacetime with a spherically symmetric space], *Soobshcheniya GAISh* **132**, 3; English translation in *Gen. Relativ. Gravit.* **33**, 2259 (2001), with an editorial note by A. Krasiński, *Gen. Relativ. Gravit.* **33**, 2255 (2001). See also Krasiński, A., Ellis, G. F. R., and MacCallum, M. A. H. (2013), pp. 397 – 438. [**183, 194, 361**]

Ohanian, H. C. and Ruffini, R. (1994). *Gravitation and Spacetime*. W. W. Norton & Company, New York – London. [**511**]

Omer, G. C. (1965). Spherically symmetric distributions of matter without pressure, *Proc. Nat. Acad. Sci. USA* **53**, 1. [**293**]

Oppenheimer, J. R. and Snyder, H. (1939). On continued gravitational contraction, *Phys. Rev.* **56**, 455. [**309**]

Ori, A. (1990). The general solution for spherical charged dust, *Class. Quant. Grav.* **7**, 985. [**378, 379, 380, 382**]

Ori, A. (1991). Inevitability of shell crossing in the gravitational collapse of weakly charged dust spheres, *Phys. Rev.* **D44**, 2278. [**378, 382**]

Padmanabhan, T. (1993). *Structure Formation in the Universe*. Cambridge University Press, Cambridge. [**261, 281, 283**]

Padmanabhan, T. (1996). *Cosmology and Astrophysics Through Problems*. Cambridge University Press, Cambridge. [**298**]

Pais, A. (1982). *Subtle is the Lord. . . The Science and the Life of Albert Einstein*. Oxford University Press, Oxford. [**512**]

Palatini, A. (1919). Deduzione invariantiva delle equazioni gravitazionali dal principio di Hamilton [Invariant deduction of the gravitational equations from Hamilton's principle], *Rendiconti del Circolo Matematico di Palermo* **43**, 203. [**133**]

Papakostas, T. (2009). A new anisotropic solution for ellipsoidal spaces, *J. Phys. Conf. Ser.* **189**, 012027. [**492**]

Partovi, H. H. and Mashhoon, B. (1984). Toward verification of large-scale homogeneity in cosmology, *Astrophys. J.* **276**, 4. [**358**]

Peebles, P. J. E. (1968). Recombination of the primeval plasma, *Astrophys. J.* **153**, 1. [**279**]

Peebles, P. J. E. (1993). *Principles of Physical Cosmology*. Princeton University Press, Princeton. [**262, 280, 281**]

Penrose, R. (1960). A spinor approach to general relativity, *Ann. Phys.* **10**, 171. [**65, 108, 112, 114**]

Penrose, R. (1964). Conformal treatment of infinity, in *Relativity, groups and cosmology*. Edited by B. and C. deWitt. Gordon and Breach, New York – London, pp. 565 – 584. Reprinted in *Gen. Relativ. Gravit.* **43**, 901 (2011), with an editorial note by H. Friedrich, *Gen. Relativ. Gravit.* **43**, 897 (2011). [**312**]

Penrose, R. (1969). Gravitational Collapse: the Role of General Relativity, *Rivista del Nuovo Cimento, Numero Speziale* **I**, 252. Reprinted in *Gen. Relativ. Gravit.* **34**, 1141 (2002), with an editorial note by A. Królak, *Gen. Relativ. Gravit.* **34**, 1135 (2002). [**336, 485**]

Penrose, R. and Rindler, W. (1984). *Spinors and Space-Time.* Cambridge University Press, Cambridge. [**108, 512**]

Penzias, A. A. and Wilson, R. W. (1965). A measurement of excess antenna temperature at 4080 Mc/s, *Astrophys. J.* **142**, 419. [**283**]

Perlmutter, S., Aldering, G., Goldhaber, G., Knop, R. A., Nugent, P., Castro, P. G., Deustua, S., Fabbro, S., Goobar, A., Groom, D. E., Hook, I. M., Kim, A. G., Kim, M. Y., Lee, L. C., Nunes, N. J., Pain, R., Pennypacker, C. R., Quimby, R., Lidman, C., Ellis, R.S., Irwin, M., McMahon, R. G., Ruiz-Lapuente, P., Walton, N., Schaefer, B., Boyle, B. J., Filippenko, A. V., Matheson, T., Fruchter, A. S., Panagia, N., Newberg, H. J. M. and Couch, W. J. (1999). Measurements of omega and lambda from 42 high-redshift supernovae, *Astrophys. J.* **517**, 565. [**270, 281, 282**]

Perlick, V. (2004). Gravitational lensing from a spacetime perspective, *Living Reviews in Relativity* **7**, article # 9. [**57, 58**]

Petrov, A. Z. (1954). Klassifikacya prostranstv opredelyaushchikh polya tyagoteniya [The classification of spaces defining gravitational fields], *Uchenye Zapiski Kazanskogo Gosudarstvennogo Universiteta im. V. I. Ulyanovicha-Lenina* [Scientific Notes of the State V. I. Ulyanovich-Lenin University] **114**(8), 55. English translation in *Gen. Relativ. Gravit.* **32**, 1665 (2000), with an editorial note by M. A. H. MacCallum, *Gen. Relativ. Gravit.* **32**, 1661 (2000). See also Krasiński, A., Ellis, G. F. R., and MacCallum, M. A. H. (2013), pp. 81 – 108. [**65**]

Pirani, F. A. E. (1957). Invariant formulation of gravitational radiation theory, *Phys. Rev.* **105**, 1089. [**65**]

Planck (2014). Planck collaboration, Planck 2013 results. XVI. Cosmological parameters, *Astron. Astrophys.* **571**, A16 (2014). [**181, 261, 270, 279, 282, 284, 347**]

Plebański, J. (1964). *Notes from Lectures on General Relativity. Part I: Mathematical Introduction.* CINVESTAV, Mexico. [Note: this is a xeroxed typescript. Part II and following probably never appeared even in this form.] [**xiii**]

Plebański, J. (1967). *On Conformally Equivalent Riemannian Spaces.* CINVESTAV, Mexico. [Note: this is a xeroxed typescript.] [**64, 86**]

Plebański, J. (1974). *Spinors, Tetrads and Forms. Part I.* CINVESTAV, Mexico 1974, Part II: CINVESTAV, Mexico 1975. [Note: these are xerox copies of handwritten notes. This work has never been published as a proper book.] [**108**]

Podurets, M. A. (1964). Ob odnoy forme uravneniy Eynshteyna dlya sfericheski simmetrichnogo dvizheniya sploshnoy sredy [On one form of Einstein's equations for spherically symmetrical motion of a continuous medium], *Astron. Zh.* **41**, 28; English translation: *Sov. Astr. A. J.* **8**, 19 (1964). [**291**]

Quevedo, H. and Sussman, R. (1995). On the thermodynamics of simple non-isentropic perfect fluids in general relativity, *Class. Quant. Grav.* **12**, 859; Thermodynamics of the Stephani Universe, *J. Math. Phys.* **36**, 1365. [**221**]

Racz, I. (1992). Note on stationary axisymmetric vacuum space-times with inside ellipsoidal symmetry, *Class.Quant.Grav.* **9**, L93. [**492**]

Racz, I. and Süveges, M. (1998). On the uniqueness of stationary axisymmetric ellipsoidal vacuum configurations, *Mod. Phys. Lett.* **A13**, 1101. [**492**]

Raine, D. J. and Thomas, E. G. (1981). Large-scale inhomogeneity in the Universe and the anisotropy of the microwave background, *Mon. Not. Roy. Astr. Soc.* **195**, 649. [**329**]

Raszewski, P. K. (1958). *Geometria Riemanna i analiza tensorowa* [*Riemann Geometry and Tensor Analysis*] (Polish translation from Russian). Państwowe Wydawnictwo Naukowe, Warszawa. [**54**]

Raychaudhuri, A. K. (1953). Arbitrary concentrations of matter and the Schwarzschild singularity, *Phys. Rev.* **89**, 417. [**183**]

Raychaudhuri, A. K. (1955). Relativistic cosmology. I *Phys. Rev.* **98**, 1123 (1955). Reprinted in *Gen. Relativ. Gravit.* **32**, 749 (2000), with an editorial note by J. Earman and A. K. Raychaudhuri, *Gen. Relativ. Gravit.* **32**, 743 (2000). [**217**]

Raychaudhuri, A. K. (1957). Singular state in relativistic cosmology, *Phys. Rev.* **106**, 172 (1957). [**217**]

Raychaudhuri, A. K. (1979). *Theoretical Cosmology.* Clarendon Press, Oxford. [**432**]

Reissner, H. (1916). Über die Eigengravitation des elektrischen Feldes nach der Einsteinschen Theorie [On the self-gravitation of the electric field according to Einstein's theory], *Ann. Physik* **50**, 106. [**168**]

Ribeiro, M. B. (1992a). On modelling a relativistic hierarchical (fractal) cosmology by Tolman's spacetime. I. Theory, *Astrophys. J.* **388**, 1. [**331, 360**]

Ribeiro, M. B. (1992b). On modelling a relativistic hierarchical (fractal) cosmology by Tolman's spacetime. II. Analysis of the Einstein–de Sitter model, *Astrophys. J.* **395**, 29. [**360**]

Ribeiro, M. B. (1993). On modelling a relativistic hierarchical (fractal) cosmology by Tolman's spacetime. III. Numerical results, *Astrophys. J.* **415**, 469. [**360**]

Riess, A. G., Filippenko, A. V., Challis, P., Clocchiatti, A., Diercks, A., Garnavich, P. M., Gilliland, R. L., Hogan, C. J., Jha, S., Krishner, R. P., Leibundgut, B., Phillips, M. M., Reiss, D., Schmidt, B. P., Schommer, R. A., Smith, R. C., Spyromilio, J., Stubbs, C., Suntzeff, N. B. and Tonry, J. (1998). Observational evidence from supernovae for an accelerating Universe and a cosmological constant, *Astron. J.* **116**, 1009. [**270, 282**]

Rindler, W. (1956). Visual horizons in world models, *Mon. Not. Roy. Astr. Soc.* **116**, 662; reprinted in *Gen. Relativ. Gravit.* **34**, 133 (2002), with an editorial note by A. Krasiński, *Gen. Relativ. Gravit.* **34**, 131 (2002). [**273, 274, 276**]

Rindler, W. (1980). *Essential Relativity. Special, General and Cosmological.* Revised 2nd ed. Springer, Berlin. [**247, 268, 513**]

Robertson, H. P. (1929). On the foundations of relativistic cosmology, *Proc. Nat. Acad. Sci. USA* **15**, 822. [**105, 256**]

Robertson, H. P. (1933). Relativistic cosmology, *Rev. Mod. Phys.* **5**, 62. Reprinted in *Gen. Relativ. Gravit* **44**, 2115 (2012), with an editorial note by G. F. R. Ellis, *Gen. Relativ. Gravit* **44**, 2099 (2012). [**105, 257**]

Robertson, H. P. and T. W. Noonan (1968). *Relativity and Cosmology.* W. B. Saunders Company, Philadelphia – London – Toronto, p. 374 – 378. [**268**]

Robinson, I. (1959). A solution of the Einstein–Maxwell equations, *Bull. Acad. Polon. Sci., Ser. Mat. Fis. Astr.* **7**, 351. [**165, 387**]

Roos, W. (1976). On the existence of interior solutions in general relativity, *Gen. Relativ. Gravit.* **7**, 431. [**493**]

Rosen, N. (1973). A bi-metric theory of gravitation, *Gen. Relativ. Gravit.* **4**, 435. [**146**]

Rothman, T. (2002). Editorial note to the Droste 1917 paper, *Gen. Relativ. Gravit.* **34**, 1541. [**168**]

Rothman, T. and Ellis, G. F. R. (1987). Metaflation? *Astronomy* **15** no 2, 6; see also the University of Cape Town preprint 85/18. [**281**]

Royal Astronomical Society Discussion (1930), *Observatory* **53**, 39. [**256**]

Ruban, V. A. (1968). T-modeli 'shara' v obshchey teorii otnositelnosti [*T*-models of a 'sphere' in general relativity theory], *Pis'ma v Red. ZhETF* **8**, 669. English translation: *Sov. Phys. JETP Lett.* **8**, 414 (1968), reprinted in *Gen. Relativ. Gravit.* **33**, 369 (2001). [**104, 385**]

Ruban, V. A. (1969). Sfericheski-symmetrichnye *T*-modeli v obshchey teorii otnositelnosti [Spherically symmetric *T*-models in the general theory of relativity], *Zh. Eksper. Teor. Fiz.* **56**, 1914. English translation: *Sov. Phys. JETP* **29**, 1027 (1969), reprinted in *Gen. Relativ. Gravit.* **33**, 375 (2001), with an editorial note by A. Krasiński, *Gen. Relativ. Gravit.* **33**, 363 (2001). [**104, 385**]

Ruban, V. A. (1972). Neodnorodnye kosmologicheskiye modeli s ploskoy i psevdosfericheskoy simmetriyami [Inhomogeneous cosmological models with flat and pseudospherical symmetries], in: *Tezisy dokladov 3-y Sovetskoy Gravitatsyonnoy Konferentsii [Theses of Lectures of the 3rd Soviet Conference on Gravitation].* Izdatel'stvo Erevanskogo Universiteta, Erevan, pp. 348 – 351. [**xvii, 384**]

Ruban, V. A. (1983). T-modeli shara w obshchey teorii otnositelnosti (II) [T-models of a sphere in the general theory of relativity (II)], *Zh. Eksper. Teor. Fiz.* **85**, 801; English translation: *Sov. Phys. JETP* **58**, 463 (1983). [**xvii, 384, 386**]

Saez, D., Arnau, J. V. and Fullana, M. J. (1993). The imprints of the Great Attractor and the Virgo cluster on the microwave background, *Mon. Not. Roy. Astr. Soc.* **263**, 681. [**330**]

Sagnac, G. (1913a). L'éther lumineux d'emontré par léffet du vent relatif d'éther dans un interféromètre en rotation uniforme [The luminous aether demonstrated by the effect of the wind relative to the aether in an interferometer in uniform rotation], *C. R. Acad. Sci. (Paris)*, **157**, 708. [**502**]

Sagnac, G. (1913b). Sur la preuve de la réalité de l'éther lumineux par l'expérience de l'interférographe tournant [On the proof of reality of the luminous aether by the experiment with the rotating interferograph], *C. R. Acad. Sci. (Paris)*, **157**, 1410. [**502**]

Sandage, A. (1962). The change of redshift and apparent luminosity of galaxies due to the deceleration of selected expanding Universes, *Astrophys. J.* **136**, 319. [**272**]

Sato, H. (1984). Voids in expanding Universe, in: *General Relativity and Gravitation.* Edited by B. Bertotti, F. de Felice and A. Pascolini. D. Reidel, Dordrecht, pp. 289 – 312. [**297**, **304**, **331**]

Sato, H. and Maeda, K. (1983). The expansion law of the void in the expanding Universe, *Progr. Theor. Phys.* **70**, 119. [**297**, **331**]

Schiff, L. I. (1960a). Possible new experimental test of general relativity theory, *Phys. Rev. Lett.* **4**, 215. [**154**]

Schiff, L. I. (1960b). Motion of a gyroscope according to Einstein's theory of gravitation, *Proc. Nat. Acad. Sci. USA* **46**, 871. [**154**]

Schneider, P., Ehlers, J. and Falco, E. E. (1992). *Gravitational Lenses.* Springer, Berlin. [**179**, **180**, **181**, **209**]

Schouten, J. A. and Struik, D. J. (1935). *Einführung in die neueren Methoden der Differentialgeometrie, Band 1* [*Introduction to the Newer Methods of Differential Geometry, Vol. 1*]. P. Noordhoff N. V., Groningen - Batavia 1935, p. 142. [**75**]

Schouten, J. A. and van Kampen, E. R. (1934). Beiträge zur Theorie der Deformation [Contributions to the deformation theory], *Prace Matematyczno-Fizyczne* **41**, 1. [Note: in citations of this paper, even by Schouten himself, the title of the journal and the date of publication are distorted. The citation here is corrected.] [**75**]

Schwarzschild, K. (1916a). Über das Gravitationsfeld eines Massenpunktes nach der Einsteinschen Theorie [On the gravitational field of a point mass according to Einstein's theory], *Sitzungsber. Preuss. Akad. Wiss.* p. 189. English translation: *Gen. Relativ. Gravit.* **35**, 951 (2003). The original editorial note contains errors and misconceptions, rectified in Senovilla (2007). [**168**]

Schwarzschild, K. (1916b). Über das Gravitationsfeld einer Kugel aus inkompressibler Flüssigkeit nach der Einsteinschen Theorie [On the gravitational field of a sphere of incompressible fluid according to the Einstein theory], *Sitzungsber. Preuss. Akad. Wiss.* p. 424. [**197**]

Sen, N. R. (1934). On the stability of cosmological models *Z. Astrophysik* **9**, 215; reprinted in *Gen. Relativ. Gravit.* **29**, 1477 (1997), with an editorial note by A. Krasiński, *Gen. Relativ. Gravit.* **29**, 1473 (1997). [**296**, **297**]

Senin, Yu. E. (1982). Kosmologiya s toroidalnym raspredeleniyem zhidkosti [Cosmology with a toroidal distribution of a fluid], in: *Problemy teorii gravitatsii i elementarnykh chastits* [*Problems of the Theory of Gravitation and Elementary Particles*], 13th issue. Edited by K. P. Stanyukovich. Energoizdat, Moscow, pp. 107 – 111. [**435**]

Senovilla, J. M. M. (1998). Singularity theorems and their consequences, *Gen. Relativ. Gravit.* **30**, 701. [**219**]

Senovilla, J. M. M. (2007). The Schwarzschild solution: corrections to the editorial note, *Gen. Relativ. Gravit.* **39**, 685. [**168**]

Serbenta, J. and Korzyński, M. (2022). Bilocal geodesic operators in static spherically-symmetric spacetimes, *Class. Quant. Grav.* **39**, 155002. [**74**]

Shikin, I. S. (1972). Gravitational fields with groups of motions on two-dimensional transitivity hypersurfaces in a model with matter and a magnetic field, *Commun. Math. Phys.* **26**, 24. [**374**, **377**]

Shikin, I. S. (1974). Issledovaniye klassa polyey tyagotyeniya dlya zaryazhennoy pylevidnoy sredy [Investigation of a class of gravitational fields of charged dust medium], *Zh. Eksper. Teor. Fiz.* **67**, 433; English translation: *Sov. Phys. JETP* **40**, 215 (1975). [**374**]

Silk, J. (1977). Large-scale inhomogeneity of the Universe, *Astron. Astrophys.* **59**, 53. [**294**, **344**]

Ślebodziński, W. (1931). Sur les équations canoniques de Hamilton [On Hamilton's canonical equations], *Bulletins de la Classe des Sciences, Acad. Royale de Belg.* (5) **17**, 864. English translation: *Gen. Relativ. Gravit.* **42**, 2529 (2010) with an editorial note by A. Trautman, *Gen. Relativ. Gravit.* **42**, 2525 (2010). See also Krasiński, Ellis and MacCallum (2013), pp. 3 – 14. [**75**]

Smoot, G. F., Bennett, C. L., Kogut, A., Wright, E. L., Aymon, J., Boggess, N. W., Cheng, E. S., de Amici, G., Gulkis, S., Hauser, M. G., Hinshaw, G., Jackson, P. D., Janssen, M., Kaita, E., Kelsall, T., Keegstra, P., Lineweaver, C., Loewenstein, K., Lubin, P., Mather, J., + 8 more (1992). Structure in the COBE differential microwave radiometer first-year maps, *Astrophys. J. Lett.* **396**, L1. [**280, 285, 302, 330**]

Sneddon, I. I. (1957). *Elements of Partial Differential Equations.* McGraw–Hill, New York – Toronto – London. [**447**]

Soldner, J. (1804). Über die Ablenkung eines Lichtstrahls von seiner geradelinigen Bewegung durch die Attraktion eines Weltkörpers, an welchem er nahe vorbeigeht [On the deflection of a light ray from its rectilinear motion by the attraction of a heavenly body, nearby which it is passing], *Berliner Astronomisches Jahrbuch* 1804, p. 161; cited after Schneider, Ehlers and Falco (1992). [**179, 209**]

Spero, A. and Szafron, D. A. (1978). Spatial conformal flatness in homogeneous and inhomogeneous cosmologies, *J. Math. Phys.* **19**, 1536. [**400, 440**]

Springel, V., Frenk, C. S. and White, S. D. M. (2006). The large-scale structure of the Universe, *Nature* **440**, 1137. [**284**]

Stephani, H. (1967a). Über Lösungen der Eisteinschen Feldgleichungen, die sich in einen fünfdimensionalen flachen Raum einbetten lassen [On solutions of Einstein's equations that can be embedded in a five-dimensional flat space], *Commun. Math. Phys.* **4**, 137. [**218, 221, 257, 285**]

Stephani, H. (1967b). Konform flache Gravitationsfelder [Conformally flat gravitational fields], *Commun. Math. Phys.* **5**, 337. [**64**]

Stephani, H. (1990). *General Relativity.* Second edition. Cambridge University Press, Cambridge. [**73, 85, 87, 141, 148, 154, 260**]

Stephani, H., Kramer, D., MacCallum, M., Hoenselaers, C. and Herlt, E. (2003). *Exact Solutions of Einstein's Field Equations.* 2nd Edition. Cambridge University Press, Cambridge. [**65, 66, 100, 103, 141, 148, 164, 165, 218, 232, 240, 257, 285, 374, 459, 487, 511, 512**]

Stoeger, R. W., Ellis, G. F. R. and Nel, S. D. (1992). Observational cosmology: III. Exact spherically symmetric dust solutions, *Class. Quant. Grav.* **9**, 509. [**294, 312, 361**]

Sussman, R. A. and Delgado Gaspar, I. (2015). Multiple non-spherical structures from the extrema of Szekeres scalars, *Phys. Rev.* **D92**, 083533. [**438**]

Suto, Y., Sato, K. and Sato, H. (1984). Expansion of voids in a matter-dominated Universe, *Progr. Theor. Phys.* **71**, 938; Nonlinear evolution of negative density perturbations in a radiation-dominated Universe, *Progr. Theor. Phys.* **72**, 1137. [**318**]

Sutter, P. M., Lavaux, G., Wandelt, B. D. and Weinberg, D. H. (2012). A public void catalog from the SDSS DR7 galaxy redshift surveys based on the watershed transform, *Astrophys. J.* **761**, 44. [**296**]

Synge, J. L. (1965). *Relativity: The Special Theory.* 2nd ed. North-Holland, Amsterdam. [**513**]

Szafron, D. A. (1977). Inhomogeneous cosmologies: New exact solutions and their evolution, *J. Math. Phys.* **18**, 1673. [**221, 391, 394, 399, 401, 435**]

Szafron, D. A. and Collins, C. B. (1979). A new approach to inhomogeneous cosmologies. II. Conformally flat slices and an invariant classification, *J. Math. Phys.* **20**, 2354. [**399, 401**]

Szafron, D. A. and Wainwright, J. (1977). A class of inhomogeneous perfect fluid cosmologies, *J. Math. Phys.* **18**, 1668. [**433**]

Szekeres, G. (1960). On the singularities of a Riemannian manifold, *Publicationes Mathematicae Debrecen* **7**, 285; reprinted in *Gen. Relativ. Gravit.* **34**, 2001 (2002), with an editorial note by P. Szekeres, *Gen. Relativ. Gravit.* **34**, 1995 (2002). See also Krasiński, A., Ellis, G. F. R., and MacCallum, M. A. H. (2013), pp. 375 – 395. [**183**]

Szekeres, P. (1975a). A class of inhomogeneous cosmological models, *Commun. Math. Phys.* **41**, 55. [**249, 403, 404, 526**]

Szekeres, P. (1975b). Quasispherical gravitational collapse, *Phys. Rev.* **D12**, 2941. [**249, 305, 414, 416, 418, 422**]

Szekeres, P. (1980). Naked singularities, in: *Gravitational Radiation, Collapsed Objects and Exact Solutions.* Edited by C. Edwards. Springer (Lecture Notes in Physics, vol. 124), New York, pp. 477 – 487. [**322, 325**]

Taub, A. H. (1951). Empty space-times admitting a three parameter group of motions, *Ann. Math.*, **53**, 472; reprinted in *Gen. Relativ. Gravit.* **36**, 2699 (2004), with an editorial note by M.A.H. MacCallum, *Gen. Relativ. Gravit.* **36**, 2689 (2004). [**94**]

Thorne, K. S., Lee, D. L. and Lightman, A. P. (1973). Foundations for a theory of gravitation theories, *Phys. Rev.* **D7**, 3563. [**144**]

Tolman, R. C. (1934). Effect of inhomogeneity on cosmological models, *Proc. Nat. Acad. Sci. USA* **20**, 169. Reprinted in *Gen. Relativ. Gravit.* **29**, 935 (1997), with an editorial note by A. Krasiński, *Gen. Relativ. Gravit.* **29**, 931 (1997). [**292, 296**]

Tomimatsu, A. and Sato, H. (1972). New exact solution for the gravitational field of a spinning mass, *Phys. Rev. Lett.* **29**, 1344. [**512**]

Tomita, K. (2000). Distances and lensing in cosmological void models, *Astrophys. J* **529**, 38. [**349**]

Trautman, A. (1972). On the Einstein–Cartan equations, *Bull. Acad. Polon. Sci., sér. sci. math., astr. et phys.* **20** Part I: 185–190, Part II: 503–506, Part III: 895–896 (1972); **21** Part IV: 345–346 (1973); On the structure of the Einstein-Cartan equations, *Symposia Mathematica* **12** 139 (1973). [**145**]

Verne, J. (1874). *Le Tour du Monde en 80 Jours (Voyages Extraordinaires)* [Around the World in Eighty Days]. Hetzel, Paris (to be had for \$ 6,000). [**488**]

Vickers, P. A. (1973). Charged dust spheres in general relativity, *Ann. Inst. Poincaré* **A18**, 137. [**374, 375**]

Wagoner, R. V. (1970). Scalar–tensor theory and gravitational waves, *Phys. Rev.* **D1**, 3209. [**145**]

Wagoner, R. V., Fowler, W. A. and Hoyle, F. (1967). On the synthesis of elements at very high temperatures, *Astrophys. J.* **148**, 3. [**283**]

Wainwright, J. (1977). Characterization of the Szekeres inhomogeneous cosmologies as algebraically special space-times, *J. Math. Phys.* **18**, 672. [**401**]

Wainwright, J. and Yaremovicz, P. E. A. (1976). Killing vector fields and the Einstein–Maxwell field equations with perfect fluid source, *Gen. Relativ. Gravit.* **7**, 345; Symmetries of the Einstein–Maxwell field equations: the null field case, *Gen. Relativ. Gravit.* **7**, 595. [**165**]

Walker A. G. (1932). Relative co-ordinates, *Proc. R. Soc. Edinburgh*, **52**, 345. [**246**]

Walker, A. G. (1935). On Riemannian spaces with spherical symmetry about a line, and the conditions for isotropy in general relativity, *Quart. J. Math. Oxford*, ser. 6, 81. [**105, 257**]

Walters, A. and Hellaby, C. (2012). Constructing realistic Szekeres models from initial and final data, *J. Cosm. Astropart. Phys.* **JCAP12**, 001. [**298, 438**]

Wambsganss, J. (2006). Gravitational microlensing. In: *Gravitational Lensing: Strong, Weak and Micro*. Saas-Fee Lectures, Springer-Verlag. Saas-Fee Advanced Courses. Vol. 33. pp. 453 – 540. [**181**]

Waugh, B. and Lake, K. (1988). Strengths of shell-focusing singularities in marginally bound collapsing self-similar Tolman spacetimes, *Phys. Rev.* **D38**, 1315. [**340**]

Weber, J. (1961). *General Relativity and Gravitational Waves*. Wiley Interscience, New York. [**511**]

Webster, A. G. (1949). *The Dynamics of Particles and of Rigid, Elastic and Fluid Bodies*. Hafner Publishing Company, Inc., New York, p. 414. [**492**]

Weinberg, S. (1972). *Gravitation and Cosmology*. Wiley, New York. [**432, 512**]

Werle, J. (1957). *Termodynamika fenomenologiczna [Phenomenological Thermodynamics]*. Państwowe Wydawnictwo Naukowe, Warszawa, pp. 78 – 80 (in Polish). [**220**]

Will, C. M. (1988). Henry Cavendish, Johann von Soldner, and the deflection of light, *Amer. J. Phys.* **56**, 413; cited after Schneider, Ehlers and Falco (1992). [**179, 209**]

Will, C. M. (2006). The confrontation between general relativity and experiment, *Living Rev. Relativity* **9**, 3. [**145, 146, 180**]

Will, C. M. (2018). *Theory and Experiment in Gravitational Physics*. 2nd edition, Cambridge University Press, Cambridge. [**2, 123, 129, 136, 144, 174, 179, 512**]

Wolf, T. (1985). *Exakte Lösungen der Einsteinschen Feldgleichungen mit flachen Schnitten [Exact solutions of the Einstein field equations with flat sections.]* Ph. D. Thesis, University of Jena. [**99**]

Yodzis, P., Seifert, H. J. and Müller zum Hagen, H. (1973). On the occurrence of naked singularities in general relativity, *Commun. Math. Phys.* **34**, 135. [**336, 338, 339**]

Yoo, C.-M., Kai, T. and Nakao, K-i (2011). Redshift drift in Lemaître – Tolman – Bondi void universes, *Phys. Rev.* **D83**, 043527. [**347**]

Zeldovich, Ya. B., Kurt, V. G., and R. A. Syunyaev, R. A. (1968). Recombination of hydrogen in the hot model of the Universe, *Zhurn. Eksper. Teor. Fiz.* **55**, 278. [**279**]

Zel'dovich, Ya. B. and Novikov, I. D. (1971). *Relativistic Astrophysics: Volume I: Stars and Relativity.* University of Chicago Press, Chicago. [**512**]

Zel'dovich, Ya. B. and Novikov, I. D. (1974). *Relativistic Astrophysics: Volume II: The Universe and Relativity.* University of Chicago Press, Chicago. [**512**]

Zsigrai, J. (2003). Ellipsoidal shapes in general relativity: general definitions and an application, *Class. Quant. Grav.* **20**, 2855. [**492**]

Index

G_3/S_2 symmetric spacetime, 367
G_3/S_2 symmetry, 401
ΛCDM model, 272, 347

Abelian group, 87
absolute
  -apparent horizon, 418–429, 521, 522
  -luminosity, 282, 347
acausal spacetime, 202
accelerating expansion of Universe, 269, 270, 272,
    273, 347–349
  mimicking-, 347, 348
acceleration, 246
  -of light ray, 229
  -transformation, 86
  -vector, 215
action
  -integral of gravitation, 131
  -of group, 99, 100
  multiply transitive-, of group, 100
  simply transitive-, of group, 100
  transitive-, of group, 100
active gravitational mass, 291, 292, 369, 385, 410
adapted coordinates, 36, 44, 64, 65, 375
affine
  -connection, 28, 30–33, 35, 42, 45, 47
  -parameter, 33, 42, 46, 57, 169, 250, 363
  -parametrisation, 45, 321, 363
age of Universe, 270, 279
AH, see apparent horizon
AH$^+$, see future apparent horizon
AH$^-$, see past apparent horizon
algebraic
  -computing, 92, 93
  -type of Weyl tensor, 65, 114
algebraically special metrics, 444
angular
  -momentum, 450, 470, 487
    -conservation of, 152
    orbital-, 470
  -velocity
    -of Earth, 501
    -of rotating fluid, 212
anomaly
  eccentric-, 505
  true-, 505

antilinear
  -index, 109
  -transformation of spinors, 108
antisymmetric
  -part of tensor, 17, 95
  -tensor (density), 17, 20, 24
antisymmetrisation, 17, 18, 22, 23, 44, 92
antisymmetry, 20, 48, 90, 92
apparent
  -horizon, 192, 193, 230, 231, 276, 305–312, 314,
      318, 326, 327, 331, 333, 334, 336, 338, 340,
      357, 363, 365, 418, 419, 422–429
    -in Friedmann models, 308
    -in L–T model, 305
    future-, 193, 230, 306–308, 310–312, 314, 315,
        317, 318, 338, 339, 356, 357, 365
    past-, 230, 277, 307–310, 312, 314, 315, 317,
        318, 356, 357, 365
  -magnitude, 282
approximation, weak-field-, 148, 149
area distance, 240
  observer-, 242, 246
  source-, 241, 277
areal radius, 290, 343, 415
associated
  -function, 18
  -mapping, 18, 19, 69, 76
asymptotically
  -Cartesian coordinates, 133, 155
  -flat spacetime, 133, 488–490
axial symmetry, 487, 488
axisymmetric spacetime, 487

back-reaction, 149
bang-time function, 291
barotropic equation of state, 221, 260, 395, 400
basis
  -in tangent space, 125
  -of Killing vector fields, 77, 78
  -of generators, 94
  -of orthonormal vector fields, 67, 68
  dual-, 27
Beidou navigation system, 499, 506
Bergmann–Wagoner theory, 145
bi-metric Rosen theory, 146
bi-tensor, 36

Bianchi
  -algebra, 106
  -basis, 106
  -classification, 94, 98, 99, 106
  -identities, 44, 55, 92, 129, 235
    contracted-, 158
  -type, 94, 96–99, 103, 106, 367
    -metrics, 255
    -models, variational principle for-, 132
    -solutions of Einstein equations, 220
    -spacetime, 94, 100, 102–104, 141, 142
Big Bang, 192, 262, 278, 280, 283, 284, 292, 293,
    301, 308, 320, 324–326, 331, 334, 342,
    346–348, 351, 352, 354–358, 360, 375–378,
    382, 402, 403, 433
  -function, 341, 342, 410
  cigar-type-, 433
  lagging cores of-, 351, 352
  pointlike-, 433
  position-dependent-, 292
  simultaneous-, 433
Big Crunch, 262, 293, 308, 310, 319, 324, 326,
    331–334, 337, 339, 340, 346, 354, 355, 357,
    358, 375–378, 382, 411, 415, 424, 425, 428
  -function, 410
  prevention of-, 375
Birkhoff theorem, 167, 169
black
  -body
    -radiation, 283
    -spectrum, 283, 327
  -hole, 47, 191, 192, 309–311, 427, 429, 449, 469,
    486, 487
    -formation, 312
    -model, 276
    dynamical-, 230
    Kerr-, 485
    object falling into-, 192
    observed-, 191
blueshift, 322, 352
  infinite-, 276, 322, 324–326
Boltzmann constant, 283
Boyer–Lindquist coordinates, 451–453, 460–462,
    464, 465, 477–480, 484, 491, 497, 498
Brans–Dicke
  -field equations, 144
  -theory, 144

Cartesian
  -coordinates, 1, 77, 78, 86, 87, 211, 225, 398, 450
    asymptotically-, 133, 155
    rectangular-, 45
  -product, 164
  -reference system, 4
  , *see also* coordinates, Cartesian
caustic, 241
centrifugal potential, 503
characteristic equation, 112
charge density, 375
charged dust, 370, 373, 378
  shell crossing in-, 381, 382
Christoffel symbols, 47, 48, 50, 72, 75, 90, 91, 124,
    125, 135, 170, 225, 443, 444

5-dimensional-, 160
  variations of-, 131
class of a Riemann space, 64
classification
  Bianchi-, 94, 98, 99, 106
  Petrov-, 65, 66, 108
closed
  -Universe, 256
  -timelike curve, 453
  -trapped surface, 230, 305
CMB, *see* cosmic microwave background
Codazzi equations, *see* Gauss–Codazzi equations
cold dark matter, 269
collineations, 84
commutator
  -of directional covariant derivatives, 35
  -of second covariant derivatives, 34, 35, 44
  -of vector fields, 36
comoving
  -coordinates, 137, 193, 257, 278, 289, 308, 327,
    331, 332, 336, 341, 391
  -observer, 276, 501
  -synchronous coordinates, 289, 366
conformal
  -Killing fields, 81, 83, 86
    finite basis of-, 81
  -Killing vector, 81, 249, 340
  -curvature, 54
    -tensor, 54
  -diagram, 199–201, 206
  -image, 67
    -of null geodesic, 200
    -of null infinity, 200
  -mapping, 53, 67
  -symmetry, 81, 82, 86, 87, 393
conformally
  -equivalent
    -Riemann spaces, 81
    -metrics, 81
  -flat
    -Riemann space, 53, 54, 64
    -hypersurface, 401
    -metric, 54, 56, 487
    -perfect fluid solutions, 257
    -solution of Einstein's equations, 285
  -related Riemann spaces, 53, 54, 67
connection
  -coefficients, 28, 30, 33, 41, 45, 47, 48
  -forms, 91
  symmetric part of-, 33
conservation
  -equation of radiation, 240
  -of angular momentum, 152
constant, *see* under specific names, e.g.
    gravitational constant
  -curvature, 99
    space of-, 87, 256
  -of motion, 466
  -time hypersurface, 214
  structure-, 76, 86, 87, 94, 95
continuity equation, 139, 155, 220
continuous medium
  -in Newtonian hydrodynamics, 211

-in relativistic hydrodynamics, 213
contraction (of a tensor over indices), 16
contrast
  curvature-, 344
  density-, 344
contravariant
  -components of a vector, 47
  -vector, 13, 24, 27, 37, 40, 47
    -density, 16
    -field, 13, 35
    -vectors, bases of-, 89
convergence of null geodesics, 240
coordinate
  -singularity, 317
  -system, 13, 65
  -transformation, 20, 46, 47, 69
    linear-, 49
  space-, 57
  time-, 57
coordinates, 13
  -adapted to Killing fields, 487
  -spherical, 78
  adapted-, 36, 44, 64, 65, 375
  Boyer–Lindquist-, 451–453
  Cartesian-, 1, 31, 41–43, 77, 78, 86, 87, 211, 398,
    450
    asymptotically-, 133
    locally-, 56
    rectangular-, 49
  comoving-, 137, 193, 257, 278, 289, 308, 327,
    331, 332, 336, 341, 391
  comoving-synchronous-, 289, 366
  curvature-, 162, 165, 184, 193, 289, 362, 378
  geodesic-, 193
  isotropic-, 193
  Lemaître-, 193
  Lemaître–Novikov-, 304, 330, 338, 339, 374, 386,
    390
  Lorentzian-, 154
  mass-curvature-, 378
  Novikov-, 193
  rotating-, 154
  spherical-, 49, 87
  synchronous-, 289
  Szekeres–Szafron-, 399
    interpretation of-, 397
Copernican principle, 223
corrected luminosity distance, 245
cosmic
  -censorship, 336, 340, 449
    -hypothesis, 336, 338, 340, 341
  -microwave background, 224, 261, 278, 280, 281,
    284, 285, 324, 327, 329, 330, 347, 352
    -temperature, 302, 355
    quadrupole anisotropy of-, 355
    temperature anisotropies of-, 280, 302
cosmological
  -constant, 140, 167, 259, 282, 341
    -parameter, 271
    -problem, 281
    value of-, 282
  -model, 276
  -principle, 223, 224

-singularity, 262
Cotton–York tensor, 54, 56, 400
covariant
  -components of a vector, 47
  -derivative, 26, 28, 30, 34, 35, 40, 41, 44, 48, 54,
    60, 91, 225, 232
    directional-, 31, 35, 60, 216
  -differentiation, 26, 34, 37, 40
  -vector, 19, 34, 37, 40, 47, 55
    -field, 13, 19, 30, 34
    -vectors, bases of-, 89
critical density, 261, 278, 281
curvature, 34, 36, 43, 48, 50, 51, 56, 67, 99
  -contrast, 344, 346
  -coordinates, 162, 165, 184, 186, 193, 289, 362,
    378
  -index, 255, 368
  -of symmetry orbit, 165
  -parameter, 271
  -scalar, 92
  -tensor, 34, 35, 43, 44, 48, 51, 101
    conformal-, 54
  conformal-, 54
  constant-, 99
  extrinsic-, 60
  normal-, 51
  quick calculation of-, 89
  Riemann-, 51
  scalar-, 99
  space of constant-, 87
curve
  null-, 57, 59
  spacelike-, 57
  timelike-, 57

dark
  -energy, 262, 269, 281, 347–349
  -matter, 261
    cold-, 269
  -star of Laplace, 192
Datt–Ruban metric, 373, 383, 389, 390, 392, 395,
  409
de Sitter metric, 85, 87, 93, 168, 285
Debever
  -method of Petrov classification, 115
  -spinor, 112–116
  -vector, 116
    degenerate-, 235, 444
  -vector field, 477
deceleration parameter, 259, 263
deflection
  -of light rays, 175
    -in a gravitational field, 177
    measurement of-, 178–180
  -of microwaves, 180
density
  -contrast, 297, 344
  -parameter, 271, 330
  -perturbations
    decreasing-, 344
    increasing-, 344
  -waves, 365
  mean mass-, 269

derivative
covariant-, 26, 28, 30, 34, 35, 40, 41, 44, 48, 54, 60, 91, 225, 232
  directional-, 35, 60, 75, 216
directional-, 13, 15, 19, 65, 77, 91
exterior-, 90, 93
Lie-, 75, 101
partial-, 91
deviation
geodesic-, 249, 335, 341
-equation, 243, 244, 249
-operator, 249
differentiable
-manifold, 12, 18, 20
-vector field, 70
differential
-form, 89, 90, 92, 93
perfect-, 31
differentiation, covariant-, 26, 34, 37, 40
dilatation, 86
dilation, time-, 504
dipole
-axis, 417
mass-, 409, 416, 417, 423
-moment, 151
direct orbit, 171, 470, 471
directional
-covariant derivative, 60, 75, 216
-derivative, 65, 77, 91
distance
-in cosmology, 240
area-, 240
corrected luminosity-, 245
luminosity-, 272, 347
observer area-, 246, 263, 347
source area-, 277, 282
divergence of a vector field, 24
Doppler effect, transverse-, 504, 507
double-null tetrad, 92, 231, 232
dragging inertial frames, 154, 453
drift
-free cosmological model, 249
-of light rays, 349, 350
position-, 248, 249, 253
redshift-, 253
Droste solution, 168
dual
-basis, 27
-electromagnetic field tensor, 157
-field of basis vectors, 89
duality, 158
-rotation, 157, 158, 166
dummy indices, 16
dust, 137, 139
charged-, 370

Earth
-surface, idealised, 503
equatorial radius of-, 502
gravitational field of-, 502
gravitational potential of-, 502
quadrupole moment of-, 499, 503, 504
eccentric anomaly, 505

eccentricity of orbit, 172
ECEF = Earth-centered, Earth-fixed reference system, 500, 501, 503
ECI = Earth-centered inertial reference system, 500–502, 504
effective
-mass, 372, 377
-potential, 503
EH, *see* event horizon
EH$^+$, *see* future event horizon
EH$^-$, *see* past event horizon
Ehlers–Kundt method of Petrov classification, 115
eigenvalue, 65
Einstein
–Cartan theory, 145
–Maxwell equations, 158, 159, 161, 166, 370, 373, 456, 458
  solutions of-, 167
  variational principle for-, 159
-Universe, 140, 261, 268–270, 285, 361
-cross, 182
-equations, 128, 129, 136, 217, 220, 235, 257, 280, 289, 290, 373, 391, 401, 458, 511, 519
  -for Robertson–Walker metric, 260
  -in 5 dimensions, 161
  -with cosmological constant, 140
  -with electromagnetic source, 160
  Hilbert's derivation of-, 129
  integration of-, 361
  linearised solution of-, 502
  Newtonian limit of-, 133, 136
  sources in-, 137
  vacuum-, 443
-tensor, 129, 391
  -for a Robertson–Walker metric, 260, 287
-theory, linearised-, 151
electric
-field, 65, 370
-part of Weyl tensor, 65, 217
electromagnetic
-energy-momentum tensor, 158, 226
-field, 225
  dual tensor of-, 157
  energy density of-, 157
  energy stream of-, 157
  energy-momentum of-, 157
  Lorentz-covariant description of-, 156
  stress tensor of-, 157
-field tensor, 156, 158, 370, 457
  -in Reissner–Nordström metric, 207
-wave, 225, 227, 500
electrovacuum metric, 374
ellipsoidal spacetime, 490–492
elliptic orbit, 2
embedding, 58, 60, 64, 80, 202, 207
  -of Schwarzschild spacetime in flat Riemann space, 187
energy
-conservation equation, 139
-density of fluid, 137
-flux, 241
-momentum tensor, 361
  -of electromagnetic field, 158, 166

-of perfect fluid, 137, 138
-of radiation, 240
electromagnetic-, 226
internal-, 220
enthalpy, 220
-density, 220
entropy, 220, 221
equation, *see also* under specific names, e.g.
Einstein equations
-of continuity, 139, 155, 220
-of geodesic deviation, 249
-of hydrostatic equilibrium, 194, 196
-of motion of charged particle, 207
-of state, 220, 221, 260, 400
barotropic-, 221, 260, 395, 400
equations of motion, 260
-of perfect fluid, 139
Euler-, 139
Newton's-, 134
equivalence problem, 66
ergosphere, 486, 487
Ernst potential, 512
Euclidean
-geometry, 124
-metric, 99
-plane, 50, 397
-space, 31, 32, 45, 49, 50, 56, 60, 77, 78, 85
Euler
−Lagrange equations, 51
-of geodesic, 464, 465
-equations of motion, 139, 154
event horizon, 192, 193, 200, 206, 230, 231,
273–277, 306, 311, 312, 314–316, 338, 339,
357, 361, 387, 388, 477, 482, 486, 487
-in Kerr metric, 459–462, 464, 473
-in Schwarzschild metric, 192, 404, 429, 460
future-, 310, 312, 314
proper time to reach-, 193
evolution
-equations of Szekeres models, 409
-of hydrodynamical tensors, 216
expansion
-of Universe, 256, 266, 269, 270, 272
accelerating-, 269, 270, 272
-of null curves, 229, 230
-of ray bundle, 254
-pattern, 369
-scalar, 212, 215, 218, 241
-of null curves, 230, 232
-of ray bundle, 254, 276
-velocity of Universe, 257
isotropic-, 212
exterior
-derivative, 90, 93
-product, 92
extreme R–N metric, maximal extension of-, 206
extrinsic curvature, 60

Fermi
−Walker derivative, 248, 253
−Walker transport, 246–248
-coordinates, 125, 126
field

electric-, 65
magnetic-, 65
vector-, 67
final singularity, 274, 275
flat
-Riemann space, 49, 50, 53, 64, 84, 85, 127
-Universe, 256
-manifold, 40–43
-metric, 64
-space, 50, 225
-spacetime, 230
asymptotically-, 133, 488–490
flatness problem, 277, 281
FLRW model, 257
see also Robertson–Walker model, 526
fluid, perfect-, 391
flux
-of energy, 241
-of radiation, 246
observed-, 240, 241
specific-, 246
focussing by curvature, 241
force, inertial-, 124
formation of structure in Universe, 297
fractal distribution of matter, 360
frame dragging, 154, 453
free motion, trajectory of-, 124
frequency shift, gravitational-, 504
Friedmann
-Lemaître models, 224, 257
-Universe, flat-, 278
-equation, 260, 266, 402
-limit, 368, 402
-of L–T model, 292, 304, 324, 361
-of Szekeres metric, 417
-model, 260, 262, 263, 265, 269, 270, 274, 276,
277, 294, 296, 312, 324, 326, 328, 360–363,
365
-matching to Schwarzschild metric, 309
-recollapsing, 268
-with $\Lambda = 0$, 262, 273
-with Λ, 266
instability of-, 296, 297
function, associated-, 18
future, 57, 58
-apparent horizon, 230, 306–308, 310–312, 314,
315, 317, 318, 338, 339, 356, 357, 365
-event horizon, 310, 312, 314
-infinity, 356
-light cone, 326
-null infinity, 338, 339
-trapped surface, 230

GAIA mission, 350
Galileo Navigation System, 499, 506
gamma-ray bursts, 352
Gauss
−Codazzi equations, 63, 64, 147
-curvature, 51
-law, 152
general relativity, Newtonian limit of-, 127
generator
-basis, 94

-of conformal symmetry, 81
-of group, 71, 99
-of invariance, 72, 76
-of rotation, 78
-of symmetry, 84, 86, 94, 104
-of transformation, 74, 77
-of translation, 99
generators
  -of $O(3)$, 77, 99
  -of symmetries
    finite basis of-, 82
geodesic
  -completeness, 482
  -coordinates, 193
  -deviation, 42, 74, 243, 249, 335, 341
    -equation, 43, 74, 243, 244, 249, 251, 252
    -operator, 249
    -vector, 249, 251
  -equation, 32, 33, 52, 67
  -equations in Kerr metric, 468
  -line, 31–33, 42, 43, 47, 51, 52, 57, 67, 124
  -motion of a perfect fluid, 139
  -vector field, 235
  null-, 57, 249, 305, 312, 314, 317, 318, 320, 321,
    328, 339, 349
    -arc, 57
    -equation, 347, 349
    -in Schwarzschild metric, 175
    radial-, 305, 321, 341, 342
  timelike-, 124, 125, 169
geodesically complete manifold, 482
geodesics
  -in Kerr metric, 468
  null-, convergence of-, 240
geoid, 503, 504
geometric optics, 224, 225
geometry
  Euclidean-, 124
  Riemann-, 123, 124
Gibbs identity, 220
Global Positioning System, 499, 500, 504, 506–509
globally naked
  -shell focussing, 339
  -singularity, 336
GLONASS = Global Navigation Satellite System,
  499, 506
GNSS = Global Navigation Satellite System, 499
Goldberg–Sachs theorem, 235, 444, 477
Goode–Wainwright representation of Szekeres
  metrics, 287, 429, 430, 433
graceful exit problem, 281
gravitation theory of Newton, 133
gravitational
  -deflection of light rays, 177
  -field, 2, 4, 5, 12, 42, 54
    -of Earth, 502
    homogeneous-, 123, 124
    spherically symmetric-, 64
  -force, 4, 5, 42
  -frequency shift, 504
  -interaction, 2
  -lens, 178, 180
    equation of-, 181

example of-, 182, 183
-mass
  -defect, 373, 385
  active-, 385, 410
-potential, 127, 503
  -of Earth, 502
-radius, 191
-redshift, 507
Great Attractor, 330
group
  -action, 99
  -of mappings, 99
  -of symmetries, 100
  -orbit, 100
  -parameter, 77, 86
  Abelian-, 87
  continuous-, of mappings, 69
  isometry-, 99
  multiply transitive action of-, 100
  simply transitive action of-, 100
  transitive action of-, 100

Haantjes transformation, 86, 87, 393
Hamilton
  –Cayley equation, 113
  -equations of geodesic, 464, 465
Hamiltonian for geodesic equations, 464–467
Hilbert's
  -axioms of gravitation theory, 130
  -derivation of Einstein equations, 129
  -variational principle, 132
history of Universe, 260
homoeoid, 492
homogeneity of Universe, 360
homogeneous
  -gravitational field, 123, 124
  -hypersurface, 102
  -space, 100
  spatially-, spacetime, 100, 103
horizon, 47, 273, 276, 336
  -problem, 277–281, 341–343
  absolute apparent-, 418–429, 521, 522
  apparent-, 192, 193, 230, 231, 276, 305–312, 314,
    318, 326, 327, 336, 338, 340, 357, 363, 365,
    418, 419, 422–429
    -in Friedmann models, 308
    -in L–T model, 305
  future-, 193, 230, 306–308, 310–312, 314, 315,
    317, 318, 338, 339, 356, 357, 365
  past-, 230, 277, 307–310, 312, 314, 315, 317,
    318, 356, 357, 365
  event-, 193, 230, 231, 273–277, 306, 311, 312,
    314–316, 338, 339, 357, 361, 387, 388, 486,
    487
    -in Schwarzschild metric, 192, 404, 429
  future-, 310, 312, 314
  particle-, 273, 276
  Schwarzschild-, 306
Hubble
  -constant, 261, 360
  -law, 258, 261, 265, 302, 360
  -parameter, 258, 259, 263, 271, 272, 282, 304,
    330, 355, 359

hydrodynamical tensors, 216
hydrostatic equilibrium, equation of-, 194, 196
hyperbolic symmetry, 367
hyperbolically symmetric
  -metric, 367, 368
  -spacetime, 367
hypersurface, 13
  -of constant time, 214
  -of simultaneous events, 214
  homogeneous-, 102
  null-, 99
  spacelike-, 99
  stationary limit-, 486, 487
  timelike-, 99

ideal observations, 361
identity
  -transformation, 75
  Gibbs-, 220
image, conformal-, 67
indefinite signature, 85
inertial
  -coordinates at infinity, 154
  -force, 124
  -frame
    local-, 123, 124, 500, 502
infinite
  -blueshift, 276
  -redshift, 276, 452
    -hypersurface, 193, 452
infinity
  future null-, 338, 339
  future-, 356
  null-, 311
  past-, 356
inflation, 280, 281, 284
inflationary models, 277, 278, 280, 281, 341
inheritance of symmetry, 84, 165, 361
initial singularity, 276
integrating factor, 220
intensity
  -of radiation, 246
  specific-, 246
interior Schwarzschild solution, 196, 197
internal energy, 220
International Space Station, 506
interval
  null-, 56
  spacelike-, 56
  timelike-, 56
invariance
  -group, 71, 83
  -of tensors, 69, 73, 75
  -transformation, 69, 71, 72, 74
invariant
  -tensor field, 69
  -vector field, 100–102, 106
ionisation temperature, 280, 283
isentropic motion, 221
isometry, 86
  -group, 77, 82, 99
  -of Riemann space, 69
isomorphic algebras, 94

isomorphism of manifold subsets, 69
isotropic
  -coordinates, 193
  -expansion, 212
  -metrics, 255
  -spacetime, 103

Jacobi
  -identity, 44, 95
  -matrix, 21, 250, 251
Jacobian, 16

Kaluza–Klein theory, 159–161
Kantowski–Sachs
  -geometry, 385, 400
  -metric, 104, 388, 409
    vacuum-, 184
  -symmetry, 80, 169, 385
Kerr
  –Newman metric, 459
  –Schild
    -form of Schwarzschild metric, 494
    -metric, 442, 444, 445, 448, 449, 477
    -vector field, 444, 477, 485
  -black hole, 485
  -manifold, maximally extended, 462, 464, 482
  -metric, 187, 442, 444, 449, 459–462, 477, 481,
      484, 487, 490–493, 511
    -derivation from separability of Klein–Gordon
        equation, 454
    -fourth first integral of geodesic equations, 465
    -in Boyer–Lindquist coordinates, 451–453
    generalisation of-, 459
    inverse-, 477
    maximal extension of-, 475, 476
    singularities of-, 453, 454
  -spacetime, 461–463
Killing
  -equations, 69, 71–75, 77, 78, 85, 99, 102–104,
      367
  -field, conformal-, 81, 83, 86
  -vector, 77, 78
    conformal-, 81, 249, 340
  -vector field, 71–74, 76, 80, 86, 87, 101–104, 159,
      366, 367, 385, 447, 448, 450, 465, 487–489
    -as geodesic deviation field, 74
    -for Kerr metric, 465
  -vector fields
    algebra of-, 76
    basis of-, 73, 74, 76–78
kinematical relativity, 266
Klein–Gordon equation, 454, 456
Kottler solution, 168, 361
Kronecker delta, 21, 26
  multidimensional-, 21, 22, 24, 26, 44
Kruskal
  –Szekeres
    -coordinates, 187
    -representation of Schwarzschild metric, 187
    -throat, 317
  -diagram, 186, 192, 199, 230, 357, 464, 478, 480

L–T, *see* Lemaître–Tolman model

lagging cores
  -of Big Bang, 351, 352
  -of expansion, 192
Lagrangian for geodesic equations, 466
Laplace
  -equation, 492
  dark star of-, 192
last scattering, 278, 279, 324, 341
  -hypersurface, 278
Lemaître
  −Novikov coordinates, 304, 330, 338, 339, 374,
     386, 390, 427
  −Tolman limit of Szekeres metrics, 406
  −Tolman model, 194, 276, 277, 291, 292, 305,
     392, 397
    -in observational coordinates, 294
    -matching to Friedmann metric, 330
    -matching to Schwarzschild metric, 330, 338,
     340
    apparent horizon in-, 305
    mass density in-, 314
    vacuum limit of-, 294
  -coordinates, 193
lens, gravitational-, 180
  equation of-, 181
  example of-, 182, 183
Lense–Thirring effect, 154
Levi-Civita symbol, 20–22, 24, 26, 54, 83, 95,
    108–111, 120
Lie
  -algebra, 76
  -derivative, 75, 77, 101
    -of general tensor field, 75
  -transport, 101, 246
light
  -cone, 57–59, 278, 308, 360, 462
    future-, 326, 462
    past-, 278, 312, 347, 348, 359–361
  -paths, repeatable-, 249
  -rays, drift of-, 349, 350
limiting focussing condition, 340, 341
linear transformation of spinors, 108
linearised
  -Einstein theory, 151
  -theory of gravitation, 149
local
  -inertial frame, 123, 124, 500, 502
  -rotational symmetry, 401
locally
  -naked singularity, 336
  -nonrotating observer, 487, 489, 490
Lorentz
  -force, 371
  -transformation, 85, 86, 89
Lorentzian
  -coordinates, 154
  -metric, 46
  -signature, 65
lowering the index, 47
luminosity
  -distance, 246, 272, 347
    -redshift relation, 272
    corrected-, 245

absolute-, 282, 347
total-, 241

Mach principle, 1–4, 145
magnetic
  -charge, 166, 370
  -field, 65, 370
  -monopole, 158, 166, 370
  -part of Weyl tensor, 65, 217, 399
magnitude, apparent-, 282
manifold, 13, 18
  differentiable-, 12, 18, 20
map, geographic-, 52
mapping, 12, 13, 18–20, 52, 69, 86
  -of Riemann spaces, 52
  associated-, 18, 19, 69, 76
  conformal-, 53, 67
  inverse-, 47, 52, 53
  Mercator-, 67
mappings
  continuous group of-, 69
  group of-, 99
mass
  -conservation equation, 261
  -curvature coordinates, 378
  -defect, 195
    gravitational-, 373, 385
    relativistic-, 292, 369
  -density, 343
    -of charged dust, 373
    mean-, 269
  -dipole, 409, 416, 417, 423
    -moment, 151
  active gravitational-, 291, 292, 369, 410
  effective-, 372, 377
matching solutions of Einstein's equations, 146
Mattig formula, 264
maximal extension
  -of Kerr metric, 475, 476
  -of R–N metric, 197, 201, 202
  -of extreme R–N metric, 206
Maxwell equations, 156, 158, 159, 166, 224, 225,
    279, 458, 495
  -from Kaluza–Klein theory, 160
  Lorentz covariance of-, 156
  variational principle for-, 158
mean mass density, 269
Mercator mapping, 67
Mercury, orbital parameters of-, 174
metric, 13, 45, 47–49, 52–54, 56, 58, 65, 80
  -form, 45, 49, 50, 60, 80, 86
    spherically symmetric-, 80
  -tensor, 45–47, 49, 50, 52, 53, 58, 67, 69, 72–74,
    78, 81, 89, 110
    limit of-, 84
  3-dimensional-, 54
  conformally flat-, 54, 56
  degenerate-, 46, 47
  Euclidean-, 52, 99
  flat-, 64
  induced-, 49
  Minkowski-, 56, 87
  positive-definite-, 46, 47

spherically symmetric-, 80, 103
Milne–McCrea equation, 266
minimal equation, 65, 114
Minkowski
  -metric, 56, 87, 109, 500, 501
  -spacetime, 57, 86, 87, 109, 116
moment, quadrupole-, -of Earth, 499, 503, 504
momenta, 464, 466
momentum
  angular-, 450, 470, 487
  orbital angular-, 470
motion, free-, trajectory of-, 124

naked
  -shell crossing, 338, 339
  -shell focussing, globally-, 339
  -singularity, 336, 338, 341, 469
    globally-, 336
    locally-, 336
    strength of-, 340
neck, 306, 307, 314, 317, 318, 356–358, 373, 521,
    see also wormhole
Newman–Penrose formalism, 232
Newton's
  -equations of motion, 134
  -gravitation theory, 133, 154
Newtonian
  -cosmology, 264
  -limit
    -of Einstein equations, 133, 136
    -of general relativity, 127
nonrotating observer, locally-, 487, 489, 490
Nordström solution, *see* Reissner–Nordström
    solution
normal
  -curvature, 51
  -section, 51
Novikov coordinates, 193
null
  -Raychaudhuri equation, 240
  -congruence, 241
  -curve, 57, 59, 228
  -geodesic, 57, 200, 227, 249, 273, 305, 312, 314,
      317, 318, 320, 321, 328, 339, 349
    -arc, 57
    -equation, 347, 349
    -in Schwarzschild metric, 175
    future-directed-, 200
    radial-, 200, 305, 321, 341, 342
  -geodesics
    bundle of-, 241, 276
    convergence of-, 240
  -hypersurface, 99
  -infinity, 200, 311, 312
    future-, 338, 339
  -interval, 56
  -orbit of a group, 99
  -relation, 57
  -tetrad, 92
  -vector, 57, 226, 228
    spinor image of-, 111
number
  -counts, 246

-density of particles, 219

observation time vector, 251
observational cosmology programme, 294
observer
  -area distance, 242, 246, 263, 347
  comoving-, 276, 501
  locally nonrotating-, 487, 489, 490
  stationary-, 453
open Universe, 256
operator of geodesic deviation, 249
optical
  -observations, 257
  -scalars, equations of propagation, 233
  -tensors, 228
optics, geometric-, 224, 225
orbit
  -of Mercury, parameters of-, 174
  -of family of mappings, 99
    dimension of-, 99
  -of group, 99, 100
    null-, 99
    spacelike-, 100
  -of planet, 2, 3, 169
  -of point, 70–72, 100
  -of symmetry group, 221, 304
  -parameter, 172
  direct-, 171, 470, 471
  elliptic-, 2
  group-, 75, 80
  parameter on-, 75
  planetary, influence of cosmic expansion on-, 303
  retrograde-, 171, 470, 471, 476, 486
orbital angular momentum, 470
orbits of $O(3)$, 77, 80
origin in Szekeres spacetimes, 409
orthogonal
  -transformations, 89
  -transitivity, 487, 488
orthonormal
  -basis of vector fields, 67, 68
  -tetrad, 92, 93

Palatini variational principle, 133
parallax free cosmological model, 249
parallel transport, 10, 11, 31–34, 36, 37, 40–42, 45,
    46, 101, 125, 247
  -along a closed curve, 32, 40
  propagator of-, 36–38, 41, 42
parameter
  -of orbit, 172
  affine-, 57, 363
  density-, 330
  group-, 77
parametrisation, affine-, 321, 363
Parametrised Post-Newtonian formalism, 136, 145
partial derivative, 91
particle horizon, 273, 276
Pascal law, 138
past, 57
  -apparent horizon, 230, 277, 307–310, 312, 314,
      315, 317, 318, 356, 357, 365
  -infinity, 356

-light cone, 278, 312, 347, 348, 359–361
-trapped surface, 230, 277
Pauli matrices, 109, 110, 116
Penrose
-diagram, 200, 388
-method of Petrov classification, 115
-process, 485
-transformation, 199, 200, 312
perfect
-differential, 31, 220
-fluid, 137, 217–219, 221, 391, 401
energy-momentum tensor of-, 137, 138
equations of motion of-, 139
geodesic motion of-, 139
single-component-, 221, 400
perihelion shift
-of Mercury, 174
-of other planets, 174
Petrov
-classification, 65, 66, 108, 116
-by Debever method, 115
-by Ehlers–Kundt method, 115
-by Penrose method, 115
-vs. Penrose's classes, 114
spinor representation of-, 111, 113
-type, 66, 68, 112, 116, 235, 399, 444
phenomenological thermodynamics, 220
photon orbit in Kerr metric, 472
Planck constant, 283
plane
-symmetric
-metric, 286, 367, 368
-spacetime, 366, 368
-symmetry, 286, 367
Euclidean-, 397
planetary orbits, influence of cosmic expansion on-, 303
Poisson
-bracket, 465, 466
-equation, 127, 129, 136
modified-, 140
solution of-, 141
position drift, 248, 249, 253
post-Newtonian approximation, 151
potential
centrifugal-, 503
effective-, 503
Ernst-, 512
gravitational-, 503
-of Earth, 502
PPN formalism, *see* Parametrised Post-Newtonian formalism
precession of gyroscope axis, 154
principal
-normal to a curve, 246, 247
-spinor of the Weyl tensor, 112
problem
cosmological constant-, 281
flatness-, 277, 281
graceful exit-, 281
horizon-, 277–281
projection, stereographic-, 397, 399, 420
projective plane, 413

propagation
-equation
-of shear, 217
-of vorticity, 217
-equations of optical scalars, 233
propagator
-of parallel transport, 36, 38, 41, 42
-along a closed curve, 37
elementary-, 37
pseudo-Riemannian manifold, 46
pullback, 18, 19, 52, 53
pushforward, 19

quadratic form, 46
quadrupole moment of Earth, 499, 503, 504
quasi
-hyperbolic Szekeres metric, 406
-plane Szekeres metric, 406, 413
-spherical Szekeres metric, 404–406

R–W metric, *see* Robertson–Walker metric
R-region, 164, 360, 361
radial null geodesic, 305, 321, 341, 342
radiation
-conservation equation, 240
-dominated Universe, 279
-intensity, 246
energy-momentum tensor of-, 240
flux of-, 240, 241, 246
radius, areal-, 290, 415
raising the index, 47
Raychaudhuri equation, 216–218, 233, 433
null-, 240, 446
reciprocity theorem, 246
recombination, 278
redshift, 226, 227, 240, 241, 246, 254, 263, 272, 282, 320, 321, 324, 358, 359
-distance relation, 258, 263, 271
-drift, 253, 272
-in R–W universe, 257
-space, 358, 359
gravitational-, 507
infinite-, 276, 324, 325, 452
-hypersurface, 193, 452
reflection (of coordinates), 171
refocussing, 241
-in Friedmann models, 264
regularity conditions at centre, 295, 380
Reissner–Nordström
-horizon, 377
-manifold, 387
-metric, 166, 168, 187, 197, 361, 374, 375, 377, 386, 388, 390, 459, 470, 476, 482, 485
electromagnetic tensor in-, 207
extension of-, 201, 202
motion of particles in-, 205
throat in-, 203
relativistic
-mass defect, 292, 369
-thermodynamics, 219
relativity, special-, 56, 86
repeatable light paths, 249
retrograde orbit, 171, 470, 471, 476, 486

Ricci
  -formula, 35, 60, 73, 81, 82, 93, 233
  -identities, 216
  -rotation coefficients, 90, 91, 231, 232, 236
  -tensor, 54, 92, 128, 129
    -for spherically symmetric metric, 167
    eigenvalues of-, 400, 401
    five-dimensional-, 160
Riemann
  -curvature, 51
  -geometry, 50, 123, 124
  -manifold, 73
  -space, 45–48, 51–54, 58, 60, 64, 69, 72, 78, 80,
      81, 83–87, 92, 124
    conformally flat-, 53, 54, 64
    conformally related-, 53, 54, 67, 81
    flat-, 49, 50, 53, 64, 74, 84, 85, 127
    signature of-, 127
    spherically symmetric-, 77
    subspace of-, 49
    symmetry of-, 69
  -surface, 462
  -tensor, 48, 50, 54, 60, 62, 73, 74, 83, 85, 91, 93,
      232, 314, 444
    -of Kerr metric, 453
    -of Szekeres metric, 410
    linearised-, 149
    scalar polynomial invariants of-, 410
    symmetries of-, 243
Robertson–Walker
  -geometry, 256, 261, 400
  -limit of Szafron metric, 399, 400
  -limit of Szekeres metric, 397
  -metric, 105, 192, 218, 220, 221, 224, 255–258,
      260, 264, 265, 271–274, 276–278, 280,
      284–287, 294, 302, 324, 355, 395, 400
    -invariant definition of-, 285, 288
    flat-, 286
    perturbations of-, 344
Rosen bi-metric theory, 146
rotating coordinates, 154
rotation, 13, 24, 366, 367
  -of null curves, 229, 230
  -of null vector field, 225
  -of ray bundle, 254
  -of vector field, 289
  -scalar, 212, 216, 232
  -tensor, 212, 215, 216
    tetrad components of-, 232
  -vector, 216, 217
  Ricci-, -coefficients, 90, 91, 231, 232, 236
rotational symmetry, local-, 401
Ruban metric, 373, 383–389
  charged-, 386, 389

Sagnac effect, 500, 502
scalar, 14, 15, 22, 23, 26, 27
  -curvature, 92, 99
    five-dimensional-, 160
  -density, 16, 27, 34, 35, 37, 40
  -field, 27, 28, 34, 37, 280
    -of Kaluza–Klein, 160
  -polynomial invariants of Riemann tensor, 410

scale factor, 218, 255, 268
Schwarzschild
  -Droste solution, 168
  -event horizon, 404, 429
  -horizon, 306, 309
  -limit of L–T model, 306
  -metric, 110, 166, 168, 182, 187, 188, 192, 193,
      195, 231, 277, 294, 309, 317, 330, 338, 339,
      357, 361, 386, 390, 452–454, 460, 493
    -as vacuum limit of L–T, 330
    -embedding in 6 dimensions, 189
    -in Novikov coordinates, 193, 194
    -matched to Szekeres metric, 427
    -matching to Friedmann metric, 303
    -matching to L–T, 338, 340
    embedding of-, 188
    generalisations of-, 193
    incompleteness of-, 187
    interior-, 196, 197
    material source for-, 196
    maximally extended-, 186, 189
    null geodesics in-, 175
    spurious singularity of-, 169, 182, 183, 194
    throat in-, 203
  -singularity, 339
screen space, 250, 253
second fundamental form, 60, 61, 64, 67, 375, 386
section, normal-, 51
self-dual spin-tensor, 111
self-similar
  -L–T model, 340
  -spacetime, 100, 340
shear, 213, 241, 417
  -in Szafron metric, 400
  -of null curves, 229, 230
  -of ray bundle, 254
  -propagation equation, 217
  -scalar, 213, 216
  -tensor, 213, 215
    eigenvalues of-, 401
    tetrad components of-, 231
shearfree vector field, 235
shearing motion, 213
shell crossing, 292, 299, 300, 306, 307, 314,
      316–320, 332, 334–343, 353, 355, 357, 359,
      363–365, 377, 378, 383, 386–389, 402, 403,
      406, 409, 412, 413, 415, 433
  -in charged dust, 381, 382
  avoiding of-, in Szekeres models, 414
  naked-, 338, 339
  pancake type-, 433
shell focussing, 331, 332, 340, 341
  globally naked-, 339
SI = Système International, 500, 502
  time unit in-, 503
signature, 46, 47, 56, 65, 92
  -of Riemann space, 127
  indefinite-, 85
  Lorentzian-, 65
simultaneous events, hypersurface of-, 214
single-component perfect fluid, 400
singularity, 218, 219, 230, 269, 276, 277, 310, 314,
      331, 332, 336, 338, 340

-of Kerr metric, 453, 454
-of Schwarzschild metric
    spurious-, 169, 182, 183
-theorems, 218, 219
coordinate-, 317
cosmological-, 262
final-, 274, 275
genuine-, 183, 185
globally naked-, 336
initial-, 276
locally naked-, 336
naked-, 336, 338, 341, 469
    strength of-, 340
permanent timelike-, 351
spurious-, 183, 193, 452
true-, 183
weak-, 316
source area distance, 241, 277, 282, 290
space
-coordinate, 57
-of constant curvature, 85, 87, 256
Euclidean-, 60
Riemann-, 52–54, 58, 60
    conformally flat-, 54
    conformally related-, 53, 54
tangent-, 12, 27, 53
Space Shuttle, 506
spacelike
-curve, 57
-hypersurface, 99
-interval, 56
-orbit of a group, 100
-relation, 57
-vector, 57
spacetime, 56, 57
-of Bianchi type, 100, 102–104
acausal-, 202
asymptotically flat-, 488–490
axisymmetric-, 487
ellipsoidal-, 490–492
isotropic-, 103
Minkowski-, 57, 86, 87
spatially homogeneous-, 94, 100, 103, 132, 141,
    194, 255, 385
static-, 487
stationary-, 487, 488
stationary–axisymmetric-, 487, 488, 490, 491
spatial distribution of matter, 358
spatially flat Universe, 269
special relativity, 57, 86
specific
-flux, 246
-intensity, 246
spherical
-coordinates, 78, 87
    generalised-, 78
-symmetry, 367
spherically symmetric
-Riemann space, 77
-gravitational field, 64
-inhomogeneous models, 289
-metric, 80, 103, 360, 367
spin-tensor, 110, 111, 117

self-dual-, 111
spinor, 108, 109
-density, 109
-image
    -of Weyl tensor, 111, 115, 116
    -of null vector, 111
    -of tensor, 109, 115, 120
-index, 109, 116
-representation of Petrov classification, 111, 113
antilinear transformation of-, 108
covariant-, 108
Debever-, 112–116
linear transformation of-, 108
principal-, of Weyl tensor, 112
spurious singularity, 193, 200
-of Kerr metric, 452
-of Reissner–Nordström metric, 205
-of Schwarzschild metric, 169, 452
standard candle, 282
state, equation of-, 220
barotropic-, 221
static spacetime, 487
stationary
-axisymmetric spacetime, 487, 488, 490, 491, 512
-limit hypersurface, 453, 459–462, 486, 487
-observer, 453
-spacetime, 487, 488
Stephani universe, 221, 285
stereographic projection, 397, 399, 420
stiff fluid, 351
stress tensor, 138
structure
-constants, 76, 86, 87, 94, 95, 106
-formation in the Universe, 297
Sun, gravitational field of-, 169
supernovae, 270, 282, 347, 349
surface
-area propagation equation, 240
-forming vector fields, 76, 77, 487
closed trapped-, 230
future-trapped-, 230
past-trapped-, 230, 277
trapped-, 230
Sylvester theorem on inertia of forms, 46
symmetric
-part of tensor, 17, 95
-spinor, 17
-tensor, 17, 65, 67
symmetrisation, 17
symmetry, 84, 90
G_3/S_2-, 401
-generator, 84, 94, 104
-group, 74, 76, 83–85, 100, 221
    dimension of-, 84
-inheritance, 84, 165, 361
-of Riemann space, 69, 72
-of manifold, 73
-transformation, 69
axial-, 487, 488
conformal-, 81, 82, 87, 393
hyperbolic-, 367
Kantowski–Sachs-, 80
local rotational-, 401

plane-, 367
spherical-, 367
synchronous coordinates, 289
Szafron metric, 221, 400
    invariant definition of-, 401
    shear in-, 400
    Weyl tensor of-, 399
Szekeres
    —Szafron coordinates, 399
        -interpretation of, 397
    —Szafron metric, 391, 401
        invariant definitions of-, 401
    -metric, 402, 406
        -in Goode–Wainwright representation, 429,
            430, 433
        -matched to Schwarzschild metric, 427
        class II-, as limit of class I, 404
        evolution equations of-, 409
        Goode–Wainwright representation of-, 287
        invariant definitions of-, 401
        origin in-, 409
        quasi-hyperbolic-, 406
        quasi-plane-, 406, 413
        quasi-spherical-, 404–406
        Riemann tensor of-, 410

T-region, 164, 360, 361, 384
tangent
    -space, 12, 13, 27, 53
        basis in-, 125
    -vector, 32, 33, 57
temperature
    -contrast, 328, 330
    -of fluid, 220
tensor, *see also* under specific names, e.g. Weyl
        tensor
    -density, 16
    -field, invariance of-, 69
    -product, 17, 26, 35–37, 75
    associated mapping of-, 19
    differentiation of-, 24
    parallel transport of-, 32, 36
test field, 225
tetrad
    -image, 91
    -metric, 89
    -of vector fields, 89
    double-null-, 92, 231, 232
    null-, 92
    orthonormal-, 92, 93
theorem, *see* under specific names, e.g. the
        reciprocity theorem
thermodynamical scheme, 220, 221, 400
    existence of-, 221
thermodynamics, 220, 221
    phenomenological-, 220
    relativistic-, 219
time
    -coordinate, 57
    -dilation, 504
    -unit in SI, 503
    hypersurface of constant-, 214
timelike

-curve, 57
    closed-, 453
-geodesic, 124, 125, 169
-hypersurface, 99
-interval, 56
-relation, 57
-vector, 57, 65
    -field, 66, 289
toroidal topology, 369
torsion, 28, 30, 46
torsion-free manifold, 34, 41–43, 48
torus, 50, 369
trace, 15, 16, 54
trajectory of free motion, 124
transformation
    acceleration-, 86
    coordinate-, 69
    symmetry-, 69
transitive
    -action of group, 100
    multiply-, action of group, 100
    simply-, -action of group, 100
transport
    Fermi–Walker-, 246–248
    parallel-, 125, 247
transverse Doppler effect, 504, 507
trapped surface, 230
    closed-, 230, 305
    future-, 230
    past-, 230, 277
true anomaly, 505
turning point
    -in Kerr spacetime, 486
    -in R–N spacetime, 208
type
    Bianchi-, 94, 96–99, 103
    Petrov-, 66, 68, 116

Universe
    closed-, 256
    flat-, 256
    open-, 256

variational principle
    -of Hilbert, 132
    -of Palatini, 133
vector, 13
    -field, 13, 19, 59, 67
        -timelike, 66
        differentiable-, 70
        tangent-, 43
        timelike-, 289
    -fields, surface-forming-, 76, 77
    -space, 12, 13, 15, 19
    contravariant-, 13–15, 18, 24, 27, 35, 37, 40, 47
    covariant-, 13–15, 19, 30, 34, 37, 40, 47, 55
    geodesic deviation-, 249
    null-, 57
    observation time-, 251
    spacelike-, 57
    tangent-, 12–14, 32, 33, 57
    timelike-, 57, 65
Virgo cluster, 330

void, 296, 297, 349, 350, 438
  -evolution, 297
  -formation, 297
  -models of acceleration, 349
vorticity propagation equation, 217

wave vector, 226
weak singularity, 316
weak-field approximation, 148, 149, 449
weight of tensor density, 16, 20–22, 24, 26, 27, 34
Weyl
  -spinor, 112

-tensor, 54, 56, 65–67, 111, 112, 120, 129, 218, 232, 444
  -of Kerr metric, 453
  -of Szafron metric, 399, 401
  algebraic type of-, 65
  algebraically special-, 233, 235
  electric part of-, 65, 217
  magnetic part of-, 65, 217, 399
  principal spinor of-, 112
  spinor image of-, 111, 115
white hole, 191, 192
wormhole, 307, 317, 415

Printed in the United States
by Baker & Taylor Publisher Services